METHANE EMISSIONS FROM MAJOR RICE ECOSYSTEMS IN ASIA

Developments in Plant and Soil Sciences

VOLUME 91

The titles published in this series are listed at the end of this volume.

Methane Emissions from Major Rice Ecosystems in Asia

Edited by

REINER WASSMANN

RHODA S. LANTIN

and

HEINZ-ULRICH NEUE

Reprinted from *Nutrient Cycling in Agroecosystems*, Volume 58, Nos. 1–3 (2000).

KLUWER ACADEMIC PUBLISHERS
DORDRECHT / BOSTON / LONDON

in cooperation with the
International Rice Research Institute

Library of Congress Cataloging-in-Publication Data

A C.I.P. Catalogue record for this book is available from the library of Congress.

ISBN 0-7923-6759-6

Published by Kluwer Academic Publishers,
P.O. Box 17, 3300 AA Dordrecht, The Netherlands

Sold and distributed in North, Central and South America
by Kluwer Academic Publishers,
101 Philip Drive, Norwell, MA 02061, U.S.A.

In all other countries, sold and distributed
by Kluwer Academic Publishers,
P.O. Box 322, 3300 AH Dordrecht, The Netherlands

Printed on acid-free paper

Printed in the Netherlands

Contents

Nutrient Cycling in Agroecosystems **58:** ix, 2000.

Foreword

This book deals exclusively with the topic of methane emission from rice fields. Interest in this subject led a decade ago to an interregional research program on methane emission from rice fields in Asia. The program started in 1993 and was completed in 1999. The Global Environment Facility (GEF) of the United Nations Development Programme (UNDP) funded the research program. The International Rice Research Institute (IRRI) coordinated the program in close collaboration with Fraunhofer Institute for Atmospheric Environmental Research (Germany) and national agricultural research systems (NARS) in China, India, Indonesia, Thailand, and the Philippines. This book presents results generated by this research program together with recent results from other research projects.

Before the onset of this interregional program, rice fields were identified as a major source of the increase in atmospheric methane. However, current understanding of the complex interactions between and variations of underlying processes and controlling factors were insufficient to accurately estimate regional and global source strength of atmospheric methane from rice fields as well as for determining potential and feasible options on reduction of methane emission rates.

The objectives of the program were

- to develop standardized measurement systems to determine methane fluxes in major rice ecosystems (irrigated, rainfed, and deepwater rice ecosystems);
- to evaluate the processes that control methane fluxes from rice fields;
- to determine the methane source strength of major rice ecologies, its controlling factors, and processes;
- to upgrade capacities of NARS to measure and analyze methane emission from rice fields; and
- to identify feasible mitigation strategies and technologies while maintaining or increasing rice productivity.

The interregional program on methane emission from rice facilitated various additional regional and international collaboration and research on methane emission issues. The program significantly contributed to the Rice-Activity within the International Global Atmospheric Chemistry (IGAC) project of the International Geosphere Biosphere Programme (IGBP).

The execution of the program required unprecedented coordination of field experiments and laboratory studies, sophisticated measurement systems, and powerful computer capabilities for data analyses and syntheses. The human, technical, and financial resources needed to implement this interregional research program successfully could only be met through interdisciplinary research and through the cooperation of scientists and policymakers from major rice-growing countries, developed countries, and international development agencies. The involvement of all groups and the spirit of equal partnership in the design and execution of the research were essential for achieving the objectives and providing a good basis for recommended policy actions and sustained follow-up research.

Simultaneously, other research groups have substantially improved the understanding of the role of rice fields in the global methane budget. Approaches vary among different research groups, but specific research activities have consistently been implemented on the basis of systematic reviews of the current state of knowledge of rice ecology systems and dynamics, climate, soil and water systems, biogeochemical dynamics, and the human dimension.

In the United Nations Framework Convention on Climate Change, signed by almost all nations of the world, climate change has been recognized as a major threat to the global environment. The ultimate goal of the convention is the stabilization of greenhouse gas concentration in the atmosphere at a level that would prevent anthropogenic interference with the climate system. The specific stipulations for the signatories are to compile a national inventory of emissions and to explore mitigation options according to their own means and potentials.

Given the expected doubling in rice production in Asia, conducting research that will help developing countries grow more rice on limited land, in ways that do not harm the environment and that benefit both farmers and consumers, will be critical. The papers presented in this book, generated through the interregional program or through other research initiatives, provide a comprehensive overview on the progress made in studying methane fluxes from rice fields for global-scale and location-specific syntheses. This book aims to document baseline emissions for major rice ecosystems (that may be used for national inventories) and to devise feasible mitigation options under a broad range of environmental and socioeconomic settings.

Ronald P. Cantrell
Director General

Facts on the project

1. Financial support

U.S. Environmental Protection Agency (1991-1995)
United Nations Development Programme/Global Environment Facility (1993-1999)
German Ministry for Economic Cooperation/Gesellschaft für Technische Zusammenarbeit (1996-1999)

2. IRRI staff (key members)

R. Wassmann (project coordinator 1996-1999)
H.U. Neue (project coordinator 1991-1996)
R.S. Lantin, K.F. Bronson, J.R.M. Arah, M.S. Aulakh, B. Wang; Y. Lu, S. Mitra, L.V. Buendia, J.B. Aduna, M.C.R. Alberto, C.S. Bueno, E.B. Abao, N. Nocon, A.M. Javellana, L.A. Lubigan

3. NARS collaborators (key members)

T. M. Corton, J. Bajita (Philippine Rice Research Institute, Philippines)
N. Chareonsilp, B. Buddhaboon (Prachinburi Rice Research Institute, Thailand)
M. Makarim, A. Setyanto (Central Research Institute for Food Crops, Indonesia)
M.C. Jain, S. Kumar (Indian Agricultural Research Institute, India)
T.K. Adhya, N. Sethunathan (Central Rice Research Institute, India)
Z. Wang, Y.X. Guo, Y.C. Xu (Institute of Crop Breeding and Cultivation, China)
W.F. Lu, W. Chen (China National Rice Research Institute, China)

4. Scientists in advanced research organizations

H.A.C. Denier van der Gon, Wageningen Agricultural University, The Netherlands
H. Rennenberg, W. Seiler, H. Papen, Fraunhofer Institute for Atmospheric Environmental Research, Germany
R.B. Matthews and J. Knox, Cranfield University, UK

Part I

Results of the interregional program on methane emissions from rice fields

Nutrient Cycling in Agroecosystems **58:** 1–12, 2000.

Characterization of methane emissions from rice fields in Asia. I. Comparison among field sites in five countries

R. Wassmann[1,2], H.-U. Neue[1,3], R.S. Lantin[1], L.V. Buendia[1] & H. Rennenberg[2]

[1] *International Rice Research Institute (IRRI), P. O. Box 3127, 1271 Makati City, Philippines;* [2] *Fraunhofer Institute for Atmospheric Environmental Research (IFU), Garmisch-Partenkirchen, Germany* [3] *UfZ-Centre for Environmental Research Leipzig-Halle, Department of Soil Sciences, Halle, Germany*

Key words: irrigated, climate, crop management, organic amendments, China, India, Thailand, Philippines, Indonesia, mitigation options

Abstract

The Interregional Research Program on Methane Emissions from Rice Fields established a network of eight measuring stations in five Asian countries. These stations covered different environments and encompassed varying practices in crop management. All stations were equipped with a closed chamber system designed for frequent sampling and long-term measurements of emission rates. Even under identical treatment—e.g., continuous flooding and no organic fertilizers—average emission rates varied from 15 to 200 kg CH_4 ha^{-1} $season^{-1}$. Low temperatures limited CH_4 emissions in temperate and subtropical stations such as northern China and northern India. Differences observed under given climates, (e.g., within the tropics) indicated the importance of soil properties in regulating the CH_4 emission potential. However, local variations in crop management superseded the impact of soil- and climate-related factors. This resulted in uniformly high emission rates of about 300 kg CH_4 ha^{-1} $season^{-1}$ for the irrigated rice stations in the Philippines (Maligaya) and China (Beijing and Hangzhou). The station in northern India (Delhi) was characterized by exceptionally low emission rates of less than 20 kg CH_4 ha^{-1} $season^{-1}$ under local practice. These findings also suggest opportunities for reducing CH_4 emission through a deliberate modification of cultural practice for most irrigated rice fields.

Introduction

Rice is the basic food for nearly half the people on earth, most of them concentrated in Asia. One hundred forty million ha of rice are harvested annually, occupying about 10% of the arable land worldwide (IRRI, 1993a). Rice production has surged over the past 30 yr, driven in the beginning by the doubling of yields and expansion of the cultivated area. Irrigated rice, which accounts for more than 75% of global rice production, has been responsible for most of this production growth (IRRI, 1993b).

Although rice production has so far kept up with population growth, new studies suggest that an additional 50-70% of the current rice supply will be needed during the 1990-2025 period (Pingali et al., 1997). While land resources are shrinking, present trends suggest that tomorrow's rice land will be under even more pressure (Greenland, 1997). Possible effects of climate change add to the problem of sustaining the natural resource base while raising production to feed more people. Uncertainties become even higher as agriculture itself has a significant effect on global warming through the release of greenhouse gases to the atmosphere such as CH_4 emissions from flooded rice fields (Neue, 1993).

The potential of rice fields to emit CH_4 has long been noted, but comprehensive field measurements were started only in the early eighties. This work was mainly driven by atmospheric science that aimed to clarify the global budget of the greenhouse gas CH_4 (Cicerone & Shetter 1981; Seiler et al., 1984). In spite of a wealth of field data on CH_4 emissions from different rice-growing environments, the available results still do not allow a conclusive estimate on the global emissions from rice. Recent estimates of the CH_4 source strength of rice fields still range from 20 to 100 Tg CH_4 yr^{-1} (IPCC, 1996; Neue & Sass., 1998). Major uncertainties are related to (1) diverging environments for growing rice resulting in pronounced spatial and temporal variation and (2) different experimental approaches, especially regarding sampling frequency and observation period, for recording CH_4 emission rates.

The interregional research program on CH_4 emissions has established a network of stations equipped with standardized measurement systems. These automated systems allowed continuous records of CH_4 fluxes over entire seasons. In some stations, emissions were recorded over 5 consecutive years. The concerted measurement program allowed clear distinction between inherent differences and those resulting from crop management.

This program on CH_4 emissions was a joint effort of the International Rice Research Institute (IRRI), the Fraunhofer Institute for Atmospheric Environmental Research (Garmisch-Partenkirchen, Germany), and agricultural research institutes in China, India, Indonesia, Philippines, and Thailand (Figure 1). The collaborating countries cover 67% of the global rice area while only two of those countries, India (42.2 million ha) and China (33.7 million ha), comprise 50% of the global rice area. The work was funded by the United Nations Development Programme/Global Environment Facility from 1993 to 1999. The overall objective was to provide baseline data for accurate estimates of regional CH_4 emissions from different rice-growing regions while fostering sustained growth in rice production in developing countries. Research has focused on quantifying CH_4 emissions from major rice ecosystems (irrigated rice, rainfed rice, and deepwater rice) in Asia, evaluating processes that control CH_4 fluxes from ricefields, and identifying mitigation technologies for CH_4 emissions that maintain or enhance rice productivity in a sustainable rice system. This work was part of a broader effort by IRRI to examine the interaction of rice and global climate change including greenhouse gas emissions and the vulnerability of rice production to a changing climate (Wassmann et al., 1998; Ziska et al., 1998; Moya et al., 1998).

The results of the project are presented comprehensively within this special issue through 16 articles—i.e., nine articles comprising detailed results from all measurement station (Table 1), a series of four articles on modeling and upscaling of emissions (Matthews et al., this issue) and a series of three articles that cut across the results of all collaborating stations. This first article of the latter series aims to describe the background, methodology, and experimental stations of the project, and to compare emissions under identical fertilizer applications as well as site-specific irrigation practices. The other articles of this series deal with the

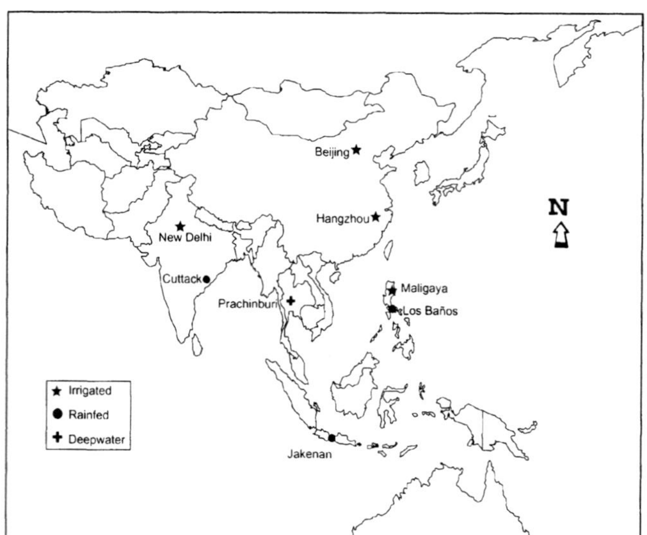

Figure 1. Stations of the Interregional Research Programme on Methane Emission from Rice Fields

Table 1. Characterization of experimental sites

Station, country	Ecosystem	Geographic coordinates	Soil properties: Texture	pH	Org C (%)	Total N (%)	Detailed information (this issue)
Beijing, China	Irrigated	39° 93′ N 116° 47′E	Silty clay loam	7.0	0.99	0.09	Wang et al.
Hangzhou, China	Irrigated	30° 23′ N 120° 20′ E	Silty clay	6.2	2.4	0.22	Lu et al.
New Delhi, India	Irrigated	20° 38′ N 70° 10′ E	Sandy clay loam	8.2	0.45	0.069	Jain et al.
Maligaya, Philippines	Irrigated	15° 67′ N 120° 88′ E	Silty clay	6.1	1.3	0.09	Corton et al.
Cuttack, India	Rainfed	20° 50′ N 86° 00′ E	Clay loam	7.0	0.54	0.048	Adhya et al.
Jakenan, Indonesia	Rainfed	6° 68′ S 111° 20′ E	Silty loam	4.7	0.48	0.05	Setyanto et al.
Los Baños, Philippines	Rainfed	14° 18′ N 121° 25′ E	Silty clay	6.3	1.5	0.14	Wassmann et al. Abao et al.
Prachinburi, Thailand	Deepwater	13° 92′ N 101° 25′ E	Clay	3.9	1.2	0.17	Chareonsilp et al.

impact of different rice ecosystems (Wassmann et al., this issue, c) and the crop management options to mitigate CH_4 emissions (Wassmann et al., this issue, b).

Background and rationale of this study

Recent observations provide compelling evidence that the global climate is changing as a direct result of human activities (IPCC, 1996). Release of chlorofluorocarbons damages the stratospheric ozone layer, which increases biologically harmful ultraviolet radiation reaching the earth. The global increase in carbon dioxide (CO_2), along with other trace 'greenhouse' gases CH_4 and nitrous oxide (N_2O), traps outgoing thermal radiation, leading to increased temperature at the earth's surface. The agricultural sector releases the greenhouse gases (CH_4) through rice cultivation and livestock and (N_2O) through intensified fertilizer use in various cropping system (GEIA, 1993).

Most of the historical and current greenhouse gas emissions have originated from developed countries (IPCC, 1996). Different nations, however, have distinct capabilities for coping with climate change, a fact recognized by the United Nations Framework Convention on Climate Change. In major rice-growing countries, rice researchers should play a crucial role in addressing the goals stipulated in the convention: conducting nationwide inventories of greenhouse gas emissions and preparing national programs for mitigating these emissions.

The tropospheric mixing ratio of CH_4, one of the main greenhouse gases, has increased from its preindustrial level of about 700 ppbv to 1720 ppbv at present (Khalil & Shearer, 1993). Although CH_4 concentrations have remained stable during the early 1990s (Dlugokencky et al., 1994), recent concentration records indicate a reestablishment of the trend of increasing CH_4 concentrations. The overall budget of atmospheric CH_4 is relatively well established, however, the strength of individual sources such as rice production is still uncertain (Rennenberg et al., 1995). The total annual source strength of all CH_4 emissions is about 500 Tg, exceeding the total sink by 37 Tg yr^{-1} (IPCC, 1996). The main sink mechanism is photochemical oxidation with the hydroxyl radicals in the troposphere. Isotopic measurements reveal that 70-80% of the atmospheric CH_4 is of biogenic origin with natural wetlands as the largest source (Khalil & Shearer, 1993). Other biological sources are related to agricultural production, namely livestock and rice.

Since the first field data from rice fields in California (Cicerone & Shetter, 1981) and southern Europe (Seiler et al., 1984; Holzapfel-Pschorn et al., 1985), extensive data sets from various rice-growing environments have indicated a pronounced variability of CH_4 emissions in space and time. The existing database on CH_4 emission from rice fields includes intensive studies conducted in Italy (Schütz et al., 1989); USA (Sass et al., 1990); China (Khalil & Rasmussen, 1991; Wassmann et al., 1993; Wang et al., 1994); India (Parashar et al., 1994), Japan (Kimura et al., 1991; Yagi et al., 1996) and Southeast Asia (Jermsawatdipong et al., 1994; Nugruho et al., 1994; Yagi et al., 1994; Neue et al., 1995; Wassmann et al., 1995; Husin et al., 1995). Global CH_4 emission from wetland rice fields is estimated to be 60 Tg yr^{-1}, with a range of 20-100 Tg yr^{-1} (IPCC, 1996). Superimposed on this uncertainty in present emission rates are rapid changes in the intensity and mode of rice production. Changes in crop management affect CH_4 emission in various ways, but the net impact of historical as well as projected progress in rice technology is difficult to assess.

While rice is preferably grown under submerged conditions, predominantly anaerobic flooded rice soils promote the production of CH_4 by anaerobic decomposition of the organic matter (native or added). The CH_4 budget of rice fields is determined by the availability of methanogenic substrate generated from organic residues, plant-borne material and, if applied, organic fertilizers. Methane emission is the interactive product of three processes (Neue et al., 1997): (1) CH_4 production controlled by Eh, pH, and mineralizable carbon and temperature; (2) CH_4 oxidation controlled by free oxygen diffusing through the rice plant, partial CH_4 pressure, and temperature; and (3) vertical transfer controlled by water depth and rice plant growth stage.

Field stations and methods

The eight field stations of this study were distributed over five countries in Asia (Figure 1) and represent a wide range of rice environments (Table 1). Four stations concentrated on irrigated rice while the rainfed and deepwater stations included irrigated rice as reference treatment. Except for Jakenan, all soils were clayey with varying proportions of silt and sand (Table 1). Chemical properties ranged from an acid sulfate soil (Prachinburi) to an alkaline soil (New Delhi) and from low concentrations of native C and N (Jakenan) to very high concentrations of these elements (Hangzhou). The different temperature regimes are schematically displayed in Figure 5.

Figure 2. Field chambers set up under dry conditions (top: Jakenan) and deepwater conditions (bottom: Prachinburi)

Methane fluxes were determined with an automated closed chamber method (Figure 2). This measurement system used in this study, a modified version of the system originally described by Schütz et al., (1989), consisted of the following components.

Field chambers

Twelve chambers made of transparent plexiglas were distributed in the field according to a complete block design (Wassmann et al., 1994). Each chamber had a basal area of 1 m^2. The height was 1.2 m in irrigated and rainfed rice (Figure 2a), while chambers in deepwater rice were 1.6 m high (Figure 2b). The chambers were placed tightly on steel frames that penetrated 20 cm into the soil. Round holes in these frames allowed water exchange during flooding, but these could be sealed for measurements during dry conditions. Chambers were equipped with hinged covers that could be opened or closed by a pneumatic system. An open stainless steel tube penetrated into the inner chamber

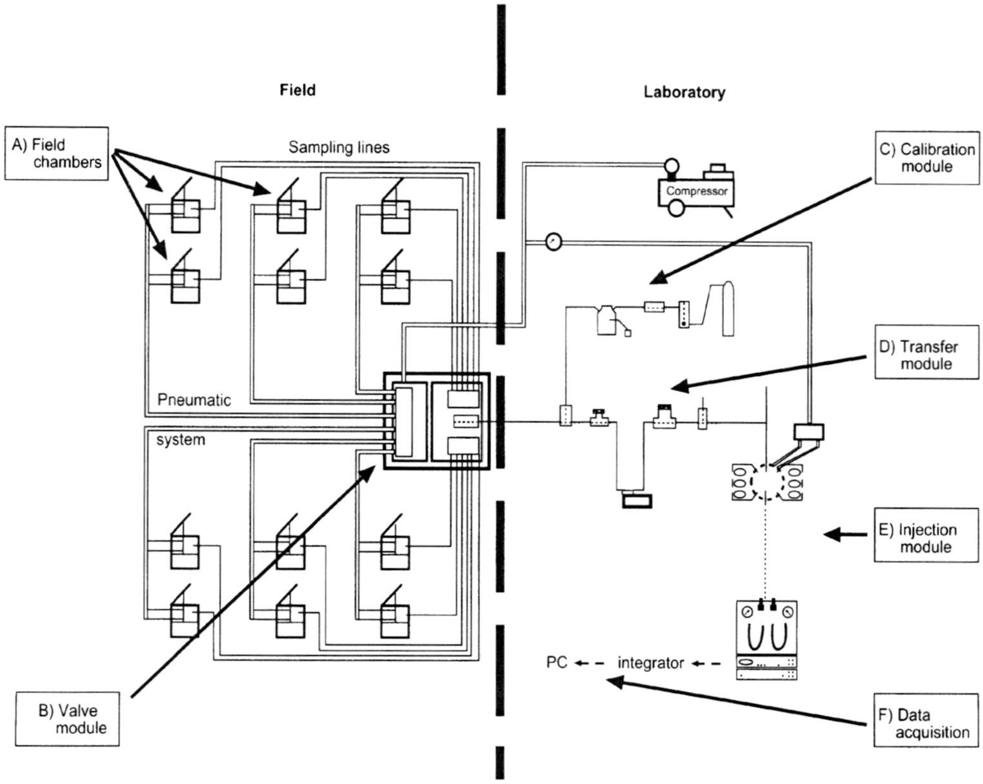

Figure 3. Schematic view of the measuring system

for sampling. Two fans inside each chamber ensured thorough mixing during enclosure and effective gas exchange with ambient air during opening.

Valve module

The valve module consisted of two valve sets—i.e., one for the pneumatic system to open and close the chambers and one for the lines connecting a pump to the inner chamber (Figue 3). Valve operations were triggered by a time control system installed in a PC. The operation sequence encompassed a 2-h cycle in which each chamber was opened for 114 min and closed for 16 min. Closing periods were staggered, so that only one pair of chambers was closed at a given moment. During closure, air was collected at 2-min intervals yielding four air samples per chamber.

Calibration module

The valve module was connected to a three-port valve that could periodically be switched to the calibration module. This module consisted of a gas cylinder filled with calibration gas and a control system that maintained ambient pressure in the lines connected to the transfer unit. During one 2-h cycle, calibration gas was tapped four times (0-2 min, 34-36 min, 68-70 min, and 102-104 min).

Transfer module

This module allowed the transfer of gas—either air from the chambers or calibration gas—to the injection module. The gas flow was driven by a pump and was controlled through electronic regulators.

Injection module

The gas was passed through a sample loop that was connected to a 10-port valve. Switching of this valve resulted in injection of a gas aliquot into the analytical device. The injection module could also be used for manual sampling without modification, e.g., during the stand-by time of the automatic system between cropping seasons. The analytical system consisted of a gas chromatograph (Shimadzu GC-8A) equipped with a Porapak column and a flame ionization detector.

Data acquisition

The signals from the gas chromatograph were converted to relative concentration values by an integrator and then logged by a computer. The computer was also equipped with the time control device that triggered all valve switches of the automatic system and a temperature acquisition system. Eight temperature sensors were distributed in the soil at 5, 10, and 15 cm depths in the floodwater and in the air.

Methane emission rates were derived from the temporal increase in CH_4 concentration inside the closed box (IAEA, 1992). The logged raw data underwent several steps of computation and quality assurance:

1) The temporal increase in CH_4 concentration was computed for each box. This procedure included a linearity test to detect possible artifacts due to leaks.
2) Flux rates were computed from the concentration increase in each chamber and were aggregated for replicate chambers for each run. After a conformity test of these replicates, the validated values for one run were compiled into 24-h cycles of emission flux rates for each treatment.
3) Occasional gaps in emission records over one 24-h cycle were recalculated by using specifically developed algorithms for diel flux patterns (Buendia et al., 1997).

Soil pH and soil Eh were measured manually at least once a week during the cropping season. Soil pH was measured with a commercially available electrode, while the Eh electrodes were manufactured using a glass tube and platinum wire. The pH electrode was exposed temporarily at 7.5 cm depth, whereas the Eh electrode remained in the soil at this depth.

Methane concentration in the soil solution was determined at weekly intervals. The solution was extracted from soil depths of 0, 5, 10, and 15 cm using a porous tube connected to a vacutainer tube (Alberto et al., 1999). Methane concentrations in the solution were derived from headspace analysis after shaking the vacutainer tube (Alberto et al., 1999).

Methane ebullition has been recorded to be equal to the total surface flux between plants. Flux rates was measured weekly by placing 40 × 15 × 20 cm chambers between rice hills (Wassmann et al., 1996). Gas samples from the inner chamber volume were collected after 24 h of exposure and were analyzed immediately for CH_4 concentration.

Results

Reference treatment

Methane emissions showed pronounced variations among sites—even under identical crop management. Figure 4 shows the results obtained for the reference treatment of this study—i.e., continuous flooding, pure mineral fertilizer, and cultivar IR72. The values for New Delhi, Cuttack, Los Baños, Jakenan, and Maligaya represent actual emission rates, whereas those for Prachinburi, Hangzhou, and Beijing had to be adjusted due to slight modifications in crop management (Chareonsilp et al., this issue; Lu et al., this issue; Wang et al., this issue). The results reflect pronounced variations from season to season. Interseasonal variations were especially large for Los Baños where different management of stubbles further amplified interseasonal differences (Wassmann et al., this issue, a).

Rice fields in New Delhi, Cuttack, and Beijing emitted less than 100 kg CH_4 ha^{-1} over one season. Emissions reached more than 200 kg CH_4 ha^{-1} for some seasons in Los Baños, Hangzhou, Jakenan, and Maligaya. The database also indicates differences in seasonal patterns of CH_4 emission, depending on temperature regime (Figure 5). With constant or increasing temperature, the bulk of CH_4 was emitted during the ripening stage of the plant. Maximum temperature in the middle of the cropping season resulted in highest emission during the reproductive stage, while a decreasing temperature trend enhanced the relative contribution of the vegetative stage. However, these emission patterns were modified by organic manure as well as drainage periods. Application of manure as well as midseason drain-

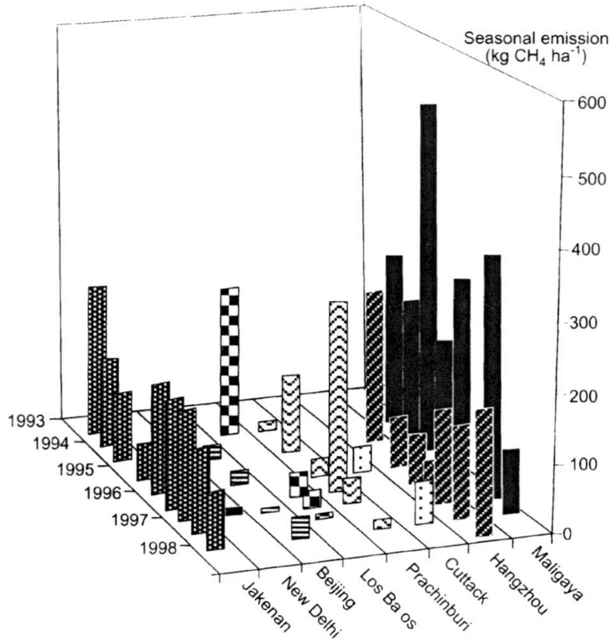

Figure 4. Seasonal emissions with mineral fertilizers under local irrigation schemes (see text for further explanation); one calendar year comprises one or two crops at the respective station

age enhanced the contribution of the vegetative stage (Wassmann et al., this issue, a,c).

Local crop management practices

Local water management practices differed among the four irrigated stations of this project. Only in Maligaya did the local practice correspond to the reference treatment (i.e., continuous flooding). At Hangzhou and Beijing, the local irrigation practice encompassed a drainage period at midseason (Lu et al., this issue; Wang et al., this issue). In New Delhi, high percolation rates on the sandy soil required continual replenishing of the floodwater, a technique referred to as intermittent irrigation (Jain et al., this issue).

Seasonal emissions with mineral fertilizers and organic manure in these four irrigated stations are illustrated in Figure 6a,b. The results with local irrigation practice and mineral fertilizer (Figure 6a) are similar to those obtained using the reference treatment for these four stations (Figure 4). Results from the four stations fall on a relatively straight line from low to high emission: New Delhi < Beijing < Hangzhou < Maligaya.

Organic amendment, however, resulted in a different picture. While emissions from New Delhi were still very low, emissions from the other three stations were increased greatly by addition of organic manures. The most notable response was recorded in Beijing, where emissions from the plots treated with organic manure were more than 10 times higher than from those receiving mineral fertilizer (Wang et al., this issue). Emission rates for organic amendments fell in similar ranges for both Chinese stations. High standard deviations with organic manure can be attributed to the different nature and quantities of the amendments—i.e., rice straw, pig manure, biogas residues, and others (Wassmann et al., this issue, b).

Discussion

Site-specific differences under identical treatments are apparently related to a combination of both climate and soil parameters. The significance of the soil can be deduced by comparing the stations in Southeast Asia. In spite of comparable temperature regimes, CH_4 emissions at Maligaya, Jakenan, Los Baños, and Prachinburi field

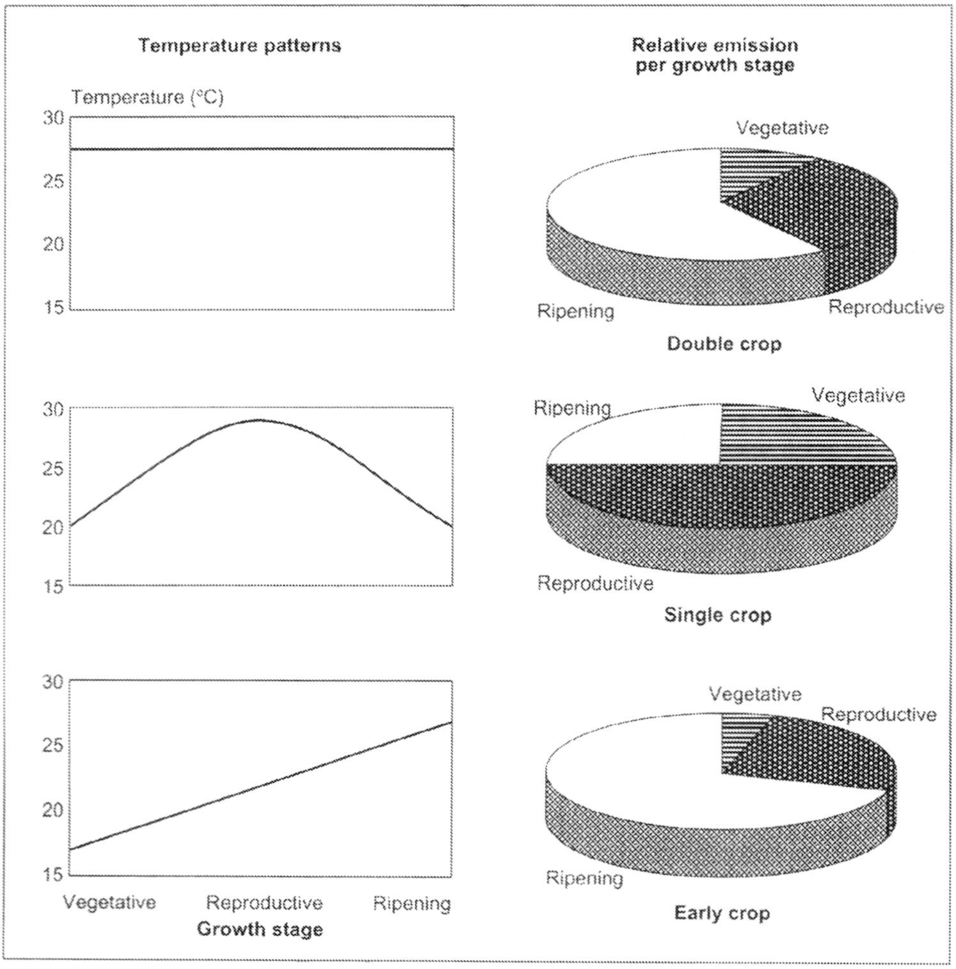

Figure 5. Schematic display of temperature and emission patterns (see text for further explanation)

stations differed over a large range (Figure 4). However, no individual soil parameter could be singled as responsible for the emission potential (Table 1). Microbial CH_4 production is affected by (1) the quality of soil organic matter and (2) the availability of alternative electron acceptors (Wassmann et al., 1998; van Bodegom et al., this issue; Matthews et al., this issue). Other soil properties such as texture may also interfere in various ways with CH_4 production, oxidation, and transport (Sass et al., this issue).

The magnitude of CH_4 emissions at the different sites also depended on crop management. The prevailing irrigation patterns differed among rice-growing regions. The four sites of irrigated rice in this study represented three different types: continuous flooding (as in the reference treatment) in Maligaya, midseason drainage in Hangzhou and Beijing, and intermittent irrigation in New Delhi. The emission potential associated with these irrigation patterns (Figure 6) was highest for continuous flooding and lowest for midseason drainage (Wassmann et al., this issue, c).

The emission potentials of the project stations also differed in their response to organic amendments (Figure 6). Again, this could be attributed to a combination

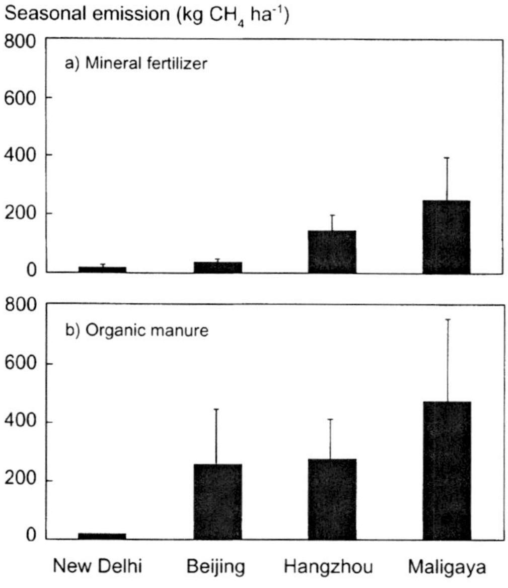

Figure 6. Seasonal emissions under local water management practice (intermittent irrigation in New Delhi, midseason drainage in Beijing and Hangzhou; continuous flooding in Maligaya) with (a) mineral fertilizer only and (b) organic manure supplemented by mineral fertilizer

of soil- and climate-related factors. Emission rates at the New Delhi site showed almost no increase with organic amendments. High percolation rates resulted in an inflow of oxygen into the soil and downward discharge of methanogenic substrate resulted in low emission rates (Yagi et al.,1990; Inubushi et al., 1992). Thus, emissions were low, irrespective of the amount of organic matter applied.

The pronounced increase due to organic amendments in Beijing could be related to seasonal pattern of the flux. The temperature regime in Beijing suppressed emissions during the late stage (Figure 5). Changes in the early stage therefore had a higher impact on the overall emissions as compared with a cropping season with high temperatures at the end. The discernible effect of organic amendments was generally limited to the early stage of the season (Wassmann et al., 1996).

Due to the common use of organic fertilizers in China, the emission rates displayed in Figure 6b represented local practices of crop management for Beijing and Hangzhou. On the other hand, farmers in the Philippines and northern India generally omit organic manure, so that the values depicted in Figure 6 for Maligaya and New Delhi correspond to the local management practices. Local management resulted in similar emission rates of approximately 300 kg CH_4 ha^{-1} in each season in Maligaya, Beijing, and Hangzhou. The station in New Delhi had distinctly low emission rates (less than 20 kg CH_4 ha^{-1} and season) under a crop management typical of northern India. Other rice-growing regions in India may have higher emissions than the site in New Delhi (Adhya et al., 1994), although the available database for Indian rice production is still not conclusive.

Spatial variations in CH_4 emissions from different rice-growing areas have previously been documented for individual countries (Parashar et al., 1994; Yagi et al., 1994). Extensive literature reviews have yielded even larger ranges of CH_4 emission rates from different sites (Wassmann et al., 1993; Neue & Sass, 1998). However, data sets compiled from different studies are only partly comparable due to different measurement techniques and field treatments; even definitions of "irrigated" rice deviated between different studies (Neue & Boonjawat, 1998). This project has, for the first time, established an interregional network with standardized measurement systems and a field design appropriate for a multilocation trial. The concerted measurement program allowed a clear distinction between inherent differences and those resulting from crop management.

Conclusion

The automatic measurement system used in this study allowed investigation of different crop management practices with high sampling frequency and long duration of the observation period. Application of a uniform reference treatment provided relative emission potentials for each station of this study. However, CH_4 emission is highly sensitive to water regime and organic inputs, so that local variations in crop management can supersede the impact of soil and climate factors. These distinct features of the rice fields can be characterized as (1) baseline and (2) actual emission potentials. In the case of the two Chinese stations of this study, baseline emissions differed by a factor of 6, whereas the actual emissions from these field sites were similar.

The site-specific identification of baseline emission and actual emission is essential for future development of mitigation strategies. Deliberate modification of agronomic practices can have the greatest impact in rice land with a large gap between baseline and actual emissions. Further investigations on the socio-economic feasibility of mitigation technologies could therefore be targeted to site-specific settings with these characteristics.

Acknowledgement

The research described was funded by the United Nations Development Programme, Global Environment Facility UNDP/GEF (GLO/91/G31) - Interregional Research Programme on Methane Emission from Rice Fields under contract no. C92379 to the International Rice Research Institute.

References

Abao EB, Bronson KF, Wassmann R & Singh U (2000) Simultaneous records of methane and nitrous oxide emissions in rice-based cropping systems under rainfed conditions Nutr Cycling Agroecosyst, this issue

Adhya TK, Rath AK, Gupta PK, Rao VR, Das SN, Parida KM, Parashar DC & Sethunathan N (1994) Methane emission from flooded rice fields under irrigated conditions. Biol Fertil Soils 18:245-248

Adhya TK, Bharati K, Mohanty SR, Mishra SR, Ramakrishnan B, Rao VR, Sethunathan N & Wassmann R (2000) Methane emissions from rice fields at Cuttack (India). Nutr Cycling Agroecosyst, this issue

Alberto MCR, Arah JRM, Neue HU, Wassmann R, Lantin RS, Aduna JB & Bronson KF (2000) A sampling technique for the determination of dissolved methane in soil solution. Global Change Sci 2:57-63

Buendia LV, Neue HU, Wassmann R, Lantin RS, Javellana AM, Arah JRM, Xu Yuchang, Lu Wanfang, Makarim AK, Corton TM & Chareonsilp N (1997) An efficient sampling strategy for estimating methane emission from rice to assess the impact of potential mitigation options. Chemosphere 36:395-407

Chareonsilp N, Buddhaboon C, Promnart P, Wassmann R & Lantin RS (2000) Methane emission from deepwater rice fields in Thailand. Nutr Cycling Agroecosyst, this issue

Cicerone RJ & Shetter JD (1981) Sources of atmospheric methane: measurements in rice paddies and a discussion. J Geophys Res 86:7203-7209

Corton TM, Bajita J, Grospe F, Pamplona R, Wassmann R & Lantin RS, Buendia LV (2000) Methane emission from irrigated and intensively managed rice fields in Central Luzon , Philippines. Nutr Cycling Agroecosyst, this issue

Dlugokencky E, Masaire K, Lang KM, Tans P, Steele P & Nisbet E (1994) A dramatic decrease in the growth rate of atmospheric methane in the northern hemisphere during 1992. Geophys Res Lett 21:45-48

GEIA — Global Emission Inventory Activity (1993) Report on the 3rd workshop, Amersford, 31 Jan -02 Feb 1993, ed. A.F. Bowman (ed) Bilthoven, The Netherlands, 83 p

Greenland, DJ (1997) The sustainability of rice farming. Wallingford, UK. 273 p

Holzapfel-Pschorn A, Conrad R & Seiler W (1985) Production, oxidation, and emission of methane in rice paddies. FEMS Microbiol Ecol 31:149-158

Husin YA, Murdiyarso D, Khalil MAK, Rasmussen RA, Shearer MJ, Sabiham S, Sunar A & Adijuwana H (1995) Methane flux from Indonesian wetland rice: the effects of water management and rice variety. Chemosphere 31: 3153-3180

Inubushi K, Muramatsu Y & Umerayasi M (1992) Influence of percolation on methane emission from flooded paddy soil. Jpn J Soil Sci Plant Nutr 63:184-189

IPCC – Intergovernmental Panel on Climate Change (1996) Climate Change 1995. The Science of Climate Change. Cambridge (UK): Cambridge University Press. 572 p

IRRI – International Rice Research Institute (1993a) IRRI Rice Almanac, Manila, Philippines, 142 p

IRRI – International Rice Research Institute (1993b) Rice Research in a Time of Change. Manila, Philippines. 79 p

Jain MC, Kumar K, Wassmann R, Mitra S, singh SD, singh JP, singh R, Yadav AK, Gupta S (2000) Methane emissions from irrigated rice fields in Northern India (New Delhi) Nutr. Cycling Agroecosyst, this issue

Jermsawatdipong P, Murase J, Prabuddham P, Hasathon P, Khomthong Y, Naklang N, Watanabe A, Haraguchi H & Kimura M (1994) Methane emission from plots with different fertilizer application. Soil Sci Plant Nutr 40: 63-71

Khalil MAK & Rasmussen RA (1991) Methane emissions from rice fields in China. Environ Sci Technol 25:979-981

Khalil MAK & Shearer MJ (1993) Atmospheric methane: sources, sinks and role in global change. Chemosphere 26:201-217

Kimura M, Miura Y, Watanabe A, Katoh K & Haraguchi H (1991) Methane emission from paddy field, part I. Effect of fertilization, growth stage and midsummer drainage: pot experiment. Environ Sci 4:265-271

Lu WF, Chen W, Duan BW, Guo WM, Lu Y, Lantin RS, Wassmann R & Neue HU (2000) Methane emission and mitigation options in irrigated rice fields in Southeast China. Nutr Cycling Agroecosyst, this issue

Moya TB, Ziska LH, Namuco OS & Olszyk D (1998) Growth dynamic and genotypic variation in tropical, field grown paddy rice rice (*Oryza sativa* L.) in response to increasing carbon dioxide and temperature. Global Change Biol (in press)

Neue HU (1993) Methane emission from rice fields. Bioscience 43:466-474

Neue, HU & Boonjawat J (1997) Methane emission from rice fields. In: Galloway J & Melillo J (eds) Asian Change in the Context of Global Change, pp.187-207, IGBP Book Series. Cambridge University Press, UK

Neue HU & Sass RL (1998) The budget of methane from rice fields. IGACtivities 12:3-11

Neue HU, Wassmann R, Kludze HK, Wang B & Lantin RS (1997) Factors and processes controlling methane emissions from rice fields. Nutr Cycling Agroecosyst 49:111-117

Neue HU, Wassmann R & Lantin RS (1995) Mitigation options for methane emission from rice fields. In: Peng S, Ingram KT, Neue HU & Ziska LH (eds) Climate Change and Rice, pp 136-144, Springer-Verlag, Berlin

Nugruho SG, Lumbanraja J, Suprapto H, Ardjasa WS, Haraguchi H & Kimura M (1994) Effect of intermittent irrigation on methane emission from an Indonesian paddy field. Soil Sci Plant Nutr 40:609-615

Parashar DC, Mitra AP, Sinha SK, Gupta PK, Rai J, Sharma RC, Singh N, Kaul S, Lal G, Chaudhary A, Ray HS, Das SN, Parida KM, Rao SB, Kanung SP, Ramasami T, Nair BU, Swamy M, Gupta SK, Singh AR, Saikia BK, Barua AKS, Pathak MG, Iyer CPS, Gopalakrishnan M, Sane PV, Singh SN, Banerjee R, Sethunathan N, Adhya TK, Rao VR, Palit P, Saha AK, Purkait NN, Chaturvedi GS, Sen SP, Sen M, Sarkar B, Banik A, Subbaraya BH, Lal S & Venkatramani S (1994) Methane budget from Indian paddyfields. In: Minami K, Mosier A & Sass RL (eds) CH_4 and N_2O: Global Emissions and Controls from Rice Fields and Other Agricultural and Industrial Sources, pp 27-39, Tsukuba, Japan: NIAES Series 2

Pingali PL, Hossain, M & Gerpacio, RV (1997) Asian rice bowls–the returning crisis. Wallingford (UK): CAB International, 341 p

Rennenberg H, Wassmann R, Papen H & Seiler W (1995) Role of methane and nitrous oxide in global change. In: Peng S, Ingram KT, Neue HU, Ziska LH (eds) Climate Change and Rice, pp 60-68, Springer-Verlag

Sass RL, Fisher FM, Harcombe PA, & Turner FT (1990) Methane production and emission in a Texas rice field. Global Biogeochem Cycles 4:47-68

Sass RL, Fisher FM, & Huang Y (2000) Modeling methane emissions from irrigated rice fields. Nutr Cycling Agroecosyst, this issue

Schütz H, Holzapfel-Pschorn A, Conrad R, Rennenberg H, & Seiler W (1989). A three-year continuous record on the influence of daytime season and fertilizer treatment on methane emission rates from an Italian rice paddy field. J Geophys Res 94:16405-16416

Seiler W, Holzapfel-Pschorn A, Conrad R & Scharffe D (1984) Methane emission from rice paddies. J Atmos Chem 1:241-268

Setyanto P, Makarim AK, Fagi AM, Wassmann R, Buendia LV (2000) Crop management affecting methane emissions from irrigated and rainfed rice in Central Java, Indonesia. Nutr Cycling Agroecosyst, this issue

Wang MX, Dai A, Shangguan X, Ren L, Shen R, H Schütz, Seiler W, Rasmussen RA & Khalil MAK (1994) Sources of methane in China. In: Minami K, Mosier A & Sass RL (eds) CH_4 and N_2O: Global Emissions and Controls from Rice Fields and Other Agricultural and Industrial Sources, pp 9-26, Tsukuba, Japan: NIAES Series 2

Wang ZY, Xu YC, Li Z, Guo YX, Wassmann R, Neue HU, Lantin RS, Buendia LV, Ding YP & Wang ZZ (2000) Methane Emissions from Irrigated Rice Fields in Northern China (Beijing). Nutr Cycling Agroecosyst, this issue

Wassmann R, Wang MX, Shangguan XJ, Xie XL, Shen RX, Papen H, Rennenberg H & Seiler W (1993) First records of a field experiment on fertilizer effects on methane emission from rice fields in Hunan Province (PR China). Geophys Res Lett 20:2071-2074

Wassmann R, Neue HU, Lantin RL, Aduna JB, Alberto MC, Andales MJ, Tan MJ, Denier van der Gon HAC, Hoffmann H, Papen H, Rennenberg H & Seiler W (1994) Temporal patterns of methane emissions from wetland ricefields treated by different modes of N application. J Geophys Res 99:16457-16462

Wassmann R, Neue HU, Lantin RS, Javellana MJ, Diego R, Lignes VE, Hoffmann H, Papen H & Rennenberg H (1995) Methane emissions from rainfed rice. In: Fragile lives in fragile ecosystems, pp 217-225, International Rice Research Institute, Manila, Philippines

Wassmann R, Neue HU, Alberto MCR, Lantin RS, Bueno C, Llenaresas D, Arah JRM, Papen H, Seiler W & Rennenberg H (1996) Fluxes and pools of methane in wetland rice soils with varying organic inputs. Environ Monit Assess 42:163-173

Wassmann R, Neue HU, Bueno C, Lantin RS, Alberto MCR, Buendia LV, Bronson K, Papen H & Rennenberg H (1998) Methane production capacities of different rice soils derived from inherent and exogenous substrates. Plant Soil 203:227-237

Wassmann R, Buendia LV, Lantin RS, Bueno CS, Lubigan LA, Umali A , Nocon NN, Javellana AM, & Neue HU (2000a) Mechanisms of crop management impact on methane emissions from rice fields in Los Baños, Philippines. Nutr Cycling Agroecosyst, this issue

Wassmann R, Lantin RS, Neue HU, Buendia LV & Corton TM, Y Lu (2000b) Characterization of methane emissions from rice fields in Asia. 3. Mitigation options and future research needs. Nutr Cycling Agroecosyst, this issue

Wassmann R, Neue HU, Lantin RS, Makarim K, Chareonsilp N, Buendia LV & Rennenberg H (2000c) Characterization of methane emissions from rice fields in Asia. 2. Differences among irrigated, rainfed and deepwater ecosystems. Nutr Cycling Agroecosyst, this issue

Yagi, K, Minami K & Ogawa Y (1990) Effects of water percolation on methane emission from paddy fields. Res. Rep. Div. Environ. Planning, National Institute of Agro-Environmental Sciences 6:105-112

Yagi K, Chairoj P, Tsurata H, Cholitkul W & Minami K (1994) Methane emission from rice paddy fields in the central plain of Thailand. Soil Sci Plant Nutr 40:29-37

Yagi K, H. Tsuruta K, Kanda & Minami, K (1996) Effect of water management on methane emission from a Japanese rice paddy field: automated methane monitoring. Global Biogeochem Cycles 10:255-267

Ziska LH, Moya TB, Wassmann R, Namuco OS, Lantin RS, Aduna JB, Abao E Jr, Bronson KF, Neue HU & Olszyk D (1998) Long-term growth at elevated carbon dioxide stimulates methane emission in tropical paddy rice. Global Change Biol 4:657-665

Nutrient Cycling in Agroecosystems **58:** 13–22, 2000.

Characterization of methane emissions from rice fields in Asia. II. Differences among irrigated, rainfed, and deepwater rice

R. Wassmann[1,2], H.U. Neue[1,3], R.S. Lantin[1], K. Makarim[4], N. Chareonsilp[5], L.V. Buendia[1] & H. Rennenberg[2]

[1] International Rice Research Institute (IRRI), P. O. Box 3127, Makati City 1271, Philippines; [2] Fraunhofer Institute for Atmospheric Environmental Research (IFU), Garmisch-Partenkirchen, Germany; [3] UfZ-Centre for Environmental Research Leipzig-Halle, Department of Soil Sciences, Halle, Germany; [4] Central Research Institute for Food Crops, Bogor, Indonesia; [5] Prachinburi Rice Research Institute, Prachinburi, Thailand

Key words: water regime, soil aeration, mineral fertilizer, rainfall, acid sulfate soil, soil pH, Indonesia, Thailand, Philippines, mitigation options

Abstract

Methane (CH_4) emission rates were recorded automatically using the closed chamber technique in major rice-growing areas of Southeast Asia. The three experimental sites covered different ecosystems of wetland rice—irrigated, rainfed, and deepwater rice—using only mineral fertilizers (for this comparison). In Jakenan (Indonesia), the local water regime in rainfed rice encompassed a gradual increase (wet season) and a gradual decrease (dry season) in floodwater levels. Emission rates accumulated to 52 and 91 kg CH_4 ha^{-1} season^{-1} corresponding to approximately 40% of emissions from irrigated rice in each season. Distinct drainage periods within the season can drastically reduce CH_4 emissions to less than 30 kg CH_4 ha^{-1} season^{-1} as shown in Los Baños (Philippines). The reduction effect of this water regime as compared with irrigated rice varied from 20% to 80% from season to season. Methane fluxes from deepwater rice in Prachinburi (Thailand) were lower than from irrigated rice but accumulated to equally high seasonal values, i.e., about 99 kg CH_4 ha^{-1} season^{-1}, due to longer seasons and assured periods of flooding. Rice ecosystems with continuous flooding were characterized by anaerobic conditions in the soil. These conditions commonly found in irrigated and deepwater rice favored CH_4 emissions. Temporary aeration of flooded rice soils, which is generic in rainfed rice, reduced emission rates due to low CH_4 production and high CH_4 oxidation. Based on these findings and the global distribution of rice area, irrigated rice accounts globally for 70-80% of CH_4 from the global rice area. Rainfed rice (about 15%) and deepwater rice (about 10%) have much lower shares. In turn, irrigated rice represents the most promising target for mitigation strategies. Proper water management could reduce CH_4 emission without affecting yields.

Introduction

The human population continues to increase by 85 million people a year; the developing world will add another 2 billion people over the next three decades. Intensification of rice cultivation to meet the demand for rice by the increasing human population is imperative, especially in Asia where approximately 90% of the rice is grown and consumed (IRRI, 1993a). Given the expected doubling in rice production in Asia, research on improving rice yield should focus on strategies that do not harm the environment. Rice fields represent globally one of the main sources of the greenhouse gas methane (CH_4) (GEIA, 1993; IPCC, 1996), but the global source strength of rice cultivation remains uncertain. The diversified conditions in crop management and environments for growing rice are not sufficiently characterized for accurate estimates (Sass et al., 1990; Rennenberg et al., 1992; Neue & Roger, 1994; Yagi et al., 1994; Byrnes et al., 1995; Wassmann et al., 1998).

Rice land is commonly differentiated into four ecosystems (IRRI, 1993a): irrigated rice (51% of global rice area) with full control of the water regime; rainfed rice (27%), which can be either drought-prone or flood-prone; deepwater rice (10%) characterized by intense inundation; and upland rice (11%). Upland rice, however, does not encompass flooding and thus, can be neglected as a CH_4 source.

This paper comprises results obtained within an interregional research project on methane emissions from rice fields—a joint effort of the International Rice Research Institute (Philippines), the Fraunhofer Institute for Atmospheric Environmental Research (Germany), and national agricultural research institutes in Asia. The project investigated various aspects of CH_4 emissions from rice fields such as the impact of crop management that are presented in this issue for each station separately. The results presented here cut across the data sets obtained in three different sites (Setyanto et al., this issue; Wassmann et al., a, this issue; Chareonsilp et al., this issue) to assess the impact of the rice ecosystem on CH_4 emission.

The distinction among irrigated, rainfed, and deepwater rice is a common feature of the available statistics of rice area (IRRI, 1997). A specific assessment of these ecosystems will therefore directly improve the accuracy of regional and global estimates of the CH_4 source strength—as opposed to uniform emission factors for all ecosystems. The IPCC guidelines for compiling national inventories of greenhouse gas emissions (IPCC, 1997) distinguish between rice fields that are (1) permanently flooded and (2) those with unstable flooding regime. Rainfed rice fields fall under the latter category, while deepwater rice is characterized by long flooding periods. For irrigated rice, a general description of the water regime is more difficult because local variations of the water management can lead to very different flooding patterns. The basic perception of irrigated rice used in this study follows the description in the rice statistics "as shallow flooded with anaerobic soil during crop growth" (IRRI, 1993b). Irrigation water is assured throughout the year but is typically only supplied when needed, i.e., during the dry season.

Materials and methods

The automatic measuring systems and the measurement protocols were identical in the three stations as described in detail in Wassmann et al.,b (this issue). In all field experiments presented in this study, rice was fertilized with mineral compounds only.

The fields in Jakenan (Indonesia) were fertilized with urea as N source (NPK=120-26-45) and were planted with IR64, a variety commonly used in rainfed rice. Irrigated plots were flooded permanently; water regimes in rainfed plots directly depended on actual precipitation (Setyanto et al., this issue). In Los Baños (Philippines), urea (NPK=120-30-30) was applied to grow IR72. Experiments in 1994 and 1996 compared different water regimes, i.e., permanent flooding representing irrigated rice vs two drainage periods (at mid-tillering and before harvest) emulating rainfed conditions (Wassmann et al., a, this issue). In the other seasons, only rainfed water regimes were investigated.

The experiment in Prachinburi (Thailand) followed local fertilizer practice for deepwater rice: burning of 12.5 t of rice straw ha^{-1} and additional urea application of 54 kg ha^{-1} (Chareonsilp et al., this issue). Fertilizer rates in deepwater rice are generally lower than in high-yielding rice systems because yields do not respond to higher doses. Rice fields were planted with local deepwater varieties (HTA60 in 1994 and 1995, PNG in 1996 and 1997). The experiment in Prachinburi also encompassed irrigated rice, but these plots were occasionally affected by technical problems in maintaining shallow water levels at the peak of the deepwater season. Different season lengths of deepwater (220 d) and irrigated rice (110 d) required a staggered cropping calendar in the dry season (only irrigated rice) and wet season (deepwater and irrigated rice) (Chareonsilp et al., this issue).

The three stations of this study have similar temperature regimes as described for Los Baños by Wassmann et al. (1994). Soils in Jakenan (pH 4.2, organic carbon 0.33%), Los Baños (pH 6.3, organic carbon 1.46%), and Prachinburi (pH 3.9, organic carbon 1.22%) showed wide ranges of acidity and organic carbon content.

Results

Emission data obtained in this experiments were compiled in Tables 1 and 2 while more detailed information on biomass, yield, and other variables can be obtained from Setyanto et al. (this issue) for Jakenan, Wassmann et al., a (this issue) for Los Baños, and Chareonsilp et al. (this issue) for Prachinburi. The high standard deviations of these experiments (Tables 1 and 2) indicated strong day-to-day fluctuations in emission

Table 1. Average CH_4 emission rates (± standard deviation) from rainfed and irrigated rice over different wet (WS) and dry seasons (DS) in Jakenan and Los Baños

Season	Jakenan ecosystem	CH_4 emission (mg CH_4 m^{-2} d^{-1})	Season	Los Baños ecosystem	CH_4 emission (mg CH_4 m^{-2} d^{-1})
1993-94 WS	Rainfed	19 (±33)	1994 DS	Rainfed	45 (±22)
	Irrigated	166 (±64)		Irrigated	227 (±126)
1994 DS	Rainfed	90 (±70)	1994 WS	Rainfed	11 (±5)
	Irrigated	134 (±102)		Irrigated	27 (±16)
1994-95 WS	Rainfed	63 (±69)	1995 DS	Rainfed	8 (±5)
	Irrigated	124 (±70)			
1995-96 WS	Rainfed	52 (±57)	1995 WS	Rainfed	8 (±7)
	Irrigated	81 (±60)			
1996 DS	Rainfed	59 (±69)	1996 DS	Rainfed	8 (±6)
	Irrigated	184 (±83)		Irrigated	10 (±9)
1996-97 WS	Rainfed	32 (±47)	1996 WS	Rainfed	34 (±11)
	Irrigated	171 (±105)		Irrigated	40 (±20)
1997 DS	Rainfed	106 (±71)	1997 DS	Rainfed	27 (±23)
	Irrigated	217 (±96)			
1997-98 WS	Irrigated	132 (±59)	1997 WS	Rainfed	14 (±8)
1998 DS	Irrigated	100 (±53)			

Table 2. Average CH_4 emission rates (± standard deviation) from deepwater and irrigated rice over different wet (WS) and dry seasons (DS) in Prachinburi

Season	Ecosystem	CH_4 emission (mg CH_4 m^{-2} d^{-1})	Season	Ecosystem	CH_4 emission (mg CH_4 m^{-2} d^{-1})
1994 WS	Deepwater	84 (±35)	1997 DS	Irrigated	43 (±57)
	Irrigated	17 (±5)			
1995 WS	Deepwater	33[a] (±28)	1997 WS	Deepwater	32 (±33)
	Irrigated	135[b] (±168)			
1996 DS	Irrigated	33 (±25)	1998 DS	Irrigated	17 (±12)
1996 WS	Deepwater	35 (±45)	1998 WS	Irrigated	144 (±154)
	Irrigated	198 (±161)			

[a] Flood damage in the middle of the season, data corresponds to 2-mo period before flood damage.
[b] Delayed season due to flood damage (rice plants were planted again after flood damage).

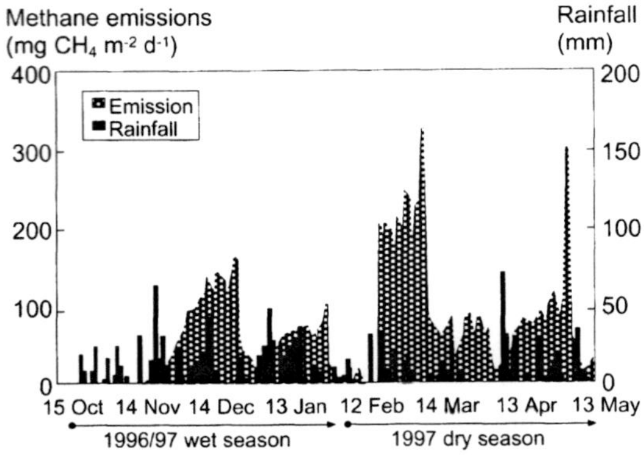

Figure 1. Methane emission rates (area) and rainfall (bars) during two consecutive seasons (1996-97) in Jakenan

rates. The conformity among replicates, i.e., chambers on different plots of an identical treatment, was ensured through rigid quality assurance protocols (Wassmann et al.,b, this issue).

Jakenan (Indonesia): rainfed vs irrigated rice

The experiment in Jakenan encompassed rainfed and irrigated rice grown in different plots within the same field and with identical cropping calendar (Setyanto et al., this issue). The cropping system in Jakenan encompassed two harvests per year as illustrated for the 1996/97 annual cycle in Figure 1. At the onset of the wet season in October, the soil was still dry and CH_4 emission rates were very low. Strong rainfall triggered high emissions in the rainfed plots while relatively dry periods resulted in lower emission rates (Figure 1). The dry season crop started in February with wet soils entailing relatively high emissions in the early phase of the rainfed crop (Figure 1). Dry season crops in Jakenan are generally subjected to drought during the maturity stage of the plants and are thus characterized by low emissions during the late stage of the rainfed crop. However, erratic rainfall as in May 1997 yielded higher emission rates in rainfed rice than was typically observed during this period (Figure 1).

Permanent flooding in the irrigated plots resulted in substantially higher emission rates than rainfed rice (Table 1). Over the course of eight consecutive seasons, CH_4 emission from rainfed rice corresponded to 37% of the emission from irrigated rice (Table 3). In most seasons, the rainfed conditions did not affect the growth of the locally used variety IR64 as compared with irrigated plots (Setyanto et al., this issue). Emission/yield indices, i.e., the ratio between cumulated emission and grain yield, were also consistently higher for irrigated rice (Table 3).

Los Baños (Philippines): rainfed vs irrigated rice

Methane flux rates obtained in the 1994 dry and wet seasons clearly demonstrated the lower emission potential of rainfed rice as compared with irrigated rice (Figure 2). Drainage occurring during the first half of the season drastically reduced emission rates; they remained low, even when the fields were reflooded. Emission rates averaged only 20% and 41% of the values for irrigated rice in the respective season. The results of the other seasons confirmed the low level of emis-

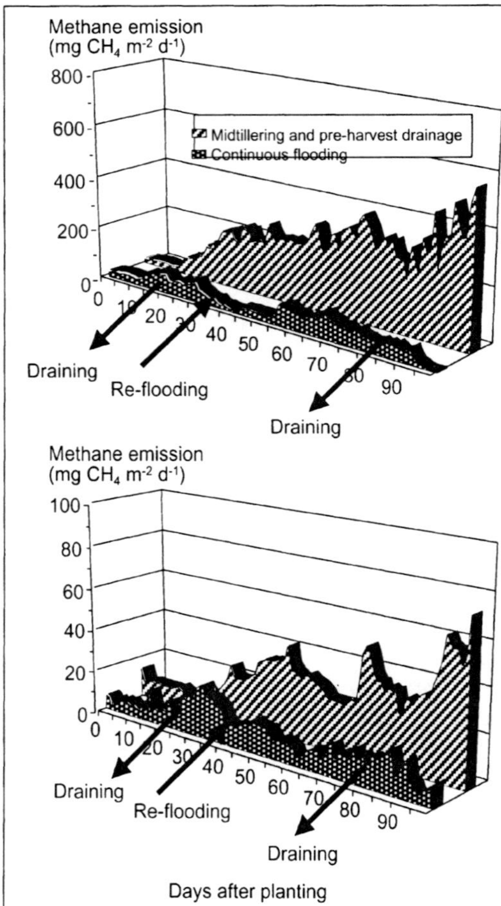

Figure 2. Methane emission rates during 1994 dry and wet season in Los Baños

Table 3. Baseline emissions and baseline emission/yield indices in different rice ecosystems at Jakenan, Los Baños, and Prachinburi

Station	Ecosystem	CH_4 emission (kg CH_4 ha^{-1} $season^{-1}$)	Emission/yield index (kg CH_4 $grain^{-1}$)
Jakenan	Rainfed	58	25
	Irrigated	137	38
Los Baños	Rainfed	40	4
	Irrigated	76	18
Prachinburi	Deepwater	81	53
	Irrigated	99	30

sion rates from this ecosystem in Los Baños. Total emission from rainfed rice during the eight seasons of this experiment was 154 kg CH_4 ha^{-1}, whereas irrigated rice released more than 200 kg CH_4 ha^{-1} in the 1994 dry season alone. In 1996, however, emissions from irrigated rice were also low and the relative impact of the dual drainage was small (Table 1). Due to equally high yields, the computed emission/yield indices for rainfed rice were generally lower than those for irrigated rice (Table 3).

Prachinburi (Thailand): deepwater rice

The seasonal pattern of the water level, pH, and CH_4 emission was displayed in Figure 3 for the 1996 wet season. The rice crop was sown in May when the field was not yet flooded. The experimental field in Prachinburi contained acid sulfate soils; pH levels were below 5 at the onset of the season (Figure 3). Methane emissions were low in the initial stage of the season and gradually increased with progressive flooding. Long-term flooding of the soil neutralized pH values until the water receded. The receding water resulted in the emergence of soil-entrapped CH_4, as can be seen from the single peak in emission rates in Figure 3.

While deepwater rice was only grown in the wet season, the experimental layout of Prachinburi station allowed an irrigated crop in wet and dry seasons (Chareonsilp et al., this issue). Deepwater rice has substantially lower average emissions than irrigated rice (Table 2), whereas the cumulated emissions over one season are similar (Table 3). This finding can be explained by different season lengths, i.e., about 110 d for irrigated and about 220 d for deepwater rice. High emission/yield indices of deepwater rice (Table 3) could primarily be attributed to low yields in this adverse ecosystem (Chareonsilp et al., this issue). Deepwater rice had the highest emission/yield indices of all rice ecosystems.

Discussion

A comprehensive overview on cumulated emission rates at the three stations is given in Figure 4. All rice ecosystems showed strong variations over time. Seasonal emissions varied not only between dry and wet seasons of a given year but also between annual cycles.

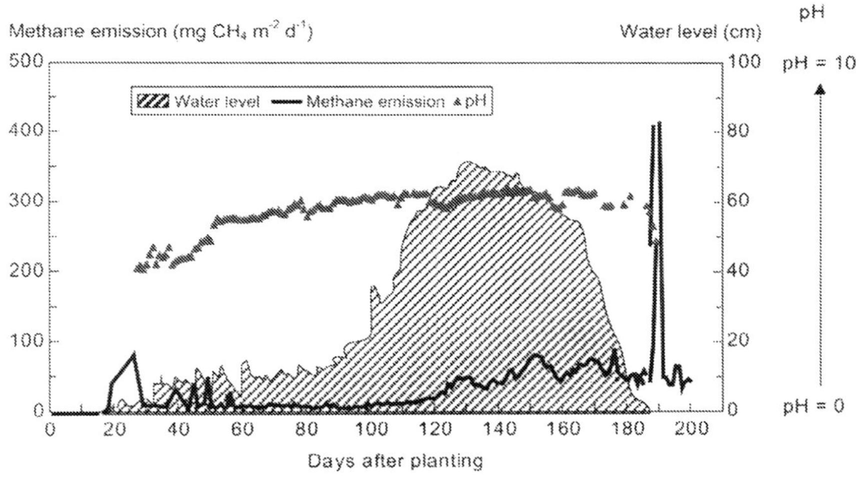

Figure 3. Methane emission rates, pH, and water level development during the 1996 season in deepwater rice (Prachinburi)

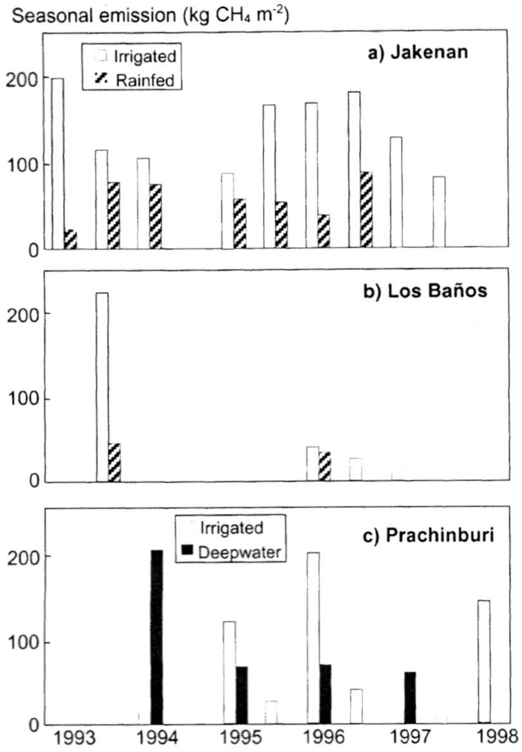

Figure 4. Seasonal emissions of irrigated and rainfed rice in Jakenan and Los Baños as well as irrigated and deepwater rice in Prachinburi

These strong interseasonal and interannual variations underscore the importance of long-term observations for assessing emission potentials of rice ecosystems. Apart from the ecosystem comparison, these results also indicated variations among irrigated sites with identical crop management. Irrigated rice had higher emission rates in Jakenan than in Los Baños (Figure 4a,b), although water regime and fertilizer were similar. Cultivar effects could be excluded because the cultivars IR64 (Jakenan) and IR72 (Los Baños) showed similar emission potentials when grown simultaneously at either site (Setyanto et al., this issue; Wassmann et al., a, this issue). Emissions from rainfed rice were also higher in Jakenan than in Los Baños.

Both stations had a similar temperature regime, so that differences may be related to soil properties. Apparently, CH_4 emission was not impeded by low organic carbon content in the soil in Jakenan as compared with the relatively high organic content of the soil in Los Baños. The soil type found in Los Baños had a high proportion of relatively inert organic material and only a small fraction that was used for methanogenesis (Wassmann et al., 1998). However, the reasons for soil-related differences will have to be clarified after more laboratory data become available.

Figure 4a,b clearly illustrated the high emission potential of irrigated rice. In Jakenan as well as in Los Baños, emission rates from this ecosystem were consistently higher than from rainfed rice. Although the water regimes in both rainfed sites were different, i.e., gradual changes in Jakenan and distinct drainage periods in Los Baños, the relative impact of the lower rainfed conditions were comparable at both stations (Table 3). Low emission potentials appeared to be a common feature of rainfed rice systems; only exceptionally high and evenly distributed precipitation may possibly result in emission potentials reaching those of irrigated systems.

Unstable water regimes affect virtually all physicochemical parameters and biological processes in rice fields (Neue, 1993). Receding floodwater—which may be induced by farmers in an irrigated system or by low precipitation in a rainfed system—triggered a short peak in CH_4 emissions due to emergence of soil-entrapped CH_4 (Wassmann et al., 1995; Denier van der Gon et al., 1996). Over the entire season, however, intermittent irrigation led to a reduction in emission. This is also shown in several field studies by other researchers (Sass et al., 1992; Yagi et al., 1996; Husin et al., 1995; Kimura et al., 1991; Kimura 1995). In the experiments of this interregional network, the redox potentials of the soil increased rapidly after the floodwater had receded (Wassmann et al., a, this issue; Lu et al., this issue; Wang et al., this issue). Oxygen input into the soil impeded CH_4 production and stimulated CH_4 oxidation.

In our experiment in Los Baños, the fields were re-flooded after a 3-wk drainage period. However, the impact of a drainage event was still detectable when the soil was fully reduced again (Wassmann et al., 1995). The decisive factor for this prolonged impact was most likely the large pool of alternative electron acceptors that became oxidized during the drainage event and impeded CH_4 production in the succeeding period. This assumption was derived independently through ecosystem modeling by van Bodegom et al. (this issue) and Matthews et al. (this issue). Drainage at the end of the growing season, however, released the fully developed CH_4 pool in the soil to the atmosphere with only a minor effect on the total amount of CH_4 emitted (Wassmann et al., 1995).

The unique properties of deepwater rice require a more distinguished consideration for a comparison of emission potentials. Deepwater rice can only be grown using specific cultivars so that impacts of water regime and cultivars could not be singled out in a comparison with other rice ecosystems. Differences in crop calendars and season lengths also impeded direct comparisons of seasonal emissions; irrigated rice can be grown in dry and wet seasons whereas deepwater rice is confined to wet seasons (Figure 4c). However, the long-term average obtained for irrigated rice in Prachinburi (Table 2) could be used as an orientation on the magnitude of CH_4 emitted from both ecosystems. Thus, the seasonal emission rates were in a similar range—i.e., 81 kg CH_4 ha^{-1} for deepwater and 99 kg CH_4 ha^{-1} for irrigated rice. In spite of high plant biomass, yields in deepwater rice reached only up to 3.2 t ha^{-1} (Chareonsilp et al., this issue). These low values of grain yield translated into high emission/yield indices (Table 3).

Considering the global distribution of rice ecosystems (Figure 5), irrigated rice has by far the highest CH_4 source strength of all rice ecologies. Based on the results of this study, irrigated rice accounts for 97% of the CH_4 emission from rice fields in East Asia and for 60% of the CH_4 emitted from South and Southeast Asian rice fields, respectively (Figure 5). Rainfed and deepwater rice are negligible for East Asia and they contribute 24% and 16%, respectively, to the CH_4 source strength of South and Southeast Asian rice (Figure 5).

These regional estimates imply considerable uncertainties. In many regions, irrigated rice typically undergoes distinct drainage periods during the cropping season. If these periods are limited to the late stage of the season, the impact on cumulative CH_4 fluxes is minor (Wassmann et al., 1995). On the other hand, emissions were substantially reduced by drainage in the middle of the season, as typically practiced in vast parts of China. Due to the small proportions of other rice ecosystems in East Asia, this practice will only marginally affect the relative dominance of irrigated rice. For South and Southeast Asia, site-specific modalities in the water management of irrigated rice could locally reduce the absolute source strength but will not alter the overall assessments. Irrigated rice contributes about 70-80% of the CH_4 emitted from rice in Asia. Since Asia comprises about 90% of the world's rice area, the contribution in the global scale will be almost identical.

The findings of this study are roughly in line with the emission factors postulated by IPCC (1997), i.e., irrigated = 1, drought-prone rainfed = 0.4, flood- prone rainfed = 0.8, and deepwater = 0.8. However, these results are contrasted by previous findings from India where Parashar et al. (1994) identified rainfed rice as the largest CH_4 source and reported only minor emis-

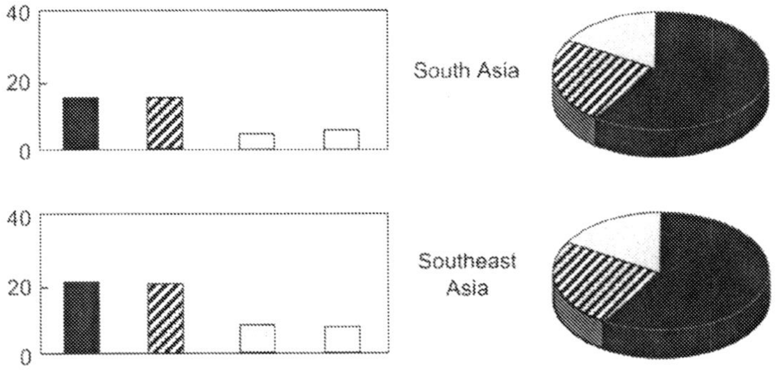

Figure 5. Area and relative emission potential per season of different rice ecosystems in East, South, and Southeast Asia

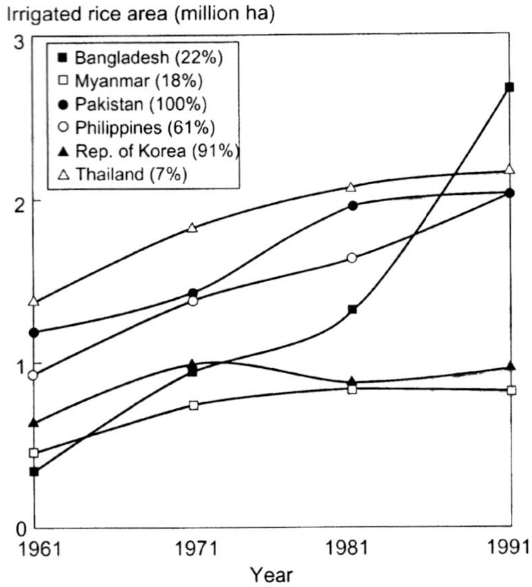

Figure 6. Historical development of irrigated rice area in selected Asian countries; percentages indicate share of irrigated rice relative to total rice area of each country as of 1991

sions from irrigated rice. In large parts of northern India, irrigated rice is grown on sandy-loamy soils with high percolation rates requiring frequent replenishment of the floodwater (Jain et al., this issue). These conditions result in a constant inflow of oxygen into the soil and thus, low emission rates in rice fields typical of this area (Jain et al., this issue). In other parts of India, however, irrigation patterns correspond more to the type described here in this study (Adhya et al., 1994), so that results obtained in the north may not be generalized for the entire country. Furthermore, Parashar et al. (1994) defined irrigated and rainfed rice in a way different from that used in common rice statistics (Neue & Boonjawat, 1998). This may also explain the big gaps between their findings and those of other studies that consistently yielded high emissions in irrigated rice (Sass et al., 1992; Husin et al., 1995; Yagi et al., 1996).

Conclusions

Agricultural production is constantly changing in response to socioeconomic pressure and technological progress. New irrigation facilities were introduced into large areas during the initial stage of the green revolution (Pingali et al., 1998). However, the trend of irrigated rice area since 1961 (IRRI 1995) showed significant differences among Asian countries (Figure 6). For example, irrigated rice area has stagnated in the Republic of Korea but has more than doubled over the last decade in Bangladesh. On the other hand, the quality of irrigation schemes has degraded substantially in recent years (Pingali et al., 1998) that may have translated into reduced emissions from a portion of the irrigated rice land. In future, this degradation process will probably be reversed due to increasing rice demand, so that the dominance of irrigated rice as a source of CH_4 should not be affected.

However, high emissions from irrigated rice should not be seen as an argument against irrigation development. Given the ever increasing food demand, advanced irrigation is one of the key elements for the agricultural sector in developing countries. Irrigated rice is not only the largest source of CH_4 but also the most promising target for mitigating CH_4 emissions from rice. Irrigation patterns could be altered to reconcile high productivity and low emissions as shown for midseason drainage in central China (Lu et al., this issue). Integrated approaches that combine crop models and process models describing carbon dynamics in the soil (Matthews et al., this issue) may yield site-specific "win-win" options for achieving these targets.

Acknowledgements

The research described was funded by the United Nations Development Programme, Global Environment Facility UNDP/GEF (GLO/91/G31)- Interregional Research Programme on Methane Emission from Rice Fields under contract no. C92379 to the International Rice Research Institute.

References

Adhya TK, Rath AK, Gupta PK, Rao VR, Das SN, Parida KM, Parashar DC & Sethunathan N (1994) Methane emission from flooded rice fields under irrigated conditions. Bio Fertil Soils 18:245-248

Byrnes BH, Austin ER & Tays BK (1995) Methane emissions from flooded rice soils and plants under controlled conditions. Soil Biol Biochem 27:331-339

Chareonsilp N, Buddhaboon C, Promnart P, Wassmann R & Lantin RS (2000) Methane emission from deepwater rice fields in Thailand. Nutr Cycling Agroecosyst (this issue)

Denier van der Gon HAC, van Breemen N, Neue HU, Lantin RS, Wassmann R, Aduna JB & Alberto MRC (1996) Release of entrapped methane from wetland rice fields upon drying. Global Biogeochem Cycles 10:1-9

Garrity DP, Oldeman LR & Morris RA (1985) Rainfed lowland rice ecosystems: characterization and distribution. In: Progress in rainfed lowland rice, pp 446, International Rice Research Institute, P.O. Box 933, Manila, Philippines

GEIA – Global Emission Inventory Activity (1993) Report on the 3rd workshop, Amersford, 31 Jan-02 Feb 1993, A.F. Bowman (ed) Bilthoven, The Netherlands, 83 p

Husin YA, Murdiyarso D, Khalil MAK, Rasmussen RA, Shearer MJ, Sabiham S, Sunar A & Adijuwana H (1995) Methane flux from Indonesian wetland rice: the effects of water management and rice variety. Chemosphere 31: 3153-3180

IPCC – Intergovernmental Panel on Climate Change (1996) Climate Change 1995. The Science of Climate Change. Cambridge (UK): Cambridge University Press, 572 p

IPCC – Intergovernmental Panel on Climate Change (1997) Guidelines for National Greenhouse Gas Inventories. Cambridge (UK): Cambridge University Press

IRRI – International Rice Research Institute (1993a) Rice Research in a Time of Change. Manila, Philippines, 79 p

IRRI – International Rice Research Institute (1993b) IRRI Rice Almanac, Manila, Philippines, 142 p

IRRI – International Rice Research Institute (1995) World Rice Statistics 1993/1994. Manila, Philippines, 260 p

Kimura M, Miura Y, Watanabe A, Katoh K & Haraguchi H (1991) Methane emission from paddy field. Part I. Effect of fertilization, growth stage and midsummer drainage: pot experiment. Environ Sci 4:265-271

Kimura M (1994) Effect of intermittent irrigation on methane emission from an Indonesian paddy field. Soil Sci Plant Nutr 40:609-615

Lu WF, Chen W, Duan BW, Guo WM, Lu Y, Lantin RS, Wassmann R & Neue HU (2000) Methane emission and mitigation options in irrigated rice fields in Southeast China. Nutr cycling Agroecosyst (this issue)

Matthews RB, Wassmann R, Buendia L & Knox J (2000) Using a crop/soil simulation model and GIS techniques to assess methane emissions from rice fields in Asia. II. Model validation and sensitivity analysis. Nutr Cycling Agroecosyst (this issue)

Neue HU (1993) Methane emission from ricefields. Bioscience 43:466-474

Neue HU & Roger PA (1994) Potential of methane emission in major rice ecologies. In: Zepp RG (ed) Climate Biosphere Interaction, New York: Wiley and Sons, p 65-93

Neue HU & Boonjawat J (1998) Methane emission from ricefields. In: Galloway J & Melillo J (eds) Asian Change in the Context of Global Change. IGBP Book Series. Cambridge University Press (in press)

Parashar DC, Mitra AP, Sinha SK, Gupta PK, Rai J, Sharma RC, Singh N, Kaul S, Lal G, Chaudhary A, Ray HS, Das SN, Parida KM, Rao SB, Kanung SP, Ramasami T, Nair BU, Swamy M, Gupta SK, Singh AR, Saikia BK, Barua AKS, Pathak MG, Iyer CPS, Gopalakrishnan M, Sane PV, Singh SN, Banerjee R, Sethunathan N, Adhya TK, Rao VR, Palit P, Saha AK, Purkait NN, Chaturvedi GS, Sen SP, Sen M, Sarkar B, Banik A, Subbaraya BH, Lal S & Venkatramani S (1994) Methane budget from Indian paddy fields. CH_4 and N_2O global emissions and controls from rice fields and other agricultural and industrial sources In: Minami K, Mosier A & Sass RL (eds) CH_4 and N_2O: Global Emissions and Controls from Rice Fields and Other Agricultural and Industrial Sources, pp. 27-39, Tsukuba, Japan: NIAES Series 2

Pingali PL, Hossain M & Gerpacio RV (1998) Asian rice bowls – the returning crisis. Wallingford (UK): CAB International, 341 p

Rennenberg H, Wassmann R, Papen H & Seiler W (1992) Trace gas emission in rice cultivation. Ecol Bull 42:164-173

Sass RL, Fisher FM, Harcombe PA & Turner FT (1990) Methane production and emission in a Texas rice field. Global Biogeochem Cycles 4:47-68

Sass RL, Fisher FM, Wang YB, Turner FT & Jund MF (1992) Methane emission from rice fields: the effect of floodwater management. Global Biogeochem Cycles 6:249-262

Setyanto P, Makarim AK, Fagi AM, Wassmann R & Buendia LV (2000) Crop management affecting methane emissions from irrigated and rainfed rice in Central Java (Indonesia). Nutr Cycling Agroecosyst (this issue)

van Bodegom PM, Leffelaar PA, Stams AJM & Wassmann R (2000) Modeling methane emissions from rice paddies: Variability, uncertainty and sensitivity analysis of processes involved. Nutr Cycling Agroecosyst (this issue)

Wassmann R, Neue HU, Lantin RL, Aduna JB, Alberto MC, Andales MJ, Tan MJ, Denier van der Gon HAC, Hoffmann H, Papen H, Rennenberg H & Seiler W (1994) Temporal patterns of methane emissions from wetland ricefields treated by different modes of N application. J Geophys Res 99:16457-16462

Wassmann R, Neue HU, Lantin RS, Javellana MJ, Diego R, Lignes VE, Hoffmann H, Papen H & Rennenberg H (1995) Methane emissions from rainfed rice. In: Fragile Lives in Fragile Ecosystems. pp 217-225, Los Baños, Philippines: International Rice Research Institute

Wassmann R, Neue HU, Bueno C, Lantin RS, Alberto MCR, Buendia LV, Bronson K, Papen H & Rennenberg H (1998) Methane production capacities of different rice soils derived from inherent and exogenous substrates. Plant Soil (in press)

Wassmann R, Buendia LV, Lantin RS, Bueno CS, Lubigan LA, Umali A, Nocon NN, Javellana AM & Neue HU (2000a) Mechanisms of crop management impact on methane emissions from rice fields in Los Baños, Philippines. Nutr Cycling Agroecosyst (this issue)

Wassmann R, Neue HU & Lantin RS, Buendia LV & Rennenberg H (2000b) Characterization of methane emissions from rice fields in Asia. 1. Comparison among field sites in five countries. Nutr Cycling Agroecosyst (this issue)

Yagi K, Chairoj P, Tsurata H, Cholitkul W & Minami K (1994) Methane emission from rice paddy fields in the central plain of Thailand. Soil Sci Plant Nutr 40:29-37

Yagi K, Tsuruta H, Kanda K & Minami K (1996) Effect of water management on methane emission from a Japanese rice paddy field: automated methane monitoring. Global Biogeochem Cycles 10:255-267

Nutrient Cycling in Agroecosystems **58**: 23–36, 2000.

Characterization of methane emissions from rice fields in Asia. III. Mitigation options and future research needs

R. Wassmann[1,2], R. S. Lantin[1], H. U. Neue[1,3], L. V. Buendia[1], T.M. Corton[4] & Y. Lu[1,5]

[1] *International Rice Research Institute, P. O. Box 3127, Makati City 1271, Philippines;* [2] *Fraunhofer Institute for Atmospheric Environmental Research (IFU), Garmisch-Partenkirchen, Germany,* [3] *UfZ-Centre for Environmental Research Leipzig-Halle, Department of Soil Sciences, Halle, Germany;* [4] *Philippine Rice Research Institute, Muñoz, Nueva Ecija, Philippines;* [5] *China National Rice Research Institute, Hangzhou, Zhejiang 310006, China*

Key words: irrigated, rainfed, deepwater, irrigation patterns, compost, biogas residues, direct wet seeding, ammonium sulfate, straw management

Abstract

Methane (CH_4) emissions from rice fields were determined using automated measurement systems in China, India, Indonesia, Thailand, and the Philippines. Mitigation options were assessed separately for different baseline practices of irrigated rice, rainfed, and deepwater rice. Irrigated rice is the largest source of CH_4 and also offers the most options to modify crop management for reducing these emissions. Optimizing irrigation patterns by additional drainage periods in the field or an early timing of midseason drainage accounted for 7-80% of CH_4 emissions of the respective baseline practice. In baseline practices with high organic amendments, use of compost (58-63%), biogas residues (10-16%), and direct wet seeding (16-22%) should be considered mitigation options. In baseline practices using prilled urea as sole N source, use of ammonium sulfate could reduce CH_4 emission by 10-67%. In all rice ecosystems, CH_4 emissions can be reduced by fallow incorporation (11%) and mulching (11%) of rice straw as well as addition of phosphogypsum (9-73%). However, in rainfed and deepwater rice, mitigation options are very limited in both number and potential gains. The assessment of these crop management options includes their total factor productivity and possible adverse effects. Due to higher nitrous oxide (N_2O) emissions, changes in water regime are only recommended for rice systems with high baseline emissions of CH_4. Key objectives of future research are identifying and characterizing high-emitting rice systems, developing site-specific technology packages, ascertaining synergies with productivity, and accounting for N_2O emissions.

Introduction

There is an increasing pressure on the rice-growing resources, especially in Asia where more than 90% of rice is grown and consumed (Blake, 1992; Becker, 1993). Rice cultivation over thousands of years has sustained Asian population, which is currently growing at 1.8% a year. Wetland rice fields, however, release the greenhouse gas methane (CH_4) and thus, affect the radiative budget of the earth (Minami & Neue, 1994; Neue & Sass, 1999). Given the expected doubling in rice production in Asia, conducting research that will help developing countries grow more rice on limited land, in ways that do not harm the environment and that benefit both farmers and consumers, will be critical. There is a need to evaluate the interaction between climate change and rice production to provide a sound basis for future decisions and technology developments by policymakers, agriculturists, environmentalists, rice producers, and rice consumers alike.

Global climate change has been recognized as a major threat for future development in the United Nations Framework Convention on Climate Change (UNFCC) in 1992. The ultimate goal of the convention is the stabilization of greenhouse gas concentration in the atmosphere at a level that would prevent anthropogenic interference with the climate system. Before target stabilization can be specified, a national greenhouse gas inventory is necessary for each country to formulate mitigation policies for international agreements. On the other hand, large portions of Asia's rice fields are located on lowlands that were natural wetlands before agricultural use. These areas have already been a source of CH_4 in their pristine state. The introduction of rice substantially enhanced the turnover of organic material and, in the next step, CH_4 emissions.

The Interregional Research Program on Methane Emissions from Rice Fields has established a network of measurement stations in China, India, Indonesia, Thailand, and the Philippines. This work was initiated by the International Rice Research Institute, Philippines, in collaboration with national agricultural research institutes and the Fraunhofer Institute for Atmospheric Environmental Research, Germany, to quantify emissions from major rice-growing systems and to identify possible strategies for mitigation. Generic strategies on mitigating CH_4 emissions have been formulated recently (Lindau et al., 1993; Wassmann et al., 1993; Neue et al., 1995; Ranganathan et al., 1995; Shin et al., 1996; Yagi et al., 1997, Minami, 1997), but information on their feasibility and efficiency in different rice-growing environments is still lacking. The immense variability in environmental and management factors in the 144 million ha of annually harvested rice fields (IRRI, 1993) demands site-specific assessments on mitigating emissions.

Flux measurements at the eight sites of this network showed a large variability in CH_4 emissions addressed in this series of articles. The first volume of this series (Wassmann et al., this issue, b) dealt with site-to-site variation under a given crop management. The second volume focused on differences among irrigated, rainfed, and deepwater rice (Wassmann et al., this issue, c). In this third volume, we evaluate crop management impacts with emphasis on possible mitigation options. Moreover, we assessed the findings of this network in the context of future research needs.

Materials and methods

Methane measurements were conducted with a standard automated closed chamber system designed and manufactured by the Fraunhofer Institute for Atmospheric Environmental Research (Wassmann et al.,this issue, b). The design of the system is suitable for continuous CH_4 flux measurements in the different rice ecosystems over long time periods. Each station in the network was supplied with 12 closed chambers, so that four treatments could be investigated with three replicates of each.

The database used in this study comprises only seven stations:

- Beijing,China (Z.Wang et al., this issue)
- Hangzhou, China (Lu et al., this issue)
- Maligaya, Philippines (Corton et al., this issue)
- Cuttack, India (Adhya et al., this issue)
- Jakenan, Indonesia (Setyanto et al., this issue)
- Los Baños, Philippines (Wassmann et al., this issue, a; Abao et al., this issue)
- Prachinburi, Thailand (Chareonsilp et al., this issue)

These stations were located in principal rice-growing areas in Asia and represented irrigated rice (Beijing, Hangzhou, and Maligaya), rainfed rice (Jakenan, Cuttack, and Los Baños) and deepwater rice (Prachinburi). Another station of this network (New Delhi) did not yield clues for mitigating emissions because baseline emissions were already very low (Jain et al., this issue; see Figure 1 for comparison). The rainfed and deepwater stations included irrigated rice for comparative purposes. Measurements were started in 1993 and ended in 1998, covering up to 10 consecutive seasons per station.

Results and discussion

Rice is grown in very diverse environments, applying a wide range of crop management practices as can be seen in GIS data in Knox et al. (this issue). Mitigation strategies can only be defined for specific target systems. According to ecosystem and crop management, five different baseline practices were distinguished as follows:

Irrigated	I_1 (continuous flooding, organic amendments)
	I_2 (midseason drainage, organic amendments)
	I_3 (continuous flooding, no organic amendments)
Rainfed:	R
Deepwater:	D

Irrigated rice accounts for 51% of the harvest rice area in Asia. For this ecosystem, water regime and organic amendments are the major determinants for the magnitude of CH_4 emissions (Wassmann et al., this issue, c). Continuous flooding and fresh organic manure resulted in highest emissions, whereas emissions were reduced in some cases by several orders of magnitude—through temporary soil aeration and omission of organic manure. Other management components such as type of inorganic fertilizer, rice cultivar, etc. had a modulating effect on emissions (i.e., they affected emissions within a range that was set by water management and organic amendment) (Wassmann et al., this issue, c). Rainfed (27% of the harvest rice area) and deepwater rice (10% of the harvest rice area) offer very limited management options and are not further split into different baseline practices.

The experiments of the network encompassed simultaneous records of different crop management options. From this database, we have extracted pairs of two management options, i.e., one representing a baseline practice (I_1, I_2, I_3, R, or D) and the other a distinct modification of this management practice termed as "mitigation practice." Figure 1 shows seasonal emissions recorded for the respective baseline practice. All

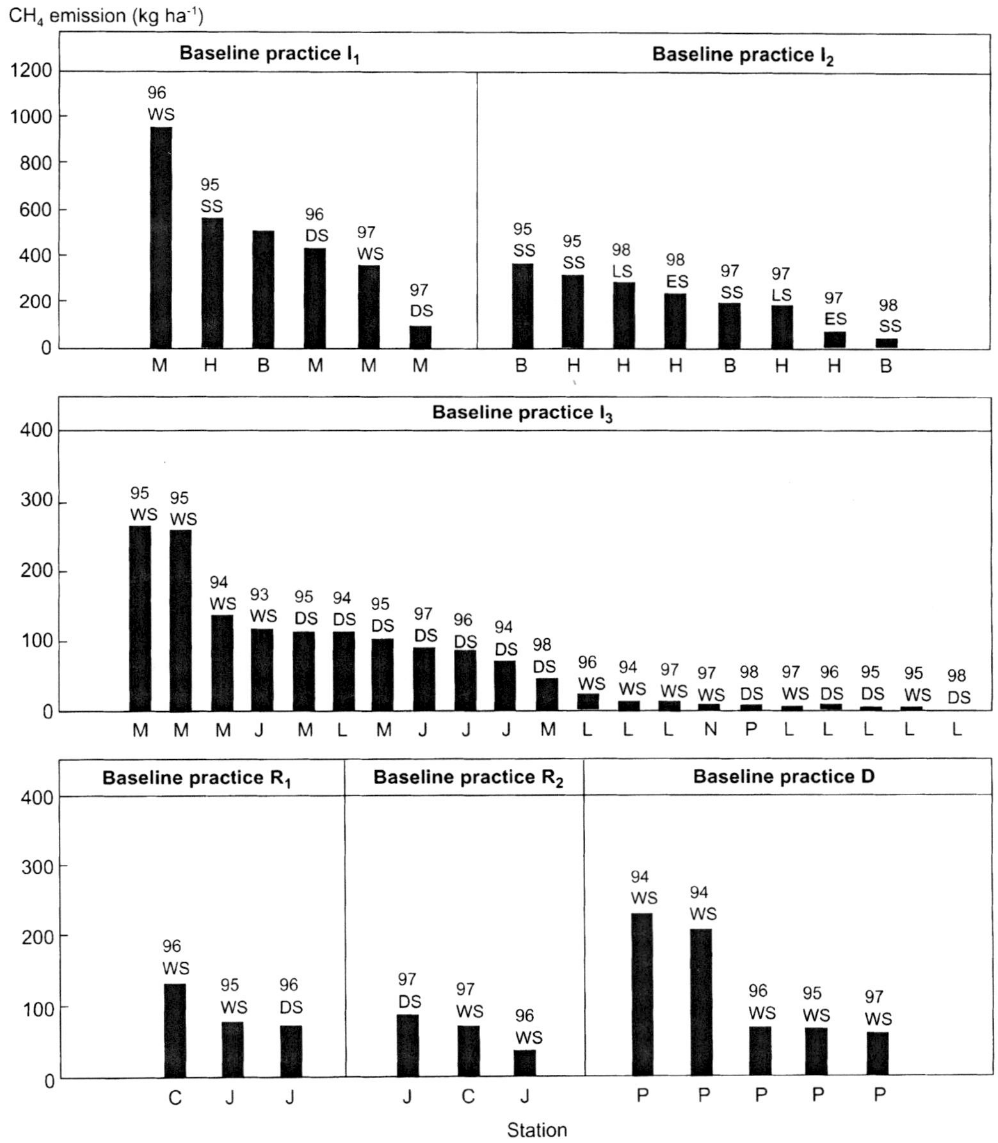

Figure 1. Seasonal CH_4 emissions of experiments with baseline practices (see text for explanation of baseline practice). Year and season of the experiment are given in labels. First letter of Hangzhou, Beijing, Maligaya, Jakenan, Los Baños, New Delhi, Cuttack, and Prachinburi indicate experimental station

Table 1. Mitigation practices for baseline practice I_1 (continuous flooding, organic amendments): seasonal emission refers to mitigation practice; reduction effect refers to emission from mitigation practice as compared with emission from baseline practice; seasons are specified as dry (DS), wet (WS), early (ES), late (LS), and single season (SS)

Mitigation practice	Seasonal emission (kg ha⁻¹)	Reduction effect: Net reduction (kg ha⁻¹)	Reduction effect: Relative reduction[§] (%)	Experiment: Station	Experiment: Season	Yield impact[a] (%)	Remark
Midseason drainage	385	118	23**	Beijing	1995 SS	14*	
	312	245	44 ns	Hangzhou	1995 SS	–1 ns	
	51	38	43**	Maligaya	1997 DS	–2 ns	
	323	25	7 ns	Maligaya	1997 WS	2 ns	
Alternate flooding/drainage	216	341	61**	Hangzhou	1995 SS	0 ns	
	207	296	59**	Beijing	1995 SS	12*	
Rice straw compost	178	242	58**	Maligaya	1996 DS	4 ns	Baseline practice: fresh rice straw
	353	599	63**	Maligaya	1996 WS	2 ns	
Phosphogypsum (3 t ha⁻¹)	308	112	27**	Maligaya	1996 DS	1 ns	
	599	353	37**	Maligaya	1996 WS	1 ns	
Direct wet seeding	272	76	22**	Maligaya	1997 WS	–28*	Baseline practice: transplanting
	75	14	16**	Maligaya	1997 DS	–15*	
Direct wet seeding and midseason drainage	48	41	46**	Maligaya	1997 DS	–19*	Baseline practice: transplanting
	150	198	57**	Maligaya	1997 WS	–36*	
Midseason drainage and no organic matter	26	477	95**	Beijing	1995 SS	–4 ns	
	239	318	57**	Hangzhou	1995 SS	–3 ns	

[a]ns = insignificant, * = significant at $p > 0.1$, ** = highly significant.

baseline practices showed pronounced variations in emission rates across sites and seasons.

The impact of the diverse mitigation practices on CH_4 emissions and rice yields are given in Tables 1 to 4 for each baseline practice separately. Results are specifically given for seasons because emissions showed pronounced interseasonal variations. This can be illustrated by the results from Maligaya where CH_4 emissions were generally high in the wet and low in the dry season (Figure 2). This interseasonal shift appears to be an influence of solar radiation and temperature difference during the early crop growth (Corton et al., this issue). Furthermore, these seasons also showed differences in the impact of mitigation practices. For example, ammonium sulfate was an efficient tool for mitigating emissions in the wet season, whereas its impact was insignificant in the dry season (Figure 2).

Baseline practice I_1 (continuous flooding, organic amendments)

The results listed in Table 1 underscore the potential of water regime to reduce CH_4 emissions from irrigated rice. Midseason drainage consistently lowered emissions as compared with a baseline practice of continuous flooding, but the reducing effect varied with both station and season. The reduction in emissions was statistically significant in seven out of the eight experiments; only the 1997 WS experiment at Maligaya showed an insignificant effect (Table 1). The reduction effect was highest when midseason drainage was supplemented by replacing organic manure with urea (Beijing, 1995 SS).

The practice of alternate flooding and drying of the field reduced emissions by about 60% as compared

Table 2. Mitigation practices for baseline practice I_2 (midseason drainage; organic amendments); see Table 1 for explanations

Mitigation practice	Seasonal emission (kg ha^{-1})	Reduction effect: Net reduction (kg ha^{-1})	Reduction effect: Relative reduction (%)	Experiment: Station	Experiment: Season	Yield impact (%)
Alternate flooding/drainage	207	178	46**	Beijing	1995 SS	-1.6
	217	95	30**	Hangzhou	1995 SS	2.8
	177	48	21**	Hangzhou	1998 ES	0.5
	215	64	23**	Hangzhou	1998 LS	4.2
Early single drainage	15	13	46**	Beijing	1998 SS	-1.7
Early dual drainage	21	7	25 ns	Beijing	1998 SS	0.3
Late dual drainage	26	2	7 ns	Beijing	1998 SS	1.2
Biogas manure	53	6	10**	Hangzhou	1997 ES	3.0
	151	28	16**	Hangzhou	1997 LS	-0.6
Rice straw winter application	200	25	11**	Hangzhou	1998 ES	-0.5
Mulching rice straw	248	31	11**	Hangzhou	1998 LS	0.0

with continuous flooding (Table 1). Alternate flooding/drying as well as midseason drainage had either no significant impact on yields or even increased rice production as compared with continuous flooding. However, temporary soil aeration could enhance emission of nitrous oxide (N_2O) (Bronson et al., 1997a,b; Abao et al., this issue), another potent greenhouse gas. When baseline levels of CH_4 emissions are high (as for I_1), midseason drainage can still be regarded as an efficient mitigation option in spite of concomitant increments in N_2O emissions (see discussion below).

The impact mechanisms of temporary soil aeration on CH_4 emissions have been discussed thoroughly in volume 2 of this series (Wassmann et al., this issue, c) as well as in various station reports (Wassmann et al., this issue, a; Lu et al, this issue; Wang et al., this issue). One rice crop requires about 1,240 mm water (Yoshida, 1981). To meet this water demand, wetland rice has evolved over the centuries as a well-adapted cultivation technique sustaining high yields. As opposed to upland systems, wetland cultivation also provides numerous advantages in terms of soil chemistry and erosion control. Large portions of Asia's rice land are on native wetlands that do not allow cultivating any crop other than rice (at least in the wet season).

The CH_4-reducing effect of midseason drainage could substantially be enhanced in conjunction with direct wet seeding (Table 1). The practice of direct wet seeding alone accounted for a reduction effect of 16-22% in the seasonal emissions as compared with the baseline practice of transplanting. In Maligaya, the difference between direct-seeded and transplanted rice occurs relatively early in the season (Figure 3). Direct-seeded rice develops high root biomass during early stages and reaches maximum root biomass soon after panicle initiation (De Datta & Nantasomsaran, 1991). Roots of transplanted rice develop slower but can penetrate into the deeper layer of the puddled soil as compared with the relatively compact soil under direct wet seeding. However, the precise mechanism involved in reducing CH_4 emissions through direct wet seeding still has to be clarified. Direct wet seeding is getting increasingly popular in major rice-growing regions. Substantial savings in labor requirements make this type of crop establishment economically viable, although yields are lower.

Application of rice straw compost significantly reduced CH_4 emissions as opposed to fresh rice straw (Table 1). The straw in the experiment in Maligaya was processed in an aerobic composter (Corton et al., this issue) so that CH_4 emissions during the composting process can be neglected. Addition of phosphogypsum also reduced CH_4 emissions. Phosphogypsum is a waste byproduct from processing of phosphate rock fertilizer and consists mainly of calcium sulfate dihydrate. The reduction effect of compost can be explained by a depletion of methanogenic substrate, whereas phosphogypsum triggers inhibition of methanogenesis through sulfate-reducing bacteria (Corton et al., this issue).

Baseline practice I_2 (midseason drainage, organic amendments)

In many rice-growing regions, the local practice of irrigation encompasses a drainage period in the early or midseason. While this drainage period itself exerts a

Table 3. Mitigation practices for baseline practice I_3 (continuous flooding, no organic amendments); see Table 1 for explanations

Mitigation practice	Seasonal emission (kg ha^{-1})	Reduction effect[a] Net reduction (kg ha^{-1})	Relative reduction (%)	Experiment Station	Season	Yield impact (%)	Remark
Preharvest drainage	251	(26)	(12*)	Los Baños	1994 DS	-6 ns	
	10	0	0 ns	Los Baños	1996 DS	-13 ns	
	28	12	30**	Los Baños	1996 WS	23 ns	
Dual drainage at midtillering and preharvest	45	180	80**	Los Baños	1994 DS	-4 ns	
	11	16	59**	Los Baños	1994 WS	-11 ns	
	8	2	20**	Los Baños	1996 DS	-9 ns	
	34	6	15**	Los Baños	1996 WS	-1 ns	
Alternate flooding/drying	14	4	22 ns	New Delhi	1997 WS	-9 ns	
Direct wet seeding	25	(10)	(67)	Thailand	1998 DS	20	
	25	(10)	(67)	Thailand	1998 DS	20	
	256	(27)	(12**)	Jakenan	1993 WS	59*	
Ammonium sulfate	230	36	14**	Maligaya	1994 WS	-2 ns	Pure ammonium
	184	20	10**	Maligaya	1995 DS	-2 ns	sulfate in Maligaya;
	327	191	37**	Maligaya	1995 WS	13 ns	Ammonium sulfate
	9	18	67**	Los Baños	1997 DS	-15 ns	Blended with urea
	7	6	46**	Los Banos	1997 WS	16 ns	
Tablet urea	104	66	39**	Jakenan	1996 WS	1 ns	
	163	18	10 ns	Jakenan	1997 DS	8 ns	
Phosphogypsum	145	83	36**	Maligaya	1995 DS	-2 ns	
	225	41	15**	Maligaya	1994 WS	-6 ns	
	241	25	9**	Maligaya	1994 WS	1 ns	
	143	388	73**	Maligaya	1995 WS	13 ns	
Rice stubbles and roots removed	14	26	65**	Los Baños	1996 WS	19 ns	
High-yielding variety (IR64)	115	26	18	Jakenan	1994 DS	10 ns	Baseline practice: IR72
New plant type	7	1	13**	Los Baños	1995 DS	-28 ns	Baseline practice: IR72
	6	2	25**	Los Baños	1995 WS	-50*	
Traditional variety	5	3	38*	Los Baños	1995 DS	-25 ns	Baseline practice: IR72
Hybrid rice	6	2	25*	Los Baños	1995 WS	0 ns	Baseline practice: IR72
	4	4	50 ns	Los Baños	1995 WS	69*	
	1	0	(12 ns)	Los Baños	1998 DS	-2 ns	
	1	0	(3 ns)	Los Baños	1998 DS	15*	
Direct wet seeding and midseason drainage	15.96	74	82	Maligaya	1998 DS	-4 ns	
Direct wet seeding and alternate flooding/drying	6.84	83	92	Maligaya	1998 DS	-11*	

[a]Figures in brackets indicate a net increase in emission

Table 4. Mitigation practices for baseline practice R (rainfed rice) and baseline practice D (deepwater rice); see Table 1 for explanations

Mitigation practice	Seasonal emission (kg ha^{-1})	Reduction effect: Net reduction (kg ha^{-1})	Reduction effect: Relative reduction (%)	Experiment: Station	Experiment: Season	Yield impact (%)	Remark
Baseline practice R							
Farmyard manure	56	22	28**	Jakenan	1995 WS		Baseline practice
	92	(19)	(26**)	Jakenan	1996 DS		Rice straw
Compost	65	67	51**	Cuttack	1996 WS	-15*	Baseline practice: Sesbania
Azolla	68	64	48**	Cuttack	1996 WS	4 ns	Baseline practice: Sesbania
Nitrate inhibitor (Nimin)	77	(7)	(10)	Cuttack	1997 WS	24*	
Nitrate inhibitor (DCN)	61	9	13	Cuttack	1997 WS	31*	
Tablet urea	45	(8)	(22**)	Jakenan	1996 WS	-2 ns	Baseline practice: prilled urea
	102	(14)	(16**)	Jakenan	1997 DS	13 ns	
Baseline practice D							
No mineral fertilizer	213	19	10**	Prachinburi	1994 WS	11 ns	Baseline practice: mineral fertilizer with burned ash
Mineral fertilizer	201	7	4 ns	Prachinburi	1994 WS	-3 ns	
	48	19	28**	Prachinburi	1995 WS	-7 ns	
Compost straw	145	(85)	(142**)	Prachinburi	1997 WS	-8	
No mineral fertilizer and RS	53	14	21 ns	Prachinburi	1995 WS	-42 ns	
Mineral fertilizer and mulching RS	127	(58)	(84**)	Prachinburi	1996 WS	-3 ns	
Mineral fertilizer and RS	619	(550)	(797**)	Prachinburi	1996 WS	24 ns	
No tillage with mulching RS	100	(40)	(67**)	Prachinburi	1997 WS	-9	

mitigation effect, CH_4 emissions may further be reduced by a number of modifications (Table 2). Alternate flooding and drying in weekly intervals decrease emissions by 21-46%. Shifting the drainage period to an early stage or adding a second drainage period can also mitigate emissions by 7-46%.

Emissions from baseline practice I_2 can also be reduced through the management of organic amendments. As compared with pig manure application, the use of cattle manure and biogas residues (both are prefermented materials) reduced emissions by 77 and 10%, respectively. Rice straw can be applied during the winter fallow or can be mulched on the soil surface; both practices reduced CH_4 emissions by 11% as compared with incorporation of fresh straw into the soil during harrowing.

Baseline practice I_3 (continuous flooding, no organic amendments)

All modifications of floodwater regime that were considered for baseline practice I_1 may also be applied for reducing emissions in this baseline practice. A late drainage period alone did not reduce emissions efficiently, whereas a preceding drainage in the early stage had a pronounced reduction effect of 15-80% (Table 3). Direct wet seeding, however, yielded a net increase for baseline practice I_3 (without organic amendment) while it had a reductive effect for baseline practice I_1 (with organic amendment). Direct-seeded rice has a higher total root biomass (De Datta & Nantasomsaran, 1991) and thus may introduce more organic material into the soil. This additional substrate for methanogenic bacteria is more significant in a soil environment with low organic inputs as opposed to a soil with high organic inputs. While this explanation may provide an initial clue for the observed discrepancies, a fully satisfactory hypothesis on the impact of direct wet seeding on CH_4 emissions is still not possible. The impact of direct wet seeding in combination with midseason drainage and alternate flooding/drying, respectively, may primarily be attributed to changes in water regime and, only to a lesser extent, to effects on plant growth.

Emissions from baseline practice I_3 could be reduced by 10-67% through application of ammonium sulfate as opposed to urea (Table 3). Competition be-

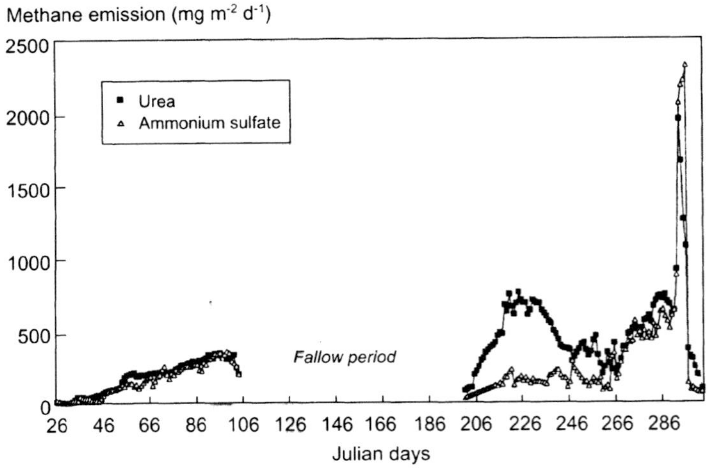

Figure 2. Methane emissions in 1995 dry and wet seasons (Maligaya) using different mineral fertilizers

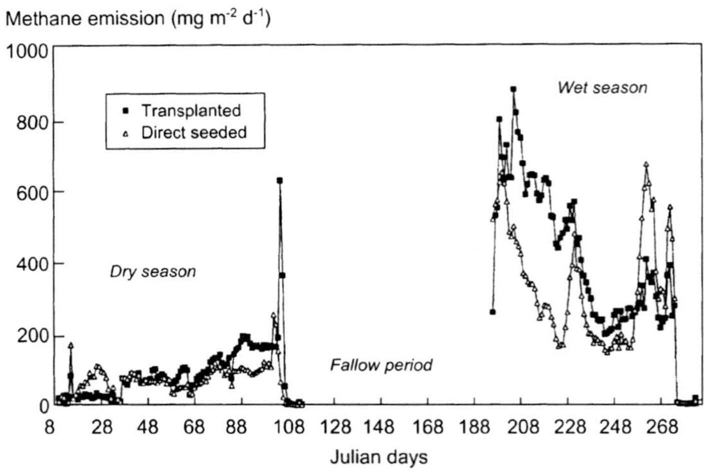

Figure 3. Methane emissions in 1997 dry and wet seasons (Maligaya) under different crop establishment methods

tween sulfate-reducing and methanogenic bacteria could be singled out to explain this effect (Corton et al., this issue). Likewise, the reductive effect of phosphogypsum could be explained by the sulfate content of this material (Corton et al., this issue). Application of tablet urea (a method for minimizing N losses) reduced CH_4 emissions by 10-39%. The experiment with stubbles and roots removed underscores the significance of residue management for CH_4 emissions, but this cannot be translated into a feasible mitigation option in farmers' field. Cultivar selection may become a mitigation option in the future, but the available database is still inconclusive to draw definite recommendations on suitable varieties.

Baseline practices R (rainfed rice) and D (deepwater rice)

Table 4 shows a limited number of CH_4 mitigation options for rainfed and deepwater rice. While farmyard manure and rice straw have a similar emission potential, compost and azolla are a preferable manure type as opposed to sesbania. Based on the available database, nitrification inhibitors and tablet urea did not represent efficient tools for mitigating CH_4 emissions in rainfed rice.

Deepwater rice also offers very limited management options to reduce CH_4 emissions. Plant biomass is substantially higher than in irrigated and rainfed rice, so that straw management plays a crucial role in controlling emissions. The common practice of straw burning reduced CH_4 emissions during the ensuing growing season but contaminated the air with aerosols and gases including CH_4 during the burning process. Mulching of rice straw entails more emissions than ash application but less emissions than direct incorporation of fresh straw. This technique may still be recommended as a preferable straw management practice in deepwater rice.

Interaction with productivity of rice systems

Productivity of rice systems can be expressed as total factor productivity (grain output divided by all inputs taken together) (Dawe & Dobermann, 1999). While most of the suggested mitigation options did not affect yields significantly (Table 1-4), their profitability is primarily determined by fertilizer and labor inputs. Economically sound doses of fertilizers may also be beneficial for the greenhouse gas budget because (i) excessive supply of N entails N_2O emissions (Smith et al., 1997; Freney, 1997) and (ii) deficiencies of nutrients such as phosphorus increases root exudation and subsequently CH_4 emissions (Lu et al., 1999).

Replacing urea with ammonium sulfate, however, may be limited by economic disadvantages. Ammonium sulfate is more expensive than urea based on N content—i.e., the costs are about twice as high in the Philippines (FADINAP, 1999). On the other hand, ammonium sulfate has gained some acceptance as fertilizer (its consumption in Indonesia corresponds to 13% of urea consumption) (FADINAP, 1999) due to easy handling, storage, and application qualities. In rice production, ammonium sulfate is used mostly in the seedbed. Phosphogypsum is a byproduct of phosphate fertilizer manufacture; distinct opportunities for this soil additive for reducing CH_4 emission can be seen in sulfur-deficient soils.

The mitigation options addressing straw management and crop establishment entail changes in labor inputs. Farmers generally prefer removal of straw from the rice land because it can exacerbate soil tillage if present in large quantities. Composting of rice straw, on the other hand, represents additional work and limits farmer acceptance. However, the use of organic amendments particularly rice straw is being promoted by most national extension services. Composting the rice straw may offer a number of benefits for soil fertility and tillage as opposed to fresh rice straw incorporation.

Biogas production could represent a low-cost source of energy for farmers, especially those with animals. In combination with rice production, biogas technology can achieve a twofold reduction of greenhouse gases (Wassmann et al., 1993): (1) prevention of CO_2 emissions by using renewable energy source and (2) reduction of CH_4 emissions from rice fields by replacing fresh manure with prefermented material. Biogas generation was successfully promoted in particular regions of China (e.g., Sichuan Province comprises about 7 million generators). Technical problems in operating small-scale biogas generators, however, have impeded their functionality and effectively stalled a further dissemination of these devices in rural areas (Wassmann et al., 1993). The beneficial greenhouse budget may become decisive arguments in favor of biogas generation in the future.

Direct wet seeding is an economically viable technique as opposed to the labor-intensive transplanting of rice plants although yields are lower. However, direct wet seeding is only recommended in systems with high organic inputs, which may also increase labor cost as compared with mineral fertilizer application.

The supply of water may incur costs for some rice farmers, e.g., through pumping. Moreover, water will become a scarce commodity in the future. Water-saving techniques can offer distinct trade-offs for mitigating CH_4 emissions as shown in this study for midseason drainage and alternate flooding/drying. While intermittent irrigation can substantially increase water use efficiency (Didiek, 1998), good timing of drainage and irrigation is essential to prevent soil compaction and subsequent water losses in reflooding the field (Tuong et al., 1996).

Midseason drainage is commonly practiced in Chinese rice fields as part of a high-yielding crop management, whereas Southeast Asian rice fields are generally not drained during the growing season. In the wet season, high precipitation constrains the effectiveness of field drainage. In the dry season, farmers are reluctant to remove water from the fields because of uncertain water supply for the remaining growing period. However, improved irrigation schemes could help in developing irrigation patterns that improve productivity and reduce CH_4 emissions.

Overall assessment of mitigation practices

The preferable mitigation options are listed in Table 5 for irrigated rice. While irrigated rice can be altered in virtually all aspects of crop management, mitigation options in rainfed and deepwater rice are very limited. However, Table 5 does not include (a) the use of chemical fertilizers, (b) straw burning, and (c) selection of rice cultivars as mitigation option for these reasons:

1. The possible mitigation effect of chemical fertilizers may be offset by CO_2 emissions through industrial N_2 fixation. One mole of ammonia fixed through the Haber-Bosch process produces 1.436 moles of CO_2 (Schlesinger, 1999). Application of 120 kg N ha^{-1}—as in the chemical fertilizer treatments of our experiments—translates into off-site emissions of 541.5 kg of CO_2. Using a conversion factor of 21 for the global warming potential of CH_4 in comparison with CO_2, (IPCC 1995), this off-site emission of CO_2 corresponds to the radiative forcing of 25.8 kg CH_4 ha^{-1}.
2. Emissions of CH_4 resulting from rice straw burning are in the range of 0.43-0.90 % of the carbon content, which is similar to the range through straw application into the soil (Miura & Kanna, 1997). Moreover, straw burning emits significant quantities of other greenhouse gases such as CO and N_2O (Miura & Kanna, 1997) and adversely affects local air quality.
3. The database on rice cultivars affecting CH_4 emissions is still inconsistent. The two important traits that determine the CH_4 emission potential of rice cultivars are (a) root exudation and (b) gas transfer through the aerenchyma (Butterbach-Bahl et al., 1997). However, the CH_4 emission potential of a given cultivar exhibits enormous variation when grown under different greenhouse and field conditions (Wassmann & Aulakh, 1999). These variations

Table 5. Mitigation matrix for different baseline practices of irrigated rice; reduction effect for each mitigation practice is given in parentheses; arrows indicate that mitigation practice can be adopted to other baseline practices although experimental results are not available

Modified crop management	Baseline practice I_1 (continuous flooding/ organic amendment)		Baseline practice I_2 (midseason drainage, organic amendment)		Baseline practice I_3 (continuous flooding, no organic amendment)
Water regime	Midseason drainage (7-44%)				Midseason drainage (15-80%)
	Alternate flooding/ drying (59-61%)		Alternate flooding/ drying (21-46%)		Alternate flooding/ drying (22%)
		←	Early/ dual drainage (7-46 %)	→	
Organic amendments	Compost (58-63%)	→			
		←	Biogas residues (10-16 %)		
Mineral amendments	Phosphogypsum (27-37%)	→			Phosphogypsum (9-73%) Ammonium sulfate (10-67%)
					Tablet urea (10-39%)
Straw management		←	Fallow incorporation (11%)	→	
		←	Mulching (11%)	→	
Crop establishment	Direct wet seeding (16-22%)	→			

complicate determination of cultivar-specific emission potentials (Aulakh et al., 1999). Therefore, at this point, it is difficult to recommend preferable rice cultivars for mitigating CH_4 emissions. Nevertheless, selection of cultivars may become an important option in the future when information on the interaction of genotype and environment in determining the respective traits become available.

Modifications of irrigation patterns are only recommended when substantial amounts of organic material is used (Table 5). Modifications of the water regime are likely to affect emissions of other greenhouse gases from rice production, namely N_2O (Bronson et al., 1997a,b; Abao et al., this issue). This greenhouse gas contributing about 6% of the anthropogenic greenhouse effect (IPCC, 1996) is generated through nitrification and denitrification occurring in soils (Rennenberg et al., 1996). When rice fields are continuously flooded during the growing season, N_2O emissions are primarily limited to the fallow period at which fields experienced alternative dryness and wetness from rainfall (Abao et al., this issue). Water regimes that encompass drainage periods stimulate nitrification (through soil drying) and denitrification (through soil wetting). Therefore, all strategies to reduce CH_4 emissions by midseason or frequent drainage may enhance N_2O emissions. Based on a global warming potential of 310 for N_2O as opposed to 21 for CH_4 (IPCC, 1995), the observed net reductions of 118 kg CH_4 ha^{-1} (Beijing) and 245 kg CH_4 ha^{-1} (Hangzhou) with midseason drainage would theoretically be compensated for by concomitant increments in N_2O emissions of 8 and 16.5 kg N_2O ha^{-1}, respectively. However, total N_2O emissions under comparable flooding regimes as in these experiments were 1.6 kg N_2O ha^{-1} in northeast China (Chen et al., 1997) and 2.4-6.2 kg N_2O ha^{-1} in central China (Cai et al., 1997); the average for the entire country under different crop management practices is given at 2.4 kg N_2O ha^{-1} (Xing & Xu, 1997). Thus, modification of water regime appears as a promising option to achieve net gains in greenhouse gas emissions when the baseline of CH_4 emissions is very high. In low CH_4-emitting rice systems, however, the net effect of modifying water regimes may in fact become negative in terms of radiative forcing of the gases emitted (Bronson et al., 1997b).

The overall aim cannot be to reduce CH_4 emissions to a zero level. A large portion of Asian rice fields are located on lowlands that would be flooded naturally (at least for some time over the year). Natural wetlands are a source of CH_4, so the net effect of growing rice on this land is less than the actual emissions. Furthermore, CH_4 emission deriving from rice is only a small driver of global warming that is mainly caused by CO_2 emitted through combustion of fossil fuels. Even for CH_4 alone, the contribution of rice fields to the global CH_4 budget ranges from 2% to 5% (Matthews et al., this issue). On a national scale, however, rice is still the prevailing CH_4 source in most of Asia. In most countries of South, Southeast, and East Asia, emissions from rice fields are too high to be ignored as a possible avenue for reduction.

Recommendations

While this study has identified possible candidates for mitigating emissions, the successful implementation of different crop management practices for reducing emissions will depend on the outcome of future research. The following objectives must be targeted:

- Identifying high CH_4-emitting rice systems
- Characterizing site-specific settings for mitigation
- Developing packages of mitigation technologies on regional bases
- Ascertaining synergies with improving productivity
- Accounting for N_2O emissions

Identifying high CH_4 -emitting rice systems

Given the spatial variability in emission rates, the most promising approach for effective mitigation is targeting those rice systems with high CH_4 emission potentials. Identification of high CH_4-emitting rice systems requires geographic data on rice ecosystems, crop management, soil, and climate. Ideally, these data should be incorporated in a geographic information system (GIS) that can be linked to a CH_4 model. An initial GIS database (Knox et al., this issue) and a CH_4 model (Matthews et al., this issue) have been developed as part of this project. The accuracy of both components, however, may be improved in the future. Incorporation of regional soil surveys in major rice- growing areas would substantially improve the accuracy of upscaling as compared with the use of global soil maps. Moreover, the current database covers only the five collaborating countries of the project and therefore should be extended to other rice-growing areas.

Characterizing site-specific conditions for mitigation

Methane emissions from rice fields cannot be reduced by using a 'blanket' strategy for the different rice-growing systems. The distinction of baseline scenarios (as done in the present study) is an initial step to classify rice systems for identifying adequate mitigation options. However, more site-specific information is needed to define the best strategy under certain natural and socioeconomic settings. Again, GIS databases could be deployed for such site characterizations.

Developing sound packages of mitigation technologies

Concepts for reducing CH_4 emissions have to consider intrinsic links between individual modifications—e.g., changes in straw treatment have to concur with appropriate soil tillage, timing of fertilizer application with irrigation pattern, etc. Furthermore, it seems unlikely that the optimum reduction effect can be accomplished by one modification only. Packages of technologies have to be based on a site-specific characterization as listed above.

Ascertaining synergies with improving productivity

As for other innovations, the success of mitigation strategies in farming practice will ultimately depend on their economic performance. Possible trade-offs beween mitigation strategies and productivity could be derived from the following relationships:

- Methane and N_2O emissions represent major pathways for energy and nutrient losses, respectively, for the rice system.
- A balanced nutrient supply prevents excessive emissions related to phosphorus deficiencies and oversupply of N.
- Modern rice plants are characterized by low root exudation leading to relatively low CH_4 emission rates.
- Temporary soil aeration reduces CH_4 emission while yields may increase and water demand may decrease depending on plant and soil type, respectively.
- The increment in CH_4 emission rates triggered by organic material can greatly be reduced by biogas techniques and applying fermented (composted) crop residues, which in turn improves soil fertility.

Accounting for N_2O emissions

It is imperative to ensure a positive net balance in greenhouse gas emissions through recommended changes in crop management. The effects on N_2O have to be elucidated further for incorporation in a decision support system.

Conclusions

The largest share of historical and current greenhouse gas emissions has come from developed countries, but different countries have distinct capabilities for coping with climate change in the widest possible cooperation (Dixon et al., 1996). These principles were acknowledged in the Framework Convention on Climate Change. With the specifications of the Kyoto Protocol, agriculture research may in the future increasingly be concerned with greenhouse gas emissions and its prevention (Smith, 1999). In countries with predominant rice cultivation, rice research could play a crucial role in accomplishing the national goals stipulated in this convention.

The achieved outputs of the Interregional Program on Methane Emission from Ricefields have opened up the possibilities to immediately develop some specific mitigation technologies for defined target areas. However, the implementation of mitigation strategies has to be seen in the context of a socioeconomically sound rice production. Increasing rice production is imperative for future generations. The challenge for rice research is to develop technologies that increase rice yields and—at the same time—reduce greenhouse gas emissions.

Acknowledgement

The research described was funded by the United nations Development Programme, Global Environment Facility UNDP/GEF (GLO/91/G31) - Interregional Research Programme on Methane Emission from Rice Fields under contract no. C92379 to the International Rice Research Institute.

References

Abao EB, Bronson KF, Wassmann R & Singh U (2000) Simultaneous records of methane and nitrous oxide emissions in rice-based cropping systems under rainfed conditions. Nutr Cycling Agroecosyst, this issue

Adhya TK Bharati K, Mohanty SR, Mishra SR, Ramakrishnan B, Rao VR, Sethunathan N & Wassmann R (2000) Methane emissions from rice fields at Cuttack (India) Nutr Cycling Agroecosyst, this issue

Becker T (1993) Asia's food challenge: to produce more with less. Dev Coop 6:15-16

Blake RO (1992) Sustainable and increased food production. Agric Syst 40:7-19

Bronson KF, Neue HU, Singh U & Abao Jr, E (1997a) Automated chamber measurements of methane and nitrous oxide flux in a flooded rice soil: I. Residue, Nitrogen, and water management. Soil Soc Am J 61:981-987

Bronson KF, Neue H, Singh U & Abao Jr, E (1997b) Automated chamber measurements of methane and nitrous oxide flux in a flooded rice soil: II. Fallow period emissions Soil Sci Soc Am J 61:988-993

Butterbach-Bahl K, Papen H & Rennenberg H (1997) Impact of gas transport through rice cultivars on methane emission from rice paddy fields. Plant Cell Environ 20: 1175-1183

Cai ZC, Xing GX, Shen GY, Xu H, Yan XY, Tsuruta, H, Yagi K & Minami K (1999) Measurements of CH_4 and N_2O emissions from rice paddies in Fengqiu, China. Soil Sci Plant Nutr 45:1-13

Cai ZC, Xing GX, Yan XY, Xu H, Tsuruta H, Yagi K, & Minami K (1997) Methane and nitrous oxide emissions from rice paddy fields as affected by nitrogen fertilizers and water management. Plant Soil 196:7-14

Chareonsilp N, Buddhaboon C, Promnart P, Wassmann R, Lantin RS & Buendia LV (2000) Methane emission from deepwater rice fields Nutr Cycling Agroecosyst, this issue

Corton TM, Bajita J, Grospe F, Pamplona R, Wassmann R & Lantin RS (2000) Methane emission from irrigated and intensively managed rice fields in Central Luzon (Philippines). Nutr Cycling Agroecosyst, this issue

De Datta SK & Nantasomsaran P (1991) Status and prospects of direct seeded flooded rice in tropical Asia. In: Direct seeded flooded rice in the tropics. pp 1-16

Dawe D & Dobermann A (1999) Defining productivity and yield. International Rice Research Institute Manila, Philippines, 13 p

Didiek SB (1998) Alternative technique of water saving through intermittent irrigation system for rice crop. Water and Land Resources Development and Management for Sustainable Use. Vol. II-A. Indonesian National Committee on Irrigation and Drainage (INACID) A-31, 8 p

Dixon RK, Sathaye JA, Meyers SP, Masera OR, Makarov AA, Toure S, Makundi W & Wiel S (1996) Greenhouse gas mitigation strategies: preliminary results from the U.S. country studies program. Ambio 25:26-32

FADINAP–Fertilizer Advisory Development Information Network for Asia and the Pacific (1999) Fertilizer statistics. Bangkok, Thailand

Freney JR (1997) Strategies to reduce gaseous emissions of nitrogen from irrigated agriculture. Nutr Cycling Agroecosyst 48:155-160

IPCC – Intergovernmental Panel on Climate Change (1996) Climate Change 1995. The Science of Climate Change. Cambridge University Press Cambridge (UK), 572 p

IRRI – International Rice Research Institute (1993) IRRI Rice Almanac, Manila, Philippines, 142 p

Jain MC, Kumar K, Wassmann R, Mitra S, Singh SD, Singh JP, Singh R, Yadav AK & Gupta S (2000) Methane emissions from irrigated rice fields in northern India (New Delhi). Nutr Cycling Agroecosyst, this issue

Knox J, Matthews RB & Wassmann R(2000) Using a crop/ soil simulation model and GIS techniques to assess methane emissions from rice fields in Asia. III. Databases. Nutr Cycling Agroecosyst, this issue

Lindau CW, Bollich PK, DeLaune RD, Mosier AR & Bronson KF (1993) Methane mitigation in flooded Louisiana rice fields. Biol Fertil Soils 15:174-178

Lu Y, Wassmann R, Neue HU & Huang C (1999) Impact of phosphorus supply on root exudation, aerenchyma formation and methane emission of rice plants. Biogeochemistry (in press)

Lu WF, Chen W, Duan BW, Guo WM, Lu Y, Lantin RS, Wassmann R & Neue HU (2000) Methane emission and mitigation options in irrigated rice fields in Southeast China. Nutr Cycling Agroecosyst, this issue

Minami K & Neue HU (1994) Rice paddies as a methane source. Climatic Change 27:13-26

Minami K (1997) Atmospheric methane and nitrous oxide: sources, sinks and strategies for reducing agricultural emissions. Nutr Cycling Agroecosyst 49:203-211

Miura Y & Kanno T (1997) Emissions of trace gases (CO_2, CO, CH_4, and N_2O) resulting from rice straw burning. Soil Sci Plant Nutr 43:849-854

Neue HU, Wassmann R & Lantin RS (1995) Mitigation options for methane emissions from rice fields. Climate Change and Rice. Springer-Verlag, Berlin, pp 136-144

Neue HU & Sass RL (1998) The budget of methane from rice fields. IGACtivities 12:3-11

Ranganathan R, Neue HU & Pingali PL (1995) Global climate change: role of rice in methane emissions and prospects for mitigation. Climate Change and Rice. Springer-Verlag, Berlin, pp 122-135

Rennenberg H, Wassmann R, Papen H & Seiler W (1992) Trace gas emission in rice cultivation. Ecol Bull 42:164-173

Sass RL & Fisher FM (1995) Methane emissions from Texas rice fields: a five-year study. In: S Peng, KT Ingram, HU Neue, L Ziska (eds) Climate Change and Rice, pp 46-59, Springer-Verlag, Berlin

Schlesinger WH (1999) Carbon sequestration in soils. Science 284:2095

Setyanto P, Makarim AK, Fagi AM, Wassmann R & Buendia LV (2000) Crop management affecting methane emissions from irrigated and rainfed rice in Central Java (Indonesia) Nutr Cycling Agroecosys, this issue

Shin YK, Yun SH, Park ME & Lee BL (1996) Mitigation options for methane emission from rice fields in Korea. Ambio 25:289-291

Smith KA (1999) After the Kyoto Protocol: can soil scientists make a useful contribution. Soil Use Manage 15: 71-75

Smith KA, McTaggart IP & Tsuruta H (1997) Emissions of N_2O and NO associated with nitrogen fertilization in intensive agriculture, and the potential for mitigation. Soil Use Manage 13:296-304

Tuong TP, Cabangon, RJ & Wopereis M C S (1996) Quantifying flow processes during land soaking of cracked rice soils. Soil Sci Soc Am J 60:872-879

Wang ZY, Xu YC, Li Z, Guo YX, Wassmann R, Neue HU, Lantin RS, Buendia LV, Ding YP & Wang ZZ (2000) Methane Emissions from irrigated rice fields in northern China (Beijing). Nutr Cycling Agroecosyst, this issue

Wassmann R, Wang MX, Shangguan XJ, Xie XL, Shen RX, Papen H, Rennenberg H & Seiler W (1993) First records of a field experiment on fertilizer effects on methane emission from rice fields in Hunan Province (PR China). Geophys Res Lett 20:2071-2074

Wassmann R, Buendia LV, Lantin RS, Bueno CS, Lubigan LA, Umali A, Nocon N, Javellana AM & Neue HU (2000 a) Mechanisms of crop management impact on methane emissions from rice fields in Los Baños, Philippines. Nutr. Cycling Agroecosys, this issue

Wassmann R, Neue HU, Lantin RS, Buendia LV & Rennenberg H (2000 b) Characterization of methane emissions from rice fields in Asia. 1. Comparison among field sites in five countries. Nutr Cycling Agroecosyst, this issue

Wassmann R, Neue HU, Lantin RS, Makarim K, Chareonsilp N, Buendia LV & Rennenberg H (2000c) Characterization of methane emissions from rice fields in Asia. 2. Differences among irrigated, rainfed and deepwater ecosystems. Nutr. Cycling Agroecosyst, this issue

Yagi K, Tsuruta H & Minami K (1997) Possible options for mitigating methane emission from rice cultivation. Nutr Cycling Agroecosyst 213-220

Yoshida S (1981). Fundamentals of rice crop science. International Rice Research Institute, Manila, Philippines, 269 p

Nutrient Cycling in Agroecosystems **58:** 37–53, 2000.

Methane emission from irrigated and intensively managed rice fields in Central Luzon (Philippines)

T.M. Corton[1], J.B. Bajita[1], F.S. Grospe[1] R.R. Pamplona[1,2], C.A. Asis Jr.[1,3], R. Wassmann[2,4], R.S. Lantin[2] & L. V. Buendia[2]

[1]Agronomy, Soils and Plant Physiology Division, Philippine Rice Research Institute, Muñoz, Nueva Ecija 3119 Philippines; [2]International Rice Research Institute, MCPO Box 3127, 1271 Makati City, Philippines; [3]National Institute for Agro-Environmental Sciences, Tsukuba, Japan; [4]Fraunhofer Institute for Atmospheric Environmental Research, Garmisch-Partenkirchen, Germany

Key words: inorganic fertilizer, ammonium sulfate, organic amendment, phosphogypsum, methane mitigation

Abstract

Methane (CH_4) emissions were measured with an automated system in Central Luzon, the major rice producing area of the Philippines. Emission records covered nine consecutive seasons from 1994 to 1998 and showed a distinct seasonal pattern: an early flush of CH_4 before transplanting, an increasing trend in emission rates reaching maximum toward grain ripening, and a second flush after water is withdrawn prior to harvesting. The local practice of crop management, which consists of continuous flooding and urea application, resulted in 79-184 mg CH_4 m^{-2} d^{-1} in the dry season (DS) and 269-503 mg CH_4 m^{-2} d^{-1} in the wet season (WS). The higher emissions in the WS may be attributed to more labile carbon accumulation during the dry fallow period before the WS cropping as shown by higher % organic C. Incorporation of sulfate into the soil reduced CH_4 emission rates. The use of ammonium sulfate as N fertilizer in place of urea resulted in a 25-36% reduction in CH_4 emissions. Phosphogypsum reduced CH_4 emissions by 72% when applied in combination with urea fertilizer. Midseason drainage reduced CH_4 emission by 43%, which can be explained by the influx of oxygen into the soil. The practice of direct seeding instead of transplanting resulted in a 16-54% reduction in CH_4 emission, but the mechanisms for the reducing effect are not clear. Addition of rice straw compost increased CH_4 emission by only 23-30% as compared with the 162-250% increase in emissions with the use of fresh rice straw. Chicken manure combined with urea did not increase CH_4 emission. Fresh rice straw has wider C/N (25 to 45) while rice straw compost has C/N = 6 to 10 and chicken manure has C/N = 5 to 8. Modifications in inorganic and organic fertilizer management and water regime did not adversely affect grain yield and are therefore potential mitigation options. Direct seeding has a lower yield potential than transplanting but is getting increasingly popular among farmers due to labor savings. Combined with a package of technologies, CH_4 emission can best be reduced by (1) the practice of midseason drainage instead of continuous flooding, (2) the use of sulfate-containing fertilizers such as ammonium sulfate and phosphogypsum combined with urea; (3) direct seeding crop establishment; and (4) use of low C/N organic fertilizer such as chicken manure and rice straw compost.

Introduction

There is an urgent need to increase rice production in the Philippines to feed a population that is growing to 70 million. Per capita consumption of rice in the Philippines is currently 103 kg. The development of reliable, efficient irrigation systems is the remaining best option as rice areas continue to decrease. Rice production in the coming years is expected to lean toward more intensification in terms of increased cropping per year and the use of high-input technologies. Expansion and intensification of the irrigated rice area could increase CH_4 emission from rice fields.

Irrigated rice fields have high potential to produce CH_4 because continuous flooding favors CH_4 production and emission. However, irrigated rice cultivation is one of the few anthropogenic sources where the management of CH_4 is possible. Thus, it becomes a critical focus of mitigation efforts. Mitigation technologies, however, must be formulated parallel to the need to increase and sustain high productivity. One major step is to identify mitigation options by investigating the influence of various factors on the processes of CH_4 production and consumption. The field experiment presented here was part of an interregional network on CH_4 emissions from rice fields (Wassmann et al., this issue). The objectives of our research were (1) to measure CH_4 fluxes in irrigated rice fields under different cultivation practices in a major rice-growing area of the Philippines; (2) to evaluate processes that control CH_4 formation; and (3) to identify mitigation options to reduce CH_4 emission from irrigated rice fields while sustaining high yield.

Materials and methods

Field site

The experimental site at PhilRice Central Experiment Station in Maligaya, Muñoz, Nueva Ecija, is located at 15° 40' 21" N latitude and 120° 53' 26" E longitude. The province of Nueva Ecija is situated in the central plain of Luzon, the top rice-producing region in the Philippines with a total irrigated land area of 300,341 ha. The central plain is a terrace in a river valley with a slope of <1% and elevation of 35 m above sea level. Annual mean precipitation is 1780 mm with distinct 4-5 mo dry season (DS) and 4-6 mo wet season (WS). The project site is fully irrigated and cropped twice in a year, one in the WS and another in the DS. The soil at PhilRice Maligaya site is derived from alluvium parent material and is poorly drained. It is classified as fine, montmorillonitic, isohyperthermic Ustic Epiaquerts (Maligaya clay). Some of its physicochemical properties are listed in Table 1.

Table 1. Some characteristics of Maligaya soil at PhilRice Central Experiment Station at Muñoz, Nueva Ecija, Philippines

pH (H_2O)	6.88
pH ($CaCl_2$)	6.36
Organic carbon (%)	1.32
Total nitrogen (%)	0.09
Ammonium nitrogen (cmol kg^{-1})	0.72
CEC (cmol kg^{-1})	34.28
Active iron (μg g^{-1})	75.02
Olsen phosphorus (mg kg^{-1})	3.10
Exchangeable potassium (cmol kg^{-1})	0.10
Available zinc (mg kg^{-1})	1.48
Available sulfate (mg kg^{-1})	13.54
% clay	43.00
% silt	51.40
% sand	5.60

Duration of experiment

Field experiments measuring CH_4 emission from irrigated Maligaya clay were conducted for nine consecutive seasons (five dry + four wet) from 1994 to 1998.

Crop management practices

The DS cropping usually starts in the second week of January and ends in late April or early May. The WS cropping starts in late June to mid-July and ends in mid- to late October. The crop was harvested leaving a 28-38 cm stubble for the next crop, except in 1996 when the crop was harvested close to the ground leaving only the roots. The amount of stubble left in the field after harvest is equivalent to 2.4-4.0 t ha^{-1} dry matter. In all experiments, the roots were incorporated to decay. The crop residues were incorporated during land preparation, which is usually 15-30 d before planting. The field was flooded 2-3 d before the start of land preparation. In 1997, the differences in date of residue incorporation between T1/T3 and T2/T4 were due to the reference dates which was the date before transplanting for T1/T3 and days before sowing for T2/T4. The use of organic amendments, using either fresh rice straw, rice straw compost, chicken manure, or commercial bio-organic fertilizers commenced in 1996. After transplanting, the field was kept moist without standing water for 7-10 d after which a 5-cm water level was kept in continuously flooded treatments. About 14 d before harvest, water was withdrawn from the plots so that the soil was dry during harvest. In 1997 and 1998, water regime treatments such as midseason drainage and intermittent irrigation treatments were imposed. Midseason drainage was done by withdrawing water for 7-10 d before the panicle initiation stage. The soil, however, was not allowed to crack. In the intermittent irrigation treatment, floodwater was left to dry out and water was introduced again when the soil started to crack. This was done continuously throughout the cropping season. Nitrogen was supplied as either urea or ammonium sulfate at 90-180 kg N ha^{-1}. The rate was 120 kg N ha^{-1} in the reference treatment (T1) both in the DS and WS of 1994-96 cropping. Rice variety IR72 was used from 1994 to 1996; IR64 in 1997; and PSBRc 28 in 1998. Fourteen-day-old seedlings were transplanted at 20- × 20-cm spacing giving a population of 25 hills m^{-2}. Seeding rates in direct-seeded rice were 140 kg ha^{-1} in 1997 and 40 kg ha^{-1} in 1998, giving a tiller density of 1,104-1,745 m^{-2}.

Experiment layout and treatments

Treatments in each cropping season are shown in Table 2. Four treatments in each season were arranged in twelve 5- × 11.6-m plots using randomized complete block design with three replications. The treatments imposed were designed to investigate CH_4 emission as influenced by 1) the amount of N application, 2) the use of sulfate (SO_4^{-2})-containing fertilizers such as phosphogypsum and ammonium sulfate, 3) the use of fresh or rice straw compost, 4) crop establishment method, 5) water management, and 6) combinations of treatments 1-5. In 1994 DS, the fertilizer rates and variety in T3 reflected the prevailing practice of farmers in Nueva Ecija. This practice was modified in T4 by balancing the amount of N, P, and K. T1 was the reference treatment across seasons and years, while T2 was the amount of fertilizer targeting high rice yield in the Maligaya site. The aim was to compare CH_4 emission under current fertilizer application practice for high yield with those under farmers' practice.

The effect of inorganic amendments on CH_4 fluxes was tested in the 1994 and 1995 experiments. The rate of N was varied from 120 to 180 kg ha^{-1}, supplied either as urea or as ammonium sulfate. Phosphogypsum, a sulfur-containing byproduct of phosphate fertilizer manufacture, was tested at the rate of 0.5-1.0 t ha^{-1} in 1994 WS, at 6.0 t ha^{-1} in 1995 WS, and at 3.0 t ha^{-1} in 1996 DS and WS. Finally, in 1998 DS, the combination of cultural practices for high yield with the least CH_4 emission—use of low C/N organic fertilizer, ammonium sulfate as N fertilizer, direct-seeding crop establishment, midseason drainage, and intermittent irrigation—were tested.

Measurements

Methane flux. CH_4 fluxes were measured continuously every 2 h from transplanting until 7 d after harvest. Continuous measurement was facilitated by chamber method — automatic sampling technique (IAEA 1993). The system used was designed by the Fraunhofer Institute for Atmospheric Environmental Research (Germany) and installed at PhilRice in September 1993. The measuring system was composed of gas-collecting plexiglas boxes installed in 12 plots connected by stainless steel and copper tubing to a field laboratory equipped with datalogger, gas chromatograph (GC), and computer for a fully automated gas sampling and analysis. Air samples trapped in plexiglas boxes were immediately pumped and flushed through the stainless steel tubing to the GC. One measurement cycle lasting 2 h started with sampling of a CH_4 gas standard followed by a series of sampling of air trapped in the boxes and ended with the CH_4 standard. During the 2-h cycle, six pairs of boxes closed successively; each pair of boxes closed for 16 min and sampled four times alternately. A datalogger program automatically controlled the closing and opening of the boxes and the timing of gas sampling and analysis. Methane concentration data were transmitted from the GC integrator to the computer after each measurement cycle. Each treatment was measured from three boxes representing three replications.

Analysis of CH_4 concentration. The concentration of CH_4 in the gas samples was analyzed in a GC (Shimadzu GC-8A) equipped with flame ionization detector and porous polymer beads Porapak N 80/100 mesh column. Analysis was performed at 60 °C column temperature and 100 °C detector temperature with N_2 as carrier gas.

Statistical analysis. The statistical analysis of mean CH_4 emission was done using the STATISTICA software. For each experiment, the daily data per treatment were evaluated as to type of distribution (i.e., normal or skewed). If the distribution is normal, t-test

Table 2. Summary treatments from 1994 dry season to 1998 dry season in PhilRice Central Experiment Station

Year/season	Treatment	T1	T2	T3	T4
1994/DS	Cultivar	IR72	IR72	IR64	IR64
	Crop establishment	Transplanted	Transplanted	Transplanted	Transplanted
	Water regime	Flooded, 5 cm	Flooded, 5 cm	Flooded, 5 cm	Flooded, 5 cm
	NPK	120-30-30	180-60-30	171-25-25	117-34-31
1994/WS	Cultivar	IR72	IR72	IR72	IR72
	Crop establishment	Transplanted	Transplanted	Transplanted	Transplanted
	Water regime	Flooded, 5 cm	Flooded, 5 cm	Flooded, 5 cm	Flooded, 5 cm
	NPK	120-30-30	120-30-30	120-30-30	120-30-30
	N source	Urea	Ammosul[a]	Urea	Urea
	Phosphogypsum	-	-	0.5 t ha^{-1}	1.0 t ha^{-1}
1995/DS	Cultivar	IR72	IR72	IR72	IR72
	Crop establishment	Transplanted	Transplanted	Transplanted	Transplanted
	Water regime	Flooded, 5 cm	Flooded, 5 cm	Flooded, 5 cm	Flooded, 5 cm
	NPK	120-40-40	120-40-40	180-40-40	180-40-40
	N source	Urea	Ammosul	Urea	Ammosul
1995/WS	Cultivar	IR72	IR72	IR72	IR72
	Crop establishment	Transplanted	Transplanted	Transplanted	Transplanted
	Water regime	Flooded, 5 cm	Flooded, 5 cm	Flooded, 5 cm	Flooded, 5 cm
	NPK	120-40-40	120-40-40	180-40-40	120-40-40
	N source	Urea	Ammosul	Urea	Urea
	Phosphogypsum	-	-	-	6.0 t ha^{-1}
1996/DS	Cultivar	IR72	IR72	IR72	IR72
	Crop establishment	Transplanted	Transplanted	Transplanted	Transplanted
	Water regime	Flooded, 5 cm	Flooded, 5 cm	Flooded, 5 cm	Flooded, 5 cm
	NPK	120-40-40	90-40-40	90-40-40	90-40-40
	Organic material	-	4 t ha^{-1} FSR[b]	2.5 t ha^{-1} RSC[c]	4 t ha^{-1} FSR[1]
	Phosphogypsum	-	-	-	3.0 t ha^{-1}
1996/WS	Cultivar	IR72	IR72	IR72	IR72
	Crop establishment	Transplanted	Transplanted	Transplanted	Transplanted
	Water regime	Flooded, 5 cm	Flooded, 5 cm	Flooded, 5 cm	Flooded, 5 cm
	NPK	120-40-40	90-40-40	90-40-40	90-40-40
	Organic material	-	4 t ha^{-1} FSR	2.5 t ha^{-1} RSC	4 t ha^{-1} FSR
	Phosphogypsum	-	-	-	3.0 t ha^{-1}
1997/DS	Cultivar	IR64	IR64	IR64	IR64
	Crop establishment	Transplanted	Direct-seeded	Transplanted	Direct-seeded
	Water regime	Continuously flooded, 5 cm	Continuously flooded, 5 cm	Midseason drained	Midseason drained
	NPK	150-60-60	150-60-60	150-60-60	150-60-60
	Organic material	300 kg ha^{-1} Commercial bio-organic fertilizer	300 kg ha^{-1} Commercial bio-organic fertilizer	300 kg ha^{-1} Commercial bio-organic fertilizer	300 kg ha^{-1} Commercial bio-organic fertilizer
1997/WS	Cultivar	IR64	IR64	IR64	IR64
	Crop establishment	Transplanted	Direct-seeded	Transplanted	Direct-seeded
	Water regime	Continuously flooded, 5 cm	Continuously flooded, 5 cm	Midseason drained	Midseason drained
	NPK	90-30-60	90-30-60	90-30-60	90-30-60
	Organic material	300 kg ha^{-1} Commercial bio-organic fertilizer	300 kg ha^{-1} Commercial bio-organic fertilizer	300 kg ha^{-1} Commercial bio-organic fertilizer	300 kg ha^{-1} Commercial bio-organic fertilizer

Table 2 continued

Year/season	Treatment	T1	T2	T3	T4
1998/DS	Cultivar	PSBRc 28	PSBRc 28	PSBRc 28	PSBRc 28
	Crop establishment	Transplanted	Transplanted	Direct-seeded	Direct-seeded
	Water regime	Continuously flooded, 5 cm	Continuously flooded, 5 cm	Midseason drained	Intermittent irrigation
	NPK	150-60-60	150-60-60	150-60-60	150-60-60
	N source	Urea	Urea	Ammosul	Ammosul
	Organic material	-	1.5 t ha^{-1} Chicken manure	2.5 t ha^{-1} Rice straw compost	2.5 t ha^{-1} Rice straw compost

[a]Ammonium sulfate. [b]Fresh rice straw. [c]Rice straw compost.

was used (parametric analysis). If the distribution is not normal, sign test was used (nonparametric analysis). The T value for t-test and the Z value for sign test were determined. Then the significance was determined from the value of probability (Table 4).

Results

The results of the 5-yr experiment were summarized by season in Table 3. In 1994 DS, flux measurement was discontinuous during the first 58 d owing to measurement system problems. Many data points during this period were actually interpolated between two actual measurements (Figure 1). IR64 with 117 kg N ha^{-1} gave slightly higher CH_4 fluxes for the period 38-83 DAT (Figure 1). This resulted in a mean emission of 114 mg CH_4 m^{-2} d^{-1} that was highest (z values = 3.2**, 7.5**, 8.2**) among the treatments. The reference treatment, IR72 with 120 kg N ha^{-1}, gave a mean emission of 90 mg CH_4 m^{-2} d^{-1}. The mean emission was lower than the reference treatment (64 mg CH_4 m^{-2} d^{-1}, z value = 8.6**) in IR72 with 180-60-30 fertilizer and in IR64 with 171-25-25 fertilizer applied (74 mg CH_4 m^{-2} d^{-1}, z value = 8.0**). An increasing trend in mean CH_4 emission was observed as rice growth progresses. The maximum was reached toward grain ripening. Two weeks before harvest, when irrigation was withdrawn, a flush of CH_4 emission occurred which was reduced to a negligible amount after 5-7 d. IR72 (T1) and (T2) treatments gave higher yields of 8.36 t ha^{-1} and 9.26 t ha^{-1}, respectively, than IR64 (T3) and (T4) with yields of 6.79 t ha^{-1} and 7.36 t ha^{-1}. Accordingly, the amount of CH_4 produced per ton of grain yield was lower in IR72. But the total aboveground biomass production did not differ among treatments.

In the 1994 WS, a distinct seasonal pattern which is an early flush of CH_4 before transplanting, followed by an increasing rate of emission reaching maximum toward grain ripening, and a second flush CH_4 after water was withdrawn before harvest was established (Figure 2). T1 (urea, 120 N) gave slightly higher fluxes starting at 30 DAT through 90 DAT, resulting in a mean emission of 266 mg CH_4 m^{-2} d^{-1}. Ammonium sulfate (T2) and (T3) urea + 0.5 t phosphogypsum (PG) ha^{-1} gave slightly lower mean emission of 232 mg CH_4 m^{-2} d^{-1}(z value = 6.6**) and 227 mg CH_4 m^{-2} d^{-1}(z value = 7.2**), respectively. The observed reduction in total seasonal CH_4 emission was about 14% in the ammonium sulfate (230 kg CH_4 ha^{-1}) compared with the urea treatment (266 kg CH_4 ha^{-1}). A 9-15% reduction of total seasonal emission was observed with application of 0.5 to 1.0 t PG ha^{-1}. The mean emission and the amount of CH_4 emitted per ton yield were the same in treatments with SO_4^{-2} (from ammonium sulfate as N fertilizer and urea plus PG). The grain yield as well as total aboveground biomass produced did not differ among the treatments.

The results of the 1995 DS also showed the distinct seasonal pattern of CH_4 emission. Measurement was discontinued after 80 DAT owing to a problem in the system. Thus, the second flush upon withdrawal of water before harvest was not observed. All treatments gave similar magnitude of CH_4 flux during the first 15 DAT (Figure 3). Starting from 25 DAT until 70 DAT, CH_4 fluxes in urea treatments were higher than those in ammonium sulfate treatments. Using ammonium sulfate in place of urea reduced mean emission from 184 mg CH_4 m^{-2} d^{-1} to 166 mg CH_4 m^{-2} d^{-1} (z value = 3.8**) at lower N level (120 kg N ha^{-1}) and from 205 mg CH_4 m^{-2} d^{-1} to 131 mg CH_4 m^{-2} d^{-1} (z value = 6.7**) at higher N level (180 kg N ha^{-1}). Increasing the amount of N applied from 120 to 180 kg ha^{-1} using urea slightly increased (z value = 3.4**) mean CH_4 emission. However, with ammonium sulfate, the higher N rate reduced

Table 3. Methane emissions from 1994 dry season to 1998 dry season in the PhilRice Central Experiment Station[a]

Year/season	Measurement	T1	T2	T3	T4
1994/DS	Mean emission (mg m^{-2} d^{-1})	90	64	74	114
	Season length (d)	105	105	91	91
	Seasonal flux (kg CH_4 ha^{-1})	95	67	67	104
	Biomass (t ha^{-1})	12.63a	13.32a	13.36a	12.79a
	Grain yield (t ha^{-1})	8.36ab	9.26a	6.79c	7.36bc
	kg CH_4 per ton yield	11.36	7.24	9.87	14.13
1994/WS	Mean emission (mg m^{-2} d^{-1})	269	232	227	243
	Season length (d)	99	99	99	99
	Seasonal flux (kg CH_4 ha^{-1})	266	230	225	241
	Biomass (t ha^{-1})	11.46a	12.54a	11.82a	11.50a
	Grain yield (t ha^{-1})	5.22a	5.10a	4.90a	5.27a
	Kg CH_4 per ton yield	50.96	45.10	45.92	45.73
1995/DS	Mean emission (mg m^{-2} d^{-1})	184	166	205	131
	Season length (d)	111	111	111	111
	Seasonal flux (kg CH_4 ha^{-1})	204	184	228	145
	Biomass (t ha^{-1})	13.79a	13.07a	14.44a	15.00a
	Grain yield (t ha^{-1})	6.54a	6.40a	6.45a	6.34a
	Kg CH_4 per ton yield	31.19	28.75	35.35	22.87
1995/WS	Mean emission (mg m^{-2} d^{-1})	503	317	516	139
	Season length (d)	103	103	103	103
	Seasonal flux (kg CH_4 ha^{-1})	518	327	531	143
	Biomass (t ha^{-1})	13.92a	13.90a	12.75a	14.33a
	Grain yield (t ha^{-1})	3.30a	3.72a	3.36a	3.78a
	Kg CH_4 per ton yield	156.97	87.90	158.04	37.83
1996/DS	Mean emission (mg m^{-2} day^{-1})	165	433	184	318
	Season length (d)	97	97	97	97
	Seasonal flux (kg CH_4 ha^{-1})	160	420	178	308
	Biomass (t ha^{-1})	15.35a	13.29ab	12.18b	10.08bc
	Grain yield (t ha^{-1})	7.30a	7.13a	7.41a	7.20a
	Kg CH_4 per ton yield	21.92	58.91	24.02	42.78
1996/WS	Mean emission (mg m^{-2} d^{-1})	272	952	353	599
	Season length (d)	100	100	100	100
	Seasonal flux (kg CH_4 ha^{-1})	272	952	353	599
	Biomass (t ha^{-1})	14.60a	14.05a	13.37a	13.05a
	Grain yield (t ha^{-1})	5.17a	5.22a	5.35a	5.27a
	Kg CH_4 per ton yield	52.61	182.38	65.98	113.66
1997/DS	Mean emission (mg m^{-2} d^{-1})	91	73	52	46
	Season length (d)	98	91	98	91
	Seasonal flux (kg CH_4 ha^{-1})	89	75	51	48
	Biomass (t ha^{-1})	12.5a	11.2a	13.5a	10.2a
	Grain yield (t ha^{-1})	7.91b	6.71a	7.74b	6.42a
	Kg CH_4 per ton yield	11.25	11.18	6.59	7.48
1997/WS	Mean emission (mg m^{-2} d^{-1})	375	323	347	178
	Season length (d)	93	84	93	84
	Seasonal flux (kg CH_4 ha^{-1})	348	272	323	150
	Biomass (t ha^{-1})	12.4a	13.4a	14.1a	11.7a
	Grain yield (t ha^{-1})	5.36b	3.84a	5.45b	3.41a
	Kg CH_4 per ton yield	64.92	70.83	59.27	43.99

Table 3 continued.

Year/season	Measurement	T1	T2	T3	T4
1998/DS	Mean emission (mg m^{-2} d^{-1})	79	80	14	6
	Season length (d)	114	114	114	114
	Seasonal flux (kg CH_4 ha^{-1})	90	91	16	7
	Biomass (t ha^{-1})	16.4b	14.7b	24.2a	23.2a
	Grain yield (t ha^{-1})	8.0ab	8.5a	7.7b	7.1c
	Kg CH_4 per ton yield	11.25	10.71	2.08	0.98

[a]In a row, numbers followed by the same letter are not significantly different at the 5% level by DMRT.

mean emission by 21% (z value = 2.9**). Grain yield and total aboveground biomass produced did not differ among treatments. The lowest amount of CH_4 (22.87 kg CH_4 t^{-1} grain yield) was observed with ammonium sulfate applied at 180 kg N ha^{-1}. In the 1995 WS, urea + PG treatment gave consistently lower CH_4 fluxes from 15 DAT until harvest as compared with the other treatments (Figure 4). No measurement was done before transplanting because of some problem in the system that started during the DS cropping. The same seasonal pattern of emission with a previous cropping was observed. Ammonium sulfate treatment also gave fluxes consistently lower than those in the urea treatments throughout the growing season. Seasonal flux in urea treatments at 180 kg N ha^{-1} (daily average of 516 mg CH_4 m^{-2}) was the same as that at 120 kg N ha^{-1} treatment (daily average of 503 mg CH_4 m^{-2}). The higher amount of N applied from urea slightly increased CH_4 emission during the 1995 DS but this was not significant during the 1995 WS. Ammonium sulfate treatment gave a mean emission of 317 mg CH_4 m^{-2} d^{-1} while urea + PG treatment gave only 139 mg CH_4 m^{-2} d^{-1}. The use of ammonium sulfate reduced seasonal CH_4 flux by 37% (z value = 8.5**), while the combination of urea + PG reduced CH_4 emission by 72% (z value = 10.0**). The lowest amount of CH_4 (37.83 kg CH_4 t^{-1} grain yield) was observed with PG addition to urea. Grain yield and total aboveground biomass produced did not differ among treatments.

In the 1996 DS, a similar distinct seasonal pattern of CH_4 emission was observed (Figure 5). Methane fluxes in the 4 t ha^{-1} fresh rice straw treatment were consistently highest among the treatments throughout the growing period. The magnitude of CH_4 fluxes was about twice as high as in the urea treatment starting at 22 DAT until 70 DAT. The CH_4 fluxes from urea-treated plots were parallel with those of fresh rice straw-treated plots starting at 70 DAT. Mean emission increased from 165 to 433 mg CH_4 m^{-2} d^{-1} (z value = 9.8**) with the addition of 4 t ha^{-1} fresh rice straw. The addition of 3 t ha^{-1} PG in rice straw-treated plots increased the mean emission to only 318 mg CH_4 m^{-2} d^{-1} (z value = 9.8**). The addition of PG in T4 with 4 t ha^{-1} fresh rice straw did not fully counteract the high CH_4 fluxes (Table 3). Methane fluxes in the rice straw compost treatment (T3) were similar to the urea treatment throughout the season. Mean CH_4 emission in compost-treated plots (184 mg CH_4 m^{-2} d^{-1}) was only slightly higher than the reference treatment (165 mg CH_4 m^{-2} d^{-1}, z value = 2.3**) where no organic amendment was added. The amount of CH_4 produced (22 kg CH_4 t^{-1} grain yield) was lowest in the urea treatment and followed by the rice straw compost treatment (24 kg CH_4 t^{-1} grain yield). The total aboveground biomass was lower when 30 kg inorganic N was replaced with organic N from fresh rice straw and rice straw compost. The grain yield, however, did not differ among treatments. In the 1996 WS, there were very high CH_4 fluxes in the fresh rice straw treatment from 15 d before transplanting until 25 DAT (Figure 6). Within 40 d from the application of fresh rice straw, CH_4 flux was high in the fresh rice straw-treated plots. Starting from 25 DAT until 100 DAT, the CH_4 fluxes in the fresh rice straw plots were parallel with those of urea and rice straw compost plots. However, the application of 4 t ha^{-1} fresh rice straw consistently gave the highest CH_4 fluxes throughout the growing season. The addition of fresh rice straw increased seasonal CH_4 flux by 250% (from 272 to 952 kg CH_4 ha^{-1}), considerably higher than the 30% increase (from 272 to 353 kg CH_4 ha^{-1}) with addition of 2.5 t ha^{-1} rice straw compost. The addition of 3 t ha^{-1} PG to fresh rice straw-treated plots decreased the seasonal CH_4 flux to almost one-half of the amount (from 952 to 599 kg CH_4 ha^{-1}) where fresh rice straw alone was added. On the other hand, CH_4 fluxes in the rice straw compost treatment were only slightly higher than those in the urea

Table 4. Results of statistical analysis of mean CH_4 emission (mg CH_4 m^{-2} d^{-1})[a]

Year/ season	Treatment no.	Treatment	Mean emission (mg CH_4 m^{-2} d^{-1})	z values[b]		
				T2	T3	T4
1994/DS	T1	120 kg N ha^{-1}: IR72	90	8.6**	8.0**	3.2**
1994/DS	T2	180 kg N ha^{-1}: IR72	64	-	3.5**	7.5**
1994/DS	T3	171 kg N ha^{-1}: IR64	74	-	-	8.2**
1994/DS	T4	117 kg N ha^{-1}: IR64	114	-	-	-
1994/WS	T1	120 kg N ha^{-1} urea	269	6.6**	7.2**	4.8**
1994/WS	T2	120 kg N ha^{-1} ammosul	232	-	0.4 ns	1.9 ns
1994/WS	T3	120 kg N ha^{-1} + 0.5 t ha^{-1} PG	227	-	-	6.6**
1994/WS	T4	120 kg N ha^{-1} + 1.0 t ha^{-1} PG	243	-	-	-
1995/DS	T1	120 kg N ha^{-1} urea	184	3.8**	3.4**	3.4**
1995/DS	T2	120 kg N ha^{-1} ammosul	166	-	4.5**	2.9**
1995/DS	T3	180 kg N ha^{-1} urea	205	-	-	6.7**
1995/DS	T4	180 kg N ha^{-1} ammosul	131	-	-	-
1995/WS	T1	120 kg N ha^{-1} urea	503	8.5**	0.8 ns	10.0**
1995/WS	T2	120 kg N ha^{-1} ammosul	317	-	8.5**	5.3**
1995/WS	T3	180 kg N ha^{-1} urea	516	-	-	10.0**
1995/WS	T4	120 kg N ha^{-1} urea + 6 t ha^{-1} PG	139	-	-	-
1996/DS	T1	120 kg N ha^{-1} urea	165	9.8**	2.3**	9.8**
1996/DS	T2	Urea + 4 t ha^{-1} rice straw	433	-	9.8**	7.9**
1996/DS	T3	Urea + 2.5 t ha^{-1} compost	184	-	-	8.9**
1996/DS	T4	Urea + 4 t ha^{-1} rice straw + 3 t ha^{-1} PG	318	-	-	-
1996/WS	T1	120 kg N ha^{-1} urea	272	9.9**	7.9**	9.9**
1996/WS	T2	Urea + 4 t ha^{-1} rice straw	952	-	9.9**	9.9**
1996/WS	T3	Urea + 2.5 t ha^{-1} compost	353	-	-	9.5**
1996/WS	T4	Urea + 4 t ha^{-1} rice straw + 3 t ha^{-1} PG	599	-	-	-
1997/DS	T1	Transplanted, continuous flooding	91	3.0**	3.2**	4.7**
1997/DS	T2	Direct-seeded, continuous flooding	73	-	4.9**	8.4**
1997/DS	T3	Transplanted, midseason drained	52	-	-	0.1 ns
1997/DS	T4	Direct-seeded, midseason drained	46	-	-	-
1997/WS	T1	Transplanted, continuous flooding	375	4.0**	0.5 ns	8.0**
1997/WS	T2	Direct-seeded, continuous flooding	323	-	1.5 ns	8.2**
1997/WS	T3	Transplanted, midseason drained	347	-	-	7.3**
1997/WS	T4	Direct-seeded, midseason drained	178	-	-	-

[a]The analysis was done using STATISTICA software. For each experiment, daily data per treatment were evaluated as to the type of distribution (i.e., normal or skewed). If distribution is normal, t-test is used (parametric analysis). If distribution is not normal, sign test is used (nonparametric analysis). T value (t-test) and Z value (sign test) were determined. The significance was determined from the value of probability. [b]Comparison is between treatment no. vs T2 or T3 or T4. Level of significance: ** = highly significant (1% level); * = significant (5% level); ns = not significant.

treatment during the period 15 d before transplanting to 25 DAT. Mean CH_4 emission amounted to 272 mg CH_4 m^{-2} d^{-1} in the urea treatment and 353 mg CH_4 m^{-2} d^{-1} (z value = 7.9**) in the rice straw compost treatment. The lowest amount of CH_4 emitted was observed in the reference treatment where no straw was added, both in the DS (22 kg CH_4 t^{-1} yield) and WS (53 kg CH_4 t^{-1} yield). In both seasons, grain yield did not differ among treatments.

In the 1997 DS, the distinct seasonal pattern of CH_4 emission was also observed (Figure 7). The first flush of CH_4 flux was observed during the early growth stage. More CH_4 flux was observed in direct-seeded rice than in transplanted rice during the early stage because there was no standing water in transplanted rice until 7-10 DAT. When irrigation water was introduced, all the treatments had the same CH_4 emission from 20 DAT until midseason drainage was introduced. After the midseason drainage, CH_4 flux was significantly reduced. The CH_4 flux increased again upon reflooding, but did not reach the same level as that in continuously flooded treatment. The second flush of CH_4 flux was observed after water was withdrawn before harvest. The magnitude of CH_4 flux during this second flush was higher in the continuously flooded treatment for both transplanted and direct-seeded rice. Direct-seeded rice

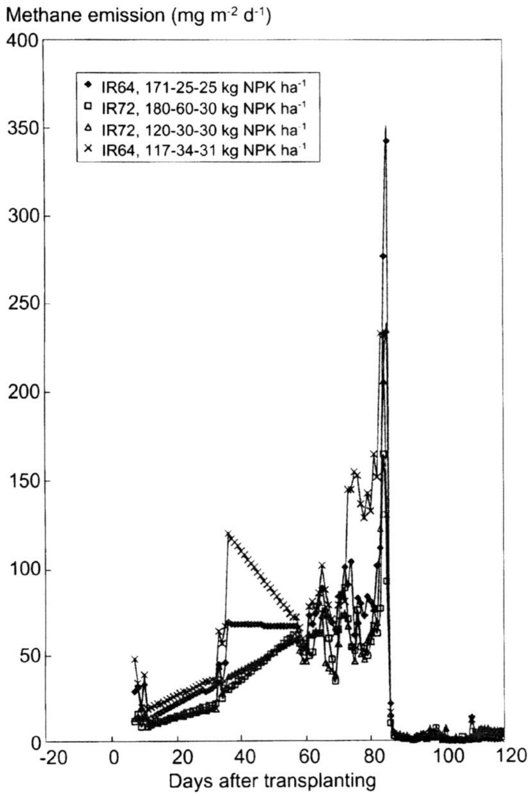

Figure 1. Effect of inorganic amendment on CH_4 emission from rice field grown to IR72 at PhilRice, Maligaya, Philippines, 1994 DS.

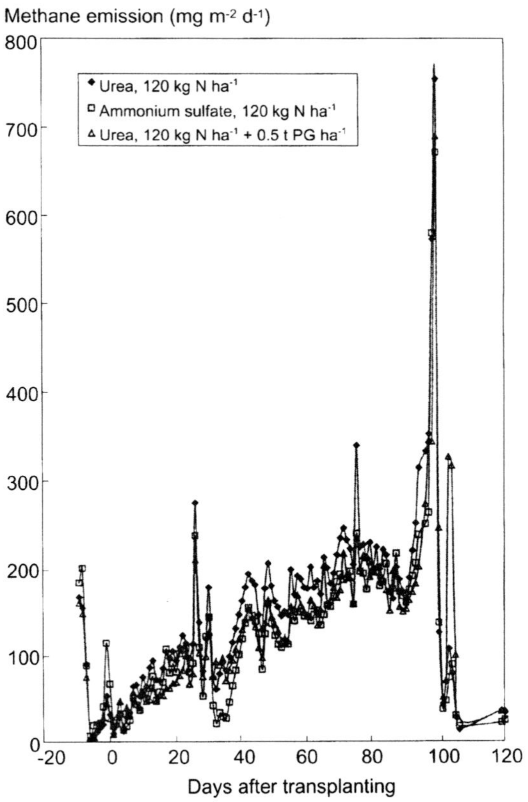

Figure 2. Effect of inorganic amendment on CH_4 emission from rice field grown to IR72 at PhilRice, Maligaya, Philippines, 1994 WS.

produced the same biomass as transplanted rice. However, grain yield of transplanted IR64 (7.7 and 7.9 t ha^{-1}) was significantly higher than that of direct seeded rice (6.4 and 6.7 t ha^{-1}). Midseason drainage significantly reduced CH_4 emission but not grain yield, hence reducing the amount of CH_4 produced from 11.3 and 11.2 kg CH_4 t^{-1} to 6.6 and 7.5 kg CH_4 t^{-1} grain yield, respectively. In 1997 WS, the CH_4 flux was high during the early vegetative growth and greater in transplanted than in direct-seeded rice (Figure 8). The reduction in CH_4 flux after midseason drainage was not distinct during the WS unlike in the DS. Water is difficult to control during the WS. The second flush of CH_4 flux before harvest was also observed. CH_4 flux was higher in continuously flooded plots than in midseason-drained plots. The final aboveground biomass in direct-seeded rice was again the same in all treatments. Also, as in the DS, the grain yield of transplanted IR64 (5.36 and 5.45 t ha^{-1}) was significantly higher than that in direct-seeded rice (3.41 and 3.84 t ha^{-1}). Midseason drainage did not reduce the mean CH_4 emission in transplanted rice (375 vs 347 mg CH_4 m^{-2} d^{-1} (z value = 0.5 ns). In direct-seeded rice, however, the mean CH_4 emission in midseason-drained plot was reduced by 45% (z value = 8.2**). The significant reduction in CH_4 emission in midseason-drained, direct-seeded rice resulted in the lowest amount (44 kg CH_4 t^{-1} grain yield) compared with the 64.9 - 70.8 kg CH_4 t^{-1} grain yield in continuously flooded rice.

In 1998 DS, the increasing trend in CH_4 emission as rice growth progresses and the flush of CH_4 before harvest were again observed, particularly in continuously flooded plots (Figure 9). Grain yield in transplanted rice was higher (8.0-8.5 t ha^{-1}) than in direct-seeded rice (7.1-7.7 t ha^{-1}), although biomass production in direct-seeded rice was much higher than that in transplanted rice. There was significant reduction in CH_4 emission of direct-seeded rice with intermittent irriga-

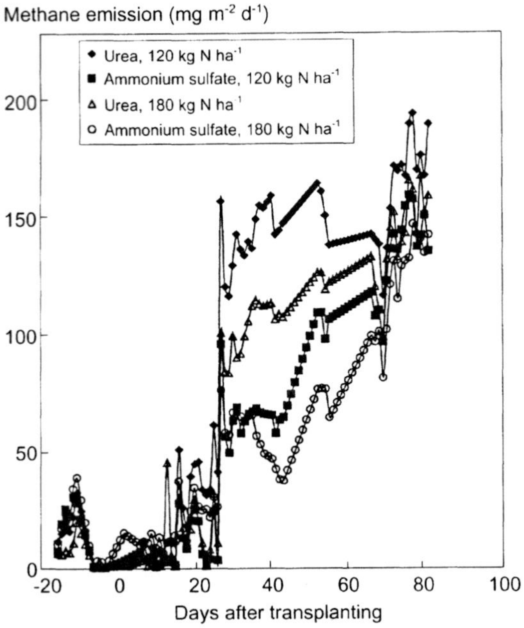

Figure 3. Effect of inorganic amendment on CH_4 emission from rice field grown to IR72 at PhilRice, Maligaya, Philippines, 1995 DS.

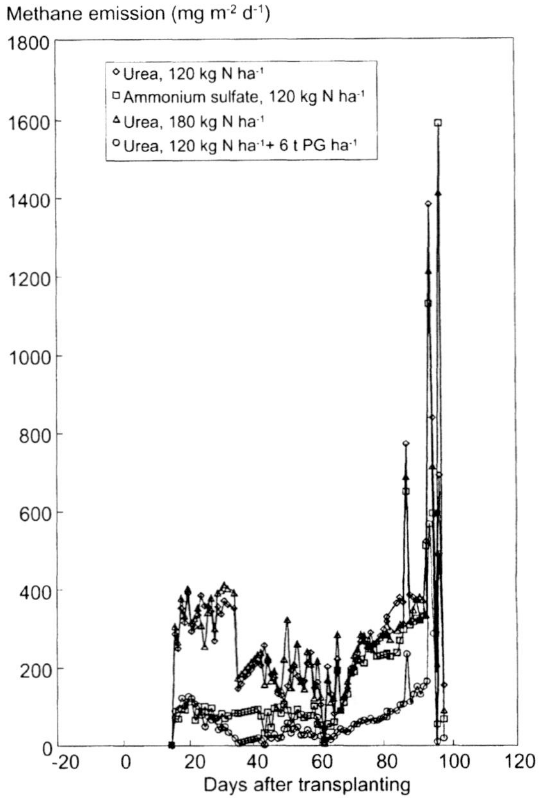

Figure 4. Effect of inorganic amendment on CH_4 emission from rice field grown to IR72 at PhilRice, Maligaya, Philippines, 1995 WS.

tion treatment (from 90 kg CH_4 ha^{-1} to 7 kg CH_4 ha^{-1}) and midseason drainage treatment (from 90 kg CH_4 ha^{-1} to 16 kg CH_4 ha^{-1}). These two treatments gave only 1 and 2.1 kg CH_4 t^{-1} grain yield as compared with 11.3 and 10.7 kg CH_4 t^{-1} grain yield in continuously flooded transplanted rice. Although intermittent irrigation resulted in negligible CH_4 emission, the yield was slightly lower.

Discussion

Effect of cropping season. The CH_4 emission at a given treatment was higher during the WS by 2 to 3 times the emission during the DS. Dry season CH_4 emissions in the reference treatment, i.e., 120 kg N ha^{-1} (urea N) were 95, 204, and 160 kg CH_4 ha^{-1} in 1994, 1995, and 1996, respectively. Wet season emissions amounted to 266, 518, and 272 kg CH_4 ha^{-1} in 1994, 1995, and 1996, respectively. Not only was there a variation between cropping season but there was also an annual variation in CH_4 emission. The T1 (reference treatment) showed wide differences in seasonal flux from year to year. This will pose problems in monitoring mitigation measures in farmers' fields. Table 5 shows that the DS months (December to April) had an average daily temperature

Table 5. Seasonal mean and range of air temperature during the 5-yr CH_4 measurement in PhilRice, Maligaya, Muñoz, Nueva Ecija, Philippines

Year	Temperature (°C)					
	Seasonal mean		Minimum		Maximum	
	DS	WS	DS	WS	DS	WS
1994	27.1	27.1	23.1	23.4	32.3	36.4
1995	26.4	27.3	22.4	25.2	29.2	29.2
1996	26.3	29.6	22.2	24.2	31.6	40.8
1997	26.7	28.4	20.9	24.9	30.3	30.8
1998	26.6	-	23.8	-	28.9	-
Mean of 5 seasons	26.62	28.10	22.48	24.42	30.46	34.30

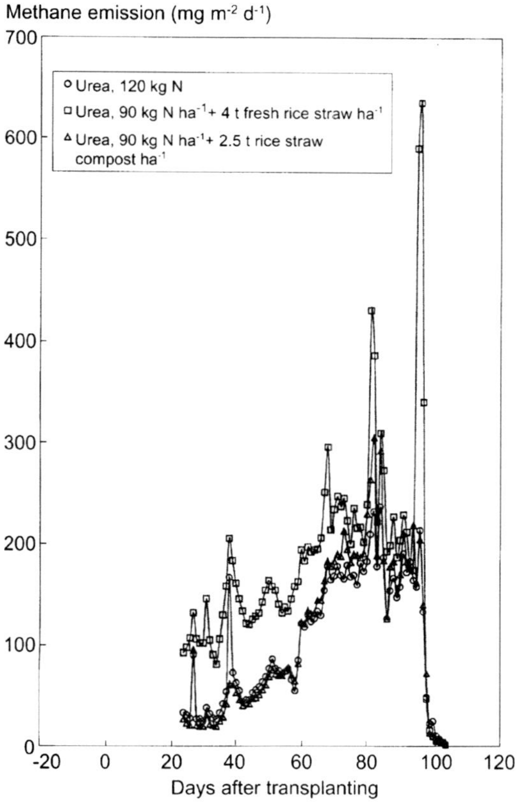

Figure 5. Effect of organic amendment on CH_4 emission from rice field grown to IR72 at PhilRice, Maligaya, Philippines, 1996 DS.

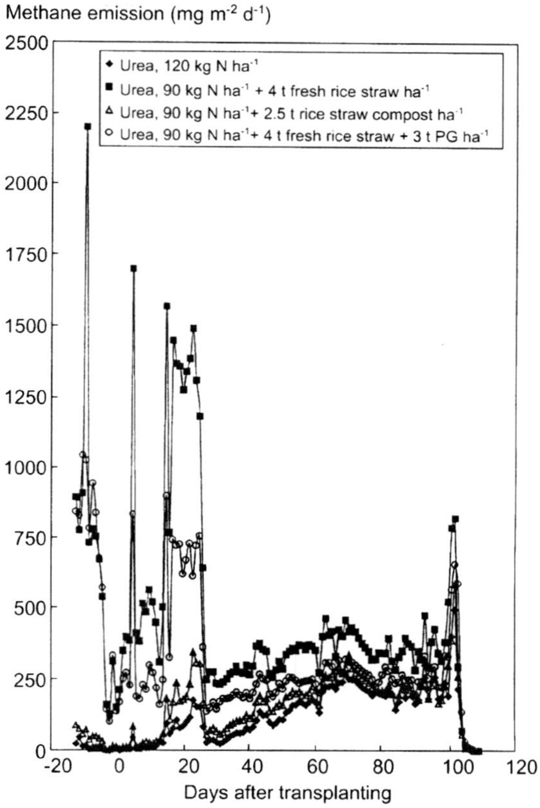

Figure 6. Effect of organic amendment on CH_4 emission from rice field grown to IR72 at PhilRice, Maligaya, Philippines, 1996 WS.

of 26.6 °C, while the WS months (June to November) had 28.1 °C. Maximum air temperature was lower in DS months by 4 and 9 °C during 1994 and 1996, respectively. But in 1995 and 1997, the maximum temperature for DS and WS were the same. Holzapfel-Pschorn and Seiler (1986) reported a marked influence of soil temperature on the CH_4 flux. Most isolates of methanogenic bacteria are mesophilic with temperature optimum of 30 °C to 40 °C (Vogel et al., 1988). The difference in daily mean temperature between the DS and the WS cropping period was, however, too small (1.6 °C) to explain the higher CH_4 emission during the WS. Temperature, theoretically, would deter or enhance the rate, not the magnitude, of emission. Contributing to this difference may be differences in labile organic carbon (OC) between the two seasons. Analysis of the OC before the 1998 DS and 1998 WS cropping showed an average of 1.15% OC before the DS cropping and 1.27% OC before the WS cropping at 0-25 cm depth. The difference was the same at the 25-50 cm depth (i.e., 0.52% OC before the DS cropping and 0.65% OC before the WS cropping). Furthermore, the field was still wet during the fallow period between WS and DS cropping. Bronson et al. (1997b) reported that CH_4 emitted from wet fallow periods is significant and should be considered when monitoring CH_4 emission from rice soils. Methane emissions during the wet fallow period during October and November before the DS crop could have resulted from decaying roots and stubble. Methane is not emitted during the April - May dry fallow period before the WS crop and accumulation of labile carbon shown by higher % OC may have resulted.

Effect of inorganic fertilizer. Most likely the SO_4^{-2} was responsible for the reduced CH_4 emission from ammonium sulfate- than from urea-treated plots. Saenjan and Wada (1990) reported that the presence of sulfate suppressed CH_4 formation. The CH_4 formation, both in flooded rice fields and in submerged soil under

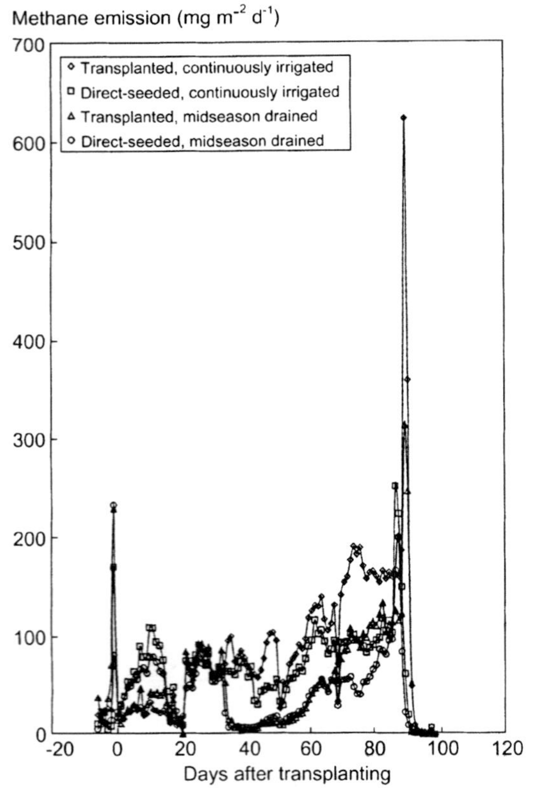

Figure 7. Effect of crop establishment and water regime on CH_4 emission from rice field grown to IR64 at PhilRice, Maligaya, Philippines, 1997 DS.

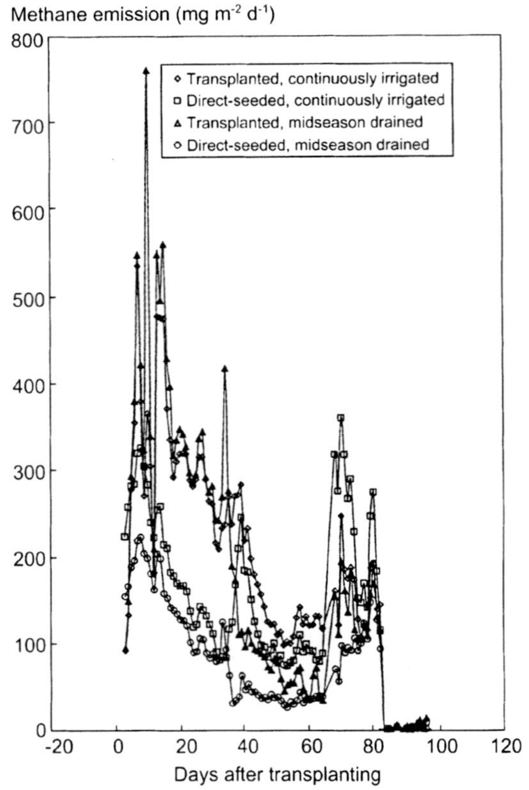

Figure 8. Effect of crop establishment and water regime on CH_4 emission from rice field grown to IR64 at PhilRice, Maligaya, Philippines, 1997 WS.

laboratory condition, is carried out largely by the transmethylation of acetic acid and by CO_2 reduction, utilizing H_2, butyric acid, etc. as hydrogen donors (Takai, 1970). Sulfate ions serve as an alternative to CO_2 as electron acceptors for the oxidation of organic matter (Delwiche & Cicerone, 1993). Differences on the effect of ammonium sulfate and urea fertilizer on CH_4 formation was reported by Wang et al. (1993) to be related to the effect on soil pH. Ammonium sulfate-treated plots had 25% to 56% less CH_4 emission averaged over the years and seasons compared with the urea-treated plots. Since addition of SO_4^{-2}-containing N fertilizers hardly changes the measured soil Eh and soil pH, the competition of SO_4^{-2}-reducing and CH_4-producing bacteria for substrates hydrogen and acetic acid, and possibly toxicity to the CH_4-producing bacteria from H_2S produced after SO_4^{-2} reduction, are likely mechanisms for the decreased CH_4 production in ammonium sulfate-treated plots. Hori et al. (1993) confirmed the possibility of competition for the usage of hydrogen between CH_4 formation and SO_4^{-2} reduction in strongly reduced rice soil. Competition for hydrogen, however, is less likely than that for acetic acid because the degree of competition for hydrogen is controlled by many factors. The added SO_4^{-2} from the ammonium sulfate fertilizer must have stimulated the SO_4^{-2}-reducing bacteria.

Increasing the rate of N from urea slightly increased CH_4 emission. Lindau et al. (1990, 1991) also reported increasing CH_4 fluxes with increasing rates of urea application. The increase in CH_4 emission with addition of higher N rate from urea could be due to the inhibitory effect of NH_4^+ on CH_4 oxidation (Conrad & Rothfuss, 1991).

Impact of phosphogypsum. Phosphogypsum (85-90% gypsum) is a waste product from the production of phosphoric acid by the wet process. The overall SO_3 content of PG is 44-46% (Alcordo & Rechcigl, 1993).

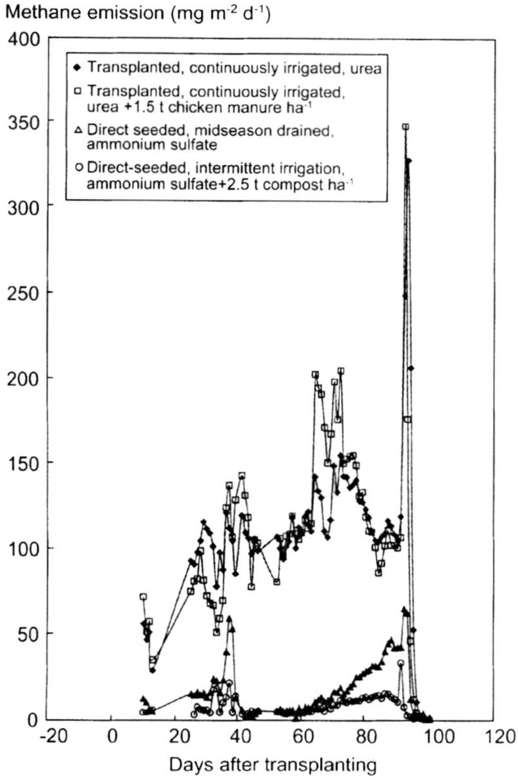

Figure 9. Combined effect of crop establishment, water regime, and inorganic and organic amendments on CH_4 emission from rice field grown to PSBRc28 at PhilRice, Maligaya, Philippines, 1998 DS.

When applied to the soil, PG solubilizes, producing Ca^{+2} and SO_4^{-2} ions. When PG was combined with urea, there was a significant reduction (z value = 10.0**) in mean CH_4 emission because of the SO_4^{-2} effect on CH_4 production. The effect of high amounts of ammonium sulfate and PG on CH_4 emission was similar. This confirms that it was the SO_4^{-2} and not the NH_4^+ that had affected the reduction in CH_4 emission. Denier van der Gon and Neue (1994) also reported a 55-70% reduction in CH_4 emission in an Aquandic Epiqualfs with amendment of 6.66 t ha^{-1} gypsum. Even with addition of green manure, the gypsum significantly decreased CH_4 emission (Denier van der Gon & Neue 1994). Addition of 3 t ha^{-1} of PG to fresh rice straw resulted in a 27% reduction in CH_4 emission compared with that from plots amended with fresh rice straw alone. The amount of carbon in fresh rice straw could be so high that the SO_4^{-2} from PG was not enough for sulfate-reducing bacteria to compete with the CH_4-producing bacteria to fully counteract the high CH_4 fluxes. However, another application of PG in succeeding cropping increased the cumulative effect of SO_4^{-2}. The succeeding cropping with same rate of PG and fresh rice straw resulted in a 37% reduction in CH_4 emission. PG is a cheaper source of SO_4^{-2} than ammonium sulfate and urea is a less expensive source of N fertilizer. Thus, the combination of urea as N fertilizer and PG as SO_4^{-2} source could be a management option to reduce CH_4 emission especially in sulfur-deficient irrigated lowland rice. Sulfate is normally reduced after the depletion of nitrate and other more energetically favorable reactions in anaerobic rice soils (Connel & Patrick, 1968, 1969; Ponnamperuma, 1972). Sulfate is reduced to H_2S which is toxic to rice at a concentration of approximately 0.07 ppm (Mitsui et al., 1951; Freney et al., 1982). However, H_2S seldom accumulates at toxic concentrations in most rice soils, since H_2S is either immediately precipitated as metallic sulfide, chiefly FeS, or is oxidized to sulfate or elemental sulfur in the rice rhizosphere by chemosynthetic microorganisms (Huang, 1991). The reoxidation of S^{-2} to SO_4^{-2} in the rhizosphere may also suppress CH_4 emission over long periods of time (Freney et al., 1982). This is the reason why addition of up to 6 t ha^{-1} PG in Maligaya clay with 75.02 μg g^{-1} active iron (Fe) did not manifest sulfide toxicity in the rice plant. In addition, PG was reported to have soil-conditioning effect in saline soils (Alcordo & Rechcigl, 1993). The annual world production of PG was estimated at 125 million Mg, and only 4% (5 million Mg) of it is used in agriculture and in gypsum board and cement industries. The remaining 120 million Mg PG accumulates annually as waste (Alcordo & Rechcigl, 1995). These could be used as soil ameliorant to decrease CH_4 emission in lowland rice.

Effect of organic amendment. Seasonal CH_4 fluxes from fresh rice straw-treated plots were 2.5 to 3.5 times greater than that from urea plots. Even the addition of PG in fresh rice straw treatment did not fully counteract these high CH_4 fluxes (Table 3). On the other hand, CH_4 fluxes in the rice straw compost treatment were similar to those in the urea treatment throughout the season.

Yagi and Minami (1990) reported that annual emission rates from plots receiving 6 t ha^{-1} of rice straw in addition to mineral fertilizer increased approximately 2 to 3 fold as compared with the mineral fertilizer plots, irrespective of soil type. Compost was also reported by Yagi and Minami (1990) to have only slightly increased emission compared with control plots. The readily mineralizable carbon (RMC) in the organic amendment was one of the principal factors affecting CH_4 emission from flooded soils (Yagi & Minami, 1990). Even without organic amendment, the readily mineralizable soil organic matter in rice soil is the main source for the fermentation products that finally drive CH_4 formation in wetland rice soils (Neue, 1993). Composting of the rice straw aerobically decreased the C/N from a range of 25-45 in fresh rice straw to a range of 6-10 in rice straw compost. This resulted in lesser carbon substrates, which in turn reduced CH_4 emission. The incorporation of rice straw during land preparation stage increased CH_4 emissions during the early vegetative stage until 30 DAT. Methane must have been produced from volatile fatty acids that were intermediate products of rice straw decomposition. In Texas, rice straw (8-12 t ha^{-1}) increased CH_4 emissions but rice yields dropped (Sass et al., 1991a,b).

Rice straw applications increased emissions 2-2.5 times but did not affect yield. Alberto et al. (1996) reported that straw incorporation increased dissolved CH_4 tenfold. Similar to the observation of Alberto et al. (1996), CH_4 emission was low in urea-and rice straw compost-treated plots 15 d before transplanting until 45 DAT and then paralleled those plots having straw treatment at later stages of rice growth (Figure 6). The early flush in CH_4 emission must have come from the decomposition of soil organic matter and added organic substrates such as rice straw. At the later stages, it is the root exudates and the decaying roots that become the major carbon source for CH_4 production (Alberto et al., 1996). Methane fluxes were slightly higher in the chicken manure treatment compared with the urea treatment at 35-45 DAT and 65-75 DAT (Figure 9). The seasonal emission, however, was the same in urea-treated plot and in chicken manure plus urea treatment. Chicken manure has a narrow C/N that is between 5 and 8. The CH_4 emission per unit of carbon from chicken manure was comparable with that of the rice straw compost that had a C/N of 6-10.

Effect of water regime and crop establishment. Methane fluxes under two water regimes (continuously flooded and midseason-drained) and two crop establishment methods (direct seeded and transplanted) were compared. The first flush of CH_4 fluxes during the early vegetative stage (Figures 7&8) could be due to decomposing stubble incorporated during land preparation and from the commercial bioorganic fertilizer applied during the final harrowing. Methane flux was reduced after midseason drainage due to aeration. This midseason drainage could be beneficial to the rice plant. The draining of rice fields for short-term periods in China at the end of tillering and before heading improved yields and reduced CH_4 emission (Wang, 1986). In Japan, the intermittent irrigation of rice fields resulted in lower CH_4 emission than those reported from western countries (Yagi & Minami, 1990). Bronson et al. (1997a) reported that midseason drainage (2-wk duration) at either maximum tillering or panicle initiation suppressed CH_4 flux. However, N_2O flux increased sharply during the drainage period, until reflooding, when it dropped back to zero. Midseason drainage as a strategy to reduce CH_4 emission should be on a short duration (7-10 d) and timed when the rice plants have used up the fertilizer N applied at basal and vegetative stages. Reflooding should be done before the application of N fertilizer at the panicle initiation stage. Intermittent irrigation, though it significantly reduced (92%) CH_4 emission, must be carefully evaluated as a mitigation strategy. Bronson (1994) reported that urea or ammonium sulfate fertilizer from irrigated rice fields have N_2O losses to a maximum of 0.1% of the applied fertilizer. With intermittent irrigation, where water regime is variable, more N_2O could be emitted as a result of higher rates of nitrification and denitrification that occur than in continuously flooded conditions. Multiple-aeration water management treatment emitted 88% less CH_4 and did not reduce yield (Sass et al., 1992). However, this intermittent drainage must be managed carefully to prevent losses of N and corresponding emission of N_2O through increased nitrification and denitrification (Neue, 1993; Bronson et al., 1997a).

Direct-seeded rice reduced CH_4 emission by 16-54% compared with transplanted rice. The mechanism explaining this difference is not yet clear. The root system of direct-seeded rice is expected to differ from that of transplanted rice. It is probable that the roots of direct-seeded rice are shallower than that of transplanted rice. With more roots present at the 0-10 cm depth, there could be more CH_4 oxidized to CO_2, thus reducing the CH_4 emission. Unfortunately, rooting characteristics of direct-seeded rice (as compared with transplanted rice) were not investigated in this experiment.

Mitigation strategies

The management practices tested in this 5-yr experiment have been primarily designed to look for mitigation strategies that are workable under Philippine conditions. It was postulated that some aspects of crop management, including the management of inorganic fertilizers, organic fertilizers, water regime, and crop establishment, could be effectively modified to mitigate CH_4 emissions from irrigated rice fields. Mitigation of CH_4 emissions, while targeting high yields, has been the prime target in using sulfur-containing inorganic amendments, in increasing N fertilizer application, in using rice straw compost, in practicing midseason drainage, and in practicing direct seeding. Whatever mitigation measure to reduce CH_4 emission has to ensure that it will not decrease grain yield. This is the most important consideration if these mitigation strategies are to be adapted by farmers. Results show significant reduction (25-36%) in CH_4 emissions with the use of ammonium sulfate as N fertilizer source instead of urea. The addition of 6 t ha^{-1} PG to urea has resulted in 72% reduction in emissions. Midseason drainage reduced CH_4 emission by 43%, while intermittent irrigation resulted in 92% reduction. Direct seeding, instead of transplanting, reduced CH_4 emission by 16-54%. Expectedly, the application of rice straw compost did not reduce emissions but rather increased it by 23-30%. But this is very small compared with the increase of 162-250% in emissions due to fresh rice straw application. Also, the use of chicken manure did not enhance CH_4 emissions in one experiment. The use of organic fertilizer and nutrient cycling from crop residues is presently being encouraged in view of soil fertility in the long term. In the last experiment (1998 DS), the different management strategies (ammonium sulfate fertilizer, rice straw compost, direct seeding, midseason drainage, and intermittent irrigation) were combined in two treatments and the result was a dramatic reduction of CH_4 emission (83-93%). This, however, needs to be verified in WS and in another DS experiment. It is important to note that these modifying treatments that successfully reduced emissions did not adversely affect grain yield. The practice of direct seeding is an exception, where grain yield was lower by 0.8-1.3 t ha^{-1} in the DS and 1.8 t ha^{-1} in the WS. Direct seeding is already widely practiced in major rice-growing areas during the DS; in central Philippines (Panay Island), 90% of the farmers are practicing direct seeding both during DS and WS cropping. Development of high yield technology for direct-seeded rice cultivation is one of the current research thrusts of PhilRice.

The workability of the above mitigation strategies under the Philippine situation needs evaluation. Results of a survey conducted in October-November 1998 showed that rice farmers in Nueva Ecija commonly use urea and complete (14-14-14) fertilizer, not ammonium sulfate because urea N is cheaper than ammonium sulfate. Ammonium sulfate is used mostly in seedbed preparation. However, 14-14-14 fertilizer also contains sulfur. Thus, the use of this fertilizer may also contribute to reduced CH_4 emissions. Farmers in the Philippines are not deliberately practicing midseason drainage or intermittent irrigation. Drainage of soils within the season is determined by the availability of rain or irrigation water. Since water is becoming scarce in many instances, farmers normally would not deliberately remove water at definite periods of the season because of the uncertainty of water availability. On the other hand, because of water becoming a limiting resource, especially during the DS, the midseason drainage practiced by farmers in China and Japan will be favorable to the Filipino farmers' management of their scarce resources. The use of organic amendments, particularly rice straw, is presently being encouraged in an effort to recycle nutrients and improve the fertility of rice soils. As a mitigation strategy, composting the rice straw aerobically must be promoted rather than fresh rice straw incorporation. A rapid rice straw composting technology is available. Most farmers, however, found composting and spreading of straw laborious. Farmers burn their rice straw instead of incorporating it into the soil so as not to encourage pests such as rats. The adoption of mitigation strategies by farmers may not be as hard as it is assumed because of the following reasons. First, our results showed that there was no real adverse effect on yield. Second, mitigation measures proposed are compatible with building soil fertility (use of rice straw compost), proper management of water (midseason drainage vs continuous flooding), and savings on labor (direct seeding vs transplanting). Third, farmers are beginning to observe the effect of global warming from longer drought (El Niño) and flood (La Niña) periods.

Conclusion

The 5-yr CH_4 measurements have established a pattern of emission common to DS and WS. The emissions, however, are magnified in the WS, and seasonal emission was found to be 2-3 times as much as that in the DS. This was partly explained by the 1.6 °C higher daily mean temperature in the WS. However, temperature theoretically would deter or enhance the rate of emission, not its magnitude. One obvious contributor to CH_4 emissions is the carbon input (Neue et al., 1994). Dry matter production and also the stubble left for the next season did not significantly differ between the two seasons. The difference in decomposable carbon between the two seasons could possibly explain this difference in WS and DS emissions. There was 0.12%

more % OC in the soil before the WS cropping than before the DS cropping. Furthermore, CH_4 from the decaying roots and stubble during the wet fallow period during October and November before the DS crop could have been emitted but was not measured. Increasing the rate of urea N from 120 to 180 kg ha^{-1} increased seasonal CH_4 emission by only ~15% in the WS. Using ammonium sulfate in place of urea at 120 kg N ha^{-1} resulted in 25% reduction in annual average of CH_4 emission. Increasing ammonium sulfate rate to 180 kg N ha^{-1} increased the reduction in annual average CH_4 emission by 36%. The effect of 0.5-1.0 t ha^{-1} PG was similar to that of ammonium sulfate at 120 kg N ha^{-1}. A significant effect of PG on CH_4 emission (72% reduction) was obtained at 6 t ha^{-1}. The residual effect of the 857 kg ammonium sulfate (180 kg N ha^{-1} yr^{-1}) and 6 t ha^{-1} yr^{-1} PG application was not clear. It is possible that one time application in a year or every 2 yr or continuous application is required to obtain the desired effect. Organic amendment such as fresh rice straw with wider C/N increased CH_4 emission to twice that of mineral fertilizer alone. Rice straw compost and chicken manure, which have narrower C/N, had little effect on CH_4 emission. Even the addition of PG with fresh rice straw could not fully counteract the high CH_4 emission.

Introduction of midseason drainage water management is one cultural practice that could be used to reduce CH_4 emission by as much as 90% compared with continuously flooded rice. This, however, has to be timed to obtain the highest N fertilizer use efficiency and minimize N_2O emissions. Another interesting result obtained was the lower mean CH_4 emission in direct-seeded than in transplanted rice. Also, direct-seeded rice had a shorter season length than transplanted rice, which could further contribute to lower seasonal CH_4 flux. This was despite the higher number of tillers per m^2 in direct-seeded rice. It would be interesting to investigate the root development, root distribution, and root characteristics of direct-seeded rice, which contributed to this lower emission.

Several management options to mitigate CH_4 emissions from irrigated rice field were identified. In terms of their effectiveness in reducing CH_4 emissions compared with the control treatment (urea fertilizer, transplanted rice, and continuously flooded), these are ranked as follows: (1) 6 t ha^{-1} PG combined with urea fertilizer, (2) midseason drainage 7-10 d before panicle initiation, (3) use of ammonium sulfate fertilizer as N source, and (4) direct seeding crop establishment. If organic fertilizer is combined with inorganic fertilizer in integrated plant nutrient management, low C/N organic fertilizers such as chicken manure and rice straw compost will not significantly increase CH_4 emission. The measurements reported here were carried out in a heavy clay soil. Whether the same results will be obtained using a different soil in a different environment remains a consideration for future measurements.

Acknowledgements

The research described in this paper was funded by the UNDP-Global Environment Facility GLO/91/G31 under agreement with the International Rice Research Institute (IRRI) and the Philippine Rice Research Institute (PhilRice)-IRRI Collaborative Project. However, UNDP did not review the paper, hence it does not necessarily reflect the views of UNDP and no endorsement should be inferred.

References

Alberto MCR, Neue HU, Capati AB, Castro RU, Bernardo LM, Aduna JB & Lantin R (1996) Effect of different straw management practices on soil fertility, rice yields and the environment. In: Attanandana T, Kheoruenromne I, Pongsakul P & Vearasilp T (eds). Proceedings of the International Symposium on Maximizing Sustainable Rice Yields Through Improved Soil and Environmental Management Vol. 2; 11-17 November 1996; Khon Kaen, Thailand. pp 721-731

Alcordo IS & JE Rechcigl (1993) Phosphogypsum in Agriculture: A Review. Adv Agron 49:55-118

Alcordo IS & Rechcigl JE (1995) Phosphogypsum and other by-product gypsums. In: Rechcigl JE (ed) Soil Amendments and Environmental QualityAgriculture and Environment Series. New York: CRC Press Inc. (Lewis Publishers) pp 365-425

Bronson KF (1994) Nitrous oxide emissions from flooded rice. Abstracts Climate Change and Rice International Symposium, 14-18 March 1994, IRRI Los Baños, Laguna, Philippines

Bronson KF, Neue HU, Singh U & Abao EB Jr (1997a) Automated chamber measurements of methane and nitrous oxide flux in a flooded rice soil. I. Residue, nitrogen and water management. Soil Sci Soc Am J 61(3):981-987

Bronson KF, Singh U & Neue HU (1997b) Automated chamber measurements of methane and nitrous oxide flux in a flooded rice soil. II. Fallow period emission. Soil Sci Soc Am J 61(3):988-993

Connel WE & Patrick WH Jr (1968) Reduction in soil: Effects of redox potential and pH. Science 159:86-87.

Connel WE & Patrick WH Jr (1969) Reduction of sulfate to sulfide in waterlogged soil. Soil Sci Soc Am Proc 33:711-715

Conrad R & Rothfuss F (1991) Methane oxidation in the soil surface layer of a flooded rice field and the effect of ammonium. Biol Fertil Soils 12:28-32

Delwiche CC & Cicerone RJ (1993) Factors affecting methane production under rice. Global Biogeochem Cycles 7:143-155

Denier van der Gon HAC & Neue HU (1994) Impact of gypsum application on the methane emission from wetland rice field. Global Biogeochem Cycles 8:127-134

Freney JR, Jacq VA & Baldensperger JF (1982) The significance of the biological sulfur cycle in rice production. In: Dommergues YR & Diem HC (eds). Microbiology of Tropical Soils and Plant Productivity. Martinus Nijhoff Publ, The Hague. pp 271-317

Holzapfel-Pschorn A & Seiler W (1986) Methane emission during a cultivation period from an Italian rice paddy. J. Geophys Res 91(D):11803-11814

Hori K, Inubushi K, Matsumoto S & Wada H (1993) Competition for hydrogen between methane formation and sulfate reduction in a paddy soil. Jpn J Soil Sci Plant Nutri 64:363-367

Huang SH (1991) Production and emission of methane from experimental paddy soils. Agricultural Improvement Station, Taiwan Provincial Taichung District, Changhua, Taiwan R.O.C. 60 p

IAEA-TECDOC-674 (1993) Manual on methane and nitrous oxide emissions from agriculture. A joint undertaking by the Food and Agriculture Organization of the United Nations and the International Atomic Energy Agency. 91 p

Lindau CW, De Laune RD, Patrick WH Jr & Bollich PK (1990) Fertilizer effects on dinitrogen, nitrous oxide, and methane emissions from lowland rice. Soil Sci Soc Am J 54:1789-1794

Lindau CW, Bollich PK, De Laune RD, Patrick WH Jr & Lau VJ (1991) Effect of urea fertilizer and environmental factors on methane emissions from a Louisiana rice field. Plant Soil 136:195-203

Minami K (1994) Methane from rice production. Fert Res 37:167-179

Mitsui S, Aso S & Kumazawa K (1951) Dynamic studies on the nutrient uptake by crop plants. I. The nutrient uptake of roots as influenced by hydrogen sulfide. J Sci Soil Manure Jpn 22:324-328

Neue HU (1993) Methane emission from rice fields. Bioscience 43:466-474

Neue HU, Lantin RS, Wassmann R, Aduna JB, Alberto MCR & Andales JF (1994) Methane emission from rice soils of the Philippines. In Minami K, Mosier A & Sass RL (eds) CH_4 and N_2O: Global Emissions and Controls from Rice Fields and Other Agricultural and Industrial Sources, Tsukuba, Japan: NIAES Series 2. pp 55-63

Neue HU & Sass RL (1994) Trace gas emissions from rice fields. In: Prinn RG (ed). Global Atmospheric-Biospheric Chemistry, Environmental Science Research Vol. 48. New York: Plenum Press. pp 119-147

Ponnamperuma FN (1972) The chemistry of submerged soils. Adv Agron 24:29-97

Saenjan P & Wada H (1990) Effects of salts on methane formation and sulfate reduction in submerged soil. Trans 14th Int Congr Soil Sci, Kyoto Japan, 2(II):244-248

Sass, RL, Fisher FM & Harcombe PA (1991a) Mitigation of methane emissions from rice fields: Possible adverse effects of incorporated rice straw. Global Biogeochem Cycles 5:275-287

Sass RL, Fischer FM, Harcombe PA & Turner FT (1991b) Methane production and emission in a Texas rice field. Global Biogeochem Cycles 4:47-68

Sass RL, Fisher FM, Wang YB, Turner FT & Lund MF (1992) Methane emission from rice fields: The effect of floodwater management. Global Biogeochem Cycles 6(3):249-262

Takai Y (1970) The mechanism of methane fermentation in flooded paddy soil. Soil Sci Plant Nutr 16:238-244

Vogel DD, Keltjens JT & Vander Drift C (1988) Biochemistry of methane production. In: Zehnder AJB (ed) Biology of Anaerobic Microorganisms. New York :Wiley. pp 707-770

Wang Z (1986) Rice based systems in subtropical China, p. 195-206. In: Juo ASR Lowe JA (eds) Wetlands and rice in Subsahara Africa. International Institute for Tropical Agriculture, Ibadan, Nigeria

Wang, ZP, De Laune RD, Lindau CW & Patrick WH Jr (1993) Methane production from anaerobic soil amended with rice straw and nitrogen fertilizers. Fert Res 33:115-1212

Wassmann R, Neue HU, Lantin RS, Buendia LV & Rennenberg H (2000) Characterization of methane emissions from rice fields in Asia. 2. Comparison among field sites in five countries. Nutr Cycling Agroecosyst (this issue)

Yagi K & Minami K (1990) Effect of organic matter application on methane emission from some Japanese paddy fields. Soil Sci Plant Nutr 36(4):599-610

Nutrient Cycling in Agroecosystems **58:** 55–63, 2000.

A four-year record of methane emissions from irrigated rice fields in the Beijing region of China

Z.Y. Wang[1], Y.C. Xu[1], Z. Li[1], Y.X. Guo[1], R. Wassmann[2,3], H.U. Neue[2,4], R.S. Lantin[2], L.V. Buendia[2], Y.P. Ding[1] & Z.Z. Wang[1]

[1]*Institute of Crop Breeding and Cultivation, Chinese Academy of Agricultural Sciences, Beijing 100081, China* [2]*International Rice Research Institute, Los Banos, Laguna, Philippines;* [3]*Fraunhofer Institute for Atmospheric Environmental Research, Garmisch-Partenkirchen, Germany;* [4]*UFZ-Center for Environmental Research, Halle, Germany*

Key words: methane emission, rice fields, water management, organic manure, rice cultivar, organic amendment, mitigation options

Abstract

Methane (CH_4) emissions from irrigated rice fields were measured using an automatic sampling-measuring system with a closed chamber method in 1995-98. Average emission rates ranged from 11 to 364 mg m^{-2} d^{-1} depending on season, water regime, and fertilizer application. Crop management typical for this region (i.e., midseason drainage and organic/mineral fertilizer application) resulted in emission of 279 and 139 mg CH_4 m^{-2} d^{-1} in 1995 and 1997, respectively. This roughly corresponds to emissions observed in other rice-growing areas of China. Emissions were very intense during the tillering stage, which accounted for 85% of total annual emission, but these were suppressed by low temperature in the late stage of the season. The local irrigation practice of drying at midseason reduced emission rates by 23%, as compared with continuous flooding. Further reduction of CH_4 emissions could be attained by (1) alternate flooding/drying, (2) shifting the drainage period to an earlier stage, or (3) splitting drainage into two phases (of which one is in an earlier stage). Emission rates were extremely sensitive to organic amendments: seasonal emissions from fields treated with pig manure were 15-35 times higher than those treated with ammonium sulfate in the corresponding season. On the basis of identical carbon inputs, CH_4 emission potential varied among organic amendments. Rice straw had higher emissions than cattle manure but lower emissions than pig manure. Use of cultivar Zhongzhuo (modern japonica) reduced CH_4 emission by 56% and 50%, in 1995 and 1997, respectively, as compared with Jingyou (japonica hybrid) and Zhonghua (tall japonica). The results give evidence that CH_4 emissions from rice fields in northern China can be reduced by a package of crop management options without affecting yields.

Introduction

Methane (CH_4) is an important greenhouse gas and it affects the chemistry and oxidation capacity of the atmosphere (Bolle et al., 1986; Rasmussen & Khalil, 1986; Thompson & Cicerone, 1986). The CH_4 concentration in the atmosphere has doubled during the last 200 yr (IPCC, 1992). Rice fields have been identified as a major source of atmospheric CH_4, contributing about 10-15% to global CH_4 emission (Neue, 1993; Sass, 1995). Owing to the rice demand of the rapidly growing population, rice cultivation and productivity will continue to increase in the coming decades. This increase in yield and harvest area of rice may further increase CH_4 emission if present practices are not changed toward lowering emission potentials.

China is the largest rice-producing country in the world. Rice harvested area in 1994 was estimated at 30.1 million ha and average rice yield was 5.83 t ha^{-1} (Agricultural Year Book of China, 1995). On the other hand, Chinese rice fields emit substantial amounts of CH_4. In recent years, research on CH_4 emission from Chinese rice fields is building up (Wang et al., 1990; Chen et al., 1993a,b; Min et al., 1993; Wassmann et al.,

1993a,b; Cai et al., 1995; Shao et al., 1996; Wassmann et al., 1996; Yao et al., 1996; Kern et al., 1997; Tao & Du, 1998).

The 4-yr study presented here was conducted within an international network of measuring stations for determining CH_4 emissions from rice fields (Wassmann et al., this issue, a). Nonvalidated data of the first two seasons of this study have been presented in a non-authorized publication (Wang et al., 1999) by a first author who was not a member of the project team.

The Beijing station represents a typical single rice cropping system in northern China and was the only network station in a temperate climate. The objectives of the studies at the Beijing station were (i) to quantify CH_4 fluxes from rice fields in northern China; (ii) to assess the impact of management practices common to this region; (iii) to evaluate effects of low temperatures in the early and late cropping seasons; and (iv) to develop mitigation strategies with low CH_4 emission in a sustainable rice system for this region.

Table 1. Some characteristics of rice soil in the Beijing experimental station (per kg dry soil)

pH	7.99
Organic carbon (g kg^{-1})	9.95
Total nitrogen (g kg^{-1})	0.91
Cation exchange capacity (cmol kg^{-1})	13.20
Olsen phosphorus (mg kg^{-1})	133.00
Exchangeable potassium (cmol kg^{-1})	0.11

Materials and methods

Field preparation

Field experiments were conducted at the experimental farm of the Institute of Crop Breeding and Cultivation, Beijing, China, for four rice seasons starting in 1995. Some characteristics of the soil (silty clay loam) are shown in Table 1. The details of field trials conducted from 1995 to 1998 are shown in Table 2. Each rice season encompassed four treatments in randomized complete block design with four replicates. Fields were flooded 1 or 2 d before transplanting for harrowing and leveling. The individual plot size was 4.5 × 5 m.

Methane emission rates

Methane emissions were monitored by an automatic sampling and measuring system (Wassmann et al., this issue, a). Methane emission was continuously measured every 2 h from each field chamber (1 m × 1 m base and 1.2 m high) during the entire growing season. Methane concentrations of air samples were measured with Shimadzu GC-8A equipped with Porapak N column and a flame ionization detector.

Table 2. Summary of modifying treatments for experiments, 1995-98

Year	Treatment no.	Water management	Ammonium sulfate		Organic manure			Variety	Date transplanted	Date harvested
			Basal (kg N ha^{-1})	Topdressing (kg N ha^{-1})	Type	Organic N (kg N ha^{-1})	Organic C (kg N ha^{-1})			
1995	1	Local irrigation practice	30	60	Pig manure	60	1783	Zhongzhuo	06/04	10/17
	2	Alternate flooding/drying	30	60	Pig manure	60	1783			
	3	Continuous irrigation	30	60	Pig manure	60	1783			
	4	Local irrigation practice	40	110	-	-	-			
1996	1	Local irrigation practice	40	80	None	-	-	Zhongzhuo	05/24	10/08
	2		40	80	-	-	-	Jingyou		
	3		40	80	-	-	-	Zhongzhua		
	4		40	80	-	-	-	IR72		
1997	1	Local irrigation practice	40	80	None	-	-	Zhongzhuo	05/21	10/06
	2		20	60	Pig manure	40	1059			
	3		27	80	Cattle manure	13	1059			
	4		31	80	Rice straw	9	1059			
1998	1	Local irrigation practice	28	60	Compost	32	1059	Zhongzhuo	05/19	10/06
	2	Dual drainage (early)	28	60		32	1059			
	3	Dual drainage (late)	28	60		32	1059			
	4	Single drainage (early)	28	60		32	1059			

Temperature, soil pH, and Eh

Temperatures of air, floodwater, and soil at 5, 10, and 15 cm depths were measured by temperature probes connected to a data logger. Soil pH and soil Eh at 7.5 cm depth were measured manually with Philips pH/Eh meter every 2 d from transplanting until harvest.

Statistical analysis of experimental data was accomplished using STATISTICA program (Statsoft, Inc. 1993). The data in each treatment were evaluated as to type of distribution. If distribution was normal, the t-test was used; when it was not, the sign test was used.

Results and discussion

Characterization of seasonal fluxes

A typical pattern of CH_4 emissions under a local crop management is shown in Figure 1 jointly with temperature, field water level, soil Eh, and pH. The fertilizers in this 1996 experiment consisted of a mineral fertilizer only (ammonium sulfate) (Table 2). Local water management included persistent flooding (at 4-cm water depth) that was interrupted by a midseason drainage. The field was dried at the end of the season.

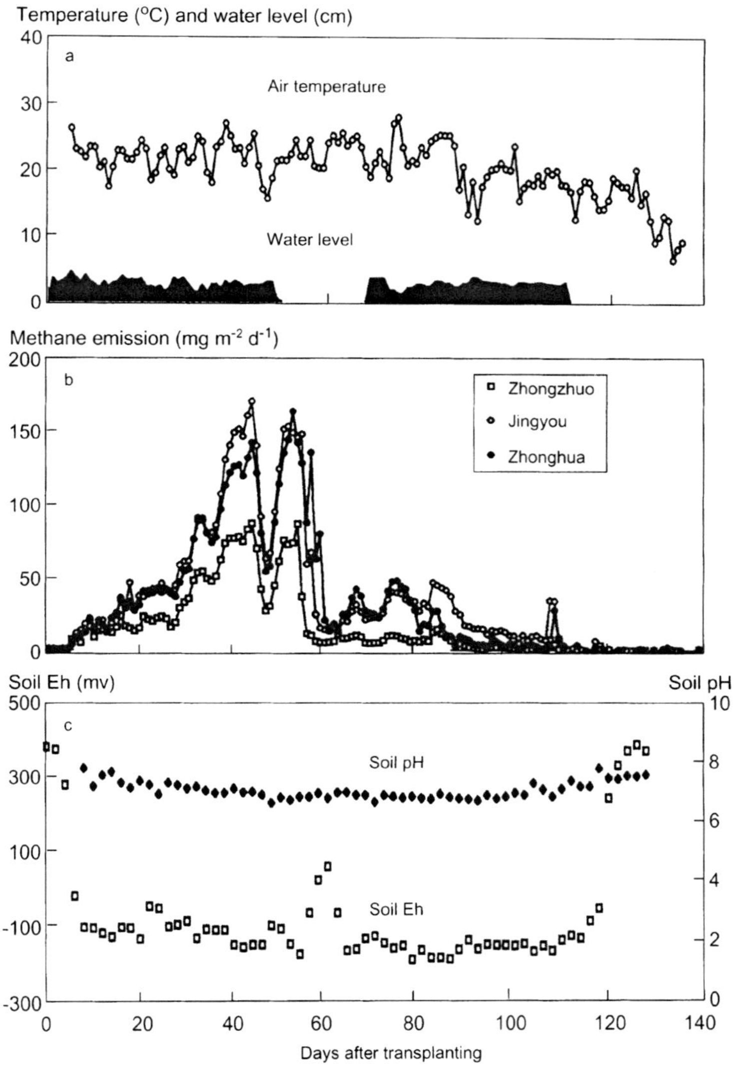

Figure 1. Seasonal pattern of (a) temperature and water level (Zhongzhuo); (b) CH_4 emission (daily averages) of three cultivars; and (c) soil pH and Eh (Zhongzhuo) in rice soil at Beijing station during the 1996 rice season

Methane emission rates rapidly increased during the first 40 d after transplanting (DAT) when temperatures were relatively high (Figure 1a). Methane emissions fluctuated strongly between 45 and 60 DAT due to a combination of climatic and management factors. Relatively cold weather between 45 and 50 DAT decreased emissions, field drainage after 51 DAT resulted in a release of entrapped CH_4 gas in the soil followed by a rapid decrease (Figure 1a,b). After reflooding at 68 DAT, CH_4 emissions remained at low levels. At the end of the season, temperatures were below 15 °C and CH_4 emission rates were virtually zero. Methane emission in the early season accounted for 85% of the total amount emitted over the season. Methane emission after midseason drainage until harvest was only a small fraction of the total CH_4 emitted from rice fields.

The seasonal pattern of CH_4 emission reflected the influence of temperature changes and midseason drainage. The pattern can be broken up into three phases (Figure 1b): (1) emission rates increase at tillering stage; (2) emissions fluctuate at reproductive stage as influenced by drainage; (3) emissions decrease at late growth stages due to temperature drop and field drainage.

The redox potential was governed by the local practice in water management. Flooding resulted in soil Eh decrease while field drying caused an increase in soil Eh (Figure 1c). Generally, soil Eh decreased from positive values to the critical value of CH_4 production (−120 to −150 mV) (Wang et al., 1993) within 1-3 wk after field flooding. Anaerobic conditions promoted CH_4 formation. Drainage resulted in a sudden increase in redox potential (Figure 1c). The effect of soil pH on CH_4 emission was negligible under field condition in Beijing. These findings were confirmed by similar trends in Eh and pH throughout the entire observation period from 1995 to 1998.

Effect of water regime

The patterns of CH_4 emission from rice fields as affected by water regime are shown in Figure 2. The experiment in 1995 compared three different water regimes: (1) local practice (field drying at 50 - 68 DAT and at 112 - 138 DAT); (2) alternate flooding/drying (7 times drying: 12 - 16, 25 - 32, 44 - 50, 59 - 64, 73 - 78, 86 - 91, and 100 - 135 DAT); and (3) continuous flooding (dry only at 32 d before harvest). All fields were fertilized with pig manure. Methane emission started to increase within the first week of flooding. For all treatments, seasonal maximum values occurred at maximum tillering stage. Continuous flooding resulted in the highest rates while alternate flooding/drying plots

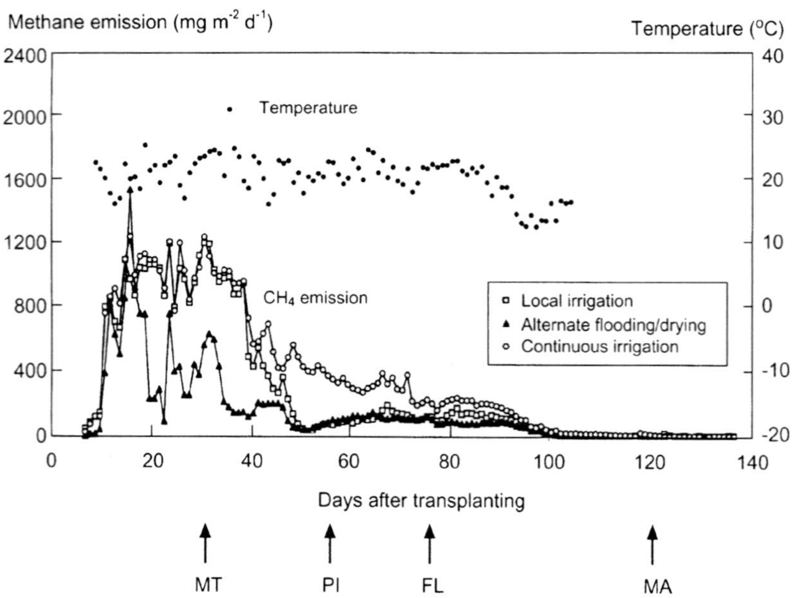

Figure 2. Seasonal patterns of temperature and CH_4 emission (daily averages) as affected by water regime, 1995 rice season. Arrows under the x axis denote the growth stages of maximum tillering (MT), panicle initiation (PI), flowering (FL), and maturity (MA)

gave the lowest CH_4 emission among the three water regimes. Low temperatures during maturity stage resulted in uniformly low emission rates.

Table 3 presents the mean and seasonal CH_4 fluxes, biomass, and grain yields for 4 yr. In 1995, CH_4 emission from local irrigation practice was 86% higher than alternate flooding/drying and 23% lower than continuous flooding. Local practice of irrigation provided highest biomass and grain yields, although only differences with continuous flooding were significant ($P < 0.05$). These results reveal that midseason drainage and alternate flooding/drying can be a promising mitigation strategy that does not affect yields.

In 1998, the field experiment included four different types of drainage (Table 2). As in previous years, local practice (T_1) encompassed late single drying at 55 - 68 DAT. Treatment T_4 represented an early single drainage (35 - 48 DAT), whereas the drainage was split into 2 separate weeks in T_3 (35 - 41 DAT and 55 - 61 DAT) and T_2 (25 - 31 DAT and 45 - 51 DAT). All fields received mineral fertilizer and compost, resulting in relatively lower emission rates even before the drying periods (Figure 3a, b). Local practice of irrigation resulted in the highest CH_4 emission that was obviously related to the relatively late onset of the drainage period at 55 DAT. Likewise, the late timing of two separate drainage periods also entailed higher emission rates. The most effective drainage period for mitigating CH_4 emissions is 35 and 48 DAT as can be seen by comparing T_2 and T_4. In the 1998 experiment, however, results are attached to strong spatial variations as can be seen in the phase before drainage was applied to T_1 and T_4 plots.

Average emission rates were 20 mg m^{-2} d^{-1} with local practice of irrigation (T_1), 19 mg m^{-2} d^{-1} with late dual drainage (T_2), 15 mg m^{-2} d^{-1} with early dual drainage (T_3), and 11 mg m^{-2} d^{-1} with early season drainage (T_4). Methane emissions in T_2 and T_3 were reduced by 5% and 25%, respectively, as compared with T_1 while similar yields were obtained (Table 3). T_4 gave 46% reduction in CH_4 emission as compared with T_1 and yields were also similar. The results indicated that the local practice of irrigation could further be optimized to reduce CH_4 emission while sustaining rice yields.

The significance of water regime for CH_4 emissions from Chinese rice fields was also shown in other field studies in China. As compared with continuous irrigation, alternate flooding/drying reduced emissions

Table 3. Mean and seasonal methane emission rates, biomass, and yields per modifying treatment, 1995 -98, Beijing

Year	Treatment no.	Modifying treatment	Mean emission (mg m^2 d^{-1})	Seasonal emission (kg ha^{-1})	Above-ground biomass (t ha^{-1})	Grain yield (t ha^{-1})
1995	1	Local irrigation practice + pig manure	279 b	385	20.66 a	6.49 a
	2	Local irrigation practice + mineral fertilizer	19 d	26	17.73 b	5.43 b
	3	Continuous irrigation + pig manure	364 a	503	18.73 b	5.61 b
	4	Alternate flooding/drying + pig manure	150 c	207	19.98 a	6.48 a
1996	1	Modern japonica (Zhongzhuo)	16 d	22	16.62 b	7.70 a
	2	Japonica hybrid (Jingyou)	36 a	49	15.06 c	6.80 b
	3	Tall japonica (Zhonghua)	32 b	44	18.14 a	6.90 b
	4	Modern indica (IR72)	23 c	32	14.74 c	4.50 c
1997	1	Pig manure/mineral fertilizer	139 a	191	15.83 a	7.74 a
	2	Cattle manure/mineral fertilizer	31 b	43	14.72 a	6.67 a
	3	Rice straw/mineral fertilizer	102 a	141	14.71 a	6.94 a
	4	Mineral fertilizer	4 c	6	15.27 a	6.94 a
1998	1	Late single drainage (local practice)	20 a	28	17.05 a	7.73 a
	2	Early dual drainage	19 a	26	17.61 a	7.82 a
	3	Late dual drainage	15 a	21	16.94 a	7.75 a
	4	Early single drainage	11 b	15	15.68 a	7.60 a

[a]Mean emission data at the same season of the same year following the same letter are not significantly different at $P < 0.05$.

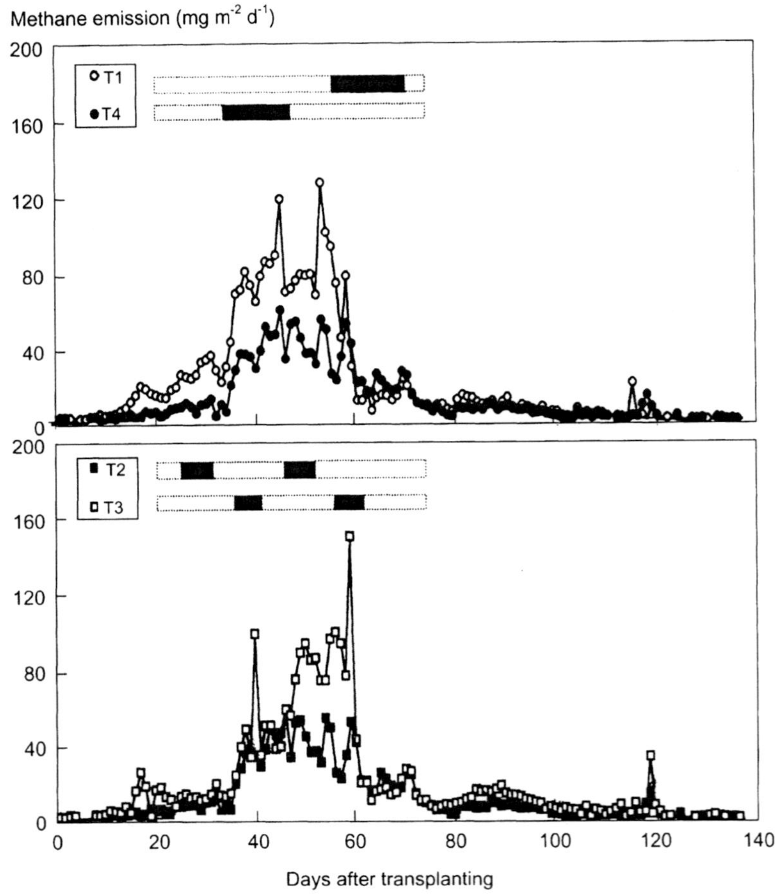

Figure 3. Effect of field drying time and duration on CH_4 emission (daily averages) during the 1998 rice season (horizontal bars [broken line] indicate drying periods [block] within 20 to 75 d after transplanting)

by 60% (Chen et al., 1993b) and 49% (Cai et al., 1994) while midseason drainage reduced emissions by 39% (Chen et al., 1993b). The reductive effect of alternate flooding/drying as compared with midseason drainage is approximately 22% as shown in a field study in Hangzhou (Zhejiang Province) conducted with the same measurement system used in this study (Lu et al., this issue).

Effect of rice cultivar

Seasonal patterns of CH_4 emission from rice cultivars are shown in Figure 1. When CH_4 emission started to increase in the second week after transplanting, rice cultivars differentiated in their CH_4 emission potential. Both Jingyou (japonica hybrid) and Zhonghua (tall japonica) gave higher CH_4 fluxes, whereas CH_4 emission from Zhongzhuo (modern japonica) was lower. Field drainage and low temperatures at the end of the season substantially reduced CH_4 emissions for all cultivars.

Average emission rates from Zhongzhuo, Jingyou, and Zhonghua were 16 mg m^{-2} d^{-1}, 36 mg m^{-2} d^{-1}, and 32 mg m^{-2} d^{-1}, respectively (Table 3). Also shown are data for IR72, a modern indica variety. However, the growth of this tropical cultivar was obviously affected by low temperatures, so that the low emission rates may be related to insufficient biomass assimilation. Among the temperate varieties, Zhongzhuo had the lowest emission rates and the highest yield. Therefore, it appears feasible and effective to maintain sustainable yield and to mitigate CH_4 emission by cultivar selection. However, results for Chinese cultivars are not yet conclusive to allow a cultivar-specific ranking of emission potentials (Lu et al., this issue).

Effect of mineral and organic fertilizers

Fertilizer impacts were investigated in the seasons of 1995 and 1997; both experiments were conducted with local farmers' irrigation practice. Organic manure greatly promoted CH_4 emissions as compared with mineral fertilizers (Table 3). Seasonal CH_4 fluxes (cumulative) in plots with pig manure exceeded those in plots with ammonium sulfate by a factor of 15 in 1995 and a factor of 35 in 1997 (Table 3). The experiment in 1997 included cattle manure and rice straw (Figure 4). Methane fluxes were low and did not differ among the four treatments during the first 7 DAT. Then, CH_4 emission increased sharply and the differences became wider after 10 DAT. The maximum CH_4 fluxes were recorded 26 DAT for pig manure, 36 DAT for rice straw, and 52 DAT for cattle manure.

The relative impact of organic manure in the Beijing station is considerably higher than in other rice-growing regions—e.g., in the tropics (Wassmann et al., this issue, b). Apparently, the soil at the Beijing station is very efficient in converting the organic amendment during the first half of the season when temperatures are relatively high. Temperatures were low at the end of the season. Lowering the temperature suppressed the CH_4 emission peak derived from plant-borne material that is commonly observed in tropical rice fields (Wassmann et al., this issue,b).

Average fluxes were 139 mg CH_4 m^{-2} d^{-1} in pig manure-treated plots, 31 mg m^{-2} d^{-1} in cattle manure-treated plots, 102 mg m^{-2} d^{-1} in rice straw-treated plots, and 4 mg m^{-2} d^{-1} in pure mineral fertilizer-treated plots (Table 3). Higher CH_4 emission rates from pig manure and rice straw were due to higher contents of easily decomposable organic carbon than in cattle manure (data not shown). In the 1997 experiments, all organic manure types resulted in similar grain yield, underscoring the potential of organic manure management as a viable mitigation option in sustainable rice production. Compost amendment in 1998 resulted in a similar range of emissions as the mineral fertilizer treatments in previous years. Therefore, composting of organic amendments can also be considered a tool for achieving low emission rates.

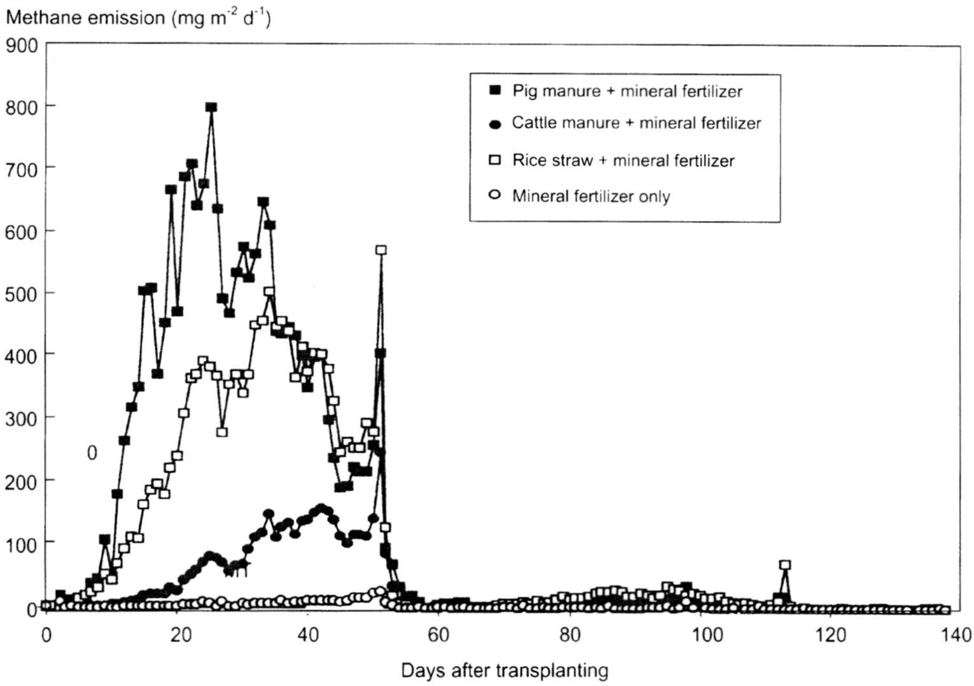

Figure 4. Impact of organic amendments on CH_4 emission (daily averages) from rice fields during the 1997 rice season

These results on fertilizer effects are in line with previous findings obtained in Chinese rice fields. Application of pure mineral fertilizer resulted in less than 50% of the CH_4 emissions that emanated from fields to which mixed organic/mineral fertilizers were applied (Chen et al., 1993b; Cai et al., 1994; Wassmann et al., 1996). The impact of compost was comparable with the impact of biogas residues reported earlier (Wassmann et al., 1993b). Both composted manure types consisted of prefermented material that had a lower emission potential than fresh organic amendments.

Impact of local crop management practices on CH_4 emissions

The common crop management in Beijing corresponds to T_1 in the experiments of 1995 and 1997 (Table 2)—i.e., mineral fertilizers mixed with pig manure, irrigation which includes midseason drainage, and a modern japonica cultivar. This practice resulted in average emission rates of 279 and 139 mg CH_4 m^{-2} d^{-1} in 1995 and 1997, respectively. These values are well within the range reported for other field measurements in China as can be seen in an extensive compilation of emission data in Cai (1997). In the Beijing area, a previous field experiment using a different type of manure (horse dung) and other irrigation schemes produced CH_4 emissions of 861 mg CH_4 m^{-2} d^{-1} with continuous flooding and 350 mg CH_4 m^{-2} d^{-1} with intermittent irrigation (Chen et al.,1993b). The unusually high application rate of organic manure in the experiments conducted by Chen et al. (1993) could explain the high emission rates.

Crop management consisting of midseason drainage and combined organic/mineral amendments was investigated in southeastern China and found to have resulted in CH_4 emissions of 259 mg m^{-2} d^{-1} (Chen et al., 1993b) and 140 mg m^{-2} d^{-1} (Cai et al., 1994). Lu et al., (this issue) presented the results of an extensive field study in Hangzhou (Zhejiang Province) conducted with the same measurement system as in this study; emissions with midseason drainage and combined mineral/organic fertilizers accumulated to 58-284 kg CH_4 ha^{-1}. The corresponding values for Beijing are similar (Table 3); it seems likely that seasonal emissions in northern China are not distinctively higher or lower than in other rice-growing areas of the country.

However, cropping systems in central, southern, and eastern China often encompass two rice seasons per year and—in many cases—have better supply of water. Continuous flooding increased CH_4 emissions by a factor of 2.3 as compared with midseason drainage (Lu et al., this issue). Wang et al. (1990) recorded 187 mg CH_4 m^{-2} d^{-1} in early rice and 672 mg CH_4 m^{-2} d^{-1} in late rice as long-term averages in continuously flooded fields in Zhejiang Province. In Hunan Province, Wassmann et al. (1993b) recorded 340 mg CH_4 m^{-2} d^{-1} in early rice and 451 mg CH_4 m^{-2} d^{-1} in late rice using mineral fertilizer plus rice straw in continuously flooded fields. For Sichuan Province, the emission rates were estimated to be in the range of 1,440 mg CH_4 m^{-2} d^{-1} (Khalil et al., 1991), but this record appears to deviate substantially from other emission data (see Cai, 1997).

Mitigation strategies

Methane mitigation options used in rice fields must both reduce CH_4 emission and sustain rice production. Water control is one of the most important factors in rice production. Midseason drainage and alternative flooding/drying management reduced CH_4 emissions by 23-59% while yield increased by 16%. Water management would be the most promising mitigation option in China where irrigation water is available and irrigation/drainage systems are established.

Application of organic manure is a common practice to maintain soil fertility but it increases CH_4 emission from rice fields. This effect on CH_4 emission may be reduced by composting manure and rice straw rather than applying the fresh material. An alternative way is rotation application of organic amendment and mineral fertilizer. Since CH_4 emissions differ among rice cultivars, variety selection may be a feasible and an effective way to combine low CH_4 emission and high rice production.

Conclusion

Methane emissions from rice fields in northern China are relatively high in spite of low temperature during the latter part of the growing season. Methane emission rates under local practice in the temperature zone of China ranged from 139 to 279 mg m^{-2} d^{-1}. Cumulated emissions per season were the range observed for other parts of China, but emission rates were more sensitive to organic amendments. These results indicate that CH_4 emissions can be reduced by a package of technologies that includes water management, composting of organic amendments, and use of selected cultivars without affecting yield.

Seasonal maxima of CH_4 emissions occurred at tillering and accounted for 85% of total seasonal flux. Therefore, it is crucial to reduce seasonal CH_4 flux by controlling CH_4 emission early in the growing season. Plant growth in the later stages can be optimized for high yields without any impact on CH_4 emissions. However, further studies are needed to convert these findings into recommendations for the farmers.

Acknowledgements

The research described in this paper was funded by the UNDP-Global Environment Facility GLO/91/G31 under agreement with the International Rice Research Institute (IRRI) and the Institute of Crop Breeding and Cultivation, Chinese Academy of Agricultural Sciences Collaborative Project. However, UNDP did not review the paper, hence it does not necessarily reflect the views of UNDP and no endorsement should be inferred.

References

Agricultural Year Book of China (1995) Agricultural Press, Beijing, China

Bolle HJ, Seiler W & Bolin B (1986) Other greenhouse gas and aerosols. In: The Greenhouse Effect. Climate Change and Ecosystems, pp 157-198, New York

Cai ZC, Xu H, Tsuruta H, Zhang HH & Jin JS (1994) Estimate of methane emission from rice paddy fields in Taihu region, China. Pedosphere 4: 297-306

Cai ZC, Yan XY, Xu H, Tsuruta H, Yagi K & Minami K (1995) Effect of nitrogen form on methane emission from rice paddy field. Acta Pedol Sin 32 (supp):136-143

Cai ZC (1997) A category for estimate of CH_4 emission from rice paddy fields in China. Nutr Cycling Agroecosyst 49:171-179

Chen DZ, Wang MX, Shangguan XJ, Huang J, Rasmussen RA & Khalil MAK (1993a) Methane emission from rice fields in southeast of China. Adv Earth Sci 8(5):47-54 (in Chinese)

Chen ZL, Li DB, Shao KS & Wang BJ (1993b) Features of methane emission from rice paddy fields in Beijing and Nanjing. Chemosphere 26:239-245

IPCC – Intergovernmental Panel on Climate Change (1992). In: Houghton JT, Callander BA & Varney SK (eds) Climate Change - The Supplementary Report to the IPCC Scientific Assessment, 200 p, Cambridge, UK

Kern JS, Gong ZT, Zhang GL, Zhuo HZ & Luo GB (1997) Spatial analysis of methane emissions from paddy soils in China and the potential for emission reduction. Nutr Cycling Agroecosyst 49:181-195

Khalil MAK, Rasmussen RA, Wang MX & Ren LX (1991) Methane emissions from rice fields in China. Environ Sci Technol 25:979-981

Lu WF, Chen W, Duan BW, Guo WM, Lu Y, Lantin RS, Wassmann R & Neue HU (2000) Methane emission and mitigation options in irrigated rice fields in Southeast China. Nutr Cycling Agroecosyst, this issue

Neue HU (1993) Methane emission from rice fields. Biosci 43:466-474

Rasmussen RA & Khalil MAK (1986) Atmospheric trace gases: Trends and distribution over the last decade. Science 32:1623-1624

Sass RL (1995) Mitigation of methane emission from irrigated rice agriculture. Global Change Newsl 22:4-5

Tao Z & Du DD (1998) Methane emission from Paddy fields in different regions and the control measures. Agric Environ Prot 17:1-7

Thompson AM & Cicerone RJ (1986) Possible perturbations to atmospheric CO, CH_4 and OH. J Geophys Res 91 (D): 10858-10864

Wang B, Xu Y, Wang Z, Li Z, Guo Y, Shao K & Chen Z (1999) Methane emissions from ricefields as affected by organic amendment, water regime, crop establishment, and rice cultivar. Environ Monit Assess 57:213-228

Wang M.X., Dai AG, Shen RX, Schütz H, Rennenberg H, Seiler W & Wu HB (1990) Methane emission from a Chinese rice paddy field. Acta Meteorol Sin 4:265-275.

Wassmann R, Schütz H, Papen H, Rennenberg H, Seiler W, Dai AG, Shen RX, Shangguan XJ & Wang MX (1993a) Quantification of methane emissions from Chinese rice fields (Zhejiang Province) as influenced by fertilizer treatment. Biogeochemistry 20:83-101

Wassmann R, Wang MX, Shangguan XJ, Xie XL, Shen RX, Papen H, Rennenberg H & Seiler W (1993b). First records of a field experiment on fertilizer effect on methane emission from rice fields in Human province (PR China). Geophys Res Lett 20:2071-2074

Wassmann R, Shangguan XJ, Tölg M, Cheng DX, Wang MX, Papen H, Rennenberg H & Seiler W (1996) Spatial and seasonal distribution of organic amendments affecting methane emission from Chinese rice fields. Biol Fertil Soils 22:191-195

Wassmann R, Neue HU, Lantin RS, Buendia LV & Rennenberg H (2000a) Characterization of methane emissions from rice fields in Asia. 1. Comparison among field sites in five countries. Nutr Cycling Agroecosyst, this issue

Wassmann R, Neue HU, Lantin RS, Makarim K, Chareonsilp N, Buendia LV & Rennenberg H (2000b) Characterization of methane emissions from rice fields in Asia. 2. Differences among irrigated, rainfed and deepwater ecosystems. Nutr Cycling Agroecosyst, this issue

Yao H, Zhuang Y & Chen ZL (1996) Estimation of methane emission from rice paddies in mainland China. Global Biogeochem Cycles 10:641-649

Nutrient Cycling in Agroecosystems 58: 65–73, 2000.

Methane emissions and mitigation options in irrigated rice fields in southeast China

W.F. Lu[1], W. Chen[1], B.W. Duan[1], W.M. Guo[1], Y. Lu[1,2], R.S. Lantin[2], R. Wassmann[2,3] & H.U. Neue[2,4]

[1]China National Rice Research Institute, Hangzhou, Zhejiang 310006, People's Republic of China; [2]International Rice Research Institute, MCPO Box 3127, Makati City, 1271, Philippines; [3]Fraunhofer Institute for Atmospheric Environmental Research, Garmisch-Partenkirchen, Germany; [4]UFZ-Center for Environmental Research, Department of Soil Sciences, Halle, Germany

Key words: midseason drainage, pig manure, rice straw, biogas residues, cultivars, winter fallow, dissolved methane

Abstract

Methane (CH_4) emissions from rice fields were monitored in Hangzhou, China, from 1995 to 1998 by an automatic measurement system based on the "closed chamber technique." The impacts of water management, organic inputs, and cultivars on CH_4 emission were evaluated. Under the local crop management system, seasonal emissions ranging from 53 to 557 kg CH_4 ha^{-1} were observed with an average value of 182 kg CH_4 ha^{-1}. Methane emission patterns differed among rice seasons and were generally governed by temperature changes. Emissions showed an increasing trend in early rice and a decreasing trend in late rice. In a single rice field, CH_4 emissions increased during the first half of the growing period and decreased during the second half. Drainage was a major modifier of seasonal CH_4 emission pattern. The local practice of midseason drainage reduced CH_4 emissions by 44% as compared with continuous flooding; CH_4 emissions could further be reduced by intermittent irrigation, yielding a 30% reduction as compared with midseason drainage. The incorporation of organic amendments promoted CH_4 emission, but the amount of emission varied with the type of organic material and application method. Methane emission from fields where biogas residue was applied was 10-16% lower than those given the same quantity (based on N content) of pig manure. Rice straw applied before the winter fallow period reduced CH_4 emission by 11% as compared with that obtained from fields to which the same amount of rice straw was applied during field preparation. Broadcasting of straw instead of incorporation into the soil showed less emission (by 12%). Cultivar selection influenced CH_4 emission, but the differences were smaller than those among organic treatments and water regimes. Modifications in water regime and organic inputs were identified as promising mitigation options in southeast China.

Introduction

Methane (CH_4) is one of the important greenhouse gases in the atmosphere (Dlugokencky et al., 1994). The increase of CH_4 in the atmosphere contributes to global warming and affects the chemical changes in the atmosphere (Cicerone & Oremland, 1988; GEIA, 1993; Khalil & Shearer, 1993; IPCC, 1996). Rice fields are one of the major atmospheric CH_4 sources (Cicerone & Shetter, 1981; Sass et al., 1990; Rennenberg et al., 1992; Neue et al., 1994; Wassmann et al., 1995; Neue & Sass, 1998; Wassmann et al., 1998). Rice plants are actively implicated in CH_4 production, oxidation, and transportation (Seiler et al., 1984; Holzapfel-Pschorn et al., 1985; Schutz et al., 1989; Neue et al., 1997). It is imperative to evaluate the contribution of rice agriculture to global CH_4 emission.

China is an important rice-producing country, accounting for 22.6% of the world rice harvested area and 36.3% of rice grain production (IRRI, 1993a,b; 1995). Rice fields in China have been considered as an important contributor to the increasing CH_4 concentration in the atmosphere (Cai et al., 1994, 1995a,b; Khalil & Rasmussen, 1991; Wang et al., 1998). The objectives of this study were to characterize and quantify

CH_4 emission from rice fields in southeast China and to develop feasible mitigation options. This study is part of an international network of measuring stations for determining CH_4 emissions from rice fields in Asia. The experimental site was located in Hangzhou, which represents a typical area of single and double rice cropping system in southeast China. The climate is subtropical and the soil and natural conditions favor rice cultivation.

Materials and methods

Experiments were conducted in a rice field at the experimental farm of the China National Rice Research Institute in Hangzhou, China. The soil has a pH of 6.2, 24.2 g organic C, 2.27 g total N, and 14.4 cmol CEC kg^{-1} soil. The details of the experiments conducted from 1995 to 1998 are shown in Table 1. The experiment in each rice season consisted of four treatments in a randomized complete block design with three replicates. Field was flooded, harrowed, and leveled 1 or 2 d before transplanting. The size of each individual plot was 5 × 5 m.

Methane emission rates were determined by an automatic measurement system based on the "closed chamber technique." The technical details of the system used in this measurement were described by Wassmann et al. (this issue, a). Sampling of gases from the chambers was done in a 2-h cycle allowing four measurements of the CH_4 inside each chamber, at 30-min intervals each measurement. Methane emission rate was calculated by regressing the four CH_4 measurements with each closing period. All sampling operations and data acquisition were controlled by a computer equipped with a timing device. Bihourly CH_4 emissions (12 readings d^{-1}) were continuously obtained during the entire growing season. Besides CH_4 emission rates, air temperature and soil temperatures at 5 and 10 cm depth were also automatically recorded continuously by the computer-regulating system at intervals of 10 min.

Dissolved CH_4 concentration at soil depths of 5, 10, and 15 cm were measured three times a week by a procedure described by Lu et al. (1999).

Statistical analysis of experimental data was accomplished using STATISTICA program (Statsoft, Inc. 1993). The data in each treatment were evaluated as to the type of distribution. If distribution is normal, t-test is used; if distribution is not normal, sign test is used.

Results and discussion

Seasonal pattern and rate of CH_4 emissions

Typical of double rice cropping pattern in southeast China, early rice is grown from April to July and late rice is grown from July to November. The treatments in the 1997 experiment consisted of urea only, urea plus pig manure, and urea plus biogas residue (Table 1). Local water management was applied, which encompassed persistent flooding (at 4 cm water depth) with 1 wk interruption at midseason. The seasonal patterns of CH_4 emission were clearly governed by both temperature change and midseason drainage (Figures 1 and 2). In early rice, air temperature increased with plant growth. Emissions increased gradually with 2 two short-period peaks at 30 d and 60 d after transplanting (DAT). These two peaks coincided with the two drainages. Methane emission rates decreased rapidly after the second short-period peak. This was due to the second drainage and the low air temperature. At the end of the season, the field was drained, and CH_4 emission rates were low. The patterns in late rice differed from those in early rice. Air temperature in late rice was high during the early growth stage, but decreased with plant growth. Correspondingly, CH_4 emission rates increased rapidly and were high after transplanting; the emissions then decreased steadily with plant growth and sharply decreased during midseason drainage (50 DAT). After 60 DAT until harvest, the emission rates remained at a low level and were virtually zero. The low air temperatures would be the main reason for the low CH_4 emissions. Methane emission at this period was only a small fraction of the total emitted CH_4 in late rice.

Besides the double rice system, local farmers also grow a single crop of rice. We measured CH_4 emission from single rice fields in 1995, 1996, and 1997. Generally, the emission patterns in single rice differed from those in early rice and late rice (Figure 3). In the continuous flooding case, CH_4 emission increased with plant growth and reached a maximum at heading stage (60 DAT); then it decreased gradually. A quick decrease in CH_4 emission rates occurred at 100 DAT. The reason for this could be the effect of drainage in some plots. Emission rates sharply decreased at 65 DAT for the treatments of local practice irrigation and intermittent irrigation. Emissions remained low during the late

Table 1. Summary of treatments and field management of experiments in Hangzhou (1995-1998)

Year	Season[a]	Treatment	Inorganic input ($kg\ N\ ha^{-1}$) Basal	Inorganic input ($kg\ N\ ha^{-1}$) Topdressed	Organic input ($kg\ ha^{-1}$)[b] Org N	Organic input ($kg\ ha^{-1}$)[b] Org C	Transplant[c]	Harvest[c]
1995[d]	Single	Local practice irrigation/green manure	60	60	38.5	600	05/30	10/10
		Intermittent irrigation/green manure	21.5	60	38.5	600	05/30	10/10
		Continuous irrigation/green manure	21.5	60	38.5	600	05/30	10/10
		Local practice irrigation/no green manure	21.5	60	n	n	05/30	10/10
1996[e]	Early	Modern indica (Zhongfu 906)	60	60	n	n	05/07	07/24
		Japonica hybrid (Jin 23a/71)	60	60	n	n	05/07	07/24
	Late	Modern japonica (Xiusui 11)	60	60	n	n	07/26	11/08
		Indica hybrid (II-you 1568)	60	60	n	n	07/26	10/30
	Single	Modern japonica (Chunjiang 06)	60	60	n	n	06/20	10/30
		Indica hybrid (Shanyou 10)	60	60	n	n	06/20	09/26
1997[e]	Early	No organic manure	60	60	n	n	05/04	07/20
		Pig manure	40	60	20	371	05/04	07/20
		Biogas residue	40	60	20	266	05/04	07/20
	Late	No organic manure	60	60	n	n	07/22	11/17
		Pig manure	40	60	20	371	07/22	11/17
		Biogas residue	40	60	20	266	07/22	11/17
	Single	Modern indica (IR72)	60	60	n	n	06/10	09/20
1998[f]	Early	Local practice irrigation/no rice straw	60	60	n	n	04/29	07/18
		Local practice irrigation/rice straw	48.4	60	11.6	600	04/29	07/18
		Intermittent irrigation/rice straw	48.4	60	11.6	600	04/29	07/18
		Local practice irrigation/rice straw incorporated in winter fallow[g]	48.4	60	11.6	600	04/29	07/18
	Late	Local practice irrigation/no rice straw	60	60	n	n	07/21	11/09
		Local practice irrigation/rice straw	48.4	60	10.7	600	07/21	11/09
		Intermittent irrigation/rice straw	48.4	60	10.7	600	07/21	11/09
		Local practice irrigation/mulched straw[h]	48.4	60	10.7	600	07/21	11/09

[a]Single crop season was from June to October, early rice in double cropping system was from May to July, and late rice was from July to November; [b]n = no organic inputs; [c]Date; [d]In 1995, modern japonica (Chujiang 06) was used for all treatments; [e]Local practice irrigation was applied for all treatments of 1996 and 1997; [f]In 1998, modern japonica (Zhongfu 906) was used for early season and modern japonica (Xiushi 11) for late season; [g]Rice straw was incorporated into the soil 5 mo before transplanting (during winter fallow); [h]Rice straw was mulched to the field at the surface (no incorporation into the soil) directly before transplanting.

growth stages, although the field was reflooded. The emission pattern of the single rice crop during early growing stages was similar to that of early rice, whereas emission rates during late growing stages were similar to those of late rice. Average total emission was 167-557 kg CH_4 ha^{-1}, which was higher than both early rice and late rice.

Wang et al. (1998) reported that the CH_4 emission from rice fields in China ranged from 28 to 206 mg m^{-2} d^{-1} for early rice, 76-526 mg m^{-2} d^{-1} for late rice, and 69-1,352 mg m^{-2} d^{-1} for single rice. Cai et al. (1994, 1995b) reported that CH_4 emission in central China ranged from 46 to 1,060 mg m^{-2} d^{-1}. Our results showed that the CH_4 emission rates in southeast China were 69-284 mg m^{-2} d^{-1} for early rice, 96-252 mg m^{-2} d^{-1} for late rice, and 87-425 mg m^{-2} d^{-1} for single rice (Table 2). The total seasonal emissions rates ranged from 53 to 225 kg ha^{-1} for early rice, 101-279 kg ha^{-1} for late rice, and 88-557 kg ha^{-1} for single rice.

Effects of water regime on CH_4 emission

The water regime of rice soil is a main factor controlling CH_4 emission (Sass et al., 1992; Adhya et al., 1994; Kimura, 1994; Neue & Sass, 1994; Husin et al., 1995; Yagi et al., 1996). In China, field drainage in the middle of the season is practiced for better growth of rice plants. The agronomic advantage of this practice is the

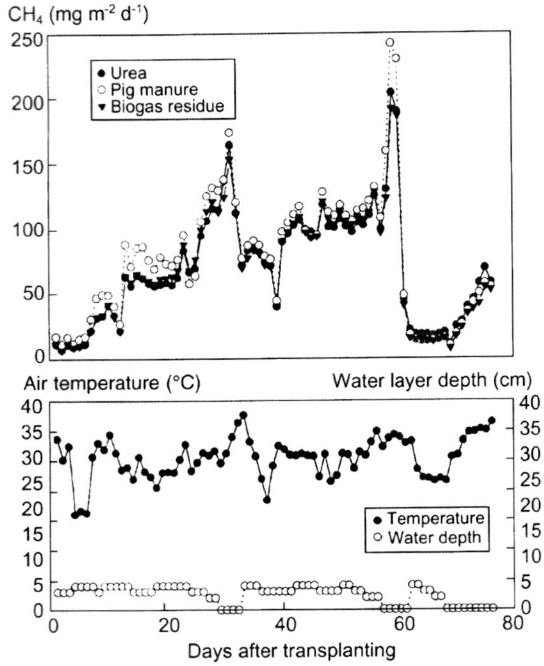

Figure 1. The seasonal patterns of CH_4 emissions, air temperature, and water layer depth in early rice season of 1997, Hangzhou

reduction of excess tillers and the promotion of root growth. The experiment with single rice in 1995 compared three different water regimes: (1) local practice, i.e., normal irrigation with midseason drainage; (2) intermittent irrigation with alternate flooding and drainage at about 10-d interval; and (3) continuous flooding. All these fields were fertilized with pig manure. The patterns of CH_4 emissions are illustrated in Figure 3. Continuous flooding resulted in highest emission, followed by local practice irrigation, while intermittent irrigation plots gave the lowest CH_4 fluxes among the three water regimes. A midseason drainage sharply decreased CH_4 emission.

Methane emission from the local practice of irrigation was 44% lower and CH_4 emission from intermittent irrigation was 61% lower than that of continuous flooding (Table 2). There were no significant differences in biomass and grain yields among the four treatments. These results revealed that proper drainage during the growing season could be a promising mitigation strategy that does not affect yields. Moreover, this finding has been corroborated by results from other stations of the network (Wassmann et al., this issue,b).

Effects of organic inputs on CH_4 emission

The impacts of organic inputs were investigated in the seasons in 1997 (pig and biogas residue) and 1998 (fresh, decomposed, and mulched rice straw). Organic amendments promoted CH_4 emissions as compared with mineral fertilizers (Table 2). In early rice (1997), pig manure increased CH_4 emission by 11% , while biogas residue did not increase CH_4 emission (Table 2 , Figure 1). In late rice, both pig manure and biogas residue significantly increased CH_4 emission, especially during the early growing season. Total emission was 26% higher than in the urea-treated plots. The application of biogas residue increased CH_4 emission slightly. The 1998 experiment was designed to test the effects of rice straw with different application methods. In the early season, rice straw was incorporated into the soil either before the winter fallow or at the time of field preparation before transplanting. Application before the winter fal-

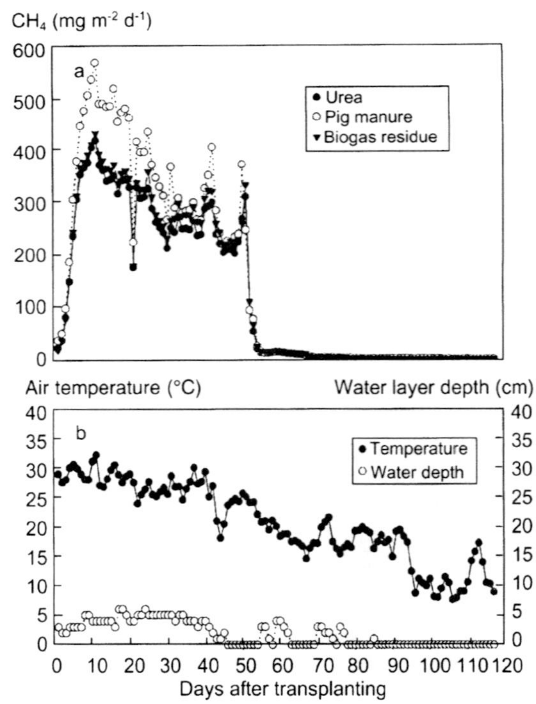

Figure 2. The seasonal patterns of CH_4 emissions, air temperature, and water layer depth in the late rice season of 1997, Hangzhou

Table 2. Mean and cumulative CH_4 emission rates, biomass, and yields per modifying treatment (1995-1998, Hangzhou)[a]

Year	Season	Modifying treatment	Mean emission (mg m^{-2} d^{-1})	Cumulative emission (kg ha^{-1})	Aboveground biomass (t ha^{-1})	Grain yield (t ha^{-1})
1995	Single	Local practice irrigation/green manure	238.3 a	312.1	14.51 a	6.49 a
		Intermittent irrigation/green manure	165.4 c	216.6	14.68 a	6.67 a
		Continuous irrigation/green manure	425.1 a	556.8	14.85 a	6.68 a
		Local practice irrigation/no green manure	182.2 b	238.7	14.72 a	6.63 a
1996	Early	Modern indica (Zhongfu 906)	123.7 a	95.2	8.54 a	5.15 a
		Japonica hybrid (Jin 23a/71	115.8 a	89.1	8.38 a	4.93 a
	Late	Modern japonica (Xiusui 11)	95.8 a	100.6	9.72 a	5.06 a
		Indica hybrid (II-you 1568)	117.0 a	112.3	8.48 a	4.85 a
	Single	Modern japonica (Chunjiang 06)	138.4 b	182.6	9.31 a	5.21 a
		Indica hybrid (Shanyou 10)	169.9 a	166.5	9.64 a	5.60 a
1997	Early	No organic manure	69.7 b	53.0	10.75 b	6.27 b
		Pig manure	77.2 a	58.6	11.01 ab	6.37 ab
		Biogas residue	69.1 b	52.5	11.36 a	6.56 a
	Late	No organic manure	121.0 c	141.6	14.01	6.02 a
		Pig manure	153.0 a	179.0	14.07 a	6.37 a
		Biogas residue	128.8 b	150.7	14.46	6.33 a
	Single	Modern indica (IR72)	86.8	87.6	14.42	6.24
1998	Early	Local practice irrigation/no rice straw	179.6 d	141.9	10.80 a	6.20 a
		Local practice irrigation/rice straw	284.2 a	224.∞	10.89 a	6.16 a
		Intermittent irrigation/rice straw	224.2 c	177.1	10.90 a	6.19 a
		Local practice irrigation/rice straw incorporated in winter fallow	253.5 b	200.3	10.74 a	6.13 a
	Late	Local practice irrigation/no rice straw	166.5 d	184.9	12.56 a	6.31 a
		Local practice irrigation/rice straw	251.7 a	279.4	12.62	6.44 a
		Intermittent irrigation/rice straw	193.3 c	214.5	13.33 a	6.71 a
		Local practice irrigation/mulched straw	223.1 b	247.7	12.99 a	6.44 a

[a]Data at the same season of the same year following the same letter are not significantly different at $P < 0.05$.

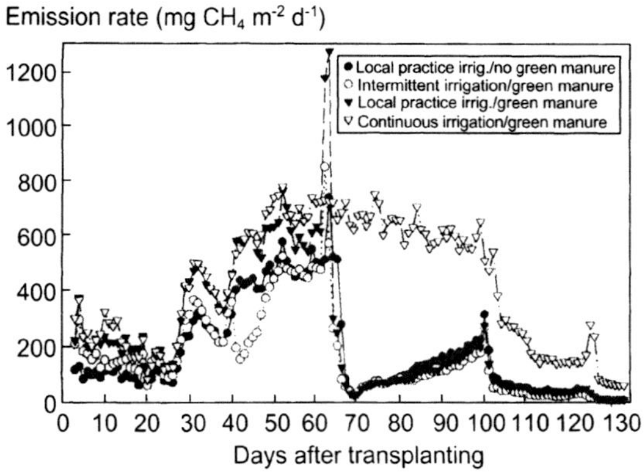

Figure 3. Effect of water regime on CH_4 emission from rice field in the single rice season of 1995, Hangzhou

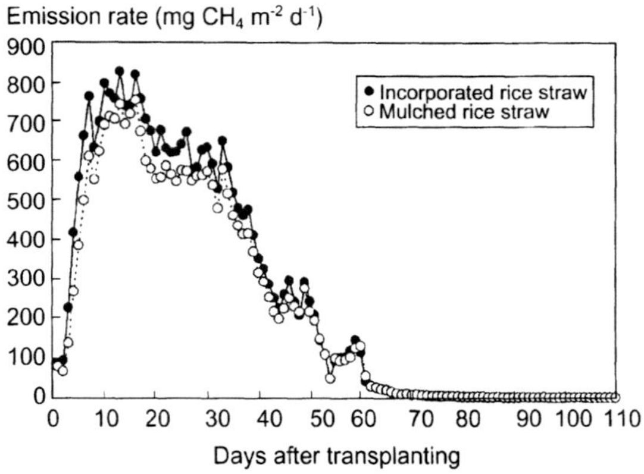

Figure 4. Methane emission as affected by mode of rice straw application in the late rice season of 1998, Hangzhou

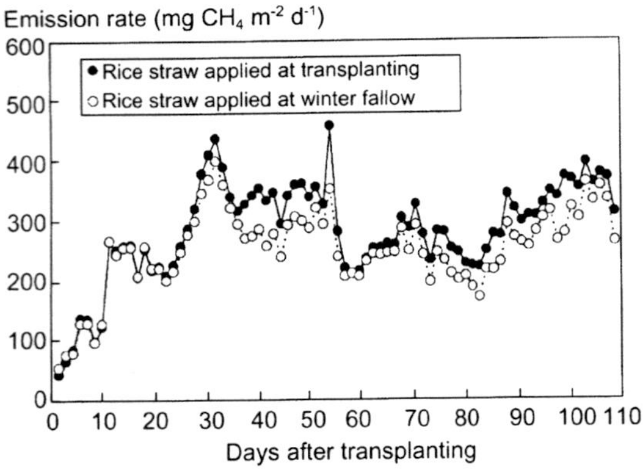

Figure 5. Methane emission as affected by time of rice straw application in the early rice season of 1998, Hangzhou

low decreased emission by 11% (Figure 5, Table 2). In the late season, rice straw was mulched on the field surface and incorporated into the soil at the time of field preparation. The plot with rice straw mulch reduced CH_4 emission by 11% compared with rice straw incorporated (Table 2).

The quality and quantity of added organic amendments and the application methods greatly affected CH_4 production and emission. The lower CH_4 emission rates from biogas residue were obviously due to the previous fermentation of the easily decomposable organic C. In the 1997 experiments, pig manure and biogas residue resulted in different CH_4 emission rates but the same grain yields, indicating that the type of organic manure is a mitigation option in a sustainable rice system. Mulching rice straw on the field surface and incorporating rice straw in the winter fallow periods promoted aerobic decomposition of rice straw, which then resulted in the reduction of CH_4 emissions. These two methods resulted in similar grain yields as the common method (i.e., incorporation of rice straw into soil at the time of field preparation), indicating that application methods of organic amendments can also be taken as important options for achieving low CH_4 emission from rice agriculture.

Effects of rice cultivars on CH_4 emission

The experiments on rice cultivars were conducted in the early, late, and single rice seasons in 1996. The accumulated CH_4 emissions of rice cultivars are shown in Figure 6. In early and late rice, the CH_4 emission rates showed slight differences between hybrid variety and conventional variety at late growth stages. In the single-rice season, accumulated CH_4 emission of hybrid rice was higher than that of conventional rice during the early growth stages, but this was reversed in the late growing season. In the late rice season, the accumulated emission of hybrid rice was constantly higher than that of the conventional variety.

The total CH_4 emissions were 92.5 kg ha^{-1} for Zhongfu 906 (indica) and 89.1 kg ha^{-1} for Jing 23 A/T1 (hybrid) in the early rice season; 100.6 kg ha^{-1} for the Xiushui 11 (japonica) and 112.3 kg ha^{-1} for the II-You 1568 (hybrid) in the late rice season; and 182.6 kg ha^{-1} for Chunjiang 06 (conventional) and 166.5 kg ha^{-1} for Shanyou 10 (hybrid) in the single rice season (Table 2). Less emission of hybrid rice in the single-rice season was due to its shorter growing period (98 d) as compared with 132 d of in-line cultivars (Table 1). The difference in CH_4 emission rates among the tested cultivars ranged from 6.8% to 11%. This difference was smaller than the difference among treatments of water regimes and organic inputs. However, screening and breeding rice cultivars with low CH_4 emission rates

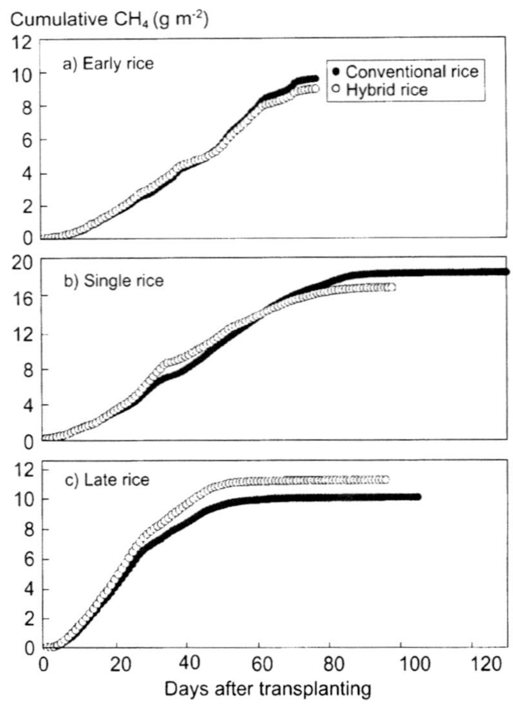

Figure 6. Comparison of accumulated CH_4 emissions between conventional rice and hybrid rice: a) early rice, b) single rice, and c) late rice seasons, 1996, Hangzhou

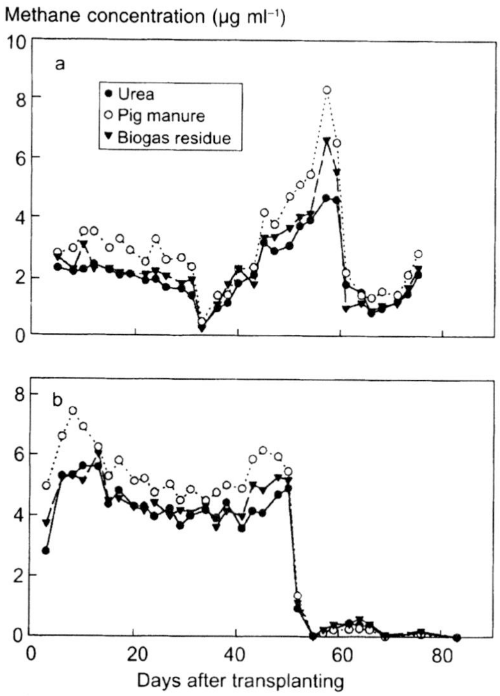

Figure 7. Seasonal patterns of dissolved CH_4 concentrations in soil solution: a) early rice and b) late rice seasons of 1997, Hangzhou

seem to deserve future research effort, considering that farmers can easily accept (without any additional input and field management) rice cultivars with low CH_4 emission rates and high yields.

Dissolved CH_4 concentration in soil solution and its relationship to CH_4 emission

The seasonal patterns of dissolved CH_4 concentrations differed between early rice and late rice. In the early rice season, dissolved CH_4 remained 2-3 μg ml^{-1} until 30 DAT. Then the concentration decreased sharply due to first midseason drainage. After reflooding, CH_4 concentration increased again and peaked at a higher level (6.58 μg ml^{-1}). Methane concentration was lower toward the end of the season because of the dryness of the soil. For late rice, temperature was high during the early stages. Dissolved CH_4 concentration was maintained at 4-6 μg ml^{-1} until 45 DAT. Methane concentration then decreased to a very low level due to midseason drainage and low temperature in the late growing season (Figure 7).

Application of organic manure triggered a rapid increase in CH_4 concentration. The average CH_4 concentrations were 2.2 μg ml^{-1} for urea, 3.09 μg ml^{-1} for urea plus pig manure, and 2.46 μg ml^{-1} for urea plus biogas residue. In the late rice season, average CH_4 concentrations were 3.07, 3.8, and 3.22 for urea, urea plus pig manure and biogas residue, respectively.

Seasonal CH_4 emissions were closely related to dissolved CH_4 concentration in soil solution at different soil depths. The highest relationship was found at the 5 cm depth. The correlations differed among different rice seasons. In early rice, r^2 was 0.68 at 5 cm depth, 0.48 at 10 cm depth, and 0.33 at 15 cm depth. For late rice, r^2 ranged from 0.88 to 0.95 for different depths. The average CH_4 concentrations were 1.9, 2.8, and 3.0 mg ml^{-1} for soil depth of 5, 10, and 15 cm, respectively, in early rice and 2.9, 3.5, and 3.7 μg ml^{-1}, respectively, in late rice. The CH_4 concentration increased with soil depth in the range of 0-15 cm depth.

Conclusions

Methane emission patterns in southeast China rice fields were the interactive results of temperature changes and irrigation regimes. Midseason drainage and intermittent irrigation sharply reduced CH_4 emission. In general, CH_4 emission rates increased with plant growth in the early rice fields; decreased with plant growth in the late rice fields; and increased during the first half growth period and decreased during the second half in single rice fields.

Organic inputs promoted CH_4 production and emission. Selecting the appropriate organic manure type (i.e., decomposed manure) and application method may reduce CH_4 emission without a yield decrease. The impact of cultivar on CH_4 emission depends on the season and growth stage. Cultivar choice may become an important mitigation option for regional and/or global CH_4 emission, but the mechanisms of varying emission potential of cultivars have to be clarified beforehand.

Southeast China is in the subtropical climatic zone and is the major rice-producing area in the country. With a growing population, China's rice production must increase to 0.6 billion over the next 30 yr. This growing demand is most likely to be met by intensifying rice production in existing rice areas. Methane emission from rice fields and mitigation options should be assessed within the overall context of rice cultivation. This will also require national and international scientific efforts and, above all, strong regulations for environment protection.

Acknowledgements

The research described in this paper was funded by the UNDP-Global Environment Facility GLO/91/G31 under agreement with the International Rice Research Institute (IRRI) and the China National Rice Research Institute Collaborative Project. However, UNDP did not review the paper, hence it does not necessarily reflect the views of UNDP and no endorsement should be inferred.

References

Adhya TK, Rath AK, Gupta PK, Rao VR, Das SN, Parida KM, Parashar DC & Sethunathan N (1994) Methane emission from flooded rice fields under irrigated conditions. Biol Fertil Soils 18:245-248

Cai ZC, Xu H, Zhang HH & Jin JS (1994) Estimate of methane emission from rice paddy fields in Tai-hu region, China. Pedosphere 4(4):297-306

Cai ZC, Yan XY, Tsuruta H, Yagi K & Minama K (1995a) Spatial variation of methane emission from rice paddy fields in hilly area. Acta Pedol Sin 32 (supp.):136-143 *(In Chinese)*

Cai ZC, Yan XY, Xu H, Tsuruta H, Yagi K & Minama K (1995b) Effect of nitrogen form on CH_4 emission from rice paddy field. Acta Pedol Sin 32 (supp.):151-159 *(In Chinese)*

Cicerone RJ & Shetter JD (1981) Sources of atmospheric methane: measurements in rice paddies and a discussion. J Geophys Res 86:7203-7209

Cicerone RJ & Oremland RS (1988) Biogeochemical aspects of atmospheric methane. Global Biogeochem Cycles 2: 299-327

Dlugokencky E, Steele P, Lang PM, Tans P & Masaire K (1994) The growth rate and distribution of atmospheric methane. J Geophys Res 99:17021-17043

GEIA—Global Emission Inventory Activity (1993) Report on the 3rd workshop, Amersford, 31 Jan -2 Feb 1993, A.F. Bowman (ed) Bilthoven, The Netherlands, 83 p

Holzapfel-Pschorn A, Conrad R & Seiler W (1985) Production, oxidation, and emission of methane in rice paddies. FEMS Microbiol Ecol 31:149-158

Husin YA, Murdiyarso D, Khalil MAK, Rasmussen RA, Shearer MJ, Sabiham S, Sunar A & Adijuwana H (1995) Methane flux from Indonesian wetland rice: the effects of water management and rice variety. Chemosphere 31: 3153-3180

IPCC – Intergovernmental Panel on Climate Change (1996) Climate Change 1995. The Science of Climate Change. Cambridge University Press, Cambridge, UK, 572 p

IPCC – Intergovernmental Panel on Climate Change (1997) Guidelines for national greenhouse gas inventories. Cambridge University Press, Cambridge, UK

IRRI – International Rice Research Institute (1993a) Rice Research in a Time of Change. Manila, Philippines, 79 p

IRRI – International Rice Research Institute (1993b) IRRI Rice Almanac. Manila, Philippines, 142 p

IRRI – International Rice Research Institute (1995) World Rice Statistics 1993/1994. Manila, Philippines, 260 p

Khalil MAK & Rasmussen RA (1991) Methane emissions from rice fields in China. Environ. Sci Technol 25:979-981

Khalil MAK & Shearer MJ (1993) Atmospheric methane: sources, sinks and role in global change. Chemosphere 26:201-217

Kimura (1994) Effect of intermittent irrigation on methane emission from an Indonesian paddy field. Soil Sci Plant Nutr 40:609-615

Lu Y, Wassmann R, Neue HU & Huang C (1999) Impacts of phosphorus supply on root exudation, aerenchyma formation and methane emission of rice plants. Biogeochemistry (in press)

Neue HU & Sass RL (1994) Trace gas emissions from rice fields. In Prinn RG (ed) Global Atmospheric-Biospheric Chemistry, pp 119-146, Plenum Press, New York

Neue HU & Sass RL (1998) The budget of methane from rice fields. IGACtivities 12:3-11

Neue HU, Lantin RS, Wassmann R, Aduna JB, Alberto MCR, & Andales MJF (1994) Methane emission from rice soils of the Philippines. In: Minami K, Mosier A & Sass RL (eds) CH_4 and N_2O: Global Emissions and Controls from Rice Fields and Other Agricultural and Industrial Sources, pp. 55-77, Tsukuba, Japan

Neue HU, Wassmann R, Kludze HK, Wang B & Lantin RS (1997) Factors and processes controlling methane emissions from rice fields. Nutr Cycling Agroecosyst 49:111-117

Rennenberg H, Wassmann R, Papen H & Seiler W (1992) Trace gas emission in rice cultivation. Ecol Bull 42:164-173

Sass RL, Fisher FM, Harcombe PA & Turner FT (1990) Methane production and emission in a Texas rice field. Global Biogeochem Cycles 4:47-68

Sass RL, Fisher FM, Wang YB, Turner FT & Jund MF (1992) Methane emission from rice fields: the effect of floodwater management. Global Biogeochem Cycles 6:249-262

Schütz H, Holzapfel-Pschorn A, Conrad R, Rennenberg H & Seiler W (1989) A three-year continuous record on the influence of daytime season and fertilizer treatment on methane emission rates from an Italian rice paddy field. J Geophys Res 94:16405-16416

Seiler W, Holzapfel-Pschorn A, Conrad R & Scharffe D (1984) Methane emission from rice paddies. J Atmos Chem 1:241-268

Wang MX, Dai A, Shangguan X, Ren L, Shen R, H Schütz, Seiler W, Rasmussen RA & Khalil MAK (1994) Sources of methane in China. In: Minami K, Mosier A & Sass RL (eds) CH_4 and N_2O: Global Emissions and Controls from Rice Fields and Other Agricultural & Industrial Sources, pp. 9-26, Tsukuba, Japan

Wang MX, Li J & Zheng XH (1998) Methane emission and mechanisms of methane production, oxidation, transportation in the rice fields. Sci Atmos Sin 22(4):600-612 *(In Chinese)*

Wassmann R, Neue HU, Lantin RL, Aduna JB, Alberto MC, Andales MJ, Tan MJ, Denier van der Gon HAC, Hoffmann H, Papen H, Rennenberg H & Seiler W (1994) Temporal patterns of methane emissions from wetland ricefields treated by different modes of N application. J Geophys Res 99:16457-16462

Wassmann R, Neue HU, Lantin RS, Javellana MJ, Diego R, Lignes VE, Hoffmann H, Papen H & Rennenberg H (1995) Methane emissions from rainfed rice. In: Fragile Lives in Fragile Ecosystems, pp 217-225, International Rice Research Institute, Manila, Philippines

Wassmann R, Neue HU, Bueno C, Lantin RS, Alberto MCR, Buendia LV, Bronson K, Papen H & Rennenberg H (1998) Methane production capacities of different rice soils derived from inherent and exogenous substrates. Plant Soil 203:227-237

Wassmann R, Neue HU, Lantin RS, Buendia LV & Rennenberg H (2000a) Characterization of methane emissions from rice fields in Asia. 1. Comparison among field sites in five countries. Nutr Cycling Agroecosys, this issue

Wassmann R, Neue HU, Lantin RS, Makarim K, Chareonsilp N, Buendia LV & Rennenberg H (2000b) Characterization of methane emissions from rice fields in Asia. 2. Differences among irrigated, rainfed, and deepwater rice. Nutr Cycling Agroecosyst, this issue

Yagi K, Tsuruta H, Kanda K & Minami K (1996) Effect of water management on methane emission from a Japanese rice paddy field: automated methane monitoring. Global Biogeochem Cycles 10:255-267

Nutrient Cycling in Agroecosystems **58**: 75–83, 2000.

Methane emissions from irrigated rice fields in northern India (New Delhi)

M.C. Jain[1], S. Kumar[1], R. Wassmann[2,3], S. Mitra[1,2], S.D. Singh[1], J.P. Singh[1], R. Singh[1,4], A. K. Yadav[1] & S. Gupta[1]

[1]*Division of Environmental Sciences, Indian Agricultural Research Institute, New Delhi 110 012, India.* [2]*International Rice Research Institute, MCPO Box 3127, Makati City 1271, Philippines;* [3]*Fraunhofer Institute for Atmospheric Environmental Research, Garmisch-Partenkirchen, Germany;* [4]*Central Ground Water Board, N.E. Region, Guwahati (India)*

Key words: water regime, intermittent flooding, rice cultivar, farmyard manure, wheat straw, automatic sampling, manual sampling

Abstract

Methane (CH_4) emission fluxes from rice fields as affected by water regime, organic amendment, and rice cultivar were measured at the Indian Agricultural Research Institute, New Delhi, using manual and automatic sampling techniques of the closed chamber method. Measurements were conducted during four consecutive cropping seasons (July to October) from 1994 to 1997. Emission rates were very low (between 16 and 40 kg CH_4 m^{-2} $season^{-1}$) when the field was flooded permanently. These low emissions were indirectly caused by the high percolation rates of the soil; frequent water replenishment resulted in constant inflow of oxygen in the soil. The local practice of intermittent flooding, which encompasses short periods without standing water in the field, further reduced emission rates. Over the course of four seasons, the total CH_4 emission from intermittently irrigated fields was found to be 22% lower as compared with continuous flooding. The CH_4 flux was invariably affected by rice cultivar. The experiments conducted during 1995 with one cultivar developed by IRRI (IR72) and two local cultivars (Pusa 169 and Pusa Basmati) showed that the average CH_4 flux from the intermittently irrigated plots without any organic amendment ranged between 10.2 and 14.2 mg m^{-2} d^{-1}. The impact of organic manure was tested in 1996 and 1997 with varieties IR72 and Pusa 169. Application of organic manure (FYM + wheat straw) in combination with urea (1:1 N basis) enhanced CH_4 emission by 12-20% as compared with fields treated with urea only. The site in New Delhi represents one example of very low CH_4 emissions from rice fields. Emissions from other sites in northern India may be higher than those in New Delhi, but they are still lower than in other rice-growing regions in India. The practice of intermittent irrigation—in combination with low organic inputs—is commonly found in northern India and will virtually impede further mitigation of CH_4 emissions in significant quantities. In turn, the results of this study may provide clues to reduce emissions in other parts of India with higher baseline emissions.

Introduction

Methane (CH_4) is a radiatively active trace gas which is present in the atmosphere and is 30 times more efficient than CO_2 in trapping heat (Ramanathan et al., 1985). Current atmospheric concentration of CH_4 is around 1.72 ppmv, but it is predicted that until the year 2100, CH_4 levels may rise to 3-4 ppmv which may have a significant effect on global warming (US-Environmental Protection Agency, 1991). The increase of CH_4 in the atmosphere contributes to global warming and affects chemical changes in the atmosphere (GEIA, 1993; Khalil & Shearer, 1993; IPCC, 1996; Cicerone & Oremland 1988). Rice fields are one of the major atmospheric CH_4 sources (Cicerone & Shetter, 1981; Neue & Sass, 1998; Neue et al., 1994; Rennenberg et al., 1992; Sass et al., 1990; Wassmann et al., 1993, 1998; Mitra et al., 1999). Soil properties, water management, organic amendment, and temperature have been reported as the major factors controlling the amount of CH_4 emitted from rice fields (Schuetz et al., 1989; Sass et al., 1991). There are few reports on the effect of rice cultivars on CH_4 emission (Parashar et al., 1991; Lindau et al., 1995; Wang et al., 1997; Mitra et al., 1999). It has also been reported that rice plants take an active part in CH_4 production, oxidation, and transportation (Holzapfel-Pschorn et al., 1985; Neue et al., 1997; Schuetz et al., 1989; Seiler et al., 1984).

India is an important rice-producing country, comprising 28.6% of world rice area (Huke & Huke, 1997). During recent years, several studies on CH_4 emission from Indian rice fields have been carried out by different researchers to study the effect of soil type, season, water regime, organic and inorganic inputs, and cultivars (Sinha, 1995; Parashar et al., 1991; Mitra, 1992; Parashar et al., 1994; Adhya et al., 1994; Mitra et al., 1999). A broad measurement campaign (1989-91) covering selected rice-growing areas of India indicated a very low source strength ranging from 3.4 to 5.4 Tg yr^{1} (Mitra, 1992). In spite of considerable improvement in the available database in recent years, large uncertainties remain.

The present study was conducted within an international network of eight measuring stations for determining CH_4 emissions from rice fields (Wassmann et al., this issue, a). The objectives of this study at the station in New Delhi were

1) to characterize and quantify the CH_4 budget under local settings;
2) to assess the effects of water management, organic amendments, and different cultivars on CH_4 emission; and
3) to develop crop management strategies with low CH_4 emission in a sustainable rice system for this region.

Materials and methods

Field preparation

Field experiments were conducted at the Indian Agricultural Research Institute (IARI) research farm, New Delhi, during rainy season (July to October), 1994-97. The soil of the experimental site is sandy loam, slightly alkaline, moderately permeable Ustochrept (old alluvium). The fertility status of the soil is medium with respect to available N, P, and K. The physicochemical characteristics of the soil are given in Table 1. Percolation rates were very high and accounted for 20 mm d^{-1}.

The field experiments were laid out with four treatments in each year (Table 2). Every year of measurement encompassed a comparison of cultivars (IR72 vs local cultivars; Table 3) and water management (continuous vs intermittent flooding). Organic amendments were introduced in 1997 and 1998. Twenty-five to thirty-day-old rice seedlings were transplanted in 5- × 5-m plots keeping 20- × 20-cm spacing among hills. Urea was applied at rates of 120 kg N ha^{-1} (as sole fertilizer) or 60 kg N ha^{-1} (in combination with organic manure) and was split in two equal doses (at 10 and 30 d after transplanting). Organic amendments consisted of farmyard manure plus wheat straw, which were added 20 d before transplanting at rates of 60 kg N ha^{-1}. Phosphorus and potassium were added to the soil of all plots as a basal dose of 50 kg P_2O_5 and 40 kg K_2O in the form of single superphosphate and muriate of potash, respectively.

Table 1. Physicochemical characteristics of IARI soil (Ustochrept, old alluvium, sandy loam)

Parameter	Value	Parameter	Value
Sand (%)	66.0	Organic carbon (%)	0.41
Silt (%)	17.1	CEC [cmol(p+)kg ha^{-1}]	7.3
Clay (%)	16.1	Available N (kg ha^{-1})	338.52
Bulk density (g cm^{-3})	1.33	Available P (kg ha^{-1})	20.04
pH (1:2 soil:water)	8.2	Available K (kg ha^{-1})	250.95
EC (dS m^{-1})	0.32	Percolation rate (mm d^{-1})	20.0

Water management

A high percolation rate of the soil required constant supply of water to maintain a water level of 5 cm (± 2 cm) in continuously flooded plots (Figure 1c). In intermittently irrigated plots, the floodwater was replenished to a level of 5-10 cm whenever the soil moisture declined near saturation level (Figure 1f) . The process of periodic flooding of the field was continued throughout the experiment.

Gas sampling and CH_4 flux measurement

Methane fluxes were monitored using automatic as well as manual sampling systems. The automatic system was set up in the vicinity of the rice fields at IARI farm in early 1995. The automatic system has been described in detail by Wassmann et al. (this issue, a). Methane measurements were carried out mostly by manual sampling (otherwise stated) and was done by using the closed chamber technique described by Hutchinson and Mosier (1981). The closed chambers (30 × 50 × 100

Table 2. Details of experiments and treatments, 1994-97

Year	Cultivar	Water management	Mineral NPK (kg ha^{-1})	Organic amendment (kg N ha^{-1})^{a}	Biomass (t ha^{-1})	Grain yield (t ha^{-1})	Methane emission (kg ha^{-1})^{b}
1994	IR72	Continuous flooding	120:50:40	-	16.8	5.1	39.8 a
	IR72	Intermittent irrigation	120:50:40	-	16.4	5.2	32.4 b
	Pusa 169	Continuous flooding	120:50:40	-	14.2	4.9	34.8 ab
	Pusa 169	Intermittent irrigation	120:50:40	-	13.9	4.8	30.0 ab
1995	IR72	Continuous flooding	120:50:40	-	17.6	5.5	22.67 a
	IR72	Intermittent irrigation	120:50:40	-	17.2	5.5	9.71 b
	Pusa 169	Intermittent irrigation	120:50:40	-	13.1	5.2	12.47 ab
	Pusa Basmati	Intermittent irrigation	120:50:40	-	21.8	4.9	13.44 ab
1996	IR72	Continuous flooding	120:50:40	-	11.8	4.4	23.00 a
	IR72	Intermittent irrigation	120:50:40	-	11.8	4.2	17.90 b
	IR72	Intermittent irrigation	60:50:40	60	11.1	4.2	22.10 ab
	Pusa 169	Intermittent irrigation	60:50:40	60	9.1	3.2	20.30 ab
1997	IR72	Continuous flooding	120:50:40	-	14.0	6.8	16.58 a
	IR72	Intermittent irrigation	120:50:40	-	12.6	6.2	12.93 a
	IR72	Intermittent irrigation	60:50:40	60	12.0	6.0	14.42 a
	Pusa 169	Intermittent irrigation	60:50:40	60	11.8	5.6	15.28 a

[a]Organic amendment = farmyard manure (50%) and wheat straw (50%) [b]Means followed by the same letter do not differ significantly at $p < 0.01$ level.

Table 3. Physiological characteristics of different rice cultivars used in the 1994-97 experiment

Character	Pusa 169	IR72	Pusa Basmati
Plant height (cm)	95	90	105
No. of panicle-bearing tillers m^{-2}	400	415	350
Av leaf area tiller^{-1} (cm^2)	100	85	95
Leaf area index	4.6	4.5	4.1
Specific leaf weight	4.5	5.4	6.1
Grain yield (t ha^{-1})	5.1	5.3	4.6
Biological yield (t ha^{-1})	10.7	12.4	11.8
Harvest index (%)	47	45	40
Days to maturity	120 ± 5	125± 5	135 ± 5

cm) were made of acrylic (Perspex) sheet and the joints were sealed with silicone grease to make them leak-proof. The acrylic chambers were placed over the aluminum jackets preinserted into the soil to a depth of 5 cm in each plot well in advance (1 d before sampling) to ensure minimum disturbance to the soil at the time of gas collection in the chambers. The water seal surrounding the acrylic chamber in a channel made the system airtight. Gas samples were drawn at 0, 10, and 20-min interval through a three-way stopcock after installation of chamber using an airtight syringe (capacity 50 ml). The mixing of the gas inside the chamber was achieved during sampling by drawing air out of the chamber head space into a syringe and releasing it back into the chamber (8-10 times) before the final sample was withdrawn. Thereafter, a little higher volume of gas samples (about 15 ml) was transferred into a preevacuated vacutainer (capacity12.5 ml) closed with an airtight rubber stopper by a hypodermic needle (26 gauge) to maintain higher pressure than the atmosphere to avoid contamination or dilution of the collected sample. The samples were analyzed by using a gas chromatograph, HP 5890 series II GC fitted with FID and Porapak N column. Column, detector, and injector temperatures were maintained at 70, 130, and 130 °C, respectively. In both GC, nitrogen was used as the carrier gas, hydrogen as the fuel gas, and zero air as the supporting gas with flow rates of 20, 30, and 250 ml min^{-1}, respectively. The CH_4 flux (F) was calculated using the following equation (Debnath et al., 1996):

$$F= [(C_t - C_0) / t] \times H \times 42.857 \text{ mg m}^{-2} \text{ h}^{-1}$$

where t is time interval (min), H is height of headspace (m), C_0 is initial concentration of CH_4 at time 0 (ppmv), and C_t is final concentration of CH_4 at time t (ppmv).

Soil pH, redox potential (Eh), and temperature measurement

The redox potential (Eh) was measured using a battery operated pH cum voltmeter (Philips). The platinum tip of the electrode was inserted into each plot under investigation at the root zone (12-15 cm depth) throughout the growing season, whereas the reference electrode (calomel) was placed at the surface only to maintain electrical contact (Ponnamperuma, 1972). Sufficient time (8-10 min) was given for the volt reading to get stabilized before recording. The pH of the submerged soil was measured using a portable pH meter (Systronics Griph D pH meter/Philips). Soil and air temperatures were measured by using the digital thermometer and soil thermometer (mercury), respectively, giving temperature values in °C. Soil temperature was measured at a depth of 10 cm.

Results and discussion

The available database consists of four consecutive seasons (1994-97). Emission measurements from 1994, 1995, and 1996 were conducted through manual sampling, whereas emission data of 1997 were based on automatic measurements.

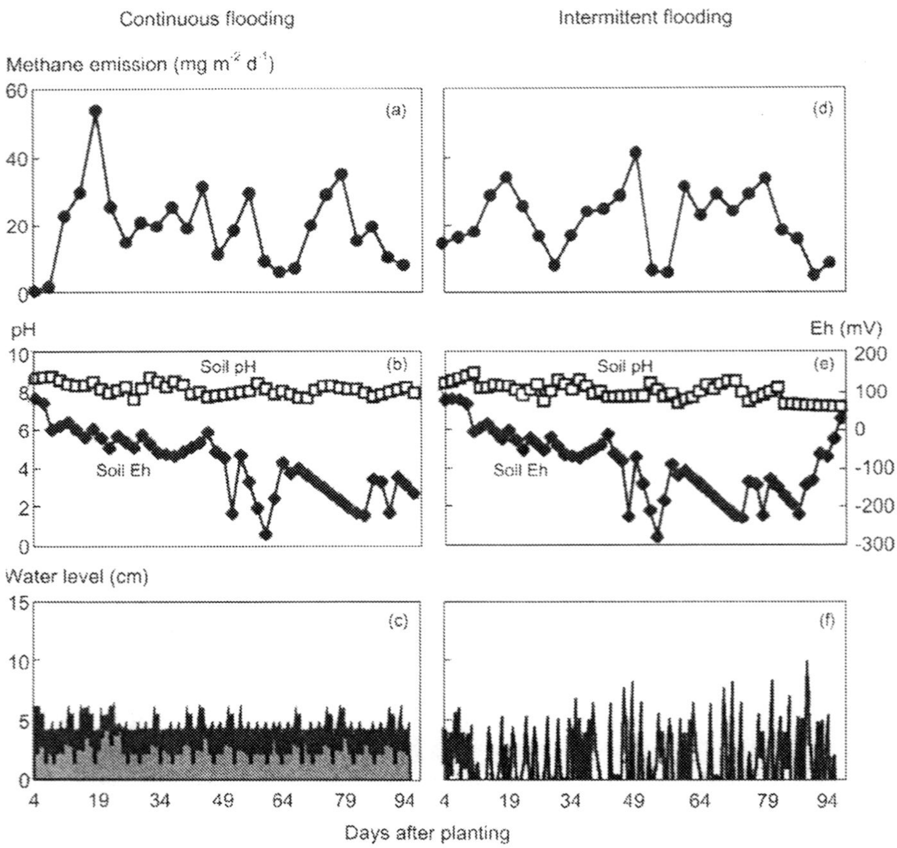

Figure 1. Seasonal patterns of CH_4 emission (a, d), soil Eh and pH (b, e), and water level (c, f) in continuously flooded and intermittently flooded rice (IR72), 1995 season

Seasonal patterns

The seasonal patterns of CH_4 emission (Figure 1a), soil Eh (Figure 1b), soil pH (Figure 1b), and water levels (Figure 1c) are depicted for the continuously flooded field in 1995. Fluctuations in water level in both 1995 (Figure 1c) and 1996 (Figure 2c) clearly illustrate the high percolation rates at this site. Irrigation water had to be added several times a week to maintain flooding of the field. Due to this constant inflow of oxygen, Eh showed a pronounced fluctuation throughout the season and reached only temporarily values below −100 mV (Figure 1b). Floodwater fluctuations had no impact on pH values that remained stable around 8.00.

Methane emission rates varied throughout the season without any distinct trend or pattern. The relationship to Eh development is evident from Figure 1: low emission rates coincided with high Eh and high emissions with low Eh. Relatively low temperatures during the final stage of the season may have caused a distinct period of low emission and high Eh.

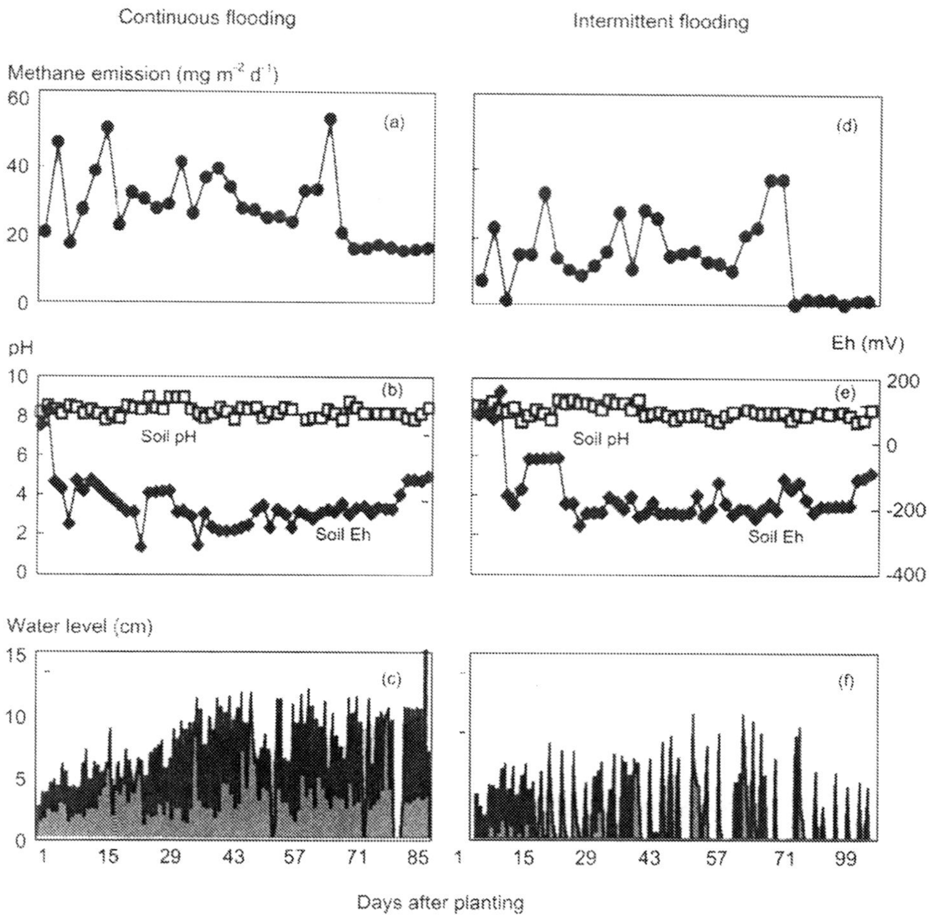

Figure 2. Seasonal patterns of CH_4 emission (a, d), soil Eh nd pH (b, e) and water level (c, f) in continuously and intermittently flooded rice (IR72), 1996 season

Effect of water regime

The impact of different water regimes on emission rates varied from year to year (Table 2). Differences in CH_4 emission were pronounced in 1 yr (1995), i.e., continuous flooding caused higher emission rate than intermittent irrigation, and only small in other years (1994, 1996, 1997). It has been observed that except for 1997, in case of cultivar IR72 on an average, intermittent irrigation significantly reduced CH_4 emission (by 29%) as compared with continuous flooding (Table 2). However, the irrigation mode of fields with continuous flooding was substantially different from rice fields in other regions because high percolation of the soil required frequent replenishment of the receding floodwater. In our experiments, the mean CH_4 emission rate during the 4 yr of experimentation amounted to 25.57 and 18.33 kg ha^{-1} under intermittent and continuous flooding, respectively, corresponding to a 28% decrease by adopting the practice of intermittent irrigation over continuous flooding. This reduction was accomplished at the expense of slightly lower (3.2%) grain yields (Table 2). Lower grain yields were also observed by Yagi et al. (1994) in intermittently irrigated rice fields in Japan.

Seasonal emission patterns under different water regimes are shown in Figures 1 and 2 jointly with Eh. The 1995 experiment (Figure 1) and 1996 experiment (Figure 2) showed different patterns of redox development. In 1995, the soil remained at redox levels of more than −100 mV throughout the first half of the growing period. High Eh values under intermittent flooding were reflected by very low emission rates from this field. In 1996, the Eh decreased immediately to the −100 mV level and the differences between treatments were small. The reasons for these divergences among seasons and treatments are not clear. At this low level of emission rates, the CH_4 budget appears to be relatively unstable so that small changes (e.g., in the fallow treatment) may have caused prominent effects. In absolute terms, however, differences among seasons and treatments are still small since all emission rates were in a comparably low range.

Intermittent flooding leads to an overall reduction of emission rates, but emissions could be enhanced during short intervals. Automatic measurements during the 1997 season documented a sudden pull of emerging CH_4 lasting for 4 d (Figure 3). This incident is similar to the emergence of gaseous CH_4 after harvest draining that is commonly observed in irrigated rice (Wassmann et al., 1994).

Effect of cultivars

The three cultivars tested in this experiment were the high-yielding cultivars IR72 and Pusa 169 as well as the tall cultivar Pusa Basmati. The physiological characteristics of these cultivars are given in Table 3. The IRRI variety, IR72, had higher emission rates than the local variety Pusa 169 in 1994 and 1996, whereas Pusa 169 had higher emissions in 1995 and 1997. The reasons for these discrepancies among cultivars and among seasons are still not clear. IR72 had a higher yield potential and higher biomass in all the experiments (Table 2). The 1994 experiment indicated that the cultivar-specific difference occurred uniformly under different water regimes (Figure 4). Both cultivars showed almost identical patterns (data not shown). Pusa Basmati showed higher emissions than IR72 in the 1995 experiment and this variety developed higher biomass which might be one of the reasons for its higher CH_4 emission potential. Some reports also show the dependence of CH_4 emission on biomass (Cicerone et al., 1983; Sinha, 1995).

Effect of organic amendment

Organic amendment inputs promoted CH_4 emissions, but total emission remained less than 25 kg CH_4 ha^{-1} (Table 2). This finding is contrasted by results from other network stations with irrigated rice where total emissions generally exceeded 100 kg CH_4 ha^{-1} after manure application (Wassmann et al., this issue, a). Previous field experiments also reported larger impacts of organic amendments in both absolute and relative terms (Neue et al., 1994; Wassmann et al., 1993; Chen et al., 1993). The low impact of organic manure in the experiment in New Delhi could be related to high percolation rates. Constant inflow of oxygen into the soil and downward discharge of methanogenic substrate resulted in low CH_4 production (Yagi et al., 1990; Inubushi et al., 1992). Thus, emissions were very low even when organic matter was applied.

The seasonal patterns of CH_4 emissions in IARI rice fields during 1996 and 1997 wet seasons as affected by organic amendment are shown in Figure 3. Differences among treatments were relatively low throughout the season. In other stations of the network, organic amendments stimulated emissions during the first half of the season (Wassmann et al., this issue,b). At the New Delhi station, no definite response pattern could be delineated on the conversion of organic amendment

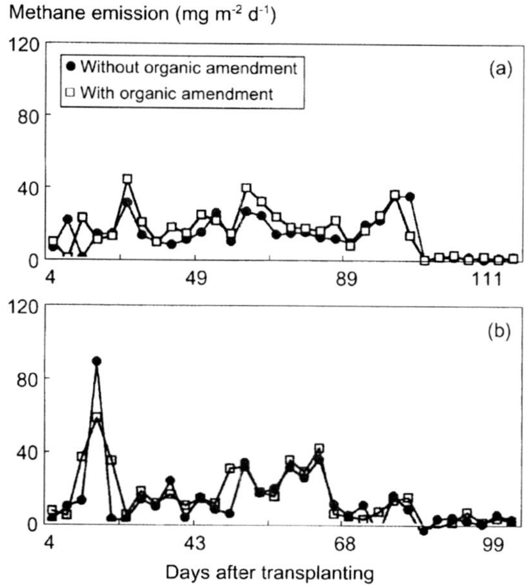

Figure 3. Effect of organic amendments on CH_4 emission during the 1996 (a) and 1997 (b) crop seasons from intermittently irrigated rice fields (IR72)

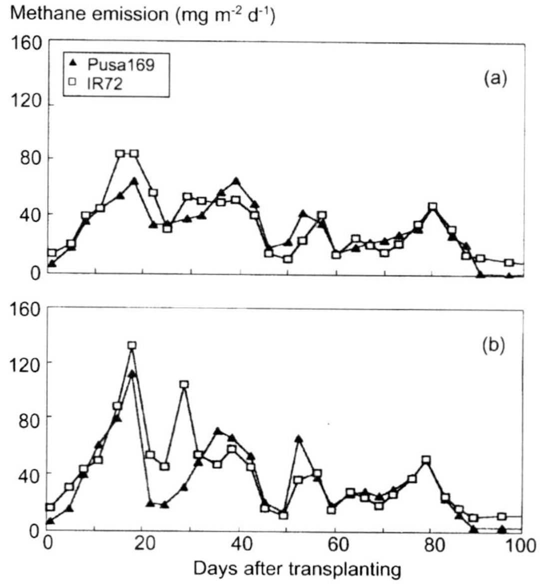

Figure 4. Seasonal patterns of CH_4 emission as affected by rice cultivar in an intermittently (a) and continuously flooded (b) rice, 1994 wet season

to CH_4 emission, which was substantially lower than the other sites of this network (Wassmann et al., this issue, a).

Conclusion and mitigation options

Emission rates at the New Delhi site were by far the lowest in the network of eight stations. The reasons for intersite differences are discussed in Wassmann et al. (this issue, a). The distinct feature of the New Delhi station was a very high percolation rate (20 mm d^{-1}) requiring periodic replenishment of floodwater through intermittent drainage. While sandy loamy soils (the abundant soil type of northern India) are characterized by relatively high percolation, the conditions at the New Delhi site appear to be rather at the extreme end for irrigated rice. Local differences in soil and crop management may slightly modify source strengths of CH_4 emission within northern India, but it seems likely that the practice of intermittent flooding will result in a relatively low level of CH_4 emission rates throughout this region. In eastern and southern India, however, environmental conditions differ largely from New Delhi and should be assessed by separate studies (e.g., the study by Adhya et al., 2000 [this issue] conducted in Cuttack [eastern India]).

In India, out of 42.3 million ha of rice cultivation area, 19.6 million ha are irrigated (Huke & Huke, 1997). In general, organic soil amendments are very low in Indian rice production systems. Even if technically possible, mitigation strategies will only have a small net effect on the CH_4 source strength as opposed to organically amended fields. On the other hand, intermittent irrigation is one of the best options for mitigating CH_4 emission in continuously flooded fields, e.g., in southern India. In our experiment, there was a net decrease of 28% in CH_4 emission (over four consecutive seasons) by applying intermittent irrigation over continuous flooding. This practice did not affect yields and may also be exploited for water saving. Site-specific adaptations will be required for an optimum effect considering rice yields, water consumption, and CH_4 emissions. In this process, appropriate selection of rice cultivars may also become a tool to control CH_4 emission from rice fields. At present, however, the database on the impact of local varieties from India as well as varieties developed at IRRI is insufficient to devise distinct recommendations on cultivar use.

Acknowledgement

The research described in this paper was funded by the UNDP-Global Environment Facility GLO/91/G31 under agreement with the International Rice Research Institute (IRRI) and the Indian Agricultural Research Institute (IARI), New Delhi, India. However, UNDP did not review the paper, hence it, does not necessarily reflect the views of UNDP and no endorsement should be inferred. The IARI center sincerely acknowledges with great appreciation the technical guidance provided by Dr. H.U. Neue, Mrs. R.S. Lantin, and other IRRI scientists. The cooperation and encouragement from Dr. R.S. Paroda, D.G., ICAR is highly acknowledged. At home, the sincere support and guidance by Prof. S.K. Sinha, national professor and former director of IARI in the formulation of the project objectives and Prof. R.B. Singh, director of IARI in the successful implementation of the project is gratefully acknowledged. The authors thank the staff of the IRRI office, New Delhi—Dr. B.P. Ghildyal, Dr. R.K. Singh and their colleagues — for their help and cooperation.

References

Adhya TK, Rath AK, Gupta PK, Rao VR, Das SN, Parida KM, Parashar DC & Sethunathan N (1994) Methane emission from flooded rice fields under irrigated conditions. Biol Fertil Soils 18:245-248

Adhya TK, Bharati K, Mohanty SR, Mishra SR, Ramakrishnan B, Rao VR, Sethunathan N (2000) Methane emissions from rice fields at Cuttack, India. Nutr Cycling Agroecosyst, this issue

Bharati K, Mohanty SR, Padmavathi PVL, Rao VR & Adhya TK (2000) Nitrification inhibitors and methane production in a flooded alluvial soil. Nutr Cycling Agroecosyst, this issue

Chen Z, Li D, Shao K & Wang B (1993) Features of CH_4 emission from rice paddy fields in Beijing and Nanjing. Chemosphere 26:239-245

Cicerone RJ & Shetter JD (1981) Sources of atmospheric methane: measurements in rice paddies and a discussion. J Geophys Res 86:7203-7209

Cicerone RJ, Shetter JD & Delwiche CC (1983) Seasonal variation of methane flux from California rice paddy. J Geophys Res 88:11022-11024

Cicerone RJ & Oremland RS (1988) Biogeochemical aspects of atmospheric methane. Global Biogeochem Cycles 2: 299-327

Debnath G, Jain MC, Kumar S, Sarkar K & Sinha SK (1996) Methane emission from rice fields amended with biogas slurry and farm yard manure. Climate Change 33:97-109

GEIA – Global Emission Inventory Activity (1993) Report on the 3rd workshop, Amersford, 31 Jan-02 Feb 1993, A.F. Bowman (ed), Bilthoven, The Netherlands, 83 p

Holzapfel-Pschorn A, Conrad R & Seiler W (1985) Production, oxidation, and emission of methane in rice paddies. FEMS Microbiol Ecol 31:149-158

Huke RE & Huke EH (1997). Rice area by type of culture: South, Southeast and East Asia. International Rice Research Institute, Los Banos, Laguna, Philippines, 15 p

Hutchinson GL & Mosier AR (1981) Improved soil cover method for field measurement of nitrous oxide flux. Soil Sci Soc Am J 45:311-316

Inubushi K, Muramatsu Y & Umerayasi M (1992) Influence of percolation on methane emission from flooded paddy soil. Jpn J Soil Sci Plant Nutr 63:184-189

IPCC – Intergovernmental Panel on Climate Change (1996) Climate Change 1995. The Science of Climate Change. Cambridge (UK): Cambridge University Press. 572 p

Khalil MAK & Shearer MJ (1993) Atmospheric methane: sources, sinks and role in global change. Chemosphere 26:201-217

Lindau CW, Bollich PK & De Laune RD (1995) Effect of rice variety on methane emission from Louisiana rice. Agric Ecosyst Environ 54:109-114

Mitra AP (1992) Greenhouse gas emission in India: 1991 Methane campaign. Science Report No. 2, Council of Scientific and Industrial Research and Ministry of Environment and Forests, New Delhi

Mitra S, Jain MC, Kumar S, Bandyopadhyay SK & Kalra N (1999) Effect of rice cultivars on methane emission. Agric Ecosyst Environ 73(3):177-183

Neue HU & Sass RL (1998) The budget of methane from rice fields. IGACtivities 12:3-11

Neue HU, Lantin RS, Wassmann R, Aduna JB, Alberto CR & Andales MJF (1994) Methane emission from rice soils of the Philippines. In: Minami K, Mosier A & Sass RL (eds) CH_4 and N_2O: Global Emissions and Controls from Rice Fields and Other Agricultural and Industrial Sources, pp 55-63, Yokendo, Tokyo, Japan

Neue HU, Wassmann R, Kludze HK, Wang B & Lantin RL (1997) Factors and processes controlling methane emissions from rice fields. Nutr Cycling Agroecosyst 49:111-117

Parashar DC, Rai J, Gupta PK & Singh N (1991) Parameters affecting methane emission from paddy fields. Indian J Radio Space Phys 20:12-17

Parashar DC, Mitra AP, Sinha SK, Gupta PK, Rai J, Sharma RC, Singh N, Kaul S, Lal G, Chaudhary A, Ray HS, Das SN, Parida KM, Rao SB, Kanung SP, Ramasami T, Nair BU, Swamy M, , Gupta SK, Singh AR, Saikia BK, Barua AKS, Pathak MG, Iyer CPS, Gopalakrishnan M, Sane PV, Singh SN, Banerjee R, Sethunathan N, Adhya TK, Rao VR, Palit P, Saha AK, Purkait NN, Chaturvedi GS, Sen SP, Sen M, Sarkar B, Banik A, Subbaraya BH, Lal S & Venkatramani S (1994) Methane budget from Indian paddy fields. In: Minami K, Mosier A & Sass RL (eds) CH_4 and N_2O: Global Emissions and Controls from Rice Fields and Other Agricultural and Industrial Sources, pp 27-39, Tsukuba, Japan: NIAES Series 2

Ponnamperuma FN (1972) The chemistry of submerged soils. Adv Agron 24:29-36

Ramanathan V, Cicerone RJ, Singh HB & Kiehl JT (1985) Trace gas trends and their potential role in climate change. J Geophys Res 90:5547-5566

Rennenberg H, Wassmann R, Papen H & Seiler W (1992) Trace gas exchange in rice cultivation. Ecol Bull 42:164-173

Sass RL, Fisher FM, Haracombe PA & Turner FT (1990) Methane production and emission in a Texas rice field. Global Biogeochem Cycles 4:47-68

Sass RL, Fisher FM, Harcombe PA & Turner FT (1991) Mitigation of methane emissions from rice fields: possible adverse effects of incorporated rice straw. Global Biogeochem Cycles 5:275-287

Schuetz H, Holzapfel-Pschorn A, Conrad R, Rennenberg H & Seiler W (1989) A three-year continuous record on the influence of daytime season and fertilizer treatment on methane emission rates from an Italian rice paddy field. J Geophys Res 94:16405-16416

Seiler W, Conrad R & Scharffe D (1984) Field studies of methane emission from termite nests into the atmosphere and measurements of methane uptake by tropical soils. J Atmos Chem 1:171-186

Sinha SK (1995) Global methane emission from rice paddies: Excellent methodology but poor extrapolation. Curr Sci 68 (6):643-646

US-Environmental Protection Agency (1991) Methane emissions and opportunities for control. EPA/400/9-90/9007, US EPA, Washington DC

Wang B, Neue HU & Samonte HP (1997) Effect of cultivar difference ('IR72', 'IR65598' and 'Dular') on methane emission. Agric Ecosyst Environ 62:31-40

Wassmann R, Neue HU, Lantin RL, Aduna JB, Alberto MC, Andales MJ, Tan MJ, Denier van der Gon HAC, Hoffmann H, Papen H, Rennenberg H & Seiler W (1994) Temporal patterns of methane emissions from wetland ricefields treated by different modes of N application. J Geophys Res 99:16457-16462

Wassmann R, Wang MX, Shangguan XJ, Xie XL, Shen RX, Papen H, Rennenberg H & Seiler W (1993) First records of a field experiment on fertilizer effects on methane emission from rice fields in Hunan Province (PR China). Geophys Res Lett 20:2071-2074

Wassmann R, Neue HU, Bueno C, Lantin RS, Alberto MCR, Buendia LV, Bronson K, Papen H & Rennenberg H (1998) Methane production capacities of different rice soils derived from inherent and exogenous substrates. Plant Soil 203:227-237

Wassmann R, Neue HU Lantin RS, Buendia LV & Rennenberg H (2000a) Characterization of methane emissions from rice fields in Asia 1. Comparison among field sites in five countries. Nutr Cycling Agroecosyst, this issue

Wassmann R, Neue HU, Lantin RS, Makarim K, Chareonsilp N, Buendia LV & Rennenberg H (2000b) Characterization of methane emissions from rice fields in Asia 2. Differences among irrigated, rainfed and deepwater ecosystems. Nutr Cycling Agroecosyst, this issue

Yagi, K, Minami K & Ogawa Y (1990) Effects of water percolation on methane emission from paddy fields. Res Rep Div Environ Planning, National Institute of Agro-Environmental Sciences 6:105-112

Yagi K, Tsuruta H & Kanda K (1994) Effect of water management on methane emission from rice paddy fields. Res Rep Div Environ Planning, National Institute of Agro-Environmental Sciences 10:61-70

Nutrient Cycling in Agroecosystems 58: 85–93, 2000.

Crop management affecting methane emissions from irrigated and rainfed rice in Central Java (Indonesia)

P. Setyanto[1], A.K. Makarim[1], A.M. Fagi[1], R. Wassmann[2,3] & L.V. Buendia[2]

[1]Central Research Institute for Food Crops, Bogor, Indonesia; [2]International Rice Research Institute (IRRI), MCPO 3127, Makati City 1271, Philippines; [3]Fraunhofer Institute for Atmospheric Environmental Research (IFU), Garmisch-Partenkirchen, Germany

Key words: closed chamber technique, rainfall, wet season crop, dry season crop, water regime, farmyard manure, straw, rice cultivars, mitigation strategies

Abstract

Methane (CH_4) emissions were determined from 1993 to 1998 using an automated closed chamber technique in irrigated and rainfed rice. In Jakenan (Central Java), the two consecutive crops encompass a gradient from low to heavy rainfall (wet season crop) and from heavy to low rainfall (dry season crop), respectively. Rainfed rice was characterized by very low emission at the onset of the wet season and the end of the dry season. Persistent flooding in irrigated fields resulted in relatively high emission rates throughout the two seasons. Average emission in rainfed rice varied between 19 and 123 mg CH_4 m^{-2} d^{-1}, whereas averages in irrigated rice ranged from 71 to 217 mg CH_4 m^{-2} d^{-1}. The impact of organic manure was relatively small in rainfed rice. In the wet season, farmyard manure (FYM) was completely decomposed before CH_4 emission was initiated; rice straw resulted in 40% increase in emission rates during this cropping season. In the dry season, intensive flooding in the early stage promoted high emissions from organically fertilized plots; seasonal emissions of FYM and rice straw increased by 72% and 37%, respectively, as compared with mineral fertilizer. Four different rice cultivars were tested in irrigated rice. Average emission rates differed from season to season, but the total emissions showed a consistent ranking in wet and dry season, depending on season length. The early-maturing Dodokan had the lowest emissions (101 and 52 kg CH_4 ha^{-1}) and the late-maturing Cisadane had the highest emissions (142 and 116 kg CH_4 ha^{-1}). The high-yielding varieties IR64 and Memberamo had moderately high emission rates. These findings provide important clues for developing specific mitigation strategies for irrigated and rainfed rice.

Introduction

Rice is grown in Indonesia in a wide range of environments comprising 10.6 million ha of harvested area (IRRI, 1995). The highly productive land, which is primarily located in Java and Sumatra, is irrigated lowland (72% of total rice area). Smaller portions of the rice land are classified as rainfed (7%) lowland, flood-prone (10%), and upland (11%). Indonesia used to import rice previously but has attained self-sufficiency since 1984 (IRRI, 1995). This success was mainly attributed to an annual production increase of 4-5 % in the 1970s and 1980s. Rice is the staple food of the Indonesian people and the major source of income in many rural areas.

However, rice production is a major source of greenhouse gases such as methane (CH_4) (Neue et al., 1990; Cicerone et al., 1992). The increase of CH_4 concentration in the atmosphere contributes to global warming (Ramanathan et al., 1985) and affects the chemistry of the atmosphere (Bolle et al., 1986; Rasmussen & Kahlil, 1986). Global annual CH_4 emissions from rice fields were estimated in very wide range—from 25 to 100 Tg (IPCC, 1996). One of the main reasons for these uncertainties is the lack of field investigations addressing the variety of cultivation techniques used. Only a few studies using manual sampling techniques are available for Indonesia (Kimura et al., 1994; Husin et al., 1995).

The modalities of rice cultivation have undergone pronounced changes in previous decades, notably in irrigation facilities, fertilizer application, and cultivar traits. Virtually all these agronomic practices affect the conditions for CH_4 production, oxidation, and transport (Neue & Roger, 1993), whereas the net result of these changes remains uncertain.

The study presented here was conducted within an international network of measuring stations for determining CH_4 emissions from rice fields (Wassmann et al., this issue, a). The station is located in an area with predominant rainfed systems while irrigation rice is scattered in the regions. The objectives of the studies at the Jakenan station were

- to quantify CH_4 fluxes from rice fields in Central Java, one of the important rice areas of Indonesia;
- to assess the impact of management practices common to this region;
- to evaluate processes that control CH_4 emission in an equatorial climate; and
- to develop mitigation strategies with low CH_4 emission in a sustainable rice system for this region.

Field site and methods

Field experiments were conducted from 1993 to 1998 at the Jakenan station located in Central Java (Indonesia). Experiments included irrigated and rainfed rice; a comparison between these two ecosystems is presented in Wassmann et al. (this issue, b). Soil properties listed in Table 1 indicate relatively high acidity, low CEC, and low content of organic material. The soil is classified as Acric Tropoqualf and has a silty loam texture.

Central Java has an equatorial climate with heavy rainfalls typically occurring from November to April. Annual precipitation varied from 950 mm to 2200 mm over the last four decades with a long-term average of 1588 mm. Farmers in this region plant two consecutive rainfed crops with a short intermediate fallow. These seasons are commonly denoted as wet season (*gogorancah*) and dry season (*walik jeramih*), although each of them encompasses a gradient from dry to wet and from wet to dry conditions, respectively. In rainfed systems, the wet season crop is dry-seeded, whereas the dry season crop is transplanted. Farmers with access to irrigation water plant two crops of irrigated rice; both crops are transplanted.

Table 1. Soil physical and chemical characteristics of Jakenan Experiment Station.

Parameters	Value
Soil texture (%)	
Sand	29
Silt	58
Clay	13
pH (H_2O)	4.7
Total N (%)	0.05
Total organic carbon (%)	0.48
Available P (ppm)	21
Exchangeable cations (meq/100 mg)	
Ca	2.13
Mg	0.44
K	0.04
Na	0.12
Al	1.45
CEC	6.14
Microelement (ppm)	
Fe	41
Mn	10
Cu	1
Zn	2

A closed chamber technique was used to determine CH_4 emission continuously through an automated system (Wassmann et al., this issue, a).

The standardized measuring systems had the following basic features: (i) three chambers per treatment distributed in the field according to a complete block design; (ii) a pneumatic system for alternate chamber closing (for 16 min to record emission) and opening (for 104 min to equilibrate with ambient air); (iii) a sampling system providing direct air transfer from the inner volume of the chambers to a sample loop and a direct injection of aliquots into the gas chromatograph (GC); and (iv) an analytical system (GC plus integrator) linked to a data acquisition device.

Results and discussion

Characterization of seasonal emissions

Figure 1 shows seasonal patterns of CH_4 emission rates for two consecutive seasons. The experiments in the 1996-97 wet season and the 1997 dry season were conducted with urea as N source in irrigated and rainfed rice (Table 2).

In irrigated rice, flooding started at 4 d before transplanting (for soaking of the soil) and ended a week

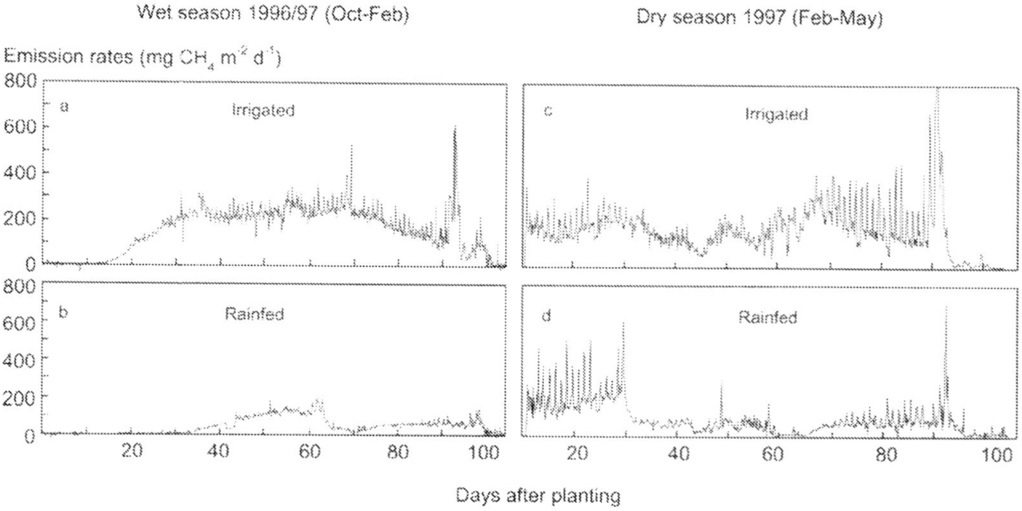

Figure 1. Seasonal patterns of CH_4 emissions in rainfed and irrigated rice during the 1996-97 wet season and 1997 dry season

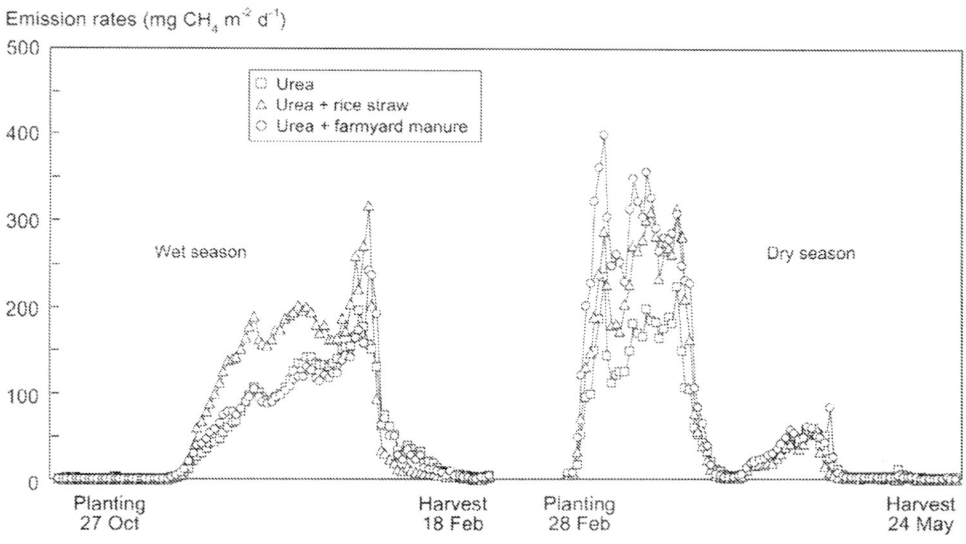

Figure 2. Methane emissions in rainfed rice as affected by amendments, 1995-96 wet season and 1996 dry season (tick marks on the x axis indicate 20-d intervals)

Table 2. Summary of treatments and management practices for the Jakenan experiments, 1993-98.

Year	Season	Treatment No.	Treatment description: Water management	Treatment description: Other management	Inorganic input (kg ha⁻¹): Type	Basal	Topdressed	Organic input (Org N)	Date transplanted	Date harvested
1993-94	Wet	1	Irrigated	IR72/transplanted/20×20	Urea	0	120		10/23	02/19
		2	Irrigated	IR72/direct seeded/20×20	Urea	0	120		10/23	02/19
		3	Rainfed	IR64/transplanted/15×20	Urea	0	120		10/02	02/16
		4	Rainfed	IR64/direct seeded/20×20	Urea	0	120		10/02	02/16
1994	Dry	1	Irrigated	IR72	Urea	0	120		03/01	06/13
		2	Rainfed	IR64	Urea	0	120		03/01	05/26
		3	Rainfed	IR72	Urea	0	120		03/01	05/26
		4	Irrigated	IR64	Urea	0	120		03/01	05/26
1994-95	Wet	1	Rainfed	Urea	Urea	0	120		11/16	03/16
		2	Rainfed	Farm manure	Urea	0	109	11	11/16	03/16
		3	Rainfed	Farm manure	Urea	0	98	22	11/16	03/16
		4	Irrigated	Urea	Urea	0	120		01/06	04/07
1995	Dry	1	Rainfed	Urea	Urea	0	120		03/22	07/17
		2	Rainfed	Farm manure	Urea	0	120	11	03/22	06/08
		3	Rainfed	Farm manure	Urea	0	120	22	03/22	06/08
		4	Irrigated	Urea	Urea	0	120		04/11	06/29
1995-96	Wet	1	Rainfed	Urea	Urea	0	120		10/27	02/18
		2	Rainfed	Rice straw	Urea	0	78.4	42	10/27	02/18
		3	Rainfed	Farm manure	Urea	0	75.4	45	10/27	02/18
		4	Irrigated	Urea	Urea	0	120		11/21	02/10
1996	Dry	1	Rainfed	Urea	Urea	0	120		02/28	05/24
		2	Rainfed	Rice straw	Urea	0	120	41	02/28	05/24
		3	Rainfed	Farm manure	Urea	0	120	45	02/28	05/24
		4	Irrigated	Urea	Urea	0	120		02/28	05/24
1996-97	Wet	1	Irrigated	Prilled urea	Prilled urea	0	120		10/30	02/06
		2	Irrigated	Tablet urea	Urea tablet	0	120		10/30	02/06
		3	Rainfed	Prilled urea	Prilled urea	0	120		10/14	02/06
		4	Rainfed	Tablet urea	Urea tablet	0	120		10/14	02/06
1997	Dry	1	Irrigated	Prilled urea	Prilled urea	0	120		02/12	05/06
		2	Irrigated	Tablet urea	Urea tablet	0	120		02/12	05/06
		3	Rainfed	Prilled urea	Prilled urea	0	120		02/12	05/06
		4	Rainfed	Tablet urea	Urea tablet	0	120		02/12	05/06
1997-98	Wet	1	Irrigated	Dodokan	Urea	0	120		12/02	03/02
		2	Irrigated	IR64	Urea	0	120		12/02	03/08
		3	Irrigated	Menberamo	Urea	0	120		12/02	03/14
		4	Irrigated	Cisadane	Urea	0	120		12/02	04/06
1998	Dry	1	Irrigated	Dodokan	Urea	0	120		05/28	08/10
		2	Irrigated	IR64	Urea	0	120		05/28	08/17
		3	Irrigated	Menberamo	Urea	0	120		05/28	08/18
		4	Irrigated	Cisadane	Urea	0	120		05/28	09/01

before harvest in each season. Therefore, CH_4 emissions increased relatively fast after transplanting and remained on a relatively high level between 100 and 200 mg CH_4 m^{-2} d^{-1} throughout the seasons (Figure 1a, c). The emission peaks at the end of the season indicate the presence of soil-entrapped CH_4 gas that is released after drainage (Wassmann et al., 1994; Denier van der Gon et al., 1996).

For rainfed rice, CH_4 emission rates were virtually zero during the first month of the wet season when the soil is still very dry. Frequent rainfall in November and December gradually increased the flooding intensity and thus, CH_4 emission rates (Figure 1b). However, emission rates were generally below 100 mg CH_4 m^{-2} d^{-1} in the wet season. High flux rates at the onset of the dry season (Figure 1d) can be attributed to persistent flooding in combination with a substrate supply derived from organic residues, i.e. stubble and roots of the preceding crop. After 20 d, emission rates were suppressed in rainfed rice to values below 100 mg CH_4 m^{-2} d^{-1}. Rainfed fields were also drained a week before harvest, but the emerging CH_4 pool from the soils was either negligible (wet season) (Figure 1b) or relatively small (dry season) (Figure 1d) as compared with irrigated rice (Figure 1a,c).

For irrigated rice, the cumulated emission computed for dry and wet seasons are in a similar range (Table 2). Rainfed rice generally had a lower emission in the wet season, although the differences varied broadly from year to year (Table 3). Apparently, dry periods at the early stage have a stronger impact on seasonal emissions than dry periods at later plant stages.

Impact of organic amendment

In the 1995-96 wet season and the ensuing 1996 dry season, the field experiments encompassed three different combinations of mineral and organic amendments in rainfed rice (Table 3). All field trials received urea; one field trial received an additional dose of rice straw and one trial an additional dose of farmyard manure (FYM) (Table 3). Organic amendment had only a minor impact at the onset of the wet season (Figure 2). Soils were still dry in this period, so that the bulk of the organic material was decomposed aerobically. Even when the fields were flooded, emissions were in an identical range in the plots without and those treated with FYM. Application of FYM had no detectable impact on CH_4 emission in the 1995-96 wet season while in the 1994-95 wet season, it even reduced CH_4 emission as compared to urea-applied plots ($P < 0.05$) (Table 3).

Application of rice straw, however, resulted in enhanced emission rates during the middle season (Figure 3). The seasonal emissions are 40% higher than the urea- and FYM-applied plots ($P < 0.05$) (Table 3). This increment in CH_4 emission can be attributed to a relatively high resilience of rice straw to aerobic decomposition. Rice straw is only partially decomposed during the first month of aerobic conditions in the soil.

The different decomposition rates of FYM and rice straw were also discernable during the dry season. Both types of amendments increased emission rates within the first month after planting, but FYM produced consistently higher emissions than rice straw during this period. Again, this difference can be attributed to a faster decomposition of FYM. In contrast to the preceding season, however, anaerobic conditions prevailed in the soil and led to a relatively faster CH_4 production at the start of the experiment.

Methane emission rates converged for all treatments after the initial stage. The overall impact of FYM in the dry season corresponded to a 72% increase in CH_4 emission. The increment triggered by rice straw (37%) roughly corresponded to the relative impact during the preceding wet season (Table 3).

Impact of rice cultivars

Four different cultivars were tested in the 1997-98 wet season and 1998 dry season (Table 2); this comparison was conducted in irrigated rice. The most distinctive feature among these cultivars was the length of cropping season (Figure 3). Dodokan was an early maturing cultivar and was harvested after 90 d in the wet season and 75 d in the dry season. The cultivars IR64 and Memberamo had similar season lengths with 98 d (wet season) and 82 d (dry season). Cisadane required the longest time for maturing with 125 d and 75 d, respectively. IR64 produced the highest yield followed by Memberamo, Dodokan, and Cisadane ($P \leq 0.05$) (Table 3).

Methane emission rates are shown in Figure 3 for Dodokan, IR64, and Cisadane; emission rates for Memberamo and IR64 were almost identical. Emissions of all cultivars were in a similar range and showed similar patterns. Preharvest drainage triggered the emergence of entrapped CH_4 from the soil, resulting in an emission peak shortly before harvest. Then, CH_4 emission rates subsided rapidly after harvest. Due to different harvesting dates, the emission peaks were staggered for the different cultivars (Figure 3).

Table 3. Summary of methane fluxes, biomass, and grain yield from the Jakenan experiments, 1993-98[a]

Year	Season	Treatment no.	Treatment description: Water management	Treatment description: Other management	Mean emission (mg m^{-2} d^{-1})	Seasonal emission (kg ha^{-1})	Biomass (t ha^{-1})	Yield (t ha^{-1})
1993-94	Wet	1	Irrigated	IR72/transplanted/20×20	166 a[†]	229	5.0 b	4.7 b
		2	Irrigated	IR72/direct seeded/20×20	152 a	256	14.3 a	7.1 a
		3	Rainfed	IR64/transplanted/15×20	28 b	59	13.2 a	6.5 a
		4	Rainfed	IR64/direct seeded/20×20	19 b	26	4.9 b	4.4 b
1994	Dry	1	Irrigated	IR72	134 a	141	6.9 b	2.6 b
		2	Rainfed	IR64	90 b	77	9.5 a	3.8 a
		3	Rainfed	IR72	66 c	69	8.7 a	3.6 a
		4	Irrigated	IR64	163 a	115	6.7 b	2.9 b
1994-95	Wet	1	Rainfed	Urea	63 b	75	11.0 a	4.9 a
		2	Rainfed	Farm manure	55 d	65	9.7 a	4.1 a
		3	Rainfed	Farm manure	58 c	69	11.9 a	5.1 a
		4	Irrigated	Urea	124 a	105	11.9 a	4.7 a
1995	Dry	1	Rainfed	Urea	-[b]	-	8.6 b	3.4 a
		2	Rainfed	Farm manure	-	-	8.2 b	3.3 a
		3	Rainfed	Farm manure	-	-	7.6 b	2.9 a
		4	Irrigated	Urea	-	-	9.5 a	3.7 a
1995-96	Wet	1	Rainfed	Urea	52 b	56	10.9 a	4.8 a
		2	Rainfed	Rice straw	73 a	78	12.2 a	5.3 a
		3	Rainfed	Farm manure	52 b	56	11.4 a	4.9 a
		4	Irrigated	Urea	81 a	87	10.5 a	4.4 a
1996	Dry	1	Rainfed	Urea	59 c	53	10.6 a	4.4 a
		2	Rainfed	Rice straw	81 c	73	11.9 a	4.6 a
		3	Rainfed	Farm manure	102 b	92	10.6 a	4.5 a
		4	Irrigated	Urea	184 a	166	10.4 a	4.6 a
1996-97	Wet	1	Irrigated	Prilled urea	171 a	170	14.9 b	7.4 a
		2	Irrigated	Urea tablet	105 b	104	14.9 b	7.4 a
		3	Rainfed	Prilled urea	32 d	37	16.3 a	6.9 b
		4	Rainfed	Urea tablet	39 c	45	15.9 a	6.8 b
1997	Dry	1	Irrigated	Prilled urea	217 a	181	-	4.6 a
		2	Irrigated	Urea tablet	197 a	163	-	5.0 a
		3	Rainfed	Prilled urea	106 c	88	-	4.2 a
		4	Rainfed	Urea tablet	123 b	102	-	4.8 a
1997-98	Wet	1	Irrigated	Dodokan	110 d	101	-	4.5 c
		2	Irrigated	IR64	132 b	128	-	7.0 a
		3	Irrigated	Memberamo	133 ac	137	-	6.0 b
		4	Irrigated	Cisadane	113 cd	142	-	4.2 c
1998	Dry	1	Irrigated	Dodokan	71 c	52	-	-
		2	Irrigated	IR64	100 b	81	-	-
		3	Irrigated	Memberamo	118 b	97	-	-
		4	Irrigated	Cisadane	121 a	116	-	-

[a]Data at the same season of the same year following the same letter are not significantly different at $P < 0.05$. [b]Missing data.

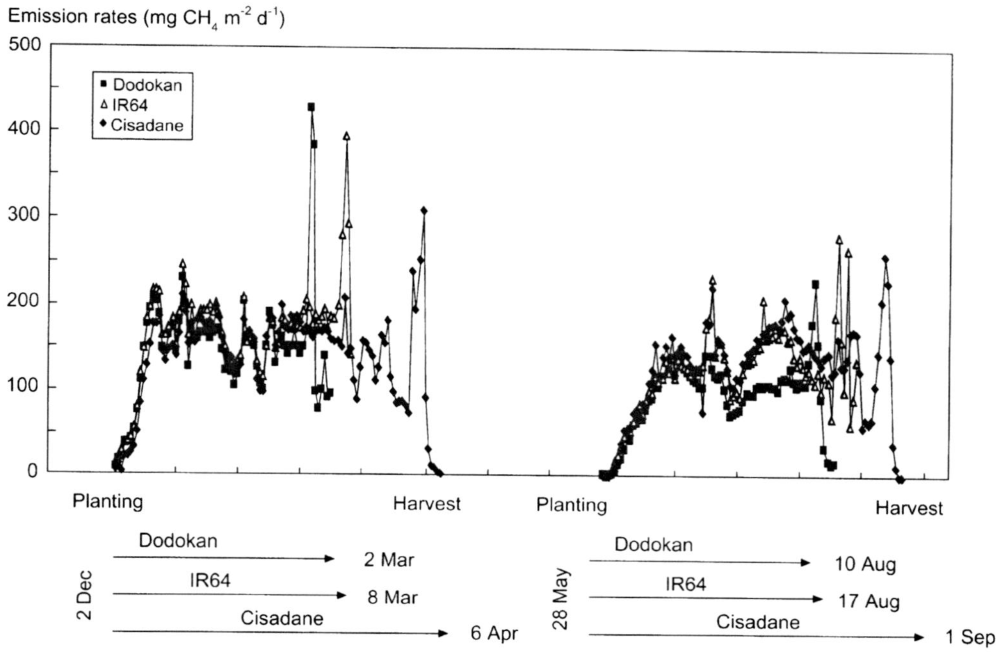

Figure 3. Methane emissions in irrigated rice as affected by cultivars, 1997-98 wet season and 1998 dry season (tick marks on x axis indicate 20-d intervals, arrows indicate length of growing season for each cultivar)

These seasonal patterns resulted in significantly different cumulative values of emission rates ($P \leq 0.05$) (Table 3). In the case of cultivars tested in this experiment, the differences in CH_4 emissions were greatly related to season length. Based on this finding, progress in breeding short-maturing cultivars should be beneficial for an environmentally sound rice production.

However, the results of the 1994 dry season indicate that cultivars with identical season length can also show significant differences in CH_4 emissions (Table 3). IR72 had a lower emission in irrigated rice (18%) and rainfed rice (17 %) than another high-yielding variety IR64 ($P < 0.05$). The reasons for this difference are not clear, but low yields of IR72 (Table 3) indicate the suboptimum growth of IR72 that may have affected emissions.

Impact of other management practices

The first season of the experiment in Jakenan (1993/94 wet season) was conceived to explore promising parameters for future field studies. The individual rice hills were spaced in one field trial in 15-cm × 20-cm distance, whereas all other trials during the entire Jakenan experiment had 20-cm × 20-cm spacing. The different spacing was conducted in rainfed rice, and CH_4 emissions in this season was on an extremely low level (average < 30 mg CH_4 m^{-2} d^{-1}) for both field trials in this season (Table 3). The observed differences between these field trials therefore can not be generalized.

In the same season, direct seeding was compared with transplanting. These modes of crop establishment were tested for irrigated rice and emission rates were relatively higher (Table 3). Direct seeding resulted in 8% reduction in CH_4 emission as compared with transplanting, but the difference was not significant. Therefore, this singular experiment also cannot be used to conclude ubiquitously valid statements on the impact of direct seeding vs transplanting.

The application mode of urea affected CH_4 emission rates in the 1996/97 wet season and 1997 dry season. In irrigated rice, deep placement of urea tablets decreased CH_4 emission rates by 39% in wet season (significant at $P < 0.05$) and 10% in the dry season as compared with broadcasting of prilled urea (Table 3). In rainfed rice, however, tablet urea caused an increase of 21% ($P < 0.05$) and 16% ($P < 0.05$) in wet season and dry season, respectively (Table 3). Yields were similar for prilled and urea tablet, so that the reasons for this diverging effect in irrigated and rainfed rice remain unclear at this point.

Conclusion

The results of this study represent the most comprehensive data set on CH_4 emission from equatorial rice systems. The range of CH_4 emission rates from Indonesian rice fields were previously reported using manual sampling techniques (Kimura et al., 1994; Husin et al., 1995). The field experiment in Jakenan allows a profound assessment of CH_4 emissions through continuous measurements over an observation period of 5 yr. The field layout encompassed virtually all agronomic parameters affecting CH_4 emissions from irrigated and rainfed rice. This article focuses on crop management while preliminary results on temporal patterns and possible mitigation strategies were presented by Buendia et al. (1997). The difference between irrigated and rainfed rice is discussed by Wassmann et al. (this issue, b).

Based on data presented in this study, organic amendments had a much lower impact on CH_4 emissions in rainfed rice than in irrigated systems (Yagi & Minami, 1990; Sass et al., 1991; Wassmann et al., 1995). Easily decomposable material such as FYM is predominantly decomposed aerobically when applied in the wet season. In the dry season, the impact is also substantially smaller than in the irrigated stations of this interregional network. Therefore, the application of organic manure can be regarded as an integral part of sustainable crop management in rainfed rice, even by considering greenhouse gas budget of rice fields as one criterion.

The results for irrigated rice provided an important clue for the selection of cultivars. Previously, different emission potentials have been related to CH_4 transport capacity of the aerenchyma and root exudation (Butterbach-Bahl et al., 1997; Wang et al., 1997). For the set of cultivars tested in Jakenan, these differences appear to be minor and emission potentials are determined by season length. This finding can be used for a preselection of cultivars to identify those with low emission potentials. Breeding attempts to shorten the season length are also beneficial for an environment-friendly rice production with less emissions of greenhouse gases.

Acknowledgements

The research described in this paper was funded by the UNDP-Global Environment Facility GLO/91/G31 under agreement with the International Rice Research Institute (IRRI) and the Central Research Institute for Food Crops-IRRI Collaborative Project. However, UNDP did not review the paper, hence it does not necessarily reflect the views of UNDP and no endorsement should be inferred.

References

Buendia LV, Neue HU, Wassmann R, Lantin RS, Javellana AM, Arah JRM, Xu Yuchang, Lu Wanfang, Makarim AK, Corton TM & Charoensilp N (1997) An efficient sampling strategy for estimating methane emission from rice to assess the impact of potential mitigation options. Chemosphere 36:395-407

Bolle HJ, Seiler W & Bollin B (1986) Other greenhouse gases and aerosols. In: The Greenhouse Effect, Climate Change and Ecosystems, pp 157-198, John Wiley & Sons, New York

Butterbach-Bahl K, Papen H & Rennenberg H (1997) Impact of gas transport through rice cultivars on methane emission from rice paddy fields. Plant Cell Environ 20: 1175-1183

Cicerone RJ, Delwiche CC, Typer SC & Zimmermann PR (1992) Methane emissions from California rice paddies with varied treatments. Global Biogeochem Cycles 6(3):233-248

Denier van der Gon HAC, van Breemen N, Neue HU, Lantin RS, Wassmann R, Aduna JB & Alberto MRC (1996) Release of entrapped methane from wetland rice fields upon drying. Global Biogeochem Cycles 10:1-9

Husin YA, Murdiyarso D, Khalil MAK, Rasmussen RA, Shearer MJ, Sabiham S, Sunar A & Adijuwana H (1995) Methane flux from Indonesian wetland rice: the effects of water management and rice variety. Chemosphere 31: 3153-3180

IPCC – Intergovernmental Panel on Climate Change (1996) Climate Change 1995. The Science of Climate Change. Cambridge (UK): Cambridge University Press. 572 p

IRRI — International Rice Research Institute (1995) World Rice Statistics 1993/1994. Manila, Philippines, 260 p

Kimura M (1994) Effect of intermittent irrigation on methane emission from an Indonesian paddy field. Soil Sci Plant Nutr 40:609-615

Neue HU, Heidmann PB & Scharpenseel HW (1990) Organic matter dynamics, soil properties, and cultural practices in rice lands and their relationship to methane production. In: Bouwman AF (ed) Soil and the Greenhouse Effect, pp 457-466, John Wiley & Sons, Chichester, England

Neue HU & Roger PA (1993) Potential of methane emission in major rice ecologies. In: Zepp RG (ed) Climate Biosphere Interaction, pp 65-93, New York: Wiley and Sons

Ramanathan V, Cicerone RJ, Singh HB & Kiehl JT (1985) Trace gas trends and their potential role in climate change. J Geophys Res 90: 5547-5566

Rasmussen RA & Khalil MAK (1986) Atmospheric trace gases: Trends and distribution over the last decade. Science 32:1623-1624

Sass RL, FM Fischer, Harcombe PA & Turner FT (1991) Mitigation of methane emissions from rice fields: possible adverse effects of incorporated rice straw. Global Biogeochem Cycles 5:275-287

Wang B, Neue HU & Samonte HP (1997) Effect of cultivar difference ('IR72', 'IR65598' and 'Dular') on methane emission. Agric Ecosyst Environ 62:31-40

Wassmann R, Neue HU, Lantin RL, Aduna JB, Alberto MC, Andales MJ, Tan MJ, Denier van der Gon HAC, Hoffmann H, Papen H, Rennenberg H & Seiler W (1994) Temporal patterns of methane emissions from wetland ricefields treated by different modes of N application. J Geophys Res 99: 16457-16462

Wassmann R, Neue HU, Alberto MCR, Lantin RS, Bueno C, Llenaresas D, Arah JRM, Papen H, Seiler W & Rennenberg H (1995) Fluxes and pools of methane in wetland rice soils with varying organic inputs. Environ Monit Assess 42:163-173

Wassmann R, Neue HU, Lantin RS, LV Buendia & Rennenberg H (2000a) Characterization of methane emissions from rice fields in Asia. 1. Comparison among field sites in five countries, (this issue)

Wassmann R, Neue HU, Lantin RS, Makarim K, Chareonsilp N, Buendia LV & Rennenberg H (2000b) Characterization of methane emissions from rice fields in Asia. 2. Differences among irrigated, rainfed and deepwater rice ecosystems (this issue)

Yagi K & Minami K (1990) Effects of organic matter applications on methane emission from Japanese paddy fields. Soil Sci Plant Nutr 36:599-610

Nutrient Cycling in Agroecosystems **58:** 95–105, 2000.

Methane emission from rice fields at Cuttack, India

T.K. Adhya[1], K. Bharati[1], S.R. Mohanty[1], B. Ramakrishnan[1], V.R. Rao[1]
N. Sethunathan[1,2] & R. Wassmann[3]

[1]Laboratory of Soil Microbiology, Division of Soil Science & Microbiology, Central Rice Research Institute, Cuttack 753006, India; [2]Division of Microbiology, Indian Agricultural Research Institute, New Delhi 110012, India; [3]International Rice Research Institute, P.O. Box 3127, Makati City 1271, Philippines, and Fraunhofer Institute for Atmospheric Environmental Research (IFU), Garmisch-Partenkirchen, Germany

Key words: organic amendment, water management, cultivar variation, nitrification inhibitors, production potential, soil type

Abstract

Methane (CH_4) emission from rice fields at Cuttack (State of Orissa, eastern India) has been recorded using an automatic measurement system (closed chamber method) from 1995-1998. Experiments were laid out to test the impact of water regime, organic amendment, inorganic amendment and rice cultivars. Organic amendments in conjunction with chemical N (urea) effected higher CH_4 flux over that of chemical N alone. Application of *Sesbania*, *Azolla* and compost resulted in 132, 65 and 68 kg CH_4 ha^{-1} in the wet season of 1996 when pure urea application resulted in 42 kg CH_4 ha^{-1}. Intermittent irrigation reduced emissions by 15% as compared to continuous flooding in the dry season of 1996. In the wet season of 1995, four cultivars were tested under rainfed conditions resulting in a range of emissions from 20 to 44 kg CH_4 ha^{-1}. Application of nitrification inhibitor dicyandiamide (DCD) inhibited while Nimin stimulated CH_4 flux from flooded rice compared to that of urea N alone. Wide variation in CH_4 production and oxidation potentials was observed in rice soils tested. Methane oxidation decreased with soil depth, fertilizer-N and nitrification inhibitors while organic amendment stimulated it. The results indicate that CH_4 emission from the representative rainfed ecosystem at the experimental site averaged to 32 kg CH_4 ha^{-1} yr^{-1}.

Introduction

Rice fields are considered to be an important anthropogenic source for methane (CH_4) (Neue et al., 1995) and contribute up to 20% or ~100 Tg CH_4 to the global budget on an annual basis (Houghton et al., 1996). With intensification of rice cultivation during the coming decades (IRRI 1999), CH_4 emission from this economically important but ecologically fragile ecosystem is anticipated to increase (Anastasi et al., 1992). Despite recent studies on identification of controlling variables (Neue et al., 1997), the uncertainty in the global CH_4 source strength estimate for rice paddies is still very high among all the established CH_4 sources (Houghton et al., 1996) due to large spatial differences (Yagi, 1997). Such uncertainty in the source strength estimate largely stems from different soil types as well as variations between crop management in space and time. Refinement in methodologies and more measurements incorporating site-specific practices are essential for an accurate assessment of the contribution of paddy ecosystem to global CH_4 budget as well as to devise methodologies for its abatement.

India produces annually 80 m t of rice on an area of 42.3 m ha corresponding to 28% of the global rice lands (Sharma et al., 1995). The rice growing areas of India can be broadly categorized into rainfed upland, rainfed lowland and irrigated medium land, representing about 15, 40 and 45% of total rice area of the country. In India, 48% of the country's rice area is irrigated while the rest is grown under rainfed situations. Extrapolating CH_4 flux measurement data from rice fields of USA and Europe, the total CH_4 emission from Indian rice fields was estimated to be 37.8 Tg CH_4 yr^{-1} (US-EPA, 1990). However, based on the actual field measurements conducted in select rice growing areas, mean CH_4 emission from Indian rice fields ranged between 2.7 and 6.4 Tg $CH_4.yr^{-1}$ (Adhya et al., 1994; Mitra, 1992, Parashar et al., 1997). Under the IRRI-UNDP collaborative project (Wassmann et al., this issue, a), we measured CH_4 flux from both rainfed and irrigated fields under the influence of different controlling factors such as organic amendments, water management, rice cultivars and chemical inhibitors. In addition, studies were also conducted to estimate the source strength of CH_4 for select Indian rice soils by measuring their CH_4 production and oxidation potentials and factors influencing these processes.

Materials and methods

Field experiments were conducted during the dry (January-May) and wet (July-December) seasons, beginning with the wet season of 1995 and continued till the wet season of 1998 in the research farm of Central Rice Research Institute, Cuttack (State of Orissa). The farm is situated at 20°25'N latitude and 85°55'E longitude. During 1995-98, the mean rainfall during dry and wet seasons was 85 and 1352 mm, respectively. The monthly mean maximum and minimum temperatures were in the range of 26.5-37.7 °C and 12.7-26.7 °C, respectively. The mean sunshine hours during dry and wet seasons were 8.1 and 5.8 h d^{-1}, respectively.

The experiment in each rice season consisted of four treatments in randomized block design with three replicates and concentrated on specific variables. The summary of the treatments for the period of 1995-98 is provided in Table 1. The soil was a typic Haplaquept (Table 2) with a percolation rate of 0.23 cm d^{-1}. The individual plot size was 5 × 5 m. The field was ploughed thoroughly and flooded 2-3 d before transplanting for puddling and leveling. Rice seedlings (21 d old) were transplanted at a spacing of 15 × 15 cm with 2 seedlings $hill^{-1}$. For the dry season crop, the field was irrigated to maintain the floodwater level between 5-10 cm during the entire period of crop growth excepting for the treatment on alternate flooding. For the alternate flooding treatment (1997 dry season), the field plots were irrigated at every 15 d interval to a maximum floodwater level of 10 cm. During the wet season, the crop was grown exclusively under rainfed conditions —floodwater level remained shallow i.e., 3-15 cm during most part of its growth.

Methane emission from field experiments was sampled and analyzed by automatic gas sampling and analysis system installed under the IRRI-UNDP Program. This system consisted of gas collection chambers made up of plexiglas (1 × 1 × 1.2 m) and the sampling system. The boxes were fitted with top covers that open and close automatically through a pneumatic system that was controlled by a microcomputer through a specific software developed by the Fraunhofer Institute for Atmospheric Sciences, Germany.

Measurement of CH_4 flux from different rice cultivars (1995 wet season) were obtained by manual measurement method (Adhya et al., 1994) at 5-d intervals from the day of transplanting till maturity. Sampling for CH_4 flux measurements were made at 0900-0930 and 1500-1530, and the average of morning and evening fluxes was used as the flux value for the day. For measuring CH_4 emission, six hills of rice plants were covered with a locally fabricated perspex box (53 cm length × 37 cm width × 71 cm height). A battery-operated air circulation pump with air displacement of 1.5 l min^{-1}, connected to polyethylene tubing was used to mix the air inside the box and draw the air samples into air-sampling bags at fixed intervals of 0, 15, and 30 min. The air samples from the sampling bags were analyzed for CH_4.

Potential CH_4 production was measured from select rice soils whose properties are listed in Table 2. The soils were collected from the plough layer (0-25 cm), air dried under shade, ground and passed through a sieve (>2 mm) and stored in glass bottles at room temperature. Twenty grams portions of air-dried soil samples were placed in 100 ml spouteless beakers and 40 ml of sterile distilled water was added to flood the soil. The beakers were closed with a rubber stopper with provisions for gas ports for headspace gas sampling, platinum electrode and placement of pH electrode assembly. Soil samples in beakers were incubated under N_2 atmosphere at 30 °C for 40 d, as described by Lantin et al. (1995). At regular intervals, headspace gas sample was analyzed by gas chromatography for quantification of CH_4 produced.

Methane oxidation potential of the soils was measured by the method of Bharati et al. (1999a). Soil samples (surface and subsurface) from different field experiments were collected with a PVC core sampler (2 cm dia) and the cores were sectioned at different depth intervals (0-5, 5-10, 10-15 cm). The profile samples of each of six cores from the same treatment, after removal of root pieces and stones, were mixed thoroughly and the moisture content of the samples was brought to approximately 60% moisture holding capacity by removing excess moisture with Whatman filter paper. Portions of the soil (10 g) were placed in 130 ml

Table 1. Summary of the treatments in 1995, 1996, 1997, and 1998

Treatment	T1	T2	T3	T4
1995 (wet season)				
Cultivar[a]	Lalat	Tulasi	Gayatri	IR72
Crop establishment	Transplanting 15 × 15 cm	Transplanting 15 × 15 cm	Transplanting 15 × 15 cm	Transplanting 15 × 15 cm
Water regime	Rainfed	Rainfed	Rainfed	Rainfed
Mineral NPK	60-30-30	60-30-30	60-30-30	60-30-30
Planting date	95/07/16	95/07/16	95/07/16	95/07/16
Harvesting date	95/11/02	95/11/18	95/11/02	95/11/02
1996 (wet season)				
Cultivar	CR 749-20-2	CR 749-20-2	CR 749-20-2	CR 749-20-2
Crop establishment	Transplanting 15 × 15 cm	Transplanting 15 × 15 cm	Transplanting 15 × 15 cm	Transplanting 15 × 15 cm
Water regime	Rainfed	Rainfed	Rainfed	Rainfed
Mineral NPK	60-30-30	40-30-30	40-30-30	40-30-30
Organic N[a]	0	Green manure (*Sesbania*) : 20	Compost : 20	Green manure *Azolla* : 20
Planting date	96/07/19	96/07/19	96/07/19	96/07/19
Harvest date	96/10/30	96/10/30	96/10/30	96/10/30
1997 (dry season)				
Cultivar	CR 749-20-2	CR 749-20-2	CR 749-20-2	CR 749-20-2
Crop establishment	Transplanting 15 × 15 cm	Transplanting 15 × 15 cm	Transplanting 15 × 15 cm	Transplanting 15 × 15 cm
Water regime[a]	Continuous flooding	Continuous flooding	Alternate flooding	Alternate flooding
Mineral NPK	60-30-30	60-30-30	60-30-30	60-30-30
Organic C	0	Rice straw (2 t ha^{-1})	0	Rice straw (2 t ha^{-1})
Planting date	97/02/25	97/02/25	97/02/25	97/02/25
Harvest date	97/05/31	97/05/31	97/05/31	97/05/31
1998 (dry season)				
Cultivar	CR 749-20-2	CR 749-20-2	CR 749-20-2	CR 749-20-2
Crop establishment	Transplanting 15 × 15 cm	Transplanting 15 × 15 cm	Transplanting 15 × 15 cm	Transplanting 15 × 15 cm
Water regime	Normal irrigation	Normal irrigation	Normal irrigation	Normal irrigation
Mineral NPK	0-40-40	120-40-40	120-40-40	120-40-40
Inorganic amendment[a]	0	0	Nimin (1% of added urea)	DCD (30 kg ha^{-1})
Planting date	98/02/12	98/02/12	98/02/12	98/02/12
Harvest date	98/05/18	98/05/18	98/05/18	98/05/18

[a]Modifying treatment focused during the season

sterile serum bottles and allowed to equilibrate with the ambient air for 3 d in the dark in an incubator at 30 ± 2 °C. Soil samples from places other than those of the field experiments were first activated in a greenhouse by putting 5 kg of each soil in earthenware pots and seedlings of rice plants (CR 749-20-2) planted to it. The soil from the pots was similarly sampled with a PVC core sampler and incubated in serum bottles. Methane oxidation was initiated by sealing the serum bottles with neoprene septa and injecting the headspace with 5 ml of pure CH_4 to provide approximately 2,100 mmol of CH_4 g^{-1} air-dried soil. Soil samples were incubated in an incubator (30 ± 2 °C) in the dark. At select intervals, headspace gas sample (5 ml) of the serum bottles was analyzed for CH_4 until 10 d. After each sampling, the headspace was replaced with an equivalent

Table 2. Physicochemical characteristics of the soils used in the study

Location	Soil type	pH	EC (dS m^{-1})	Organic C (%)	Total N (%)	CEC (meq 100 g^{-1} soil)	Soil separates		
							Clay (%)	Silt (%)	Sand (%)
Balasore	Alluvial	6.69	1.70	1.36	0.14	16.40	30.50	39.30	30.20
Bhubaneswar	Laterite	5.89	0.82	0.71	0.06	11.63	9.00	11.20	79.80
Cochin	Sandy loam	6.10	0.77	1.36	0.09	7. 83	16.18	15.82	68.00
Cuttack	Alluvial	6.16	0.50	0.86	0.09	15.00	25.90	21.60	52.50
Hyderabad	Laterite	7.90	14.57	0.60	0.07	51.50	50.00	26.00	24.00
Kalahandi	Black	6.90	3.78	0.52	0.04	9.00	18.60	21.50	59.90
Khuntuni	Laterite	5.87	0.16	0.21	0.08	15.20	35.00	33.30	31.70
Pokkali	Acid sulfate	3.90	5.01	4.86	0.21	19.20	40.60	49.60	9.80
Sukinda	Laterite	6.87	1.10	0.62	0.04	6.00	14.60	10.60	74.80

Table 3. Methane emission from a rainfed alluvial field planted to rice (cv CR 749-20-2) under the influence of urea N in combination with different organic amendments (1996 wet season)

Treatments[a]	Mean emission[b] (mg m^2 d^{-1})	Seasonal flux (kg ha^{-1})	Grain yield[c] (t ha^{-1})	Kg CH_4 t^{-1} grain yield	% change
Urea N	41.06 ± 41.47	42.30	3.02 ± 0.82	14.00	-
Sesbania + Urea N	128.12 ± 115.35	131.97	3.50 ± 1.12	37.70	212
Compost + Urea N	63.53 ± 65.33	65.44	2.98 ± 0.52	21.96	55
Azolla + Urea N	65.74 ± 69.30	67.71	3.65 ± 0.45	18.55	60

[a]All organic amendments made on equal N basis (20 kg N ha^{-1}) with urea to provide a total of 60 kg N ha^{-1}. [b]Mean of daily observations in a cropping season (n = 103) ± SD. [c]Av of three replicate observations ± SD.

amount of high purity Ar to maintain the pressure equilibrium.

Methane concentrations in samples collected from field and laboratory experiments were analyzed with a Shimadzu GC-8A gas chromatograph (GC) equipped with FID and a Porapak N column. The column and detector were maintained at 70 and 110 °C, respectively. The gas samples from field experiments were automatically injected through a sample loop (3 ml) with the help of an on-column injector using a multiport valve. Samples of CH_4 from production and oxidation experiments were analyzed by injecting the gas samples through a secondary injection port. The GC was calibrated before and after each set of measurement using 5.38, 9.03, and 10.8 ml CH_4 ml^{-1} in N_2 as primary standard and 2.14 ml CH_4 ml^{-1} in air as secondary standard. Under these conditions, the retention time of CH_4 was 0.65 min and the minimum detectable limit was 0.5 ml ml^{-1}.

Results and discussion

Methane emission from rice fields

Organic matter amendment. Methane flux from flooded plots planted to rice (CR 749-20-2) under different organic amendments, was monitored during the wet season of 1996. Seasonal flux of CH_4 was high following the application of fertilizer-N and organic amendments further enhanced it (Table 3). All the organic treatments in combination with urea effected higher CH_4 flux over that of chemical-N (urea) alone. Organic amendment affected an immediate increase in emission values (up to 400 mg CH_4 m^{-2} d^{-1}). After 10 days emission rates decreased to less than 100 mg CH_4 m^{-2} d^{-1} but remained consistently on a higher level than the other treatments (Figure 1). Over the season, the ranking in emission from these four treatments was *Sesbania* (212% increase as compared to urea alone) > *Azolla* (61% increase) >

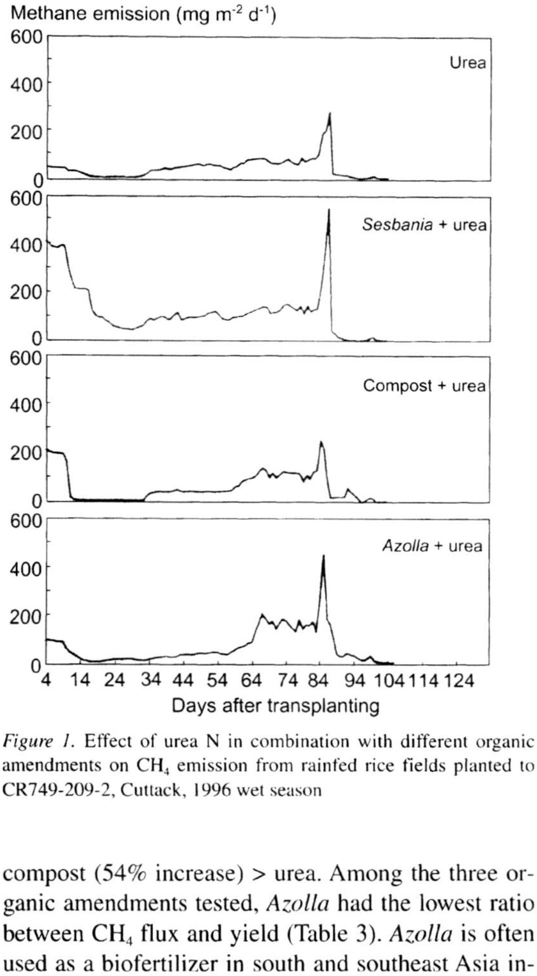

Figure 1. Effect of urea N in combination with different organic amendments on CH_4 emission from rainfed rice fields planted to CR749-209-2, Cuttack, 1996 wet season

compost (54% increase) > urea. Among the three organic amendments tested, *Azolla* had the lowest ratio between CH_4 flux and yield (Table 3). *Azolla* is often used as a biofertilizer in south and southeast Asia including China (Liu & Zheng, 1992), India (Singh & Singh, 1997), Bangladesh (Islam et al., 1984) and Vietnam (Lumpkin & Plucknett, 1982) to improve the N balance of paddy fields and is either incorporated as green manure at the beginning of the cropping season or grown as a dual crop alongwith rice, in the standing water of flooded fields. Our studies indicate that amendment with *Azolla* although increased CH_4 flux over that of chemical-N alone, the effect was compensated for by higher grain yield.

Organic matter amendment to flooded soil increased CH_4 production and emission (Cicerone et al., 1992; Sass et al., 1991; Wassmann et al., 1996; Yagi & Minami, 1990). Readily mineralizable soil organic matter is the main source of fermentation products in flooded soils and sediments that are driven to CH_4 by strict anaerobic bacteria (methanogens) (Ferry, 1992). Results obtained from this study indicate substantial increase in CH_4 efflux from rainfed paddy following amendment with organic sources.

Water management. Flooding the soil creates anaerobiosis and conditions favorable for CH_4 production and emission. Thus, floodwater regime can have a strong influence on CH_4 emission rates from rice fields (Minami, 1994; Wassmann et al., 1995; Yagi et al., 1996) and a single midseason drainage is considered to reduce seasonal CH_4 rates by about 50% (Sass et al., 1992). In a controlled experiment during the dry season of 1997, seasonal CH_4 flux as influenced by continuous flooding *vis-à-vis* alternate flooding (intermittent irrigation) was investigated.

Mean CH_4 emission was lowest (13.80 mg m^{-2} d^{-1}) in field plots that were alternately flooded as compared to continuously flooded (16.32 mg m^{-2} d^{-1}) field plots (Table 4) leading to a 15% reduction in seasonal CH_4 flux. Amendment with rice straw at 2 t ha^{-1} significantly increased CH_4 production under both continuously flooded and intermittently flooded field plots with the maximum increase under the continuously flooded conditions (Figure 2). However, grain yield was higher under rice straw-amended, intermittently flooded field plots, resulting in the least amount of CH_4 t^{-1} grain yield.

Table 4. Methane emission from an irrigated alluvial field planted to rice (CR749-20-2) as affected by water regime and straw amendment, 1997 dry season

Treatment[a]	Mean emission[b] (mg m^{-2} d^{-1})	Seasonal flux (kg ha^{-1})	Grain yield[c] (t ha^{-1})	Kg CH_4 t^{-1} grain yield	% change
Continuously flooded	16.32 ± 27.61	18.61	3.21 ± 0.93	5.80	-
Continuously flooded + rice straw	31.73 ± 51.61	36.18	3.52 ± 1.28	10.28	94
Alternately flooded	13.80 ± 18.89	15.73	3.47 ± 0.82	4.53	-15
Alternately flooded + rice straw	23.81 ± 42.05	27.14	3.11 ± 1.61	8.73	46

[a]Rice straw was added at 2 t ha^{-1}; for alternate flooding treatment, the field plots were flood-irrigated at 15-d intervals. [b]Mean of daily observations in a cropping season (n = 114) ± SD. [c]Av of three replicate observations ± SD.

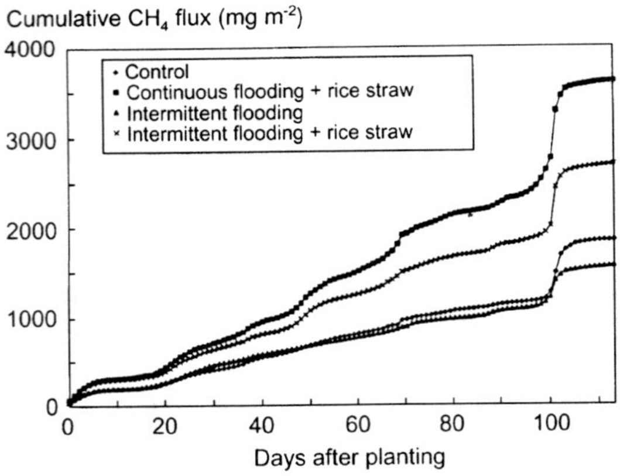

Figure 2. Cumulative CH_4 efflux from an irrigated alluvial field planted to rice under the influence of water management and rice straw amendment, 1997 dry season

In a greenhouse study, with soil from this field site, intermittent flooding resulted in distinctly less CH_4 than continuous flooding (Mishra et al., 1997). In rainfed rice ecosystem, drying and wetting of soil occurs naturally and frequently with alternate drought and rainy periods. While such situations would automatically reduce CH_4 flux from a rainfed ecosystem, efficient water management in areas with effective drainage facility would further limit CH_4 flux.

Cultivar variation. Rice plants serve as the major conduit for the transfer of CH_4 from the reduced soil layer to the atmosphere and more than 90% of CH_4 fluxes from paddy soils are mediated by the rice plants (Denier van der Gon & Neue, 1996; Holzapfel-Pschorn et al., 1985; Schutz et al., 1989). In view of the inherent variability in plant architecture, metabolic activity and gas transport potential among different rice cultivars (Neue & Sass, 1994), cultivar variation in CH_4 efflux from rice has attracted attention (Satpathy et al., 1998; Wang et al., 1997).

The role of rice cultivar on CH_4 emission from flooded fields was investigated in a field experiment in the wet season of 1995. Among the four modern improved rice cultivars tested, *cv.* Lalat gave the highest seasonal CH_4 flux (44.41 kg ha^{-1}) and the degree of CH_4 efflux followed the order of Lalat > IR 72 > Gayatri > Tulasi. Cultivars Gayatri and Tulasi had lower CH_4 flux (Table 5), thereby producing –13% and –22% CH_4 over that of IR72. Wide variations among rice cultivars tested with regard to CH_4 flux opens up possibilities for breeding rice cultivars with low CH_4 emission potential.

Nitrification inhibitors. Nitrification inhibitors are being increasingly recommended for intensive agriculture to regulate fertilizer N losses (Prasad & Power, 1995) from flooded paddy. In addition to their acknowledged role in controlling various processes of N losses, nitrification inhibitors like acetylene (wax coated calcium carbide) and nitrapyrin have been shown to inhibit CH_4 emission from flooded soil planted to rice (Bronson & Mosier, 1991; Keertisinghe et al., 1993). In a field experiment during the dry season of 1998, the effect of two nitrification inhibitors, dicyandiamide (DCD) and Nimin (alcoholic extract of *Azadirachta indica*) on CH_4 efflux from flooded paddy was investigated.

Seasonal flux of CH_4 (Figure 3) increased by 94% following application of fertilizer-N (urea). Among the nitrification inhibitors tested DCD reduced CH_4 emission by 13%, while Nimin, at the concentration used in this study, increased CH_4 flux by 9.6% over that of urea-N alone (Table 6). Nitrification inhibitors are known to inhibit CH_4 oxidation and CH_4-oxidizing microbial population (Hanson & Hanson, 1996), but their exact role in CH_4 emission is not clear. Inhibition of CH_4 production in DCD-amended alluvial soil was related to high redox potential, low pH, Fe^{2+} and readily mineralizable carbon content as well as lower population of methanogenic bacteria and their activity (Bharati et al., 1999b). It is possible that low source strength of

Table 5. Methane emission from a rainfed alluvial field planted to different rice cultivars under uniform conditions, 1995 wet season

Rice cultivar	Mean emission[a] (mg m^{-2} d^{-1})	Seasonal flux (kg ha^{-1})	Grain yield[b] (t ha^{-1})	Kg CH_4 t^{-1} grain yield	% change
IR72	23.36 ± 17.62	25.84	2.37 ± 0.80	10.90	-
Gayatri	19.89 ± 14.55	22.58	3.15 ± 1.05	7.16	-13
Tulasi	17.85 ± 11.89	20.21	3.28 ± 1.45	6.16	-22
Lalat	39.58 ± 26.41	44.41	3.85 ± 0.72	11.53	72

[a]Mean of observations in a cropping season (n = 22) ± SD. [b]Av of three replicate observations ± SD.

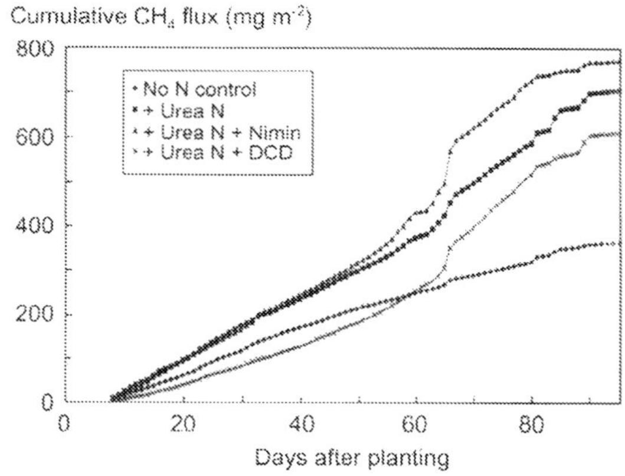

Figure 3. Cumulative CH_4 efflux from an irrigated alluvial field planted to rice as influenced by urea N and nitrification inhibitors, 1998 dry season

CH_4 due to inhibition of CH_4 production, resulted in a low emission of CH_4 in DCD-amended plots. Two nitrification inhibitors tested in this study significantly increased grain yield over that of urea-N alone, probably due to better N-use efficiency (Prasad, 1998). The results with DCD amendment have applied significance in view of low CH_4 flux with increased N use efficiency and higher grain yield.

Methane production in tropical rice soils

CH_4 production rates of the nine soils studied showed appreciable differences among themselves and were of lower magnitude in almost all the soils except that of Balasore soil (Figure 4). Temporal pattern of production rates during 50 d of incubation indicated three different classes of production patterns, namely (I) suppressed (Kalahandi, Pokkali, Sukinda, and Bhubaneswar), (II) delayed (Cuttack, Khuntuni, Cochin, and Balasore) and (III) immediate (Hyderabad). Kalahandi soil was not microbially active as indicated by a slow reduction of the soil following flooding. However, in spite of fast reduction and near neutral pH, CH_4 production was low in acid sulfate soil (Pokkali) and could be due to the presence of sulfate and volatile sulfides in the soil. Methane production rates were low throughout the incubation period for the soils classified in category I. On the contrary, CH_4 production in category II soils was low during the first 10 d followed by an increase around 30 d of incubation. Interestingly, in Hyderabad soil, CH_4 production reached its peak within first 10 d of incubation after which it declined. In a laboratory incubation study, Wang et al. (1993) classified the soils in two groups, the first group where CH_4 production was inhibited until 10 d while in the other group CH_4 production was not inhibited. Among the soils used in the present study, only Hyderabad soil falls in the second group.

A correlation analysis of different soil characters and CH_4 production rates is indicated in Table 7. Methane production was significantly correlated only with soil CEC over 10 d incubation while no significant correlation existed between any of the soil characters and CH_4 production for incubation period of 50 d. In a study on CH_4 production capabilities of eleven Philippine rice soils, Wassmann et al. (1998) indicated significant positive relationship between concentrations of organic C and organic N for soils incubated over 8 wk period.

While soil physicochemical properties are known to affect CH_4 production through various pathways, soils used in the present study did not reveal any such effect. The scope of the present study involving incubation of select native rice soils is of limited nature and probably can not be extended to field situations where growing rice plants will affect CH_4 production by providing exogenous substrates through root exudates and dead and decaying roots. However, the results indicate the inherent spatial variability among different rice soils and further studies with a wider range of soils and different

Table 6. Methane emission from an irrigated field planted to rice (CR749-20-2) as affected by urea N and nitrification inhibitors, 1998 dry season

Treatment[a]	Mean emission[b] (mg m^{-2} d^{-1})	Seasonal flux (kg ha^{-1})	Grain yield[c] (t ha^{-1})	Kg CH_4 t^{-1} grain yield	% change
No N control	41.09 ± 19.47	36.15	2.49 ± 1.23	14.5	-
+ Urea N	79.66 ± 47.87	70.10	3.43 ± 2.05	20.4	94
+ Urea N + Nimin	87.32 ± 83.79	76.84	4.25 ± 2.43	18.1	113
+ Urea N + DCD	68.98 ± 60.63	60.69	4.48 ± 1.51	13.5	68

[a]Urea N was added at 120 kg N ha^{-1}. Nimin was applied at 1% of urea-N and DCD was applied at 30 kg ha^{-1}. [b]Mean of daily observations in a cropping season (n = 96) ± SD. [c]Av of three replicate observations ± SD.

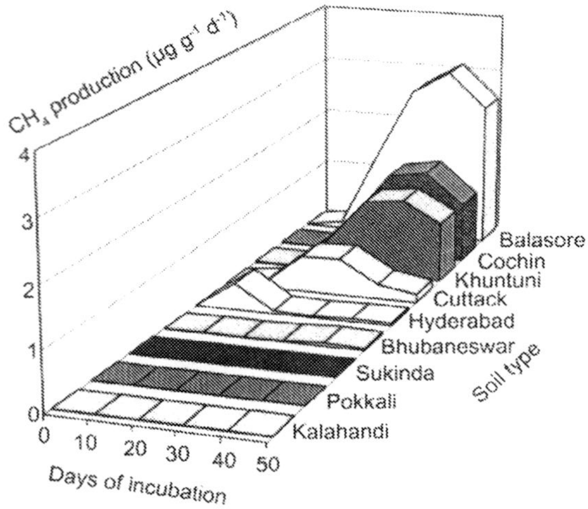

Figure 4. Methane production potential of selected rice soils of India

amendments would probably help in explaining the basic mecahnisms of variability of CH_4 production and emission from these soils.

Methane oxidation in flooded rice soils

In submerged rice paddies, the oxic surface soil-water interface modulates the CH_4 flux to the atmosphere through microbial CH_4 oxidation. About 80% of CH_4 produced in anaerobic soil is oxidized to CO_2 in the aerobic thin surface layer and rice rhizosphere (Conrad & Rothfuss, 1991). The biochemical process carried out exclusively by a group of autotrophic bacteria, the methanotrophs, is unique in their ability to utilize methane as a sole carbon and energy source (Conrad, 1996).

In studies on CH_4 oxidation in select rice soils, CH_4 oxidation potential varied widely among different soils. Alluvial soils from Cuttack and Balasore had high oxidation potential while laterite soils of Bhubaneswar, Khuntuni and Sukinda had low oxidation potential (Table 8). The oxidation potential also varied depending upon the growth stage of the rice plant.

Organic amendment increased the CH_4 oxidation potential of the field soil while fertilizer N inhibited the process (Table 9). CH_4 oxidation potential also decreased depending upon the depth. In an earlier study from this laboratory, Methane oxidation was low at deeper layers (Kumaraswamy et al., 1997). Application of N fertilizers, especially NH_4^+-containing compounds inhibit the process of CH_4 oxidation (Conrad & Rothfuss, 1991). In the present study, CH_4 oxidation was inhibited in soils amended with urea N. Application of nitrification inhibitors DCD and Nimin with urea N further inhibited the CH_4 oxidation process (Table 10).

Approximately 95% of the CH_4 produced in flooded soils is oxidized to CO_2 before it's release to the environment and thus CH_4 oxidation plays an important role in the biogeochemical cycling of CH_4. Our studies indicate high spatial variability in CH_4 oxidation in different soils. The process of CH_4 oxidation is also subject to several agricultural processes including organic amendment and application of inorganic fertilizers and nitrification inhibitors. While accelerating CH_4 oxidation can be a feasible approach to mitigate CH_4 emission, detailed studies both under greenhouse and field conditions are essential before this process can be developed as a field-scale technology.

Table 7. Correlation analyses of physicochemical soil properties (independent variables) and incubation results (dependent variables)

Parameter	pH	EC	CEC	Organic C	Total N	Clay	Silt	Sand
				10-d capacity of CH_4 production				
r	0.336	0.107	0.954	0.199	0.078	0.342	0.095	0.425
a	–0.420	0.100	–0.088	0.109	0.103	–0.116	0.057	0.204
b	0.081	–0.009	0.010**	–0.021	–0.214	0.007	0.001	–0.002
				50-d capacity of CH_4 production				
r	0.078	0.250	0.119	0.071	0.292	0.080	0.354	0.232
a	0.153	0.761	0.662	0.589	0.113	0.399	–0.051	0.919
b	0.061	–0.132	–0.007	–0.043	4.640	0.005	0.023	–0.008

r = correlation coefficient; *a* = intercept; *b* = slope; ** = significant at 1% level.

Table 8. Methane[a] oxidation potential of selected rice soils[b] planted to rice (CR749-20-2) under greenhouse conditions

Soil type	Plant growth stage					
	Tillering		Panicle initiation		Maturity	
	k	$t_{1/2}$ (d)	k	$t_{1/2}$ (d)	k	$t_{1/2}$ (d)
Balasore	0.161	4.30	0.183	3.78	0.152	4.55
Bhubaneswar	0.021	32.96	0.199	3.47	0.043	16.10
Cochin	0.108	6.41	0.169	4.09	0.144	4.80
Cuttack	0.340	2.03	0.294	2.35	0.352	1.96
Khuntuni	0.030	23.07	0.047	14.72	0.029	23.87
Sukinda	0.056	12.36	0.032	21.63	0.041	16.88

[a]Concentration of CH_4 added to headspace air was 2100 mmol g^{-1} air-dried soil. [b]Soils collected from 0-5 cm depth of planted pots.

Table 9. Methane oxidation potential of an alluvial soil at different depths from a flooded field planted to rice (*cv.* CR 749-20-2) under the influence of urea N in combination with different organic amendments

Treatments[b]	Soil depth					
	0-5 cm		5-10 cm		10-15 cm	
	k	$t_{1/2}$ (d)	k	$t_{1/2}$ (d)	k	$t_{1/2}$ (d)
Urea N	0.619	1.11	0.340	2.03	0.156	4.43
Sesbania + urea N	0.771	0.89	0.663	1.04	0.292	2.37
Compost + urea N	0.683	1.01	0.672	1.03	0.614	1.12
Azolla + urea N	0.621	1.11	0.603	1.14	0.578	1.19

[a]Concentration of CH_4 added to headspace air was 2100 mmol g^{-1} air-dried soil.[b]All organic amendments made on an equal N basis (20 kg N ha^{-1}) with urea to provide a total of 60 kg N ha^{-1}.

Table 10. Methane[a] oxidation potential of a flooded alluvial soil at different depths under the influence of fertilizer N and nitrification inhibitors

Treatments[b]	Soil depth					
	0-5 cm		5-10 cm		10-15 cm	
	k	$t_{1/2}$ (d)	k	$t_{1/2}$ (d)	k	$t_{1/2}$ (d)
No N control	0.409	1.69	0.104	6.66	0.076	9.11
+ Urea N	0.318	2.18	0.062	11.17	0.037	18.71
+ Urea N + Nimin	0.238	2.91	0.059	11.73	0.018	38.46
+ Urea N + DCD	0.096	7.21	0.032	21.63	0.036	19.23

[a]Concentration of CH_4 added to headspace air was 2100 mmol g^{-1} air-dried soil. [b]Urea added at 120 kg N ha^{-1}. Nimin applied at 1% of urea and DCD applied at 30 kg ha^{-1}.

Conclusion

Flooded paddy is one of the most important anthropogenic source of atmospheric CH_4. Research worldwide indicates that organic amendments, water management, fertilizer management and candidate rice cultivars affect the flux of CH_4 from this economically important ecosystem. Studies conducted under the IRRI-UNDP Interregional Research Program using automatic measurement system have clearly indicated that (I) although organic amendment increased CH_4 flux under rainfed conditions, application of *Azolla* resulted in a lower CH_4

flux per ton of grain yield; (II) CH_4 emission was reduced by 15% when intermittent irrigation was practiced during the dry season; and (III) nitrification inhibitor DCD distinctly inhibited CH_4 flux.

Acknowledgements

The research described in this paper was funded by the UNDP-Global Environment Facility GLO/91/G31 under agreement with the International Rice Research Institute (IRRI) and the Central Rice Research Institute, Cuttack Collaborative Project. However, UNDP did not review the paper, hence it does not necessarily reflect the views of UNDP and no endorsement should be inferred. Support rendered by Dr. H.U. Neue, former coordinator nd Dr. R. Wassmann, coordinator, IRRI-UNDP Project, is gratefully acknowledged.

References

Adhya TK, Rath AK, Gupta PK, Rao VR, Das SN, Parida KM, Parashar DC & Sethunathan N (1994) Methane emission from flooded rice fields under irrigated conditions. Biol Fertil Soils 18:245-248

Anastasi C, Dowding M & Simpson VJ (1992) Future CH_4 emissions from rice production. J Geophys Res 97:7521-7525

Bharati K, Mohanty SR, Adhya TK, Banerjee A, Rao VR & Sethunathan N (1999a) Influence of a commercial formulation of tridemorph on methane production and oxidation in a tropical rice soil. Chemosphere 39:933-943

Bharati K, Mohanty SR, Padmavathi PVL, Rao VR & Adhya TK (1999b) Nitrification inhibitors and methane production in a flooded alluvial soil. Nutr Cycling Agroecosyst, this issue

Bronson KF & Mosier AR (1991) Effect of encapsulated calcium carbide on dinitrogen, nitrous oxide, methane and carbon dioxide emissions from flooded rice. Biol Fertil Soils 11:116-120

Cicerone R, Delwiche CC, Tyler SC & Zimmerman PR (1992) Methane emissions from California rice paddies with varied treatments. Global Biogeochem Cycles 6:233-248

Conrad R (1996) Soil microorganisms as controllers of atmospheric trace gases (H_2, CO, CH_4, OCS, N_2O and NO). Microbiol Rev 60:28-32

Conrad R & Rothfuss F (1991) Methane oxidation in the soil surface layer of a flooded rice field and the effect of ammonium. Biol Fertil Soils 12:28-32

Denier van der Gone HAC & Neue HU (1996) Oxidation of methane in the rhizosphere of rice plants. Biol Fertil Soils 22:359-366

Ferry JG (1992) Biochemistry of methanogenesis. CRC Crit Rev Biochem Mol Biol 27:473-503

Hanson RS & Hanson TE (1996) Methanotrophic bacteria. Microbiol Rev 60:439-471.

Holzapfel-Pschorn A, Conrad R & Seiler W (1985) Production, oxidation and emission of methane in rice paddies. FEMS Microbiol Ecol 31:343-351

Houghton JT, Meira Filho LG, Callander BA, Harris N, Katterberg A & Maskell K (1996) IPCC report on climate change: The science of climate change. WG1 contribution to the IPCC second assessment report on methane emission from rice cultivation, Cambridge, UK, Cambridge University Press

IRRI – International Rice Research Institute (1999) Sustaining Food Security Beyond the Year 2000 - A Global Partnership for Rice Research : Medium-term Plan 2000-2002, pp 130, International Rice Research Institute, Los Baños, Philippines

Islam A, Molla AL & Haque S (1984) *Azolla* and blue-green algae as alternative source of nitrogen for rice and their mineralization in soils of Bangladesh. Ind J Agric Sci 54:1056-1060

Keerthisinghe DG, Freney JR & Mosier AR (1993) Effect of wax-coated calcium carbide and nitrapyrin on nitrogen loss and methane emission from dry-seeded flooded rice. Biol Fertil Soils 16:71-75

Kumaraswamy S, Ramakrishnan B, Satpathy SN, Rath AK, Mishra S, Rao VR & Sethunathan N (1997) Spatial distribution of methane oxidizing activity in a flooded rice soil. Plant Soil 191:241-248

Lantin RS, Aduna JB & Javellana AM (1995) Methane measurements in rice fields : Instruction manual on methodologies. International Rice Research Institute, Los Baños, Philippines

Liu ZZ & Zheng WW (1992) Nitrogen fixation of *Azolla* and its utilization in agriculture in China. In : Hong GF (ed) The Nitrogen Fixation and its Research in China, pp 526-537. Berlin, Germany, Springer-Verlag

Lumpkin TA and Plucknett DL (1982) *Azolla* as a green manure: use and management in crop production. Colorado, USA, West View Press

Minami K (1994) Methane from rice production. Fert Res 37:167-180

Mishra S, Rath AK, Adhya TK, Rao VR & Sethunathan N (1997) Effect of continuous and alternate water regimes on methane efflux from rice under greenhouse conditions. Biol Fertil Soils 24:399-405

Mitra AP (1992) Greenhouse gas emission in India : 1991 Methane Campaign. Scientific Report No. 2, pp. 91. New Delhi, Council of Scientific and Industrial Research

Neue HU & Sass RL (1994) Trace gas emission from rice fields. In: Prinn RG (ed) Global Atmospheric-Biospheric Chemistry, Environmental Science Research, Vol 48, pp 119-148, New York, Plenum Press

Neue HU, Ziska LH, Matthews RB & Dai Q (1995) Reducing global warming - the role of rice. GeoJournal 35:351-362

Neue HU, Wassmann R, Kludze HK, Wang BJ & Lantin RS (1997) Factors and processes controlling methane emissions from rice fields. Nutr Cycling Agroecosyst 49:111-117.

Parashar DC, Gupta PK & Bhattacharya S (1997) Recent methane budget estimates from Indian rice paddy fields. Ind J Radio Space Phys 26:237-243

Prasad R (1998) Fertilizer urea, food security, health and the environment. Curr Sci 75:677-683

Prasad R & Power JF (1995) Nitrification Inhibitors for agriculture, health and environment. Adv Agron 54:233-281

Sass RL, Fisher FM, Harcombe PA & Turner FT (1991) Mitigation of methane emission from rice fields: possible adverse effects of incorporated rice straw. Global Biogeochem Cycles 5:275-287

Sass RL, Fisher FM, Wang YB, Turner FT & Jund MF (1992) Methane emission from rice fields: the effect of floodwater management. Global Biogeochem Cycles 6:249-262

Satpathy SN, Mishra S, Adhya TK, Ramakrishnan B, Rao VR & Sethunathan N (1998) Cultivar variation in methane efflux from tropical rice. Plant Soil 202:223-229

Schutz H, Holzapfel-Pschorn A, Conrad R, Rennenberg H & Seiler W (1989) A 3-year continuous record on the influence of daytime, season and fertilizer treatment on methane emission rates from an Italian rice paddy. J Geophys Res 94:16405-6416

Sharma AR, Moorthy BTS, Rao KS & Venkateswarlu B (1995) Improvement of rice production technology in the conventional areas of eastern India. In: Deb DL (Ed.) Sustaining Crop and Animal Productivity - The Challenge of the Decade, pp 157-178, New Delhi, Associated Publishing Co

Singh PK & Singh DP (1997) *Azolla-Anabaena* symbiosis. In: Dadarwal KR (ed) Biotechnological Approaches in Soil Microorganisms for Sustainable Crop Production, pp 93-107, Jodhpur, India, Scientific Publishers

US-EPA – United States-Environmental Protection Agency (1990) Overview of methane's contribution to global warming. In: Methane Emission and Opportunities for Control. US-EPA/Air and Radiation, EPA 1400/9-90/007, pp 2-24, Washington

Wang B, Neue HU & Samonte HP (1997) Effect of cultivar difference (IR 72, IR 65598 and Dular) on methane emission. Agric Ecosyst Environ 62:31-40

Wang ZP, Lindau CW, DeLaune RD and Patrick WH Jr (1993) Methane emission and entrapment in flooded rice soils as affected by soil properties. Biol Fertil Soils 16:163-168

Wassmann R, Neue HU, Lantin RS, Javellana MJ, Diego R, Lignes VE, Hoffmann H, Papen H & Rennenberg H (1995) Methane emission from rainfed rice fields. In: Fragile Lives in Fragile Ecosystems, pp 218-225, Manila, Philippines, International Rice Research Institute

Wassmann R, Shangguan XJ, Tolg M, Cheng DX, Wang MX, Papen H, Rennenberg H & Seiler W (1996) Spatial and seasonal distribution of organic amendments affecting methane emission from Chinese rice fields. Biol Fertil Soils 22:191-195

Wassmann R, Neue HU, Bueno C, Lantin RS, Alberto MCR, Budendia LV, Bronson K, Papen H & Rennenberg H (1998) Methane production capacities of different rice soils derived from inherent and exogenous substrates. Plant Soil 203:227-237

Wassmann R, Neue HU, Lantin RS, Buendia LV & Rennenberg H (2000a) Characterization of methane emissions from rice fields in Asia. 1. Comparison among field sites in five countries. Nutr Cycling Agroecosyst, this issue

Yagi K (1997) Methane emissions from paddy soils. Bull Natl Inst Agro-Environ Sci 14:96-210

Yagi K & Minami K (1990) Effect of organic matter application on methane emission from some Japanese paddy fields. Soil Sci Plant Nutr 36:599-610

Yagi K, Tsuruta H, Kanda K & Minami K (1996) Effect of water management on methane emission from a Japanese rice paddy field: automated methane monitoring. Global Biogeochem Cycles 10:255-267

Nutrient Cycling in Agroecosystems **58:** 107–119, 2000.

Mechanisms of crop management impact on methane emissions from rice fields in Los Baños, Philippines

R. Wassmann[1,2], L.V. Buendia[1], R.S. Lantin[1], C.S. Bueno[1], L.A. Lubigan[1], A. Umali[1], N.N. Nocon[1], A.M. Javellana[1], & H.U. Neue[1,3]

[1]International Rice Research Institute, P. O. Box 3127, Makati City 1271, Philippines; [2]Fraunhofer Institute for Atmospheric Environmental Research (IFU), Garmisch-Partenkirchen, Germany; [3]UfZ-Centre for Environmental Research, Leipzig-Halle, Department of Soil Sciences, Halle, Germany

Key words: rainfed, irrigated, midtillering drainage, organic amendments, green manure cultivars, ebullition, dissolved methane, mitigation options

Abstract

This article comprises 4 yr of field experiments on methane (CH_4) emissions from rice fields conducted at Los Baños, Philippines. The experimental layout allowed automated measurements of CH_4 emissions as affected by water regime, soil amendments (mineral and organic), and cultivars. In addition to emission records over 24 h, ebullition and dissolved CH_4 in soil solution were recorded in weekly intervals. Emission rates varied in a very wide range from 5 to 634 kg CH_4 ha^{-1}, depending on season and crop management. In the 1994 and 1996 experiments, field drying at midtillering reduced CH_4 emissions by 15-80% as compared with continuous flooding, without a significant effect on grain yield. The net impact of midtillering drainage was diminished when (i) rainfall was strong during the drainage period and (ii) emissions were suppressed by very low levels of organic substrate in the soil. Five cultivars were tested in the 1995 dry and wet season. The cultivar IR72 gave higher CH_4 emissions than the other cultivars including the new plant type (IR65597) with an enhanced yield potential. Incorporation of rice straw into the soil resulted in an early peak of CH_4 emission rates. About 66% of the total seasonal emission from rice straw-treated plots was emitted during the vegetative stage. Methane fluxes generated from the application of straw were 34 times higher than those generated with the use of urea. Application of green manure (*Sesbania rostrata*) gave only threefold increase in emission as compared with urea-treated plots. Application of ammonium sulfate significantly reduced seasonal emission as compared with urea application. Correlation between emissions and combined dissolved CH_4 concentrations (from 0 to 20 cm) gave a significant R^2 of 0.95 (urea + rice straw), and 0.93 (urea + *Sesbania*), whereas correlation with dissolved CH_4 in the inorganically fertilized soils was inconsistent. A highly significant correlation (R^2 =0.93) existed between emission and ebullition from plots treated with rice straw. These findings may stimulate further development of diagnostic tools for easy and reliable determination of CH_4 emission potentials under different crop management practices.

Introduction

Global mean temperature of the earth's surface has risen by about 0.3-0.6 °C due to anthropogenic interference, namely the emissions of greenhouse gases (Hadley Center, 1998). By 2100, increases in temperature between 1 and 3.5 °C are expected to take place due to global warming because human-induced warming of the atmosphere is likely to continue. The most important greenhouse gases are carbon dioxide (CO_2), methane (CH_4), nitrous oxide (N_2O), and halocarbons. Carbon dioxide accounted for 70-72% of the additional greenhouse effect accumulated since industrialization while CH_4 has contributed 21-22% (Oberthur & Ott, 1999). Except for a brief episode in 1992/1993, the atmospheric concentration of CH_4 is consistently showing an upward trend (Tyler et al., 1999).

Wetland rice fields are an important source of CH_4, but uncertainties in the source strength remain high (Bachelet & Neue, 1993). Methane emissions from rice fields are governed by a complex set of parameters that link the physical and biological characteristics of flooded soil environments with specific agricultural management practices. In particular the impact of different management practices has been addressed in several field studies over recent years. Methane emissions are influenced by water regime (Kimura, 1992; Sass et al., 1992; Buendia et al., 1997), cultivars (Neue et al., 1996; Wang et al., 1997; Butterbach-Bahl et al., 1997), and application of organic and inorganic amendments (Schütz et al., 1989; Yagi & Minami, 1990; Sass et al., 1991; Cicerone et al., 1992). Overall, the interaction of these controlling factors makes it difficult to arrive at better prediction and estimates of CH_4 emission from rice fields. Variations within a 24-h cycle, one season, and a multiyear observation period demand long-term records with high temporal resolution that can best be accomplished by automated systems.

The pressure on Asia's land resources to produce more rice will aggravate in the coming years due to increasing population and demand for food. Rice cultivation practices have to adjust to facilitate higher yield. Future technologies will rely on the adoption of high-yielding cultivars, efficient water management, and increased use of fertilizers. Some production practices may promote CH_4 emissions while others may infer a net decrease of the CH_4 source strength. The extent to which different rice ecosystems and currently employed technologies contribute to CH_4 emission is not known. A mechanistic understanding of crop management impacts is pivotal in achieving an environmentally sound future rice production in the future.

The Philippines comprises 3.4 million ha of rice land, of which 61% is irrigated and 32% is rainfed (IRRI, 1997). The CH_4 source strength of irrigated rice land in the Philippines has been addressed in several publications (Wassmann et al., 1994; Corton et al., this issue). This study describes the amplitude in CH_4 emissions as affected by different water management practices, rice cultivars, and organic amendments to cover the scope of CH_4 emissions found in the region. The specific objectives of this field study conducted within an international network of measuring stations (Wassmann et al., this issue,b) at the station in Los Baños were

1) to quantify CH_4 fluxes as affected by a wide range of management practices;
2) to evaluate processes that control CH_4 emissions; and
3) to identify mitigation strategies for CH_4 emission in a sustainable rice system.

Methods and materials

Experiments were conducted in a rice field at the experimental farm of the International Rice Research Institute in Los Baños, Laguna, Philippines (14° 09'N, 121° 15'E). Los Baños is located in the warm humid tropics (FAO-AEZ classification) with annual rainfall of 2027 mm, mean solar radiation of 16.1 MJ m^{-2} d^{-1}, mean temperature of 26.8 °C (highest at 36.0 °C in May) and mean rainy days of 155 in a year. The soil is classified as Aquandic Epiaqualf with soil pH of 6.6, 1.2% organic C, 14% total N, 2.8% active Fe, 19 mg kg^{-1} available P , and 0.92 mg kg^{-1} available K. The soil has silt-clay texture (44% silt and 43% clay).

Details of the experiments in the dry seasons (DS) and wet seasons (WS) from 1994 to 1997 were summarized in Table 1. Three aspects of rice cultivation were evaluated as to their influences on CH_4 emission: water regime (1994 and 1996); cultivar (1995), and organic/inorganic amendments (1997). The treatments of water regime consisted of continuous flooding (maintaining 5 cm of floodwater throughout the season), preharvest drainage (drainage at 14 d before harvest until harvest), and dual drainage (drainage at mid-tillering for 20 d and drainage at 14 d before harvest). The 1997 experiment encompassed amendments of rice straw, green manure (*Sesbania rostrata*), urea, and ammonium sulfate. Three cultivars were evaluated in 1995 DS and four cultivars in 1995 WS.

Table 1. Summary of modifying treatments for 1994-97 experiments, Los Baños

Year	Season	Water management[a]	Organic amendment: Input Type	Organic amendment: Input Rate (kg N ha^{-1})	Organic amendment: Plant residues	Inorganic amendment: Type[b]	Inorganic amendment: Rate (kg N ha^{-1})	Cultivar	Transplant[c]	Harvest[c]
1994	Dry	Continuous flooding	None	0	Incorporated	Urea	120	IR72	13/01	22/04
		Preharvest drainage	None	0	Incorporated	Urea	120	IR72	13/01	22/04
		Dual drainage	None	0	Incorporated	Urea	120	IR72	13/01	22/04
	Wet	Continuous flooding	None	0	Removed	Urea	120	IR72	14/07	22/10
		Preharvest drainage	None	0	Removed	Urea	120	IR72	14/07	22/10
		Dual drainage	None	0	Removed	Urea	120	IR72	14/07	22/10
1995	Dry	Dual drainage	None	0	Removed	Urea	120	IR72	11/01	16/04
		Dual drainage	None	0	Removed	Urea	120	IR65597	20/01	16/04
		Dual drainage	None	0	Removed	Urea	120	Dular	20/01	16/04
	Wet	Dual drainage	None	0	Removed	Urea	120	IR72	04/06	11/10
		Dual drainage	None	0	Removed	Urea	120	IR65597	04/06	03/10
		Dual drainage	None	0	Removed	Urea	120	PSBRc14	04/06	03/10
		Dual drainage	None	0	Removed	Urea	120	Magat	04/06	03/10
1996	Dry	Continuous flooding	None	0	Incorporated	Urea	120	IR72	09/01	18/04
		Preharvest drainage	None	0	Incorporated	Urea	120	IR72	09/01	18/04
		Dual drainage	None	0	Incorporated	Urea	120	IR72	09/01	18/04
	Wet	Continuous flooding	None	0	Incorporated	Urea	120	IR72	09/07	17/10
		Preharvest drainage	None	0	Incorporated	Urea	120	IR72	09/07	17/10
		Dual drainage	None	0	Incorporated	Urea	120	IR72	09/07	17/10
		Continuous flooding	None	0	Removed	Urea	120	IR72	09/07	17/10
1997	Dry	Dual drainage	None	0	Incorporated	Urea	150	IR72	10/01	20/04
		Dual drainage	None	0	Incorporated	Urea + AS	150	IR72	10/01	20/04
		Dual drainage	Rice straw	60	Incorporated	Urea	90	IR72	10/01	20/04
		Dual drainage	Sesbania	60	Incorporated	Urea	90	IR72	10/01	20/04
	Wet	Dual drainage	None	0	Incorporated	Urea	150	IR72	09/07	14/10
		Dual drainage	None	0	Incorporated	Urea + AS	150	IR72	09/07	14/10
		Dual drainage	Rice straw	60	Incorporated	Urea	90	IR72	09/07	14/10
		Dual drainage	Sesbania	60	Incorporated	Urea	90	IR72	09/07	14/10

[a]Dual drainage = midtillering and preharvest drainage. [b]AS = ammonium sulfate. [c]Date (dd/mm).

Twelve experimental plots (4 m × 5 m) were laid out using a randomized complete block design. One season experiment comprised 3 or 4 treatments with four or three replicates, respectively (Table 1). Rice plants (21 d old) were transplanted at 20- × 20-cm spacing. Plots were applied with 120-30-30 kg ha^{-1} of N (as urea), P_2O_5, and K_2O, respectively, except for studies on organic and inorganic amendments where N applications were modified (Table 1).

Methane emission rates were determined by an automatic system based on the "closed chamber technique." The technical details of the measurements and data acquisition were described by Wassmann et al (this issue,b). Emission rates were determined in 2-h intervals; four records of the CH_4 concentrations inside each chamber were used for regression analysis.

Methane ebullition was quantified using small plexiglas chambers (l:40 cm, w:20 cm, h:20 cm) placed for 24 h between plant rows (Wassmann et al., 1996). These measurements were conducted once a week with four chambers per treatment; ebullition records covered the entire flooding period (before and after transplanting) and also included the drainage periods during the growing season. Weekly records were interpolated to compute cumulative ebullition rates. Dissolved CH_4 was determined in soil solution that was sampled through tubing of porous ceramic (Alberto et al., 2000). Measurements were conducted in weekly intervals.

Statistical analysis of experimental data was accomplished using STATISTICA program (Statsoft, Inc. 1993). Significant differences among treatment means, for fluxes, were determined as to the type of distribution. T test is used for normal distribution and sign test is used for non-normal distribution.

Results and discussion

Effect of water regime

Different water regimes did not affect biomass and grain yields in 1994 and 1996 seasons (Table 2) in either season. In 1994 DS, field drying for 20 d starting at midtillering (19 d after transplanting [DAT]) resulted in a substantial reduction in CH_4 emission (Figure 1a). While the emission rates from the flooded plots showed a steady increase, draining in the early phase of the growing season resulted in constant levels of emission rates. A late drainage between 85 and 99 DAT triggered a short-term spike in emissions followed by a deep plunge that lasted until harvest (Figure 1a). This practice did not have a significant impact on overall emissions (Table 2). Seasonal patterns of ebullition rates in flooded conditions were closely related to the seasonal emission patterns (Figure 1b). Ebullition rates were reduced when field drying was imposed and showed only a moderate increase during the second half of the growing season.

In the two seasons of 1996, however, the reductive effect of drainage at midtillering was not as pronounced as in 1994. Different reasons may be singled out for the relative similarity among the different treatments in each of these seasons. In the 1996 DS, CH_4 emission was generally on a very low level (Figure 2a). Even the preharvest drainage did not stimulate any release of en-

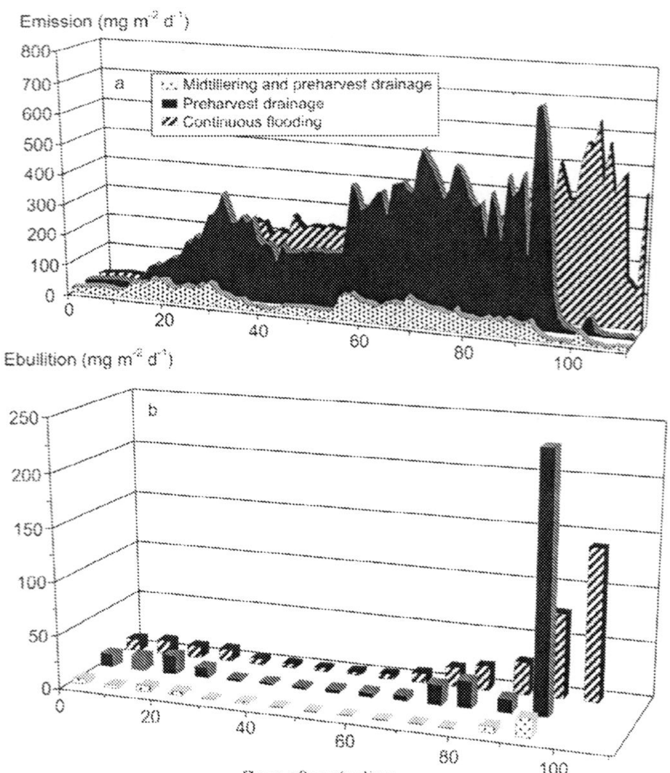

Figure 1. Effect of water regime on CH_4 emission (a) and ebullition (b) in 1994 DS

Table 2. Mean CH_4 emission rates, cumulative CH_4 emission (before and after transplanting), and relative contribution by ebullition (in relation to cumulative emission), aboveground biomass, and yield per modifying treatments (1993-97)

Year	Season	Modifying treatment	After transplanting Mean emission (mg m-2d-1)	Cumulative emission (kg ha-1)	Ebullition (%)	Before transplanting Cumulative emission (kg ha-1)	Above-ground biomass (t ha-1)	Grain yield (t ha-1)
1994	Dry	Continuous flooding	227 b	225	11	n.d	7.9 a	5.3 a
		Preharvest drainage	254 a	251	10	n.d.	8.3 a	5.0 a
		Dual drainage	45 c	45	7	n.d.	8.0 a	5.1 a
	Wet	Continuous flooding	27 b	27	20	n.d.	7.5 a	3.8 a
		Preharvest drainage	35 a	35	25	n.d	6.9 a	4.1 a
		Dual drainage	11 c	11	27	n.d	7.0 a	3.4 a
1995	Dry	IR72	8 a	8	8	1	6.1 b	5.5 a
		IR65597	7 b	7	22	1	8.0 a	4.0 a
		Dular	6 b	5	15	0	7.0 ab	4.1 a
	Wet	IR72	8 a	8	11	0	7.9 a	3.1 b
		IR65597	6 b	6	11	0	7.9 a	1.5 c
		PSBRc14	6 b	6	12	0	7.1 ab	3.1 bc
		Magat	4 c	4	18	0	6.3 b	5.1 a
1996	Dry	Continuous flooding	10 a	10	5	0	7.1 a	4.6 a
		Preharvest drainage	10 b	10	9	0	6.6 a	4.0 a
		Dual drainage	8 c	8	9	0	7.1 a	4.2 a
	Wet	Continuous flooding	40 a	40	20	6	7.0 a	3.0 a
		Preharvest drainage	28 b	28	17	8	7.3 a	3.6 a
		Dual drainage	34 b	34	26	10	7.8 a	2.9 a
		Continuous flooding	14 c	14	7	3	7.1 a	3.5 a
1997	Dry	Urea	27 c	27	15	0	9.0 a	5.4 a
		Urea + ammonium sulfate	9.0 d	9	6	1	6.5 ab	4.6 ab
		Urea + rice straw	634 a	634	55	26	4.5 b	3.5 b
		Urea + green manure	119 b	119	12	17	7.6 a	4.8 a
	Wet	Urea	14 c	13	37	3	6.0 a	3.0 a
		Urea + ammonium sulfate	7 d	7	15	1	6.9 a	3.5 a
		Urea + rice straw	621 a	602	52	30	5.5 a	3.0 a
		Urea + green manure	42 b	40	45	7	7.2 a	3.7 a

[a]Data following the same letter in the same season of the same year are not significantly different at $P < 0.05$.

trapped CH_4, indicating a low CH_4 production over the course of the flooding period. In the preceding three seasons, plant residues were completely removed from the field (Table 1). Plant residues were incorporated before 1996 DS, but depletion in soil organic matter was apparently not yet compensated for and this constrained CH_4 production in the soil. Methane emissions in the succeeding 1996 WS (Figure 2b) were back on the level found 2 yr earlier. However, heavy rainfalls in the drainage period have reversed the potential impact of the midtillering drainage. The amount of rainfall during midtillering drainage of 1996 WS was 153 mm, whereas the corresponding amount in the other drainage experiments in 1994 was less than 2 mm.

Dual drainage gave 80% and 59% reduction in seasonal emission as compared with continuous flooding in 1994 DS and WS (Table 2). This proportional reduction is higher than the effect observed by Sass et

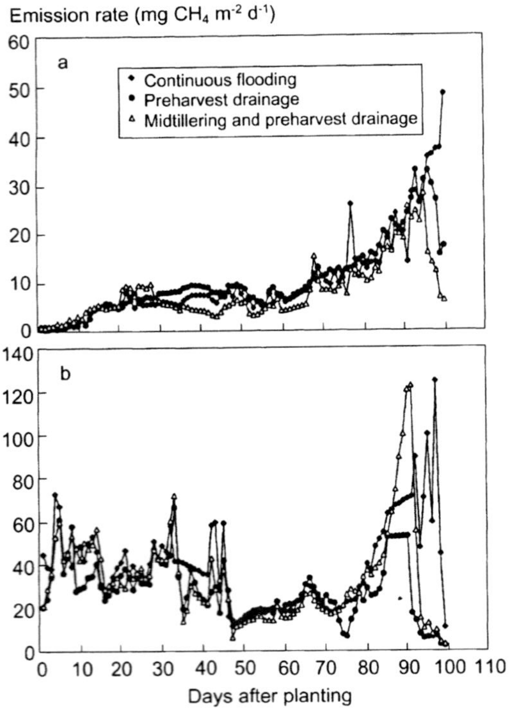

Figure 2. Methane emission as affected by different water regimes in 1996 DS (a) and WS (b)

al. (1992) in a Texas rice field. Their study reported a 50% reduction in emission rates in plots treated with normal water management imposing a single midseason drainage. In a previous experiment in Los Baños, midseason drainage at either midtillering or panicle initiation was suppressing CH_4 emission up to 60% (Bronson et al., 1997).

In 1996, low levels of soil organic matter and high precipitation drastically diminished the reductive effect of drainage on CH_4 emissions. Field studies in Indonesia (Nugroho et al., 1997) and Japan (Ishibashi et al., 1997) also reported low or inconsistent net effects of drainage events on CH_4 emissions, but timing and frequency of drainage periods differed in these field studies. Due to large variations in net effects obtained in different locations and under different climatic conditions, field drying cannot be recommended as a blanket strategy for mitigating CH_4 emissions. Moreover, drainage events stimulate N_2O emissions that may offset possible gains in CH_4 emissions (Bronson et al., 1997). However, an early drainage may still be considered a mitigation strategy for specific baseline practices, i.e. those with high organic inputs (Wassmann et al., this issue,b).

The impact mechanism of field drying on CH_4 emission may be derived from the seasonal courses of dissolved CH_4 in soil solution (Figure 3a,b). Methane concentrations were very low in the dry season and were further reduced by midtillering drainage (Figure 3b). This reduction is due to inflow of oxygen-inhibited CH_4 production and oxidized CH_4 dissolved in the soil solution. In rice field with high levels of CH_4 emission, midseason drainage also released entrapped gaseous CH_4 as soil pore spaces started to open (Lu et al., this issue). In Los Baños, this peak in emissions was only observed (i) during late drainage periods and (ii) in the early drainage period of the 1997 experiments following high inputs of organic manure. In the other experiments, an early drainage event caused low emissions throughout the remaining season.

Effect of residue management

Methane emissions are extremely sensitive to incorporation of plant residues. This can be illustrated by the results of 1996 WS when the removal of plant residues resulted in a 65% reduction of emissions under continuous flooding (Table 2). Ebullition rates and dissolved CH_4 were high at the early growth stage due to decomposing plant stubbles incorporated during land preparation (Figure 3c). Removal of plant residues from the field resulted in low levels of ebullition rates and dissolved CH_4 in the early phase of the growing season (Figure 3d). As the plants developed, root exudates and decomposing roots provided substrates for CH_4 production which resulted in similar level of dissolved CH_4 in plots with and without stubbles (Figures 3c and 3d). However, ebullition rates were still higher in plots with incorporated residues than those without residue incorporation. This prolonged effect may be attributed to pockets of gaseous CH_4 residing in the soil for longer time spans.

Effect of organic and inorganic amendments

In the 1997 experiment, incorporation of rice straw resulted in an early peak of CH_4 emissions. About 66% of the total seasonal emission from plots treated with these amendments were emitted during the vegetative stage of the dry WS (Figure 4a,b). However, soil drying at midtillering abruptly terminated this period of high emission rates. After reflooding, CH_4 emissions

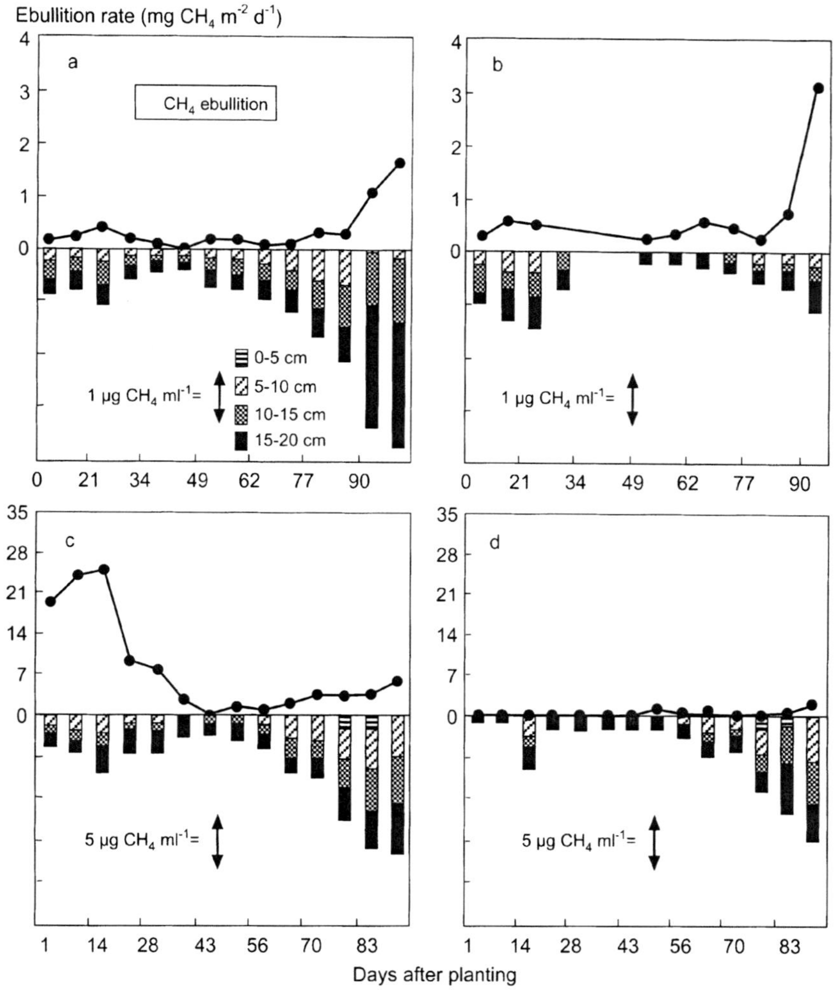

Figure 3. Concentration of dissolved CH_4 and ebullition in plots with different irrigation schemes and residue management in 1996 wet and dry seasons: continuous flooding/ with residues/ dry season (a); dual drainage/ with residues/ dry season (b); continuous flooding/ with residues/ wet season (c); and continuous flooding/ without residues/ wet season (d)

in the rice straw plots were higher than in the other plots, but did not return to the high initial values. Green manure triggered a lower increment in emissions than rice straw (Table 2); its impact was virtually offset after the drainage event (Figure 4a,b).

The differences between organic and inorganic treatments were also reflected in dissolved CH_4 concentrations in soil solution (Table 3). The drainage events (19-39 DAT) drastically reduced dissolved CH_4 concentrations in all treatments, but the ranking among the treatments (rice straw > green manure > urea > ammonium sulfate) basically remained throughout the season. Mean values for rice straw treatment were 19 times (DS) and 15 times (WS) higher than for urea treatment. The corresponding value for the green manure treatment exceeded the urea treatment by factors of 3.9 (DS) and 6.6 (WS). Ammonium sulfate, on the other hand, had significantly lower values than urea (factors of 0. 2 and 0.7 for DS and WS, respectively).

In 1997 DS, CH_4 emission rates were significantly correlated to dissolved CH_4 concentrations (across 0 to 20 cm depth); R^2 values were 0.65 (urea), 0.81 (urea +

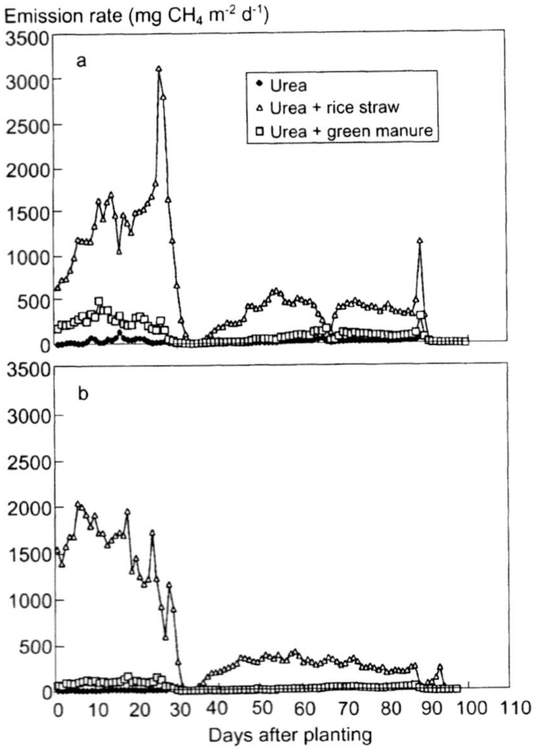

Figure 4. Methane emission as affected by inorganic and organic amendments, 1997 DS (a) and WS (b)

Table 3. Means of CH_4 concentrations (μl CH_4 ml soil solution^{-1}) across the soil column (0-20 cm) at different days after transplanting (DAT) in 1997 dry and wet seasons; letters indicate $P < 0.05$ significance level (DMRT) for given DAT; data following the same letter in the same season of the same year are not significantly different at $P < 0.05$.

DAT	Urea	Urea + ammonium sulfate	Urea + rice straw	Urea + green manure
		Dry season		
3	0.19 c	0.09 d	4.47 a	1.92 b
10	0.52 c	0.14 d	14.45 a	3.68 b
13	0.62 c	0.15 d	14.39 a	3.88 b
18	0.69 c	0.16 d	13.08 a	3.27 b
20	0.70 c	0.13 d	11.68 a	3.25 b
25	0.56 c	0.15 d	7.70 a	2.26 b
27	0.55 c	0.16 d	5.20 a	1.32 b
39	0.04 c	0.01 d	2.56 a	0.11 b
41	0.12 b	0.01 d	3.19 a	0.09 c
46	0.10 b	0.02 c	4.60 a	0.09 b
48	0.08 b	0.02 c	4.47 a	0.08 b
53	0.06 b	0.02 d	3.14 a	0.08 c
55	0.09 c	0.03 d	5.91 a	0.15 b
60	0.08 c	0.01 d	3.40 a	0.14 b
67	0.14 c	0.03 d	2.69 a	0.23 b
69	0.28 c	0.05 d	4.17 a	0.51 b
74	0.35 c	0.13 d	3.64 a	0.66 b
81	0.27 c	0.05 d	2.26 a	0.59 b
87	0.65 c	0.13 d	3.78 a	1.35 b
Ø	0.32 c	0.08 d	6.04 a	1.25 b
		Wet season		
6	0.87 c	0.71 c	19.28 a	12.84 b
12	0.73 c	1.08 d	10.69 a	5.43 b
19	0.93 c	0.53 d	12.41 a	3.77 b
27	0.51 c	0.08 d	1.19 a	0.76 b
34	0.03 b	0.00 c	0.11 a	0.02 b
42	0.03 c	0.01 d	1.41 a	0.07 b
51	0.06 c	0.01 d	1.88 a	0.14 b
55	0.07 c	0.01 d	1.82 a	0.21 b
63	0.07 c	0.01 d	1.82 a	0.21 b
70	0.40 c	0.14 d	4.11 a	1.13 b
Ø	0.37 c	0.26 d	5.47 a	2.46 b

ammonium sulfate), 0.95 (urea + rice straw), and 0.93 (urea + green manure). In 1997 WS, significant correlations were restricted to the straw plots ($R^2 = 0.94$) and green manure plots ($R^2 = 0.99$) only. The average concentration across the soil column from 0 to 20 cm (Table 3) showed better correlation to emission rates than any of the individual depth layers (data not shown).

Organic amendments were applied at 14 d before transplanting and resulted in high CH_4 release of CH_4 before transplanting (Table 2). High emissions in the preseason indicate that both rice straw and green manure contain sizable amounts of readily decomposable substances which favored CH_4 production. During the growing period, ebullition from the rice straw treatment remained on a high level; ebullition rates were significantly correlated to emissions in the rice straw plots:

$$\text{Emission} = 211.13 + 1.65 * \text{ebullition}$$
$$R^2 = 0.93^{**}$$

However, there was no significant correlation observed for the other treatments as seasonal patterns of ebullition and emissions were distinct.

Ebullition contributed to more than 50% to the overall emission in the rice straw plots which is a higher percentage than in the other treatments (Table 2). Due to the sturdy structure of straw, soil puddling results in a patchy distribution of straw aggregates representing clusters of high organic contents in the bulk soil. A fraction of the CH_4 produced in these clusters will force its way through the soil pore spaces and floodwater in the form of emerging gas bubbles. However, concentrations of CH_4 in the solution of the bulk soil are also enhanced (Table 3), so that more CH_4 will diffuse to the rhizosphere followed by emission through the aerenchyma. However, given a limitation of the CH_4 transport capacity of rice plants (Aulakh et al., this issue), such high CH_4 concentrations in the soil solution would, in turn, also intensify the passage through the water column, namely ebullition.

Computed over two seasons, CH_4 fluxes from the straw treatment were 34 times higher than those from urea treatment (Table 2). This increment appears very high as compared with observations made in the United States (Sass et al., 1991; Lauren et al., 1994), Japan (Yagi & Minami 1990; Watanabe et al., 1995), Indonesia (Nugroho et al., 1994) and India (Rath et al., 1999). However, a recent study in northern China found a similar increment in emission rates caused by straw application (Wang et al., this issue).

Site-to-site differences in the response to identical organic amendments were attributed to a combination of soil- and climate-related factors (Wassmann et al., this issue, b). The net effect on emission depended on soil type; Subadiyasa et al., (1997) observed a 36-41% increase in an Alfisol and a 45-48% increase in an Inceptisol using identical amounts of rice straw. Periodic straw application over several years generally resulted in high emissions, e.g. a fivefold increase in total CH_4 emissions was observed in California rice plots in which rice straw had been incorporated each for 4 consecutive years (Bossio et al., 1999). Moreover, the increment in emission rate depended on timing of straw application. An early incorporation of straw 3 mo before transplanting reduced CH_4 emissions by 23% as compared with the common practice of incorporation at the end of the fallow period (Shin et al., 1996). Nugroho et al. (1997) observed that the relative increment in emissions triggered by straw application varied from 23 to 98% in fields planted with different cultivars.

Green manure application gave only threefold increase in emission as compared with urea-treated plots which is in line with previous findings (Lauren et al., 1994; Bronson et al., 1997). Sulfate application significantly reduced seasonal emission due to competition between CH_4-producing and methanogenic bacteria (Denier van der Gon & Neue, 1996). In 1997 DS, use of straw as additional source of N resulted in a significant reduction in grain yield. No significant differences in yield were observed in the 1997 wet season.

Effect of cultivars

Field experiments of the 1995 DS and WS encompassed five different cultivars (Table 1): the modern cultivars IR72 (DS and WS) and PSBRc14 (WS), the new plant type IR65597 (DS and WS), the traditional cultivar Dular (DS), and the hybrid Magat (WS). There were no significant differences in grain yields among cultivars in the 1995 DS (Table 2). In the 1995 WS, the hybrid tested in this experiment (Magat) produced more grain yield than other cultivars.

For both seasons, IR72 consistently gave the highest seasonal CH_4 emission. However, no specific plant trait could be singled out as determinant of the high emission potential of this cultivar (Table 4). Plants of PSBRc14 have morphological features similar to those of IR72. In contrast, rice plants of Dular are very tall and have a low number of tillers and a low root biomass. Dular plants are characterized by a delayed development that appears to limit emissions in the early stage (Table 4). Plants of Magat are taller and have a higher yield potential than IR72.

IR65597 deserves special attention because it belongs to the advanced lines currently under development at IRRI. This new plant type should be able to raise the existing yield barrier by 25% based on the following features: (i) low number of tillers as can be seen in Table 4, (ii) large panicles, (iii) and a vigorous root system (IRRI, 1997). The new plant type had similar emission rates as IR72 during most parts of both seasons, but maximum values of IR72 exceeded those of the new plant type in 1995 DS and WS (Figure 5). This temporary divergence in emission patterns resulted in 24% lower emissions for IR65597 over both seasons. It should be noted, however, that emissions were generally on a very low level during these seasons due to the removal of plant residues from the field. The comparison among cultivars may deviate from these findings when background levels of CH_4 production and emission are high.

In 1995 DS, the relative contribution of ebullition was very high in the IR65597 plots (Table 2) indicating, in turn, a low contribution of the plant-mediated transport to overall emission. In the succeeding WS, however, the ebullition in the IR65597 plots had a similar contribution as IR72.

The different plant parameters were statistically analyzed for correlation with cumulative fluxes recorded in 1995 DS and WS. About 86% in the change of cumulative emission (EM_{cum}) was explained by the

Table 4. Means of plant parameters, cumulative CH_4 emission (from transplanting to day of measurement), and contribution of cumulative ebullition (related to cumulative emission) at different days after transplanting (DAT) in 1995 DS and WS; data following the same letter in the same season of the same year are not significantly different at P < 0.05 (not tested for cumulative emission and ebullition)

DAT	Cultivar	Plant height (cm)	Tillers (no m^{-2})	Root length (cm)	Root weight (g m^{-2})	Aboveground biomass (g m^{-2})	Cumulative emission (mg m^{-2})	Contribution of ebullition (%)
				Dry season				
29	IR72	52.0 b	300 a	15.5 b	8.0 b	50.5 a	267	8
	IR65597	57.2 a	275 a	19.0 a	14.8 a	56.3 a	203	37
	Dular	52.8 b	175 b	18.2 ab	3.5 b	21.3 b	136	28
57	IR72	65.6 b	425 a	20.6	227.3 a	487.0	423	8
	IR65597	72.8 b	300 b	19.6	159.3 ab	375.5	359	27
	Dular	121.3a	300 b	20.2	104.8 b	444.8	263	20
71	IR72	78.2 b	350	nd	nd	nd	575	7
	IR65597	84.6 b	300	nd	nd	nd	495	23
	Dular	140.5a	325	nd	nd	nd	416	15
				Wet season				
28	IR72	58.3	368.8 a	18.8 a	329.9 a	137.1 a	135	8
	IR65597	64.9	241.7 b	17.0 b	206.0 b	101.6 b	111	9
	PSBRc14	63.5	356.2 a	17.5 a	297.7 a	139.1 a	146	12
	Magat	64.1	412.5 a	17.7 a	294.1 a	140.1 a	150	8
56	IR72	93.4	425.0	16.7	209.5 b	744.5 ab	272	13
	IR65597	99.8	300.0	17.9	252.7 ab	619.5 b	255	11
	PSBRc14	93.1	400.0	14.6	338.2 a	827.2 a	270	11
	Magat	103.8	425.0	15.8	307.5 a	879.2 a	292	11
84	IR72	103.3 c	400.0	14.8	123.5	1248.2	456	16
	IR65597	117.7 ab	275.0	16.0	108.7	1153.7	386	15
	PSBRc14	109.6 bc	400.0	13.5	145.7	1150.2	409	15
	Magat	118.9 a	425.0	14.9	118.5	1364.2	416	15

combined effect of plant height (PHT), tiller number (TNO), root length (RLT), root weight (RWT), and biomass (BIO). The effect of RLW, RLT, RWT, and BIO were more stable (t values equal 2.7, 2.7, 4.2, respectively) as compared with PHT and TNO. The equation is

$$EM_{cum} = -178.5 - 1.92*PHT + 0.38*TNO + 22.10*RL - 0.40*RW + 0.32*BM \; ; R_2=0.86$$

The relative contribution of ebullition was negatively correlated to tiller number ($R^2 = 0.26$) and root weight ($R^2 = 0.40$). No significant correlation was determined between cumulative emission and percent contribution of ebullition ($R^2 = 0.001$). Apparently, the interaction of different plant traits in determining CH_4 emission and ebullition rates is very complex. Huang et al. (1997) found that daily CH_4 emission was correlated to aboveground vegetative biomass and to root biomass; total seasonal CH_4 emission was positively correlated to rice aboveground biomass. In spite of statistical relationships found in this field and other field experiments, a mechanistic understanding is still lacking at this point.

A screening of 10 cultivars yielded variations in CH_4 emissions by 440% (Satpathy et al., 1997). Cultivars grown in the same regions show distinct variations, indicating a possible reduction of regional source

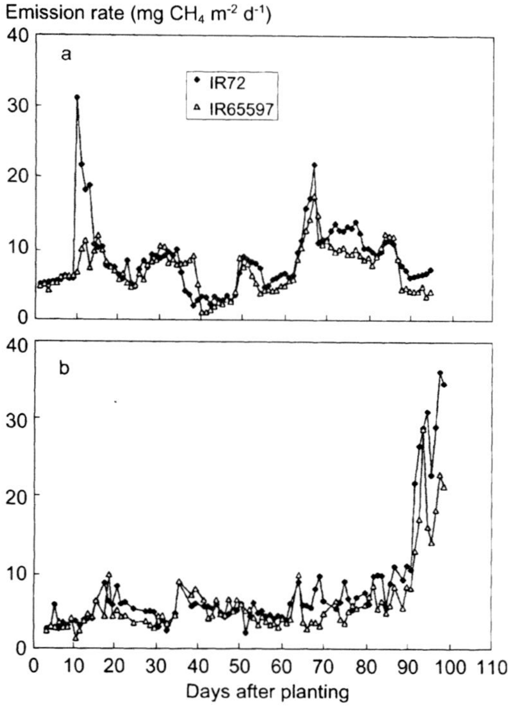

Figure 5. Methane emission as affected by different cultivars IR72 and IR65597), 1995 DS (a) and WS (b)

strengths through cultivar selection. Cultivars commonly found in China differed in their emission potential by 9-56% (Shao & Li, 1997) and 19% (Cai et al., 1994); four Indian cultivars differed by 1-42.6% (Mitra et al, 1999), two Italian cultivars by 24-31% in different growing seasons (Butterbach-Bahl et al., 1997), and two Texan cultivars by 47% (Sigren et al., 1997). According to Lindau et al. (1995), semidwarfs emitted 38% less than tall cultivars, but our results indicated a 33% higher emission potential for the semidwarf IR72 than the tall cultivar Dular.

The two decisive functions of rice plants in regulating CH_4 emissions are (i) gas transfer through the aerenchyma and (ii) root exudation (Wassmann & Aulakh, 2000). However, an assessment of cultivar-specific emission potentials is compounded by the fact that these two traits show an enormous plasticity under field conditions (Aulakh et al., 2000). The aerenchyma formation and root exudation are affected by cultivar and soil parameters such as nutrient availability (Lu et al., 1999), physical impedance (Marschner 1996), and redox potential (Kludze et al., 1993) that may supersede possible differences between cultivars. Therefore, the results of this experiment should be taken as an initial step to accomplish a thorough understanding of CH_4 emissions as affected by different cultivars.

Conclusions and recommendations

The available data set covering 4 yr of field experiments at Los Baños, Philippines, clarified some important issues on measurement approaches, extrapolation, and mitigation of CH_4 from rice fields. Ebullition and dissolved CH_4 in soil solution can be used to diagnose emission potentials of given rice fields—as long as specific conditions are met. These two parameters can be taken as fairly good indicators for emission rates under two prerequisites: (i) fields are continuously flooded and not drained during the growing season, and (ii) sizeable amounts of organic matter are applied to the soil. Ebullition rates and dissolved CH_4 concentrations are easy to record—as compared with emission data over sufficient time spans—and may be used for screening of CH_4 emission potentials in a large number of rice fields. In particular, the localization of 'hot-spots,' i.e. rice fields with very intense CH_4 release, would be a promising application for this screening approach. A systematic screening for high-emitting systems could corroborate and eventually correct regional CH_4 budgets derived from upscaling of local measurements. High-emitting systems also represent prime targets for implementing mitigation measures. For low-emitting systems, however, these parameters have limited diagnostic values and records of emission rates are indispensable for assessment of emission potentials.

The results of this study underscore the significance of organic inputs for emission rates. The removal of plant stubbles from the preceding crop displaced an essential starting substrate for methanogenesis in field trials at Los Baños. Many resource-limited areas still depend on organic manure as a primary source of nutrients. Straw and green manure provide readily mineralizable carbon sources that enhance the reductive capacity of soils which finally drive CH_4 formation and emission. Substituting organic manures by mineral fertilizers reduced CH_4 emissions but entailed emissions of CO_2 and N_2O during fertilizer production and application, respectively (Wassmann et al., this issue, a).

Drainage periods during the cropping season are generic to rainfed rice. Irrigated rice fields can also encompass distinct drainage periods, either caused by water shortage in the irrigation scheme or as part of the local management practice (Wassmann et al., this issue,a). The impact of field drying showed large season-to-season variations but still exerted a considerable reduction effect on overall emissions over 2 yr. As a consequence of these large variations, CH_4 source strengths of rainfed rice with unstable water should ultimately show large interannual variations than CH_4 released from rice fields with continuous water supply. In view of mitigation options in irrigated rice, modified water regimes may effectively be ruled out for seasons with high precipitation.

The results of this study also clarified that there is no inextricable link between grain yield and CH_4 emission. In turn, this would allow developing rice cultivars with less CH_4 emissions but higher grain yield. However, the available data base on cultivar effects does not yet allow a clear guidance on the preferable plant traits to be incorporated by breeding.

Acknowledgements

The research described was funded by the United Nations Development Programme, Global Environment Facility UNDP/GEF (GLO/91/G31) - Interregional Research Programme on Methane Emission from Rice Fields under contract no. C92379 to the International Rice Research Institute.

References

Alberto, MCR, Arah JRM, Neue HU, Wassmann R, Lantin RS & Aduna JB (2000) A sampling technique for the determination of dissolved methane in soil solution. Chemosphere - Global Change Sci 2:57-63

Aulakh MS, Wassmann R, & Rennenberg H (2000) Methane production and emission from rice fields—significance, measurement, role of management, and mitigation options. Adv Agron (in press)

Bachelet D & Neue HU (1993) Methane emissions from wetland rice areas of Asia. Chemosphere 26(1-4):219-237

Bossio DA, Horwath WR, Mutters RG & van Kessel C (1999) Methane pool and flux dynamics in a rice field following straw incorporation. Soil Biol Biochem 31:1313-1322

Bronson KF, Neue HU, Singh U & Abao EB Jr (1997) Automated chamber measurement of methane and nitrous oxide flux in a flooded rice soil: I. Residue, nitrogen, and water management. Soil Sci Soc Am J 61:981-987

Buendia LV, Neue HU, Wassmann R, Lantin R S, Javellana A M, Xu Y, Makarim A K, Corton T M & Chareonsilp N (1997) Understanding the nature of methane emission from rice ecosystems as basis of mitigation strategies. Appl Energy, 56:433-444

Butterbach-Bahl K, Papen H & Rennenberg H (1997) Impact of gas transport through rice cultivars on methane emission from rice paddy fields. Plant Cell Environ 20: 1175-1183

Cai ZC, Xu H, Zhang HH & Jin JS (1994) Estimate of methane emission from rice paddy fields in Taihu Region, China. Pedosphere 4:297-306

Cicerone RJ, Delwiche CC, Tyler SC & Zimmermann PR (1992) Methane emissions from California rice paddies with varied treatments. Global Biogeochem Cycles 6: 233-248

Corton TM, Bajita J, Grospe F, Pamplona R, Wassmann R & Lantin RS (2000) Methane emission from irrigated and intensively managed rice fields in Central Luzon (Philippines). Nutr Cycling Agroecosyst (this issue)

Denier van der Gon HAC & Neue HU (1994) Impact of gypsum application on the methane emission from a wetland rice field. Global Biogeochem. Cycles 8:127-134

Hadley Center (1998) Climate change and its impacts. UK Meteorological Office, Bracknell

Huang Y, Sass RL & Fisher FM Jr (1997) Methane emission from Texas rice paddy soils. 2. Seasonal contribution of rice biomass production to CH_4 emission. Global Change Biol 3:491-500

IRRI – International Rice Research Institute (1997) Rice Almanac. 2nd edition, Los Baños, Philippines, 181 p

Ishibashi E, Akai N, Itoshima Y, Kawanaka K & Yanai M (1997) Effect of soil type and water management on methane emission from paddy fields in Okayama Prefecture. Jpn J Soil Sci Plant Nutr 68:417-422

Kimura M (1992) Methane emission from paddy soils in Japan and Thailand. In: World Inventory of Soil Emission Potentials. Batjes N H & Bridges E M (eds), pp 43-79, WISE Report 2, ISRIC, Wageningen

Kludze HK, DeLaune RD & Patrick WH Jr (1993) Aerenchyma formation and methane and oxygen exchange in rice. Soil Sci Soc Am J 57:386-391

Lauren JG, Pettygrove GS & Duxbury JM (1994) Methane emissions associated with a green manure amendment to flooded rice in California. Biogeochemistry 24:53-65

Lindau CW, Bollich PK & DeLaune RD (1995) Effect of rice variety on methane emission from Louisiana rice. Agric Ecosyst Environ 54:109-114

Lu Y, Wassmann R, Neue HU & Huang C (1999) Impact of phosphorus supply on root exudation, aerenchyma formation and methane emission of rice plants. Biogeochemistry (in press)

Lu WF, Chen W, Duan BW, Guo WM, Lu Y, Lantin RS, Wassmann R & Neue HU (2000) Methane emission and mitigation options in irrigated rice fields in Southeast China. Nutr Cycling Agroecosyst (this issue)

Marschner, H (1996) Mineral nutrition of higher plants. Academic Press, London

Mitra S, Jain MC, Kumar S, Bandyopadhyay SK & Kalra N (1999) Effect of rice cultivars on methane emission. Agric Ecosyst Environ 73:177-183

Neue HU, Wassmann R, Lantin RS, Alberto MCR, Aduna J B & Javellana AM (1996) Factors affecting methane emission from rice fields. Atmos Environ 30(10/11): 1751-1754

Nugroho SG, Lumbanraja J, Suprapto H, Sunyoto, Ardjasa WS, Haraguchi H & Kimura M (1994) Methane emission from an Indonesian paddy field subjected to several fertilizer treatments. Soil Sci Plant Nutr 40:275-281

Nugroho SG, Sunyoto, Lumbanraja J, Suprapto H, Ardjasa WS & Kimura M (1997) Effect of rice variety on methane emission from an Indonesian paddy field. Soil Sci Plant Nutr 43:799-809

Oberthur S & Ott H (1999) The Kyoto Protocol: International Climate Policy for the 21st Century. Springer-Verlag, Berlin, Heidelberg

Rath AK, Mohanty SR, Mishra S, Kumaraswamy S, Ramakrishnan B & Sethunathan N (1999) Methane production in unamended and rice-straw-amended soil at different moisture levels. Biol Fertil Soils 28:145-149

Rasmussen, RA & Khalil MAK (1986) Atmospheric trace gases: trends and distributions over the last decade. Sci 232:1623-1624

Sass RL, Fisher FM, Harcombe PA & Turner F T (1991) Mitigation of methane emissions from rice fields: possible adverse effects of incorporated rice straw. Global Biogeochem Cycles 5:275-287

Sass RL, Fisher FM, Wang YB, Turner FT & Jund MF (1992) Methane emission from rice fields: the effect of floodwater management. Global Biogeochem Cycles 6:249-262

Satpathy SN, Mishra S, Adhya TK, Ramakrishnan B, Rao, VR, Sethunathan N (1998) Cultivar variation in methane efflux from tropical rice. Plant Soil 202: 223-229

Schütz H, Holzapfel-Pschorn A, Conrad R, Rennenberg H & Seiler W (1989) A three-year continuous record on the influence of daytime, season, and fertilizer treatment on methane emission: rates from an Italian rice paddy field. J Geophys Res 94(16):405-416

Shao KS & Li Z (1997) Effect of rice cultivars and fertilizer management on methane emission in a rice paddy in Beijing. Nutr Cycling Agroecosyst 49:139-146

Shin YK, Yun SH, Park ME & Lee BL (1996) Mitigation options for methane emission from rice fields in Korea. Ambio 25:289-291

Sigren LK, Byrd GT, Fisher FM & Sass RL (1997) Comparison of soil acetate concentrations and methane production, transport, and emission in two rice cultivars. Global Biogeochem Cycles 11:1-14

Subadiyasa N, Arya N & Kimura M (1997) Methane emissions from paddy fields in Bali Island, Indonesia. Soil Sci Plant Nutr 43:387-394

Tyler SC, Ajie HO, Gupta ML, Cicerone RJ, Blake DR & Dlugokencky EJ (1999) Stable carbon isotopic composition of atmospheric methane: a comparison of surface level and free tropospheric air. J Geophys Res (in press)

Wang B, Neue HU & Samonte H P (1997) Effect of rice plant on seasonal methane emission patterns. Acta Agron Sin 23:271-279

Wang ZY, Xu YC, Li Z, Guo YX, Wassmann R, Neue HU, Lantin RS, Buendia LV, Ding YP & Wang ZZ (2000) Methane emissions from irrigated rice fields in northern China (Beijing). Nutr Cycling Agroecosyst (this issue)

Wassmann R & Aulakh MS (2000) The role of rice plants in regulating mechanisms of methane emissions. Biol Fertil Soils 31:20-29

Wassmann R, Neue HU, Lantin RS, Aduna JB, Alberto MCR, Andales MJ, Tan MJ, Denier van der Gon HAC, Hoffmann H, Papen H, Rennenberg H & Seiler W (1994) Temporal patterns of methane emissions from wetland ricefields treated by different modes of N application. J Geophys Res 99:16457-16462

Wassmann R, Neue HU, Alberto MCR, Lantin RS, Bueno C, Llenaresas D, Arah JRM, Papen H, Seiler W & Rennenberg H (1996) Fluxes and pools of methane in wetland rice soils with varying organic imputs. Environ Monit Assess 42:163-173

Wassmann R, Lantin RS, Neue HU, Buendia LV, Corton TM & Lu YH (2000a) Characterization of methane emissions from rice fields in Asia. 3. Mitigation options and future research needs. Nutr Cycling Agroecosyst (this issue)

Wassmann R, Neue HU, Lantin RS, Buendia LV & Rennenberg H (2000b) Characterization of methane emissions from rice fields in Asia. 1. Comparison among field sites in five countries. Nutr cycling agroecosyst (this issue)

Watanabe A, Satoh Y & Kimura M (1995) Estimation of the increase in CH_4 emission from paddy soils by rice straw application. Plant Soil 173:225-231

Yagi K & Minami K (1990) Effects of organic matter application on methane emission from some Japanese paddy fields. Soil Sci Plant Nutr 36:599-610

Nutrient Cycling in Agroecosystems **58:** 121–130, 2000.

Methane emission from deepwater rice fields in Thailand

N. Chareonsilp[1], C. Buddhaboon[1], P. Promnart[1], R. Wassmann[2,3] & R.S. Lantin[2]

[1]*Prachinburi Rice Research Center, Rice Research Institute, Department of Agriculture, Thailand;* [2]*International Rice Research Institute, MCPO Box 3127, Makati City 1271, Philippines;* [3]*Fraunhofer Institute for Atmospheric Environmental Research, Garmisch-Partenkirchen, Germany*

Key words: acid sulfate soil, crop management, straw application, mitigation strategy, ebullition

Abstract

Field experiments were conducted in the Prachinburi Rice Research Center (Thailand) from 1994 to 1998. The major objective was to study methane (CH_4) emission from deepwater rice as affected by different crop management. Irrigated rice was investigated in adjacent plots, mainly for comparison purposes. The 4-yr average in CH_4 emission from deepwater rice with straw ash (burned straw) treatment was 46 mg $m^{-2} d^{-1}$ and total emission was 98 kg $ha^{-1} yr^{-1}$. For irrigated rice, the average emission rate and total emission for the straw ash treatment was 79 mg m^{-2} d^{-1} and 74 kg $ha^{-1} yr^{-1}$, respectively. Low emission rates may partially be related to acid sulfate soil of the experimental site. Without organic amendment, the seasonal pattern of CH_4 emission from deepwater rice was correlated with an increase in biomass of rice plants. Emission rates from deepwater rice depend on the production of biomass and the straw management as well. Methane emission was greatest with straw incorporation, followed by straw compost incorporation, zero-tillage with straw mulching, and least with straw ash incorporation. The seasonal pattern of CH_4 ebullition in deepwater rice was consistent with seasonal emission, and total ebullition corresponded to 50% of total emission. Dissolved CH_4 concentrations in the surface soil (0-5 cm) were similar to those in the subsoil (5-15 cm), and the seasonal fluctuation of dissolved CH_4 was also consistent with the seasonal CH_4 emission. Increase in plant density and biomass of irrigated rice grown by pregerminated seed broadcasting enhanced CH_4 emission as compared with transplanting.

Introduction

The atmospheric concentration of greenhouse gas methane (CH_4) has been increasing rapidly over the past 300 years (Cicerone & Oremland, 1988; Khalil & Rasmussen, 1989). While CH_4 is generated biologically from the decomposition of organic matter under anaerobic conditions, flooded rice field has been identified to be one of the agricultural sources of CH_4 (Bouwman, 1990; Bartlett & Harriss, 1993; Wassmann et al., 1993; Neue & Sass, 1994; IPCC, 1995). Recent global estimates of emission rate from wetland rice fields range from 20 to 100 Tg yr^{-1} (IPCC, 1992). Methane fluxes from irrigated and rainfed lowland rice fields were well documented by several field studies (Yagi & Minami, 1990; Sass et al., 1991; Cicerone et al., 1992; Denier van der Gon & Neue, 1995; Bronson et al., 1997; Watanabe et al., 1998). Little is known about CH_4 flux from deepwater rice.

The deepwater rice area in the world is about 9 million ha, of which 60% is in the Indian subcontinent, 35% in Southeast Asia (mainly in Myanmar, Thailand, Vietnam, and Cambodia) and 5% in West Africa (Catling, 1992). In Thailand, the area of deepwater rice is approximately 0.5 million ha (Charoendham et al., 1994) corresponding to 2% of the total rice land. Rice production in Thailand is about 22 million t, most of which was consumed locally, with only 25% exported (IRRI, 1995).

Deepwater rice fields in Thailand reach water depths of 0.7-2 m during the peak of the flooding season in October/November. Deepwater rice requires elongating plants which produce from 9.4 to 12.5 t residues ha^{-1} which, in turn, are left on the fields. In traditional practice, plant residues remain on the field throughout the fallow period. This practice causes difficulties in plowing the land for the next growing season. Hence residues are commonly burned in February to March to facilitate land preparation. Seeds of deepwater rice are broadcast immediately after the first or second tillage. Harvest time is from late December to mid-January and average grain yield is 2 t ha^{-1}.

The Prachinburi Rice Research Center (PRRC) participated in the Interregional Research Program on Methane Emission from Rice Fields (1994-98) coordinated by IRRI (Wassmann et al., this issue, a). The specific objectives of this PRRC field study were (i) to quantify CH_4 emissions from deepwater rice ecosystems, (ii) to evaluate processes that control CH_4 fluxes from rice fields, and (iii) to develop mitigation technologies while maintaining or enhancing rice productivity in a sustainable deepwater rice system.

Materials and methods

Soil and field management

The PRRC soil is clayey, acid sulfate soil. Some of the physicochemical properties of PRRC soil are shown in Table 1. Deepwater rice was seeded in late May to early June. The dry land was plowed twice and dry seeds were broadcast directly onto the soil at 94 kg ha^{-1}. Chemical fertilizers were applied in two doses: basal application 30 d after seed germination at 25-31-0 kg NPK ha^{-1} and topdressing when water depth was 30-40 cm at 29-0-0 kg NPK ha^{-1}. Nitrogen was applied as urea. Application rate of straw (applied as fresh, compost, or straw ash) was 12.5 t ha^{-1} fresh weight corresponding to 54 kg N ha^{-1}. Deepwater rice was harvested in late December.

Irrigated rice was transplanted except for the 1996 and 1998 wet season experiments when seeds were broadcast. Stubble was removed from the irrigated plot before land preparation. Chemical fertilizers for irrigated rice were applied in three doses: basal at 40-30-30 kg NPK ha^{-1}, midtillering, and panicle initiation at 40-0-0 kg NPK ha^{-1}. Mineral N was applied as urea.

Table 1. The physical and chemical characteristics of soil in Prachinburi Experiment Station

Soil property	Analysis
pH	3.93
Organic matter (%)	1.93
N (%)	0.18
K (%)	0.04
P (ppm)	4.5
Clay (%)	62.9
Silt (%)	26.7
Sand (%)	10.4
CEC (cmol) (+) kg^{-1})	21.8
Fe (%)	1.2
Mn (mg kg^{-1})	33.6
SO_4^{-2} (mg kg^{-1})	371

Experimental layout and treatment arrangement

Methane emissions from rice fields were monitored from 1994 to 1998 with an automatic system (Wassmann et al., this issue). The deepwater rice field was divided into nine plots measuring 7 m × 7 m arranged in randomized complete block design with three replications. While the water level in these nine plots followed the pattern of deepwater rice in the area, three adjacent plots (5 m × 7 m) were separated from the other area by a dam. In these plots, water levels were controlled to simulate irrigated rice fields. This field layout with one chamber in each plot facilitated simultaneous records of CH_4 emissions from deepwater and irrigated rice fields during the wet season. In the 1997 and 1998 dry seasons, each of these irrigated plots was used for one treatment and was equipped with three chambers.

Methane emission measurement

The principles of sampling and analytical procedure were described by Schütz et al. (1989); technical details in the system applied in this network were described by Wassmann et al. (this issue). The automatic measurement system was programmed to monitor CH_4 flux continuously for 16 min every 2 h (12 measurements a day) from planting until a week after harvest.

Methane ebullition was determined weekly by capturing gas bubbles emerging from the water surface (Wassmann et al.,1996). Plexiglas boxes were installed on the soil surface between rice hills. After 24 h, gas sample was withdrawn from each box and analyzed for CH_4 concentration using a gas chromatograph. Floodwater height inside each box was also determined for headspace calculation (Neue & Sass,1993). Dissolved CH_4 in the soil were measured weekly according to the techniques described by Wassmann et al. (1996).

Auxiliary data measurement

Amount of rainfall, water depth, soil pH, and redox potential (Eh) were measured daily. Growth of plants was evaluated monthly by harvesting 10 plants from each plot for biomass and height determination. Grain yield was determined after harvest.

Statistical analysis of experimental data was accomplished using STATISTICA by Statsoft. The data in each treatment were evaluated according to the type of distribution. When the distribution was normal, t test was used; when it was not normal, sign test was used.

Results and discussion

Seasonal patterns in deepwater rice

The distribution and amount of rainfall in Prachinburi is shown for the 1994 wet season (Figure 1). Average annual rainfall was 1750 mm. The period of strong rainfall began in April/May and stopped in October/ November. The field was initially flooded in June. Water levels typically rose at the rate of 2-5 cm d^{-1}, and reached the maximum depth of 70-80 cm in October (Figure 1). In 1995, however, the water level rose rapidly to 145 cm by mid-September. Water level started to recede in November and the field was dry by mid-December (Figure 1). Temperatures in air, water, and soil in deepwater rice field varied between 22-35 °C, 25-29 °C, and 26-27 °C, respectively (data not shown).

The pH of dry soil was 3.8-4.0 and increased gradually after flooding to values of 4.3-6.5 (Figure 2). Soil Eh decreased after flooding and remained below −150 mV for most of the season (Figure 2).

The local practice of applying burned rice straw resulted in very low emission rates during the early growth stage (Figure 3). Emission rates gradually increased at flowering stage and reached maximum at ripening before harvest (December). After harvest, emission rates declined sharply and leveled off. The increase of emission rates with plant growth was correlated with the continuous increase in biomass of deepwater rice. Due to the absence of organic amendments, the methanogenic material could either come from root exudation, decaying roots, or aquatic biomass. Emission rates showed pronounced fluctuations at the end of the season when the field dried out.

Effect of crop management on CH_4 emissions in deepwater rice

Different crop management options for deepwater rice were evaluated from 1994 to 1998 (Table 2). Grain yields and biomass of deepwater rice were rated uniform, irrespective of crop management, except for higher grain yields through fresh straw application in 1996 (Table 2). Under favorable conditions in 1994 and 1996 wet seasons, average grain yield of deepwater rice was 3 t ha^{-1}. However, plants were heavily damaged by flooding in 1995 and partially damaged by insect disease in 1997 and yields were lower.

In 1994, CH_4 emissions were about 200 kg ha^{-1} in all treatments, i.e. chemical fertilizer, burned straw, and without fertilization (Table 2). Apparently, CH_4 emission was limited in all fields by low organic carbon levels for CH_4 production as previously described in other experiments without organic amendments (Schütz et al., 1989; Yagi & Minami, 1990).

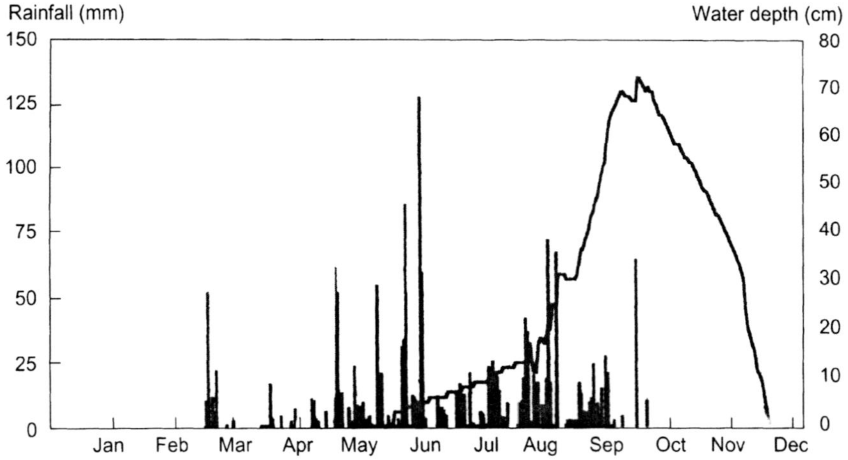

Figure 1. Distribution of rainfall (column) and water depth (line) in a deepwater rice field in PRRC, 1994 wet season

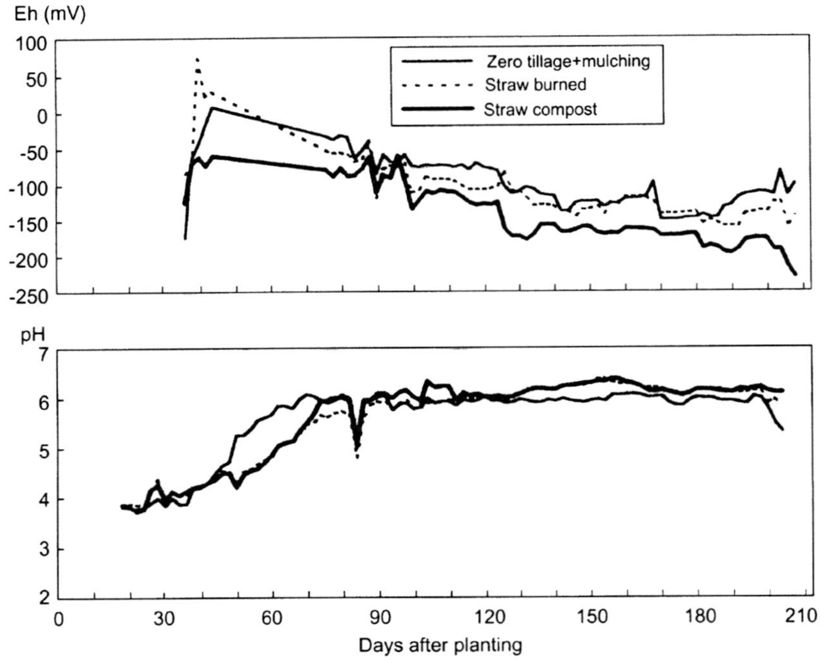

Figure 2. Seasonal patterns of soil Eh and pH in deepwater rice field with different straw management, 1997 wet season

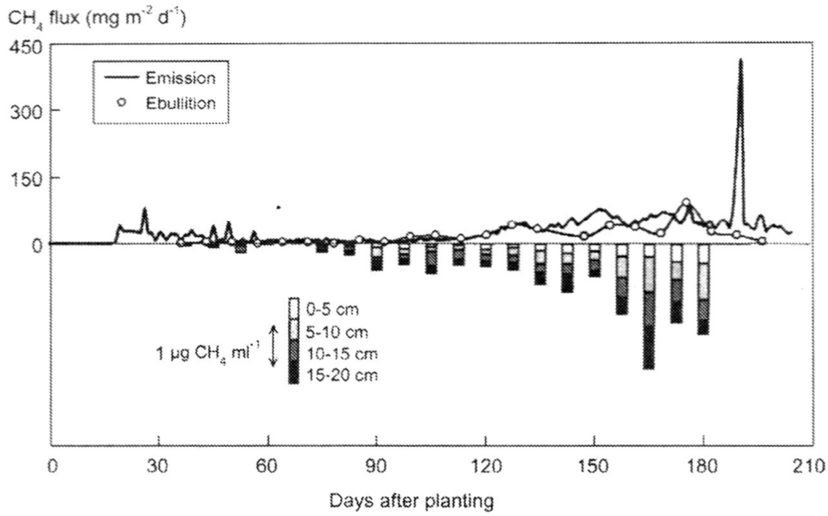

Figure 3. Methane flux rate (emission and ebullition) and dissolved CH_4 concentrations (at 0-5, 5-10, 10-15, and 15-20 cm soil depths) in deepwater rice field with treatment of burned straw application, 1996 wet season

Table 2. Methane emission rates, biomass, and grain yields of rice in 1994-98 wet season

Year	Ecosystem/modifying treatment	Cultivar	Mean CH_4 emission ($mg\ m^{-2} d^{-1}$)[a]	Cumulative CH_4 emission ($kg\ ha^{-1}$)	Biomass ($t\ ha^{-1}$)	Grain yield ($t\ ha^{-1}$)
1994	Deepwater/urea	HTA60	87±42	201	26.9	2.99
	Deepwater/straw ash	HTA60	84±35	194	27.3	3.07
	Deepwater/no N	HTA60	92±48	213	29.3	3.42
	Irrigated/urea	IR72	17±5	18	-	3.42
1995	Deepwater/urea	HTA60	24±20	48 (10)[b]	25.6	0.75
	Deepwater/straw ash	HTA60	33±28	67 (12)[b]	27.8	0.81
	Deepwater/fresh straw incorporated	HTA60	26±20	83 (18)[b]	1.3	0.47
	Irrigated/urea	RD25[c]	135±168	119	-	3.6
1996	Deepwater/straw mulching	PNG	64±55	127	16.6	2.82
	Deepwater/straw ash	PNG	35±45	69	19.8	2.91
	Deepwater/fresh straw incorporated	PNG	311±145	619	19.8	3.60
	Irrigated/urea	SPR2	198±161	289	20.2	3.14
1997	Deepwater/straw mulching	PNG	54±36	100	9.7	1.22
	Deepwater/straw ash	PNG	32±33	60	10.8	1.34
	Deepwater/straw compost	PNG	78±62	145	16.7	1.23
	Irrigated/urea	KDML 105	22±26	22	13.1	-
1998	Deepwater/straw mulching	PCR1[d]	-	-	17.6	3.31
	Deepwater/straw ash	PCR1[d]	-	-	17.8	3.16
	Deepwater/straw compost	PCR1[d]	-	-	17.3	3.08
	Irrigated/urea	KDML 105	144±154	188	8.0	1.66

[a]± = standard deviation of mean. [b]Flood damage, accumulated emission data in parenthesis computed for 2-mo period before flood damage. [c]RD25 was retransplanted after flood crisis. [d]Emission data of deepwater rice in 1998 not completed.

The experiments included a variety of different straw treatments: (i) straw burned, (ii) straw incorporation, (iii) straw mulching on zero-tillage field, and (iv) straw compost (Table 2). In 1996, incorporation of fresh straw into the soil strongly enhanced CH_4 emission (Figures 3-5). The seasonal emissions were highest for rice straw incorporation (619 kg ha^{-1}), moderately high for zero tillage (127 kg ha^{-1}), and low for burned straw (69 kg ha^{-1}) (Table 2, Figures 3-5). In the 1995 wet season, the data were limited to the initial 2 mo when the characteristic difference between those treatments had not yet fully evolved.

The experiment in 1997 included straw compost as an additional treatment (Figure 6). Seasonal CH_4 flux in plots with rice straw compost (145 kg ha^{-1}) was higher than with mulching (100 kg ha^{-1}) and burned straw (60 kg ha^{-1}) (Table 2). In comparison with the straw ash treatment of the respective year (1996 and 1997), composted straw increased emissions by a factor of 2.4, fresh straw by a factor of 9, and mulching by 1.7-1.8. The decomposition of straw during the composting process reduced potential precursors of CH_4, hence CH_4 production after compost application was relatively low.

Methane ebullition and dissolved CH_4 in deepwater rice

Methane ebullition was evaluated in the 1996 experiments. Seasonal patterns of CH_4 ebullition rates were consistent with those of emission rates (Figures 3-5).

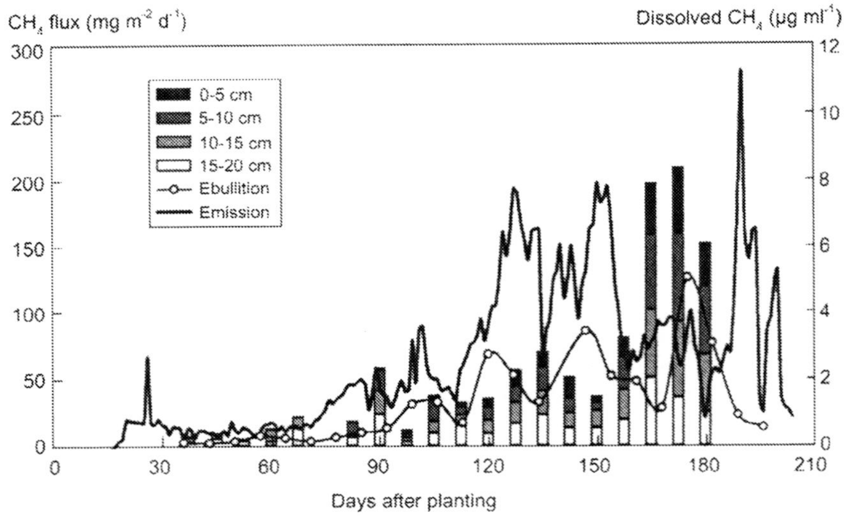

Figure 4. Methane flux rate (emission and ebullition) and dissolved CH_4 concentrations (at 0-5, 5-10, 10-15, and 15-20 cm soil depths) in deepwater rice field with treatment of zero tillage plus mulching straw, 1996 wet season

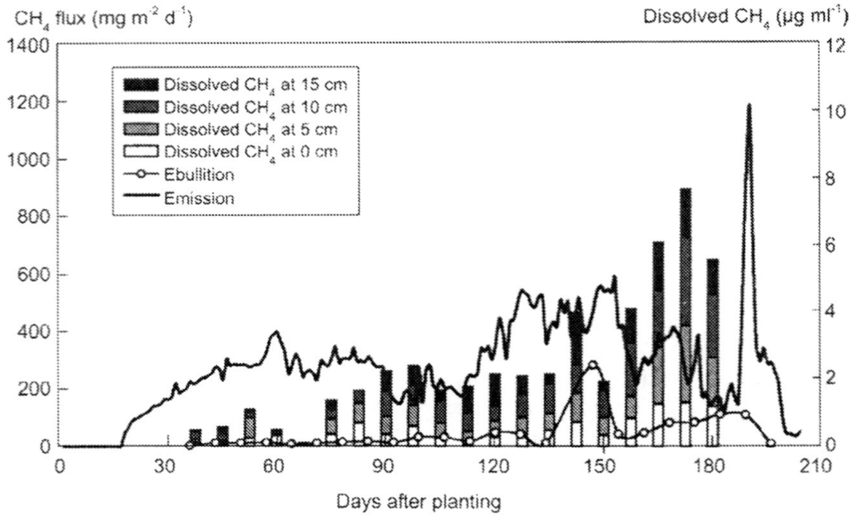

Figure 5. Methane flux rate (emission and ebullition) and dissolved CH_4 concentrations (at different soil depths) in deepwater rice field with treatment of straw incorporation, 1996 wet season

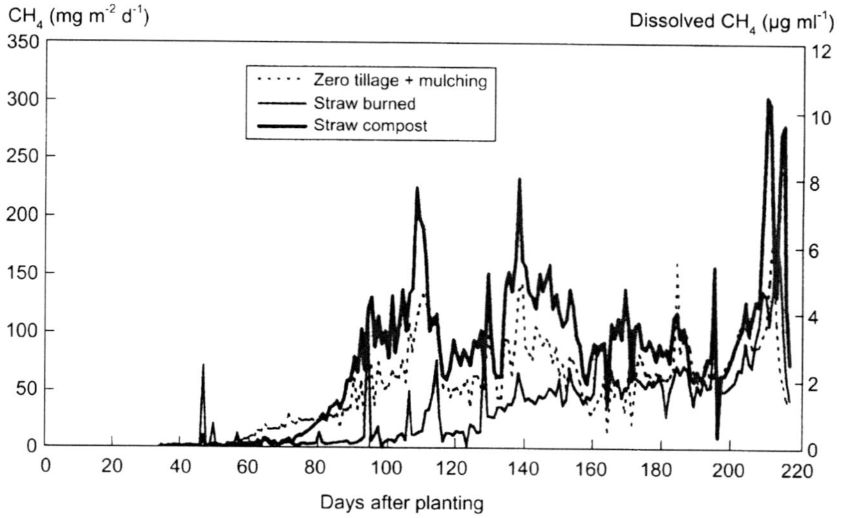

Figure 6. Seasonal CH_4 emissions from deepwater rice field with different straw management, 1997 wet season

Addition of rice straw enhanced ebullition in absolute terms but decreased its relative contribution to overall emission. The total CH_4 from ebullition in the treatments of straw incorporation, zero tillage, and straw burned corresponded to 14%, 47%, and 59% of total emission, respectively. When CH_4 production in soil was high, CH_4 was primarily emitted through aerenchyma of rice plants rather than through ebullition.

Seasonal patterns of dissolved CH_4 were relatively uniform among the different straw treatments in 1996 (Figures 3-5). The stimulation of emissions by straw incorporation was not reflected by high CH_4 concentrations in soil solution. Apparently, the bulk of CH_4 produced in the soil escaped rapidly to the atmosphere—without longer storage in aqueous media. Concentrations of dissolved CH_4 at the surface soil (0-5 cm depth) did not significantly differ from those at the 5-20 cm depth (Figures 3-5), indicating relative homogeneity of the soil in the vertical direction.

Possible technology for mitigating CH_4 emissions from deepwater rice

Mitigation options in deepwater rice can be assessed through experiments in 1994, 1996, and 1997 (Table 2). Due to the lack of water control, deepwater rice offers limited options to modify crop management. Burned straw incorporation, instead of fresh straw incorporation, gives 89% reduction in CH_4 emission. Burning of straw, however, causes local air pollution and is therefore not recommended as a mitigation option. Zero tillage and mulching also reduced emissions as compared with fresh straw incorporation. Although yields were lower than with incorporation of fresh straw, yield levels were still in the same range as for burned straw application. Hence, zero tillage and mulching would be the most promising mitigation technology for deepwater rice in Thailand. There are, however, some problems such as crop establishment, weed control, and land preparation that may have to be addressed.

Methane emissions from irrigated rice

Methane emission rates in irrigated rice were determined in the wet and dry seasons from 1994 to 1998 (Tables 4-5). The experiments were conducted for purposes of comparison, i.e., to determine the emission potential of deepwater rice in comparison with irrigated fields; only the dry season experiments of 1997 and 1998 encompassed a comparison of treatments.

Emission rates in irrigated rice in Prachinburi reflected pronounced variations over time. In the dry seasons, emissions were generally in a low range, between 15 and 42 kg ha^{-1}. Low emissions were attributed to high acidity of the soil. Soil pH remained below pH 6 for 60 d after flooding, i.e., half of the veg-

Table 3. Methane emission and biomass of irrigated rice cultivar SPR2 grown in three soils under pot and field cultivation, 1996 dry season

Soil location	Soil texture	Soil pH	Organic matter (%)	Mean CH_4 emission (mg m^{-2} d^{-1})	Cumulative CH_4 emission (kg ha^{-1})	Biomass (t ha^{-1})
PRRC (pot)	Clay	3.9	1.93	19 ± 9	17	5.9
Ayutthaya (pot)	Clay	4.9	1.10	5 ± 3	5	6.6
Hinsorn (pot)	Sandy loam	5.3	0.77	100 ± 45	91	6.4
PRRC (field)	Clay	3.9	1.93	33 ± 25	28	7.0

Table 4. Methane emissions, biomass, and grain yields of three high-yielding irrigated rice cultivars in 1997 dry season

Cultivar	Mean CH_4 emission (mg m^{-2} d^{-1})	Cumulative CH_4 emission (kg ha^{-1})	Biomass (t ha^{-1})	Grain yield (t ha^{-1})
Poe-Thong	43±60	41	22.9	3.9
SPR1	43±59	41	21.7	2.8
SPR60	44±57	42	21.0	3.2

Table 5. Methane emissions, plant densities, biomass, and grain yields of high-yielding irrigated rice KLG1 as affected by crop establishment in 1998 dry season

Planting method	Spacing/ seed rate	Plant density (tillers m^{-2})	Mean CH_4 emission (mg m^{-2} d^{-1})	Cumulative CH_4 emission (kg ha^{-1})	Biomass (t ha^{-1})	Grain yield (t ha^{-1})
Transplanting	20×20 cm	444	17±12	15	11.3	1.5
Pregerminated seed broadcasting	94 kg seed ha^{-1}	657	26±20	25	16.0	1.8
Pregerminated seed broadcasting	188 kg seed ha^{-1}	645	26±15	25	15.4	1.8

etation period. In the wet season, cumulative emissions exceeded 100 kg m^{-2} in 1995, 1996, and 1998, while emissions in 1994 and 1997 were below 50 kg ha^{-1} (Table 6). One explanation for the different emissions could be that the experimental setup did not ensure full hydrological isolation of the irrigated plots. Limited control of water levels required an adjustment in the timing of the irrigated crop to allow shallow water levels during crop establishment. While the dam around the plots prevented high flooding, soil pH was apparently affected by seepage from the adjacent deepwater plots even before irrigation started. In the wet seasons with high emissions, the pH levels were relatively favorable (pH > 5; data not shown).

In the 1997 dry season, CH_4 emissions were determined in three high-yielding varieties: Poe-Thong, SPR1, and SPR60. There were no significant differences among three cultivars, with total CH_4 emissions being 41-42 kg ha^{-1} (Table 4). This is probably the result of the similar production of biomass among three cultivars although the grain yield was relatively higher for Poe-Thong.

In the 1998 dry season, CH_4 flux was determined in irrigated rice with different planting methods: (i) transplanted with spacing of 20 × 20 cm, (ii) pregerminated seed broadcast at 94 kg ha^{-1}, and (iii) pregerminated seed broadcast at 188 kg ha^{-1}. Total CH_4 emissions in plots using seed broadcasting were 25 kg ha^{-1} (for both seed rates) which were significantly higher than that in transplanted plot (15 kg ha^{-1}) (Table 5). This variance was apparently related to a difference in plant growth. Broadcasting pregerminated seeds at 94 and 188 kg ha^{-1} yielded 657 and 645 tillers m^2, whereas transplanting resulted in only 444 tillers per m^2. Simi-

Table 6. Average methane emission from deepwater rice with straw burned treatment (A), irrigated rice (B), and rainfed rice (C) from 1994 to 1998

Year/season	Cultivar	Mean CH_4 emission (mg m^{-2} d^{-1})	Cumulative CH_4 emission (kg ha^{-1})
(A) Deepwater rice			
1994 wet season	HTA60	84	194
1995 wet season	HTA60	33	67
1996 wet season	PNG	35	69
1997 wet season	PNG	32	60
	Av	**46**	**98**
(B) Irrigated rice			
1994 wet season	IR72	17	18
1995 wet season	RD35	135	119
1996 dry season	SPR2	33	28
1996 wet season	SPR2	298	289
1997 dry season	PT	43	41
1997 dry season	SPR1	43	41
1997 dry season	SPR60	44	42
1998 dry season	KLG1	17	15
	Av	**79**	**74**
(C) Rainfed rice			
1997 wet season	KDML 105	22	22
1998 wet season	KDML 105	144	188
	Av	**83**	**105**

larly, the grain yields and biomass were significantly higher for seed broadcasting than for transplanting (Table 5). Apparently, high plant density and biomass enhanced CH_4 emission from rice field. In the later growth stage, rice plant in all plots was approximately 50% damaged by disease and insect pests, hence grain yields in this experiment were relatively lower than those in other year experiments (Table 5). Consequently, the emission rate observed in this experiment was also relatively lower than in other years.

Conclusion

The results of this study showed that deepwater rice had low CH_4 emission rates, but due to long season lengths, seasonal emission rates accumulate to relatively high levels. The 4-yr observation of CH_4 emission from deepwater rice with the burned straw treatment corresponded to a daily average of 46 mg m^{-2} d^{-1} and a seasonal average of 98 kg ha^{-1} yr^{-1}. Average emission from irrigated rice was 79 mg m^{-2} d^{-1} and 74 kg ha^{-1} yr^{-1}, respectively, and that from rainfed rice was 83 mg m^{-2} d^{-1} and 105 kg ha^{-1}, respectively (Table 6).

However, the comparison between deepwater and irrigated rice may be affected by site-specific conditions (Wassmann et al., this issue, b). In the acid sulfate soil of Prachinburi, pH of the soil reached a neutral range within 2 mo of flooding. While this pattern will drastically reduce emissions from irrigated rice (with 110 d growing period), the soil impact on deepwater rice is less severe due to its long growing period.

The only practical option for reducing CH_4 emissions in deepwater rice is proper straw management. Mulching of straw in zero-tillage fields slightly enhanced CH_4 emission as compared with burned straw application but significantly reduced emissions as compared with fresh straw incorporation. Apparently, the straw on the surface of the soil was partially decomposed during the fallow period. However, further research is needed to integrate these findings into an overall strategy of sound crop management for high yields and low emissions.

References

Bartlett KB & Harriss RC (1993) Review and assessment of methane emissions from wetlands. Chemosphere 26(1-4):261-320

Bouwman AF (1990) Exchange of greenhouse gases between terrestrial ecosystems and the atmosphere. In: AF Bouwman AF (ed) Soil and the Greenhouse Effect. John Wiley, New York. pp 61-127

Bronson KF, Neue HU, Singh U & Abao Jr EB (1997) Automated chamber measurement of methane and nitrous oxide flux in flooded rice soil: I. Residue, nitrogen, and water management. Soil Sci Soc Am J 61:981-987

Catling D (1992) Rice in Deepwater. International Rice Research Institute. Macmillan Press Ltd, London. 542 p

Charoendham P, Palaklang W, Niyomwit L, Promnart P, Chareonsilp N & Puckridge DW (1994) Deepwater Rice Situation of Thailand in 1992/1993. Proc. 6th Deepwater Rice Annual Meeting, Prachinburi Rice Research Center, 11 March 1994

Cicerone RJ & Oremland RS (1998) Biogeochemical aspects of atmospheric methane. Global Biogeochem Cycles 2: 299-327

Cicerone RJ, Delwiche CC, Tyler SC & Zimmermann PR (1992) Methane emissions from California rice paddies with varied treatments. Global Biogeochem Cycles 6: 233-248

Denier van der Gon HAC & Neue HU (1995) Influence of organic matter incorporation on the methane emission from a wetland rice field. Global Biogeochem Cycles 9: 11-22

IPCC – Intergovernmental Panel on Climate Change (1992) Climate Change. The supplementary report on the IPCC Scientific assessment. Houghton J T, Callender B A, Varney S K. (eds) Cambridge University Press, UK. 200 p

IPCC – Intergovernmental Panel on Climate Change (1995) Climate Change: The supplementary report on the IPCC Scientific Assessment. Cambridge University Press, UK. 365 p

IRRI – International Rice Research Institute (1995) Rice Almanac, International Rice Research Institute, Manila, Philippines

Khalil MAK & Rasmussen RA (1989) Climate induced feedback for the global cycles of methane and nitrous oxide. Tellus 41B:554-559

Neue HU & Sass R (1993) Trace gas emissions from rice fields. Paper presented at the Methane Emission at the Ricefield Workshop held at IRRI-Philippines, 30 Aug-3 Sep 1993

Neue HU & Sass R (1994) Trace gas emissions from rice fields. In Global Atmospheric-Biospheric Chemistry Prinn RG (ed) Plenum Press , New York. pp 119-146

Sass RL, Fisher FM, Turner FT & Jund MF (1991) Methane emission from rice fields as influenced by solar radiation, temperature, and straw incorporation. Global Biogeochem Cycles 5(4):335-350

Schütz H, Holzapfel-Pschorn A, Conrad R, Rennenberg H & Seiler W (1989) A three-year continuous record on the influence of daytime season and fertilizer treatment on methane emission rates from an Italian rice paddy field. J Geophys Res 94:16405-16416

Wassmann R, Neue HU, Lantin RS, Buendia LV & Rennenberg H (2000a) Characterization of methane emissions from rice fields in Asia. 1. Comparison among field sites in five countries. Nutr Cycling Agroecosyst, this issue

Wassmann R, Neue HU, Lantin RS, Makarim K, Chareonsilp N, Buendia LV & Rennenberg (2000b) Characterization of methane emissions from ricefields in Asia. 2. Differences among irrigated, rainfed, and deepwater rice. Nutr Cycling Agroecosyst, this issue

Wassmann R, Neue HU, Alberto MCR, Lantin RS, Bueno C, Llenaresas D, Arah JRM, Papen H, Seiler W & Rennenberg H (1996) Fluxes and pools of methane in wetland rice soils with varying organic inputs. Environ Monit Assess 42:163-173

Wassmann R, Papen H & Rennenberg H (1993) Methane emission from rice paddies and possible mitigation. Atmosphere 26:201-217

Watanabe A, Yoshida M & Kimura M (1998) Contribution of rice straw carbon to CH_4 emission from rice paddies using ^{13}C-enriched rice straw. J Geophys Res 103:8237-8242

Yagi K & Minami K (1990) Effect of organic matter application on methane emission from some Japanese paddy fields. Soil Sci Plant Nutr 36:599-610

Nutrient Cycling in Agroecosystems **58:** 131–139, 2000.

Simultaneous records of methane and nitrous oxide emissions in rice-based cropping systems under rainfed conditions

E.B. Abao Jr. [1], K.F. Bronson[1,2], R. Wassmann[1,3] & U. Singh[1,4]

[1] International Rice Research Institute, P. O. Box 3127, Makati City 1271, Philippines; [2] Texas A&M University, Agricultural Experiment Station, Route 3, Box 219, Lubbock, TX 87401, USA; [3] Fraunhofer Institute for Atmospheric Environmental Research (IFU), Garmisch-Partenkirchen, Germany; [4] International Fertilizer Development Center, Muscle Shoals, AL 35662 USA

Key words: automated closed chamber method, wheat, cowpea, slow-release nitrogen fertilizer, residue management, denitrification, methane sink, rainfall

Abstract

Rainfed rice (*Oryza sativa* L.)-based cropping systems are characterized by alternate wetting and drying cycles as monsoonal rains come and go. The potential for accumulation and denitrification of NO_3^- is high in these systems as is the production and emission of CH_4 during the monsoon rice season. Simultaneous measurements of CH_4 and N_2O emissions using automated closed chamber methods have been reported in irrigated rice fields but not in rainfed rice systems. In this field study at the International Rice Research Institute, Philippines, simultaneous and continuous measurements of CH_4 and N_2O were made from the 1994 wet season to the 1996 dry season. During the rice-growing seasons, CH_4 fluxes were observed, with the highest emissions being in organic residue-amended plots. Nitrous oxide fluxes, on the other hand, were generally nonexistent, except after fertilization events where low N_2O fluxes were observed. Slow-release N fertilizer further reduced the already low N_2O emissions compared with prilled urea in the first rice season. During the dry seasons, when the field was planted to the upland crops cowpea [*Vigna unguiculata* (L.) Walp] and wheat (*Triticum aestivum* L.), positive CH_4 fluxes were low and insignificant except after the imposition of a permanent flood where high CH_4 fluxes appeared. Evidences of CH_4 uptake were apparent in the first dry season, especially in cowpea plots, indicating that rainfed lowland rice soils can act as sink for CH_4 during the upland crop cycle. Large N_2O fluxes were observed shortly after rainfall events due to denitrification of accumulated NO_3^-. Cumulative CH_4 and N_2O fluxes observed during this study in rainfed conditions were lower compared with previous studies on irrigated rice fields.

Introduction

Rainfed rice-based production systems make up 25% of the world's area of harvested rice (IRRI, 1998). These systems are characterized by a monsoon season in which rice is grown in the wet season and various upland crops are grown in the dry season without irrigation (Tripathi et al., 1997). At any time of the year, rains can flood the soil, resulting in denitrification and leaching of accumulated NO_3^- (Buresh et al., 1989; George et al., 1993).

Production and emission of CH_4, a "greenhouse gas" about 30 times more radiatively active than CO_2, is an important feature in the cycle of C in flooded rice soils. Methane and CO_2 are the final products of organic matter decomposition under anaerobic conditions. Emission of CH_4 from rice fields makes up about one-fifth of all sources of CH_4 emitted to the atmosphere globally (IPCC, 1992). Nitrous oxide (N_2O) is about 300 more radiatively active than CO_2 (mass basis, considering residence time in the atmosphere (Rodhe, 1990). Agriculture is the main source of most N_2O emissions. Nitrous oxide is produced from soil processes as an intermediate product of microbial nitrification and denitrification (Granli & Bockman, 1994). The potential of N_2O emission increases when the amount of N available for microbial transformation is enhanced through fertilizer application (Eichner, 1990), cropping of legumes, return to soil of manures and crop residue (Aulakh et al., 1991), and mineralization of soil biomass and other forms of soil organic matter. In previous work, we found that residue incorporation had no effect on N_2O emissions in fallow rice fields (Bronson et al., 1997b) but could reduce N_2O fluxes during a rice growing season with midseason drainage (Bronson et al., 1997a).

Previous research by our team involved measurements of CH_4 and N_2O emissions in irrigated rice fields using automated chambers from double-cropped irrigated rice fields (Bronson et al., 1997a) and the short rainfed fallow periods (Bronson et al., 1997b). This study represents a continuation of those studies in which we hypothesized that CH_4 and N_2O emissions will be of different magnitude and pattern in rainfed rice-upland cropping systems compared with double-cropped irrigated rice.

Materials and methods

Experimental site and field design

The field studies were conducted at the International Rice Research Institute, Los Baños, Philippines on Maahas clay soil (pH 7.0, 1.2 g N kg^{-1}, CEC of 17.2 cmol(+) kg^{-1}). The experiments covered two cropping cycles with wet and dry seasons and the fallow periods in between. Rice was grown under rainfed lowland conditions in the wet seasons while wheat and cowpea was grown in the dry seasons.

The treatments during the 1994 wet/rice season were

1. Prilled urea (90 kg N ha^{-1} applied in three equal splits at final harrowing, midtillering, and flowering)
2. Polyon 12, a slow-release N fertilizer urea (90 kg N ha^{-1} applied at final harrowing)

In the 1995 dry season, the treatments/crops were

1. Weed-free fallow
2. Cowpea (30 kg urea N ha^{-1} applied pre-plant) planted in previous prilled urea plots
3. Cowpea (30 kg urea N ha^{-1} applied pre-plant) planted in previous slow-release N plots
4. Wheat (60 kg urea N ha^{-1} applied pre-plant)

In the 1995 wet/rice season, the treatments were

1. Urea (90 kg N ha^{-1} applied in three equal splits at final harrowing, midtillering, and flowering) in weed-free fallow plots
2. Urea (90 kg N ha^{-1} applied in three equal splits at final harrowing, midtillering, and flowering) with cowpea residue removed
3. Urea (30 kg N ha^{-1} applied in three equal splits at final harrowing, midtillering, and flowering) and 3 t ha^{-1} dry cowpea residue incorporated at final harrowing
4. Urea (90 kg N ha^{-1} applied in three equal split applications at final harrowing, midtillering, and flowering) with 3 t ha^{-1} dry wheat residue incorporated at final harrowing

During the 1996 dry season, the treatments/crops were

1. Weed-free fallow
2. Cowpea (30 kg N ha^{-1} applied pre-plant in plots with previous cowpea residue removed)
3. Cowpea (30 kg N ha^{-1} applied pre-plant in plots with previous cowpea residue incorporated)
4. Wheat (90 kg N ha^{-1} applied pre-plant in plots with previous wheat residue incorporated)

Measurement of CH_4 and N_2O fluxes

An automated chamber system which operated for 24 h a day was used to measure CH_4 and N_2O fluxes. The details of the system were described in Bronson et al. (1997a). Fluxes were measured from all plots every 2 h. Two-hour flux rates were averaged over 12-h daytime and 12-h night time periods for each treatment. Cumulative fluxes for each season were also calculated. This measurement system was used continuously from the 1994 wet season to the 1996 dry season.

Grain yield determination

Harvesting was done on a 2- × 2-m area in the middle of each experimental plot. The crops were cut at ground level and put in cloth bags and dried. After drying, the grains were threshed and weighed. Grain yields were adjusted to 14% moisture. For cowpea, the pods were collected and the seeds separated, dried, and weighed.

Statistical analysis

Analysis of variance was done using SAS (SAS, 1987) on 12-hourly and seasonal CH_4 and N_2O fluxes. Duncan's multiple range test was used at P= 0.05 level of probability to distinguish treatment differences.

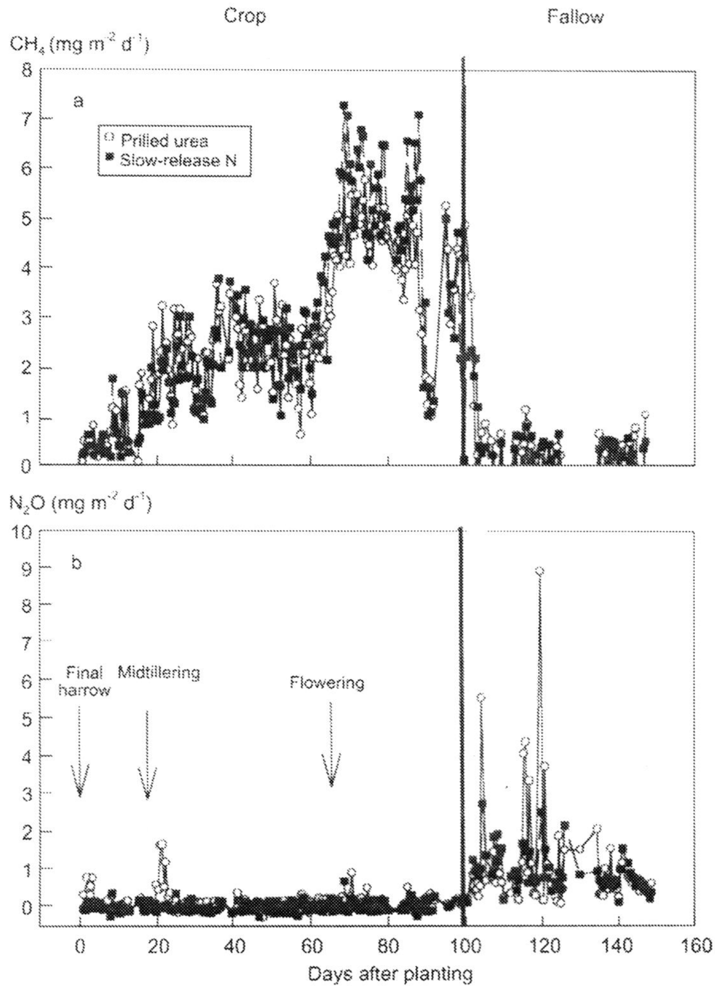

Figure 1. Methane (a) and N_2O (b) fluxes during the 1994 wet season (rice crop and fallow period)

Results and discussion

1994 wet season

Methane fluxes for both prilled urea and slow-release N showed the same pattern during the entire season wherein two peaks were observed (Figure 1a). The first major peak of CH_4 activity was at 40 d after transplanting (DAT) or maximum tillering when CH_4 fluxes rose to about 4 mg CH_4-C $m^{-2} d^{-1}$. The second peak was observed at 70 DAT where CH_4 fluxes increased to 7 mg CH_4-C $m^{-2} d^{-1}$. The two distinct peaks of CH_4 flux observed may be attributed to increase in tillers which serve as CH_4 channels and decomposing roots which provide C source for CH_4-producing bacteria (Neue et al., 1994). There was no significant difference in cumulative CH_4 fluxes between the two N fertilizer sources, slow-release and prilled urea (Table 1). Seasonal fluxes of CH_4 were lower than those reported by Bronson et al. (1997a) for a nearby irrigated site of higher soil organic matter content.

During the fallow period after the 1994 wet season, CH_4 fluxes drastically decreased shortly after harvest to less than 1 mg CH_4-C $m^{-2} d^{-1}$ for both treatments (Figure 1a) until the end of the fallow period.

Nitrous oxide fluxes were generally less than 1 mg N_2O-N m^{-2} d^{-1} during the entire rice-growing season. In the prilled urea treatment, low but distinct N_2O

Table 1. Grain yields of rice, cowpea, and wheat, and cumulative CH_4 and N_2O fluxes as affected by crop and residue management under rainfed conditions during 1994-96 dry and wet seasons.[a]

Year/ season	Treatment	Cropping period			Fallow	
		CH_4 emission (mg C m^{-2})	N_2O emission (mg N m^{-2})	Yield (t ha^{-1})	CH_4 emission (mg C m^{-2})	N_2O emission (mg N m^{-2})
1994/WS	Rice, prilled urea	230 a	9.7 a	6.0	11.1 a	48.6 a
	Rice, polyon 12	220 a	0.3 b	5.9	1.5 a	41.2 a
1995/DS	Cowpea, urea (after urea)	–67.8 b	31.9 b	0.99	12.6 a	34.7 a
	Cowpea, urea (after polyon)	–37.8 b	38.3 b	1.11	19.8 a	67.1 a
	Fallow, (weed-free)	2.8 a	36.8 b	–	14.5 a	42.2 a
	Wheat, urea	4.4 a	64.5 a	1.41	24.7 a	59.2 a
1995/WS	Rice, urea, no residue	530 b	24.9 a	5.2	10.2 a	40.0 a
	Rice, urea, cowpea residue	1560 a	23.2 a	5.3	3.9 a	56.0 a
	Rice, urea	560 b	24.5 a	5.4	29.8 a	59.1 a
	Rice, urea, wheat residue	2580 a	11.5 a	5.1	40.7 a	40.6 a
1996/DS	Cowpea, urea (after no residue)	–15.3 a	10.6 b	0.9		
	Cowpea, urea (after cowpea residue)	–15.1 a	27.7 b	1.0		
	Fallow (weed-free)	2.4 a	28.5 b	–		
	Wheat, urea	1.8 a	61.2 a	1.1		

[a]Values in the same season of the same year followed by the same letter in a column are not significantly different by Duncan's multiple range test at P = 0.05.

fluxes appeared shortly after fertilizer applications at final harrow, midtillering, and flowering. Low N_2O emissions with small peaks after N fertilization events and high CH_4 emissions in rice have been observed by other workers (Bronson et al., 1997a; Cai et al., 1997). A maximum flux of 1.7 mg N_2O-N m^{-2} d^{-1} was observed at midtillering. Slow-release N resulted in very low N_2O flux rates throughout the season and showed no distinct peaking pattern (Figure 1b). Cumulative seasonal N_2O fluxes were significantly higher in prilled urea than in slow-release N (Table 1). In the fallow period, N_2O fluxes were also generally higher in prilled urea than in slow-release N fertilizer (Figure 1b). This is one of the first reports of N_2O emissions from slow-release N fertilizer in rice. Minami (1994) first reported that slow-release N fertilizer in carrots can reduce N_2O emissions compared with ammonium sulfate. Delgado and Mosier (1996) reported N_2O flux measurements using polyolefin-coated urea in an upland crop—spring barley. They reported initial mitigation of N_2O fluxes with coated urea compared with prilled urea, but the opposite result was observed in the latter part of the growing season. The amounts of N_2O seasonal emission in our study were much smaller than those reported by Bronson et al. (1997a) on the same soil with higher soil organic matter under irrigated condition.

1995 dry season

Starting in the 1995 dry season, the field experiments encompassed four treatments per season. Cumulative flux results of all treatments are shown in Table 1 while the respective figures on seasonal patterns show only two out of four treatments to allow a visual distinction among the graphs (Figure 2a,b).

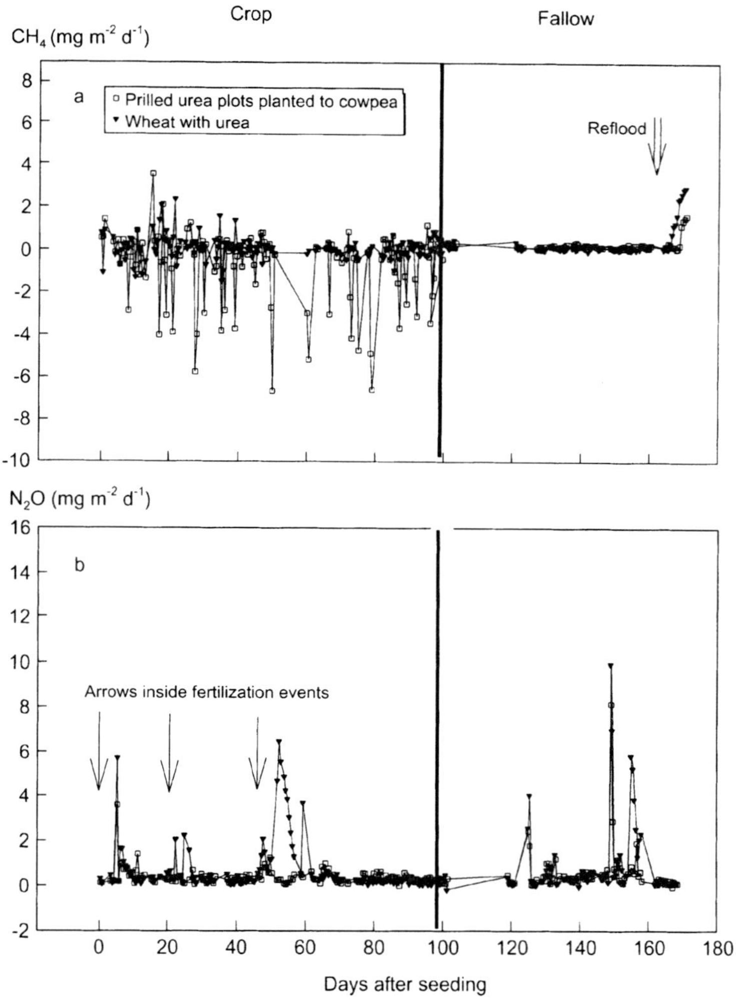

Figure 2. Methane (a) and N_2O (b) fluxes during the 1995 dry season (upland crop and fallow period)

Methane fluxes during the entire dry season crop were generally very low and ranged from -7 to 5 mg CH_4-C m^{-2} d^{-1} (Figure 2a). For all treatments, CH_4 uptake by the aerobic soil was evident throughout the season particularly in the cowpea plots. Only plots planted to cowpea showed net cumulative CH_4 uptake for the season (Table 1). Methane uptake or consumption in soil is a result of CH_4 oxidation by methanotrophic bacteria (Lidstrom & Stirling, 1990). Methane uptake has been reported in temperate native grasslands and in fertilized cropped fields (Bronson & Mosier, 1993) and in tropical forests and agricultural soils (Keller et al., 1990). Only recently have reports been made of CH_4 consumption in rice soils (Singh et al., 1998; 1999). It is not clear why the cowpea plots exhibited the highest CH_4 uptake rates. Nitrogen fertilizer addition in the wheat plots may have inhibited CH_4 uptake (Bronson & Mosier 1994; Singh et al., 1999), but this would not explain the similar result for the unfertilized fallow treatment.

In the fallow period after the 1995 dry season, CH_4 fluxes were generally below detection limit for all treatments. Methane fluxes as high as 3 mg CH_4-C m^{-2} d^{-1} appeared about a week after the imposition of a permanent flood prior to rice cultivation (Figure 2a).

Nitrous oxide fluxes appeared shortly after seeding and 25 and 55 d after seeding of cowpea and wheat, events which coincided with the time of fertilizer application. Fluxes of N_2O fluxes were generally low (mean <2 mg N_2O-N m^{-2} d^{-1}) during the entire season. Plots planted to wheat showed the highest cumulative fluxes (Table 1) since these plots received the highest amount of N (90 kg N ha^{-1}). Nitrous oxide fluxes were generally low during the ensuing fallow period except at 52 d after harvest where N_2O fluxes as high as 9 (Figure 2b) mg N_2O-N m^{-2} d^{-1} appeared after a large rainfall event. Smaller N_2O fluxes also appeared after the imposition of a permanent flood prior to rice transplanting (Figure 2b). These trends of N_2O fluxes were similar to the report of Bronson et al. (1997b) for a rainfed fallow, although the magnitude of the fluxes was lower.

1995 wet season

Methane fluxes appeared shortly after transplanting in all treatments, but the residue-amended plots had higher CH_4 emissions than the unamended plots (Figure 3a). Initially, CH_4 fluxes were higher in cowpea residue-added plots than wheat-residue plots (data not shown). Thereafter, wheat residue-amended plots showed higher CH_4 fluxes. Cowpea had more easily decomposable C than wheat, but more C on a dry-weight basis was added as wheat straw. Wheat straw-amended plots showed the highest cumulative CH_4 fluxes followed by cowpea residue-added plots (Table 1). The maximum cumulative CH_4 flux of 2.6 g CH_4-C m^{-2} with wheat residue was lower than those reported by Bronson et al. (1997a) with similar amounts of straw addition on an irrigated soil. Plots that were weed-free in the previous fallow and those that did not receive any residue had the same magnitude of CH_4 fluxes (Table 1). Stimulation of CH_4 fluxes in rice following organic amendments have been reported extensively (Yagi & Minami, 1990; Sass et al., 1990; Neue et al., 1994).

In the fallow period after the 1995 wet season, there was a rapid decline of CH_4 fluxes after harvest especially with residue-amended plots as CH_4 entrapped in the soil was completely released. Thereafter, CH_4 fluxes remained at a lower level of <10 mg CH_4-C m^{-2} d^{-1} (Figure 3a).

Nitrous oxide fluxes were again low during the rice-growing season except shortly after transplanting and at 65 d after transplanting which corresponded to fertilization applications where N_2O fluxes rose to as much as 2.5 mg N_2O-N m^{-2} d^{-1} (Figure 3b). Fluxes of N_2O continued at a low level (<2 mg N_2O-N m^{-2} d^{-1}) after harvest (Figure 3b). Rainfall events during this fallow period resulted in increased N_2O emission to as high as 8 mg N_2O-N m^{-2} d^{-1} (Figure 3b).

1996 dry season

During the 1996 dry season, CH_4 fluxes were generally insignificant with values ranging from –4 to 4 mg CH_4-C m^{-2} d^{-1} (Figure 4a). Unlike in the previous 1995 dry season, negative CH_4 fluxes were few and small (Table 1). Again, as in the 1995 dry season, cowpea plots without residue added had the highest cumulative CH_4 uptake (–12.6 mg CH_4-C m^{-2} d^{-1}) during the entire fallow period. The reasons for the much lower CH_4 uptake levels in this dry season than in the previous one are not clear, but this was probably related to the less frequent rains. Soil moisture is one of the main controlling factors in CH_4 uptake in rice soils (Singh et al., 1999).

Nitrous oxide fluxes appeared right after seeding for all treatments with residue-amended plots showing the highest N_2O fluxes. Nitrous oxide emissions, however, remained low (<2 mg N_2O-N m^{-2} d^{-1}) during the entire season except during fertilizer application where small (<4 mg N_2O-N m^{-2} d^{-1}) but significant N_2O fluxes appeared. Particularly, after a big rainfall event of >3 cm, a dramatic increase in N_2O fluxes was observed from plots with wheat straw amended in the previous season. Nitrous oxide flux rose to as high as 16 mg N_2O-N m^{-2} d^{-1} in these plots. Similar to the 1995 dry season, plots planted to wheat had the highest seasonal flux of N_2O (Table 1).

Crop yields

Rice grain yields were similar between treatments of a given season (Table 1). Rice yields were very low in dry seasons due to water stress under rainfed conditions (Table 1). Cowpea seed yields were stable at about 1 t ha^{-1} regardless of season or treatment (Table 1). Wheat yields were low as expected in a tropical environment.

Conclusions

The results from this study revealed that positive CH_4 fluxes were evident during the rice-growing season but not during the fallow periods or dry seasons except when the field was subjected to submergence prior to rice transplanting. Addition of residues such as cowpea,

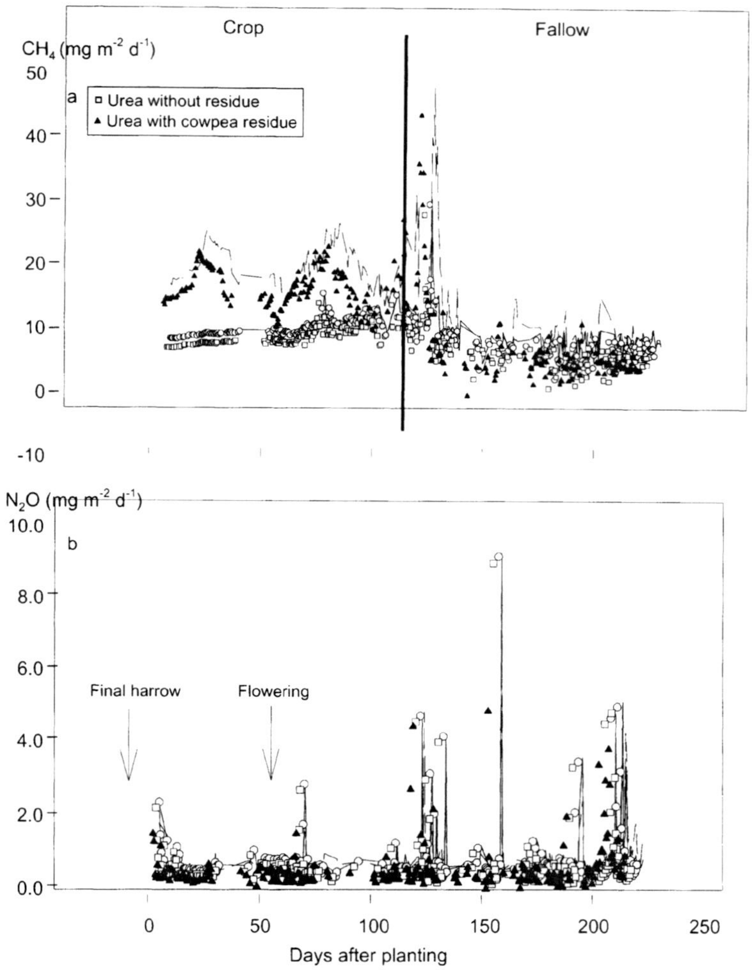

Figure 3. Methane (a) and N_2O (b) fluxes during the 1995 wet season (rice crop and fallow period)

wheat, or rice straw enhanced CH_4 emissions. Methane uptake was observed during the first dry season particularly in cowpea plots, apparently due to the activities of CH_4-oxidizing bacteria. Nitrous oxide fluxes were insignificant during the rice-growing period except after fertilization events where low but significant N_2O peaks were observed. During the fallow periods, larger N_2O fluxes were seen shortly after large rainfalls (>2 cm), apparently due to denitrification of accumulated NO_3. The use of slow-release N fertilizer reduced N_2O emissions, although the emissions from prilled urea were already low. These findings in rainfed rice-upland crop systems are similar to our previous studies in irrigated double-cropped rice fields, with the important exception that these rainfed studies showed lower CH_4 and N_2O emissions and some CH_4 uptake.

Acknowledgements

The research described in this paper was funded by the UNDP-Global Environment Facility (GLO/91/G31 under agreement with the International Rice Research Institute and the International Fertilizer Development Center (IFDC).

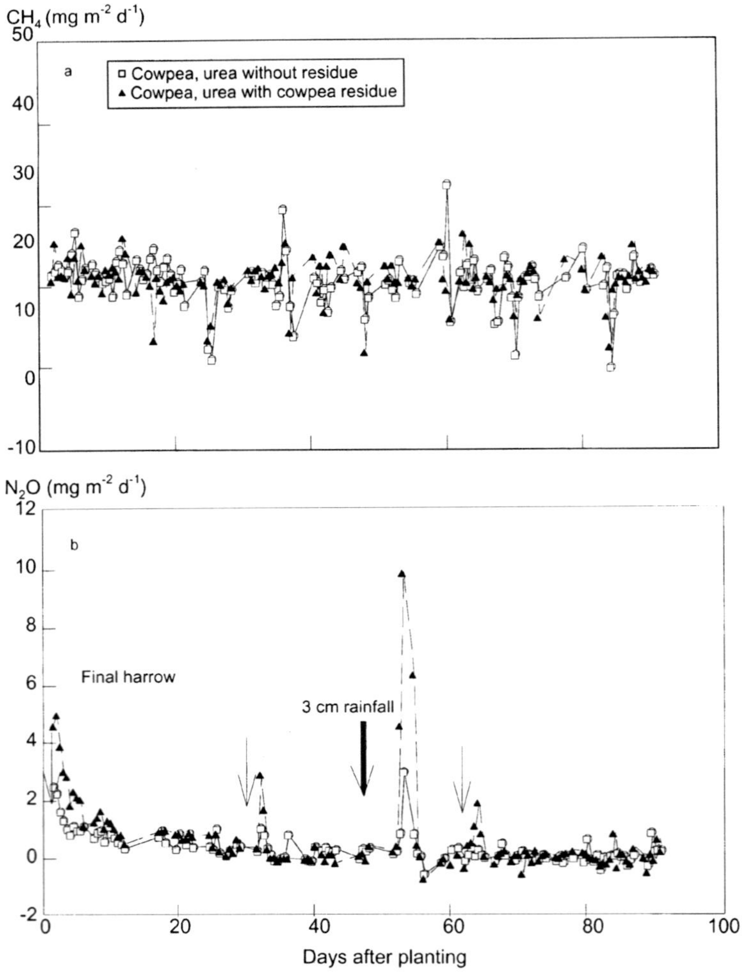

Figure 4. Methane (a) and N_2O (b) fluxes during the 1996 dry season (cowpea crop and fallow period)

References

Aulakh MS, Doran JW, Walters DT, Mosier AR & Francis DD (1991) Crop residue type and placement effects on denitrification and mineralization. Soil Sci Soc Am J 55:1010-1025

Bronson KF & Mosier AR (1993) Nitrous oxide emissions and methane consumption in wheat and corn-cropped systems in northeastern Colorado. p 133-144. In: L. Harper (ed) Agroecosystem Effects on Trace Gases and Global Climate Change. American Society of Agronomy, Madison, Wisconsin

Bronson KF & Mosier AR (1994) Suppression of methane oxidation in aerobic soil by nitrogen fertilizers, nitrification inhibitors, and urease inhibitors. Biol Fertil Soils 17:263-268

Bronson KF, Neue H, Singh U & Abao Jr, E (1997a) Automated chamber measurements of methane and nitrous oxide flux in a flooded rice soil: I. Residue, nitrogen, and water management. Soil Soc Am J 61:981-987

Bronson KF, Neue H, Singh U & Abao Jr, E (1997b) Automated chamber measurements of methane and nitrous oxide flux in a flooded rice soil: II. Fallow period emissions. J Soil Sci Soc Am 61:988-993

Buresh RJ, Woodhead T, Shepherd KD, Flordelis E & Cabangon R (1989) Nitrate accumulation and loss in mungbean lowland rice cropping system. Soil Sci Soc Am J 53:477-482

Cai ZC, Xing GX, Yan XY, Xu H, Tsuruta H, Yagi K & Minami K (1997) Methane and nitrous oxide emissions from rice paddy fields as affected by nitrogen fertilizers and water management. Plant Soil 196:7-14

Delgado JA & Mosier AR (1996) Mitigation alternatives to decrease nitrous oxide emissions and urea-nitrogen loss and their effect on methane flux. J Environ Qual 25:1105-1111

Eichner MJ (1990) Nitrous oxide emissions from fertilized soils: Summary of available data. J Environ Qual 19:272-280

George T, Ladha JK, Buresh RJ & Garrity DP (1993) Nitrate dynamics during the aerobic phase in lowland rice-based cropping systems. Soil Sci Soc Am J 57:1526-1532

Granli T & Bockman OC (1994) Nitrous oxide from agriculture. Norwegian J Agric Sci ( ):7-16

IPCC—Intergovernmental Panel on Climate Change (1992) Climate change. The supplementary report to the IPCC scientific assessment. Cambridge University Press, New York.

IRRI—International Rice Research Institute (1998) IRRI Rice Almanac, (2nd ed) Manila, Philippines, 181 p

Keller M, Mitre ME & Stallard RF (1990) Consumption of atmospheric methane in soils of Central Panama: effects of agricultural development. Global Biogeochem Cycles 4:21-27

Lidstrom ME & Stirling DI (1990) Methanotrophs: genetics and commercial applications. Annu Rev Microbiol 44:27-58

Minami K (1994) Effect of nitrification inhibitors and slow-release fertilizer on emissions of nitrous oxide from fertilized soils. pp 187-196. In: CH_4 and N_2O: Global Emissions and Controls from Rice Fields and Other Agricultural and Industrial Sources, National Institute of Agro-Environmental Sciences, Tsukuba, Japan

Neue, H, Lantin RS, Wassmann R, Aduna JB, Alberto CR & Andales MJF (1994) Methane emission from rice soils off the Philippines. pp 55-63. In: K. Minami, A. Mosier, & R. Sass (eds) CH_4 and N_2O - Global Emissions and Controls from Rice Fields and Other Agricultural and Industrial Sources. National Institute of Agro-Environmental Sciences, Tsukuba, Japan

Rodhe H (1990) A comparison of the contribution of various gases to the greenhouse effect. Science 248:1217-1219

SAS Institute Inc (1987) SAS user's guide: Statistics. Cary, NC

Sass R, Fischer FM, Wang YB, Turner FT & Jund MF (1990) Methane emission from rice fields as influenced by solar radiation, temperature, and straw incorporation. Global Biogeochem Cycles 5:335-350

Singh JS, Raghubanshi AS, Reddy VS, Singh S & Kashyap AK (1998). Methane flux from irrigated paddy and dryland rice fields, and from seasonally dry tropical forest and savanna soils of India. Soil Biol Biochem 30:135-139

Singh S, Singh JS & Kashyap AK (1999) Methane consumption by soils of dryland rice agriculture: Influence of varieties and N-fertilization. Chemosphere 38:175-189

Tripathi BP, Ladha JK, Timsina J & Pascua SR (1997) Nitrogen dynamics and balance in intensified rainfed lowland rice-based cropping systems. Soil Sci Soc Am J 61:812-821

Yagi K & Minami K (1990) Effect of organic matter application on methane emission from some Japanese rice fields. Soil Sci Plant Nutr 36:599-610

Nutrient Cycling in Agroecosystems **58:** 141–159, 2000.

Using a crop/soil simulation model and GIS techniques to assess methane emissions from rice fields in Asia. I. Model development

R.B. Matthews[1], R. Wassmann[2,3] & J. Arah[2,4]

[1]*Institute of Water and Environment, Cranfield University, Silsoe, Bedfordshire MK45 4DT, United Kingdom;* [2]*International Rice Research Institute, MCPO Box 3127, Makati 1271, Philippines;* [3]*Fraunhofer Institute for Atmospheric Environmental Research, Garmisch-Partenkirchen, Germany;* [4]*AAT Consultants, 15 Clerk Street, Edinburgh EH8 9JH, United Kingdom*

Key words: methane, rice, *Oryza sativa*, anaerobic, model, simulation, carbon dynamics

Abstract

The development of the MERES (*M*ethane *E*missions in *R*ice *E*co*S*ystems) model for simulating methane (CH_4) emissions from rice fields is described. The CERES-Rice crop simulation model was used as a basis, employing the existing routines simulating soil organic matter (SOM) decomposition to predict the amount of substrate available for methanogenesis. This was linked to an existing submodel, described elsewhere in this volume (Arah & Kirk, 2000), which calculates steady-state fluxes and concentrations of CH_4 and O_2 in flooded soils. Extra routines were also incorporated to simulate the influence of the combined pool of alternative electron acceptors in the soil (i.e., NO_3^-, Mn^{4+}, Fe^{3+}, SO_4^{2-}) on CH_4 production. The rate of substrate supply is calculated in the SOM routines of the CERES-Rice model from (a) the rate of decomposition of soil organic material including that left from the previous crop and any additions of organic matter, (b) root exudates (modified from the original CERES-Rice model using recent laboratory data), and (c) the decomposition of dead roots from the current crop. A fraction of this rate of substrate supply, determined by the concentration of the oxidized form of the alternative electron acceptor pool, is converted to CO_2 by bacteria which outcompete the methanogenic bacteria, thereby suppressing CH_4 production. Any remaining fraction of the substrate supply rate is assumed to be potentially available for methanogenesis. The CH_4 dynamics submodel uses this potential methanogenesis rate, along with a description of the root length distribution in the soil profile supplied by the crop model, to calculate the steady-state concentrations and fluxes of O_2 and CH_4. The reduced form of the alternative electron acceptor pool is allowed to reoxidize when soil pores fill with air if the field is drained. The MERES model was able to explain well the seasonal patterns of CH_4 emissions in an experiment involving mid- and end-season drainage and additions of organic material at IRRI in the Philippines.

Introduction

Methane (CH_4) is one of the principal greenhouse gases and has been estimated to account for 15-20% of current radiative forcing (Bouwman, 1991). Rice field soils, characterized by O_2 depletion, high moisture, and relatively high organic substrate levels, offer an ideal environment for the activity of methanogenic bacteria, and are one of the major anthropogenic CH_4 sources. Global emission estimates for this source range from 20 to 100 Tg yr^{-1} (Sass & Fisher, 1997), which may be 4-30% of the total anthropogenic contribution to the atmosphere, making it one of the CH_4 sources with the largest uncertainty. Precise estimates have been difficult due to the large spatial and temporal variability in CH_4 measured at different sites due to differences in climate, soil properties, duration and pattern of flooding, rice cultivars and crop growth, organic amendments, fertilization, and cultural practices. Spatial information on these factors along with mechanistic modeling of CH_4 fluxes would help to improve these estimates, but the use of geographical information systems coupled with ecosystem models has so far been limited (e.g., Bachelet & Neue, 1993).

It has been estimated that rice production must almost double by the year 2020 in order to meet the demand of an increased population, which may *increase* CH_4 production by up to 50% (Bouwman, 1991). However, the Intergovernmental Panel on Climate Change has recommended immediate *reductions* of 8% in anthropogenic emissions of CH_4 to stabilize atmospheric concentrations at current levels (IPCC, 1996). The only feasible way in which these two opposing requirements can be met are by using crop management practices that reduce CH_4 emissions without affecting crop yields. Manipulation of some or all of the factors causing variability in CH_4 emission rates mentioned above may offer a way in which this reduction target is met.

To address these issues, a multinational project, coordinated by the International Rice Research Institute (IRRI) in collaboration with selected national agricultural research systems in major rice-growing countries of Asia, was established in 1993. The aims of the project were (a) to provide more accurate estimates of CH_4 emission rates and (b) to develop strategies that would mitigate CH_4 emissions from rice fields without sacrificing crop yields. Experimental data on CH_4 emissions and the factors influencing them were collected from eight sites in five Asian countries, namely India, China, Indonesia, Thailand, and the Philippines. An important part of the project was the use of these experimental data to develop a simulation model describing the processes involved in CH_4 emission. This model could then be used, together with databases of weather, soils, and crop management, to provide estimates of current CH_4 emissions and to evaluate potential mitigation strategies.

This modeling component of the project is summarized in this series of papers. In this first paper, the development of the process-based simulation model is described. Subsequent papers in the series describe validation and sensitivity analysis of the model (Matthews et al., 2000a), the databases used (Knox et al., 2000), and the extrapolation of the experimental data to the national and regional levels (Matthews et al., 2000b).

Previous CH_4 models

A number of models have been developed in recent years to predict the rate of emission of CH_4 from rice fields. Early models used regression relationships between rates of emission and either the crop biomass (e.g., Aselmann & Crutzen, 1990; Taylor et al., 1991; Bachelet & Neue, 1993; Bachelet et al., 1995; Kern et al., 1997) or grain yield (e.g., Anastasi et al., 1992). These relationships were based on the assumption that the higher the biomass production of the crop, the more substrate would be available for CH_4 production, either from increased crop residues or from higher rates of rhizodeposition.

As our knowledge of the processes involved in CH_4 emission from flooded soils has increased, however, subsequent models have gradually replaced this empiricism with more mechanistic descriptions. For example, Nouchi et al. (1994) describe a model in which CH_4 emissions for the first 73 d of the crop are calculated using a function dependent on the leaf area of the rice crop, the concentration of dissolved CH_4 in the soil water, and a constant crop conductance calculated from previous work. After 73 d, crop conductance was expressed as an empirical function of temperature.

Cao et al. (1995) present a more mechanistic model describing CH_4 production and oxidation in rice fields. In this model, soil organic carbon was assumed to be partitioned between three main pools based on their rate of decomposition. The carbon in these pools was assumed to be released by decay according to first-order reactions, which, together with that released from the growing rice plants as root exudates and dead root tissue, was available as substrate for methanogens. Methane production was calculated as a function of this substrate, modified by factors accounting for the influence of the soil redox potential (Eh), pH, temperature, floodwater depth, and addition of mineral fertilizers. The seasonal pattern of Eh was a required input of the model. The fraction of the CH_4 produced that was oxidized by methanotrophs was calculated using an empirical function based on the dry matter of the crop. Methane emission rate was then calculated as the difference between rate of production and rate of oxidation.

Huang et al. (1998) used two pools in their model to represent soil organic matter, with different potential decomposition rates for each; these could be modified by multipliers representing the influence of soil texture and temperature. Variations in soil water content were not accounted for. Rhizodeposition rate was calculated as a function of aboveground biomass on a given day, account being also taken of varietal differences and soil texture effects. The amount of substrate available for CH_4 production was taken as the sum of that from SOM decomposition and rhizodeposition. As with the Cao et al. (1995) model, CH_4 production was affected directly by soil Eh, although this was simulated by a negative power function rather than as a model input. The fraction of CH_4 produced that is oxidized by methanotrophs was described by an empirical function related to the aboveground biomass of the crop as a proportion of the maximum biomass reached at the end of the season. Aboveground biomass was calculated with a logistic growth equation whose parameters were empirically related to final grain yield.

Other approaches have focused on individual processes involved in CH_4 emissions. Lu et al. (2000) developed a model for CH_4 production derived from incubation studies, while Cai et al. (1996) present a model describing CH_4 oxidation in incubations. The mechanistic basis for modeling concentrations and fluxes of O_2 and CH_4 in real systems was set out by Lassiter & Plis (1994), and was implemented first for peat lands (e.g. Walter et al., 1996; Arah & Stephen, 1998). This last model was subsequently developed further to describe profiles of CH_4 and O_2 in rice soils (Arah & Kirk, 2000), a version of which is also used in the work we describe in this paper.

None of these models just described explicitly simulate the effect of the alternative electron acceptors in the soil (i.e. NO_3^-, Mn^{4+}, Fe^{3+}, and SO_4^{2-} ions), the quantity of which strongly influence the time CH_4 production begins after initial flooding. In the models of Cao et al. (1995) and Huang et al. (1998), for example, this was accounted for by the pattern of decline of Eh. However, Eh is a difficult variable to simulate for reasons discussed by van Bodegom et al. (2000), who then make an attempt to simulate directly the behavior of these alternative electron acceptors and their influence on CH_4 production in rice soils by considering the effects of NO_3^-, Mn^{4+}, Fe^{3+}, and SO_4^{2-} ions separately.

Although all of these models marked major steps forward in the simulation of CH_4 dynamics in rice soils, they all have limitations in some way or another. For example, in most, the growth of the crop is described in an empirical way, limiting their ability to describe the effects of various management practices on both crop performance and substrate C supply from the crop. Similarly, not all can simulate the processes involved in the intermittent draining and reflooding of rice fields, while in several, CH_4 oxidation rate is calculated as a fraction dependent on crop status rather than as a function of the processes involved. There is a clear need, therefore, to bring together into one model routines describing

- crop growth and rhizodeposition over the season;
- soil organic matter decomposition under anaerobic conditions;
- the effect of alternative electron acceptors in the soil such as NO_3^-, Mn^{4+}, Fe^{3+}, and SO_4^{2-} ions;
- a mechanistic description of CH_4 oxidation and fluxes of CH_4 from the soil; and
- the influence of crop management practices such as water management and application of organic and inorganic fertilizers.

We have attempted to do this by using the CERES-Rice crop simulation model (Ritchie et al., 1998) as a basis (Figure 1). The advantage of using this model is that it already includes soil organic matter decomposition routines, along with routines describing the relevant crop management options such as water management and applications of organic and inorganic fertilizers, allowing us to evaluate the effects of varying any of these on both crop yields and CH_4 emissions. We have incorporated a subroutine describing the effect of alternative electron acceptors on CH_4 production and have linked all these to the model of Arah & Kirk (2000) describing the interaction between CH_4 and O_2 in the soil. We have also improved the calculations of root exudation with data recently obtained from laboratory experiments.

Methods

Much of the development of the model described in this paper is derived from data from an experiment carried out at IRRI in the 1997 dry season, referred to hereafter as the IRRI-1997DS experiment. A full description of the experimental methodology is given by Wassmann et al. (2000), but a brief summary is included here for convenience. Relevant treatments were (a) no organic amendments, (b) 10 t dry matter (DM) ha^{-1} of rice straw was added to the field 14 d before planting, and (c) 3 t DM ha^{-1} of green manure was added 14 d before planting. The plots were planted with IR72 on 10 January and harvested on 20 April 1997. All treatments received 150 kg N ha^{-1} as urea in addition to the organic amendments. Stubble from the previous crop was cut to ground level and removed before plowing, but any dead root material was left remaining in the soil. All plots were drained in the middle of the season from 23-40 d after planting (DAP) (i.e., for 17 d) and again at the end of the season from 85 DAP until harvest (for 15 d). Bihourly measurements of CH_4 emitted from each treatment were made using the methodology described by Wassmann et al. (2000) and integrated to give daily emission rates.

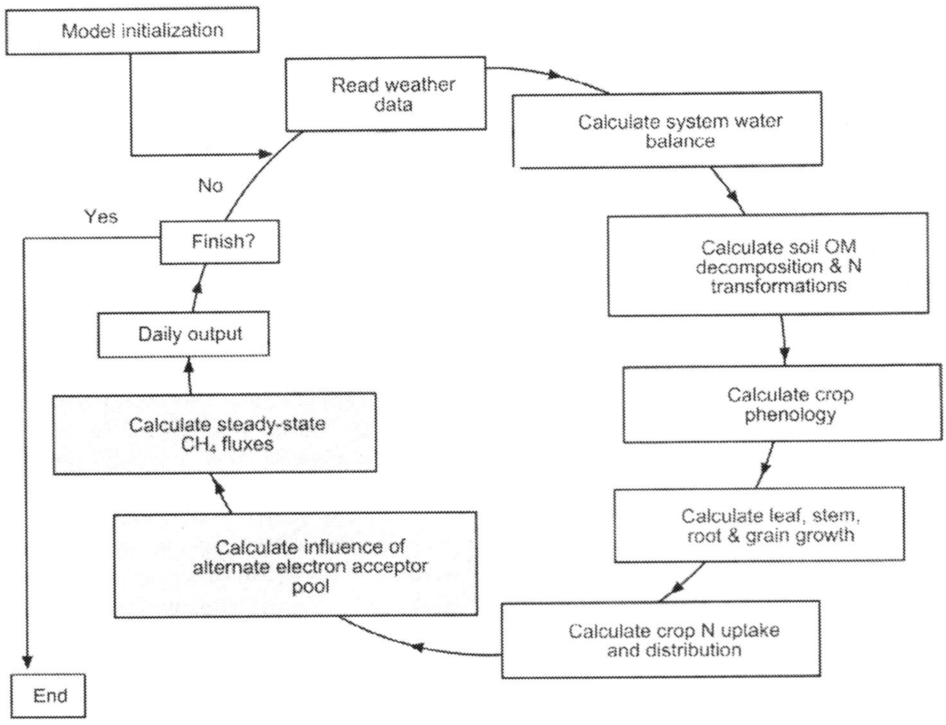

Figure 1. Schematic diagram of the structure of the MERES crop/soil model. Components on the loop are calculated on a daily basis. Shaded components represent the modifications made to the basic CERES-Rice model to take into account the effect of the alternative electron acceptor pool and the calculation of the steady-state fluxes and concentrations of CH_4 and O_2. Root death and exudation rates are calculated in the root growth routines. Water management options are executed in the water balance calculations and organic and inorganic fertilizer management in the OM and N dynamic routines

Model development

Background

Methanogenesis is the last stage in the mineralization of organic matter under anaerobic conditions. Carbon as a substrate for methanogenic microorganisms is assumed to come from three sources: the decay of organic matter (both freshly added and humus), the death of root tissue from the crop, and carbohydrate exudates from living root tissue. Depending on the pathway followed, the breakdown of organic matter (CH_2O) can result in the production of H_2 and CO_2 or acetate (CH_3COO^-) (Conrad, 1989). *Methanogenic* bacteria can then produce CH_4 either from the H_2 and CO_2 (i.e., $CO_2 + 4\ H_2 \rightarrow CH_4 + 2\ H_2O$), or from the acetate (i.e., $CH_3COO^- + H^+ \rightarrow CO_2 + CH_4$). Whichever route is followed, the summary reaction can be written as

$$2(CH_2O) \rightarrow CO_2 + CH_4 \qquad (1)$$

Thus, a maximum of 50% of the carbon present in organic matter can be converted to CH_4, a value which has been confirmed by laboratory measurements (Tsutsuki & Ponnamperuma, 1987). The actual amount depends on the soil pH, temperature, and the presence in the soil of other ions (e.g., NO_3^-, Fe^{3+}, Mn^{4+}, SO_4^{2-}) which can act as electron acceptors for microbial respiration, resulting in the production of CO_2 rather than CH_4.

A certain proportion of the CH_4 that is produced can be oxidized to CO_2 by *methanotrophic* bacteria if it happens to pass through an O_2-rich environment:

$$CH_4 + 2\ O_2 \rightarrow CO_2 + 2\ H_2O \qquad (2)$$

Such environments may occur in the thin layer of topsoil interfacing with the floodwater, and in the rhizosphere where CH_4 and O_2 gradients overlap due to diffusion of O_2 from the atmosphere down through the aerenchyma to the roots. The rates of diffusion of

O_2 downward and of CH_4 upward through the plant and floodwater are dependent on the concentration gradients of the respective gases between atmosphere and soil and the conductance of the routes followed.

Simulation of all these processes, therefore, requires calculations of (a) the rate of production of substrate available for methanogenesis, (b) the rate of production of CH_4 from this substrate, (c) the rate of oxidation of this CH_4, and (d) the rates of CH_4 flux from soil to atmosphere through the rice plant, ebullition, or diffusion through the floodwater. For (a), we have made use of the routines in the CERES-Rice crop model describing root death, root exudation, and organic matter decomposition, and have added new routines describing the effect of the alternative electron acceptors on CH_4 production. For (b), (c), and (d), we have used the model simulating the steady state concentrations and fluxes of CH_4 and O_2 described by Arah & Kirk (2000). Each of these components is described in more detail below.

The CERES-Rice crop simulation model

CERES-Rice (Ritchie et al., 1998) is a process-based, management-oriented model simulating the growth and development of rice. We decided to use it as it has been relatively well tested in a range of environments (e.g., Bachelet et al., 1993) and already has routines describing the main crop components involved in CH_4 dynamics, i.e., organic matter decomposition, root growth and death, and root exudation. Interestingly, Cao et al. (1995) use the CERES-Rice model in their approach, but only for the calculation of crop dry matter production for estimating rhizodeposition and the fraction of CH_4 oxidized —the existing routines in CERES describing organic matter decomposition and root processes were not used.

A full description of the CERES-Rice model is given by Ritchie et al. (1998). Briefly, the model operates on a daily time-step (Figure 1) and calculates biomass production, which is then partitioned to the leaves, stems, roots and grain, depending on the phenological stage of the plant. Submodels calculate the water balance and N transformations in the soil, and crop uptake of water and N. Under fully irrigated conditions, the height of the surrounding bund and the initial floodwater depth can be specified—subsequent floodwater depth is simulated taking into account inputs from rainfall or irrigation and losses from evapotranspiration, percolation, and runoff over the bund. In the N submodel, mineralization of N is linked to the routines describing the decomposition of organic matter, described in more detail in the next section. The soil profile is characterized by its initial organic matter and N content, water-holding properties, and texture. Differences between genotypes are accounted for through the use of a set of coefficients specific to each genotype. The user is able to specify various crop management options such as sowing and/or planting dates, water management (e.g., dates and amounts of irrigations), fertilizer management (dates, amounts, incorporation depth, and types of fertilizers applied), organic matter management (dates, amounts, incorporation depth, and types of organic amendments applied), and crop harvest dates.

Decomposition of soil organic matter

The approach used in the CERES-Rice model to simulate soil carbon dynamics (Godwin & Jones, 1991) is to assume two types of organic matter—these are the fresh organic matter (FOM) pool, which includes crop residues and green manure, and a more stable organic or humic pool (HUM). The FOM pool is further divided into three arbitrary pools corresponding approximately to the carbohydrate, cellulose, and lignin fractions. It is assumed that initially any fresh organic matter is distributed as 20% carbohydrate, 70% cellulose, and 10% lignin. The model requires as input data the amount of straw added, its C/N, and its depth of incorporation, along with an estimate of the amount of root residue from the previous crop, all of which are used to initialize the FOM pools. Initialization of the HUM pool is calculated from the soil organic carbon as specified in the soil data file.

Each of the three FOM pools is assumed to have a different potential relative rate of decay—under nonlimiting conditions, the decay constants ($R_{p(max)}$, d^{-1}) as reported by Seligman & van Keulen (1981) are 0.2, 0.05 and 0.0095 d^{-1} for pool p ($p \in$ (carbohydrate, cellulose, and lignin)) respectively. However, these potential relative rates of decay are usually limited by soil temperature, soil moisture, and the C/N of the decaying material. Thus, actual decay rates (dO_p/dt, kg C ha^{-1} d^{-1}) are calculated as

$$dO_p/dt = O_p R_{p(max)} \cdot f(T_s) \cdot g(\theta_s) \cdot h(\kappa_s) \tag{3}$$

where O_p (kg ha^{-1}) is the amount of organic matter remaining in the pool p on the day in question, and $f(T_s)$, $g(\theta_s)$ and $h(\kappa_s)$ are dimensionless multipliers for soil temperature (T_s, °C), soil moisture (θ_s, m^3 water m^{-3} soil),

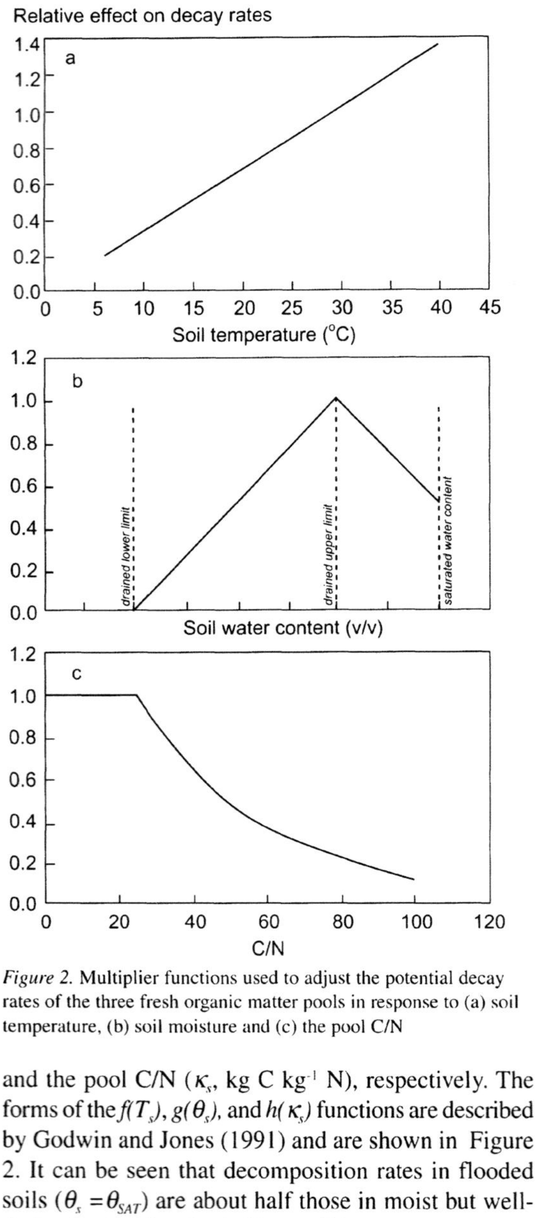

Figure 2. Multiplier functions used to adjust the potential decay rates of the three fresh organic matter pools in response to (a) soil temperature, (b) soil moisture and (c) the pool C/N

and the pool C/N (κ_s, kg C kg^{-1} N), respectively. The forms of the $f(T_s)$, $g(\theta_s)$, and $h(\kappa_s)$ functions are described by Godwin and Jones (1991) and are shown in Figure 2. It can be seen that decomposition rates in flooded soils ($\theta_s = \theta_{SAT}$) are about half those in moist but well-drained soils ($\theta_s = \theta_{DUL}$) (Figure 2b).

A similar procedure is used to estimate the rate of decay of the humus pool (dO_H/dt, kg C ha^{-1} d^{-1}), except that the pool C/N multiplier κ_s is not used and the potential relative rate of decomposition ($R_{H(max)}$, d^{-1}) is much slower, with a value of 0.000085 d^{-1}. The total amount of carbon released by decay of organic matter on a given day (R_{Cdecay}, kg C ha^{-1} d^{-1}), and therefore available as substrate for methanogenic microorganisms, is the sum of the decay rates of the individual pools:

$$R_{Cdecay} = dO_H/dt + \sum_{p=1}^{p=3} (dO_p/dt) \tag{4}$$

Rhizodeposition: root exudates and root death

The contribution to CH_4 production of organic matter originating from living rice plants through root exudates and root death, collectively referred to as rhizodeposition, was first recognized by Seiler et al. (1984). The peak in emission rates commonly observed toward the end of the growing season was ascribed by these authors to be due to the increase in decaying root tissue or root exudates after flowering. Watanabe and Roger (1985) suggest that the amount of carbon released by rhizodeposition over a growing season can exceed that contained in the root biomass by a factor of four. Cao et al. (1995) refer to a number of studies on annual crops indicating that rhizodeposition accounts for 35-60% of carbon transferred to roots.

In the model of Cao et al. (1995), both of these sources were treated as one. As a way to understanding the underlying processes, we have attempted to treat them separately.

Root exudates. Root exudates contain high-molecular-weight substances such as mucilage and ectoenzymes, and low-molecular-weight substances (LWS) such as organic acids, phenols, and amino acids. The total amount of carbon exuded has been shown to be closely related to root dry weight (r^2 = 0.919) and aboveground DM production (r^2 = 0.954) (Wang et al., 1997). We have used data from Lu et al. (1999) to estimate the rate of exudation of organic compounds per unit of root biomass (Figure 3). The relationship was relatively linear at about 0.6 mg C (g root)$^{-1}$ d^{-1} up until the time of flowering, after which it increased to an average of 1.6 mg C (g root)$^{-1}$ d^{-1}. The rate of root exudation (g C m^{-2} d^{-1}), therefore, is calculated as the product of these values (depending on the crop growth stage) and the root weight in each soil layer which is simulated in another part of the CERES-Rice model.

Root death. Very little information exists on the rate of root death in rice. Root death in the CERES-Rice model is assumed to be a constant 2% of existing root dry weight (W_{root}, kg DM ha^{-1}) in each soil layer per day, i.e.,

$$R_{Croots} = 0.4 \times 0.02 \times W_{root} \tag{5}$$

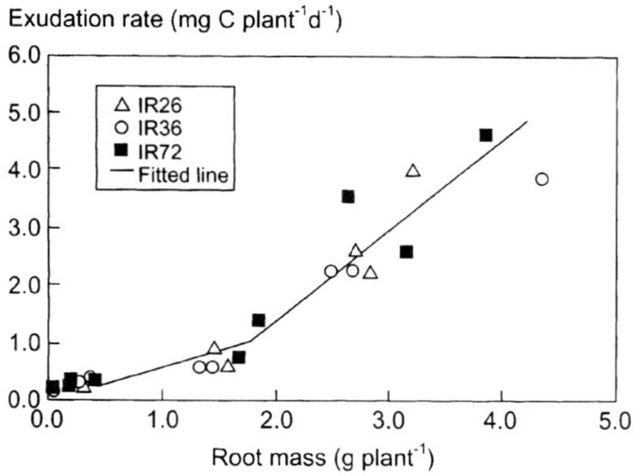

Figure 3. Relationship between exudation rate and root mass in three rice cultivars, IR26, IR36, and IR72 (See Lu et al. [1998] for details)

Dry matter lost from the plant in this way is assumed to enter the FOM pools with the same proportion of 0.2:0.7:0.1 allocated to the carbohydrate, cellulose, and lignin pools as described previously. The carbon in this DM therefore becomes available as substrate for methanogens according to the decomposition rate of each of the three pools described in the previous section.

Using these parameters for root exudation and death gave total rhizodeposition figures of about 18% of the aboveground biomass at final harvest, toward the top end of the range of 5-20% obtained by Shamoot et al. (1968) in a greenhouse study with 11 plant species (but not including rice). Our value also agrees closely with the 17% predicted by the model of Huang et al. (1998) after 110 d.

The effect of alternative electron acceptors on CH_4 production

As long as O_2 is present in the soil, it acts as the sole electron acceptor for microbial respiration. However, after a rice soil is flooded, O_2 dissolved in the floodwater and soil is consumed rapidly. The need for electron acceptors by anaerobic organisms results in the reduction of a number of other oxidized species of ions in the soil. Reductions of NO_3^- to NO_2^-, N_2O to N_2, Mn^{4+} to Mn^{2+}, Fe^{3+} to Fe^{2+}, SO_4^{2-} to S^{2-} all resulting in the production of CO_2, and finally CO_2 to CH_4, occur sequentially, provided available carbon sources exist (Patrick Jr & Delaune, 1977). Thus, CH_4 production will not occur until most of the NO_3^-, N_2O, Mn^{4+}, Fe^{3+}, and SO_4^{2-} ions in the soil have been reduced. Methanogenesis is thought to be inhibited by the presence of alternative electron acceptors because bacteria using these electron acceptors outcompete methanogens for substrate. This is particularly so in the case of O_2 where aerobic bacteria are able to maintain concentrations of catabolic substances so low that methanogens cannot compete.

The models of Cao et al. (1995) and Huang et al. (1998) accounted for this sequence of events by relating CH_4 production rate to the soil Eh—CH_4 production was switched on at an Eh value of –200 mV and switched off at values above this. In both models, Eh was described statically, being required as a model input in the first case, and as a negative power function in the second. The rate of Eh decline after flooding at the start of the season, however, is dependent on several factors, including the type of soil (i.e., the quantity of alternative electron acceptors present), and the amount of fresh organic matter present at the start of the season (e.g., from previous crop residues and organic amendments). It is difficult to simulate Eh using these factors (van Bodegom et al., 2000), and so we decided to take a different approach.

We have assumed the presence in the soil of a pool of alternative electron acceptors in oxidized form (AEA_{ox}), which reacts with the substrate C from decomposition to form CO_2, becoming reduced in the process (AEA_{red}). For simplicity, we have not differentiated between any of the species of ions and have specified the quantity of *AEA* present in moles of C equivalents m^{-3}, assuming a 1:1 stoichiometric relationship between substrate C and the *AEA*, i.e.,

$$(CH_2O) + AEA_{ox} \rightarrow CO_2 + AEA_{red} + 2\,H^+ \qquad (6)$$

Methane production can occur when there are still some alternative electron acceptors remaining in the bulk soil, for various reasons. While methanogenic bacteria are generally outcompeted completely by the nitrate-, iron- and manganese-reducing bacteria, they are only partially so by the sulfate-reducing bacteria (e.g., van Bodegom et al., 2000). Similarly, heterogeneity in the soil system (i.e., in microenvironments which have exhausted all their electron acceptors), and the ability of methanogens to use specific organic substrate molecules (e.g., methyl amides, methyl sulfides) not used by the other bacteria can also result in CH_4 production before the AEA_{ox} are completely reduced.

Deriving the relationship between CH_4 production and the concentration of the alternative electron acceptor pool in the soil ($[AEA_{ox}]$, mol C_{eq} m^{-3}) is not a

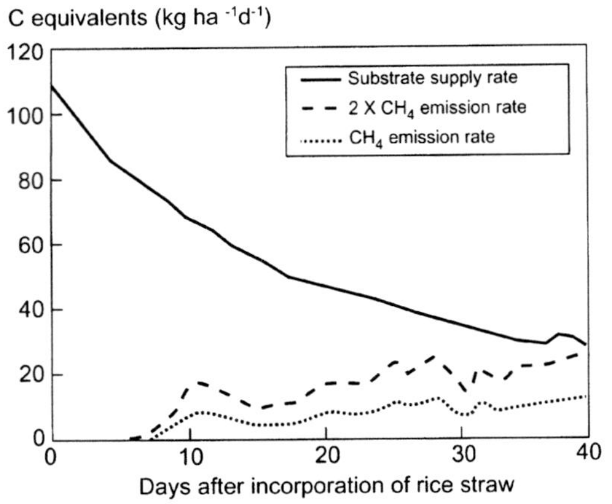

Figure 4. Estimation of the alternative electron acceptor pool size in the IRRI soil. Substrate supply rate (solid line) is that predicted by the organic matter decomposition routines of the CERES-Rice model after addition of 10 t ha^{-1} rice straw, CH_4 emission rate (dotted line) is that measured in the rice-straw treatment of the IRRI-1997DS experiment. The dashed line represents the amount of carbon that is *not* oxidized by the AEA_{ox} pool (i.e. produced according to Equation 1). The area between the solid line and the dashed line represents the size of the AEA_{ox} pool in kg (C equivalents) ha^{-1}

straightforward task due to a paucity of measured data. However, a first approximation can be made from measured CH_4 emissions when a large amount of organic matter is applied to the field such that most of its decomposition occurs before there is appreciable oxidation of CH_4 produced as a result of O_2 being introduced through rice plant aerenchyma. Such a dataset is provided by the IRRI-1997DS experiment described above.

The first step is to estimate the initial concentration of the AEA_{ox} pool of the IRRI soil. Assuming a carbon content of 0.31 kg C (kg DM)$^{-1}$, 10 t DM ha^{-1} of rice straw represents the addition of 3,100 kg C ha^{-1}. Figure 4 shows the rate of substrate production (kg C ha^{-1} d^{-1}) predicted by the organic matter decomposition routine of the CERES-Rice model (solid line) for this treatment. Assuming that for every mole of carbon released as CH_4, there must have also been one mole of carbon released as CO_2 (equation 1) and assuming that there is negligible reoxidation of CH_4 produced, we can calculate that the amount of carbon <u>not</u> reacting with the AEA_{ox} pool is twice the measured CH_4-C emission rate, and by difference, the amount reacting with the AEA_{ox} pool can be calculated. The point at which these two curves intersect represents the stage at which the AEA_{ox} pool has been completely converted into a reduced form (i.e., all the substrate is then being metabolized into CO_2 and CH_4 according to equation 1). This appears to be about 39 d after incorporation of the rice straw, or about 25 d after planting (Figure 4). Integrating the amount of carbon that was calculated to have reacted with the AEA_{ox} pool from the date of incorporation until 39 d later (the area between the solid and dashed lines in Figure 4) gives a value of 1,590 kg C ha^{-1}, which can be taken as the size of the AEA_{ox} pool in C equivalents (C_{eq}). Converting this to a mean concentration, assuming a soil depth of 50 cm, gives a value of around 26.5 mol C_{eq} m^{-3}.

Knowing the initial size of the AEA_{ox} pool, it is then possible to calculate its size as a function of time between the date of rice straw incorporation and 39 d later, by reiteratively subtracting the amount of carbon reacting with the AEA_{ox} pool each day (the difference between the solid and dashed lines in Figure 4) from the size of the pool on the preceding day. The same procedure was also followed for the treatments with no organic amendments and with 3 t ha^{-1} green manure added, assuming the initial AEA_{ox} pool size of 1,590 kg C ha^{-1} calculated from the rice straw treatment. The measured rates of CH_4 emissions on each day during this period can then be plotted against the size of the AEA_{ox} pool (converted to a concentration (mol C_{eq} m^{-3})) for each treatment (Figure 5). The relationship shows clearly that as the concentration of the AEA_{ox} pool declines, CH_4 production increases. A regression line can be fitted through the data with the equation $y = 0.2\ [1.0 - x/24.0]$ ($r = 0.87$, $n = 120$, $P < 0.01$). We have, therefore, assumed a two-stage process in the relationship between potential CH_4 production (P_{CH4}*, mol C m^{-3} s^{-1}) and $[AEA_{ox}]$ (mol C_{eq} m^{-3}):

$$\begin{array}{lll} \text{for} & [AEA_{ox}] > [AEA_{ox}]^* & P_{CH4}^* = 0.0 \\ & [AEA_{ox}]^* > [AEA_{ox}] > 0.0 & P_{CH4}^* = \min(0.2 * (1 - [AEA_{ox}]/[AEA_{ox}]^*), S) \\ & [AEA_{ox}] = 0.0 & P_{CH4}^* = S \end{array} \tag{7}$$

where $[AEA_{ox}]^*$ is the critical concentration of the oxidized alternative electron acceptor pool (mol C_{eq} m^{-3}) above which no CH_4 production occurs (taken as 24.0 mol C_{eq} m^{-3} from the line in Figure 5), and *S* is the rate of substrate-C production (mol C_{eq} m^{-3} s^{-1}). The rate of change of the oxidized alternative electron acceptor pool ($d[AEA_{ox}]/dt$, mol C_{eq} m^{-3} s^{-1}) is given by

$$d[AEA_{ox}]/dt = S - 2.0 \times P_{CH4}^* \tag{8}$$

and the rate of change of the reduced form of the alternative electron acceptor pool ($d[AEA_{red}]/dt$, mol C_{eq} m^{-3} s^{-1}) by

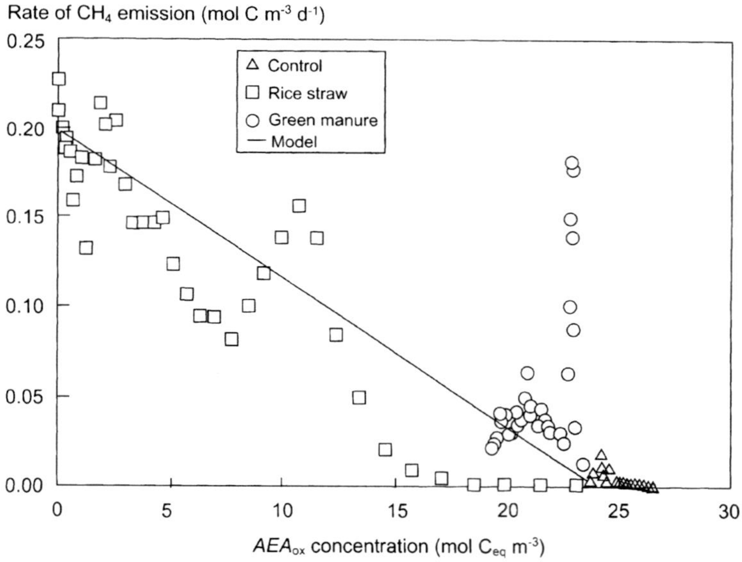

Figure 5. The relationship between rates of CH_4 emissions measured in the three treatments of the IRRI-1997DS experiment and the corresponding estimated concentrations of the AEA_{ox} pool (see text for details of method of calculation). Transient peaks are assumed to be due to the heterogeneous distribution of the added organic matter resulting in microenvironments in which AEA_{ox} pool is exhausted allowing CH_4 production to occur. The line ($y = 0.2\ (1.0 - x/24.0)$) indicates the relationship used in the model

$$d[AEA_{red}]/dt = -d[AEA_{ox}]/dt \qquad (9)$$

Account also needs to be taken of reoxidization of the *AEA* pool in the case of midseason drainage when air reenters the soil profile. For this, we assumed that this oxidation rate ($d[AEA_{red}]/dt$, mol C_{eq} m^{-3} d^{-1}) is related to the air-filled porosity (ε, m^3 air m^{-3} soil), i.e.

$$d[AEA_{red}]/dt = k \times \varepsilon / \varepsilon^* \times [AEA_{red}] \qquad (10)$$

where k is a rate constant (units: d^{-1}), ε^* is the maximum air-filled porosity (m^3 air m^{-3} soil) of the soil, and $[AEA_{red}]$ is the concentration of the reduced form of the alternative electron acceptor pool (mol C_{eq} m^{-3}). The two air-filled porosity values can be calculated as $\varepsilon = (1.0 - \rho/2.65 - \theta)$ and $\varepsilon^* = (1.0 - \rho/2.65 - \theta_L)$, where ρ is the bulk density (g cm^{-3}) of the soil, and θ and θ_L (m^3 water m^{-3} soil) are respectively the actual soil water content and the soil water content at the drained lower limit of the soil. Trial and error indicated that $k = 0.06\ d^{-1}$ gave realistic results, with complete reoxidization of the *AEA* pool occurring in about 2 wk. This value is comparable with that of $7.6 \times 10^{-7}\ s^{-1}$ ($= 0.068\ d^{-1}$) for FeS used by van Bodegom et al. (2000) in their model. It is assumed that all of the *AEA* pool remains in either its oxidized or reduced forms — i.e., that losses by leaching, denitrification, etc. are negligible—and that there is no diffusion or mass flow of the *AEA* pool between soil layers.

Steady-state concentrations and fluxes of CH_4 and O_2

To simulate the interactions between O_2 and CH_4 throughout the soil profile, we have used the submodel described by Arah & Kirk (2000) elsewhere in this volume. For convenience, a brief description of this submodel is included here along with details of how it links to the main CERES-Rice model. Concentration profiles of nonadsorbed materials (O_2 and CH_4 in our case) can be described by the differential equation:

$$\frac{\partial y}{\partial t} = \frac{\partial}{\partial z}\left(D\frac{\partial y}{\partial z}\right) - \frac{\partial}{\partial z}(Ly_w) + O + P - Q - R - E; (0 \le z \le Z) \qquad (11)$$

where z is the depth below the surface (m), D is the coefficient of diffusion of the material through the bulk matrix ($m^2\ s^{-1}$), L is the rate of leaching ($m^3\ s^{-1}$); O is the root-mediated influx ($m^3\ s^{-1}$), P is the rate of production of the material ($m^3\ s^{-1}$), Q is the rate of consumption of the material ($m^3\ s^{-1}$), R is the root-mediated efflux ($m^3\ s^{-1}$), and E is the rate of ebullition ($m^3\ s^{-1}$). Diffusion depends on the bulk concentration $y\ (z, t)$, leaching and consumption on the solution-phase concentration y_w (z, t), and root-mediated efflux and ebullition on the gas-phase concentration y_a (z, t). Root-mediated influx and production are assumed independent of y, y_w and y_a, though they may of course depend on other properties of the system (surface concentrations, concentrations of other substrates, root density profiles).

Methane production. The Arah & Kirk (2000) submodel requires the potential rate of CH_4 production (P_{CH4}*, mol C m^{-3} s^{-1}) as an input, which we have assumed is that calculated in equation 7 after the effects of the AEA_{ox} pool have been taken into account. However, the presence of O_2, even in small concentrations, affects the enzyme mechanisms of the methanogenic bacteria, so that the actual rate of CH_4 production can be considerably less than this potential rate. Actual CH_4 production (P_{CH4}, mol m^{-3} s^{-1}) in a given soil layer is therefore calculated as

$$P_{CH4} = P^*_{CH4}/(1 + \eta\,[O_2]) \qquad (12)$$

where η is a parameter (units: m^3 mol^{-1}) representing the sensitivity of methanogenesis to the concentration of O_2 ($[O_2]$, mol m^{-3}). Thus, when there is no O_2 present, the CH_4 production rate is equivalent to its potential rate, but this rapidly decreases to near zero as O_2 enters the system. A value of 400 m^3 mol^{-1} was used for η (Arah & Stephen, 1998). We have assumed that the size of the microbial population does not limit CH_4 production during the growing season (Schütz et al., 1989).

Methane oxidation. The rate of CH_4 consumption (Q_{CH4}, mol m^{-3} s^{-1}) by the methanotrophic bacteria (see equation 2) in a soil layer is given by the Michaelis-Menten equation

$$Q_{CH4} = P^*_{CH4}\,\frac{[CH_4]\qquad[O_2]}{(k_1 + [CH_4])\,(k_2 + [O_2])} \qquad (13)$$

where P^*_{CH4} is the potential rate of methanogenesis defined previously, $[CH_4]$ and $[O_2]$ are the concentrations of CH_4 and O_2 (mol m^{-3}) respectively, and k_1 and k_2 are Michaelis-Menten constants (units: mol m^{-3}) for a dual-substrate reaction. Oxygen consumption rate (Q_{O2}, mol m^{-3} s^{-1}) consists of a component due to this CH_4 oxidation and also a component due to aerobic respiration of the substrate:

$$Q_{O2} = 2Q_{CH4} + 2P^*_{CH4}\,\frac{[O_2]}{(k_3 + [O_2])} \qquad (14)$$

where $2P^*_{CH4}$ represents the maximum rate of aerobic respiration (mol O_2 m^{-3} s^{-1}) when O_2 is not limiting. Values of 0.33, 0.44, and 0.22 mol m^{-3} were used for k_1, k_2, and k_3, respectively.

Plant-mediated gaseous transport. Rice, like many other wetland plants, possesses channels (aerenchyma) within its stem and roots which have evolved to allow O_2 to diffuse from the atmosphere to the roots to allow aerobic respiration by the root cells in an otherwise anoxic environment. Both O_2 and CH_4, therefore, are able to be transported between atmosphere and soil via this route, usually in opposite directions. On a seasonal basis, transport of CH_4 through the aerenchyma is probably the most important pathway by which emissions reach the atmosphere. The contribution of plant-mediated transport may exceed 90% at given moments (Seiler et al., 1984), but over the season this contribution typically ranges from 38 to 85% (Wassmann et al., 1996). Nouchi et al. (1990) have described the process. Dissolved CH_4 in the soil water surrounding the roots diffuses through to the root cortex via the water in the cortex cell walls (the apoplastic pathway) driven by the gradient in concentration. Methane is gasified within the root cortex and transported to the shoots via the aerenchyma, where it is eventually released through the micropores in the leaf sheaths at the base of the leaf, not the stomata. A very small amount may be carried in the transpiration stream.

In the original Arah & Kirk (2000) model, the fluxes of O_2 and CH_4 through the plant were separated into inward fluxes (O, mol m^{-3} s^{-1}) and outward fluxes (R, mol m^{-3} s^{-1}). These fluxes were expressed as a function of the conductance of the pathway through the plant (λ, m air m^{-3} soil), the diffusivity of the respective substance through air (D_a, m^2 s^{-1}), and the concentration difference (mol m^{-3}) of the substance between source and sink. We have combined these fluxes into one ($F = O - R$), in which the sign of the flux denotes its direction. As in the original model, we have assumed that the conductance of the plant pathway is proportional to the root length density (L_v, cm root cm^{-3} soil) present in each soil layer, such that $\lambda = \lambda_r L_v$, where λ_r represents the specific conductivity (units: m air (m root)$^{-1}$) of the root system. Thus, the flux (F, mol m^{-3} s^{-1}) for each substance (O_2 or CH_4) is given by

$$F = \lambda_r\,(L_v \times 10^4)D_a\,(y_{a0} - y_a) \qquad (15)$$

where y_{a0} is the concentration of the respective substance in the atmosphere (O_2: 7.76 mol m^{-3}; CH_4: 7.5 × 10^{-5} mol m^{-3}), and y_a is its concentration in the gaseous phase in each soil layer. A positive value for F represents flux from atmosphere to soil, and a negative value vice versa. A value of 3.0 × 10^{-4} m air (m root)$^{-1}$ was used for λ_r. Diffusion constants (D_a) of O_2 and CH_4 in air were taken as 2.02×10^{-5} m^2 s^{-1} and 1.06×10^{-5} m^2 s^{-1}, respectively.

Diffusion. Only minor amounts of CH_4 are transported by diffusion across the air-water interface (Shearer & Khalil, 1993). Calculation of the diffusion rate between layers in the soil-water-atmosphere continuum is the same as in the original Arah & Kirk (2000) model, as described by the first term in equation 11.

Leaching. Again, the method of calculating movement of O_2 and CH_4 by leaching is the same as in the original Arah & Kirk (2000) model. Percolation rate (L in equation 11) is calculated in the water balance part of the main CERES-Rice model, and used as an input to the Arah & Kirk submodel after conversion to the appropriate units (i.e. mm d^{-1} to m^3 water m^{-2} s^{-1}).

Ebullition. We have modified the algorithm describing ebullition rate from that in the original Arah & Kirk (2000) model by expressing the rate of ebullition (E, mol m^{-3} s^{-1}) as a function of the concentration of the substance in solution (y_w, mol m^{-3})

$$E = \max\,[0, (y_w - y_w^*)/k_e] \qquad (16)$$

where y_w* is the solubility (mol m^{-3}) of the substance in water, and k_e is a constant (units: s) equal to the time-step of the simulation. Thus, if y_w exceeds y_w*, ebullition occurs, but if y_w is less than y_w*, there is no ebullition. As the time-step in the CERES-Rice model is 1 d, k_e takes a value of 8.64×10^4 s. Although this approach allows for ebullition of O_2, in practice this does not occur as O_2 concentrations never reach the y_w* value. Methane lost by ebullition is assumed to travel straight to the surface to be released into the atmosphere, with no oxidation by methanotrophs occurring en route. Rates of loss of CH_4 from the system through ebullition on an areal basis (i.e., kg CH_4-C ha^{-1} d^{-1}) are therefore calculated by summing the ebullition rates from each layer.

Values of y_w* (at 25 °C) used were 1.23 mol m^{-3} and 1.31 mol m^{-3} for O_2 and CH_4, respectively. Currently, there is no temperature dependence of y_w* included in the model, but this could be incorporated in future versions.

Stored CH_4. It is commonly observed on draining a rice field that there is a sharp peak in CH_4 emissions immediately following the drainage, which is generally ascribed to the release of entrapped and dissolved CH_4 in the soil water. To simulate this peak, we have assumed that if the floodwater drops to zero, 50% of the existing total CH_4 stored in the soil (in both gaseous and aqueous forms as calculated by the Arah & Kirk submodel) is released as emissions on each day. Allowing only 50% to be released per day rather than the total amount gives a lower but wider peak matching more closely to that observed.

Implementation of the CH_4 dynamics submodel. In the Arah & Kirk (2000) submodel, the floodwater and soil profile are divided into approximately 1 cm layers, and equations 12 to 16 solved for [O_2] and [CH_4] for each layer using the reiterative Newton-Raphson technique, to give the steady-state concentrations of O_2 and CH_4. As changes in the rates of methanogenesis dynamics occur over time periods of much less than a day (typically 10^{-3}-10^{-5} d^{-1}) and the CERES-Rice model operates on a daily time step, we feel that it is valid to assume steady-state conditions on a daily basis. Currently, it is assumed that each layer is homogeneous; no attempt is made to subdivide each layer into rhizosphere and bulk soil compartments. The original Arah & Kirk submodel was translated from Turbo Pascal into Fortran for compatibility with the CERES-Rice model. An 'interface' subroutine passes data from CERES-Rice to this submodel and receives data back from the submodel for use in the main model, in each case making the appropriate conversions for units and resolution of soil layers. Reflecting its parentage, we have called the combined crop/soil model MERES (*M*ethane *E*missions from *R*ice *E*co*S*ystems) (Figure 1).

Effect of inorganic fertilizers

It was assumed that the carbon in urea fertilizer, $(NH_2)_2CO$, does not contribute to CH_4 production. On application, urea undergoes hydrolysis to form NH_4^+ ions and HCO_3^- ions, the latter of which establish an equilibrium with CO_2 production depending on pH. Methanogens can use CO_2 but require a source of H_2 in order to do so. As most free H_2 in the soil has been produced from the decay of organic matter, it is stoichiometrically related to the carbon from the same source, and so there is no excess H_2 to combine with the carbonate-C. Urea as a source of carbon for CH_4 production can therefore be ignored.

In the case of ammonium sulfate (AS) fertilizer, the SO_4^{2-} is added to the oxidized alternative electron acceptor pool. In the CERES model, any fertilizer applied is partitioned between the floodwater and soil according to a 'mixing efficiency' depending on the method of application specified in the input file. Where fertilizer is broadcast onto flooded soil, for example, this mixing efficiency is such that about 15% enters the soil, and the rest is dissolved in the floodwater. It was assumed that on application of AS, the partition-

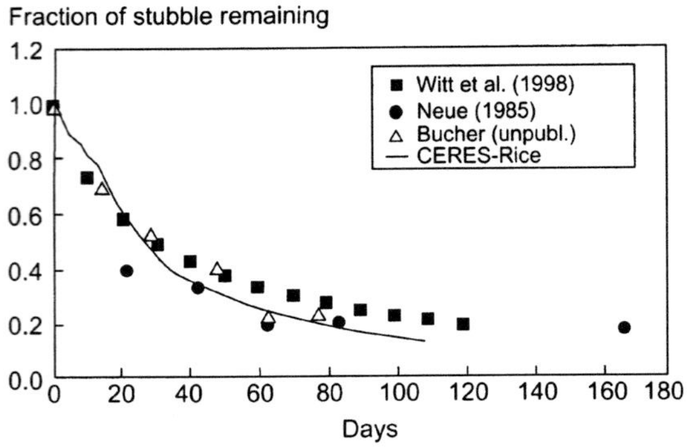

Figure 6. Comparison of simulated (solid line) and observed (symbols) fractions of rice stubble organic material remaining as a function of time since incorporation into flooded soils in field experiments at IRRI. Observed data are from Witt et al. (1998), Neue (1985), and S. Bucher (unpubl. data)

ing of SO_4^{2-} between floodwater and soil was the same as for the NH_4^+ ions, and that there is subsequently no transfer of SO_4^{2-} between floodwater and soil.

In the original version of the CERES-Rice model, it is only possible to make one application of fertilizer on a given day. In some of the treatments, however, both urea and AS were applied at the same time; the model code was therefore modified to allow this.

Results

Decay of organic matter

A comparison of the simulated decay of rice stubble and observed data from three studies is shown in Figure 6. Agreement is good, although there is a tendency for the model to overestimate the rate of decomposition in the latter part of the season. Possible reasons for this are discussed later.

Dynamics of the AEA pool

Predicted changes in the concentration of the pool of alternative electron acceptors in the top 10 cm of the soil for the rice-straw treatment of the IRRI-1997DS experiment are shown in Figure 7. It was assumed that this pool was in the fully oxidized state at the start of the simulation when flooding of the field occurred at 21 d before planting (-21 DAP). Addition of 10 t ha^{-1} of rice straw was 7 d later, which resulted in rapid reduction of these electron acceptors and release of organic carbon as CO_2, so that by 10-12 DAP, all of the *AEA* pool in this layer was in its reduced form. At this point, CH_4 production rate was limited only by the amount of substrate available.

At 23 DAP, the field was drained for a period of 17 d, as can be seen from the simulated floodwater level (Figure 7). As air entered the soil profile during the drainage period, the *AEA* pool was slowly reconverted from its reduced form to the oxidized form, although with the rate constants used in the model, complete reoxidation did not occur. On reflooding, conditions once more became anoxic and the *AEA* pool was again converted into its reduced form, although at a slower rate due to slower substrate production. The field was drained for a second time around 2 wk before harvest, when a similar pattern of behavior of the *AEA* pool was predicted.

Methane fluxes

The seasonal patterns of the various CH_4 fluxes in the rice-straw treatment of the IRRI-1997DS experiment are shown in Figure 8. The CH_4 production rate rises rapidly after the addition of the rice-straw 14 d before planting, to a maximum around the time of planting. Most of the CH_4 produced during this time is emitted through ebullition, due to the absence of plants. As the crop grows from planting onward, the fraction of CH_4 emitted through ebullition declines gradually, with an increasing fraction being transported through the plants, so that by about 70 DAP, almost all of the CH_4 emitted is through the plant and ebullition rates are almost negligible. Over the season, 24% of the total CH_4 emitted was through the plant, and 76% was through ebullition.

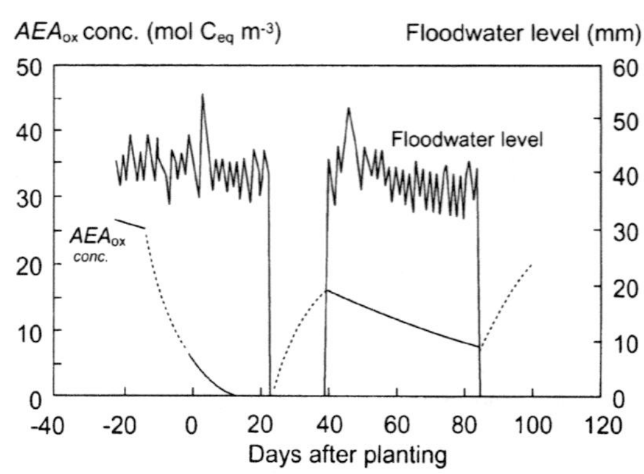

Figure 7. Simulated AEA_{ox} concentration in the top 10 cm of the soil profile for the rice-straw treatment of the IRRI-1997DS experiment. Rice straw was added at the rate of 10 t ha^{-1} 14 d before planting. The field was drained from 23-40 DAP and from 85 DAP until harvest, as indicated by the simulated floodwater level

The rate of oxidation also increases steadily over the season, as the plant conductance to gaseous transfer increases and more O_2 can diffuse into the rhizosphere. However, the proportion of CH_4 produced that is then oxidized is never large, and constitutes only some 7% of the seasonal total. The rates of CH_4 loss by diffusion are also negligible. As might be expected, all CH_4 fluxes drop to zero during the two drainage periods when CH_4 production ceases.

Comparison of simulated and measured CH_4 emissions

Comparison of the predicted and measured rates of CH_4 production over the season for the rice-straw and green manure treatments of the IRRI-1997DS experiment is shown in Figure 9. There is generally good agreement, although in the rice straw treatment the model overpredicted the plume of CH_4 at the second drainage just before harvest. In the green manure treatment, the initial plume of CH_4 emission immediately after incorporation could not be captured by the model, and the plume predicted by the model to occur on draining the field was not evident in the observed data.

Discussion

The soil organic matter decomposition routines of the CERES-Rice model appear able to match measured data well (Figure 6) with no modification or calibration from the original. The slight deviation of observed and simulated values from about 70 d onwards, which corresponds mainly to the decomposition of the lignin pool, may be due to the resistance of lignin to anaerobic degradation, such that it does not decompose at all in anoxic habitats. Currently, the decomposition rate of this pool in the model is influenced by soil water content in the same way as for the two other FOM pools (Figure 2b), but future improvements may include a modification to the multiplier function so that at the saturated water content, this rate is zero for the lignin pool only. Cao et al. (1995) have approached this problem by assuming the water content multiplier is 0.4 at the saturated water content for all pools (i.e., slower decomposition) rather than the 0.5 used in the CERES-Rice model.

The temperature multiplier used by the CERES-Rice model (Figure 2a), linear from 5 °C and above, may also need to be treated with some caution — bacteria usually have clearly defined optima. Cao et al. (1995) use a function with an optimum between 30 and 40 °C, declining below and above these values, respectively, the shape of which at least seems more realistic. For the time being, we have decided to leave the existing function unaltered, as no sites used in the upscaling exercise, described in Part IV of this series, experience soil temperatures in excess of 40 °C.

Unlike Cao et al. (1995), we have assumed that the pH of the soil has little effect on CH_4 production. Although the optimum pH for CH_4 production has been shown to be 7.0, with none below 5.7 or above 8.5 (Wang et al., 1993), pH is strongly linked to changes in the soil redox potential (Eh) so that at the Eh at which CH_4 production occurs, the soil pH is usually close to 7.0 anyway, regardless of its starting point (Wassmann et al., 1998). Similarly, we have also not included a separate effect of floodwater depth as Cao et al., (1995) have done—we have assumed that any effect of this is taken account of in the calculation of O_2 flux into the floodwater/soil profile by the Arah & Kirk submodel.

The ways in which environmental factors affect CH_4 production still need to be clarified. We have assumed that the main effect of temperature and soil water content is on the rate of production of substrate as just discussed and not on the rate of CH_4 generation from this substrate. This is supported by studies that have shown that the syntropic microbial processes involved in supplying substrate were more sensitive to temperature than methanogenesis itself (e.g., Conrad et al., 1987), and also follows the approach taken by both Cao et al. (1995) and Huang et al. (1998) in their models. Available evidence suggests that the effect of temperature on oxidation rate is small (Dunfield et al., 1993).

The submodel we have used to simulate the effect of the pool of alternative electron acceptors on CH_4 production is essentially a simplified version of that described by van Bodegom et al. (2000). The main difference between the two approaches is that in the latter, the ion species (i.e., NO_3^-, Mn^{4+}, Fe^{3+} and SO_4^{2-}) making up the pool are considered individually. While this approach is undoubtedly the more rigorous, we consider that the uncertainties of our knowledge of the processes involved justify the simpler approach we have adopted, particularly when used for upscaling exercises with a paucity of available data at the global scale on the concentrations of these ions in soils (see Part IV). Concentrations higher than AEA_{ox}* (see Figure 5), when no CH_4 is produced, correspond to the activity of the NO_3^- and Fe^{3+} reducing bacteria in the van Bodegom (2000) model, while concentrations between 0.0 and

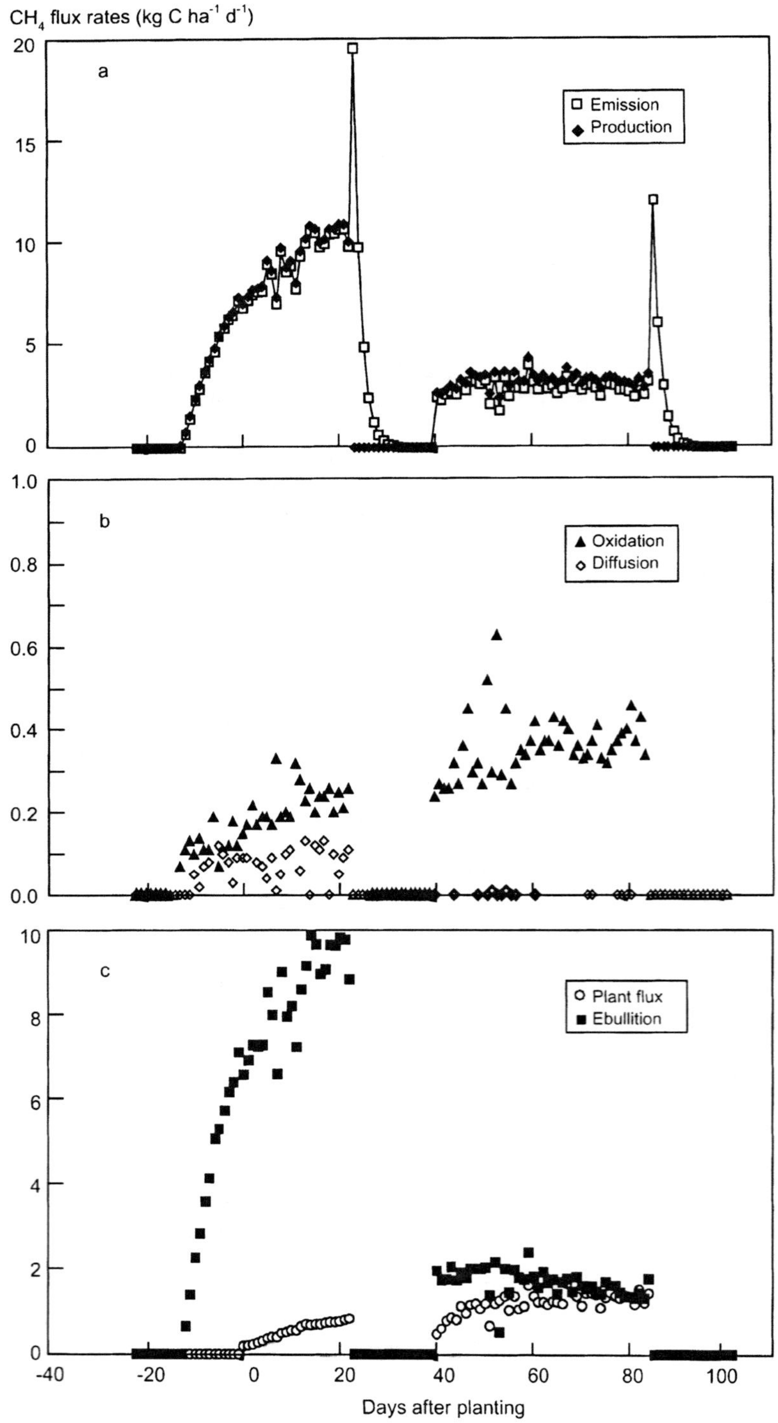

Figure 8. Predicted seasonal patterns of various CH_4 fluxes for the treatment in which 10 t ha^{-1} of rice straw was added 14 d before planting (IRRI-1997DS experiment). The field was drained from 23-40 DAP and from 85 DAP until harvest. (a) emission and production, (b) oxidation and diffusion, and (c) plant and ebullition fluxes. Note differences in y-axis scales

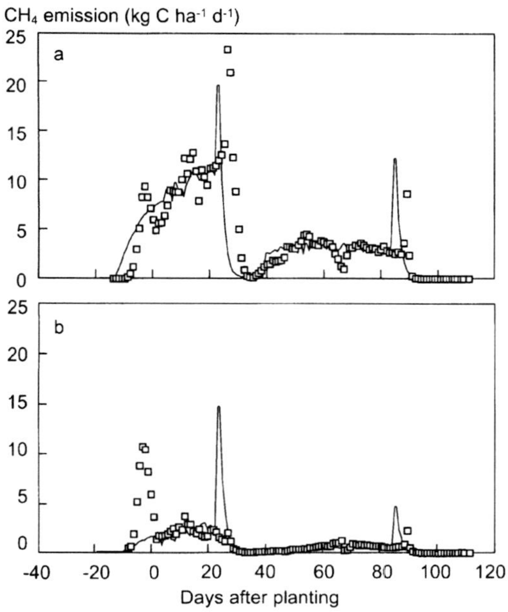

Figure 9. Comparison of observed and simulated seasonal patterns of CH_4 emissions in treatments of the IRRI-1997DS experiment with additions of (a) 10 t ha^{-1} of rice straw and (b) 3 t ha^{-1} of green manure. Both were added 14 d before planting

AEA_{ox}* correspond to the activity of the sulfate-reducers only partially competing with the methanogens for substrate. However, the relation between CH_4 production and AEA_{ox} is also governed by the heterogeneity of the organic matter distribution in the soil and the presence of other organic substrates such as methyl amides and methyl sulfides that can be used by the methanogens but not by the other bacteria. Thus, the relationship we have used in the current model is a 'blanket' relationship that takes into account all of these factors.

Estimating the size of the *AEA* pool in different soils is clearly a problem that needs to be addressed. The accuracy of the method we have used to determine the size of the *AEA* pool depends on the validity of the assumption that the amount of CH_4 produced that was reoxidized was negligible, which seems reasonable in a period where the rice crop was small and therefore not likely to transport significant quantities of O_2 through the aerenchyma. However, any errors in this assumption would underestimate the amount of CH_4 produced and therefore overestimate the size of the buffer. Similarly, the procedure is only possible if large quantities of carbon have been added as organic amendments such that the *AEA* pool is completely exhausted in a short time. In this case, 10 t ha^{-1} of rice straw was required to achieve this, but in most experiments, much less, if any, organic material is added.

As an alternative way to estimating the size of the *AEA* pool, we have used soil analysis data in which the concentrations of the main ion species of the pool were measured for the Maahas soil at IRRI (Yao et al., 1999). Converting the Fe^{3+}, Mn^{4+}, SO_4^{2-}, and NO_3^- concentrations to C equivalents according to the stoichiometry of the summary reactions gives a total *AEA* pool concentration of about 63.3 mol C_{eq} m^{-3}. This provides an upper limit to the estimate, but the 'effective' concentration is likely to be less than this due to a proportion of the ions in the *AEA* pool being unavailable for oxidation of organic matter due to occlusion, fixation, or their general insolubility—the latter particularly in the case of Fe^{3+}. Comparison of this value just calculated with the value of 26.5 mol C_{eq} m^{-3} calculated earlier would suggest that, in the Maahas soil at least, only about 42% of the measured *AEA* pool is actively involved in reacting with the substrate. Further work is clearly needed to see if this proportion is a general one for all soils.

The proportion of CH_4 produced that was oxidized by methanotrophs was predicted to be only 7% in the data set we have used. This is considerably lower than previous estimates of 50-80% (e.g., Sass et al., 1991). Similarly, Neue and Roger (1993) reported laboratory studies showing that 50-90% of the CH_4 produced in the soil can be oxidized before it reaches the surface. On the other hand, Frenzel et al. (1992) calculated that 50-90% of CH_4 *transported* to the rhizosphere is oxidized, which would suggest that a lower fraction of the total CH_4 *produced* was oxidized. Comparing seasonal totals, however, may be misleading, particularly in our case where a large proportion of the total CH_4 emitted is from the large organic matter supply early in the season before the crop reaches a stage where it is able to significantly influence oxidation rates. Where little or no organic matter is added, it might be expected that the oxidized/produced ratio be much higher. Calculation of this ratio on predicted instantaneous fluxes toward the end of the season (84 DAP), when most of the initial carbon source has decomposed, indicate that the rate of oxidation is about 20% of the production rate.

In this first version of the MERES model, we have not included the effect of soil texture on CH_4 emissions noted by some authors (e.g., Sass & Fisher, 1995; van Bodegom et al., 2000) as the mechanisms involved are far from clear. Cao et al. (1995) use the relationship

$M = 0.25 + 0.75\ S$, where M is the value of the zero-to-unity multiplier and S is the sand fraction of the soil. M is used to modify the decomposition rate of only the recalcitrant and lignin FOM pool. Huang et al. (1998), use a different relationship in their model, $M = 0.325 + 2.25\ S$, to modify decomposition rates of both OM pools and the rate of rhizodeposition. Initial simulations evaluating a number of possible mechanisms suggest that the effect on seasonal totals of CH_4 emissions may be small anyway (van Bodegom et al., 2000). Nevertheless, for completeness, we intend to survey our existing data sets on incubation studies and include a function in the next version of the MERES model to modify CH_4 production rate according to soil texture.

For the time being, we have assumed that the specific conductivity (λ_r) of the plant to gaseous transfer is constant throughout the growing season, so that changes in the conductance of the plant to gaseous transfer are due to changes in root length density only. This approach seems to work well in predicting the pattern of the plant-mediated CH_4 flux over the season (Figure 8). However, there is evidence that the conductance of the rice plants to CH_4 transport decreases as the plant ages (e.g., Nouchi et al., 1990), particularly toward the end of the season. Nouchi et al. (1990) suggested that this was due to reduced permeability of the root epidermis during ageing, but it can also be explained by root death reducing the size of the root system. Certainly, changes in root porosity due to continued exposure to low Eh values have been reported (Kludze et al., 1993). However, it is not certain to what extent this contributes to overall plant conductance. Butterbach-Bahl et al. (1997), for example, consider that the main site of resistance to gaseous movement is the transition from root to stem — they found that a high CH_4 transport capacity was associated with an increase in the relative pore diameter in this zone. Similarly, Ueckert et al. (1990) found that the size of aerenchyma was the main plant parameter that controlled O_2 transport through the plant to the rhizosphere. If this is the case, relating plant conductance to cross-sectional stem area (e.g., tiller number) rather than root length density as we have done, may be a better approach. However, in view of the uncertainty in this relationship, we consider for the time being that the assumption of a constant relationship between conductance and root length is justified, particularly as cross-sectional stem area and root length are likely to be strongly correlated over a season anyway. Nevertheless, if better ways of estimating plant conductance are discovered, these can be easily incorporated into the model. The effect of varying λ_r on overall CH_4 emissions, particularly in relation to differences between varieties (e.g., Butterbach-Bahl et al., 1997) also needs to be explored.

The assumption that ebullition rate is proportional to the difference between the aqueous concentration of CH_4 in the soil and its solubility concentration is obviously a simplification from reality, but appears to work adequately in describing the seasonal pattern of CH_4 flux lost from the soil in this way (Figure 8). Certainly, the relative effects of plant-mediated and ebullition fluxes match observed patterns well, with most emissions early in the season due to ebullition and an increasing dominance of plant-mediated flux as the season progresses. The seasonal total of 76% of emissions through ebullition predicted by the model for the rice-straw treatment of the IRRI-1997DS experiment is high, but is comparable with the 70% reported by Crill et al. (1988) from peat lands. Similarly, Bartlett et al. (1988) measured values between 49 and 64% for the Amazonian floodplain. A value of 60% has been reported for rice fields in studies where ebullition rates were high at the beginning of the season due to additions of organic matter (e.g., Denier van der Gon & Neue, 1995; Wassmann et al., 1996). In rice fields where little or no organic material is added, ebullition normally contributes only 10-20% to the seasonal CH_4 emission (e.g., Schütz et al., 1989; Nouchi et al., 1994). There is clearly a wide range in the estimates of the contribution made by ebullition to overall CH_4 emissions, depending mainly on the balance between substrate supply and the presence of plants to act as a conduit.

Walter et al. (1996) use a method similar to ours of calculating ebullition rates in their model, but assume that bubble formation occurs when the aqueous CH_4 concentration exceeds 0.5 mol m^{-3} rather than the 1.31 mol m^{-3} we have used. Their value takes into account a mixing ratio of 25% of CH_4 in the bubble with the remaining 75% being inert N_2. However, using this value would have the effect of increasing ebullition rates still further and could not explain the high ebullition fractions predicted by our model. Clearly, further work is necessary to simulate the processes involved in bubble formation more mechanistically. Consideration could be given in future versions of MERES to describing the formation and release of bubbles in terms of when the partial pressure of entrapped CH_4 within the soil exceeds the hydrostatic pressure (Wang et al., 1995). Similarly, the effects of soil temperature and solar radiation on increasing ebullition rates that some workers have reported (e.g., Nouchi et al., 1990) also need to be looked into.

Despite some of its limitations just discussed, we conclude that the MERES model describes the basic features of CH_4 emissions from rice fields with reasonable accuracy. Furthermore, it contains the crucial components required for adequately evaluating (a) the effects of altered crop management practices on CH_4 emissions, and (b) upscaling experimental measurements to national and regional levels using the weather, soils, crop management, and rice-growing area data described in Part III of this series (Knox et al., 2000). Further papers deal with the validation and sensitivity analysis of the model (Matthews et al., 2000a), and the results of the upscaling exercise itself (Matthews et al., 2000b).

Acknowledgements

The authors would like to thank the staff in the Soil and Water Sciences Division at the International Rice Research Institute for their tireless work in collecting and collating the data used in this study, particularly Mr. Leandro Buendia. This study was conducted with funding from the United Nations Development Programme Global Environment Facility.

References

Anastasi C, Dowding M & Simpson VJ (1992) Future CH_4 emission from rice production. J Geophys Res 97:7521-7525

Arah JRM & Kirk GJD (2000) Modelling rice-plant-mediated methane emission. Nutr Cycling Agroecosyst (this issue)

Arah JRM & Stephen KD (1998) A model of the processes leading to methane emission from peatland. Atmos Environ 32:3257-3264

Aselmann I & Crutzen DJ (1990) A global inventory of wetland distribution and seasonality, net primary productivity, and estimated methane emissions. In: AF Bouwman (ed) Soils and the Greenhouse. Soil Biol Biochem 25:321-326

Bachelet D & Neue H-U (1993) Methane emissions from wetland rice areas of Asia. Chemosphere 26:219-237

Bachelet D, van Sickle J & Gay CA (1993) The impacts of climate change on rice yield: evaluation of the efficacy of different modelling approaches. In: FWT Penning de Vries, PS Teng & K Metselaar (eds), Systems Approaches for Agricultural Development. Kluwer Academic Publishers, Dordrecht, The Netherlands, pp 145-174

Bartlett KB, Crill PM, Sebacher DI, Hariss, RC, Wilson JO & Melack JM (1988) Methane fluxes from the central Amazonian floodplain. J Geophys Res 93:1571-1582

Bouwman AF (1991) Agronomic aspects of wetland rice cultivation and associated methane emissions. Biogeochemistry 15:65-88

Butterbach-Bahl K, Papen H & Rennenberg H (1997) Impact of gas transport through rice cultivars on methane emission from rice paddy fields. Plant Cell Environ 20:1175-1183

Cai Z & Yan XY (1996) Simulation of methane oxidation by paddy soils in a closed system. Pedosphere 6(4):313-320

Cao M, Dent JB & Heal OW (1995) Modelling methane emissions from rice paddies. Global Biogeochem Cycles 9:183-195.

Conrad R (1989) Control of methane production in terrestrial ecosystems. In: MO Andreae & DS Schimel (eds) Exchange of Trace Gases between Terrestrial Ecosystems and the Atmosphere, pp 39-58, J. Wiley & Sons, New York

Conrad R, Schütz H & Barbble M (1987) Temperature limitation of hydrogen turnover and methanogenesis in anoxic paddy soil. FEMS Microbiol Ecol 45:281-289

Crill PM, Bartlett KB, Harriss RC, Verry ES, Sebacher DI, Madzar L & Sanne W (1988) Methane flux from Minnesota peatlands. Global Biogeochem Cycles 2:371-384

Denier van der Gon HAC & Neue H-U (1995) Influence of organic matter incorporation on the methane emission from a wetland rice field. Global Biogeochem Cycles 9:11-22

Dunfield P, Knowles R, Dumont R & Moore TR (1993) Methane production and consumption in temperate and subarctic peat soils. Soil Biol Biochem 25:321-326

Godwin DC & Jones CA (1991) Nitrogen dynamics in soil-plant systems. In: Hanks J & Ritchie JT (eds) Modeling Soil and Plant Systems, pp 287-322, Agron Soc Am, Madison

Huang Y, Sass RL & Fisher FM Jr (1998) A semi-empirical model of methane emission from flooded rice paddy soils. Global Change Biol 4(3):247-268

IPCC — Intergovernmental Panel on Climate Change (1996) XII. Summary for policymakers. In: JT Houghton, LG Meira-Filho, BA Chancellor, A Kattenberg & K Maskell (eds) Climate Change 1995: The Scientific Basis of Climate Change. Cambridge University Press, Cambridge, UK, 572 p

Kern JS, Gong Z, Zhang G, Zhuo H & Luo G (1997) Spatial analysis of methane emissions from paddy soils in China and the potential for emissions reduction. Nutr Cycling Agroecosyst 49:181-195

Kludze HK, DeLaune RD & Patrick WH (1993) Aerenchyma formation and methane and oxygen exchange in rice. Soil Sci Soc Am J 57:386-391

Knox J, Matthews RB & Wassmann R (2000) Using a crop/soil simulation model and GIS techniques to assess methane emissions from rice fields in Asia. III. Databases. Nutr Cycling Agroecosyst (this issue)

Lassiter RR & Plis YM (1994) Descriptive basis for a model of soils biogeochemistry. In: R.G. Zepp (ed) Climate Biosphere Interaction: Biogenic Emissions and Environmental Effects of Climate Change. pp 223-240, John Wiley & Sons, Inc

Lu Y, Wassmann R, Neue H-U & Huang C (1999) Impact of phosphorus on root exudation, aerenchyma formation, and methane emission of rice plants. Biogeochemistry 47:203-218

Matthews RB, Wassmann R, Buendia LV & Knox J (2000a) Using a crop/soil simulation model and GIS techniques to assess methane emissions from rice fields in Asia. II. Model validation and sensitivity analysis. Nutr Cycling Agroecosyst (this issue)

Matthews RB, Wassmann R, Knox JW & Buendia LV (2000b) Using a crop/soil simulation model and GIS techniques to assess methane emissions from rice fields in Asia. IV. Upscaling to national levels. Nutr Cycling Agroecosyst (this issue)

Neue HU (1985) Organic matter dynamics in wetland soils. In: Wetland Soils: Characterization, Classification and Utilization, pp 109-122, International Rice Research Institute, Los Baños, Philippines

Neue HU & Roger PA (1993) Rice agriculture: factors controlling emissions. In: MAK Khalil (ed) Atmospheric Methane: Sources, Sinks, and Role in Global Change, pp 254-298, NATO ASI Series. Springer-Verlag, Berlin

Nouchi I, Hosono T, Aoki K & Minami K (1994) Seasonal variation in methane flux from rice paddies associated with methane concentration in soil water, rice biomass and temperature, and its modelling. Plant Soil 161(2):195-208

Nouchi I, Mariko S & Aoki K (1990) Mechanism of methane transport from the rhizosphere to the atmosphere through rice plants. Plant Physiol 94:59-66

Patrick Jr WH & Delaune RD (1977) Chemical and biological redox systems affecting nutrient availability in the coastal wetlands. GeoScience Manage 18:131-137

Ritchie JT, Singh U, Godwin DC & Bowen WT (1998) Cereal growth, development and yield. In: G Tsuji, G Hoogenboom & P Thornton (eds) Understanding Options for Agricultural Production. Systems Approaches for Sustainable Agricultural Development, pp 79-98, Kluwer Academic Publishers, Dordrecht, The Netherlands

Sass RL & Fisher FM (1995) Methane emissions from Texas rice fields. In: S Peng, KT Ingram, HU Neue & LH Ziska (eds) pp 46-59, Climate Change and Rice, Springer-Verlag, Berlin, Germany

Sass RL & Fisher FM (1997) Methane emissions from rice paddies: a process study summary. Nutr Cycling Agroecosyst 49:119-127

Sass RL, Fisher FM, Turner FT & Yund MF (1991) Methane emission from rice fields as influenced by solar radiation, temperature, and straw incorporation. Global Biogeochem Cycles 5(4):335-350

Schütz H, Seiler W & Conrad R (1989) Processes involved in formation and emission of methane in rice paddies. Biogeochemistry 7:33-53

Seiler W, Holzapfel-Pschorn A, Conrad R & Scharffe D (1984) Methane emission from rice paddies. J Atmos Chem 1:241-268

Seligman NG & van Keulen H (1981) PAPRAN: a simulation model of annual pasture production limited by rainfall and nitrogen. In: M. Frissel & J. van Veen (eds), Simulation of Nitrogen Behaviour in Soil-Plant Systems pp 192-221, PUDOC, Wageningen, The Netherlands

Shamoot S, McDonald L & Bartholomew WV (1968) Rhizodeposition of organic debris in soil. Soil Sci Soc Am Proc 32:817-820

Shearer MJ & Khalil MAK (1993) Rice agriculture: emissions. In: MAK Khalil (ed), Atmospheric Methane: Sources, Sinks, and Role in Global Change, pp 230-253, NATO ASI Series, Springer-Verlag, Berlin

Taylor JA, Brasseur GP, Zimmerman PR & Cicerone RJ (1991) A study of the sources and sinks of methane and methyl chloroform using a global three-dimensional Lagrangian tropospheric tracer transport model. J Geophys Chem 96:3013-3044

Tsutsuki K & Ponnamperuma FN (1987) Behaviour of anaerobic decomposition in submerged soils: effects of organic material amendment, soil properties and temperature. Soil Sci Plant Nutr 33:13-33.

Ueckert J, Hurek I, Fendrik I & Niemann EG (1990) Radial gas diffusion from roots of rice and kallargrass (*Leptochloa fusca* L. Kunth), and effects of inoculation with *Azospirillium brasilense* Cd. Plant Soil 112:59-65

van Bodegom PM, Leffelaar PA, Stams AJM & Wassmann R (2000) Modelling methane emissions from rice paddies: variability, uncertainty and sensitivity analyses of processes involved. Nutr Cycling Agroecosyst (this issue)

Walter BP, Heimann M, Shannon RD & White JR (1996) A process-based model to derive methane emissions from natural wetlands. Geophys Res Lett 23(25):3731-3734

Wang B, Neue HU & Samonte HP (1997) Effect of cultivar difference ('IR72', 'IR65598', and 'Dular') on methane emission. Agric Ecosyst Environ 62:31-40

Wang Z, Kludze, H, Crozier CR & Patrick WH Jr (1995) Soil characteristics affecting methane production and emission in flooded rice. In: S Peng, KT Ingram, HU Neue & LH Ziska (eds) Climate Change and Rice, pp 80-90, Springer Verlag, Berlin, Germany

Wang ZP, DeLaune RD, Masscheleyn PH & Patrick WH (1993) Soil redox and pH effects on methane production in a flooded rice soil. Soil Sci Soc Am J 57:382-385

Wassmann R, Neue HU, Alberto MCR, Lantin RS, Bueno C, Llenaresas D, Arah JRM, Papen H, SeilerW & Rennenberg H (1996) Fluxes and pools of methane in wetland rice soils with varying organic inputs. Environ Monit Assess 42:163-173

Wassmann R, Neue HU, Bueno C, Lantin RS, Alberto MCR, Buendia LV, Bronson K, Papen H & Rennenberg H (1998) Inherent properties of rice soils determining methane production potentials. Plant Soil 203:227-237

Wassmann R, Buendia LV, Lantin RS, Bueno CS, Lubigan LA, Umali A, Nocon NN, Javellana AM, Neue HU (2000) Mechanisms of crop management impact on methane emissions from rice fields in Los Baños, Philippines. Nutr Cycling Agroecosyst (this issue)

Watanabe A & Roger PA (1985). Ecology of flooded rice fields. In: Wetland Soils: Characteristics, Classification and Utilization. International Rice Research Institute, Manila, Philippines, pp 229-243

Witt C, Cassman KG, Ottow JCG & Biker U (1998) Soil microbial biomass and nitrogen supply in an irrigated lowland rice soil as affected by crop rotation and residue management. Biol Fertil Soils 28:71-80

Yao H, Conrad R, Wassmann R & Neue HU (1999) Effect of soil characteristics on sequential reduction and methane production in sixteen rice paddy soils from China, the Philippines, and Italy. Biogeochemistry 47:267-293

Nutrient Cycling in Agroecosystems 58: 161–177, 2000.

Using a crop/soil simulation model and GIS techniques to assess methane emissions from rice fields in Asia. II. Model validation and sensitivity analysis

R.B. Matthews[1], R. Wassmann[2,3], L.V. Buendia[2] & J.W. Knox[1]

[1]*Institute of Water and Environment, Cranfield University, Silsoe, Bedfordshire MK45 4DT, United Kingdom;* [2]*International Rice Research Institute, MCPO Box 3127, Makati City 1271, Philippines;* [3]*Fraunhofer Institute for Atmospheric Environmental Research, Garmisch-Partenkirchen, Germany*

Key words: methane, rice, *Oryza sativa*, anaerobic, model, simulation, carbon dynamics

Abstract

The MERES (Methane Emissions from Rice EcoSystems) simulation model was tested using experimental data from IRRI and Maligaya in the Philippines and from Hangzhou in China. There was good agreement between simulated and observed values of total aboveground biomass, root weight, grain yield, and seasonal methane (CH_4) emissions. The importance of the contribution of the rice crop to CH_4 emissions was highlighted. Rhizodeposition (root exudation and root death) was predicted to contribute about 380 kg C ha^{-1} of methanogenic substrate over the season, representing 37% of the total methanogenic substrate from all sources when no organic amendments were added. A further 225 kg C ha^{-1} (22%) was predicted to come from previous crop residues, giving a total of around 60% originating from the rice crop, with the remaining 41% coming from the humic fraction of the soil organic matter (SOM). Sensitivity analysis suggested that the parameter representing transmissivity to gaseous transfer per unit root length (λ_r) was important in determining seasonal CH_4 emissions. As this transmissivity increased, more O_2 was able to diffuse to the rhizosphere, so that CH_4 production by methanogens was reduced and more CH_4 was oxidized by methanotrophs. These effects outweighed the opposing influence of increased rate of transport of CH_4 through the plant, so that the overall effect was to reduce the amount of CH_4 emitted over the season. Varying the root-shoot ratio of the crop was predicted to have little effect on seasonal emissions, the increased rates of rhizodeposition being counteracted by the increased rates of O_2 diffusion to the rhizosphere. Increasing the length of a midseason drainage period reduced CH_4 emissions significantly, but periods longer than 6-7 d also decreased rice yields. Organic amendments with low C/N were predicted to be more beneficial, both in terms of enhancing crop yields and reducing CH_4 emissions, even when the same amount of C was applied. This was due to higher rates of immobilization of C into microbial biomass, removing it temporarily as a methanogenic substrate.

Introduction

Methane (CH_4) is one of the principal greenhouse gases and has been estimated to account for 15-20% of current radiative forcing. Rice soils, characterized by O_2 depletion, high moisture, and relatively high organic substrate levels, offer an ideal environment for the activity of methanogenic bacteria and are one of the major anthropogenic CH_4 sources. Precise estimates of source size have been difficult because of the large spatial and temporal variability in CH_4 emission rates measured at different sites due to differences in climate, soils, rice cultivars used, and crop management practices. Representation and integration of these factors within a geographical information system, coupled with the development of mechanistic models describing the processes involved in CH_4 production and emission, is a logical way forward.

Part I of this series (Matthews et al., 2000a) describes the development of a process-based model of CH_4 dynamics in rice fields, in which the CERES-Rice crop simulation model was linked to a submodel (Arah & Kirk, 2000) calculating the steady-state concentrations of CH_4 and O_2 in flooded soils. Routines to account for the influence on CH_4 production of the pool of alternative electron acceptors in the soil were also developed. The model was able to simulate well the seasonal pattern of CH_4 emissions from a rice field in the Philippines in which rice straw had been incorporated as an organic amendment.

Most of the parameters in the model were physical constants for which the values were known precisely. However, for four of the parameters (i.e., the active fraction of the alternative electron acceptor pool, the root death constant (δ_r), the specific root exudation rate (ε_r), and the transmissivity of the roots to gaseous transfer (λ_r), values were not known exactly, and reasonable estimates had to be made. It is the purpose of this paper to validate the model against observed data from a number of experiments carried out as part of the UNDP-funded project described elsewhere in this volume, and to evaluate the sensitivity of the model to various parameters, including the four mentioned above.

Methods

Model validation

Data from three sets of experiments within the UNDP project were used for testing and validating the model. These were the ones carried out at the International Rice Research Institute (IRRI) at Los Baños (latitude 14.18° N; longitude 121.25° E; altitude 21 m) in the Philippines, those from PhilRice at Maligaya (latitude 15.75° N, longitude 120.93° E, altitude 48 m), also in the Philippines, and those at Hangzhou (latitude 30.23° N; longitude 120.20° E; altitude 45 m) in China. A summary of the experiments and their treatments is shown in Table 1.

IRRI experiments

These experiments were conducted in both the dry and wet seasons of each year from 1994 to 1997, with treatments including frequency and timing of drainage, different rice genotypes, and amounts and types of inorganic fertilizers and organic amendments. Each plot was sealed with a plastic sheet to exclude lateral seepage as well as percolation. At the beginning of the drainage period, floodwater was allowed to flow out of the field, and no irrigation water was applied throughout the drainage period. During this period, the soil was allowed to dry out, and cracks were observed. Total aboveground biomass and yield were determined at final harvest, and root biomass was determined about 2 wk before this.

In all of these experiments, stubble from the previous crop was removed to ground level, and in three experiments (1994 WS, 1995 DS, 1995 WS), root residues were also removed by sieving the soil. In the remaining experiments, the contributions of C from the previous crop residues were estimated in the following manner. From the data of Bronson et al. (1998), it was calculated that 1 cm of stubble represents 163 kg dry matter (DM) ha^{-1} for high N levels (190 kg N ha^{-1}), and 93 kg DM ha^{-1} for zero N. It was also assumed that there was little decomposition of stubble remaining above ground between seasons. Root residues underground, however, did decompose. Their contribution to the soil organic matter (OM) pools was estimated from measurements of root weights taken in the previous crop just prior to harvest. The mean of these measurements across all experiments and treatments was 1450 kg DM ha^{-1}, measured on average about 14 d before the final harvest. This represented a root-shoot ratio (i.e., root dry weight to aboveground dry weight) of 11.8%. Although aboveground biomass was on average 11.5% higher in the DS than in the WS, there were no significant differences between absolute root weights at the end of the two seasons. To determine the quantity of this root biomass remaining at the start of the following season, the model was run from the dates of the root measurements until the date that the field was reflooded. For the period following the DS crop, this was from early-April to mid-June (72 d) and from early October to mid-December (79 d) following the WS crop. During this time, it was assumed that there was no crop present and no standing water, so the model was simulating only soil processes under an aerobic fallow. The contribution of any weeds was ignored. It was also assumed that the initial root biomass was partitioned between the three fresh organic matter (FOM) pools in the ratio of 20% carbohydrate, 70% cellulose, and 10% lignin. These simulations predicted a mean value of 395 kg DM ha^{-1} (27% of the original total) remaining at the end of the fallow period before the WS (i.e., April-June). However, the distribution of this remaining OM between the respective pools had changed to 2%, 71%, and 27%, reflecting their relative speeds of decay. The corresponding value for the fallow pe-

Table 1. Main cultural details of experimental treatments used in testing the MERES model. Roots are estimated in terms of dry weight remaining from the previous crop at the start of the season. Drains are number of times per season the field was drained of water

Dataset ID	Date of planting	Date of harvest	Variety	Roots (kg ha^{-1})	Stubble (kg ha^{-1})	Straw (kg ha^{-1})	Manure (kg ha^{-1})	Drains (no.)
IRRI								
I94DS-T1	13 Jan 94	22 Apr 94	IR72	330	0	0	0	0
I94DS-T2	13 Jan 94	22 Apr 94	IR72	330	0	0	0	1
I94DS-T3	13 Jan 94	22 Apr 94	IR72	330	0	0	0	2
I94WS-T1	14 Jul 94	22 Oct 94	IR72	0	0	0	0	0
I94WS-T2	14 Jul 94	22 Oct 94	IR72	0	0	0	0	1
I94WS-T3	14 Jul 94	22 Oct 94	IR72	0	0	0	0	2
I95DS-T1	11 Jan 95	16 Apr 95	IR72	0	0	0	0	2
I95DS-T2	11 Jan 95	16 Apr 95	IR65597	0	0	0	0	2
I95DS-T3	20 Jan 95	16 Apr 95	Dular	0	0	0	0	2
I95WS-T1	4 Jul 95	11 Oct 95	IR72	0	0	0	0	2
I95WS-T2	4 Jul 95	11 Oct 95	IR65597	0	0	0	0	2
I95WS-T3	4 Jul 95	11 Oct 95	PSBRc14	0	0	0	0	2
I95WS-T4	4 Jul 95	3 Oct 95	Magat	0	0	0	0	2
I96DS-T1	9 Jan 96	18 Apr 96	IR72	330	0	0	0	0
I96DS-T2	9 Jan 96	18 Apr 96	IR72	330	0	0	0	1
I96DS-T3	9 Jan 96	18 Apr 96	IR72	330	0	0	0	2
I96WS-T1	9 Jul 96	17 Oct 96	IR72	395	0	0	0	0
I96WS-T2	9 Jul 96	17 Oct 96	IR72	395	0	0	0	1
I96WS-T3	9 Jul 96	17 Oct 96	IR72	395	0	0	0	2
I96WS-T4	9 Jul 96	17 Oct 96	IR72	395	0	0	0	0
I97DS-T1	10 Jan 97	20 Apr 97	IR72	330	0	0	0	2
I97DS-T2	10 Jan 97	20 Apr 97	IR72	330	0	0	0	2
I97DS-T3	10 Jan 97	20 Apr 97	IR72	330	0	10,000	0	2
I97DS-T4	10 Jan 97	20 Apr 97	IR72	330	0	0	3,000	2
Maligaya								
M96DS-T1	22 Dec 95	15 Apr 96	IR72	370	0	0	0	1
M96DS-T2	22 Dec 95	15 Apr 96	IR72	370	0	4,000	0	1
M96DS-T3	22 Dec 95	15 Apr 96	IR72	370	0	2,500	0	1
M96DS-T4	22 Dec 95	15 Apr 96	IR72	370	0	4,000	0	1
M96WS-T1	28 May 96	4 Oct 96	IR72	490	0	0	0	1
M96WS-T2	28 May 96	4 Oct 96	IR72	490	0	4,000	0	1
M96WS-T3	28 May 96	4 Oct 96	IR72	490	0	2,500	0	1
M96WS-T4	28 May 96	4 Oct 96	IR72	490	0	4,000	0	1
Hangzhou								
H95S-T1	30 May 95	10 Oct 95	Chunjiang 06	0	820	0	0	2
H95S-T2	30 May 95	10 Oct 95	Chunjiang 06	0	820	0	2,000	4
H95S-T3	30 May 95	10 Oct 95	Chunjiang 06	0	820	0	2,000	2
H95S-T4	30 May 95	10 Oct 95	Chunjiang 06	0	820	0	2,000	1
H96E-T2	7 May 96	24 Jul 96	Zhongyou 906	92	820	0	0	3
H96E-T4	7 May 96	24 Jul 96	Jin 23A/71	92	820	0	0	3
H96S-T1	20 Jun 96	30 Oct 96	Chunjiang 06	92	820	0	0	4
H96L-T2	26 Jul 96	8 Nov 96	Xiu-shui 11	980	820	0	0	4
H96S-T3	20 Jun 96	26 Sep 96	Shan-you 10	92	820	0	0	4
H96L-T4	26 Jul 96	30 Oct 96	II-yiu 1568	980	820	0	0	4

riod before the DS (i.e., October-December) was 330 kg DM ha^{-1} (23% of the original total) remaining, of which 1%, 69%, and 30%, respectively, was in the three FOM pools. These estimated values were used to initialize the FOM pools at the start of each treatment in which the roots or stubble had not been removed. As the original CERES-Rice model assumes a distribution of 20:70:10 for crop residue organic matter regardless of its degree of decomposition, the model code was modified to read the appropriate distributions as inputs from the crop management data input file (the X-file). For all pools, it was assumed that the C content of the DM was 0.31 kg C (kg DM)$^{-1}$ (Bronson et al., 1998).

In the IRRI 1997DS experiment, organic amendments were added – in treatment 3, 10 t ha^{-1} of rice straw were applied, and in treatment 4, 3 t ha^{-1} of green manure in the form of *Sesbania rostrata* prunings were applied, both at 14 d before planting. It was assumed that the C content in the DM was 0.31 kg C (kg DM)$^{-1}$ in both cases, and that the C/N were 50 and 15 kg C (kg N)$^{-1}$, respectively (Bronson et al., 1998). These organic amendments, therefore, both contribute about 60 kg N ha^{-1}.

The model was set to maintain a constant depth of floodwater throughout the season, except for the times that the field was drained. This was achieved in the model by setting the irrigation mode to automatic so that water was added whenever the floodwater level fell below a prescribed depth. Drainage during and at the end of the season was simulated by setting both the bund height and floodwater depth to zero on the day of draining, and to 10 cm and 5 cm, respectively, on the day when the field was reflooded. During these times of drainage, no water was added, although any rain that fell was taken into account and would influence the soil water balance.

In all the experiments with the exception of 1997DS, urea was applied at the rate of 120 kg N ha^{-1} in four equal splits throughout the season. For all of these, it was assumed that the fertilizer was broadcast onto flooded soil with 30% of it being incorporated into the soil.

Maligaya experiments

Eight experiments with four treatments each were carried out at PhilRice at Maligaya in the Philippines from the 1994 DS to the 1997 WS. Due to difficulties in setting up equipment in several of the experiments, reliable measurements of seasonal CH_4 emissions were obtained in only two of these, the 1996DS and 1996WS. Both of these experiments investigated the effects of the addition of rice straw (both fresh material and compost) and the use of phosphogypsum (K_2SO_4, so named as it is a byproduct during manufacture of phosphoric acid). Variety IR72 was used in all treatments of both experiments, and the field was drained one week before harvest. N fertilizer was applied in the form of urea in three splits such that the total N applied (including that in the organic amendments) was 120 kg N ha^{-1}. No stubble was left remaining from the previous crop.

Root residues from the previous crop were left in the soil in each experiment, with a similar methodology being used to calculate the quantity and quality remaining at the start of the next season as that done for the IRRI experiments. Root weight measurements at the end of the season were not made, but using the same root-shoot ratio of 11.8% measured in the IRRI experiments, end-of-season root weights were calculated from the total aboveground biomass figures as 1,525 and 1,540 kg DM ha^{-1} for the DS and WS, respectively. These mean values for each season were used for initialization of the relevant SOM pools because although there may have been differences in final root weights between treatments in each season, it was not possible to link these treatments with the ones in the following season that were in the same part of the field. Mean dates of harvest were 25 Apr and 16 Oct for each season, with the next season starting on average on 16 Jun and 22 Dec, respectively. Using the mean weather data, the model predicted 490 kg DM ha^{-1} (0.012:0.737:0.251) and 370 kg DM ha^{-1} (0.010:0.690:0.299) of root residues remaining at the start of the WS and DS, respectively.

Soil parameters were obtained for the Maligaya soil from Wassmann et al. (2000), and weather data recorded at PhilRice over the period of the experiments were used to run the model.

Hangzhou experiments

At the Hangzhou site, experiments were carried out in 1995 and 1996 to evaluate the effect of various drainage regimes and different varieties, including hybrid rice. In the 1995 experiment, all roots and stubble biomass were removed before the start of the season, so that there was no input of carbon from previous crop residues. In the 1996 experiment, treatments 1 and 3 were left fallow until early June, when they were planted with single rice. Treatments 2 and 4 were planted in early May with an early rice crop, and again in late July with a late rice crop. Root measurements were not made at this site, so the root mass remaining at the end of the season was estimated from the final biomass using a method similar to that used in the Maligaya experiments (assuming the same root-shoot ratio of 11.8%). This gave a value of 1,530 kg root DM ha^{-1} at the end of October. Model simulations predicted that about 92 kg root DM ha^{-1} (6% of the original total) was remaining at the start of early June (T1 and T3) with a distribution of 0%, 23%, and 77% between the three pools. A similar procedure gave 980 kg root DM ha^{-1} (20%:70%:10%) for treatments 2 and 4. For all four treatments, it was estimated that 820 kg DM ha^{-1} (= 5 cm) of stubble was remaining from the previous crop.

In the 1995 experiment, treatment 1 received 120 kg N ha^{-1} of urea, while treatments 2, 3, and 4 received 2000 kg DM ha^{-1} of *S. rostrata* green manure and 82 kg N ha^{-1} as urea. In the 1996 experiment, all treatments received 120 kg N ha^{-1} of urea. Irrigation was set to maintain a floodwater depth of 5 cm automatically, except during the times of drainage. In all treatments, the field was drained for a short time before harvest, and in addition, various midseason drainage regimes, ranging from one to three periods, were imposed.

Genotype parameters for the varieties Chunjiang 06, Jin 23A/71, Zhongyou 906, Xiu-shui 11, Shan-you 10, and Il-yiu 1568 used in the experiments were estimated by adjusting the length of the basic vegetative period to match observed phenological dates. Weather data recorded at the site over the period of the experiments were used to run the model.

Sensitivity analysis

The model was used to evaluate the sensitivity of seasonal CH_4 emissions to changes in various parameters. In each case, the 1996 weather data at IRRI, Maahas soil parameters, and IR72 genotype parameters were taken as the standard conditions. In most cases, to avoid complications due to excess water from rainfall, simulations were made for the DS only, with irrigation being set to automatically maintain the floodwater depth at 5 cm throughout the growing season until harvest. Fertilizer was applied at the rate of 120 kg N ha^{-1} as urea in four equal splits. For both seasons, the quantity of stubble and root biomass left from the previous crop at the start of the simulation was standardized at 500 and 400 kg DM ha^{-1}, respectively.

For the sensitivity analysis, the model was run several times, with the parameter being evaluated varying in a number of steps over a predefined range, with all other parameters being held constant at the standard value. We recognize that this 'one factor at a time' approach has limitations in that it does not explore all of the input space and does not account for interactions between the input variables (Saltelli, 1999), but we consider that the approach is adequate for our purposes, particularly as we are more interested in investigating the response surface of CH_4 emissions generated by variations in particular inputs rather than the rank of input variables in terms of degree of sensitivity.

Influence of the crop

The presence of the crop can influence seasonal CH_4 emissions in two ways. First, it is a source of organic material through the loss of dead root material and exudation of carbon-containing compounds from the live roots (collectively referred to as rhizodeposition). Second, the aerenchyma in the stem can act as a conduit for gaseous exchange, allowing O_2 from the atmosphere to reach the anaerobic soil and CH_4 to be transported easily from the soil to the atmosphere. These two influences will tend to counteract each other — an increase in the size of the root system should result in higher rates of rhizodeposition, thereby increasing the substrate available for methanogenesis. On the other hand, increasing the size of the rhizosphere so that more O_2 can reach the soil will result in greater inhibition of CH_4 production and a greater fraction of that which is produced being oxidized by methanotrophs. However, the enhanced transport of CH_4 from the soil to the atmosphere via the aerenchyma reduces its residence time in the soil, thereby lessening the chance that it will be oxidized. The actual influence of the root system size on seasonal CH_4 emissions, therefore, depends on the balance between these different processes.

To start with, we evaluated the sensitivity of seasonal CH_4 emissions to changes in the parameters representing the two components of rhizodeposition, the root death coefficient (∂_r, d^{-1}) and the specific root exudation rate (ε_r, mg C (g root)$^{-1}$ d^{-1}). Values of ∂_r were varied from 0 to 0.05 d^{-1}, and from 0 to 5 mg C (g root)$^{-1}$ d^{-1} for ε_r. The values of ε_r spanned those of 0.6-1.2 mg C (g root)$^{-1}$ d^{-1} reported by Lu et al. (1999) and the 5 mg C (g root)$^{-1}$ d^{-1} of Wang et al. (1997). While each parameter was being analyzed, the other was held at its standard value; these were 0.02 d^{-1} and 0.913 mg C (g root)$^{-1}$ d^{-1} for ∂_r and ε_r, respectively.

We then investigated how changes in transmissivity to gaseous transfer per unit root length (λ_r) influence seasonal CH_4 fluxes in rice soils, by running the model with values of the λ_r parameter varying from 0 to 3×10^{-5} m air (m root)$^{-1}$, using weather data for the 1996 dry season at IRRI. A second set of simulations was then made varying the size of the root system by modifying root growth rate calculated in the model by a multiplier factor within the range of 0.5-2.0. This gave a fourfold range of root-shoot values from 0.05 to 0.22, allowing an evaluation of the possible plant breeding strategy of selecting for genotypes with different root sizes as a mitigation option.

Initial size of oxidized alternative electron acceptor pool

To investigate the sensitivity of seasonal CH_4 emissions to the initial size of the oxidized *AEA* pool, the model was run using weather and crop management data for the dry season, with the irrigation option set to maintain the floodwater depth at 5 cm. For simplicity, it was assumed that there was no midseason or end-of-season drainage. Soil analysis data from Yao et al. (1999) was used to determine the range of likely values of the effective *AEA* pool size from 8 to 40 mol C_{eq} m^{-3}. For comparison, the Maahas soil at IRRI was estimated to have an initial *AEA* pool size of 26.5 mol C_{eq} m^{-3} (Matthews et al., 2000a), about midway within the range. Two scenarios were evaluated—that when a large amount of organic material (i.e., 10 t ha^{-1} rice straw) was added to the soil at the start of the season and that when no organic material was added.

Seasonal temperature

To investigate how CH_4 production over the season is influenced by mean seasonal temperature, the model was run using weather data for the 1996 DS and WS. To vary the temperature throughout the season, the 'Environmental Modifications' facility of the CERES-Rice model was used—the recorded daily maximum and minimum temperatures were adjusted by amounts ranging from −5 °C to +5 °C in 0.5 °C increments.

Floodwater depth

The model of Cao et al. (1995), based on field data from subarctic conditions (Sebacher et al., 1986), assumes a linear relationship between the rate of CH_4 emission and floodwater depth up to a depth of 10 cm beyond which there is no further increase. To investigate this further, we ran the model for the WS and DS at IRRI with the irrigation option set to maintain the floodwater at specified 'nominal' depths ranging from 0.0 to 5.0 cm. In the DS, these nominal depths were accurately maintained, but in the WS, floodwater depths exceeded the nominal depths on occasions when rainfall was high, as excess floodwater was not drained from the field.

Length of midseason drainage period

Midseason drainage of rice fields has been proposed as a possible mitigation option that farmers could practice to reduce CH_4 emissions. As this could potentially reduce the yields they obtain and therefore reduce the likelihood of their adopting the practice, it is useful to evaluate the likely effect of duration of drainage period on both CH_4 production and crop yields. For this, we ran the model for both the WS and DS at IRRI, with the field being drained at 20 d after planting for varying lengths of time ranging from 0 up to 30-d duration. For simplicity, we also assumed no drainage before harvest at the end of the season. When the field was not drained, the model's automatic irrigation facility maintained the floodwater level at 5 cm.

Type of organic amendments

To investigate the effect of the type of organic amendments applied on CH_4 emissions and crop performance, two sets of simulations were made — one in which the amount of C applied was constant, but the amount of N varied, and the second on which the amount of C applied varied, but the amount of N was constant. In the first set of simulations, the model was run with 3000 kg DM ha^{-1} of organic material of different C/N being applied. A range of C/N from 5 to 100 kg C (kg N)$^{-1}$ was generated by assuming that C concentration was 0.31 kg C (kg DM)$^{-1}$ with the N concentration of the material varying from 0.003 to 0.062 kg N (kg DM)$^{-1}$. In the second set of simulations, the same C and N concentrations as in the first set were used, but the amount of applied organic material varied from 500 to 10,000 kg DM ha^{-1} in such a way as to maintain the same amount of organic N applied in each case. In all simulations, the organic amendments were applied 22 d before planting when the field was first flooded. As previously, the quantity of stubble and root biomass left from the previous crop at the start of the simulation was standardized at 500 and 400 kg DM ha^{-1}, respectively.

Sulfate fertilizers

Sulfate ions are among the alternative electron acceptors (*AEA*) that can be used in the oxidation of organic carbon compounds, in this case by the sulfate-reducing bacteria which compete with the methanogenic bacteria for substrate. Addition of SO_4^{2-} ions in fertilizer, therefore, has the potential to increase the size of the *AEA* buffer (see Matthews et al., 2000a for details), thereby decreasing the proportion of organic material being reduced to CH_4. Examples of commonly used sulfate-containing fertilizers are ammonium sulfate and phosphogypsum (PG) (K_2SO_4).

To evaluate the effect of the level of applied PG on seasonal emissions of CH_4, the model was run using standard values for all parameters, but varying the amount of PG applied as fertilizer from 0 to 10,000 kg PG ha^{-1}. The model takes into account the addition of SO_4^{2-} ions by assuming that 1 mole of SO_4^{2-} is used to oxidize 2 moles of substrate carbon (i.e., 96 kg SO_4^{2-} oxidizes 24 kg C) — the amount of applied SO_4^{2-} is therefore converted to the *AEA* units of C equivalents (Matthews et al., 2000a), by multiplying by 24/96. The proportion of SO_4^{2-} by weight in PG is 55%, of which it is assumed (similarly to other fertilizers taken account of in the CERES-Rice model) that a somewhat arbitrary 30% of that applied enters the soil and is available for both uptake by the plant roots and participation in the *AEA* pool dynamics, the remainder being dissolved in the floodwater and not available.

Percolation rate

The sensitivity of seasonal CH_4 emissions to rates of loss of CH_4 by leaching beyond the soil profile was investigated by running the model with the rate of percolation of floodwater through the profile set at values ranging from 0 to 10 mm d^{-1}.

Results

Model performance

A comparison of the observed and predicted aboveground biomass values is shown in Figure 1. In general, there was good agreement, although there were three outlying points representing treatments in the DS in which there was a midseason drainage. These are discussed in more detail later. A comparison of the observed and predicted root biomass values and grain yield values is shown in Figures 2 and 3, respectively. Again, agreement was good, with the exception of the same three treatments.

A comparison of the observed and predicted seasonal CH_4 emission values is shown in Figure 4. There was some scatter, but agreement was generally good.

Influence of the crop

The predicted contributions to the total amount of substrate available for methanogenesis over the season

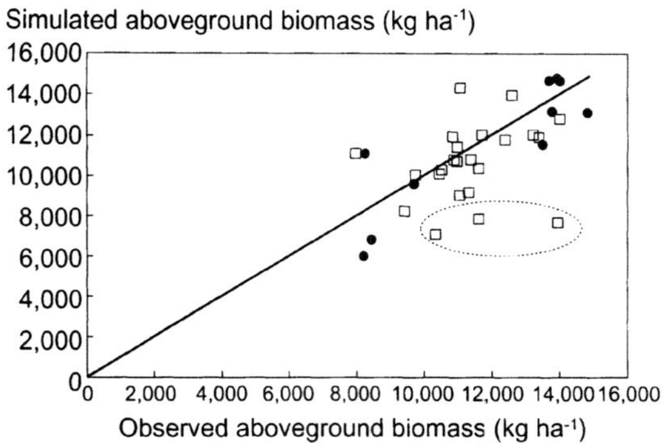

Figure 1. Comparison between observed and simulated total aboveground biomass values for the IRRI (open squares) and Hangzhou (filled circles) experiments. The dotted line encloses the three experiments at IRRI in which there was a midseason drainage during the dry season (see text for discussion)

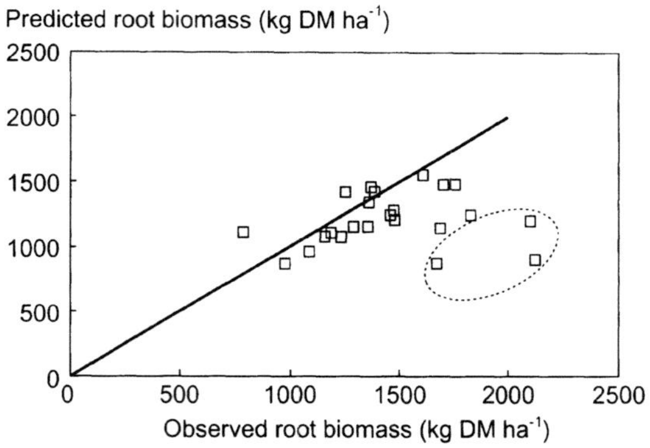

Figure 2. Comparison between observed and simulated root biomass values for the IRRI experiments. Root measurements were not made at the other sites. The dotted line encloses the three experiments in which there was a midseason drainage during the dry season (see text for discussion). Straight line indicates the 1:1 line

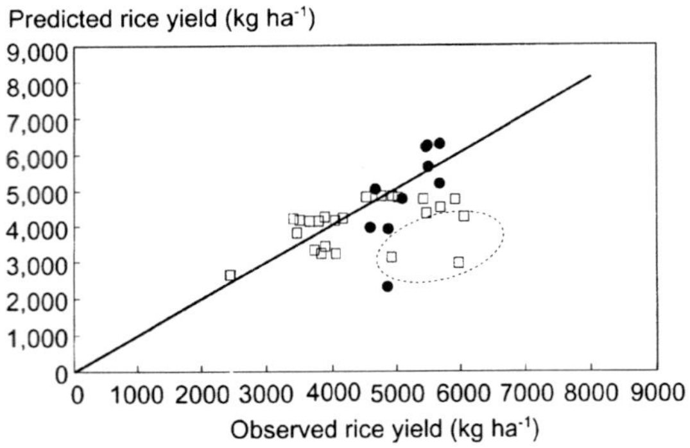

Figure 3. Comparison between observed and simulated grain yield values for the IRRI (open squares) and Hangzhou (filled circles) experiments. The dotted line encloses the three experiments at IRRI in which there was a midseason drainage during the dry season (see text for discussion). Straight line indicates the 1:1 line

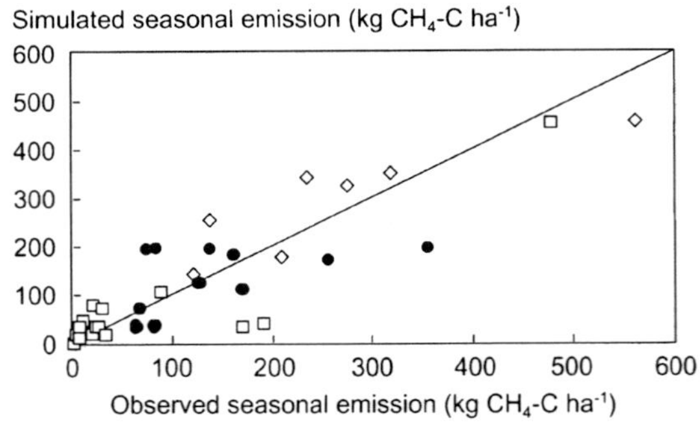

Figure 4. Comparison between observed and simulated seasonal CH_4 emissions for the IRRI (open squares), Maligaya (open diamonds), and Hangzhou (filled circles) experiments. The 1:1 line is also shown

by various sources are shown in Table 2. For this, the model was run for the 1996 DS at IRRI, with a) no organic amendments added and b) 3,000 kg DM ha^{-1} of green manure added. Values of all other parameters were set at standard values described above. With no organic amendments, rhizodeposition (i.e., root exudates + dead root tissue) contributed about 37% of the total substrate, previous crop residues a further 22%, with the remaining 41% coming from long-lived SOM (humus). Around 59%, therefore, originated from the rice crop in one way or another. With 3,000 kg DM ha^{-1} of green manure added, the absolute quantities from each of the above sources remained the same, but the proportions fall to 21%, 12%, and 25% for rhizodeposition, previous crop residues, and humus, respectively, with the remaining 42% coming from green manure.

The relative sensitivity of seasonal CH_4 emissions to the root death coefficient (∂_r) and the specific root exudation rate (ε_r) are shown in Figure 5. Of the two parameters, emissions were most sensitive to ∂_r, as indicated by the steeper gradient of the relative response curve. Errors in the estimation of this parameter, therefore, could have a significant influence on seasonal CH_4 emission estimates. For ε_r, the response was much less sensitive, so that even with the fivefold difference in estimates of ε_r from the studies of Lu et al. (1999) and Wang et al. (1997), seasonal CH_4 emissions differed by only 32%.

The influence of changes in the root transmissivity parameter (λ_r) on the different seasonal CH_4 fluxes is shown in Figure 6. The effect of increasing λ_r was to decrease overall seasonal emissions, but this was mainly due to a decrease in the amount of CH_4

Table 2. Predicted contributions (kg C ha^{-1} season^{-1}) from various sources to total methanogenic substrate in rice fields growing in the dry season at IRRI with (a) no organic amendments and (b) 3,000 kg DM ha^{-1} of green manure added 20 d before planting. Figures in parentheses represent percentage of total substrate

	Residues	Humus	Exudates	Dead roots	OM amendments	Total
No OM added	227 (22%)	418 (41%)	88 (9%)	285 (28%)	0 (0%)	1018 (100%)
3,000 kg DM ha^{-1} added	223 (12%)	452 (24%)	88 (5%)	290 (16%)	793 (43%)	1845 (100%)

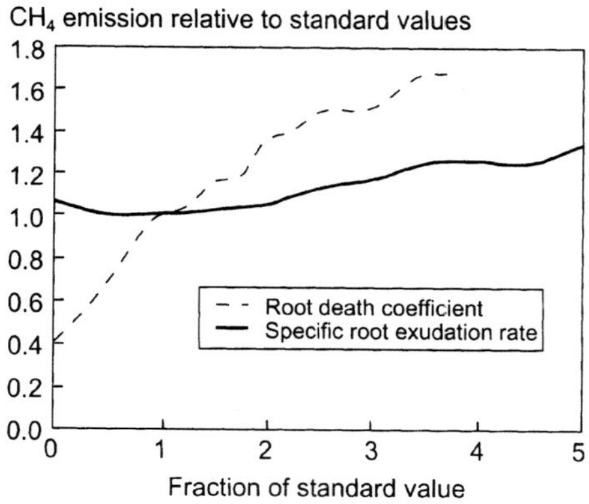

Figure 5. Sensitivity analysis of the two components of rhizodeposition, root death coefficient and specific root exudation rate. Standard values of these two parameters are 0.02 d^{-1} and 0.913 mg C (g root)$^{-1}$ d^{-1}, respectively

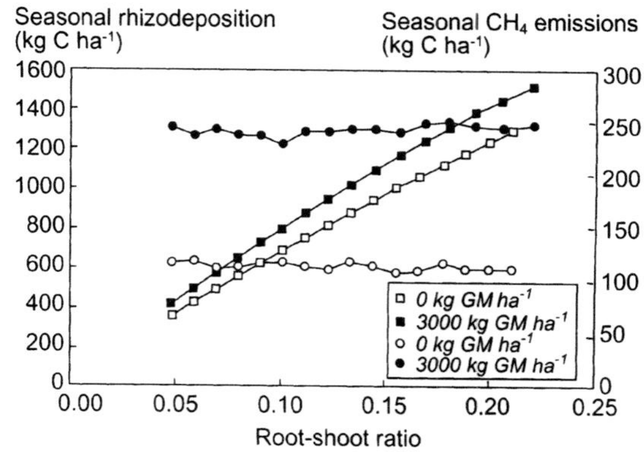

Figure 7. Predicted effect of changing the root-shoot ratio of the crop on seasonal CH_4 emissions (circles) and on total rhizodeposition (root exudates and dead roots) (squares) over the season. Standard root-shoot ratio is 0.1

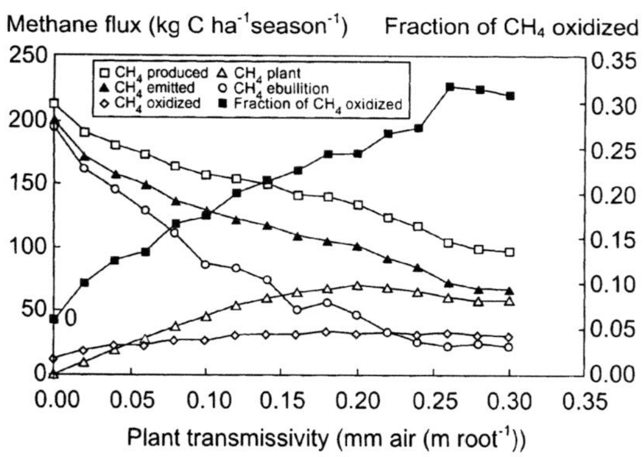

Figure 6. Effect of changes in root transmissivity parameter (λ_r) on seasonal CH_4 fluxes in the dry season at IRRI

produced due to the inhibitory effect of increased O_2 concentrations on methanogenic activity and, to a lesser extent, on an increase in the amount of CH_4 oxidized to CO_2 by increased methanotrophic activity. Increasing λ_r also increased the fraction of CH_4 emitted through the plants but reduced the fraction through ebullition.

The effects of varying the root-shoot ratio of the crop on seasonal CH_4 emissions and on total rhizodeposition over the season is shown in Figure 7. Despite there being around a fourfold range in the amount of organic material from rhizodeposition available for methanogenesis, seasonal CH_4 emissions hardly varied over this range due to the opposing effects of reduced CH_4 production and an increase in the fraction of this CH_4 produced that is oxidized by methanotrophs.

Initial size of oxidized alternative electron acceptor pool

Results of the sensitivity analysis of the initial size of the oxidized *AEA* are shown in Figure 8. In both cases, there was a steady decline in seasonal CH_4 emission as the size of the oxidized *AEA* pool size increased, declining to nearly zero when no organic material was added. Even when a large quantity of rice straw was added, seasonal CH_4 emissions declined by 54% over the range of *AEA* pool sizes considered. It would therefore seem that the initial size of the oxidized *AEA* pool is a major factor in determining the emission of CH_4 from different soils, suggesting that for accurate estimation of CH_4 emission from rice soils, accurate estimates of this pool are essential.

Seasonal temperature

The predicted effect of mean seasonal temperature on CH_4 emissions is shown in Figure 9. As the temperature rose from 20 °C, emissions were predicted to decrease until about 30 °C was reached, beyond which they began to rise again. Except at the lower temperatures, there was close agreement between the two seasons. Closer examination showed that the pattern of the response to temperature was almost entirely explained by the effect on crop duration—the crop matured fastest at 30 °C, the optimum temperature for development, but at temperatures on either side of this value, maturity was progressively delayed. The longer the crop was in the ground, the more time there was available for CH_4 production. There was little effect of mean sea-

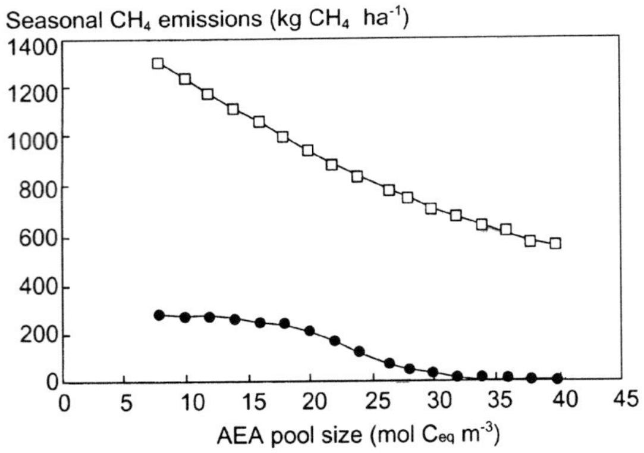

Figure 8. Sensitivity of seasonal CH_4 emissions to changes in size of the alternative electron acceptor (*AEA*) pool. The open squares represent the response when 10 t ha^{-1} of rice straw was added at the start of the season, while the filled circles represent the response when no organic amendments were added

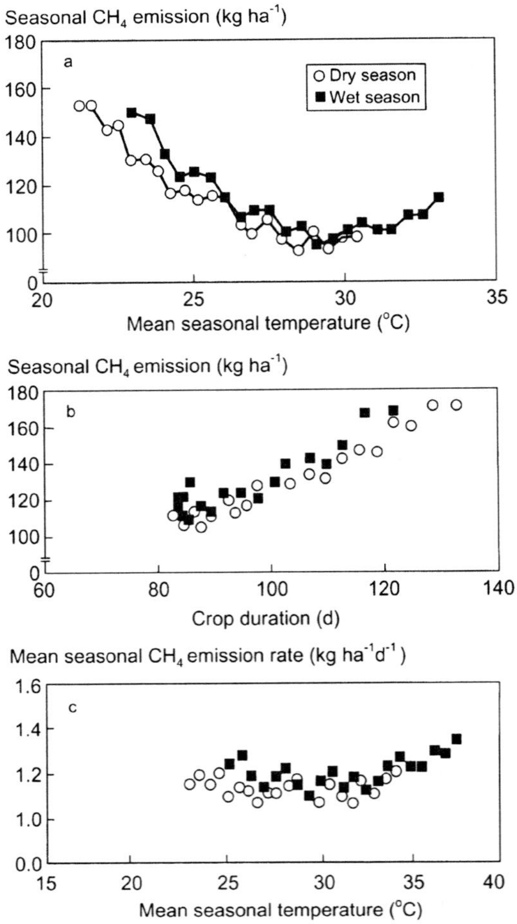

Figure 9. (a) Predicted effect of mean seasonal temperature on CH_4 emissions from rice fields in the wet and dry seasons at IRRI. (b) Relationship between crop duration and seasonal CH_4 emissions. (c) Mean seasonal CH_4 emissions plotted against mean seasonal temperature

sonal temperature predicted on the average rate of CH_4 production over the season.

Floodwater depth

The predicted relationships between the 'nominal' floodwater depth and seasonal CH_4 emissions for the WS and DS are shown in Figure 10. In all cases, there was an increase in emissions as the depth of water increased from 0 to 2 cm, but there was no further increase in emissions beyond this depth. Closer examination showed that 2 cm was the depth of water that caused the O_2 concentration in the top layer of the soil to fall to almost zero, thereby causing the soil to become anaerobic and favorable for CH_4 production. Differences between the WS and DS relationships were due to the fact that the actual floodwater depth in the WS was sometimes in excess of the 'nominally' maintained depth in periods of high rainfall as excess floodwater was not drained from the field; the effective depth of water was greater than the 'nominal' value in such cases, resulting in higher CH_4 production and emission.

Length of drainage period

The predicted effects of duration of midseason drainage period on seasonal CH_4 emissions and grain yields are shown in Figure 11. There was a steady decline in CH_4 emissions in both WS and DS as the duration of the drainage period increased and the proportion of time the soil was under anaerobic conditions decreased. However, the effect of this drainage on crop yields depended on the season. In the WS, there was little effect on yields with drainage periods up to 30 d in length mainly because rain during this period was able to maintain soil water status at a level sufficient to meet crop water requirements, but at the same time there being sufficient air in the soil profile to reduce the amount of CH_4 production. In the DS, however, at drainage durations longer than about 6 d, there was a decline in yields to about 50% of the fully irrigated value when the field was drained for 30 d.

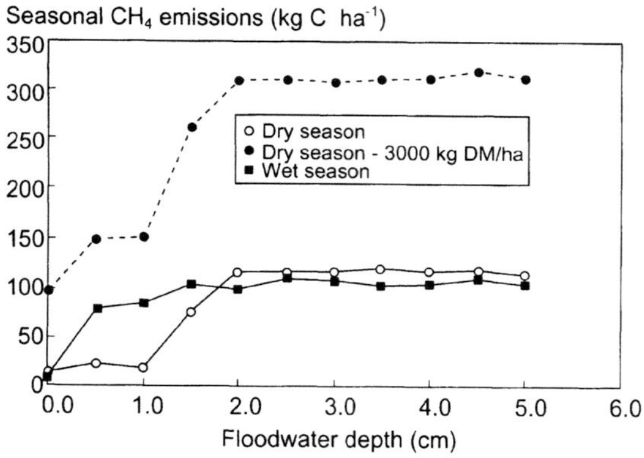

Figure 10. Predicted relationships between 'nominal' floodwater depth and seasonal CH_4 emissions. The two solid lines represent the wet and dry seasons at IRRI with only crop residues present at the start of the season. The dashed line represents the dry season with 3000 kg DM ha^{-1} of rice straw added 18 d before planting

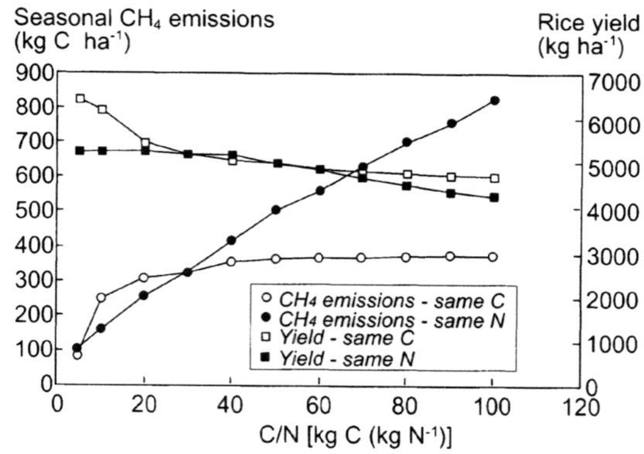

Figure 12. Predicted response of seasonal CH_4 emissions and grain yield to application of 3,000 kg DM ha^{-1} of organic amendments with varying C/N ratios. Filled circles represent CH_4 emissions; open squares represent grain yield

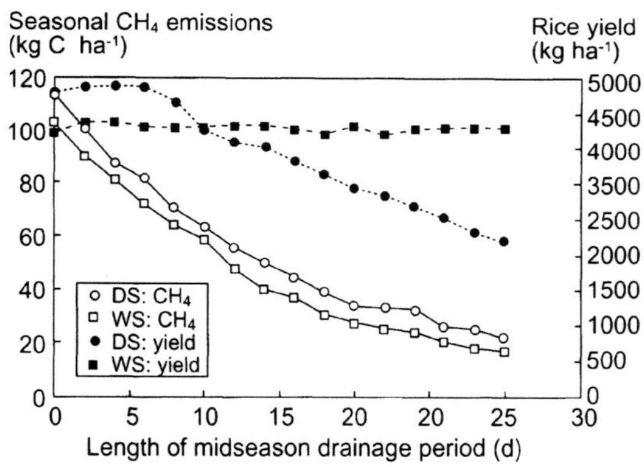

Figure 11. Predicted effect of length of midseason drainage on seasonal CH_4 emissions and grain yields in the dry and wet seasons at IRRI

Type of organic amendments

The predicted effects of organic amendments with different C/N are shown in Figure 12. In the first set of simulations (where the amount of C applied was the same in each case), as the C/N of the added organic material increased, there was a rapid increase in the predicted seasonal CH_4 emissions until a C/N of around 40 kg C (kg N)$^{-1}$ was reached, beyond which there was a leveling off. When the amount of C applied varied but the amount of N applied remained the same, seasonal CH_4 emissions increased almost linearly in response to changes in C/N. In both sets of simulations, there was a general decline in grain yields predicted with an increase in C/N, the decline being steeper in the constant C simulations at lower C/N.

In the case where the amount of C being applied was the same, the reasons for the predicted response of seasonal CH_4 emissions to changes in C/N are of interest. Closer examination showed that the lower CH_4 emissions at the lower C/N were due to more of the C in the applied organic matter being immobilized by microbial activity stimulated by the higher levels of N present. Although this C started to be released later in the season through death of microbial biomass, it was not soon enough for all to become available, so that by the end of the season, much was still locked up and therefore not able to contribute to methanogenesis. In the second set of simulations, as the amount of organic C being applied was increasing proportionally to C/N, the predicted linear response of CH_4 emissions is to be expected.

The decline in crop yields at the higher C/N in the first set of simulations was due to lower quantities of organic N being supplied through the amendments. In the second set, although the amount of organic N applied was the same at each C/N, the proportion of this N being mineralized and becoming available for use by the crop declined at the high C/N due to the influence of C/N on mineralization rate incorporated into the model (see Figure 2 in Part I of this series).

At the lower C/N, the model also predicted higher rates of root exudation and root death due to increased

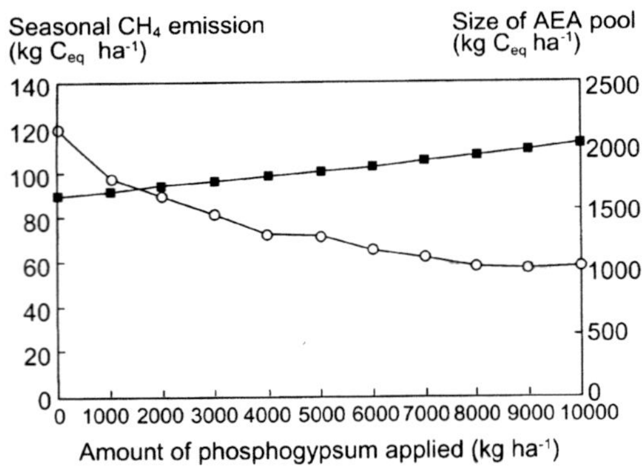

Figure 13. Predicted effects of applying varying amounts of phosphogypsum (K_2SO_4) on seasonal CH_4 emissions (open circles) and on the size of the *AEA* pool (filled squares)

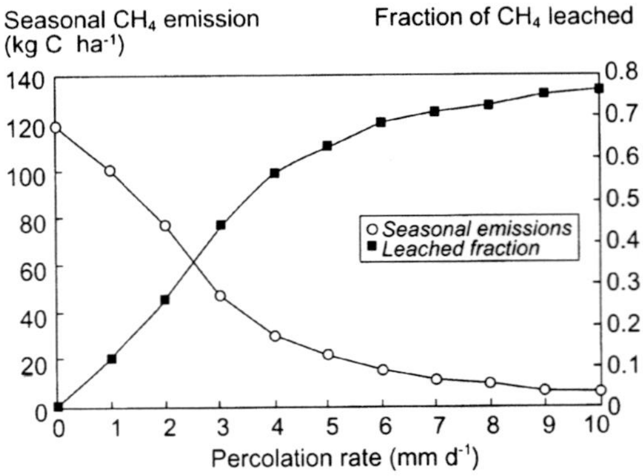

Figure 14. Predicted effects of percolation rate on seasonal CH_4 emissions and on the fraction of the CH_4 produced by methanogenesis that is lost by leaching

crop growth, but these were not of sufficient magnitude to significantly offset the effect of the immobilized C or the reduced supply of C.

Sulfate fertilizers

The effect of varying the amount of applied phosphogypsum from 0 to 10,000 kg PG ha^{-1} on seasonal CH_4 emission rates is shown in Figure 13. There was an initial rapid decline in emissions as the application rate increased to about 4,000 kg PG ha^{-1} (~1800 kg SO_4^{2-} ha^{-1}), after which the response leveled off. As would be expected, there was a linear increase in the size of the *AEA* pool from about 1,600 kg C_{eq} ha^{-1} to 2,000 kg C_{eq} ha^{-1} over the range.

Percolation rate

Seasonal CH_4 emissions were predicted to be highly sensitive to percolation rates between 0 and 4 mm d^{-1}, dropping to about 25% of their initial value as percolation rates increased over this range (Figure 14). There was a leveling off in seasonal emissions predicted at higher percolation rates. The decrease in seasonal emission rates was reflected in the increasing proportion of the CH_4 produced by methanogenesis that was lost by leaching.

Discussion

In general, there was good agreement between the simulated and observed values of aboveground biomass, root biomass, and grain yield of the crops, although there were consistent discrepancies in some cases (Figures 2, 3, & 4). These corresponded to treatments at IRRI with a midseason drainage in the DS, for which the model predicted a decline in biomass and final yield as a result of water stress suffered by the crop during this period, whereas the measurements show no effect. In each case, the drainage period was around 21 d. The amount of plant-extractable water (PESW) held in the soil at the start of this period is difficult to estimate due to the presence of the plastic sheet preventing free percolation, but it probably lies within the range of the 48 mm held between the drained upper limit (DUL) and the drained lower limit (DLL), and the 83 mm held between the saturated water content (SAT) and the drained lower limit (to a depth of 50 cm in each case). The model actually calculates 83 cm. Taking the maximum of these two estimates and using the potential evaporation calculated using the Penman-Monteith formula, it can be calculated that all of the available water would be gone after 15 d. It is therefore difficult to see how the crop did not suffer from water stress in the last week of the drainage period with a resulting decline in biomass and yield, unless the presence of the plastic sheet altered the hydrological characteristics of the soil (e.g., pooling of water at the bottom) so much that the model cannot describe it. Certainly, lower grain yields were observed by Yagi et al. (1994) in intermittently irrigated rice fields in Japan.

Agreement between observed and simulated values of seasonal CH_4 emissions was good, particularly as it was across three different rice-growing environments, and gives some confidence in the use of the model for upscaling experimental results to national and regional levels in Part IV of this series (Matthews et al., 2000b).

The results show that the influence of the crop on seasonal CH_4 emissions is considerable, perhaps even more so than the soil itself, contributing some 600 kg C ha^{-1} $season^{-1}$ with a large part of this coming from rhizodeposition by the current crop. This value is within the range obtained by Cao et al. (1996). The transmissivity of the rice plant to gaseous transfer can also have a large effect on the amount of this substrate that is actually converted into CH_4 and on the fraction of this CH_4 that is oxidized to CO_2. Variation in these characteristics, therefore, offers scope for varietal selection to reduce CH_4 emissions from rice cultivation. Indeed, various studies have reported differences in CH_4 emission potential between rice genotypes (e.g., Parashar et al., 1990; Lindau et al., 1995; Watanabe et al., 1995a; Mitra, 1999), and even more importantly, that low emission potential can be achieved while still maintaining a high yield potential (Wang et al., 1997). Compared with other mitigation strategies such as intermittent drainage, which require substantial changes in farmer practice, new varieties may be adopted much more readily by farmers.

Rhizodeposition was predicted by the model to contribute about 37% of the total substrate, a proportion that agrees closely with the 30-40% estimated by Cao et al. (1996). High rates of rhizodeposition not only increase the amount of substrate available for methanogenesis but also represent a loss of assimilates for the crop and can therefore be detrimental to yields. Reduction of the rates of rhizodeposition, therefore, would likely be beneficial to both yields and CH_4 emissions. Unfortunately, the quantification of the two component rates of crop rhizodeposition, root exudation and root death, is the part with the largest uncertainty.

The state of knowledge on exudation from rice plants is rudimentary, but recent studies have reported differences between genotypes in the amount of C lost by root exudation (Wang et al., 1997; Lu et al., 1999), which, within each study, seems to be more due to the quantity of roots present rather than the exudation rate per unit length of root, or specific exudation rate. However, specific exudation rates varied considerably between these two studies — the data of Wang et al. (1997) indicate a value of around 5 mg C (g root)$^{-1}$ d^{-1} while that of Lu et al. (1999) varies from 0.6 to 1.6 mg C (g root)$^{-1}$ d^{-1}, depending on the stage of growth of the crop. The sensitivity analysis described above showed that this difference is significant — there is a 34% increase in the total substrate available (assuming other sources remain constant), with a similar rise of 32% in the seasonal CH_4 emissions (Figure 5). Clearly, further work is required to clarify these rates and also the factors that affect them. It is well known that mechanical impedance, presence of toxic elements (e.g., Pb, Cd, and Al), nutrient deficiencies, water status of the growing medium, and nitrogenase activity can all affect the amount and composition of root exudates (Wassmann & Aulakh, 2000). Similarly, it is not known how variation in the constituents of root exudates affects rates of methanogenesis. Lin and You (1989) noted that root exudates from rice contained varying amounts of organic acids, carbohydrates, and amino acids. Among the organic acids, citric was highest, followed by malic, succinic, and lactic acid, although there was a large variation in components and contents of root exudates of different varieties.

Estimates of root death rates are even more uncertain. To our knowledge, there have been no studies on rice to measure the amount of C lost over a season in this way. We have used a value for the relative root death rate (∂_r) of 0.02 d^{-1}, which produces reasonable behavior in terms of the CH_4 dynamics. However, this value is based only on the figure for total rhizodeposition being 5-20% of the aboveground biomass at final harvest obtained by Shamoot et al. (1968) in a greenhouse study with 11 plant species which did not include rice. To some extent, errors in the estimation of root death rates can be offset by negatively correlated errors in the root exudation rates, so long as the total rate of rhizodeposition is not affected significantly. For example, the possible higher specific exudation rate obtained by Wang et al. (1997), discussed previously, may suggest that the root death rates are lower than 0.02 d^{-1}. Clearly, more research in this area is required to be more certain of the relative contribution of each source of methanogenic substrate.

Our results suggest that the transmissivity of the plant to gaseous transfer may also be of considerable importance (Figure 6)—increasing the root transmissivity parameter (λ_r) has the effect of both reducing the amount of CH_4 produced due to the toxic effect of O_2 on the enzyme systems of the methanogens, and increasing the proportion of CH_4 produced that is oxidized to CO_2 by the methanotrophs. The model predictions also suggest that it is the first of these two effects that is the greatest—CH_4 production was reduced to a much greater extent than the increase in the amount of CH_4 oxidized (Figure 6).

Selecting for genotypes that have a greater conductance to gaseous transfer, therefore, would seem to be a strategy to follow to reduce the amount of CH_4 emitted. Increasing the number of tillers may be one way of achieving this, although some studies (e.g., Mariko et al., 1991; Wang et al., 1997) have found that CH_4 emission rates increase as tiller number increases. However, most of these studies have measured the rates of CH_4 emission through plants only and, as shown in Figure 6, the flux of CH_4 through the plant increases (up to a plateau) even though total CH_4 emissions (which include ebullition) are decreasing. There is some doubt, therefore, whether these pot experiments can be reliably extrapolated to field conditions. The reciprocal pattern of behavior of the plant and ebullitive fluxes suggest that any CH_4 that is produced and not oxidized will be emitted somehow, either through the plant or by ebullition.

Our results also suggest that there is little gain to be made in selecting genotypes with differently sized root systems alone as a plant breeding strategy to reduce CH_4 emissions. While reducing the size of the root system was predicted to reduce the amount of rhizodeposition over a season (Figure 7) and therefore the amount of substrate available for methanogenesis, it also reduced the size of the conduit for O_2 to enter the soil to both inhibit the production of CH_4 and increase the fraction that is oxidized by methanotrophs. These two opposing influences, therefore, seem to cancel each other out, resulting in the stable emission response seen in Figure 7. This presupposes, however, that the value of the root transmissivity parameter (λ_r) remains constant, and indeed, that the overall conductivity of the plant to gaseous transfer is determined by the quantity of roots present. Certainly, the porosity of the roots to gaseous diffusion may vary—Kludze et al. (1993), for example, found that root porosity was increased threefold in flooded plants compared with nonflooded or drained plants. This enhanced the transport of O_2 to the roots, which increased by more than a factor of three. Whether similar variations exist between genotypes needs to be clarified—Wassmann et al. (1998) have suggested it is possible, and Wang et al. (1997) did find differences in the proportion of air spaces in the roots of three rice cultivars during the heading and ripening stages, although these differences were not evident earlier. Root air space differences did not correlate with the oxidation potential of the roots in this study, however.

It may also be that the main site of resistance to gaseous movement is the transition from root to stem (Butterbach-Bahl et al., 1997) and not the quantity of roots present, so that the main effect of a larger root system would be on increased rhizodeposition rates. This may explain the observation by Lindau et al. (1995) that tall genotypes emit more CH_4 than semidwarf varieties, but as unfortunately no plant biomass data were presented in this paper, this must remain conjecture. Similarly, Wang et al. (1997) found that genotypes with the highest root biomass also had the highest CH_4 emission potential. Clearly, these various uncertainties need to be explored with the model and further experimentation. Cultivar selection may be crucial for mitigating CH_4 emissions—a thorough understanding of the mechanisms involved is required, therefore, to direct efforts toward developing high-yielding rice plants with a limited emission potential.

The main contributor to the predicted decrease in seasonal emissions with increased temperature was the shortening of crop duration, with mean emission rates over the season not being greatly affected. These effects are similar to those observed experimentally; although diel emission rates are strongly correlated to temperature, mean seasonal rates are only poorly so (Kimura & Minami, 1995). Closer examination showed that although higher temperatures brought about higher decomposition rates, and hence higher CH_4 emission rates early in the season from previous crop residues, these rates fell close to those at lower temperatures once the rapidly decomposable FOM pools had disappeared. Thus, mean seasonal emission rates were influenced more by the total amount of C in the system, which did not vary much as a result of the higher temperatures.

Results from our study suggest that less depth of floodwater is required to ensure near-anaerobic conditions than was assumed previously (e.g., Cao et al., 1995). No further increase in CH_4 emissions were found after about 2 cm of water, compared with the 10 cm observed in subarctic conditions (Sebacher et al., 1986). Experimental work is required to confirm this value for rice fields in tropical environments.

Midseason drainage of rice fields has been proposed as a possible mitigation option that farmers could practice to reduce CH_4 emissions, but as this could potentially reduce their yields, the effect of duration of drainage period on both CH_4 production and crop yields is of interest. Our results suggest that midseason drainage is a viable practice in the WS when there is likely to be sufficient rainfall to meet crop water requirements without the field being flooded, with a subsequent reduction in the amount of CH_4 produced. In the DS, however, except in the case of relatively short drainage periods (less than 6 d), crop yields are likely to decline. Nevertheless, even if the field is drained for 6 d, the model suggests that there could be around a 25% reduction in CH_4 emissions with no loss in yield. Planners and policymakers, therefore, could use this information to decide to what extent farmers might have to be compensated for lost crop revenue, if CH_4 emissions from rice fields are to be reduced to a specified level.

Much work has been done on the effect on seasonal CH_4 emissions of incorporating varying amounts of rice straw to the field before planting (e.g., Sass et al., 1991; Nouchi et al., 1994), the results of which have been summarized by Denier van der Gon and Neue (1995) and Watanabe et al. (1995b). In general, adding rice straw leads to an increase in CH_4 emissions, as might be expected from the addition of more C to an anaerobic system. Similarly, Lindau et al. (1995) found much higher emissions in a ratoon rice crop due to the residues of the first crop being left in the field.

Rice straw, however, has a relatively high C/N of around 50 kg C (kg N)$^{-1}$ (Bronson et al., 1998). The strong links between C and N dynamics in the soil raise the question of whether the addition of organic material of different qualities has any effect on the emission of CH_4. Our results predict that material with a lower C/N (i.e., < 40 kg C (kg N)$^{-1}$) results in less CH_4 being emitted even though the amount of C being applied remains the same, the main reason, according to the model, being the greater immobilization of C in microbial biomass stimulated by the larger quantities of N present. Of course, this C would be emitted later as the microbes die, but a greater proportion of this will be after the crop is harvested when conditions are aerobic and would be emitted as CO_2 rather than CH_4. Bouwman (1991) summarized the literature on the effect of management practices on CH_4 emissions and concluded that although increasing the amount of organic fertilizers applied increased emissions, composted materials (with lower C/N) tended to cause a smaller increase. Similarly, in greenhouse experiments, Mariko et al. (1991) found that additions of rice straw compost resulted in a sixfold reduction in CH_4 emissions compared with uncomposted straw. Data on the C and N contents of various organic amendments are summarized by Kern et al. (1995) from which the following C/N can be calculated: animal manure 100 kg C (kg N)$^{-1}$, rice straw 51 kg C (kg N)$^{-1}$, compost 12 kg C (kg N)$^{-1}$, green manure 10 kg C (kg N)$^{-1}$, and rapeseed cake 8.7 kg C (kg N)$^{-1}$. Thus, applying green manure rather than rice straw would appear to be desirable, as not only is there likely to be a response in grain yield, but the increase in CH_4 emissions would also be less. Animal manure would appear to be the worst option in terms of reducing CH_4 emissions.

The use of sulfate fertilizers has been suggested as a way to reduce CH_4 emissions by increasing the size of the soil pool of alternative electron acceptors (Wassmann et al., 1993). The model predicts a significant effect of adding SO_4^{2-} to the soil, emissions being reduced by 50% when 10,000 kg ha^{-1} phosphogypsum (4500 kg SO_4^{2-} ha^{-1}) is added (Figure 13). This compares with a value of 43% reduction in emissions with addition of sulfate fertilizer at a rate of 685 kg SO_4^{2-} ha^{-1} obtained by Schütz et al. (1989) The model of van Bodegom et al. (1999) predicts a 3% reduction in CH_4 emissions when 400 kg ha^{-1} of ammonium sulfate (290 kg SO_4^{2-} ha^{-1}) is added. The MERES model differs from the van Bodegom model in that the effect of SO_4^{2-} is not simulated explicitly and instead is part of the general soil pool of alternative electron acceptors. As such, the partial competition of the sulfate-reducing bacteria with methanogens for C substrate is not taken into account. Similarly, we have assumed in MERES that the mixing ratio of added SO_4^{2-} between the floodwater and soil is 30%, although this figure is quite subjective. To some extent, these two assumptions will cancel each other out so that differences between the two modeling approaches are not likely to be large.

The model predicts that overall seasonal CH_4 emissions are quite sensitive to percolation rates in the range from 0 to 4 mm d^{-1}. Extremely high percolation rates of around 28 mm d^{-1} have been reported in northern India (Mitra, 1999) which probably explain the low CH_4 emission rates measured there (~25 kg CH_4 ha^{-1} season^{-1}). For comparison, average percolation (including seepage) rates in Philippine rice fields are about 2-4 mm d^{-1} (Wickham & Singh, 1978) depending on season. High percolation rates and the necessary high frequency of irrigation could influence CH_4 emission rates either by increasing the flux of O_2 dissolved in the irrigation water into the soil or by transporting CH_4 produced downward into groundwater, thereby preventing it from being emitted from the rice field into the atmosphere. It is also possible that the rapid flux of water through the profile transports the methanogenic substrate away before it can be acted upon by the methanogens (Yagi & Minami, 1990; Inubushi et al., 1992), although this is not currently accounted for in the model.

The results presented in this paper, therefore, indicate that the MERES model is capable of exploring quantitatively the major aspects of CH_4 production and emissions from rice fields. In the final paper in this series (Matthews et al., 2000b), we use the model together with the spatial databases described in Part III (Knox et al., 2000) to upscale experimental measurements of CH_4 emissions to national levels and to evaluate various mitigation options on the overall emission of CH_4 from each of the countries in the study.

Acknowledgements

The authors would like to thank the staff of the Soil and Water Sciences Division at the International Rice Research Institute for their tireless work in collecting and collating the data used in this study. The study was conducted with funding from the United Nations Development Programme's Global Environment Facility.

References

Arah JRM & Kirk GJD (2000) Modelling rice-plant-mediated methane emission. Nutr Cycling Agroecosyst (this issue)

Bouwman AF (1991) Agronomic aspects of wetland rice cultivation and associated methane emissions. Biogeochemistry 15:65-88

Bronson KF, Cassman KG, RWassman, Olk DC, van Noordwijk M. & Garrity DP (1998) Soil carbon dynamics in different cropping systems in principal ecoregions of Asia. In: R Lal, J Kimble, RF Follett & BA Stewart (eds) Management of Carbon Sequestration in Soil, pp 35-57, CRC Press, Boca Raton

Butterbach-Bahl K, Papen H & Rennenberg H (1997) Impact of gas transport through rice cultivars on methane emission from rice paddy fields. Plant, Cell & Env 20:1175-1183

Cao M, Dent JB & Heal OW (1995) Modelling methane emissions from rice paddies. Global Biogeochem Cycles 9:183-195

Cao M, Gregson, Marshall S, Dent B & Heal OW (1996) Global methane emissions from rice paddies. Chemosphere 55(5):879-897

Denier van der Gon HAC & Neue HU (1995) Influence of organic matter incorporation on the methane emission from a wetland rice field. Global Biogeochem Cycles 9:11-22

Inubushi K, Muramatsu Y & Umerayasi M (1992). Influence of percolation on methane emission from flooded paddy soil. Japanese J Soil Sci Plant Nutr 63:184-189

Kern JS, Bachelet & Tölg M (1995). Organic matter inputs and methane emissions from soils in major rice-growing regions of China. In: R Lal, J Kimble, E Levine & B.A. Stewart (eds) Soils and Global Change. Adv Soil Sci 189-198

Kimura M & Minami K (1995) Dynamics of methane in rice fields: emissions to the atmosphere in Japan and Thailand. In: S Peng, KT Ingram, HU Neue & LH Ziska (eds) Climate Change and Rice, pp 30-45. Springer Verlag, Berlin, Germany

Kludze HK, DeLaune RD & Patrick WH (1993). Aerenchyma formation and methane and oxygen exchange in rice. Soil Sci Soc Am J 57:386-391

Knox J, Matthews RB & Wassmann R (2000) Using a crop/soil simulation model and GIS techniques to assess methane emissions from rice fields in Asia. III. Databases. Nutr Cycling Agroecosyst (this issue)

Lin M & You C (1989) Root exudates of rice (*Oryza sativa* L.) and its interaction with *Alcaligenes faecalis*. Sci Agric Sin 22(6):6-12

Lindau CW, Bollich PK & DeLaune RD (1995) Effect of rice variety on methane emission from Louisiana rice. Agric Ecosyst Environ 54:109-114.

Lu Y, Wassmann R, Neue HU & Huang C (1999) Impact of phosphorus on root exudation, aerenchyma formation, and methane emission of rice plants. Biogeochemistry 47:203-218

Mariko S, Harazono Y, Owa N & Nouchi I (1991) Methane in flooded soil water and the emission through rice plants to the atmosphere. Environ Exp Bot 31:343-350

Matthews RB, Wassmann R & Arah J (2000a) Using a crop/soil simulation model and GIS techniques to assess methane emissions from rice fields in Asia. I. Model development. Nutr Cycling Agroecosyst (this issue)

Matthews R, Wassmann R, Knox JW & Buendia LV (2000b) Using a crop/soil simulation model and GIS techniques to assess methane emissions from rice fields in Asia. IV. Upscaling to national levels. Nutr Cycling Agroecosyst (this issue)

Mitra S (1999) Effect of rice cultivars on methane emission. Agric Ecosyst Environ 73(3):177-183

Nouchi I, Hosono T, Aoki,K & Minami K (1994) Seasonal variation in methane flux from rice paddies associated with methane concentration in soil water, rice biomass and temperature, and its modelling. Plant Soil 161(2):195-208

Parashar D, Rai CJ, Gupta PK & Singh N (1990) Parameters affecting methane emission from rice paddies. Indian J Radio Space Physics 20:12-17

Saltelli A (1999) Sensitivity analysis: could better methods be used? J Geophys Res 104(D3):3789-3793.

Sass RL, Fisher FM, Harcombe PA & Turner FT (1991) Mitigation of methane emissions from rice fields: possible adverse effects on incorporated rice straw. Global Biogeochem Cycles 5:275-287

Schütz H, Holzapfel-Pschorn A, Conrad R & Rennenberg,H (1989) A three-year continuous record on the influences of daytime, season and fertiliser treatment on methane emission rates from an Italian rice paddy. J Geophys Res 94:405-416

Sebacher DI, Harriss RC, Bartlett KB & Sebacher SM (1986) Atmospheric methane sources: Alaskan tundra bogs, an alpine fen and a subarctic boreal marsh. Tellus 38B:1-10

Shamoot S, McDonald L & Bartholomew WV (1968) Rhizodeposition of organic debris in soil. Soil Sci Soc Am Proc 32:817-820

van Bodegom PM, Leffelaar PA, Stams AJM & Wassmann R (2000) Modelling methane emissions from rice paddies: variability, uncertainty and sensitivity analyses of processes involved. Nutr Cycling Agroecosyst (this issue)

Wang B, Neue HU & Samonte HP (1997) Effect of cultivar difference ('IR72', 'IR65598', and 'Dular') on methane emission. Agric Ecosyst Environ 62:31-40

Wassmann R & Aulakh MS (2000) The role of rice plants in regulating mechanisms of methane emissions. Biol Fertil Soils 31:20-29

Wassmann R, Neue HU, Bueno C, Lantin RS, Alberto MCR, Buendia LV, Bronson K, Papen H & Rennenberg H (1998) Inherent properties of rice soils determining methane production potentials. Plant Soil 203:227-237

Wassmann R, Schütz H, Papen H, Rennenberg H, Seiler W, Dai AG, Shen RX, Shangguan XJ & Wang MX (1993) Quantification of methane emissions from Chinese rice fields (Zhejiang Province) as influenced by fertilizer treatment. Biogeochemistry 11:83-101

Watanabe A, Kajiwara M, Tashiro T & Kimura M (1995a) Influence of rice cultivar on methane emission from rice paddy fields. Plant Soil 176:51-56

Watanabe A, Satoh Y & Kimura M (1995b) Estimation of the increase in CH_4 emission from paddy soils by rice straw application. Plant Soil 173(2):225-231

Wickham TH & Singh VP (1978). Water movement through wet soils. In: Soils and Rice, International Rice Research Institute, Los Baños, Philippines, pp 337-358

Yagi K & Minami K (1990) Effect of organic matter application on methane emission from some Japanese paddy fields. Soil Sci Plant Nutr 36:599-610

Yagi K, Tsuruta H & Kanda K (1994) Effect of water management on methane emission from rice paddy field, Research Report. Division of the Environment of the National Institute of Agro-Environmental Science (in Japanese), pp 61-70

Yao H, Conrad R, Wassmann R & Neue HU (1999) Effect of soil characteristics on sequential reduction and methane production in sixteen rice paddy soils from China, the Philippines, and Italy. Biogeochemistry 47:267-293

Nutrient Cycling in Agroecosystems **58:** 179–199, 2000.

Using a crop/soil simulation model and GIS techniques to assess methane emissions from rice fields in Asia. III. Databases

J.W. Knox[1], R.B. Matthews[1] & R. Wassmann[2,3]

[1]Institute of Water and Environment, Cranfield University, Silsoe, Bedfordshire, MK45 4DT, United Kingdom; [2]International Rice Research Institute, MCPO Box 3127, Makati City 1271, Philippines; [3]Fraunhofer Institute for Atmospheric Environmental Research, Garmisch-Partenkirchen, Germany

Key words: climate change, databases, GIS, methane, rice, soils, weather

Abstract

As part of a series of papers describing the use of a simulation model to extrapolate experimental measurements of methane (CH_4) emissions from rice fields in Asia and to evaluate the large-scale effect of various mitigation strategies, the collation and derivation of the spatial databases used are described. Daily weather data, including solar radiation, minimum and maximum temperatures, and rainfall were collated from 46 weather stations from the five countries in the study, namely China, India, Indonesia, Philippines, and Thailand. Quantitative soil data relevant to the input requirements of the model were derived by combining data from the World Inventory of Soil Emissions (WISE) database, the ISIS database, and the FAO Digital Soil Map of the World (FAO-DSMW). These data included soil pH; organic carbon content; sand, silt, and clay fractions; and iron content for top and subsoil layers, and average values of bulk density and available water capacity for the whole profile. Data on the areas allocated to irrigated, rainfed, upland, and deepwater rice at the province or district level were derived from the Huke & Huke (1997) database developed at IRRI. Using a geographical information system (GIS), a series of geo-referenced data sets on climate, soils, and land use were derived for each country, at the province or district level. A summary of the soil-related derived databases is presented and their application for use in global change modeling discussed.

Introduction

At the global level, methane (CH_4) is the second most important greenhouse gas because of its strong ability to absorb infrared radiation. Its concentration in the atmosphere has been rising in recent years (Houghton et al., 1992), which has led to concerns about its role in global warming. Emissions of CH_4 from rice fields are an important contributor to levels of the gas in the atmosphere, with the current best estimate of its source strength being ~60 Tg CH_4 yr^{-1} with a range of 20-150 Tg CH_4 yr^{-1} (Houghton et al., 1992). This wide range, making it the most uncertain of all the CH_4 sources, is due to the large variation between sites in measured CH_4 fluxes—such measurements, often limited in number and obtained only for short periods, may not necessarily be representative of average emission rates at the national and regional levels to which they are frequently extrapolated. However, in view of the significance of rice cultivation as a source of CH_4 and of the fact that world rice production must increase by an estimated 70% in the next few decades to meet the demands of an increasing populaton (IRRI, 1993), estimates of the magnitude of the contribution of rice cultivation to global CH_4 emissions need to be refined and the mechanisms involved better understood. This, in turn, should enable the development of mitigation options that could reduce emissions while still allowing the required increases in rice yield.

Much of the uncertainty in the current estimates of CH_4 emissions from rice cultivation is due to the large spatial variation between sites in the controlling factors—climate, soil type, and cultural practices such as water management, fertilizer, and organic matter applications. Representation and integration of these factors within a geographical information system (GIS) framework, coupled with the development of mechanistic models describing the processes involved in CH_4 production and emission, have been suggested as the way forward (Bliss, 1990; Matthews, 1993; Batjes & Bridges, 1995; Neue et al., 1995). Up until now, a major limitation to this approach has been the lack of high-quality databases of climate, soils, and rice-growing areas for use with the mechanistic models (Shearer & Khalil, 1993).

Several geographic databases relevant to CH_4 emissions from rice lands have been developed and published in recent years. The FAO soil map of the world (FAO, 1974) provides the basis for several of these — Zobler (1986), for example, used it to create a 1° (latitude × longitude) digital soil data set. More recently, it has been used as the basis of the World Inventory of Soil Emissions (WISE) database of soil qualitative characteristics (Batjes, 1997) which is described in more detail later in this paper. For climate, Leemans and Cramer (1990) developed a database of global monthly air temperature and precipitation. Similarly, for rice production statistics, Matthews et al. (1991) developed a database indicating the location and the harvested area of rice cultivation, derived from an earlier database of land use (Matthews, 1983). Several of these databases have since been used to estimate CH_4 emissions at the national or regional scales (e.g., Bachelet & Neue, 1993; Bachelet et al., 1995; Kern et al., 1997).

In Part I of this series (Matthews et al., 2000a), a mechanistic model, MERES (Methane Emissions in Rice EcoSystems), simulating the main processes involved in CH_4 production and emission in rice fields was described. This model is based on the CERES-Rice crop simulation model (Godwin et al., 1990) with additional routines describing the influence of alternative electron acceptors in the soil and steady-state profiles of O_2 and CH_4 resulting from CH_4 production, oxidation, diffusion, leaching, and flux through plant aerenchyma (Arah & Kirk, 2000). Previously, the use of such mechanistic models for analysis of CH_4 emissions was limited by the lack of daily weather data they require to run; however, such a database is now available for most of the rice-growing areas in Asia (Centeno et al., 1995).

This paper describes the derivation of geo-referenced databases for input into MERES for upscaling of experimental measurements of CH_4 emissions at a number of sites in Asia. The model output, aggregated from province/district to regional and national levels, is described in a subsequent paper in this series (Matthews et al., 2000b).

Methods

For modeling CH_4 emissions, spatial and temporal information on a range of parameters are required, including

- soils
- climate
- land use (rice cropped areas and current production)
- national and administrative boundaries

This study involved the collation, integration, and preprocessing of data from various sources, using a combination of methodologies including GIS techniques, computer program routines, and spreadsheet analyses. The GIS represented the principal database management and visualization tool, working in conjunction with existing databases as an integrated part of the overall model. A schematic representation of the methodological framework is given in Figure 1. A brief description of the data sources and procedures used for deriving the databases is given.

Data sources

The original FAO-UNESCO Soil Map of the World, published between 1974 and 1978, has since been computerized to produce a digital soil map of the world (FAO-DSMW). The database is available in raster or vector format and subdivided into 10 regions of the world. The scale of the original map is 1:5,000,000. The basic level of classification used in the database is the soil unit. The database comprises an estimated 4,930 mapping units. Where a mapping unit is not homogeneous, it is composed of a dominant soil unit and component soil units. The latter are further categorized into associated soils (covering at least 20% of the area) and inclusions (important soils covering less than 20% of the area). The FAO-DSMW is supplied with a second relational database (termed the 'expansion' file) which contains qualitative and quantitative attribute data for each of the 4,930 mapping units, detailing the proportions of dominant and component soil units in each mapping unit, together with information on slope and soil texture class (FAO, 1995).

A global data set of derived quantitative soil characteristics, classified by FAO-UNESCO soil units, has been produced by Batjes (1997). These data were de-

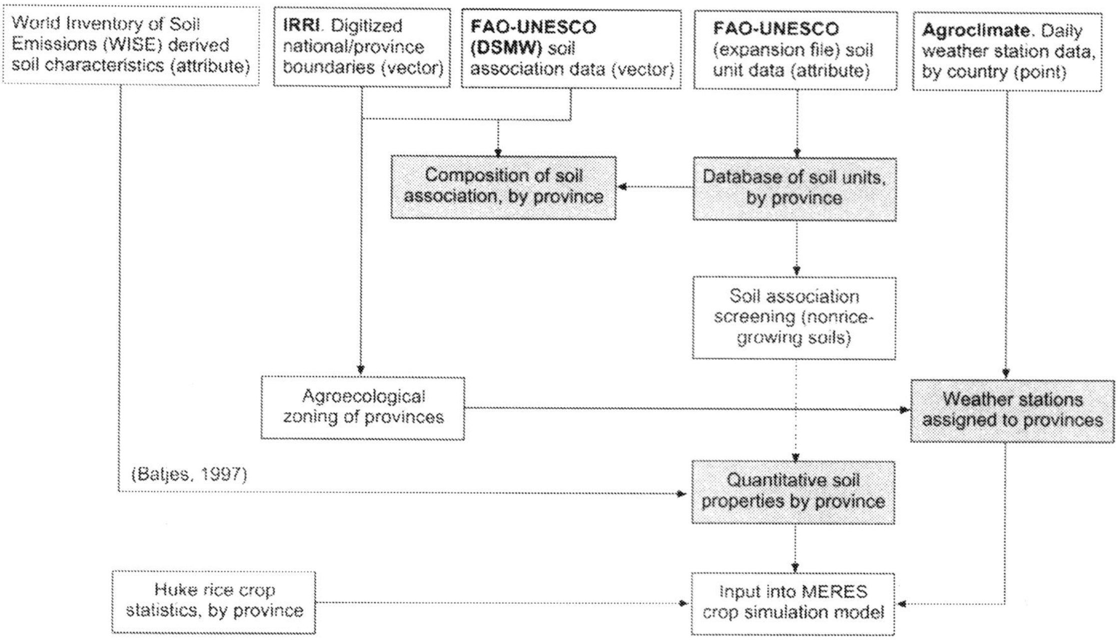

Figure 1. Schematic representation showing database integration (gray boxes denote derived databases)

rived from statistical analyses of the 4,353 soil profiles held in the WISE database. Median values by soil unit for soil properties including pH, organic carbon content (OC), bulk density (BD), and available water capacity (AWC) were used in this study.

Soil particle distribution (i.e., percent sand, silt, and clay) and soil iron content, required for estimating the soil water release characteristics and initial size of the oxidized alternate electron acceptor pool (see Part IV, Matthews et al., 2000b), were obtained from the ISIS database (van de Ven & Tempel, 1994). These data were in the form of measurements at a number of depths in each soil profile, but for consistency with the data in the WISE database, a program was written to aggregate the ISIS data into median values for topsoil (0-30 cm) and subsoil (>30 cm) for each FAO soil unit. These aggregated data were then merged with the main WISE database for input into the MERES model.

The location and extent of rice-growing areas in the five countries were obtained from the Huke and Huke (1997) database developed at the International Rice Research Institute (IRRI). This revised database, covering South, Southeast, and East Asia, was developed largely on the basis of a wide range of official data published by various government agencies and data collated by field research teams from IRRI and the national agricultural research systems (NARS). The database provides detailed statistics, at the province or district level, on rice production and cropped area under the four main rice ecosystems (irrigated, rainfed, upland, and deepwater systems). For selected countries, additional information on administrative regions, population size, and hybrid rice production are also included. For most countries, the database relates to 1990.

To complement the published Huke and Huke (1997) database, staff of the GIS laboratory at IRRI have digitized the national and province boundaries for China, Indonesia, the Philippines, and Thailand, and the national, state, and district boundaries for India. These vector data sets for each country were used as the basic or 'polygon' level for modeling CH_4 emissions. For China, Indonesia, the Philippines, and Thailand, these polygons typically represented a single province or subunit of a province. For India, on the other hand, each polygon represented a district or a subunit of a district. Subunits occurred in provinces or districts containing discontinuous components, such as islands in coastal regions. Thus, provinces or districts can be made up of one or more polygons. Sample output given in the Results section therefore represent data derived at the polygon level but aggregated to either province or state level depending on the country.

Daily weather data were collected from various rice-growing areas in Asia, either by the national weather bureau in the different countries, by the Climate Unit at IRRI (10 stations in the Philippines), or by the participants of the IRRI-WMO Rice-Weather Project based at IRRI from 1984 to 1986 (Oldeman et al., 1987). Subsequently, five institutes collaborating in a project funded by the US Environmental Protection Agency (EPA) modeling the effects of climate change on rice production in Asia (Matthews et al., 1995) continued to collect and supply weather data after the end of the Rice-Weather project in 1986. The current database contains 10 or more years of historic weather data from 87 stations from rice-growing areas in 11 countries in Asia, of which 46 were used for the five countries in the present study. The CLICOM system (CLImate COMputing system developed by the World Meteorological Organization, [WMO] 1989) was used for data storage and data validation. Database management is undertaken by the Climate Unit at IRRI. A description of the database, including the procedures used to ensure data quality and to estimate missing data, is given by Centeno et al. (1995), and a map showing the location of the weather stations is shown in Matthews et al. (1995). A computer program was developed to convert these data into the appropriate format for input into the MERES model.

To associate each polygon with the most appropriate weather data (as described later), weather stations were classified according to the agroecological zone (AEZ) in which they were located (Table 1). The zoning system used was that developed by the FAO, based on climatic conditions and landforms that determine relatively homogeneous crop-growing environments (IRRI, 1993). The classification distinguishes between tropical regions, subtropical regions with winter or summer rainfall, and temperate regions. These major regions are further subdivided into rainfed moisture zones, lengths of the growing period, and thermal zones based on the temperature regime that prevails during the growing season. Most of the countries in the study fell within a single agroecological zone, although China and India spanned several zones (Figure 2). Details of each weather station and a summary of the long-term averages of the key climate variables are given in Table 2. Dates of sowing and transplanting were, in general, supplied by the collaborating institutions along with the weather data. Where this information was not provided, transplanting dates were obtained from IRRI (1991), and date of sowing in the seedbed assumed to be 25 d prior to this. Where a range of transplanting dates was given, generally a date near the start of the range was used. In some countries, second, and even third, crops are grown in the same year; these were also simulated. Sowing dates and ages at transplanting for each season at each site are shown in Table 2. These agree well with the dates used by Jansen (1990) with the exception of those in Indonesia; both, however, are within the range given in IRRI (1991).

Table 1. Description of the FAO-defined agroecological zones (AEZ) used in this study (IRRI, 1993)

AEZ	FAO description
1	Warm arid and semiarid tropics
2	Warm subhumid tropics
3	Warm humid tropics
5	Warm arid and semiarid subtropics with summer rainfall
6	Warm subhumid subtropics with summer rainfall
7	Warm/cool humid subtropics with summer rainfall
8	Cool subtropics with summer rainfall

It is recognized that these dates may sometimes be somewhat arbitrary and not always a reflection of actual planting dates in a given region. Published crop calendars for a number of regions are available, but there is often disagreement between these even for the same regions, thereby limiting their use. Transplanting dates depend on the decisions of individual farmers, which are influenced by actual weather conditions, economic considerations, and other factors. Often, transplanting in a region can take place over extended periods, particularly in tropical regions; in the higher latitudes, planting date is generally constant.

Derived spatial databases

Using the procedures described earlier but relying predominantly on GIS techniques, a series of databases were derived. SPANS GIS v7.0 software (TYDAC, 1994) was used due to its comprehensive functionality and strong data integration and modeling capabilities.

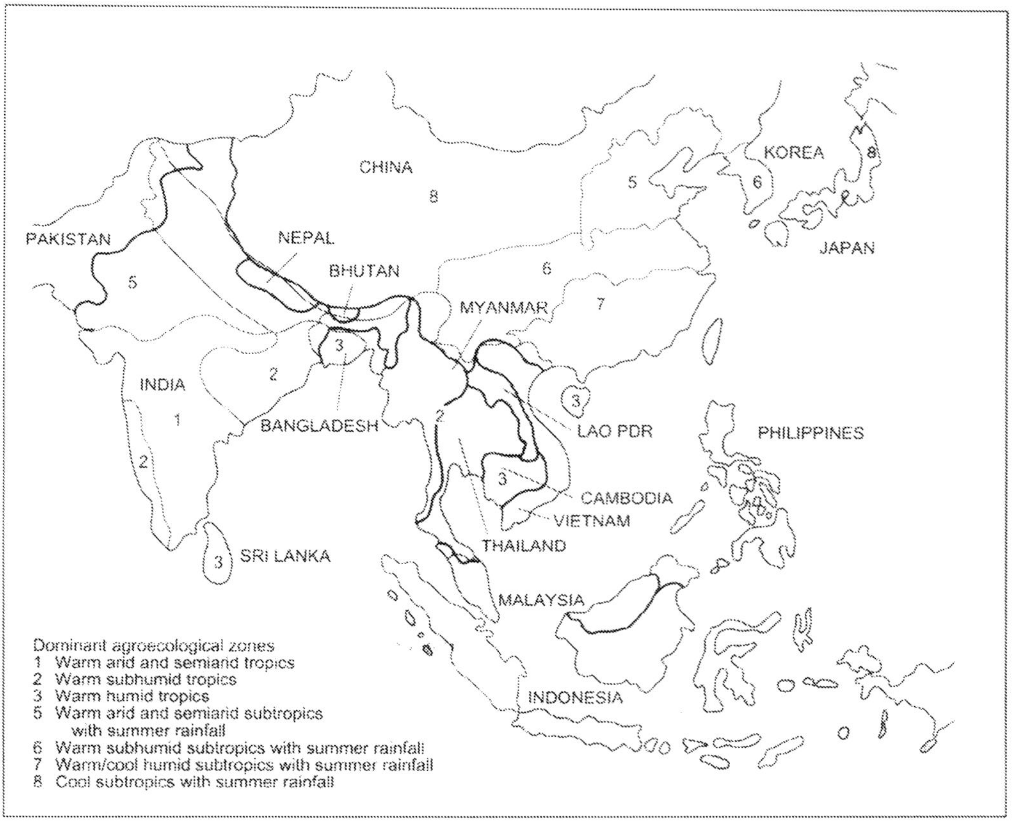

Figure 2. Regional agroecological zones in Asia (IRRI, 1993)

Creating a database of quantitative soil characteristics for each province/subprovince

As described earlier, the basic mapping unit in the FAO-DSMW database is generally a soil association, a group of soil units occurring in close proximity to each other. The WISE database, on the other hand, contains soil quantitative information at the soil unit level. Linking these two databases, therefore, requires an expansion of the soil associations of the FAO-DSMW into their constituent soil units. A schematic representation of the procedures developed are shown in Figure 1.

To begin with, the digitized province/subprovince layers for each country were overlaid onto the FAO-DSMW. The soil associations contained within each polygon were then extracted into an intermediate database containing the name of each association and the fraction (f_a) of the total area of the polygon it occupied. The soil units and the proportion (f_u) they occupy in each of the soil associations are described in the separate expansion file, a relational database linked to the DSMW by a sequential code number representing the association. The second step, therefore, was to replace each of the soil associations in each polygon with its constituent soil units. In many cases, the same soil units occur in different soil associations of the polygon, so all fractions of the same unit in each polygon were pooled to give a single fraction for that unit. This was achieved using a separate program which identified each polygon in turn taking each soil association and expanding it into its constituent soil units. The proportion of each soil unit in the association was multiplied by the proportion of the soil association in the polygon to give the contribution of each soil unit to its overall proportion in the polygon.

The third step was to calculate the weighted average of each of the quantitative soil characteristics for each polygon. However, not all soil units are suitable

Table 2. Details of weather stations used in the study showing location, longitude, latitude, and elevation of the site; long-term averages of main climatic variables; and times of sowing (SOW) and transplanting (TPLT) for each growing season (adapted from Centeno et al., 1995)[a]

Site	Longitude (°E)	Latitude (°N)	Elevation (m)	AEZ	Rainfall (mm)	PET (mm)	Solar radn (MJ m^{-2} d^{-1})	Max temp (°C)	Min temp (°C)	Av temp (°C)	Av humidity (%)	VP (kPa)	VPD (kPa)	Windspeed (m s^{-1})	Main season		2nd season		3rd season	
															SOW	TPLT	SOW	TPLT	SOW	TPLT
China																				
Beijing	116.5	39.9	55	5	541	841	14.7	17.8	7.1	12.5	55	1.10	0.70	2.4	121	35	*	*	*	*
Chansa	113.1	28.2	45	7	1360	676	12.0	21.4	13.8	17.5	78	1.75	0.56	2.1	90	30	196	25	*	*
Chendu	104.0	30.7	506	6	887	513	9.7	19.9	12.8	16.3	83	1.71	0.35	1.1	91	40	*	*	*	*
Fuzhou	119.3	26.1	85	7	1290	753	12.2	24.2	16.6	20.4	76	2.01	0.62	2.4	86	30	176	25	*	*
Guangzhou	113.3	23.1	18	7	1742	710	12.4	26.2	18.8	22.5	77	2.30	0.62	1.7	79	30	196	25	*	*
Guiyang	106.7	26.6	1074	6	1030	627	10.7	19.6	12.2	15.9	76	1.52	0.49	1.9	107	40	*	*	*	*
Hangzhou	120.2	30.2	45	7	1470	651	12.8	20.4	12.9	16.6	77	1.72	0.47	2.3	90	30	175	25	*	*
Nanjing	118.8	32.1	9	7	1043	648	13.1	19.9	11.1	15.5	77	1.62	0.44	2.5	126	41	*	*	*	*
Shenyang	123.4	41.8	42	5	718	733	13.5	13.9	3.6	8.8	61	0.99	0.48	2.8	121	35	*	*	*	*
Wuhan	114.1	31.0	23	7	1367	632	13.4	20.7	12.7	16.7	78	1.73	0.48	2.0	85	30	175	25	*	*
India																				
Aduthurai	79.5	11.0	19	1	1116	2022	19.2	32.7	23.7	28.3	82	3.19	0.72	2.1	161	25	350	25	*	*
Bijapur	75.8	16.8	594	1	662	2027	18.9	33.2	20.8	27.0	m	m	m	m	160	25	350	25	*	*
Coimbatore	77.0	11.0	431	1	651	1911	18.2	31.6	21.0	26.3	78	2.40	0.17	m	161	25	350	25	*	*
Cuttack	86.0	20.5	23	2	1569	1675	17.0	31.8	22.1	27.0	80	2.78	0.90	1.5	213	25	15	25	*	*
Hyderabad	78.4	17.4	545	1	706	1270	18.4	32.1	19.6	25.9	86	2.35	0.42	2.1	160	25	350	25	*	*
Kapurthala	75.9	30.9	247	6	767	1477	17.1	29.8	15.2	22.5	88	2.11	0.39	1.1	191	25	*	*	*	*
Madurai	78.0	8.5	147	1	824	2133	20.7	34.0	23.7	28.8	82	3.26	0.79	1.4	161	25	350	25	*	*
Patancheru	78.4	17.5	25	1	898	2122	18.8	31.8	20.0	25.9	55	1.83	1.62	2.9	160	25	350	25	*	*
Pattambi	76.2	10.8	25	2	2744	1643	17.4	31.6	22.9	27.2	93	2.85	0.22	0.5	120	25	300	25	*	*
Indonesia																				
Bandung	107.5	–6.9	791	3	2173	1188	15.1	28.0	17.9	23.0	79	2.23	0.60	0.7	270	20	80	20	*	*
Ciledug	106.8	–6.3	26	3	2312	1611	16.8	31.7	22.7	27.2	82	3.00	0.65	0.8	270	20	80	20	*	*
Cimanggu	106.7	–6.6	240	3	4738	1112	14.5	30.5	22.8	26.6	79	2.80	0.38	1.0	270	20	80	20	*	*
Cipanas	107.0	–6.8	1100	3	2756	1050	*	24.0	17.1	20.5	83	2.03	0.41	m	270	20	80	20	*	*
Maros	119.5	–5.0	5	3	3252	1718	17.7	31.0	22.3	26.6	91	3.22	0.31	m	270	20	140	20	*	*
Muara	106.7	–6.8	260	3	4215	1395	14.0	31.0	21.4	26.2	69	2.36	0.91	0.9	270	20	80	20	*	*
Pacet	107.0	–6.8	1138	3	3232	979	12.2	24.8	16.8	20.8	85	2.11	0.38	m	270	20	80	20	*	*

Table 2 continued.

Site	Longitude (°E)	Latitude (°N)	Elevation (m)	AEZ	Rainfall (mm)	PET (mm)	Solar radn (MJ m^{-2} d^{-1})	Max temp (°C)	Min temp (°C)	Av temp (°C)	Av humidity (%)	VP (kPa)	VPD (kPa)	Windspeed (m s^{-1})	Main season SOW	Main season TPLT	2nd season SOW	2nd season TPLT	3rd season SOW	3rd season TPLT
Philippines																				
Albay	123.7	13.4	*	3	4995	1566	16.2	30.9	22.4	26.7	90	3.35	0.37	1.2	135	20	15	20	258	20
Banaue	121.1	16.9	1040	3	3608	1380	15.1	25.3	17.3	21.3	89	2.21	0.28	0.5	135	20	15	20	258	20
Betinan	123.5	7.9	45	3	2891	1652	16.0	31.0	22.2	26.6	92	2.91	0.33	0.8	135	20	15	20	258	20
Butuan	125.7	11.0	60	3	2070	1681	16.6	32.1	22.5	27.3	88	3.15	0.43	1.3	135	20	15	20	258	20
Cavinti	121.5	14.3	305	3	4196	1517	16.4	28.1	20.7	24.4	92	2.81	0.25	2.0	135	20	15	20	258	20
CLSU	120.9	15.7	76	3	1944	1928	18.9	31.8	22.4	27.1	85	2.93	0.55	2.4	135	20	15	20	258	20
CSAC	123.3	13.6	36	3	2162	1528	*	31.7	22.7	27.2	88	3.13	0.44	1.2	135	20	15	20	258	20
Guimba	120.8	15.7	66	3	1656	1986	21.0	32.6	22.0	27.3	89	2.90	0.36	1.6	135	20	15	20	258	20
IRRI Wetland	121.3	14.2	21	3	2027	1741	16.1	30.6	23.2	26.8	88	3.05	0.44	1.5	135	20	15	20	258	20
La Granja	122.9	10.4	84	3	2589	1539	13.7	32.3	22.0	27.1	86	3.05	0.50	1.8	135	20	15	20	258	20
MMSU	124.3	8.0	854	3	1876	1765	18.4	32.1	21.9	27.0	87	2.95	0.46	0.9	135	20	15	20	258	20
Muñoz	120.9	15.8	48	3	1581	*	20.8	31.4	23.0	27.2	89	2.90	0.37	1.9	135	20	15	20	258	20
PNAC	118.6	9.4	7	3	1636	1772	18.4	31.2	23.2	27.2	90	3.23	0.37	1.1	135	20	15	20	258	20
Solana	121.7	17.7	21	3	1422	1755	18.5	31.7	22.5	27.1	92	2.88	0.27	0.9	135	20	15	20	258	20
UPLB	121.3	14.2	21	3	2035	1581	16.9	31.6	22.6	27.1	87	3.05	0.48	1.2	135	20	15	20	258	20
VES	122.6	10.8	14	3	2200	1809	18.3	31.4	23.5	27.5	91	3.26	0.35	1.4	135	20	15	20	258	20
Thailand																				
Chiang Mai	99.0	18.8	313	2	1140	1424	18.5	32.0	20.6	26.3	71	2.46	1.04	*	150	28	350	20	*	*
Khon Kaen	102.8	16.4	165	2	1228	1554	18.8	32.7	22.0	27.3	70	2.62	1.11	*	180	28	350	20	*	*
Nakhon S.	100.2	15.8	28	2	1104	1689	19.0	34.0	23.2	28.6	70	2.80	1.21	*	*	*	*	*	*	*
Ubon R.	104.9	15.3	123	2	1652	1583	19.1	32.5	21.8	27.2	72	2.65	1.04	*	180	28	350	20	*	*

[a]AEZ = agroecological zone, PET = potential evapotranspiration, VP = vapor pressure, VPD = vapor pressure deficit. Asterisks indicate data not available. SOW is day of the year on which sowing into the nursery takes place, TPLT is days after sowing that transplanting into the field occurs.

for growing rice — even where cultivation is possible, certain soil factors (e.g., slope, texture, soil depth, and stoniness) as well as agroclimatic conditions may preclude successful cultivation. Using a combination of information from the literature (IRRI, 1978) and from expert consultation (Batjes, pers. comm., 1998), only soil units suitable for rice cultivation in each polygon were selected. In particular, Acrisols, Cambisols, Fluvisols, Luvisols, Histosols, Vertisols, Planisols, and most Gleysols were included. For each valid soil unit present in each polygon, the quantitative soil characteristics (pH, OC, BD, AWC, etc.) were extracted from the combined data set described above, using the FAO soil unit code (e.g., G = gleysol) as the common field. The mean value (V_p) of each soil characteristic in each polygon was then obtained by summing each variable across all valid soil units contained in the polygon weighted by the proportion of each valid soil unit in that polygon, i.e.,

$$V_p = \frac{1}{N}\left[\sum_{u=l}^{N} (V_u \bullet f_u \bullet f_a)\right] \qquad (1)$$

where N is the number of valid soil units in the polygon, and V_u is the value of a particular soil characteristic for the soil unit *u*. For each country, a database listing the mean values for the selected soil characteristics, by polygon, was derived. These data are summarized in Tables 3-7.

Assigning weather stations to each polygon

As individual polygons were to be the basic level for simulation and because the MERES crop/soil simulation model requires daily weather data as an input, it was necessary to associate each polygon with a representative weather station. This was done by selecting the nearest station within the same AEZ to the center of each polygon.

First, to determine the AEZ into which each polygon fell, the boundaries of each AEZ (IRRI, 1993) were overlaid onto the map containing the province/subprovince boundaries. Using the GIS, the latitude and longitude of the geometric centroid of each polygon were determined. These centroid coordinates were then used to identify which AEZ each polygon is located in. In the cases of Indonesia, the Philippines, and Thailand, the whole country falls into a single AEZ, namely 3 (warm humid tropics). China, however, spans four AEZs (5, 6, 7, and 8), while India spans six (1, 2, 3, 5, 6, and 8). In the latter two countries, no attempt was made to partition a polygon between two AEZs if the AEZ boundary bisected it, as the resolution of the original AEZ map was such that this would represent false accuracy. The AEZ associated with each polygon, therefore, was based only on the position of its centroid.

Second, the nearest weather station in the same AEZ to this polygon centroid position was determined using in-built nearest-neighbor procedures. This was achieved by creating a layer containing the coordinates (latitude and longitude) of the geometric centroid of each polygon and overlaying this onto another layer containing the weather station coordinates. Weather stations were assigned to each polygon by comparing the distance between each polygon centroid and weather station locations and allocating the station with the shortest distance, provided it was in the same AEZ.

For each country, columns were then added to the database of quantitative soil characteristics described in the previous section to include this newly derived information defining the AEZ and nearest weather station for each polygon.

Results and discussion

Derived data sets

A summary of the derived mean values for selected soil properties are given in Tables 3-7. For convenience (i.e., to reduce the number of individual records), data are aggregated from the polygon level to the province level (or state level in the case of India) for each country. This was undertaken by weighting the appropriate values from each polygon by the fraction of the total province/district area occupied by that polygon. In most cases, as mentioned previously, each province/district was typically represented by only one polygon. Polygons for which FAO soil data were unavailable (e.g., small off-shore islands) were removed from the analysis. These were always of insignificant area in comparison with the whole province/district and are unlikely to contain any significant rice-growing area.

Limitations

The databases described here were derived through integration and analysis of existing spatial databases, primarily within a GIS framework. The spatial accuracy of both existing and derived data sets are, however, a potential source of error; this is a common problem for any GIS-based analysis. For simplicity, a number of

Table 3. Derived mean values for selected soil properties, by province, for China. Province locations are shown in Figure 3. Values are means of rice-growing soils only and not necessarily of all soils in the province[a]

Province	Code	Area	pH_t	pH_s	OC_t	OC_s	BULK	AWC	SA_T	SA_B	SI_T	SI_B	CL_T	CL_B	Fe_T	Fe_B
Anhui	1	146748	6.38	6.79	1.17	0.31	1.44	119	25.5	23.1	38.6	34.8	30.1	31.8	0.94	1.05
Beijing	2	16078	6.88	7.18	1.12	0.34	1.40	127	19.4	19.1	34.6	33.4	40.4	40.1	0.95	1.06
Fujian	3	128877	5.31	5.46	0.88	0.33	1.41	116	54.1	42.1	24.5	21.4	23.9	34.1	1.61	1.79
Gansu	4	408357	7.03	7.23	1.28	0.61	1.40	128	26.9	26.6	34.6	33.5	37.0	35.3	1.08	1.24
Guangdong	5	189492	5.53	5.59	0.90	0.34	1.45	116	47.6	31.7	23.5	21.3	26.4	37.7	2.26	1.99
Guangxi	6	250055	5.15	5.21	0.81	0.28	1.42	116	53.6	39.3	21.9	19.6	24.3	35.9	1.54	1.60
Guizhou	7	171144	5.16	5.28	0.82	0.27	1.38	115	56.2	44.5	24.4	21.1	22.8	32.9	2.30	2.16
Hainan	8	34129	5.42	5.46	0.87	0.32	1.43	110	51.4	38.3	22.1	19.3	25.7	35.8	2.74	2.46
Hebei	9	189737	6.91	7.25	1.28	0.35	1.42	125	19.1	20.4	38.3	35.1	37.8	36.7	2.87	3.00
Heilongjiang	10	479702	6.48	6.95	1.74	0.96	1.46	116	28.3	30.0	44.3	36.0	23.3	28.9	2.08	2.55
Henan	11	165545	6.84	7.26	1.12	0.30	1.45	122	16.7	18.2	44.0	41.5	35.4	33.0	2.09	2.28
Hubei	12	186828	6.09	6.37	1.08	0.32	1.41	120	32.4	27.1	31.8	29.1	31.5	35.0	1.02	1.13
Hunan	13	215165	5.54	5.74	0.94	0.33	1.44	117	48.2	37.2	25.0	22.3	26.8	35.7	2.17	2.27
Jiangsu	14	102139	6.65	7.14	1.17	0.26	1.42	118	16.6	17.6	46.9	41.3	30.1	29.8	2.07	1.95
Jiangxi	15	172236	5.52	5.69	0.93	0.33	1.44	117	48.7	37.0	23.6	21.4	27.3	36.5	2.00	2.08
Jilin	16	200430	6.86	7.22	1.43	0.69	1.45	120	26.5	27.5	39.0	34.3	30.7	32.5	1.94	2.09
Liaoning	17	159948	6.91	7.24	1.22	0.36	1.43	125	20.7	21.1	36.5	34.4	37.5	36.9	1.09	1.26
Nei Mongol Zizhiqu	18	1159931	6.58	7.07	1.42	0.33	1.47	116	23.8	27.4	48.6	34.6	23.1	27.8	1.33	1.53
Ningxia	19	51744	6.88	7.23	1.04	0.32	1.44	121	19.5	17.9	41.2	31.8	34.5	33.1	2.00	2.21
Qinghai	20	732507	4.81	4.91	25.43	26.51	0.58	377	40.1	40.4	45.5	36.9	32.3	27.8	1.53	1.78
Shaanxi	21	202136	6.94	7.13	0.97	0.35	1.40	126	22.9	21.8	31.4	31.8	40.6	39.4	1.04	1.19
Shandong	22	153224	6.75	7.06	1.24	0.37	1.44	119	25.0	24.6	39.0	34.6	31.2	33.2	1.35	1.68
Shanghai	23	6584	6.61	7.06	1.35	0.33	1.46	117	18.3	17.9	44.0	35.9	30.9	32.1	1.43	1.75
Shanxi	24	154585	7.09	7.37	1.01	0.36	1.38	127	14.9	15.3	32.7	33.0	47.2	45.3	1.53	1.86
Sichuan	25	574975	6.05	6.27	1.85	1.11	1.36	127	33.7	29.6	33.6	30.3	32.1	35.2	1.55	2.01
Taiwan	26	38630	6.02	6.12	0.93	0.38	1.44	117	40.9	27.3	25.3	24.4	30.5	38.9	1.88	2.32
Tianjin	27	11674	6.75	7.20	1.75	0.39	1.42	125	16.2	19.6	43.6	37.2	35.0	34.0	1.38	1.82
Xinjiang Uygur	28	1715787	6.79	6.96	4.37	3.81	1.29	148	23.6	32.0	52.8	43.8	29.7	25.1	1.65	2.14
Xizang Zizhiqu	29	1196089	5.59	5.73	7.25	6.20	1.10	185	38.7	40.3	43.4	36.3	30.6	27.5	1.70	2.25
Yunnan	30	399655	5.57	5.59	0.87	0.32	1.42	120	44.8	34.2	24.0	22.1	29.0	37.2	1.77	2.36
Zhejiang	31	105128	5.60	5.81	0.95	0.32	1.43	118	48.1	37.7	25.9	23.0	26.3	34.9	1.65	2.41

[a]The abbreviation _T refers to topsoil (0-30 cm) and _S to the subsoil (30-100cm). pH: median soil pH, measured in water. OC: median organic carbon (%). BULK: median bulk density (g cm^{-3}). AWC: available water capacity in mm to a depth of 100 cm (or less if soil depth is shallower), for range pF2.5 to pF4.2 (Ψ_s = 0.033 to 1.5 MPa). 2SA : % sand, SI: % silt, CL: % clay, Fe: % iron. Medians for soil pH and AWC are taken from Batjes (1995a, 1996), respectively. (Adapted from Batjes, 1997)

Table 4. Derived mean values for selected soil properties, by state, for India. State locations are shown in Figure 4. Values are means of rice-growing soils only and not necessarily of all soils in the state[a]

State	Code	Area	pH_t	pH_s	OC_t	OC_s	BULK	AWC	SA_T	SA_B	SI_T	SI_B	CL_T	CL_B	Fe_T	Fe_B
Andaman & Nicobar	1	7449	5.12	5.31	1.60	0.44	1.26	131	40.2	36.9	28.7	27.1	30.7	30.6	1.07	1.23
Andhra Pradesh	2	279036	6.85	7.10	0.80	0.42	1.53	112	24.8	20.2	28.7	29.6	40.7	43.9	1.98	1.85
Arunachal Pradesh	3	84824	5.10	5.22	1.48	0.44	1.34	123	42.2	35.0	27.0	23.2	31.4	34.0	1.21	1.27
Assam	4	81667	5.33	5.46	1.18	0.48	1.40	118	44.6	35.6	26.6	23.4	26.9	32.6	1.96	1.74
Bihar	5	165307	6.68	6.89	0.80	0.32	1.47	108	29.7	28.2	36.8	36.4	29.3	32.0	1.43	1.33
Chandigarh	6	127	7.49	7.59	0.62	0.32	1.46	113	20.7	30.7	52.7	42.8	26.0	24.9	1.57	1.34
Dadra & Nagar Haveli	7	591	7.32	7.53	0.77	0.41	1.61	122	20.8	17.7	28.0	30.4	47.0	48.7	1.96	1.69
Daman and Diu	8	737	7.89	8.01	0.59	0.37	1.46	119	15.8	25.8	45.6	42.0	37.8	31.5	1.97	2.80
Delhi	9	1398	6.30	6.70	0.63	0.26	1.55	91	30.3	31.9	55.9	35.5	19.8	31.2	1.00	1.10
Goa	10	2958	5.75	6.02	0.93	0.30	1.46	109	36.6	30.3	35.0	41.4	22.5	26.2	1.76	2.36
Gujarat	11	191208	7.24	7.45	0.77	0.40	1.53	123	21.5	22.2	31.4	33.2	41.8	40.9	2.01	2.48
Haryana	12	43360	6.61	6.90	0.67	0.29	1.52	100	28.9	30.7	49.9	36.1	23.1	30.5	1.83	2.73
Himachal Pradesh	13	54127	6.05	6.28	1.39	0.41	1.39	123	34.7	32.7	36.1	30.5	28.6	30.7	1.72	2.29
Jammu and Kashmir	14	219780	6.18	6.45	0.97	0.34	1.40	124	63.2	54.7	19.2	15.9	17.0	27.5	1.80	2.04
Karnataka	15	193842	6.64	6.83	0.79	0.40	1.56	111	31.1	23.3	25.3	27.2	38.8	43.5	1.98	2.07
Kerala	16	39350	5.10	5.26	1.43	0.64	1.43	121	50.4	38.2	25.3	28.3	22.0	28.6	1.50	2.13
Madhya Pradesh	17	451656	6.90	7.10	0.71	0.38	1.56	111	28.1	22.0	24.8	28.3	42.8	45.7	2.17	3.02
Maharashtra	18	316331	7.13	7.39	0.78	0.42	1.56	123	20.4	17.8	25.4	29.7	50.2	49.1	1.99	1.64
Manipur	19	22889	5.00	5.17	2.09	0.58	1.28	127	41.4	37.1	31.7	26.3	33.9	32.9	2.03	2.78
Meghalaya	20	24820	5.24	5.40	1.20	0.50	1.38	119	46.7	37.8	26.8	21.9	25.7	32.1	1.71	2.26
Mizoram	21	21800	5.04	5.18	2.11	0.53	1.25	129	35.5	34.4	34.6	28.0	32.4	30.2	1.90	2.62
Nagaland	22	16064	4.93	5.07	1.40	0.43	1.33	120	48.9	39.9	26.0	21.7	29.3	33.9	2.05	1.90
Orissa	23	141762	6.19	6.30	0.69	0.31	1.50	98	43.9	32.4	24.0	25.7	27.8	37.9	1.19	1.30
Pondicherry	24	537	6.17	6.28	1.23	0.77	1.37	115	27.8	24.6	40.5	45.0	21.0	25.9	2.11	1.98
Punjab	25	52838	7.06	7.29	0.60	0.30	1.50	102	23.6	31.8	55.6	40.5	22.7	26.5	2.15	2.33
Rajasthan	26	353055	7.14	7.29	0.72	0.34	1.50	114	25.0	26.4	38.5	35.6	32.8	33.9	2.00	2.19
Sikkim	27	7516	5.11	5.27	2.50	0.59	1.22	134	35.8	36.3	38.9	32.4	31.3	26.3	1.73	3.51
Tamil Nadu	28	127395	6.62	6.84	0.82	0.38	1.52	107	27.0	22.4	30.2	31.4	35.6	39.3	1.40	2.12
Tripura	29	11889	5.12	5.23	1.33	0.34	1.35	122	38.7	34.0	30.9	27.4	25.4	28.9	2.24	2.55
Uttar Pradesh	30	290922	6.44	6.70	1.01	0.36	1.45	115	28.2	29.2	39.2	34.8	27.2	29.8	1.85	2.62
West Bengal	31	88529	6.32	6.49	1.45	0.97	1.44	114	33.8	28.6	37.1	34.3	26.0	31.3	2.29	2.53

[a]See Table 3 for meaning of abbreviations.

Table 5. Derived mean values for selected soil properties, by province, for Indonesia. Province locations are shown in Figure 5. Values are means of rice-growing soils only and not necessarily of all soils in the province[a]

Province	Code	Area	pH_t	pH_s	OC_t	OC_s	BULK	AWC	SA_T	SA_B	SI_T	SI_B	CL_T	CL_B	Fe_T	Fe_B
Bali	1	5521	6.75	6.96	0.93	0.36	1.45	121	26.4	26.1	33.1	33.1	31.8	34.2	1.60	2.10
Bengkulu	2	26373	5.11	5.24	3.11	1.98	1.27	140	44.2	40.3	30.7	26.3	26.8	27.5	1.62	2.13
Daerah Istimewa Aceh	3	67437	4.92	5.01	2.14	1.25	1.32	131	52.1	45.8	25.1	22.0	20.9	24.7	1.38	1.63
Irian Jaya	4	67801	5.52	5.66	2.90	2.13	1.35	125	46.6	39.1	28.3	26.8	26.1	30.7	1.74	2.26
Jambi	5	53551	4.92	5.01	7.12	6.72	1.18	182	53.7	46.0	24.7	21.9	22.5	27.7	1.73	2.32
Jawa Barat	6	49982	5.30	5.43	1.25	0.41	1.39	119	51.2	44.6	25.5	23.1	22.4	27.3	1.59	2.23
Jawa Tengah	7	36715	6.54	6.79	1.10	0.58	1.47	111	24.3	24.2	29.6	28.9	39.2	41.1	1.89	2.38
Jawa Timur	8	47365	6.55	6.77	1.22	0.50	1.45	107	25.4	24.3	31.6	31.8	38.5	39.4	2.07	2.59
Kalimantan Barat	9	145686	4.95	5.05	7.29	6.79	1.16	184	42.8	38.2	25.3	22.1	28.5	29.6	1.63	1.94
Kalimantan Selatan	10	36533	4.81	4.86	10.14	9.89	1.09	215	40.8	33.8	20.8	19.2	32.9	33.7	1.72	1.49
Kalimantan Tengah	11	162279	4.92	5.01	5.78	5.10	1.22	168	47.0	40.4	23.4	21.3	26.9	29.0	2.09	2.52
Kalimantan Timur	12	198973	5.20	5.34	2.49	1.60	1.33	132	38.6	33.7	26.8	22.7	30.9	32.2	2.16	2.18
Lampung	13	36583	5.28	5.38	2.83	1.95	1.33	134	44.1	36.8	26.4	24.4	26.9	31.1	2.35	1.28
Maluku	14	59490	5.60	5.73	1.29	0.55	1.41	114	40.6	33.7	26.0	24.2	27.7	32.5	1.97	2.82
Nusa Tenggara Barat	15	19173	6.88	7.12	1.01	0.53	1.52	106	16.4	16.2	29.5	29.0	46.5	48.5	2.28	3.49
Nusa Tenggara Timur	16	43943	6.77	6.99	0.81	0.37	1.52	97	29.3	24.9	27.8	34.5	34.8	35.5	1.94	2.84
Riau	17	92117	4.74	4.79	15.43	15.82	0.93	270	57.9	49.1	22.1	20.5	19.3	25.5	1.94	2.91
Sulawesi Selatan	18	60865	5.63	5.85	1.62	0.78	1.39	122	33.1	31.1	29.8	25.7	31.3	32.8	1.99	2.68
Sulawesi Tengah	19	57509	5.22	5.38	1.63	0.54	1.35	121	33.5	28.3	24.7	22.0	37.9	36.4	2.29	3.62
Sulawesi Tenggara	20	32956	6.19	6.41	1.47	0.54	1.39	113	33.2	31.4	32.8	33.2	30.7	31.1	2.43	2.92
Sulawesi Utara	21	26148	5.17	5.32	1.62	0.53	1.35	124	33.4	28.0	24.0	21.0	38.0	36.4	2.25	3.53
Sumatera Barat	22	48644	5.91	6.23	2.09	1.53	1.42	116	32.9	36.6	45.9	32.7	20.2	26.8	2.03	2.64
Sumatera Salatan	23	115319	4.85	4.89	5.06	3.95	1.28	157	72.0	60.5	18.6	21.9	10.2	13.6	2.14	2.73
Sumatera Utara	24	85122	4.92	5.01	2.20	1.30	1.32	132	51.9	45.6	25.1	22.0	21.1	24.9	1.71	1.30
Timor Timur	25	13476	6.97	7.24	0.72	0.30	1.49	105	29.3	29.6	25.6	40.8	34.6	28.3	2.57	3.49
Yogyakarta	26	3054	5.90	6.05	1.37	0.44	1.46	116	37.8	33.7	24.4	24.5	31.9	33.4	2.66	4.28

[a]See Table 3 for meaning of abbreviations.

Table 6. Derived mean values for selected soil properties, by province, for the Philippines. Province locations are shown in Figure 6. Values are means of rice-growing soils only and not necessarily of all soils in the province[a]

Province	Code	Area	pH_t	pH_s	OC_t	OC_s	BULK	AWC	SA_T	SA_B	SI_T	SI_B	CL_T	CL_B	Fe_T	Fe_B
Abra	1	3739	5.38	5.55	1.10	0.33	1.39	118	42.2	35.8	30.6	25.6	27.0	32.1	2.04	2.24
Agusan del Norte	2	2843	5.72	5.96	0.96	0.31	1.43	115	36.3	31.6	35.8	30.9	27.4	32.8	1.74	2.19
Agusan del Sur	3	7575	5.67	5.94	1.12	0.34	1.41	121	30.9	27.0	34.6	29.2	30.8	33.6	1.73	1.92
Aklan	4	1607	5.59	5.82	1.02	0.34	1.42	117	40.5	34.2	31.7	25.5	27.9	33.6	1.61	2.21
Albay	5	2299	6.06	6.35	0.83	0.28	1.47	110	33.1	29.9	40.6	35.5	25.7	32.0	1.69	1.76
Antique	6	3038	5.98	6.31	1.05	0.36	1.45	121	26.2	22.7	36.4	31.4	34.7	37.4	1.67	2.33
Aurora	7	2511	5.37	5.53	1.04	0.32	1.39	115	45.4	38.5	30.5	24.9	24.8	31.1	1.89	1.88
Babuyan Island	8	187	6.19	6.51	0.71	0.26	1.49	103	32.9	30.3	44.6	38.3	23.0	30.8	1.63	2.31
Basilan	9	1249	5.70	5.90	0.77	0.27	1.44	111	47.1	40.2	34.1	27.2	21.4	30.9	1.86	2.08
Bataan	10	1200	6.28	6.54	1.29	0.38	1.35	124	54.6	53.5	21.4	16.1	18.8	16.7	1.67	2.00
Batangas	11	3169	6.52	6.91	0.87	0.36	1.50	118	24.7	21.9	39.2	33.0	33.4	37.6	1.58	2.13
Benguet	12	2686	5.19	5.33	1.19	0.35	1.37	119	44.4	37.5	28.2	23.4	26.9	31.6	1.47	1.57
Bohol	13	3868	6.12	6.44	0.74	0.27	1.49	106	34.6	31.1	42.0	35.3	24.2	31.9	1.64	2.45
Bukidnon	14	7648	5.79	6.10	1.16	0.41	1.43	122	31.2	25.9	31.6	24.8	35.2	38.0	1.80	2.05
Bulacan	15	2178	6.54	6.84	0.86	0.35	1.53	119	24.1	21.5	36.9	32.7	36.5	40.2	2.05	2.19
Burias Island	16	410	6.19	6.51	0.71	0.26	1.49	103	32.9	30.3	44.6	38.3	23.0	30.8	2.05	1.97
Cagayan	17	8951	5.68	5.93	1.05	0.34	1.42	119	34.8	29.9	33.9	29.0	30.0	34.3	1.59	1.66
Calamian Group	18	944	5.23	5.38	1.12	0.33	1.38	117	46.2	38.9	28.6	23.3	25.5	31.2	1.89	1.71
Camarines Norte	19	2400	5.77	6.00	0.85	0.28	1.44	114	40.7	35.1	35.5	29.6	24.8	32.2	1.20	1.50
Camarines Sur	20	5129	6.15	6.50	0.85	0.33	1.48	113	31.8	27.7	38.4	31.3	29.4	35.5	1.58	2.02
Camotes Island	21	79	6.19	6.51	0.71	0.26	1.49	103	32.9	30.3	44.6	38.3	23.0	30.8	1.48	1.61
Capiz	22	2656	5.96	6.28	0.94	0.35	1.45	118	34.1	29.1	34.9	27.0	29.9	35.5	2.02	1.99
Catanduanes	23	1455	5.44	5.61	0.97	0.31	1.40	115	46.6	39.5	31.0	25.0	23.7	31.1	2.14	2.08
Cavite	24	1371	6.45	6.88	0.85	0.36	1.51	114	24.3	21.5	40.8	33.3	33.2	38.0	1.32	1.14
Cebu	25	4479	6.08	6.38	0.72	0.26	1.48	104	36.0	32.5	42.4	35.9	22.7	30.8	1.81	1.59
Cotabato North	26	6310	5.92	6.22	0.96	0.30	1.45	114	30.9	27.1	38.8	32.5	28.4	33.1	1.52	1.16
Cotabato South	27	7558	5.32	5.48	1.21	0.35	1.37	119	43.0	36.8	28.3	24.1	27.2	30.9	1.82	1.76
Davao (N)	28	8195	5.54	5.74	1.04	0.32	1.41	116	39.2	33.8	33.3	28.4	27.0	32.2	1.80	2.35
Davao del Sur	29	5743	5.44	5.63	1.11	0.33	1.39	118	39.3	33.5	31.6	27.1	28.2	32.7	1.92	2.35
Davao Oriental	30	4881	5.62	5.83	0.98	0.31	1.42	116	39.6	34.1	33.9	28.8	26.4	32.4	1.61	2.18
Dinagat Island	31	937	5.70	5.90	0.77	0.27	1.44	111	47.1	40.3	34.1	27.2	21.4	30.8	1.64	2.05
Dumaran	32	349	4.96	5.07	1.33	0.37	1.34	121	45.6	38.1	25.4	21.1	27.9	31.4	1.66	2.59
East Samar	33	3934	5.48	5.67	1.05	0.32	1.40	115	40.8	35.0	32.6	27.5	26.4	31.8	1.52	1.50
Ifugao	34	1982	5.33	5.51	1.13	0.35	1.39	118	43.4	36.5	29.1	23.6	27.3	32.5	1.75	1.73
Ilocos Norte	35	3654	5.69	5.94	1.08	0.33	1.42	120	32.3	28.0	34.7	30.6	31.0	34.5	1.41	1.33
Ilocos Sur	36	3161	5.82	6.12	1.05	0.36	1.44	120	31.7	27.0	34.1	28.5	32.4	36.2	1.67	1.55
Iloilo	37	4175	6.00	6.33	0.94	0.36	1.46	118	33.4	28.1	34.6	27.0	31.5	36.9	1.78	1.75
Isabela	38	9854	5.69	5.96	1.11	0.37	1.42	120	33.9	28.7	32.4	26.8	32.0	35.7	1.34	1.22
Kalinga Apayao	39	6188	5.15	5.28	1.19	0.35	1.37	118	45.9	38.5	27.6	22.7	26.4	31.3	1.57	1.88
La Union	40	1884	6.02	6.36	1.01	0.37	1.46	120	28.1	24.0	35.7	29.6	34.2	37.8	1.34	1.22

Table 6 continued.

Province	Code	Area	pH_t	pH_s	OC_t	OC_s	BULK	AWC	SA_T	SA_B	SI_T	SI_B	CL_T	CL_B	Fe_T	Fe_B
Laguna	41	1430	6.22	6.54	0.87	0.30	1.48	113	27.6	25.2	41.2	36.7	29.1	33.9	1.42	1.34
Lanao del Norte	42	3067	5.77	5.98	0.78	0.27	1.45	111	44.3	38.2	35.5	28.9	22.3	31.1	1.68	1.74
Lanao del Sur	43	3990	6.02	6.29	0.99	0.31	1.44	121	28.3	25.3	37.2	34.3	31.3	34.6	1.69	1.79
Leyte	44	5245	5.83	6.13	1.10	0.32	1.42	121	28.1	23.2	36.3	28.9	32.2	34.8	1.68	2.07
Maguindanao	45	5817	5.95	6.26	1.00	0.31	1.44	116	29.8	26.3	38.2	30.9	28.8	32.6	1.93	2.07
Manila	46	799	6.05	6.36	0.99	0.34	1.46	120	25.9	23.0	37.6	33.5	33.8	36.9	1.66	2.12
Marinduque	47	866	5.91	6.19	0.86	0.29	1.45	113	34.5	30.5	38.5	33.3	26.7	32.8	1.34	1.22
Masbate	48	3297	5.97	6.24	0.74	0.26	1.47	106	39.2	34.7	40.0	33.4	22.3	30.8	1.38	1.28
Mindoro Occidental	49	6118	5.44	5.64	1.14	0.34	1.39	119	37.4	32.0	31.9	27.7	29.3	33.1	1.37	1.27
Mindoro Oriental	50	3452	5.33	5.50	1.17	0.34	1.38	119	40.2	34.2	30.3	26.0	28.4	32.4	1.34	1.22
Misamis Occidental	51	1625	5.96	6.28	1.08	0.33	1.44	122	21.2	19.4	39.2	36.8	35.3	36.6	1.76	2.34
Misamis Oriental	52	3623	5.91	6.20	0.97	0.31	1.45	114	29.3	26.4	39.3	35.1	29.5	33.6	1.84	2.00
Mountain Province	53	2175	5.22	5.36	1.13	0.34	1.37	117	46.1	38.9	28.4	23.2	25.7	31.2	1.41	1.48
Negros Occidental	54	8333	5.76	6.01	0.99	0.31	1.43	115	34.0	29.9	36.6	32.1	28.1	32.9	1.46	1.54
Negros Oriental	55	4547	5.34	5.52	1.19	0.35	1.38	118	39.4	33.7	30.7	26.5	28.5	32.2	1.55	1.86
North Samar	56	3376	5.79	6.03	0.88	0.29	1.44	112	38.3	33.5	36.7	31.2	25.3	32.0	1.92	2.00
Nueva Ecija	57	5167	6.15	6.48	0.98	0.35	1.47	121	24.5	21.6	37.7	33.2	34.9	37.9	1.60	2.12
Nueva Vizcaya	58	3204	5.80	6.06	0.90	0.33	1.44	115	39.9	33.8	33.6	26.6	27.2	34.1	1.84	1.92
Palawan	59	11757	5.55	5.78	1.06	0.32	1.41	115	37.9	32.9	34.1	28.8	26.9	31.8	1.52	1.50
Pampanga	60	2130	6.25	6.61	1.14	0.35	1.41	124	32.3	29.8	33.0	24.0	28.4	28.5	1.74	1.72
Pangasinan	61	5828	6.10	6.43	0.94	0.32	1.45	116	27.7	27.5	39.0	34.7	27.5	31.2	1.52	1.50
Quezon	62	7257	5.81	6.07	0.88	0.29	1.45	109	38.2	33.7	37.8	32.0	24.3	31.2	1.55	1.51
Quirino	63	2558	5.06	5.19	1.26	0.36	1.35	120	45.4	38.0	26.5	21.8	27.4	31.5	2.30	2.08
Rizal	64	1187	5.83	6.09	0.93	0.30	1.44	117	33.9	29.6	36.7	32.1	28.5	33.8	1.70	1.81
Romblon	65	719	5.70	5.90	0.77	0.27	1.44	111	47.1	40.3	34.1	27.2	21.4	30.8	1.73	1.81
Siargao Island	66	506	6.10	6.40	0.72	0.26	1.48	104	35.4	32.1	42.8	36.4	22.7	30.8	1.84	2.35
Siquijor	67	337	6.19	6.51	0.71	0.26	1.49	103	32.9	30.3	44.6	38.3	23.0	30.8	1.70	2.29
Sorsogon	68	1957	5.80	6.04	0.86	0.29	1.44	114	39.2	34.0	36.0	30.4	25.4	32.5	1.85	2.26
South Leyte	69	2107	5.44	5.64	1.09	0.33	1.40	116	40.4	34.6	32.3	27.2	26.8	31.8	1.74	2.07
Sultan Kudarat	70	5047	5.44	5.64	1.21	0.35	1.38	121	39.7	34.5	29.6	25.6	28.6	31.3	1.77	1.97
Sulu	71	865	5.76	5.99	0.84	0.28	1.44	114	41.2	35.5	35.2	29.4	24.6	32.2	1.75	2.00
Surigao del Norte	72	994	5.80	6.05	0.94	0.30	1.44	115	35.0	30.7	36.9	32.0	27.5	33.0	1.38	1.27
Surigao del Sur	73	5290	5.61	5.83	1.00	0.31	1.42	115	38.6	33.4	34.3	29.2	26.7	32.3	1.52	1.50
Tarlac	74	2464	5.77	6.02	1.04	0.33	1.42	119	32.7	29.3	35.2	31.2	29.4	33.1	1.85	1.92
Tawi Tawi	75	685	5.70	5.90	0.77	0.27	1.44	111	47.1	40.3	34.1	27.2	21.4	30.8	1.84	1.97
Ticao Island	76	292	6.19	6.51	0.71	0.26	1.49	103	32.9	30.3	44.6	38.3	23.0	30.8	1.94	1.99
West Samar	77	5292	5.80	6.06	0.92	0.30	1.44	113	36.1	31.6	37.3	31.8	26.3	32.3	1.43	1.84
Zambales	78	4235	5.64	5.87	1.02	0.32	1.42	116	36.6	31.8	34.9	30.2	27.7	32.6	2.52	2.49
Zamboanga del Norte	79	7560	5.28	5.43	1.09	0.33	1.38	117	46.3	39.0	29.1	23.7	25.1	31.2	1.99	2.31
Zamboanga del Sur	80	7346	5.71	5.94	0.93	0.30	1.43	116	38.2	33.1	35.1	29.9	26.6	32.7	1.34	1.22

[a]See Table 3 for meaning of abbreviations.

Table 7. Derived mean values for selected soil properties, by province, for Thailand. Province locations are shown in Figure 7. Values are means of rice-growing soils only, and not necessarily of all soils in the province[a]

Province	Code	Area	pH_t	pH_s	OC_t	OC_s	BULK	AWC	SA_T	SA_B	SI_T	SI_B	CL_T	CL_B	Fe_T	Fe_B
Ang Thong	1	1019	5.85	6.09	1.13	0.44	1.39	120	34.7	32.4	32.2	31.7	21.0	25.4	1.55	1.85
Bangkok	2	1609	5.49	5.39	1.91	0.99	1.30	125	29.7	9.9	37.0	32.9	28.9	38.1	2.20	6.48
Buri Rum	3	10865	5.14	5.25	1.08	0.30	1.36	119	42.2	35.5	28.6	27.1	24.3	30.6	2.21	6.54
Chachoengsao	4	5355	5.05	4.90	1.63	0.78	1.31	123	47.0	29.1	27.2	27.9	24.3	34.9	1.67	3.73
Chai Nat	5	2612	5.54	5.73	0.92	0.32	1.41	121	41.0	33.6	26.6	26.4	26.2	31.7	2.00	5.59
Chaiyaphum	6	12659	5.02	5.11	0.97	0.30	1.37	118	50.1	39.3	24.2	22.4	24.6	32.3	1.42	1.53
Chanthaburi	7	6268	5.31	5.44	1.03	0.35	1.39	119	44.6	34.1	26.1	22.2	27.8	33.9	1.55	2.14
Chiang Mai	8	23845	5.21	5.35	1.02	0.33	1.38	117	49.1	40.1	25.5	22.5	25.1	31.5	1.17	1.38
Chiang Rai	9	12270	5.23	5.39	0.94	0.30	1.39	115	47.2	38.0	27.8	25.0	24.5	31.7	1.40	2.09
Chon Buri	10	4639	5.03	5.06	1.09	0.40	1.37	120	49.9	36.5	23.8	22.2	25.1	33.7	1.59	2.20
Chumphon	11	6199	5.28	5.43	1.01	0.35	1.39	119	44.8	33.2	27.3	22.0	27.1	34.2	1.74	3.44
Kalasin	12	7078	5.06	5.15	0.96	0.29	1.38	119	48.7	37.4	24.2	23.6	24.8	32.0	1.65	2.41
Kamphaeng Phet	13	9013	5.15	5.27	0.97	0.30	1.38	119	48.0	37.9	24.4	23.2	24.8	31.6	1.57	2.22
Kanchanaburi	14	20190	5.82	5.93	0.88	0.34	1.42	112	43.5	34.6	24.8	24.1	28.8	34.4	1.80	3.87
Khon Kaen	15	11409	5.09	5.19	1.03	0.30	1.37	118	46.2	37.5	26.7	25.3	24.1	30.9	1.60	2.44
Krabi	16	4821	5.43	5.54	1.10	0.43	1.40	122	41.2	28.4	28.1	23.3	28.3	35.5	1.70	2.26
Lampang	17	12411	5.14	5.25	0.98	0.32	1.39	117	51.0	40.4	24.4	21.5	25.3	32.7	1.45	2.56
Lamphun	18	4340	5.63	5.84	0.91	0.31	1.43	116	43.6	37.3	29.2	26.8	23.1	29.6	1.74	2.30
Loei	19	11233	5.03	5.13	1.04	0.32	1.37	118	51.2	40.9	24.1	20.9	25.2	32.4	1.73	3.65
Lop Buri	20	6657	6.60	6.92	1.07	0.42	1.47	124	23.5	24.4	32.6	31.7	34.5	34.4	1.59	2.25
Mae Hong Son	21	14095	5.01	5.11	1.04	0.33	1.37	117	52.2	41.7	23.6	20.2	25.4	32.5	1.50	2.18
Maha Sarakham	22	5476	5.15	5.27	1.05	0.29	1.37	118	44.2	36.3	28.6	28.2	23.1	29.2	1.60	2.21
Mukdahan	23	3382	4.97	5.04	0.97	0.30	1.38	119	51.6	40.0	22.7	20.5	24.9	33.1	1.28	1.82
Nakhon Nayok	24	2113	4.99	4.77	1.78	0.90	1.29	123	47.5	27.7	27.7	30.3	23.5	35.1	1.32	1.91
Nakhon Pathom	25	2020	5.17	5.01	1.74	0.89	1.30	123	41.6	21.9	30.6	32.3	24.6	35.2	1.71	2.70
Nakhon Phanom	26	7085	5.02	5.08	0.94	0.29	1.39	119	50.2	38.2	23.2	22.1	24.6	32.9	1.59	2.10
Nakhon Ratchasima	27	20604	5.20	5.30	1.10	0.31	1.36	119	42.1	35.9	28.2	25.8	25.3	31.5	1.44	2.01
Nakhon Sawan	28	9641	5.59	5.80	1.00	0.33	1.40	120	40.0	34.6	28.1	26.8	25.3	29.9	1.51	2.18
Nakhon Si Thammarat	29	11131	5.20	5.31	1.05	0.36	1.39	118	47.6	36.0	27.0	23.8	24.8	31.9	1.63	2.26
Nan	30	12448	4.99	5.09	1.07	0.33	1.37	118	52.4	42.2	24.0	20.2	25.2	32.2	1.50	1.89
Narathiwat	31	4876	5.05	5.08	4.38	4.00	1.28	154	52.9	40.0	24.1	22.5	24.1	32.9	1.55	2.29
Nong Khai	32	7845	5.02	5.07	0.94	0.29	1.40	118	51.1	38.9	23.4	22.2	23.7	32.7	1.43	1.75
Nonthaburi	33	624	4.92	4.36	2.56	1.52	1.20	127	45.0	10.7	31.6	40.2	21.4	41.2	1.58	2.10
Pathum Thani	34	1433	4.91	4.34	2.57	1.53	1.20	127	45.4	10.7	31.5	40.4	21.2	41.3	1.58	2.07
Pattani	35	2112	5.46	5.61	1.12	0.43	1.39	120	39.3	29.7	31.7	30.0	25.2	30.5	1.53	2.14
Phanakhon si Ayuthaya	36	2478	5.35	5.07	2.14	1.21	1.26	125	38.2	18.0	33.4	38.2	20.4	34.2	1.58	2.33
Phangnga	37	3839	5.01	4.94	1.33	0.60	1.35	120	52.5	36.6	24.5	23.9	23.8	34.4	1.52	2.13
Phatthalung	38	3356	4.98	4.98	1.16	0.46	1.36	119	51.7	37.4	24.2	24.0	23.4	32.6	1.32	1.72
Phayao	39	7059	5.38	5.58	0.92	0.29	1.41	113	43.6	35.4	30.2	27.7	24.6	31.1	1.60	2.14
Phetchabun	40	12203	5.44	5.57	0.96	0.32	1.40	120	46.1	38.0	25.0	22.6	25.8	32.3	1.60	2.28

Table 7 continued.

Province	Code	Area	pH_t	pH_s	OC_t	OC_s	BULK	AWC	SA_T	SA_B	SI_T	SI_B	CL_T	CL_B	Fe_T	Fe_B
Phetchaburi	41	6897	5.13	5.25	1.03	0.33	1.38	117	50.0	39.5	25.4	21.9	25.2	32.1	1.65	2.24
Phichit	42	4539	5.57	5.80	0.96	0.31	1.40	119	40.2	35.0	29.0	28.8	22.5	27.2	1.37	1.86
Phitsanulok	43	9958	5.21	5.35	0.94	0.30	1.38	119	48.3	39.1	24.9	22.9	23.9	31.2	1.60	2.05
Phrae	44	5640	5.14	5.27	0.98	0.31	1.39	117	48.6	38.6	25.6	22.8	25.3	32.4	1.59	2.04
Phuket	45	864	5.27	5.32	0.80	0.30	1.44	119	55.2	41.9	21.3	19.5	25.0	34.7	1.62	2.14
Prachin Buri	46	12350	5.10	5.17	1.19	0.41	1.36	120	48.2	37.6	25.2	23.2	25.5	32.2	1.58	1.97
Prachup Khiri Khan	47	7258	5.00	5.10	0.97	0.31	1.37	117	52.7	41.7	23.6	20.7	24.7	32.6	1.64	2.12
Ranong	48	3114	5.01	5.02	1.19	0.47	1.36	119	53.1	39.7	23.9	21.7	24.5	33.6	1.70	2.14
Ratchaburi	49	5033	5.20	5.34	1.00	0.33	1.38	120	46.9	36.9	25.1	21.6	25.5	32.6	1.67	2.03
Rayong	50	4105	5.09	5.18	0.96	0.32	1.38	121	47.6	35.3	23.2	20.1	27.2	34.9	1.51	1.85
Roi Et	51	7915	5.15	5.27	1.11	0.30	1.37	117	43.8	37.8	30.3	28.5	21.8	28.3	1.47	1.84
Sakon Nakhon	52	9602	5.13	5.19	0.94	0.29	1.41	119	50.3	36.8	22.8	22.0	24.5	33.4	1.57	2.22
Samut Prakan	53	838	5.81	5.98	1.54	0.69	1.36	124	23.2	9.5	39.2	29.8	32.1	36.7	1.55	1.93
Samut Sakhon	54	940	6.02	6.36	1.29	0.50	1.40	124	19.5	9.3	40.5	28.1	33.9	36.0	1.57	2.01
Samut Songkham	55	504	5.80	6.08	1.21	0.44	1.40	121	27.5	17.1	37.4	28.2	30.4	33.7	1.58	1.95
Saraburi	56	2986	5.65	5.76	1.07	0.41	1.40	118	40.8	32.1	27.5	28.5	26.8	31.7	1.59	2.07
Satun	57	2720	5.50	5.69	1.08	0.39	1.40	120	39.8	28.6	30.0	23.2	28.6	34.5	1.59	2.05
Si Sa Ket	58	9047	5.13	5.24	1.06	0.30	1.37	118	44.7	37.0	28.2	27.0	23.2	29.8	1.60	2.03
Sing Buri	59	858	6.29	6.69	1.01	0.35	1.42	120	27.8	32.6	35.1	32.8	20.4	22.5	1.64	2.23
Songkla	60	8076	5.07	5.15	1.04	0.35	1.38	118	51.1	39.5	25.2	23.2	23.8	31.6	1.55	2.18
Sukhothai	61	7210	5.12	5.22	0.98	0.30	1.38	119	48.3	37.9	24.3	22.7	25.3	32.2	1.52	2.10
Suphan Buri	62	5133	5.11	5.09	1.22	0.52	1.36	119	48.0	33.3	25.0	26.2	24.8	33.5	1.44	1.95
Surat Thani	63	13906	5.14	5.21	1.01	0.35	1.39	119	49.7	37.7	24.3	21.6	25.5	33.4	1.43	2.04
Surin	64	9132	5.14	5.24	1.08	0.30	1.37	118	43.6	36.9	28.9	27.2	23.2	29.8	1.54	2.33
Tak	65	16150	5.34	5.44	0.95	0.33	1.40	115	49.2	38.9	23.8	21.7	26.6	33.5	1.47	2.16
Trang	66	4777	5.51	5.68	1.04	0.38	1.41	120	40.7	29.1	29.4	23.0	28.4	34.8	1.65	2.92
Trat	67	3581	5.07	5.19	0.99	0.33	1.37	118	49.8	38.4	25.0	20.6	25.6	33.6	1.59	2.49
Ubon Ratchathani	68	19771	5.06	5.14	0.94	0.29	1.38	119	48.7	37.3	24.2	23.9	24.5	31.9	1.50	1.85
Udon Thani	69	16111	5.03	5.11	0.92	0.29	1.39	119	50.4	38.2	23.4	22.9	24.5	32.4	1.59	2.36
Uthai Thani	70	6813	5.16	5.26	1.00	0.32	1.38	117	48.7	38.3	24.0	22.1	26.0	32.6	1.60	2.60
Uttaradit	71	8731	4.95	5.04	1.01	0.31	1.37	118	52.3	41.1	23.2	20.4	25.1	32.6	1.49	2.17
Yala	72	4612	4.97	5.04	1.05	0.36	1.37	118	52.9	41.0	23.6	20.6	24.7	32.9	1.44	2.00
Yasothon	73	4172	5.11	5.21	1.05	0.30	1.37	118	44.9	37.1	27.8	26.3	23.4	30.3	1.57	2.16

[a]*See Table 3 for meaning of abbreviations.*

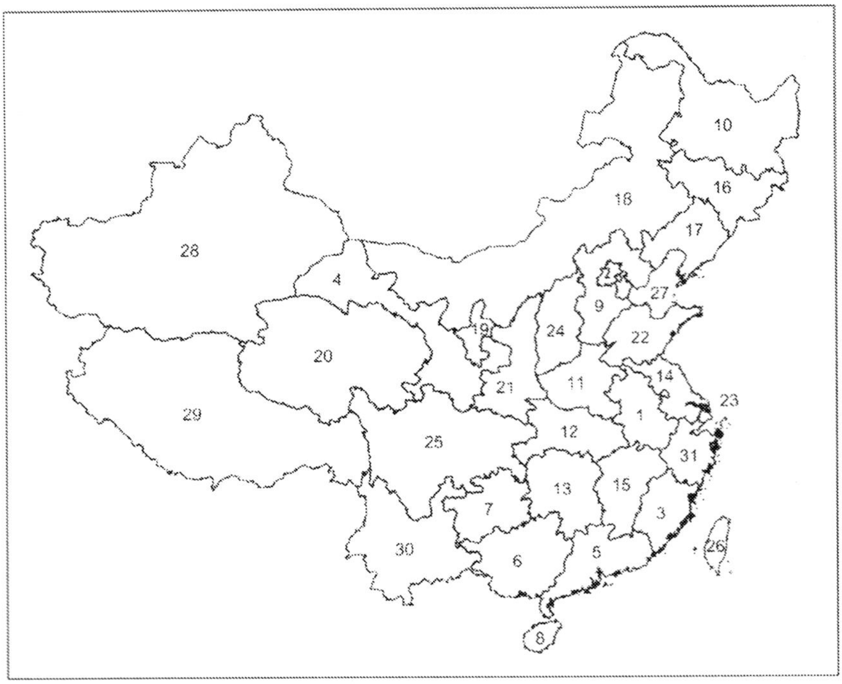

Figure 3. Province codes for China

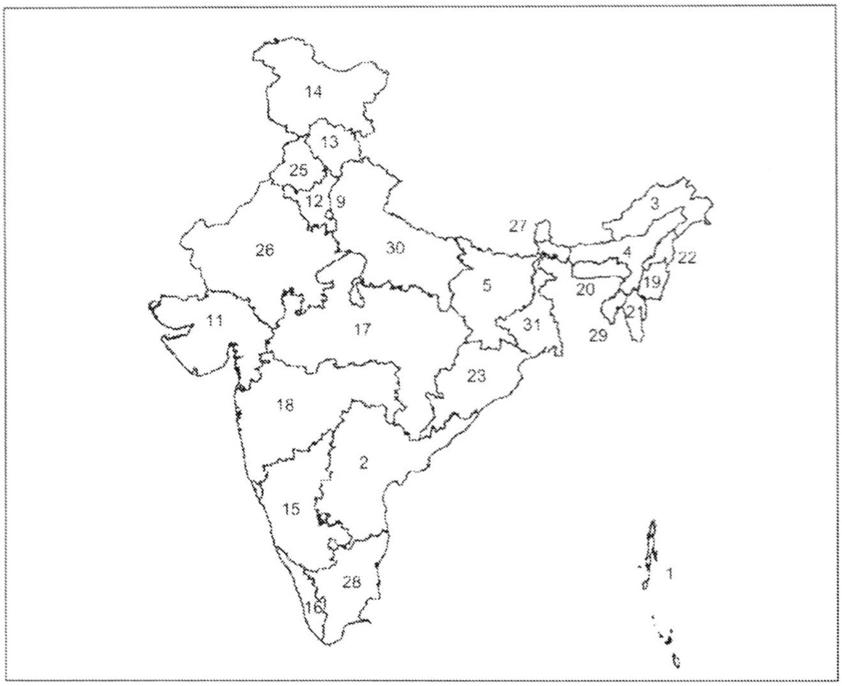

Figure 4. State codes for India

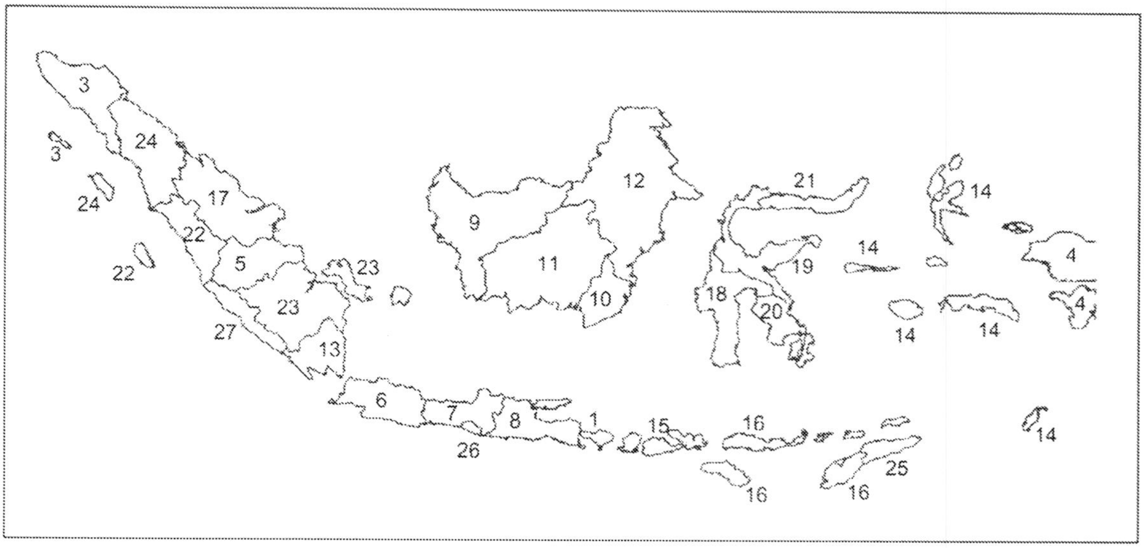

Figure 5. Province codes for Indonesia

assumptions were made in the methodology and their limitations should be recognized.

The areal coverage, resolution, level of pre-processing required, cost, and availability remain the key criteria when determining the suitability of data sets for modeling and integration within a GIS. Some of the spatial inaccuracies associated with the original FAO-DSMW have been discussed by Bachelet and Neue (1993) and Zobler (1986), the latter of whom identified limitations when reclassifying the FAO-DSMW at a 1 × 1° resolution. Other limitations, such as locally out-of-date information on soil geographic patterns, have been documented by Sombroek (1990) and Bouwman (1990). However, the latest FAO database release has addressed many of these problems, including errors in the original digitized version of the maps and consistency errors in the expansion file (FAO, 1995). Indeed, Richter and Babbar (1991) consider the FAO-DSMW as the best summary of global scale soil taxonomic data even though (a) it is based on a wide range of primary sources which consisted of mostly surveys and few actual soil data, (b) the quality of the mapping varies between regions especially in the tropics, and (c) it is not a complete soil classification since it only includes two to three levels of organization. For this study, the FAO-DSMW data set was considered appropriate for quantifying spatial soil variability at the polygon level, and when linked to the WISE database via the FAO soil unit code, provided a sound basis for extrapolating quantitative soil characteristics to the province/district level. For localities where soil information were missing (for example, small island regions), these polygons were ignored, since they were typically of minor significance with regard to rice production and CH_4 emissions.

The accuracy of the soil data held in the WISE database has been discussed by Batjes (1995). Even within a given soil unit, there may be considerable variation in measured values of a particular characteristic (e.g., soil carbon [Batjes, 1997]) which is represented by a single median value in the database. Nevertheless, a high degree of quality control over data collation and recording, coupled with the definition of stringent criteria for accepting data into the WISE database, have ensured that the spatial and soil profile integrity of the information in this database has been maintained.

The method we have used for estimating the mean values of the soil properties for each polygon is also a potential source of error. We have calculated the proportion of each soil unit in a polygon from the propor-

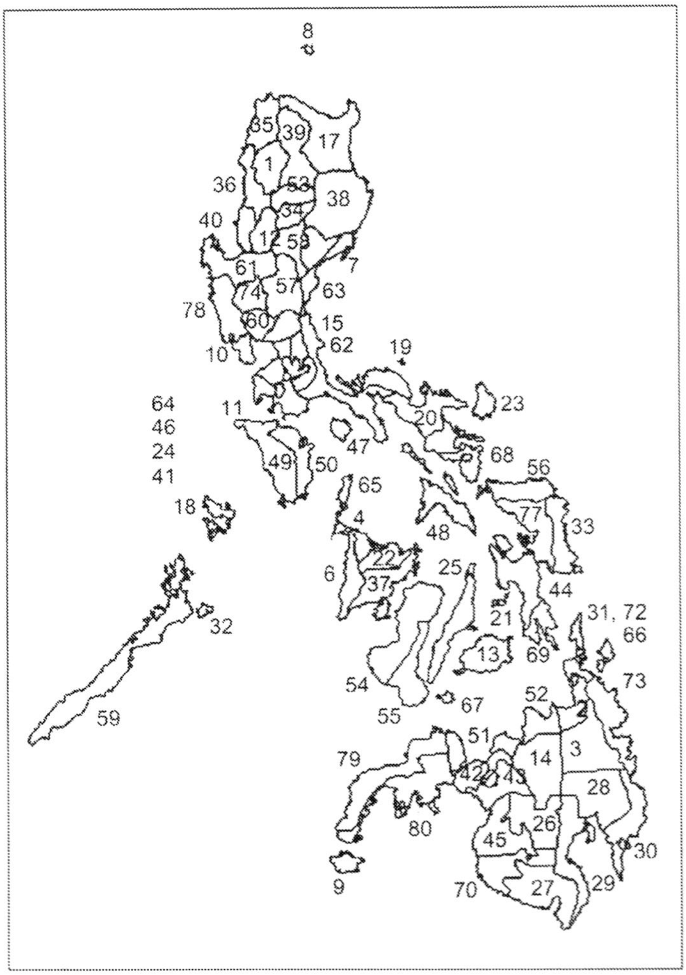

Figure 6. Province codes for the Philippines

tion it occupies in a soil association (f_u), which is a typical figure, but may also vary with geographical location, and the area fraction (f_a) of each soil association in the polygon. The method assumes that the distribution of each soil unit in each association is uniformly distributed, which may not be the case. If, for example, a particular soil unit is located in one corner of an association, it may be entirely outside the polygon in question in the case of an association that spans two or more adjacent polygons. In this case, the calculation would assume that the polygon contains a proportion of the soil unit, whereas in actual fact, it may be totally in a neighboring polygon. Alternatively, the whole soil unit may be in the polygon and not at all in any of its neighbours.

The allocation of AEZs to provinces/districts is a potential source of error. In particular, the scale and accuracy of the base map (IRRI, 1993) used for defining AEZs was not of high quality. Difficulties were encountered when digitizing AEZ boundaries, especially in regions where the base map provided little distinction between an AEZ and country boundary. Furthermore, for some provinces/districts, AEZs were defined for which no weather station was available. In this case, the nearest weather station with a similar AEZ was allocated.

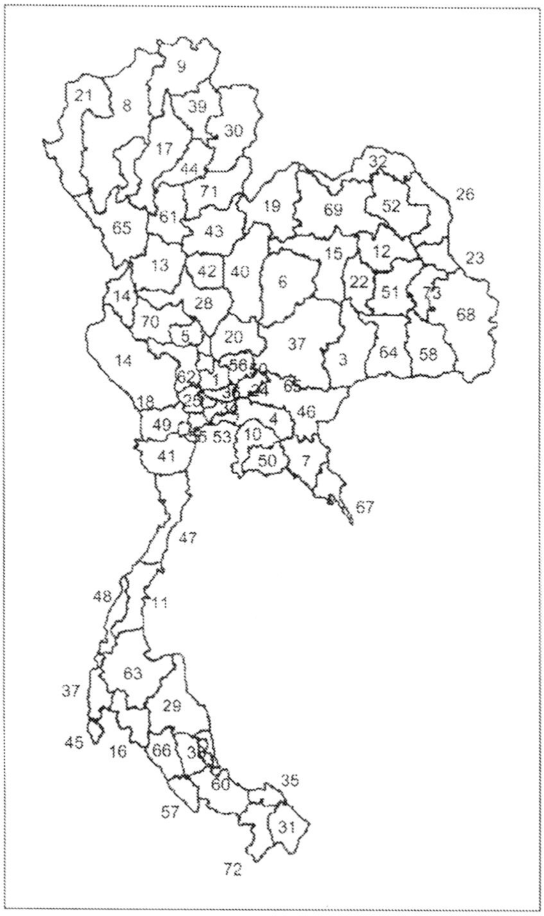

Figure 7. Province codes for Thailand

The use of geometric centroids represented one of several potential approaches to allocating weather stations to each province/district. An alternative approach for analyzing proximity to point features would have been to use 'graded buffering,' a technique used to define 'zones of influence' away from a particular polygon. This approach would have worked well for reasonably symmetrical areas, but for nonsymmetrical, especially elongated provinces/districts, a weather station allocated using this procedure may well have been unrepresentative of the region as a whole. The preferred approach of using geometric centroids overcomes this problem by assigning the weather station to a point which is representative of the majority of the province/district. However, where a province/district spans two AEZs, the shape of the province/district is clearly critical when locating the geometric centroid. Furthermore, the use of centroids ignores local topographic variation (e.g., elevation) across a province/district.

Another source of uncertainty in the results lies in the sparseness of weather data sites in some countries; areas in both India and China, for example, are represented by only a few stations, although fortunately for our analysis, little rice is grown in these regions anyway. While an attempt was made to stratify these areas into AEZs, it is unknown to what extent weather conditions are homogeneous within a zone. Comparison of weather stations in countries within the same AEZ (e.g., Thailand; AEZ 2) would suggest that there could be significant variability in climate within a designated AEZ. However, until further high-quality weather data become available to enable a more detailed coverage, estimates based on the current data cannot be more accurate.

The rice database compiled by Huke and Huke (1997) represents the most comprehensive statistics available on rice area by type. The areal extent of rice ecosystems was previously published in the form of multicolored maps for South, Southeast, and East Asia (Huke, 1982). These data were constantly updated for incorporation in *World rice statistics* (IRRI, 1994) and computerized for use in a GIS. Obviously, the criterion for classifying rainfed rice into groups with less and more than 30 cm water depth may not always be that distinct. Furthermore, the changing of political boundaries has, in some instances, complicated the data collection (Huke & Huke, 1997). These considerations, however, have not impeded the use of this database in the context of this study that was aiming to provide a broad assessment for Asia.

Application

The final stage of the project involves combining statistical data from the Huke and Huke (1997) database on rice cropped areas in each province/district, for each country, with predicted regional emission rates estimated from MERES. A data-bridge approach was adopted to pass the spatially derived data from the GIS to the crop simulation model, then convert the results back to further analyze and display the data within the GIS. Although this approach lacks flexibility and speed when compared with a fully integrated GIS model with embedded code, for combination methodologies where external models such as MERES require regular updating and modification, this approach is preferable. The final output includes tables and maps showing the estimated CH_4 emission at the province/district level, then aggregated to national levels for each country (Matthews et al., 2000b).

Conclusions

A series of derived databases relating to soils and climate at province/district level have been produced for five countries in Asia. These databases provide the geographical basis for generating improved estimates of CH_4 emission from rice fields in Asia and evaluating options for mitigation of these emissions. Additionally, the databases will supplement the world data set of derived soil properties described by Batjes (1997) for use in further GIS-based studies of soil gaseous emission potentials.

Acknowledgements

We are grateful to L.R. Oldeman, D.V. Seshu, F.B. Cady, and their colleagues in the IRRI Rice-Weather Project, for making their weather data available to form the basis of the current database. We also thank the collaborators in the IRRI-EPA project for providing subsequent data after the end of the Rice-Weather Project. The assistance of N.H. Batjes at the International Soils Reference and Information Centre (ISRIC) in providing the ISIS database, giving advice on using the WISE database, and in helping to identify rice-growing soils, is also gratefully acknowledged.

References

Arah JRM & Kirk GJD (2000) Modeling rice-plant-mediated methane emission. Nutr Cycling Agroecosyst (this issue)

Bachelet D, Kern J & Tolg M (1995) Balancing the rice carbon budget in China using spatially distributed data. Ecol Model 79(1-3):167-177

Bachelet D & Neue H-U (1993) Methane emissions from wetland rice areas of Asia. Chemosphere 26:219-237

Batjes NH (1997) A world dataset of derived soil properties by FAO-UNESCO soil unit for global modelling. Soil Use Manage 13:9-16

Batjes NH & Bridges EM (1995) World inventory of soil emission potentials (WISE): geographic quantification of soil factors that control fluxes of greenhouse gases. In: S Zwerver, RSAR van Rompaey, MTJ Kok & MM Berk (eds) Climate Change Research: Evaluation and Policy Implications. Elsevier Science, pp 645-650

Bliss NB (1990) A hierarchy of soil databases for calibrating models of global climate change. In: AF Bouwman (ed) Soils and the Greenhouse Effect, pp 523-529, John Wiley & Sons, Ltd,, New York

Bouwman AF (1990) Global distribution of the major soils and land cover types. In: AF Bouwman (ed) Soils and the Greenhouse Effect, pp 33-59, John Wiley & Sons, Ltd, New York

Centeno HGS, Aclan AP, Balbarez AD, Pascual ZP, Penning de Vries FWT & Kropff MJ (1995) The rice-weather data base. In: RB Matthews, MJ Kropff, D Bachelet & HH van Laar (eds) Modeling the Impact of Climate Change on Rice Production in Asia, pp 289, IRRI/CAB International, Wallingford, UK

FAO (Food and Agriculture Organization of the United Nations) (1974) Soil Map of the World, Volumes 1-10, Paris

FAO (Food and Agriculture Organization of the United Nations) (1995) The Digital Soil Map of the World. Notes, Version 3.5. Rome, Italy

Godwin DC, Singh U, Buresh RJ & De Datta SK (1990) Modelling of nitrogen dynamics in relation to rice growth and yield. In: Proceedings of the 14th International Congress of Soil Science Transactions, pp. 320-325, August 1990, International Society of Soil Science, Kyoto, Japan

Huke R (1982) Rice area by type and culture: South, Southeast, and East Asia. International Rice Research Institute, Manila, Philippines, 32 p

Huke RE & Huke, EH (1997) Rice area by type of culture: South, Southeast, and East Asia. A revised and updated data base. International Rice Research Institute, Manila, Philippines, 59 p

IRRI (International Rice Research Institute) (1978) Soils and Rice. Manila, Philippines.

IRRI (International Rice Research Institute) (1991) World Rice Statistics 1990. Manila, Philippines, 320 p

IRRI (International Rice Research Institute) (1993) IRRI Rice Almanac 1993-1995. Manila, Philippines, 142 p

IRRI (International Rice Research Institute) (1994) World Rice Statistics 1993. Manila, Philippines, 320 p

Jansen DM (1990) Potential rice yields in future weather conditions in different parts of Asia. Neth J Agric Sci 38:661-680

Kern JS, Gong Z, Zhang G, Zhuo H & Luo G (1997) Spatial analysis of methane emissions from paddy soils in China and the potential for emissions reduction. Nutr Cycling Agroecosyst 49:181-195

Leemans R & Cramer W (1990) The IIASA database for mean monthly values of temperature, precipitation and cloudiness of a global terrestrial grid. WP-41, International Institute of Applied Systems Analyses, Laxenburg, Austria

Matthews E (1983) Global vegetation and land use: new high-resolution data bases for climate studies. J Clim Appl Meteorol 22:474-487

Matthews E (1993) Global geographical databases for modelling trace gas fluxes. Int J Geog Inf Syst 7:124-142

Matthews E, Fung I & Lerner J (1991) Methane emission from rice cultivation: geographic and seasonal distribution of cultivated areas and emissions. Global Biogeochem Cycles 5:3-24

Matthews RB, Kropff MJ, Bachelet D & van Laar HH (1995) Modeling the Impact of Climate Change on Rice Production in Asia. IRRI/CAB International, Wallingford, UK, 289 p

Matthews RB, Wassmann R & Arah J (2000a) Using a crop/soil simulation model and GIS techniques to assess methane emissions from rice fields in Asia. I. Model development. Nutr Cycling Agroecosyst (this issue)

Matthews RB, Wassmann R, Knox JW & Buendia LV (2000b) Using a crop/soil simulation model and GIS techniques to assess methane emissions from rice fields in Asia. IV. Upscaling to national levels. Nutr Cycling Agroecosyst (this issue)

Neue HU, Wassmann R & Lantin RS (1995) Mitigation options for methane emissions from rice fields. In: S. Peng, KT Ingram, HU Neue & LH Ziska (eds), Climate Change and Rice, pp 136-144, Springer Verlag, Berlin, Germany

Oldeman LR, Seshu DV & Cady FB (1987) Response of rice to weather variables. In: Weather and Rice. pp 5-39. International Rice Research Institute, Manila, Philippines

Richter DD & Babbar LI (1991) Soil diversity in the tropics. In: M Begon, AH Fitter & A MacFadyen (eds) Advances in Ecological Research, pp 315-389, Academic Press Ltd, London

Shearer MJ & Khalil MAK (1993) Rice agriculture: emissions. In: MAK Khalil (ed) Atmospheric Methane: Sources, Sinks, and Role in Global Change, pp 230-253, NATO ASI Series. Springer-Verlag, Berlin

Sombroek WG (1990) Geographic quantification of soils and their changes in properties. In: A.F. Bouwman (ed) Soils and the Greenhouse Effect, pp 225-244, John Wiley & Sons, Ltd, New York

TYDAC (1994) SPANS GIS Modeling Handbook. INTERA TYDAC Technologies Inc, Ontario, Canada

van de Ven T & Tempel P (1994) ISIS 4.0, ISRIC Soil Information System: User Manual (revised edition). Technical Paper 14, ISRIC, Wageningen

WMO (World Meteorological Organization) (1989) CLICOM Project (Climate Data Management System). World Climate Data Programme No 299, Geneva

Zobler L (1986) A world soil file for global climate modelling. #87802, National Aeronautical and Space Administration, USA

Nutrient Cycling in Agroecosystems **58:** 201–217, 2000.

Using a crop/soil simulation model and GIS techniques to assess methane emissions from rice fields in Asia. IV. Upscaling to national levels

R.B. Matthews[1], R. Wassmann[2,3], J.W. Knox[1] & L.V. Buendia[2]

[1]*Institute of Water and Environment, Cranfield University, Silsoe, Bedfordshire MK45 4DT, United Kingdom;* [2]*International Rice Research Institute, MCPO Box 3127, Makati 1271, Philippines;* [3]*Fraunhofer Institute for Atmospheric Environmental Research, Garmisch-Partenkirchen, Germany*

Key words: rice, model, simulation, carbon dynamics, China, India, Indonesia, Philippines, Thailand, estimates

Abstract

The process-based crop/soil model MERES (*M*ethane *E*missions from *R*ice *E*co*S*ystems) was used together with daily weather data, spatial soil data, and rice-growing statistics to estimate the annual methane (CH_4) emissions from China, India, Indonesia, Philippines, and Thailand under various crop management scenarios. Four crop management scenarios were considered: (a) a 'baseline' scenario assuming no addition of organic amendments or field drainage during the growing season, (b) addition of 3,000 kg DM ha^{-1} of green manure at the start of the season but no field drainage, (c) no organic amendments but drainage of the field for a 14-d period in the middle of the season and again at the end of the season, and (d) addition of 3,000 kg DM ha^{-1} of green manure and field drainage in the middle and end of the season. For each scenario, simulations were made at each location for irrigated and rainfed rice ecosystems in the main rice-growing season, and for irrigated rice in the second (or 'dry') season. Overall annual emissions (Tg CH_4 yr^{-1}) for a province/district were calculated by multiplying the rates of CH_4 emission (kg CH_4 ha^{-1} yr^{-1}) by the area of rice grown in each ecosystem and in each season obtained from the Huke and Huke (1997) database of rice production. Using the baseline scenario, annual CH_4 emissions for China, India, Indonesia, Philippines, and Thailand were calculated to be 3.73, 2.14, 1.65, 0.14, and 0.18 Tg CH_4 yr^{-1}, respectively. Addition of 3,000 kg DM ha^{-1} green manure at the start of the season increased emissions by an average of 128% across the five countries, with a range of 74-259%. Drainage of the field in the middle and at the end of the season reduced emissions by an average of 13% across the five countries, with a range of -10% to -39%. The combination of organic amendments and field drainage resulted in an increase in emissions by an average of 86% across the five countries, with a range of 15-176%. The sum of CH_4 emissions from these five countries, comprising about 70% of the global rice area, ranged from 6.49 to 17.42 Tg CH_4 yr^{-1}, depending on the crop management scenario.

Introduction

Methane (CH_4) is an important greenhouse gas whose concentration has more than doubled over the past 200 yr (Pearman et al., 1986), a phenomenon causing some concern in view of its equivalent warming effect being some 32 times higher than carbon dioxide (CO_2). Flooded rice fields, with their abundant organic matter, warm temperatures, and anaerobic conditions, provide an ideal environment for methanogenic activity, and due to the significant areas under cultivation, are a major anthropogenic source of CH_4. Methane concentrations remained stable for a brief period in 1992-93, but have returned to increasing at an annual rate of 8 ppbv since then (IPCC, 1996). This is of particular concern as rice production has been estimated to have to increase by 270 million t, or by 60%, by the year 2020 to keep pace with projected population increases (Hossain, 1998). Fortunately, irrigated rice fields are one of the few sources of atmospheric CH_4 in which options are available to reduce emissions, in this case through crop management. Estimates of the contribution of rice cultivation to the total global budget have varied widely, ranging from as much as 280 Tg CH_4 yr^{-1} (Ehhalt &

Schmidt, 1978) to as low as 12 Tg CH_4 yr^{-1} (Minami, 1993), although more recent estimates have narrowed this range to 25-54 Tg CH_4 yr^{-1} (e.g., Sass & Fisher, 1997).

Initial approaches of estimation were to use emission rates measured in field experiments and extrapolate these to the global scale. Thus, Holzapfel-Pschorn & Seiler (1986) and Schütz et al. (1989) used measurements from rice fields in Italy to obtain estimates of 120 (70-170) Tg CH_4 yr^{-1} and 96 (47-145) Tg CH_4 yr^{-1}, respectively. The emission rates they used (3.6-10 kg CH_4 ha^{-1} d^{-1} and 2.4-8.0 kg CH_4 ha^{-1} d^{-1}), however, were somewhat higher than those measured in most rice fields, illustrating the dangers inherent in extrapolating from field experiments at one location. This problem was addressed by Khalil and Shearer (1993), who developed an inventory of direct flux measurements from a number of studies and modified the information from Matthews et al. (1991) on the duration of growing seasons to estimate global and regional annual emission rates. They arrived at a figure for the global emission rate of 66 Tg CH_4 yr^{-1}.

A second approach to CH_4 estimation assumed a constant fraction of net primary productivity (NPP) being converted into CH_4. For example, Aselmann and Crutzen (1990) estimated the fraction of the area in 2.5° latitude by 5° longitude boxes occupied by irrigated and rainfed rice cultivation, and the NPP of these areas from published yield data. Methane emissions were calculated using values of the CH_4-NPP ratio ranging from 3 to 7%. A similar approach was used by Taylor et al. (1991) assuming a CH_4-NPP ratio of 5%. Neue et al. (1990) estimated NPP from rice production statistics (using constants for grain-shoot and root-shoot ratios), taking into account aquatic biomass and weed biomass. They assumed that 15% of this was returned to the soil, of which 30% was converted into CH_4, giving a CH_4-NPP ratio of 4.5%, close to those used above. Matthews et al. (1991) refined the approach of Aselmann and Crutzen (1990) by estimating the fraction of rice cultivation area in 1° × 1° cells and calculated detailed rice crop calendars indicating the months of cultivation of rice by country, each state for India, and each province for China. A mean daily emission rate of 5 kg CH_4 ha^{-1} d^{-1} was calculated by assuming that the global emission was 100 Tg CH_4 yr^{-1}. Although this approach could give some idea of the relative CH_4 emissions between the different countries, states, or provinces, it obviously could not be used to give a global estimate. Bachelet and Neue (1993) subsequently modified the approaches of Taylor et al. (1991), Neue et al. (1990), and Matthews et al. (1991) by taking into account the CH_4 production potential of soils in the rice-growing regions using Zobler's (1986) digitized soil map, reducing each previous estimate by about 26%.

A third approach was the use of process-based simulation models using data held in spatial databases as input to the model. Cao et al. (1996) used their methane emission model (MEM) together with an updated version of the data set of rice geographical and seasonal distribution of Matthews et al. (1991). Temperature data were obtained from the IIASA Terrestrial Climate Data set of Leemans and Kramer (1990) and soil information (texture, OC%, and pH) from the digitized FAO soil map (Zobler, 1986). In the absence of crop management information, it was assumed that CH_4 was emitted at the potential rate.

As far as estimates of CH_4 emissions from individual countries are concerned, China seems to have received the most attention, probably on the basis that it has the largest area of rice cultivation, accounting for around 22% of the world rice area (Cao et al., 1995a). Data from Huke and Huke (1997) show that the annual area of harvested rice fields in China was 31 × 10^6 ha, of which 95% was irrigated. Khalil et al. (1989) extrapolated field measurements to the whole area to estimate the total CH_4 emission from China as 55 Tg CH_4 yr^{-1}, but this was soon revised downward to 30 Tg CH_4 yr^{-1} (Khalil et al., 1991), and again to 23 Tg CH_4 yr^{-1}, (Khalil et al., 1993) on the basis of further field experiments. Bachelet and Neue (1993) produced estimates ranging from 9.1 to 14.9 Tg CH_4 yr^{-1}, using the approaches of Matthews et al. (1991), Taylor et al. (1991), and Neue et al. (1990), but taking the emission characteristics of soils into account. Lin (1993) estimated 11 Tg CH_4 yr^{-1}. Cao et al. (1995a) calculated a value of 16.2 Tg CH_4 yr^{-1} by using a simplified version of their process-based CH_4 model and a geo-referenced database of soil and weather data for a number of homogeneous agroecological zones. Although the types of organic matter applied in rice cultivation in China is rather diverse, they reasoned that about 30% of the crop biomass was returned to the soil in one way or another, either as crop residues or as human and animal manures. Inorganic fertilizer applications were assumed to reduce CH_4 emissions by 20%.

Kern et al. (1995) used data from published studies to develop regression equations relating CH_4 emission rates to C and N inputs and the duration of the growing season. The latter was calculated using a 'growing degree-days' concept and temperature data from the IIASA database (Leemans & Cramer, 1990). Areas of rice cultivation were calculated from a vegetation map produced by the Chinese Academy of Sciences. The total annual CH_4 emission from rice cultivation in China was estimated in this manner to be 10 Tg CH_4 yr^{-1}. Bachelet et al. (1995), using the same databases and calculating CH_4 emission rates using the methods of Neue et al. (1990) and Taylor et al. (1991) in addition to that of Kern et al. (1995), arrived at total annual emission figures ranging from 7 to 16 Tg CH_4 yr^{-1}.

Kern et al. (1997) revised the estimates of Kern et al. (1995) for China by taking into account management practices (intermittent drainage and fertilizer inputs) and also new estimates of organic matter addition, soil drainage, and rice-growing areas. Methane emissions from continuously flooded fields were calculated using similar regression equations to those used in their previous paper. Intermittent drainage was assumed to reduce CH_4 emission rates to 50% and rainfed conditions to 40% of these values. Assuming adoption of intermittent drainage on 33% of poorly drained soils, CH_4 emission estimates were reduced by 10% to 8.9 Tg CH_4 yr^{-1}. Reduction of organic inputs by 50% resulted in a reduction of only 3% to 9.6 Tg CH_4 yr^{-1}, and the combination of 33% adoption of intermittent drainage and 50% reduction in organic inputs gave a 12% reduction to 8.7 Tg CH_4 yr^{-1}. The relatively small responses to intermittent drainage was due to the area of rice soils suitable for drainage being not large; on better drained rice soils, reflooding is difficult, while on very poorly drained soils, draining is difficult.

The problem throughout has been the correct association of CH_4 flux rates with the areas of rice production they represent. Flux rates can vary widely even between geographically close areas because of climate, soil properties, duration and pattern of flooding, rice cultivars and crop growth, organic amendments, fertilization, and cultural practices (Neue et al., 1995). The approach of using an empirical ratio between CH_4 emission and NPP is limited—apart from the crop itself, soil organic matter and applied manure are also important sources of substrate for CH_4 production. Moreover, CH_4 emission is not dependent on substrate availability alone—other factors such as the alternative electron acceptor pool, temperature, and soil water may also have significant influences. Similarly, simple extrapolation from limited field experiments is also risky, as shown by the rapid downward revision of estimates for China as more experimental evidence became available (Khalil et al., 1989; 1991; 1993). Attempts to take account some of the factors causing variation in CH_4 emission rates by multiple regression (Kern et al., 1995; 1997) using experimental data are a step in the right direction, but they do not describe the processes involved in CH_4 production and emission, and therefore have limitations in extrapolation to other countries and regions.

Simulation models based on knowledge of processes and factors that control CH_4 emissions, coupled with spatial databases in a GIS environment, have been suggested as the way forward (Neue et al., 1995), and, indeed, the model of Cao et al. (1995b) represents a useful contribution in this direction. The work described in this current series of papers builds on this progress by developing a detailed CH_4 dynamics model to integrate influences of climate, soil, agricultural management, and rice growth on CH_4 flux rates. In earlier papers in the series (Matthews et al., 2000a; b), we describe the development and testing of this model, which is based on the CERES-Rice crop simulation model (Alocilja & Ritchie, 1988) and includes a submodel calculating the steady-state fluxes and concentrations of CH_4 and O_2 in flooded rice soils (Arah & Kirk, 2000). A third paper (Knox et al., 2000) describes the development of spatial databases of variables for input into the model. In this fourth paper, we describe how the model and the databases were used together to predict overall CH_4 emissions from the countries involved in the United Nations Development Programe project, China, India, Indonesia, the Philippines, and Thailand, under different crop management scenarios.

Methodology

Part III of this series (Knox et al., 2000) describes the development of a database containing representative values of soil pH, soil organic carbon, soil iron content, soil texture, soil water release characteristics, and soil bulk density, for the polygons making up the five countries included in the study (i.e., China, India, Indonesia, Philippines, and Thailand). In general, each polygon represented the provinces (or districts in the case of India) in each of the countries, although in some cases, particularly in coastal areas including islands, a province/district may have been made up of more than one polygon. This database was merged with rice production statistics (areas cultivated and rice production) for each province/district obtained from the Huke and Huke (1997) database, with each record representing information for a single polygon. To link this data to the MERES crop simulation model, a 'driver' program was written to take information for each polygon one at a time, convert this into a form that the model could use, run the model as an external stand-alone program, and collect and collate the relevant data from the output files produced by the model. The following sections describe the way in which the information available for each polygon was converted into a form that the crop model could use.

Weather station information and dates of sowing and planting

Data on weather stations and sowing and transplanting dates associated with each weather station were contained in a separate database, with each record representing one station. A field in the polygon database, described above, referenced each polygon to a single weather station as determined by the 'nearest neighbor' procedure described in Part III. The 'driver' program matched this field to the appropriate station in the weather station database, extracted the required station-related data and inserted these into the appropriate part of the model input file.

Size of the alternative electron acceptor pool

The MERES model requires an estimation of the initial concentration of the oxidized alternative electron acceptor (AEA_{ox}, mol C_{eq} m^{-3}) pool in order to calculate the quantity of carbon from organic matter decomposition that is converted to CO_2 before conditions have reached a redox potential (Eh) sufficient for CH_4 production to occur. The concentrations of the ions involved (NO_3^-, Fe^{3+}, Mn^{4+}, and SO_4^{2-}) are not contained in the standard WISE database (Batjes, 1997), although Fe^{3+} concentrations were extracted from the ISIS database (van de Ven & Tempel, 1994) and merged with the WISE database as described in Part III of this series. The problem remains of how to estimate the AEA_{ox} pool concentration from this information.

Data on concentrations of the four species of ions in 16 rice-growing soils are given by Yao et al. (1999). From these and from a knowledge of the stoichiometric relationship between each ion and the quantity of CO_2 released, it is possible to calculate the potential concentration of the AEA_{ox} pool in C equivalents per unit weight of soil (mol C_{eq} g^{-1}). We assumed that the relevant bacteria would use either 0.5 mole of NO_3^-, 4 moles of Fe^{3+}, 2 moles of Mn^{4+}, or 0.5 mole of SO_4^{2-} to produce 1 mole of C in the form of CO_2 from the organic substrate. Using these values, the estimated concentrations of the potential AEA_{ox} pool range from 26 to 117 μmol C_{eq} g^{-1} (Figure 1, *y*-axis), with a mean of 53.4 μmol C_{eq} g^{-1}.

The largest contributor to this potential AEA_{ox} pool is iron, as shown by the strong relationship between these two variables (Figure 1), with 94% of the variation of the AEA_{ox} pool concentration being explained by variation in iron concentration. Thus, if iron concentration (x, μmol Fe g^{-1}) of the soil is known, it is possible to use the regression equation $y = 0.3015\,x$ (Figure 1) to estimate the potential AEA_{ox} pool concentration (y, μmol C_{eq} g^{-1}).

Although this gives the 'potential' AEA_{ox} pool concentration, the 'effective' pool concentration is likely to be a proportion of this. Due to lack of any other estimates of the value of this proportion, we have assumed that 42% of the potential AEA_{ox} pool is effective in acting as alternative electron acceptors for decomposition of organic C (see Part I [Matthews et al., 2000a] for derivation of this value). We recognize that this fraction is based only on Maahas soil at IRRI, but until more accurate information on how this proportion may vary between soils on which rice is grown, we feel justified in using a single value.

The initial concentration of the oxidized *AEA* pool (AEA_{ox}, mol C_{eq} m^{-3}) was therefore estimated from the iron concentration (Fe, g kg^{-1}) of each soil using the equation

$$AEA_{ox} = 0.3015 \times (Fe/M_{Fe}) \times \rho \times 0.42 \times \eta_{Fe} \qquad (1)$$

where M_{Fe} is the molecular weight of iron (55.8 g mol^{-1}), ρ is the bulk density (kg m^{-3}) of the soil, and η_{Fe} is a dimensionless normalization coefficient.

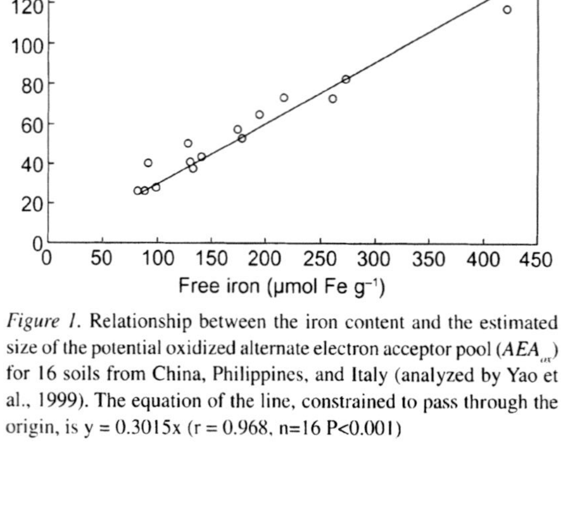

Figure 1. Relationship between the iron content and the estimated size of the potential oxidized alternate electron acceptor pool (AEA_{ox}) for 16 soils from China, Philippines, and Italy (analyzed by Yao et al., 1999). The equation of the line, constrained to pass through the origin, is y = 0.3015x (r = 0.968, n=16 P<0.001)

The coefficient η_{Fe} is used to normalize the Fe values from the ISIS database to those measured by Yao et al. (1999) upon which the model was calibrated. A value of 0.54 for η_{Fe} was calculated from the slope of the regression between the free Fe value of each of the 14 sites in China and the Philippines analyzed by Yao et al. (1999) and the corresponding mean value (calculated as described in Part III of this series) for the province in which each of these sites fell.

Soil water release characteristics

The saturated soil water content (θ_{SAT}), field capacity (θ_{DUL}, drained upper limit), and permanent wilting point (θ_{DLL}, drained lower limit) values for each soil were calculated from the sand (SA, %), silt (SI, %) and clay (CL, %) fractions using the pedotransfer functions given by Cosby et al. (1984):

$$\theta_{SAT} = 50.5 - 0.142\,SA - 0.037\,CL \quad (2)$$

$$\theta_{DUL} = \theta_{SAT} * (0.03/\Psi_s)^{-1/b} \quad (3)$$

$$\theta_{DLL} = \theta_{DUL} - AWC/1000 \quad (4)$$

where Ψ_s is the soil matric potential (MPa) at saturation, b is the slope of the ln (Ψ)/ln(θ) relationship, and AWC is the available water content (mm m^{-1}) obtained from the WISE database (Batjes, 1997). θ_{DUL} is assumed to occur at -0.03 MPa. The parameters Ψ_s and b were calculated as

$$\Psi_s = \exp(1.54 - 0.0095\,SA + 0.0063\,SI)/1000 \quad (5)$$

$$b = 3.10 + 0.157\,CL - 0.003\,SA \quad (6)$$

Creating soil profile data for input into the MERES model

The MERES model requires soil data to be input in the form of a soil profile, i.e., values for each parameter at specific soil depths. We have assumed in each case that the soil depth is 50 cm and that parameter values are provided at 10-cm intervals down to this depth. Values for each of the parameters, pH, % organic carbon, % silt, % clay, θ_{SAT}, θ_{DUL}, θ_{DLL}, and AEA_{ox}, are either stored as, or are calculated from, mean values of variables for the topsoil (0-30 cm) and the subsoil (>30 cm) in the WISE database. Thus, we have assigned the topsoil values to the profile depths of 10 cm, 20 cm, and 30 cm, and the subsoil values to the 40 cm and 50 cm depths. In the case of bulk density (ρ) and available water content (AWC) data which are stored in the WISE database as single values for the whole profile rather than as values for topsoil and subsoil, we have assumed that these apply to all depths throughout the soil profile.

Genotype parameters

Parameters for the indica genotype IR72 were used for all areas in India, Philippines, Indonesia, and Thailand. In China, parameters for japonica genotypes were used – Chunjiang 06 for latitudes less than an arbitrary line at 30.5° N (southern China) and Zhongzhuo 93 for latitudes greater than 30.5° N (northern China). Values of these parameters are shown in Table 1.

Description of scenarios simulated

For each polygon, a total of 16 different simulations were made (Table 2). These included four scenarios: two levels of organic amendments (0 and 3,000 kg DM ha^{-1}) and two levels of field drainage (either none at all, or drainage in the midseason and at end of season). For each scenario, two seasons each year (the main planting season and the dry season) and two rice ecosystems (irrigated rice and rainfed rice) were simulated.

Irrigated rice ecosystems obviously have the highest potential to produce and emit CH_4 because of assured and controlled flooding, high fertilization, and good rice growth. Irrigated rice was simulated by using the automatic irrigation option in the MERES model, which adds water when required in order to maintain the floodwater level at a specified value, in

Table 1. Genotype parameters of the varieties used in the simulations. See Part I of this series (Matthews et al., 2000a) for a description of each parameter

Genotype	P1	P2R	P5	P2O	G1	G2	G3	G4
IR72	548	0	390	12.0	46	.0250	1.0	1.0
Chunjiang 06	600	140	380	12.0	46	.0250	1.0	1.0
Zhongzhuo 93	400	60	430	12.0	46	.0250	1.0	1.0

Table 2. Description of the 16 simulations made for each polygon

	Growing season	Rice ecosystem	Organic amendments (kg DM ha^{-1})	Drainage regime
1a	Main	Irrigated	0	None
2a	Main	Irrigated	3000	None
3a	Main	Irrigated	0	Mid- and end-of-season
4a	Main	Irrigated	3000	Mid- and end-of-season
1b	Main	Rainfed	0	None
2b	Main	Rainfed	3000	None
3b	Main	Rainfed	0	Mid- and end-of-season
4b	Main	Rainfed	3000	Mid- and end-of-season
1c	Second	Irrigated	0	None
2c	Second	Irrigated	3000	None
3c	Second	Irrigated	0	Mid- and end-of-season
4c	Second	Irrigated	3000	Mid- and end-of-season
1d	Second	Rainfed	0	None
2d	Second	Rainfed	3000	None
3d	Second	Rainfed	0	Mid- and end-of-season
4d	Second	Rainfed	3000	Mid- and end-of-season

this case, 5 cm. Rice production and CH_4 emission in rainfed rice ecosystems vary widely in space and time, with rainfall within the watershed primarily controlling floodwater regimes with periods of droughts and floods common during the growing season. To simulate rainfed rice, we assumed that the field was flooded at the start of the season, and that any water added after that time during the season was from rainfall only. Although rainfed scenarios in the second season (normally the dry season) were simulated (scenarios 1d-4d in Table 2), in all cases there was little or no crop yield and no CH_4 emissions due to lack of water. In the subsequent analysis, therefore, these scenarios were ignored.

The two levels of organic amendments were chosen to represent the two extremes likely to be applied by farmers to give 'best-' and 'worst-' case scenarios. In China, recent studies have estimated a mean application of organic fertilizers of around 1,000 kg C ha^{-1} $season^{-1}$ (Kern et al., 1997), representing about 3,300 kg DM ha^{-1} $season^{-1}$. Most rates of organic amendments applied by farmers in the other countries would be likely to be less than this value. The relationship between level of organic amendments and CH_4 emission rates were examined in more detail in Part II (Matthews et al., 2000b) of this series.

Similarly, the two drainage levels were again chosen to represent the extremes of likely farmer practice to examine the sensitivity of overall CH_4 emission rates on the amount of drainage. Midseason drainage was assumed to occur from 20 d after transplanting for the following 14 d, and end-of-season drainage was assumed to occur for 14 d before harvest. Again, the effect of timing and duration of these drains was examined in more detail in Part II.

For simplicity, it was assumed that there was 5 cm of stubble (= 820 kg DM ha^{-1}) and 350 kg DM ha^{-1} of root material left from the previous crop in each case. The soil was assumed to be at the drained upper limit at the start of the simulation, with incorporation of previous crop residues and any organic amendments and flooding of the field occurring on the first day also. Transplanting occurred 20-25 d later as determined from the data for each weather station shown in Part III (Knox et al., 2000). Fertilizer in the form of urea was applied at a rate of 120 kg N ha^{-1} in four equal splits of 30 kg N ha^{-1} at 2 d before transplanting, and 15, 48, and 59 d after transplanting.

Calculating overall emissions for each country

The model simulations predicted CH_4 emission rates for each polygon in kg C ha^{-1} $season^{-1}$. These were first aggregated into mean values for each province or district by summing the predicted emission rates of each polygon weighted by its area and then dividing by the total area of the province/district.

Overall annual emissions from each province/district were then calculated using these mean emission values and the data compiled by Huke and Huke (1997) on areas of rice production in each of the main rice ecosystems (i.e., irrigated, rainfed, deepwater, and upland rice). Upland rice was assumed to produce no CH_4

at all because it is never flooded for a significant period (Neue et al., 1995). Tidal wetlands and deepwater rice comprise less than 10% of the total rice-growing area, and their CH_4 emission potential may be low because of salinity and deepwater, respectively (Neue et al., 1995). As the mechanisms involved in CH_4 emissions from these ecosystems are not well understood, a constant emission rate of 98 kg CH_4 ha^{-1} $season^{-1}$ obtained from the field experiments in Thailand was assumed. Thus the annual emission (E_A, Tg CH_4 yr^{-1}) from a province/district was calculated as

$$E_A = (E_{MI}A_{MI} + E_{MR}A_{MR} + E_{2I}A_{2I} + 98A_{DW}) \times 16/12 \times 10^{-9} \quad (7)$$

where E_{MI}, E_{MR}, and E_{2I} are the mean emission rates (kg C ha^{-1} $season^{-1}$) calculated above for irrigated rice in the main season, rainfed rice in the main season, and irrigated rice in the second season, respectively, and A_{MI}, A_{MR}, A_{2I} and A_{DW} are the areas (ha) of main season irrigated rice, main season rainfed rice, second season irrigated rice, and deepwater rice, respectively. The 16/12 is to convert kg C into kg CH_4, while 10^{-9} converts kg into Tg.

The annual emissions from each province/district were then summed to give an overall value for the whole country.

Results

Spatial distributions of CH_4 emissions under the different scenarios

Maps showing the predicted spatial distribution of mean CH_4 emission rates in the five countries under the four different scenarios are shown in Figures 2-6. Data are the total estimated annual CH_4 emission from the whole province/district divided by its total land area. Provinces or districts, therefore, which have little rice growing in them but have relatively large areas, will have low average emission rates.

It can be seen that large areas of all countries have relatively low mean emission rates, less than 50 kg CH_4 ha^{-1} yr^{-1}. The areas with higher CH_4 emission rates generally reflect the areas where the most rice is grown. As might be expected, the addition of 3,000 kg DM ha^{-1} of green manure increases the areas with relatively higher CH_4 emission rates, while field drainage reduces them. The combination of organic amendments and field drainage generally cancel each other out, resulting in CH_4 emissions close to those in the baseline scenario.

Overall

The summary of annual emissions from the five countries is shown in Table 3. The largest emissions are from China and India due to their large areas of rice grown, followed by Indonesia, with lesser rates from the Philippines and Thailand. Addition of 3,000 kg DM ha^{-1} of green manure at the start of the season increased emissions by an average of 128% (Table 3) although in individual countries this ranged from 74% to 259%. On the other hand, drainage of the field during the middle of the season and again at the end of the season reduced national emissions by an average of 13%, with a range between individual countries of -10% to -40%. The combination of adding green manure and draining the field together resulted in an average of 86% increase in CH_4 emissions over the baseline, although this varied from 15% to 176% between countries.

Discussion

Our estimates of total CH_4 emissions from China, India, Indonesia, Philippines, and Thailand are generally a little lower than most previous estimates from each of these countries (Table 4), although they do agree closely with those of Sass & Fisher (1997). This is discussed in more detail in the following paragraphs.

For China, our predicted emission value for the baseline scenario of 3.73 Tg CH_4 yr^{-1} is considerably lower than other estimates in recent years (Table 4). However, significant amounts of organic manures are added to Chinese rice fields, and in many areas drainage during the season is practiced to restrict the numbers of unproductive tillers. A more realistic estimate, therefore, is probably between scenarios 3 and 4, in which emissions were calculated to be 8.64 and 7.22 Tg CH_4 yr^{-1}, respectively. These are only a little lower than the value of 10 Tg CH_4 yr^{-1} arrived at by Kern et al (1995; 1997), who used regression equations relating CH_4 emission rates to C and N inputs and the duration of the growing season from five field experiments. However, there is considerable uncertainty in the average rates of application of organic manures - Kern et al. (1995) assume organic additions of 1,000 kg C ha^{-1} $season^{-1}$ on 25% of the rice fields, giving an average of only 250 kg C ha^{-1} $season^{-1}$, a figure somewhat lower than the value of 980 kg C ha^{-1} $season^{-1}$ we have used in scenarios 2 and 4. On the other hand, in a followup study, these authors obtained similar emission values of 9.8 Tg CH_4 yr^{-1} when they assumed an average of 1,070 kg C ha^{-1} $season^{-1}$ was ap-

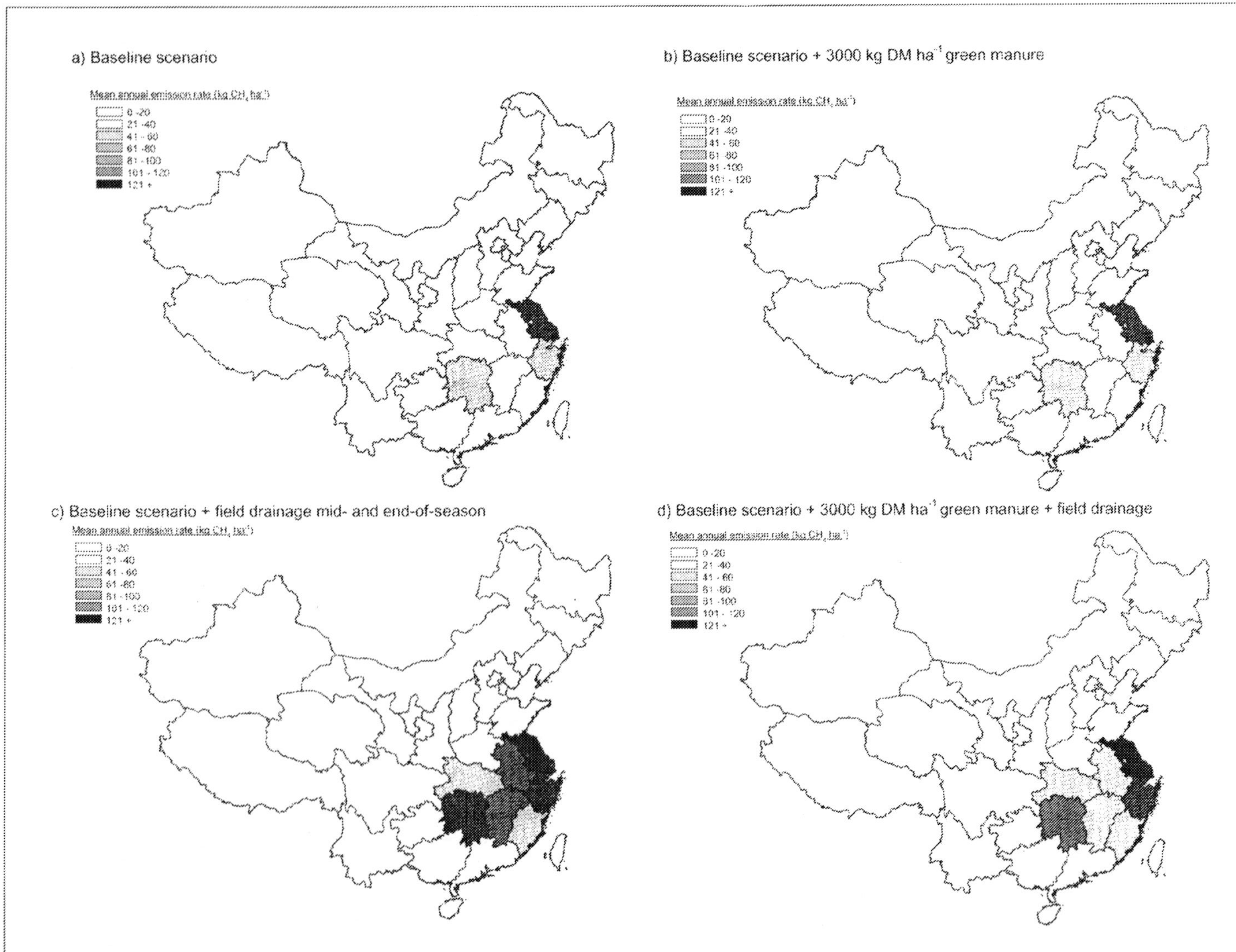

Figure 2. Map showing the distribution of predicted emissions (kg CH_4 ha^{-1} yr^{-1}) for the provinces of China under the four scenarios

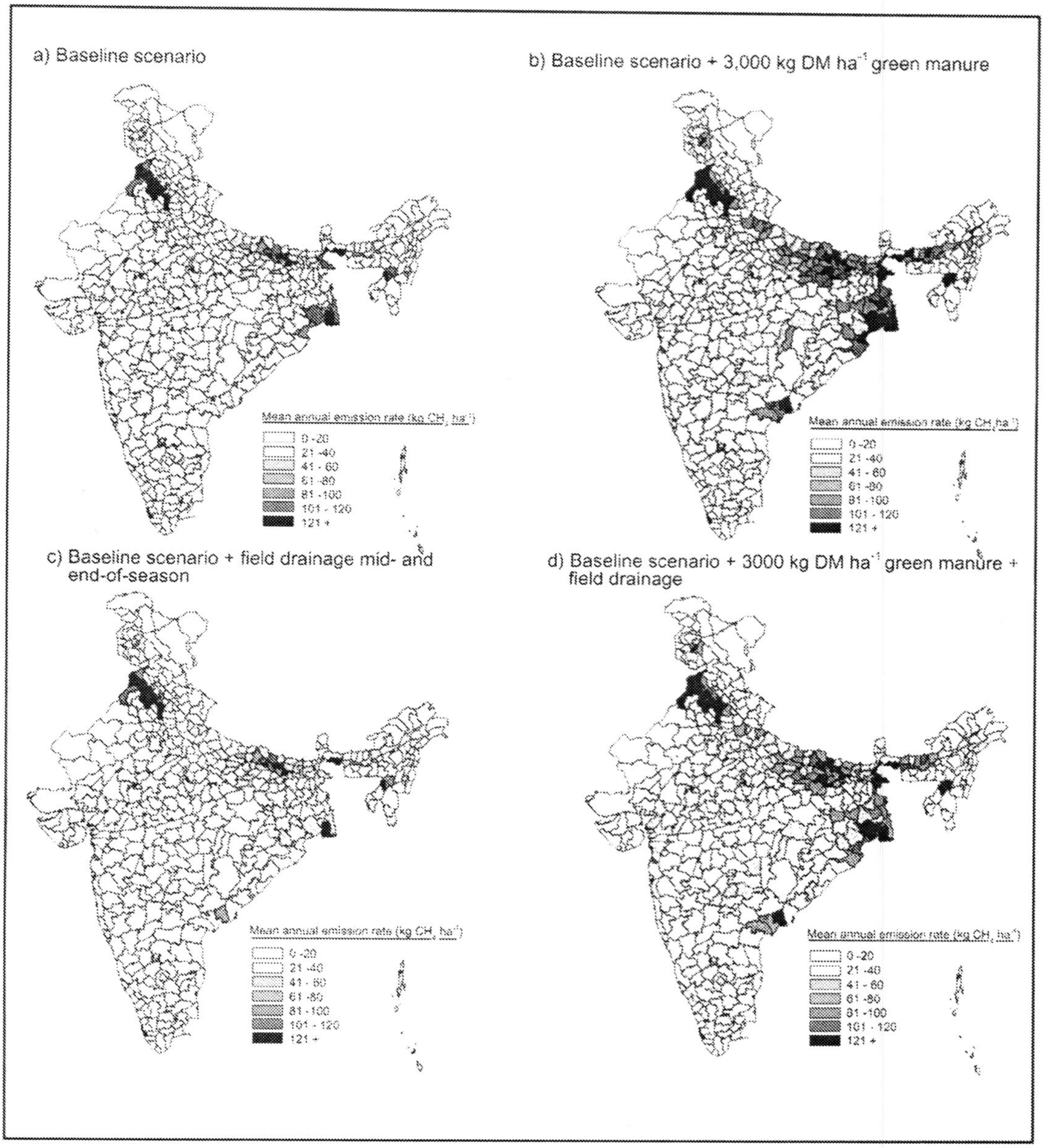

Figure 3. Map showing the distribution of predicted emissions (kg CH_4 ha^{-1} yr^{-1}) for the districts of India under the four scenarios

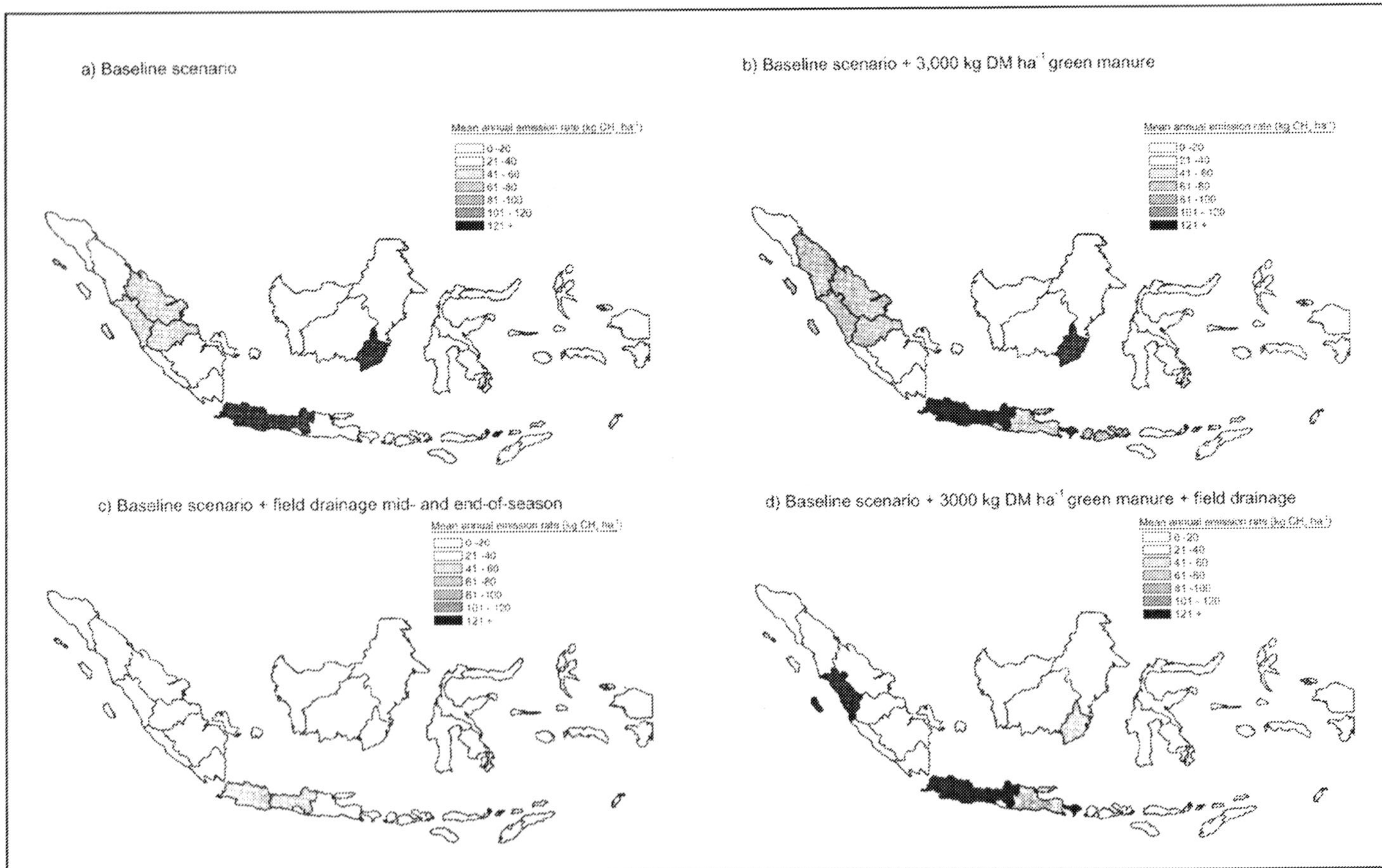

Figure 4. Map showing distribution of the predicted emissions (kg CH_4 ha^{-1} yr^{-1}) for the provinces of Indonesia under the four scenarios

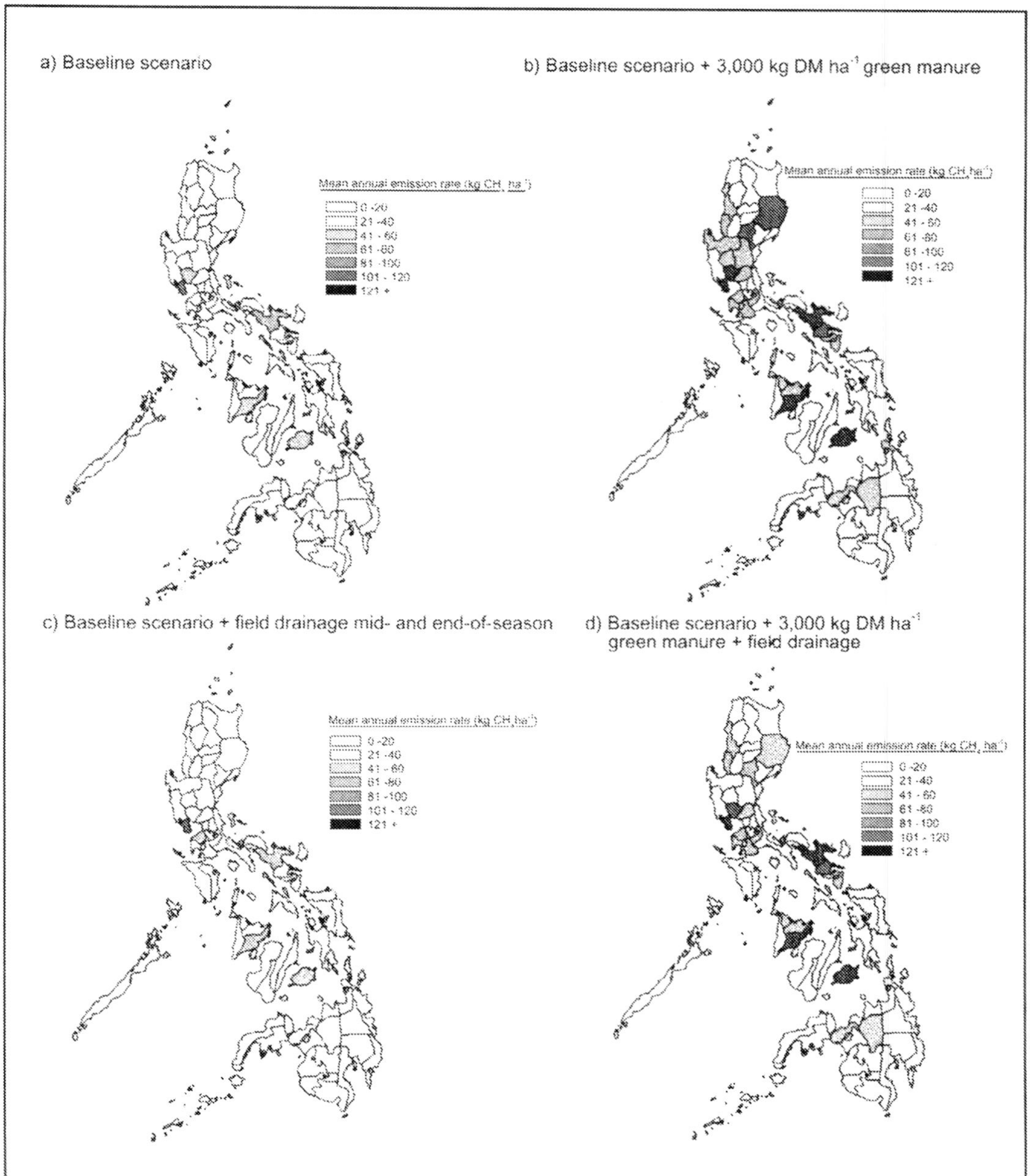

Figure 5. Map showing the distribution of predicted emissions (kg CH_4 ha^{-1} yr^{-1}) for the provinces of the Philippines under the four scenarios

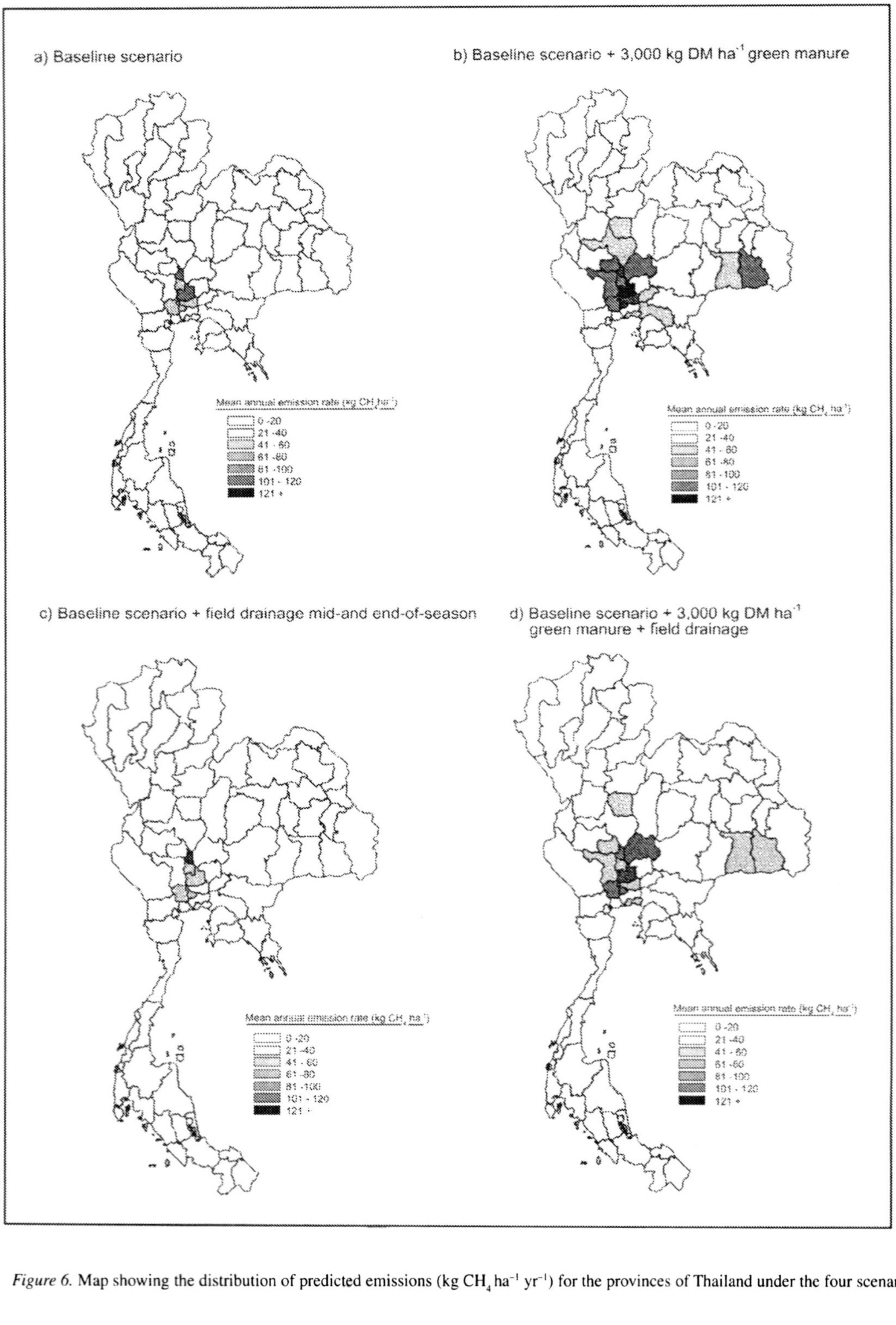

Figure 6. Map showing the distribution of predicted emissions (kg CH_4 ha^{-1} yr^{-1}) for the provinces of Thailand under the four scenarios

plied as organic fertilizer (Kern et al., 1997), similar to our value. Clearly, the constraint to more accurate estimates of CH_4 emissions from China is the shortage of detailed and accurate information on rates of application of organic manures, a factor which needs to be addressed in future studies.

For India, our baseline estimate of 2.14 Tg CH_4 yr^{-1} is again considerably lower than many of the previous estimates, but is comparable with that of Sass and Fisher (1997). Their estimate was based on results from a broad measurement campaign from 1989 to 1991 covering selected rice-growing areas of India, which indicated very low emission rates ranging from 3.4 to 5.4 Tg CH_4 yr^{-1} (Mitra, 1992). This close agreement between our simulated values and their measured values suggests that the extremely high percolation rates of around 28 mm d^{-1} reported at some sites in northern India (Mitra, 1999) are not typical of all rice-growing areas throughout the whole country. Our current simulations with MERES have assumed the percolation rate to be zero in all five countries due to the lack of spatial information on this parameter. The influence of percolation and seepage on CH_4 emissions are discussed in more detail in Part II of this series in (Matthews et al., 2000b). However, as with China, a large uncertainty in the estimates for India is in the rates of application of organic material.

Table 3. Predicted annual CH_4 emissions (Tg yr^{-1}) from each of the five countries in the study. Scenarios are (1) baseline scenario: continuous flooding and no organic amendments, (2) continuous flooding + 3000 kg DM ha^{-1} as green manure, (3) field drainage and no organic amendments, (4) field drainage + 3000 kg DM ha^{-1} green manure. Details of each scenario are given in the text.

Country	Rice area (km^2)	Scenario			
		1	2	3	4
China	323,910	3.73	8.64	3.35	7.22
India	424,947	2.14	4.99	1.88	4.07
Indonesia	110,088	1.65	2.87	1.00	1.90
Philippines	36,205	0.14	0.50	0.12	0.39
Thailand	96,442	0.18	0.42	0.14	0.32
TOTAL	991,591	7.83	17.42	6.49	13.90
% change from baseline			128	-13	86

The figures for Indonesia also deserve further mention. The emissions predicted in the current study are generally lower than in previous estimates (see Table 4), with the exception of that of Matthews et al., (1991). This was despite differences in the rice-growing areas used in some cases – for example, Bachelet and Neue (1993) use an area of 79,440 km^2 compared with the figure of 110,000 km^2 we have used from the Huke and Huke (1997) database. Closer examination indicated that several regions in Indonesia had significant fractions of peaty soils — Histosols with around 35% organic carbon (OC) and Andosols with 10% OC. Many of the Histosols are in the low-lying coastal plains of Sumatra (Bridges, 1997), Kalimantan, and other islands. The high %OC in these soils would suggest that emission rates should be high due to the greater supply of methanogenic substrate from mineralization of this peaty organic matter. However, the average iron content of these soils was also the highest of all the five countries (see Part III, Knox et al., 2000), indicating the presence of a large AEA_{ox} pool which would offset the effect of the higher %OC levels in terms of CH_4 production. Our predicted mean rate of CH_4 emissions for the irrigated main season under each scenario of 99.5-299.3 kg CH_4 ha^{-1} $season^{-1}$ is well within the range

Table 4. Comparison of annual CH_4 emissions (Tg yr^{-1}) from China, India, Indonesia, Philippines, and Thailand predicted in recent studies. Areas shown are from the Huke & Huke (1997) database and may vary slightly from published values in individual studies. Estimates indicated as Matthews et al. (1991), Taylor et al. (1991) and Neue et al. (1990) are those modified by Bachelet & Neue (1993) to account for soil CH_4 emission potential.

Reference	China	India	Indonesia	Philippines	Thailand
Rice area (km^2)	321,449	428,545	79,439	25,464	92,366
Matthews et al. (1991)	14.92	21.68	2.90	0.99	4.10
Taylor et al. (1991)	13.46	18.35	4.81	1.14	4.73
Neue et al. (1990)	14.71	14.54	3.54	0.82	2.24
Khalil & Shearer (1993)	23.0	15.3	6.2	1.2	4.7
Cao et al. (1996)	12.3	14.4	4.7	-	2.9
Sass & Fisher (1997)	15.0	4.2	3.5	0.51	4.62
Current study	3.35-8.64	1.88-4.99	1.00-2.87	0.12-0.50	0.14-0.32

of 90-440 kg CH_4 ha^{-1} $season^{-1}$ values measured by Husin et al. (1995). Estimates of CH_4 emission rates in previous studies—445 kg CH_4 ha^{-1} $season^{-1}$ (Bachelet & Neue, 1993) and 520 kg CH_4 ha^{-1} $season^{-1}$ (Cao et al., 1996)—would appear to be somewhat high.

For Thailand, our estimates of 0.14-0.42 Tg CH_4 yr^{-1} were lower than previous estimates, including those of Sass and Fisher (1997). Closer examination showed that this was due to the large area there of main-season rainfed rice—about 84% of total rice area (Huke & Huke, 1997). The predicted emission rates from these areas were very low (scenario averages: 4.0-20.6 kg CH_4 ha^{-1} $season^{-1}$) because of intermittent rainfall during this season allowing aeration of the soil between rainfall events. The iron contents of soils in Thailand are not excessively high (Knox et al., 2000), so these low emission rates do not seem to be due to the presence of a large AEA_{ox} pool. Previous studies have not been able to explicitly take into account the effect of intermittent rainfall on CH_4 production and may have overestimated CH_4 emission rates from these areas.

The current study provides a new approach in that it uses a crop model to estimate several of the components of the CH_4 flux—the contribution of the rice plant itself (rhizodeposition), which previously was estimated from aboveground biomass data alone (e.g., Huang et al., 1998), and on the length of the growing season, previously estimated from crop calendars (e.g., Matthews et al., 1991). Nevertheless, the results of the present simulation study depend on the many assumptions built into the model used and the quality of the data used as input. In the case of the MERES model, many of the relationships describing the behavior of the processes involved in CH_4 emissions have been derived from a limited number of experiments, some in laboratory conditions, and are, therefore, not fully tested, particularly for field conditions. The rate of root exudation, for example, is based on one laboratory experiment and needs further testing under a wider range of conditions, including in the field if possible. There is also considerable uncertainty in the root death rate —this is currently estimated as a constant 2% d^{-1} of the root biomass present, but despite it giving reasonable estimates of rhizodeposition, little measured data exist to support this value. We have also assumed that the rate of substrate supply for the methanogens from fermentation is not a limiting factor (i.e., that all substrate available on a given day is consumed within that day). While this assumption does seem to produce realistic behavior in terms of the pattern of CH_4 emissions over the season, independent confirmation is needed. The transmissivity of the plants to gaseous movement of CH_4 and O_2 is also an estimate and is assumed in the current model to remain constant throughout the season, although there is evidence to suggest that this is not the case (Nouchi et al., 1990).

For the soil, a major uncertainty is in the estimation of the initial oxidized alternative electron acceptor pool. There does seem to be a strong correlation between the free iron content of the soil and the potential size of this AEA pool, but it is unclear what fraction of this potential size is active in accepting electrons from the breakdown of organic matter, thereby suppressing CH_4 production. We have used a value of 42% based on estimates from the Mahaas soil at IRRI in the Philippines, but further work is clearly needed to see if this is a general one for all soils. The large variability in many of the quantitative soil characteristics (e.g., soil organic matter levels, discussed by Batjes, 1997) is also another source of uncertainty in the soil data. Nevertheless, despite these uncertainties, we feel that this approach of estimating the influence of the soil quantitatively is an improvement on that of Bachelet & Neue (1993) in which the CH_4 emission potential of different soils was estimated somewhat subjectively.

Another major area of uncertainty is the quantity of organic fertilizer applied to rice fields. In terms of green manure, this is probably only of significance in China, but previous crop residues may also be an important source of C for methanogenesis, such as in some areas of the Philippines where 30-40 cm of stubble may be left and subsequently incorporated. The main problem is that data on organic amendments to rice fields are scarce, although globally the trend appears to be declining (Neue et al., 1990). Wen (1984) estimated the average use of rice straw in Chinese rice agriculture at 3,300 kg DM ha^{-1} yr^{-1} and the sum of pig, cattle, and human wastes at about 5,000 kg DM ha^{-1} yr^{-1}. Using the % C data provided, this represents the application of a total of about 3,400 kg C ha^{-1} yr^{-1}, or for two crops a year, about 1,700 kg C ha^{-1} $season^{-1}$. Cao et al. (1995a) estimated that about 30% of the crop biomass was returned to the soil in some way or another, either through straw or animal or human excreta. Assuming an aboveground biomass production of around 15,000 kg DM ha^{-1} $season^{-1}$, this represents about 1,800 kg C ha^{-1} $season^{-1}$, similar to the value calculated from Wen (1984). Kern et al. (1995) used national N fertilizer production to estimate the likely requirements for N from organic fertilizer and calculated that a mean value of only 250 kg C ha^{-1} $season^{-1}$ needed to be added, considerably lower than other estimates.

Bachelet et al. (1995) use the same approach. The low values for organic amendments used in these two studies may explain the lower overall CH_4 emissions (~10 Tg CH_4 yr^{-1}) they obtained for China compared with others. Kern et al. (1997) present estimates of organic amendments ranging from 465 to 2,075 kg C ha^{-1} $season^{-1}$, with an average across provinces of 1,075 kg C ha^{-1} $season^{-1}$. Thus, there is a considerable range in the estimated level of organic fertilizers applied. The value of 3,000 kg DM ha^{-1} $season^{-1}$ (~1,000 kg C ha^{-1} $season^{-1}$) we have used in the present work is midway between the two extremes described and close to the mean of the Kern et al. (1997) estimates.

In our study, we have used the province or district as the unit of resolution as this was the level at which rice growing statistical data were available (Huke & Huke, 1997). Similarly, we have aggregated all of the input data to this level, despite some of them existing at finer levels of resolution. The quantitative soil data, for example, are available for individual soil units, of which there were many in a province. It can be argued that it might have been preferable to simulate CH_4 emissions from each soil unit, and aggregate the model output to the province/district level rather than aggregating the input data, but due to the large increase in computing time required for this approach, we feel justified in our approach. Current simulations for all four scenarios, two seasons, and two rice ecosystems require nearly 50 h of continuous running on a 300 MHz desktop computer. Simulating at the soil unit level would require some 20 times this amount.

The sparseness of weather data sites in some countries is also cause for some concern; large areas in both India and China, for example, are represented by only a few stations. While an attempt was made to stratify these areas into agroecological zones, it is not known to what extent weather conditions are homogeneous within a zone. Comparison of changes in different countries but the same AEZ (e.g., zone 8 in both China and India) suggests that there could be significant variability in climate within a designated zone. However, until more high-quality weather data become available to enable a more detailed coverage, estimates based on current data cannot be more accurate.

Nevertheless, despite these limitations, we feel that the current study marks significant progress in the estimation of CH_4 emissions from rice fields in the Asian region. In total, the source strength of the five countries, which comprise about 70 % of the global rice area, ranged from 6.49 to 17.42 Tg CH_4 yr^{-1}, depending on the crop management scenario used. There has been a general decline in the size of the estimated emissions from a high value of 280 Tg CH_4 yr^{-1} in 1978 (Ehhalt & Schmidt, 1978). As noted by IPCC (1992), there was clearly an overestimation of the source strength of rice fields in the early studies.

An important output of the project has been the synthesis from other existing databases of an extensive database for the region of quantitative soil characteristics important in influencing CH_4 emissions. It is also the first study of its kind to employ a detailed process-based model integrating the crop and soil processes important in the production and emission of CH_4. This approach allows an evaluation at the field, national, and regional levels of the effects of various crop management strategies on mitigation of CH_4 emissions, of which we have only considered two—the use of organic amendments and of field drainage during the season. However, the use of simulation models and spatial databases to upscale measurements made in field experiments to higher levels in this way is an evolving science, and we hope that this study can be used as a baseline for future studies, in which some of the current limitations are addressed, so that increasingly better predictions can be made.

Acknowledgements

The authors would like to thank the staff in the Soil and Water Sciences Division at the International Rice Research Institute for their efforts in collecting and collating the data used in this study. The study was conducted with funding from the United Nations Development Programme's Global Environment Facility.

References

Alocilja EC & Ritchie JT (1988) Upland rice simulation and its use in multicriteria optimization. Report 01, International Benchmark Sites Network for Agrotechnology Transfer

Arah JRM & Kirk GJD (2000) Modeling rice-plant-mediated methane emission. Nutr Cycling Agroecosyst (this issue)

Aselmann I & Crutzen DJ (1990) A global inventory of wetland distribution and seasonality, net primary productivity, and estimated methane emissions. In AF Bouwman (ed) Soils and the Greenhouse Effect, pp 441-449. John Wiley, New York

Bachelet D, Kern J & Tölg M (1995) Balancing the rice carbon budget in China using spatially-distributed data. Ecol Modelling 79(1-3):167-177

Bachelet D & Neue HU (1993) Methane emissions from wetland rice areas of Asia. Chemosphere 26:219-237

Batjes NH (1997) A world dataset of derived soil properties by FAO-UNESCO soil unit for global modelling. Soil Use Manage 13:9-16

Bridges EM (1997) World Soils. Cambridge University Press, Cambridge, UK, 170 p

Cao M, Dent JB & Heal OW (1995a) Methane emissions from China's paddyland. Agric Ecosyst Environ 55:129-137

Cao M, Dent JB & Heal OW (1995b) Modelling methane emissions from rice paddies. Global Biogeochem Cycles 9:183-195

Cao M, Gregson K, Marshall S, Dent JB & Heal O (1996) Global methane emissions from rice paddies Chemosphere 55(5):879-897

Cosby BA, Hornberger GM, Clapp RB & Ginn TR (1984) A statistical exploration of the relationship of soil moisture characteristics to the physical properties of soil. Water Resour Res 20:682-690

Ehhalt DH, Schmidt U (1978) Sources and sinks of atmospheric methane. Pure Appl Geophys 116:452-464

Holzapfel-Pschorn A, Conrad R & Seiler W (1986) Effects of vegetation on the emission of methane by submerged paddy soil. Plant Soil 92:223-233

Hossain M (1998) Sustaining food security in Asia: economic, social and political aspects. In: NG Dowling, SM Greenfield & KS Fischer (eds) Sustainability of Rice in the Global Food System. Pacific Basin Study Centre, Davis, California, and the International Rice Research Institute, Manila, Philippines, 404 p

Huang Y, Sass RL & Fisher FMJr (1998) A semi-empirical model of methane emission from flooded rice paddy soils. Global Change Biol 4(3):247-268

Huke RE & Huke EH (1997) Rice area by type of culture: South, Southeast, and East Asia. A revised and updated data base. International Rice Research Institute, Manila, Philippines, 59 p

Husin YA, Murdiyarso D, Khalil, MAK, Rasmussen R.A, Shearer MJ, Sabiham S, Sunar A & Adijuwana H (1995) Methane flux from Indonesian wetland rice, the effect of water management and rice variety. Chemosphere 31:3153-3180

IPCC (Intergovernmental Panel on Climate Change) (1992) Climate Change: The Supplementary Report to the IPCC Scientific Assessment, Houghton JT, Callander BA & Barney SK (eds) Cambridge University Press, Cambridge

IPCC (Intergovernmental Panel on Climate Change) (1996) XII. Summary for policymakers. In: Houghton JT, Meira-Filho LG, Chancellor BA, Kattenberg A & Maskell K (eds) Climate Change 1995: The Scientific Basis of Climate Change. Cambridge University Press, Cambridge, UK, 572 p

Kern JS, Bachelet D & Tölg M (1995) Organic matter inputs and methane emissions from soils in major rice-growing regions of China. In: Lal R, Kimble J, Levine E & Stewart BA (eds) Soils and Global Change, Advances in Soil Science, CRC Lewis Publishers, Boca Raton, pp 189-198

Kern JS, Gong Z, Zhang G, Zhuo H & Luo G (1997) Spatial analysis of methane emissions from paddy soils in China and the potential for emissions reduction Nutr Cycling Agroecosyst 49:181-195

Khalil MAK, Rasmussen RA, Wang MX & Ren L (1989) Sources of methane in China: rice fields, biogas pits, cattle, urban areas, and wetlands. In: Proceedings of the International Conference on Global and Regional Environmental Atmospheric Chemistry, 3-10 May 1989, Beijing, China, pp 1-10

Khalil MAK, Rasmussen RA, Wang M X & Ren L (1991) Methane emissions from rice fields in China. Environ Sci Technol 25:979-981

Khalil MAK & Shearer MJ (1993) Methane emissions from rice agriculture. Cited in Shearer MJ, Khalil MAK (1993) Rice agriculture: emissions. In: Atmospheric Methane: Sources, Sinks, and Role in Global Change. Berlin, Springer-Verlag 113:250-253

Khalil MAK, Shearer MJ & Rasmussen RA (1993) Methane sources in China: historical and current emissions. Chemosphere 26:127-142

Knox J, Matthews RB & Wassmann R (2000) Using a crop/soil simulation model and GIS techniques to assess methane emissions from rice fields in Asia. III. Databases. Nutr Cycling Agroecosyst (this issue)

Leemans R & Cramer W (1990) The IIASA database for mean monthly values of temperature, precipitation and cloudiness of a global terrestrial grid. WP-41, International Institute of Applied Systems Analyses, Laxenburg, Austria

Lin E (1993) Agricultural techniques: factors controlling methane emissions. In: Gao L, Wu L, Zheng D & Han X (eds) Proceedings of the International Symposium on Climate Change, Natural Disasters, and Agricultural Strategies, pp 120-126. China Meteorological Press, Beijing, China

Matthews E, Fung I & Lerner J (1991) Methane emission from rice cultivation: geographic and seasonal distribution of cultivated areas and emissions. Global Biogeochem Cycles 5:3-24

Matthews RB, Wassmann R, & Arah J (2000a) Using a crop/soil simulation model and GIS techniques to assess methane emissions from rice fields in Asia. I. Model development. Nutr Cycling Agroecosyst (this issue)

Matthews RB, Wassmann R, Buendia LV & Knox J (2000b) Using a crop/soil simulation model and GIS techniques to assess methane emissions from rice fields in Asia. II. Model validation and sensitivity analysis. Nutr Cycling Agroecosyst (this issue)

Minami K (1993) Methane from rice production. In: AR van Amstel (ed) Methane and Nitrous Oxide, pp 143-162. Proceedings of the IPCC (Intergovernmental Panel for Climate Change) Workshop, Bilthoven, The Netherlands

Mitra AP (1992) Greenhouse gas emission in India: 1991 Methane campaign. Council of Scientific and Industrial Research and Ministry of Environment and Forests, New Delhi, India

Mitra S (1999) Effect of rice cultivars on methane emission. Agric Ecosyst Environ 73(3):177-183

Neue HU, Becker-Heidmann P & Scharpenseel HW (1990) Organic matter dynamics, soil properties and cultural practices in rice lands and their relationship to methane production. In: A.F. Bouwman (ed) Soils and the Greenhouse Effect, pp 457-466, Wiley, New York

Neue HU, Wassmann R & Lantin RS (1995) Mitigation options for methane emissions from rice fields. In: Peng S, Ingram KT, Neue HU & Ziska LH (eds) Climate Change and Rice, pp 136-144, Springer Verlag, Berlin, Germany

Nouchi I, Mariko S & Aoki K (1990) Mechanism of methane transport from the rhizosphere to the atmosphere through rice plants Plant Physiol 94:59-66

Pearman GI, Etheridge D, de Silra F & Fraser PJ (1986) Evidence of changing concentrations of atmospheric CO_2, N_2O and CH_4 from air bubbles in Arctic ice. Nature 320:248-350

Sass RL & Fisher FM (1997) Methane emissions from rice paddies: a process study summary. Nutr Cycling Agroecosyst 49:119-127

Schütz H, Holzapfel-Pschorn A, Conrad R & Rennenberg H (1989) A three-year continuous record on the influences of daytime, season and fertiliser treatment on methane emission rates from an Italian rice paddy. J Geophys Res 94:405-416

Taylor JA, Brasseur GP, Zimmerman PR & Cicerone RJ (1991) A study of the sources and sinks of methane and methyl chloroform using a global three-dimensional Lagrangian tropospheric tracer transport model. J Geophys Chem 96:3013-3044

Van de Ven & Tempel P (1994) ISIS 4.0, ISRIC Soil Information System: User Manual (revised ed) Technical Paper 14, ISRIC, Wageningen, The Netherlands

Wen QX (1984) Utilization of organic materials in rice production in China. In: Organic Matter and Rice. International Rice Research Institute, Manila, Philippines, 631 p

Yao H, Conrad R, Wassmann R & Neue HU (1999) Effect of soil characteristics on sequential reduction and methane production in sixteen rice paddy soils from China, the Philippines, and Italy. Biogeochemistry 47:267-293

Zobler L (1986) A world soil file for global climate modelling #87802, National Aeronautical and Space Administration, USA

Part II

Selected modeling approaches, field and laboratory experiments

Nutrient Cycling in Agroecosystems **58:** 221–230, 2000.

Modeling rice plant-mediated methane emission

J.R.M. Arah[1,2] & G.J.D. Kirk[1]

[1]*International Rice Research Institute, MCPO Box 3127, Makati City 1271, Philippines;* [2]*AAT Consultants, 15 Clerk Street, Edinburgh EH8 9JH, UK*

Abstract

Late-season methane (CH_4) emissions from flooded ricefields appear to be fueled by root exudation and death and to be transmitted to the atmosphere largely through the plant. We present a general transport-reaction model which accommodates these phenomena, together with a simplified ("cartoon") version intended to reproduce the salient features of most plant-dominated CH_4-emitting systems. Our cartoon model is capable of reproducing measured concentration profiles and fluxes. Sensitivity analysis suggests that cultivars with high specific root transmissivity may, other things being equal, reduce rather than enhance net emission. Simulations assuming exponential growth of the root system followed by Gaussian die-back resemble measured flux trajectories and also point to great variability in the fraction of CH_4 oxidized before it reaches the atmosphere. Air entry on drainage reduces simulated CH_4 fluxes and the fractions of those fluxes mediated by plants. It also increases the fraction of CH_4 oxidized.

Introduction

Methane (CH_4) emissions from flooded ricefields typically show a late-season peak around the time of grain filling (Holzapfel-Pschorn et al., 1986; Schütz et al., 1989a; Wassmann et al., 1996; Neue, 1997). The organic substrate from which that CH_4 is derived presumably comes from root exudation and death (Wang 1995; H. Kludze, IRRI, 1996, pers. commun.), and most of the CH_4 emitted reaches the atmosphere via aerenchymatous roots (Nouchi et al., 1990; Denier van der Gon & van Breemen, 1993; Kludze et al., 1993), part of it being oxidized en route (Schütz et al., 1989b; Denier van der Gon & Neue, 1996). Supply by and transport through roots must therefore be taken into account when setting up a model to account for the late-season peak. We present and discuss such a model, with a view to suggesting management options and cultivar properties which might minimize plant-derived and plant-mediated CH_4 emission.

We first describe a general transport-reaction scheme governing the behavior of any nonadsorbed substance which simultaneously moves through and reacts in an effectively homogeneous soil-plant system. We then abstract, from what little information is known about the controlling variables of our model, a simplified description—a "cartoon" model—of a soil-plant system which we hope captures the important features of the plant-dominated latter period of the growing season. We examine the behavior of this model, explore its sensitivity to the various parameters which define it, and discuss links with dynamic crop models. The cartoon model, in which substrate supply and root transmissivity are both taken to be proportional to root length density, applies only to late-season emissions; the general model from which it is derived applies also to the early season, where incorporated residues are the main source of oxidizable organic substrate.

Model

Concentration profiles of nonadsorbed substrates in an areally homogeneous system are governed by the following differential equation (Arah & Stephen, 1998), where depth z is zero at the surface:

$$\frac{\partial y}{\partial t} = \frac{\partial}{\partial z}\left(D\frac{\partial y}{\partial z}\right) - \frac{\partial}{\partial z}\left(Ly_w\right) + O + P - Q - R - S; \quad (0 \le z \le Z) \qquad (1)$$

D represents diffusion through the bulk matrix; L, leaching; O, root-mediated influx; P, production; Q, consumption; R, root-mediated efflux; and S, ebullition. Temperature (T) is an implicit variable in Equation 1, influencing the instantaneous rates of all transport and reaction processes, which nevertheless remain constrained by the equation. Diffusion depends on the bulk concentration y (z, t), leaching and consumption on the solution-phase concentration y_w (z, t), root-mediated efflux and ebullition on the gas-phase concentration y_a (z, t). Root-mediated influx and production are independent of y, y_w and y_a, though they may of course depend on other properties of the system (surface concentrations, concentrations of other substrates, root density profiles). D, L, O, P, Q, R, S and y are effective areal averages at depth z and time t: they subsume within themselves any areal heterogeneity present in the real system.

Boundary conditions at the surface ($z = 0$) are

$$y_0(t) = y(0, t) \text{ for volatiles} \quad (2a)$$

$$\frac{\partial y}{\partial z} = 0 \text{ for involatiles} \quad (2b)$$

and at the lower boundary ($z = Z$)

$$\frac{\partial y}{\partial z} = 0 \text{ for all substrates} \quad (3)$$

Equation 2 simply states that the concentration at the surface is known for volatiles, and that the flux is zero for involatiles. Equation 3 states that the concentration gradient at the lower boundary of the active layer is zero. All symbols are defined in Table 1.

Phase conversion

The concentrations y, y_w and y_a are easily interconverted assuming equilibrium between solution and gas phases (there is no surface-adsorbed phase):

$$y_w = \alpha y_a \quad (4)$$

where α is the solubility constant. Bulk concentration is the volume-weighted sum of the phase concentrations:

$$y = \varepsilon y_a + \theta y_w \quad (5)$$

where ε (z, t) is the air-filled porosity and θ (z, t) the volumetric moisture constant. Hence,

$$y_a = \left(\frac{y}{\varepsilon+\alpha\theta}\right) \Rightarrow \left(\frac{\alpha y_a}{\alpha y}\right) = \left(\frac{1}{\varepsilon+\alpha\theta}\right) \quad (6)$$

$$y_w = \left(\frac{\alpha y}{\varepsilon+\alpha\theta}\right) \Rightarrow \left(\frac{\alpha y_w}{\alpha y}\right) = \left(\frac{\alpha}{\varepsilon+\alpha\theta}\right) \quad (7)$$

Whatever the forms of D, L, O, P, Q, R, and S, Equation 1 be may solved numerically by finite-difference approximation. Both transient and steady-state solutions are available, the latter being particularly attractive for volatiles, where concentration profiles may be expected to adjust so rapidly to changes in the driving variables as to be effectively decoupled from them.

How are the input variables $D, L, O, P, Q, ,R$, and S to be generated? On what do they depend?

Table 1. Symbols used in the equations

Symbol	Meaning	Control	Units
α	Solubility constant		mol m^{-3} water (mol m^{-3} air)$^{-1}$
ε	Air-filled porosity	z t	m^3 air m^{-3}
θ	Volumetric moisture content	z t	m^3 water m^{-3}
κ	Root transmissivity	z, t	m air m^{-3}
D	Diffusion coefficient	z t	m^2 s^{-1}
L	Leaching rate	z, t	m^3 water m^{-2} s^{-1}
O	Root-mediated influx	z, t	mol m^{-3} s^{-1}
P	Production rate	z, t	mol m^{-3} s^{-1}
Q	Consumption rate	y, z, t	mol m^{-3} s^{-1}
R	Root-mediated efflux	y, z, t	mol m^{-3} s^{-1}
S	Ebullition rate	y, z, t	mol m^{-3} s^{-1}
t	Time		s
y	Concentration	z, t	mol m^{-3}
y_a	Gas-phase concentration	z t	mol m^{-3} air
y_s	Solution concentration	z t	mol m^{-3} water
z	Depth		m
Z	Depth of active layer		m

Diffusion constant D

The diffusion constant D employed above is that for diffusion through the bulk medium, which is the concentration-weighted sum of the gas- and solution-phase diffusivities. It is conventionally calculated (Stephen et al., 1998b) as

$$D = \left(\frac{1}{\tau}\right)\left(\frac{\varepsilon D_a + \alpha\theta D_w}{\varepsilon+\alpha\theta}\right) \tag{8}$$

where D_a is the diffusion coefficient in air and D_w that in water. We take the tortuosity factor τ in an originally puddled soil to be equal to unity.

Leaching L

We assume that water in the system is in a state of pseudo-equilibrium (i.e., there is no change in storage):

$$L = \lambda\left(\frac{\theta_z}{\theta}\right) \tag{9}$$

where λ is the rate at which water supplied by irrigation escapes through the lower boundary of the system.

Root-mediated influx O

This may be represented by some form of exchange in which only the gas phase moves:

$$O = \kappa D_a y_a (0, t) \tag{10}$$

where the transmission constant κ (z, t) is a portmanteau variable which depends on root length density, root tip permeability, aerenchyma conductivity, and root configuration. It is difficult to specify exactly what factors enter into κ (z, t), but the property κ (z) is measurable at arbitrary time t by monitoring the rate at which argon (Ar) moves through the system when the headspace is replaced (Stephen et al., 1998a). Where this has been done, κ (z) has been found to be roughly proportional to root length density ρ (z) of aerenchymatous plants.

Root-mediated efflux R

Similarly,

$$R = \kappa D_a y_a (z, t) \tag{11}$$

The separation of root-mediated transport into an influx term (O) and an efflux term (R) is essentially a computational convenience. It should be clear that *net* root-mediated transport depends on the difference between the gas-phase concentration y_a (z,t) at depth z and that y_a $(0,t)$ at the surface.

Ebullition S

The rate at which a particular substance is lost from depth z through ebullition presumably depends on its gas-phase concentration y_a (z, t), so we can write

$$S = \sigma y_a \tag{12}$$

where σ (z, t) is an ebullition rate constant. No field data exist which unambiguously point to the importance of ebullition as a transport process, especially during the plant-dominated later stages of the growing season. Since it is these stages we are primarily concerned with, we take σ (z, t) to be equal to zero.

Transformation processes

Everything so far has been quite general. With appropriate values for α, D_a, D_w, l_Z, y_0 and the depth profiles ε, θ, λ and σ, Equations 1 to 12 apply whatever nonsurface-adsorbed substance is under consideration. We cannot retain this degree of generality when discussing specific substances and transformations. In the case we set out to simulate here, that of CH_4 production, transport, oxidation and emission, we need to consider at least two mobile substances (oxygen - O_2 - and CH_4) and at least three reactions (oxic respiration, CH_4 production, and CH_4 oxidation):

$CH_2O + O_2 \rightarrow CO_2 + H_2O$; *respiration* (13)

$CH_2O + CH_2O \rightarrow CO_2 + CH_4$; *methanogenesis* (14)

$CH_4 + 2O_2 \rightarrow CO_2 + 2H_2O$; *oxidation* (15)

Oxidizable organic matter is represented in these equations as CH_2O, oxygen as O_2, and methane as CH_4. There are strong reasons for seeking to include other reactions (iron and sulfate reduction in particular) in a comprehensive treatment of soil CH_4, but stronger reasons (lack of data, computational economy) exist for leaving them out. We adopt here the minimal defensible treatment (Watson et al., 1996; Arah & Stephen, 1998). Work currently in progress examines the consequences of introducing these complications.

Production P

Methanogenesis (Equation 14) is inhibited by solution-phase O_2:

$$P_{CH4} = IV_M \tag{16}$$

where $V_M\,(z, t)$ is the CH_4 production potential and $I\,(z, t)$ is an inhibition function which we take to be

$$I = \left(\frac{1}{1 + \eta y_{wO2}}\right) \tag{17}$$

where y_{wO2} is the solution-phase O_2 concentration and η is an inhibition efficiency constant.

No reaction produces O_2:

$$P_{O2} = 0 \tag{18}$$

Consumption Q

Methane is consumed by oxidation (Equation 15), which follows dual-substrate Michaelis-Menten kinetics:

$$Q_{CH_4} = V_O \left(\frac{y_{wCH4}}{K_{CH4} + y_{wCH4}}\right)\left(\frac{y_{wO2}}{K_{O2} + y_{wO2}}\right) \tag{19}$$

where $V_O\,(z, t)$ is the oxidation potential and K_{CH4} and K_{O2} are Michaelis constants.

Oxygen is consumed by respiration (Equation 13) and oxidation (Equation 15), the latter requiring two molecules of O_2 per molecule of CH_4. We assume Michaelis-Menten kinetics:

$$Q_{O2} = V_R\left(\frac{y_{wO2}}{K_R + y_{wO2}}\right) + 2Q_{CH_4} \tag{20}$$

where $V_R\,(z, t)$ is the respiration potential and K_R ,a Michaelis constant.

Reaction potentials

The reaction potentials $V_M\,(z, t)$, $V_O\,(z, t)$, and $V_R\,(z, t)$ are the rates at which methanogenesis, CH_4 oxidation, and aerobic respiration would proceed in situ were all enzymes saturated with the necessary substrates. They depend on in situ enzyme concentrations and thus on in situ microbial populations. They change over time. Our cartoon model seeks to represent these changes as simply as possible.

Table 2. Model parameter values

Symbol	Value	Units
α	0.03	mol m^{-3} water (mol m^{-3} air)$^{-1}$
D_a	10^{-5}	m^2 air s^{-1}
D_w	10^{-9}	m^2 water s^{-1}
η	400	m^3 water mol^{-1}
K_{CH4}	0.44	mol m^{-3} water
K_{O2}	0.33	mol m^{-3} water
K_R	0.22	mol m^{-3} water
$y_{aCH4}\,(0,t)$	7.5×10^{-5}	mol m^{-3} air
$y_{aO2}\,(0,t)$	8.9	mol m^{-3} air

Methane-oxygen model

Given the constant parameters of Table 2 (adapted from Arah & Stephen, 1998, ignoring differences in solubility and diffusion constant between CH_4 and O_2 in the interest of simplicity), equations 1-20 can be solved to provide transient or steady-state O_2 and CH_4 concentration profiles y, reaction rates P and Q, and surface fluxes J for any combination of the controlling variables ε, θ, κ, λ, σ, V_M, V_O, and V_R. Steady-state surface fluxes J are simply equal to the difference between production P and consumption Q integrated over the depth (0-Z) of the system (minus any losses due to leaching, here set equal to zero); root-mediated fluxes are equal to the difference between efflux R and influx O again integrated over 0-Z. Nonsteady-state (transient) surface fluxes are not reported here.

Where, as is usual, one or more of the controlling variables may be further simplified, approximated, or neglected, process-based simulation of CH_4 emission becomes possible using a relatively limited set of input data.

Cartoon model

In abstracting our cartoon model system, we assume the following:

1. The soil is saturated and air-filled porosity external to roots is negligible ($\varepsilon = 0$ m^3 air m^{-3});
2. Moisture content is uniform with depth ($\theta = 0.8$ m^3 water m^{-3});
3. Leaching is negligible ($l_Z = 0$ m^3 water m^{-2} s^{-1});
4. Root transmissivity is proportional to root length density $\rho\,(z, t)$ with proportionality constant k_T (*i.e.*, $\lambda = k_T\,\rho$; Stephen et al., 1998a);
5. Ebullition is negligible ($\sigma = 0$ s^{-1});
6. Oxidation potential is constant ($V_O = 5\times10^{-5}$ mol m^{-3} s^{-1}; unpubl. data, IRRI 1996);

Table 3. Standard cartoon model parameter values

Symbol	Value	Units
ρ_{max}^{0}	10^{4}	m root m^{-3}
z_{max}^{0}	0.1	m
k_T^{0}	10^{-6}	m air m^{-1} root
k_V^{0}	10^{-8}	mol m^{-1} root s^{-1}

7. Methane production potential is proportional to respiration potential ($V_M = V_R / 50$; unpubl. data, IRRI, 1996);
8. Respiration potential is proportional to root length density ρ with proportionality constant k_V (i.e., $V_R = k_V \rho$; unpubl. data, IRRI 1996);
9. Root length density ρ (m root m^{-3}) is normally distributed with depth, with maximum value ρ_{max} at depth z_{max}, and standard deviation equal to $z_{max} / 2$:

$$\rho = \rho_{max} \exp\left(-2 \left(\frac{z - z_{max}}{z_{max}} \right)^2 \right) \tag{21}$$

"Standard" values, denoted by superscript 0, of the parameters ρ_{max}, z_{max}, k_T and k_V are given in Table 3. These values are defined in order subsequently to explore the consequences of departure from them.

Assumptions 1-9 are merely ad hoc simplifications introduced in order to define a standard system with characteristics we can explore. Where possible, they are founded on experimental data (largely conducted at IRRI, otherwise at the Institute of Terrestial Ecology Edinburgh). We do not take them to be universally applicable. They can and should be overridden wherever measured data are available. Our aim here is to examine the behavior of one type of system defined by Equations 1-20, that type being characterized by assumptions 1-9, not to lay claim to a general description of what must occur in all rice fields. Some of the assumptions (1-3, 5) are insignificant or relatively uncontroversial; others (7-9) depend on an underlying supposition that root-mediated processes dominate. The specific values of assumption 4 and Table 3 derive from incubations conducted at IRRI (data not shown) and experiments (on root transmissivity in peat) conducted at ITE and elsewhere (Stephen et al., 1998a; b). We have no grounds for assuming these values to be general. Most of what follows explores the consequences of their not being so.

Results

Snapshot

Figure 1 illustrates steady-state CH_4 and O_2 concentration profiles and Figure 2 shows reaction rates within the standard soil-plant system defined by equations 1-21 and assumptions 1-9.

Sensitivity analysis

What if the assumed transmissivity factor k_T and the substrate supply factor k_V are allowed to vary? Figure 3 indicates the consequences for simulated steady-state CH_4 flux J_{CH4} of altering these factors while holding everything else constant; the abscissa is $k'_T = k_T / k_T^{0}$, the ordinate $k'_V = k_V / k_V^{0}$. Figure 4 shows the same thing for the plant-mediated flux fraction ϕ_{CH4} (plant-mediated flux / total flux), Figure 5 for the CH_4 oxidation fraction ξ (CH_4 oxidized/CH_4 produced) and Figure 6 for the maximum simulated CH_4 concentration y_{maxCH4}.

Time course

We represent the development of the rice root system over the course of a 100-d growing season as follows:

$$\left(\frac{z_{max}}{z_0}\right) = \left(\frac{z_f}{z_0}\right)^{(t/t_f)} ; (t \le t_f); \quad z_{max} = z_f ; (t > t_f) \tag{22}$$

$$\left(\frac{\rho_{max}}{\rho_0}\right) = \left(\frac{\rho_f}{\rho_0}\right)^{(t/t_f)^2} ; (t \le t_f); \quad \left(\frac{\rho_{max}}{\rho_f}\right) = \exp\left(-\frac{(t - t_f)^2}{\sigma_t}\right) ; (t > t_f) \tag{23}$$

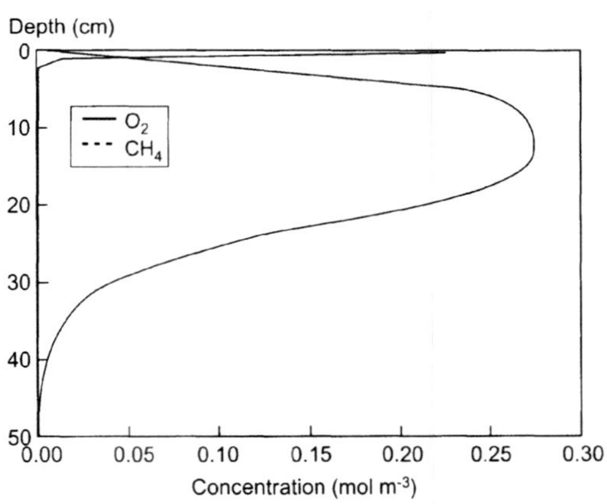

Figure 1. Simulated steady-state CH_4 and O_2 concentrations in "standard" cartoon model

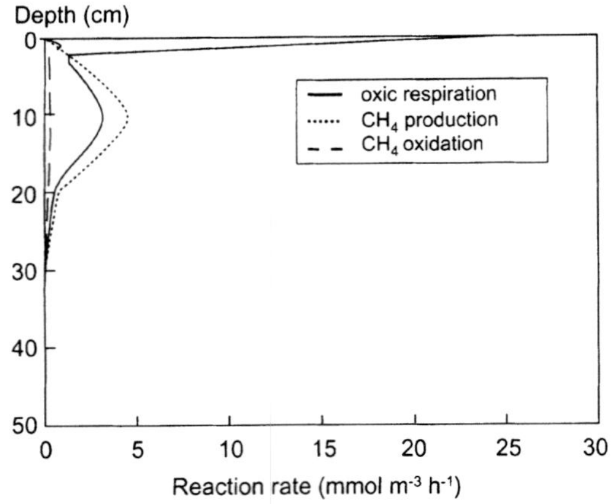

Figure 2. Simulated steady-state CH_4 and O_2 reaction rates in "standard" cartoon model

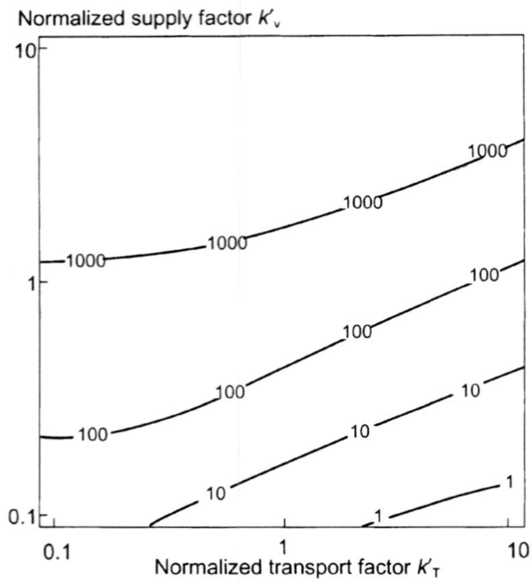

Figure 3. Sensitivity analysis of cartoon model: effect on steady-state CH_4 flux J_{CH4} (μmol m^{-2} h^{-1}) of varying supply factor k_V and transport factor k_T. Normalized factors k'_V and k'_T are divided by the standard values indicated in Table 3

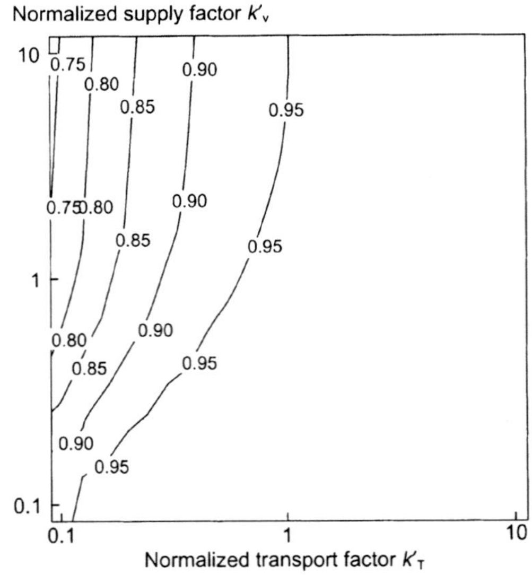

Figure 4. Sensitivity analysis of cartoon model: effect on steady-state plant-mediated fraction CH_4 flux fraction ϕ_{CH4} of varying supply factor k_V and transport factor k_T about their "standard" values

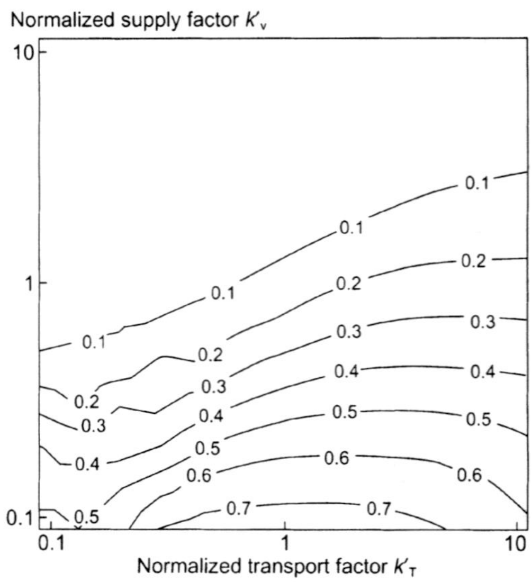

Figure 5. Sensitivity analysis of cartoon model: effect on CH_4 fraction oxidized x of varying supply factor k_V and transport factor k_T. Normalized factors k'_V and k'_T are divided by the standard values indicated in Table 3.

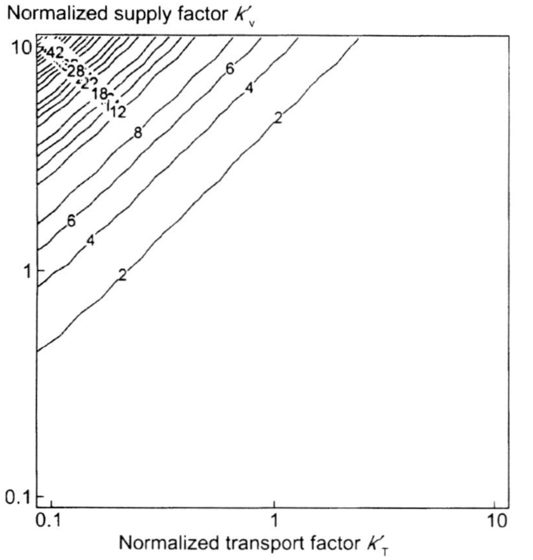

Figure 6. Sensitivity analysis of cartoon model: effect on steady-state maximum CH_4 concentration y_{maxCH_4} (mol m^{-3}) of varying supply factor k_V and transport factor k_T. Normalized factors k'_V and k'_T are divided by the standard values indicated in Table 3

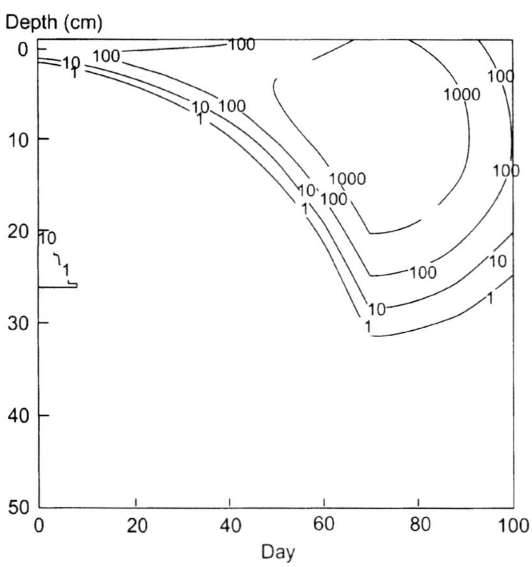

Figure 7. Root-length density profiles ρ (m root m^{-3}) generated by developmental model

Table 4. Developmental model parameter values.

Symbol	Value	Units
r_0	10^2	m root m^{-3}
r_f	10^4	m root m^{-3}
z_0	0.01	m
z_f	0.1	m
t_f	70	d
s_t	200	d

Equation 22 represents an exponential increase in modal rooting depth (z_{max} in assumption 9) from a starting value of z_0 to a final value z_f at time t_f. Equation 23 represents a similar but faster increase in modal root length density ρ_{max} over the same period, followed by a Gaussian-type decline as roots senesce and are lost. Both z_{max} and ρ_{max} increase over time from starting values z_0 and ρ_0 to maxima $z_f = z_{max}{}^0$ and $\rho_f = \rho_{max}{}^0$ at time t_f after which they decline. Values for the developmental parameters z_0, z_f, ρ_0, ρ_f, t_f and σ_t are given in Table 4. Again, we make no claim for the generality of Equations 22-23 or the parameter values in Table 4; the idea is merely to provide a simple description of root growth which looks reasonable and allows us to explore the properties of the system thereby defined.

Figure 7 illustrates the development of the root-length density profile ρ defined by Equations 21-23. Reaction potentials V_M and V_R and root transmissivity λ are all proportional to ρ. Figure 8 illustrates the corresponding steady-state CH_4 concentration profiles y_{CH4} calculated assuming the standard values of the substrate supply and root transmissivity factors k_V and k_T given in Table 3.

Figure 9 illustrates steady-state CH_4 fluxes J_{CH4} calculated for a range of k_V and k_T values indicated on the graph, and Figure 10 the corresponding CH_4 oxidation fractions ξ.

Figures 11-12 illustrate the impact of a small degree of air entry on the system properties illustrated in Figures 9-10. In these simulations, the air-filled porosity ε is set at 0.01 m^3 air m^{-3} throughout; nothing else is changed.

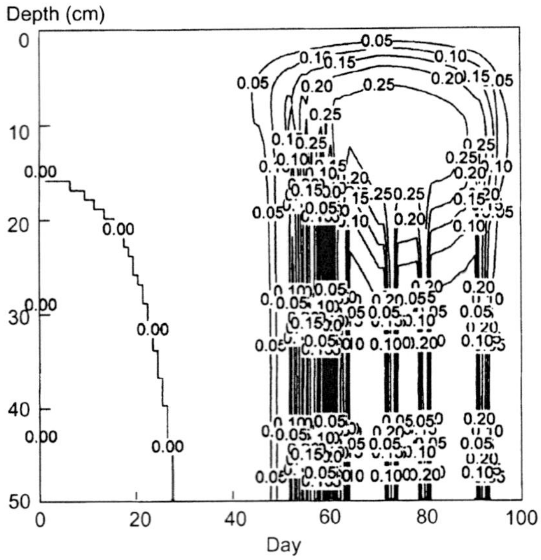

Figure 8. CH_4 concentration profiles y_{CH_4} (mol m^{-3}) generated by developmental model using standard values of k_V and k_T

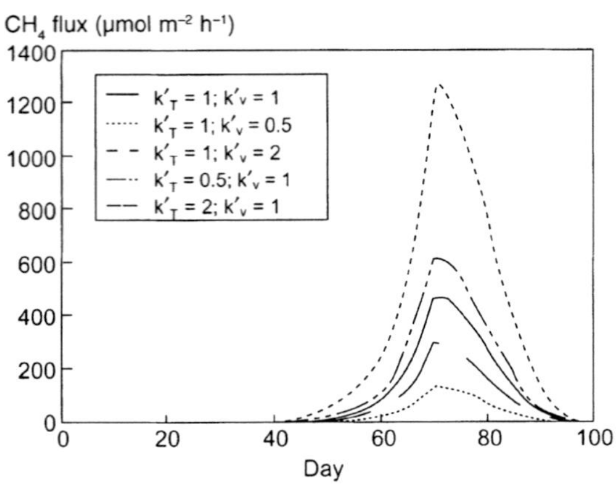

Figure 9. Steady-state CH_4 fluxes J_{CH_4} (μmol m^{-2} h^{-1}) generated by developmental model using values of k_V and k_T indicated; air-filled porosity $\varepsilon = 0$ m^3 air m^{-3}. Normalized factors k'_V and k'_T are divided by the standard values indicated in Table 3

Discussion

Simulated concentration profiles and reaction rates in the standard cartoon model (Figures 1 and 2) seem reasonable. Net O_2 flux (consumption) J_{O_2} is 460 μmol m^{-2} h^{-1}, net CH_4 flux (emission) J_{CH_4} is 480 μmol m^{-2} h^{-1}, plant-mediated O_2 flux fraction ϕ_{O2} is 0.84, plant-mediated CH_4 flux fraction ϕ_{CH_4} is 0.97, and the fraction ξ of CH_4 oxidized prior to emission is 0.13. These are all credible numbers.

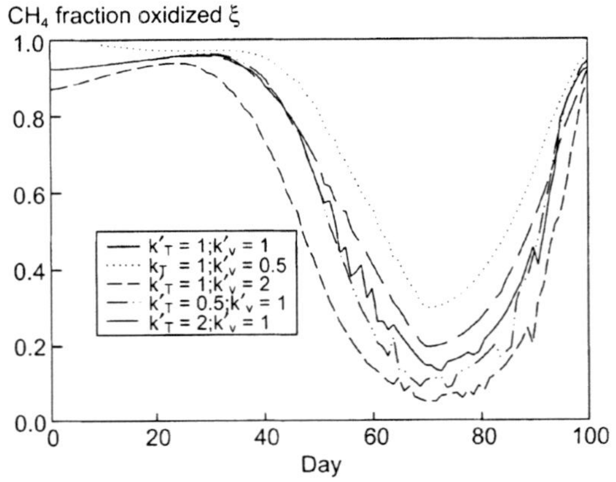

Figure 10. CH_4 fraction oxidized ξ generated by developmental model using values of k_V and k_T indicated; air-filled porosity $\varepsilon = 0$ m^3 air m^{-3}. Normalized factors k'_V and k'_T are divided by the standard values indicated in Table 3

Figure 3 indicates that, everything else being equal, increasing the substrate supply factor k_V leads to an increased CH_4 flux J_{CH4}, while increasing the transmissivity factor k_T reduces J_{CH4}. This latter, perhaps counter-intuitive, effect reflects the fact that transport through roots allows O_2 into the system as well as CH_4 out. Enhanced O_2 concentrations in the rhizosphere inhibit methanogenesis and promote oxidation, and the combined effect of these two processes more than compensates for the greater ease with which CH_4 can escape.

Figure 4 indicates, unsurprisingly, that the fraction ϕ_{CH4} of the CH_4 flux transmitted through the plants increases as root transmissivity increases and decreases as substrate supply increases.

Figure 5 shows that the fraction ξ of CH_4 production which is oxidized before reaching the atmosphere is a sensitive function of k_V and k_T. Increasing k_V reduces ξ, presumably because the oxidation potential V_O is held constant in these simulations and increased production simply overwhelms the oxidation capacity; increasing k_T enhances ξ where transmissivity is low and reduces it where transmissivity is high, presumably reflecting the intricate balance between the twin effects of O_2—inhibiting CH_4 production and promoting CH_4 oxidation (and thereby anaerobiosis, and thereby CH_4 production).

Figure 6 is included to indicate those regions in which the cartoon model becomes untenable. Methane saturation occurs at around 1 mol m^{-3}, implying that simulations in the upper left-hand corners of Figures 3-5 are physically implausible. What must occur under

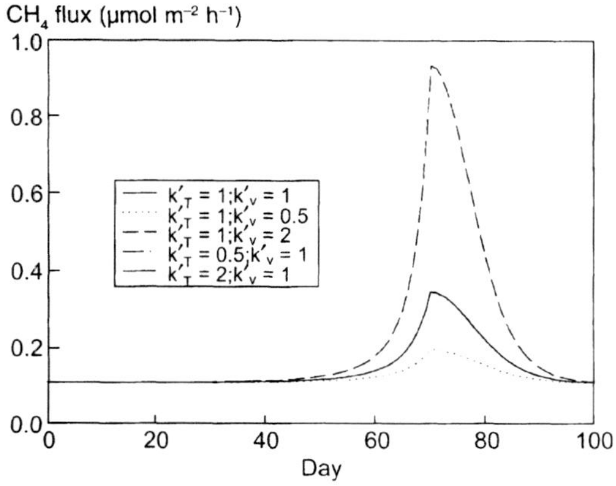

Figure 11. Steady-state CH_4 fluxes J_{CH_4} (µmol m^{-2} h^{-1}) generated by developmental model using values of k_V and k_T indicated; air-filled porosity $\varepsilon = 0.01$ m^3 air m^{-3}. Normalized factors k'_V and k'_T are divided by the standard values indicated in Table 3

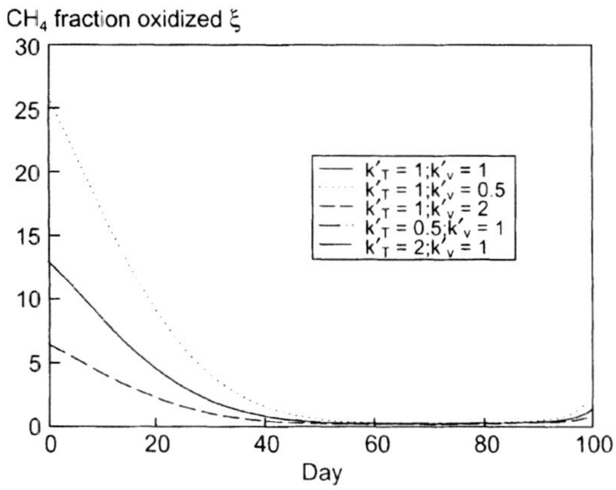

Figure 12. CH_4 fraction oxidized ξ generated by developmental model using values of k_V and k_T indicated; air-filled porosity $\varepsilon = 0.01$ m^3 air m^{-3}. Normalized factors k'_V and k'_T are divided by the standard values indicated in Table 3

such conditions (high k_V, low k_T) is ebullition, which we earlier (assumption 5) arbitrarily set equal to zero. This assumption can easily be altered.

Figure 7 shows how root length density ρ develops according to Equations 21-23 and the parameters of Table 4. The general pattern looks credible; actual values may readily be substituted where available. In any case, what really matters to the model is not ρ but the reaction potential profiles V_R and V_M and the transmissivity profile λ, all of which must change in some way as rooting patterns develop. In these simulations, we adopt the simplest assumption that V_R, V_M and λ are all proportional to ρ, but other relationships are possible.

Figure 8 shows simulated steady-state CH_4 concentration profiles driven by the developing root length density profiles illustrated in the previous figure and the standard values for k_V and k_T. The details of the figure are confusing, indicating the limitations of the graphical interpolation routine. Nevertheless, the general pattern is clear. There is an increase in y_{CH_4} in the root zone around t_f, but the increase is much less marked than that for ρ itself (compare the contour-line scales in Figures 7 and 8). Simulated CH_4 concentrations do not approach saturation; increased transmissivity mitigates the effects of increased substrate supply.

The CH_4 fluxes J_{CH4} illustrated in Figure 9, for ρ as in Figure 7 and various values for k_V and k_T, indicate that the cartoon model presented here can reproduce a wide range of fluxes. They also demonstrate the nonlinear nature of the system described by Equations 1-20. Increasing substrate supply (k_V) by a factor of 2 can lead to a CH_4 flux J_{CH4} enhanced by a factor of 5 or more; increasing transmissivity (k_T) leads to CH_4 emissions reduced roughly proportionally.

Figure 10 illustrates the variability over the season of the fraction ξ of CH_4 oxidized prior to emission. This suggests that it may not be acceptable to assume a constant value for ξ, as is common.

The degree of air entry invoked in Figures 11-12 ($\varepsilon = 0.01$ m^3 air m^{-3}) is undetectable by normal methods. Nevertheless, even so small a gas phase has significant (and explicable) effects on J_{CH4}, ϕ_{CH4} and ξ. Increased O_2 penetration into the system inhibits CH_4 production and promotes CH_4 oxidation, leading to much lower simulated CH_4 fluxes and higher oxidation fractions (ξ can exceed unity where the system as a whole consumes more CH_4 than it produces). As diffusive transport through the bulk medium becomes faster, the plant-mediated route becomes less significant. All these trends become more marked as ε increases (simulations not shown).

Conclusions

Our cartoon model of CH_4 production, transport, oxidation, and emission seems to be able to simulate observed late-season CH_4 emission events satisfactorily (e.g., Neue, 1997). It takes as input the soil physical properties θ (volumetric moisture content) and ε (air-filled porosity), both routinely simulated in ecosystem models, and relates its other driving variables (V_R, V_M,

λ) to the root length density ρ (which is also simulated within a number of ecosystem—or crop—models). Optimization of the proportionality constants k_V and k_T against suitable databases should allow this model to be incorporated within larger scale models.

We do not need to optimize k_V and k_T against any particular data set, however, in order to draw the following conclusions from our model: for any given root length density profile, (i) cultivars with high specific substrate supply rates will lead to enhanced CH_4 emissions; (ii) cultivars with high specific transmissivities will reduce CH_4 emissions; and (iii) drainage leading to even so small an air-filled porosity as 0.01 m^3 air m^{-3} can reduce CH_4 emissions practically to zero. Further, the fraction ξ of CH_4 oxidized before it reaches the atmosphere is not a constant—it depends critically on the root length density, and thus varies throughout the season.

The late-season plant-mediated peak in CH_4 emission does not always dominate. There is often an early-season peak dominated by ebullition as a transport process and by transient consumption of a finite pool of incorporated residue as a substrate supply (e.g., Yagi & Minami, 1990; Wassmann et al., 1996). Such phenomena can readily be incorporated within our general modeling scheme, especially where an intermediate oxidant (nitrate, soluble ferric iron, or sulfate) is introduced. This work is in hand (Matthews et al., this issue).

Acknowledgements

We thank everyone who worked at IRRI on CH_4 and C.T. Hoanh for a very careful review. Much of the work presented here was started under the UK Overseas Development Agency Environment Programme grant ER5305, and continued under the Department for International Development Natural Resources Systems Programme Grant NRSP R6750.

References

Arah JRM & Stephen KD (1998) A model of the processes leading to methane emission from peatland. Atmos Environ 32: 3257-3264

Denier van der Gon HAC & van Breemen N (1993) Diffusion-controlled transport of methane from soil to atmosphere as mediated by rice plants. Biogeochemistry 21: 177-190

Denier van der Gon HAC & Neue HU (1996) Oxidation of methane in the rhizosphere of rice plants. Biol Fertil Soils 22: 359-366

Holzapfel-Pschorn A, Conrad R & Seiler W (1986) Effects of vegetation on the emission of methane from submerged paddy soil. Plant Soil 92: 233-233

Kludze HK, DeLaune RD & Patrick WH Jr (1993) Aerenchyma formation and methane and oxygen exchange in rice. Soil Sci Soc Am J 57: 386-391

Matthews RB, Wassmann R, Arah JRM & Knox J (2000) Upscaling of experimental measurements of CH_4 emissions from rice fields in South East Asia using a process-based crop/methane simulation model within a GIS environment. Nutr Cycling Agroecosyst (this issue)

Neue HU (1997) Fluxes of methane from rice fields and potential for mitigation. Soil Use Manage 13: 258-267

Nouchi I, Mariko S & Aoki K (1990) Mechanism of methane transport from the rhizosphere to the atmosphere through rice plants. Plant Physiol 94: 59-66

Schütz H, Holzapfel-Pschorn A, Conrad R, Rennenberg H & Seiler W (1989a) A three-year continuous record on the influence of daytime, season and fertilizer treatment on methane emission rates from an Italian rice paddy field. J Geophys Res 94:16405-16416

Schütz H, Seiler W & Conrad R (1989b) Processes involved in formation and emission of methane in rice paddies. Biogeochemistry 7: 33-53

Stephen KD, Arah JRM, Daulat W & Clymo RS (1998a) Root-mediated gas transport in peat determined by argon diffusion. Soil Biol Biochem 30: 501-508

Stephen KD, Arah JRM, Thomas KL, Benstead J & Lloyd D (1997b) Gas diffusion coefficient profile in peat determined by modelling mass spectrometric data: implications for gas phase distribution. Soil Biol Biochem 30: 429-431

Wang B (1995) Effects of rice cultivars on diel and seasonal methane emissions. PhD thesis, University of the Philippines Los Baños, Philippines

Wassmann R, Neue HU, Alberto MCR, Lantin RS, Bueno C, Llenaresas D, Arah JRM, Papen H, Seiler W & Rennenberg H (1996) Fluxes and pools of methane in wetland soils with varying organic inputs. Environ Mon Assess 42:163-173

Watson A, Stephen KD, Nedwell DB & Arah JRM (1997) Oxidation of methane in peat: kinetics of CH_4 and O_2 removal, and the role of plant roots. Soil Biol Biochem 29:1257-1267

Yagi K & Minami K (1990) Effect of organic matter application on methane emission from some Japanese paddy fields. Soil Sci Plant Nutr 36: 599-610

Nutrient Cycling in Agroecosystems **58:** 231–248, 2000.

Modeling methane emissions from rice fields: variability, uncertainty, and sensitivity analysis of processes involved

P.M. van Bodegom[1,2,3], P.A. Leffelaar[3], A.J.M. Stams[2] & R. Wassmann[4,5]

[1]Present address: Free University Amsterdam, Department of Systems Ecology, de Boelelaan 1087, The Netherlands; [2]Laboratory of Microbiology, Wageningen University and Research Centre (WUR), P.O. Box 430, 6700 AK Wageningen, The Netherlands; [3]Laboratory of Theoretical Production Ecology, WUR; [4]International Rice Research Institute, MCPO Box 3127, Makati City 1271, Philippines; and [5]Fraunhofer Institute for Atmospheric Environmental Research, Garmisch-Partenkirchen, Germany

Abstract

Estimates of global methane (CH_4) emissions, to which rice cropping systems contribute significantly, are uncertain. The variability and uncertainty of variables governing emission rates and the sensitivity of emissions to these variables determine the accuracy of CH_4 emission estimates. A good tool for quantification of sensitivities is a process-based model. This paper describes a model that has been validated previously by experimental data. Variability and uncertainty in processes and variables underlying CH_4 emissions are reviewed and the sensitivities of modeled CH_4 emission estimates for process variables are tested. The sensitivity analysis is carried out for two sites in the Philippines at which CH_4 emissions have been measured for several years. The sensitivities of the model are compared with measured sensitivities, both as a function of input parameters. The model sensitivity analysis shows that the system is not sensitive to mechanisms of CH_4 production or the pathway of gas transport through the plant. Methane emissions are very sensitive, however, to the description of substrate supply (both from the soil and from organic fertilizers). Unfortunately, this description also represents a main uncertainty. Uncertainty in CH_4 emission estimates will thus remain large as long as this process is not well quantified.

Introduction

Methane (CH_4) is one of the principal greenhouse gases and accounts for 15-20% of the radiative forcing added to the atmosphere (Houghton et al., 1996). Rice fields contribute 9-30% to global CH_4 emissions (Houghton et al., 1996; Matthews et al., 1991). Estimates of global CH_4 emissions from rice fields differ largely depending on approaches, techniques, and databases used for extrapolation. Lelieveld et al. (1998) estimated 80 ± 50 Tg yr^{-1} using atmospheric chemistry models and tropospheric CH_4 distribution. Upscaling of field measurements generally indicate lower source strengths, in the range of 50 ± 20 Tg yr^{-1} (Neue, 1997).

One of the principal causes for uncertainties in global estimates results from the large intrinsic spatial and temporal variability in CH_4 emissions. Over the past 15 yr, numerous field experiments identified magnitude, temporal pattern, and controlling factors of CH_4 emissions from rice fields (this issue; Denier van der Gon and Neue, 1995a; Nouchi et al., 1994; Wassmann et al., 1996). The large number of data that has become available from these experiments is of great value for improved understanding of the variability in CH_4 emissions. The data show, among other things, that the variability in CH_4 emissions cannot be described by a simple relationship between CH_4 emissions and environmental variables (Walter et al., 1996). This is attributed to the dynamic (diurnal and seasonal) and non-linear interactions between the processes underlying CH_4 emissions.

It is therefore beneficial to link the available data on CH_4 emissions to knowledge of the underlying processes, i.e. through a mathematical model. In recent years, a number of models of CH_4 emissions from rice fields have been published. Some models are empirical (Hosono & Nouchi, 1997; Huang et al., 1998), which can be problematic in view of the nonlinear interactions and number of fitted parameters, leading to a loss in extrapolation reliability. Other models (Arah & Stephen, 1998; van Bodegom et al., 2000; Cao et al., 1995) are process-based models. They vary in the purposes for which they were developed and in the degree of mechanistic detail included.

Although most models can reproduce the patterns of CH_4 emissions at one experimental site with reasonable accuracy, their potential for simulating emissions at other sites remains unknown. This potential depends on the variability and uncertainty of variables and processes on a process level and, in the next step, on the sensitivity of real systems and of models for those variable or uncertain processes.

The objectives of this paper are i) to review the variability and uncertainty in processes and variables underlying CH_4 emissions, ii) to quantify the sensitivity of a model for such variability and to compare the model sensitivity with the real sensitivity where possible, and iii) to determine the uncertainties in the range of CH_4 emissions.

The sensitivity analysis is based on a process-based model, fully described below. Arguments for this model, validation with field experiments and analysis on model structure are in van Bodegom et al. (2000). Other process-based models that easily link parameters with measured entities can be used as well for such an analysis and will be mentioned when relevant. For the model sensitivity analysis, two sites, Maligaya (MA94) and Los Baños (LB97), both in the Philippines, were chosen to determine the responses to variable changes. At both sites, CH_4 emissions have been measured for several years (this issue; Corton et al., 1995; Wassmann et al., 1994; 1996) and soil characteristics, management and temperature are known (Table 1). The analysis comprises two steps. First, a sensitivity analysis of CH_4 emissions varying one single variable independently is presented. The effects of a variable on emissions are analyzed and compared with literature. Secondly, the relative sensitivity of modeled CH_4 emissions for individual variable changes is determined for both sites. The accuracy to which the model is able to reproduce and predict encountered CH_4 emissions at different experimental sites can be assessed from a combination of the relative importance and information on uncertainty in variables.

Table 1. Site characteristics of experimental stations in the case study

	Los Baños	Maligaya
Soil organic carbon content (%)	1.86	1.21
Dithionite extractable iron (%)	2.27	1.15
Clay (%)	43	59
Silt (%)	44	33
Average seasonal temperature (°C)	26.5	29.6
Rice cultivar	IR72	IR72
Yield (t ha^{-1})	5.4	5.2
Fertilizer addition (kg ha^{-1})	Urea 150	Urea 120, solophos 40, KCl 40

Model description

Methane emissions from rice fields are strongly influenced by the presence of the root system. The model incorporates this explicitly and distinguishes a rhizosphere and a bulk soil compartment. The processes involved in emissions—described from the moment of flooding onward—take place independently in both compartments (a flow diagram is given in Figure 1). In the mathematical description of the processes, it is attempted to combine simple process descriptions, while maintaining the most important characteristics of the processes. This is done to avoid excessively high data demand and to allow a future linkage to geographic information systems for scenario analysis.

Dynamics of compartment contribution

The model calculates the extent of the rhizosphere compartment in time from actual root length density (*RLD*) (in m m^{-3}), which is empirically related to maximum root length density (RLD_{max}) (in m m^{-3}) based on data by Beyrouty et al (1988), Drenth et al. (1991), Kang et al. (1995), Slaton et al. (1990), and Teo et al. (1995):

$$RLD = \frac{RLD_{max}}{1+K \cdot e^{-rgr \cdot time'}} \qquad for\ RLD < RLD_{max}$$

$$RLD = \frac{RLD_{max}}{1+K \cdot e^{-rgr \cdot time'_{mor}}} \cdot e{-}k_{mor}(time' - time'_{mor}) \qquad for\ RLD \geq RLD_{max} \quad (1)$$

in which the *time'* is the relative time (time divided by the length of the growing season) and *rgr* (dimensionless relative growth rate), *K* $((RLD_{max}-RLD_{t=0})/RLD_{t=0})$, k_{mor} (dimensionless relative mortality rate of the roots) and $time'_{mor}$ (relative time at which roots start to die) are empirical constants (Table 2). The logarithm of RLD_{max} is empirically related to the logarithm of aboveground

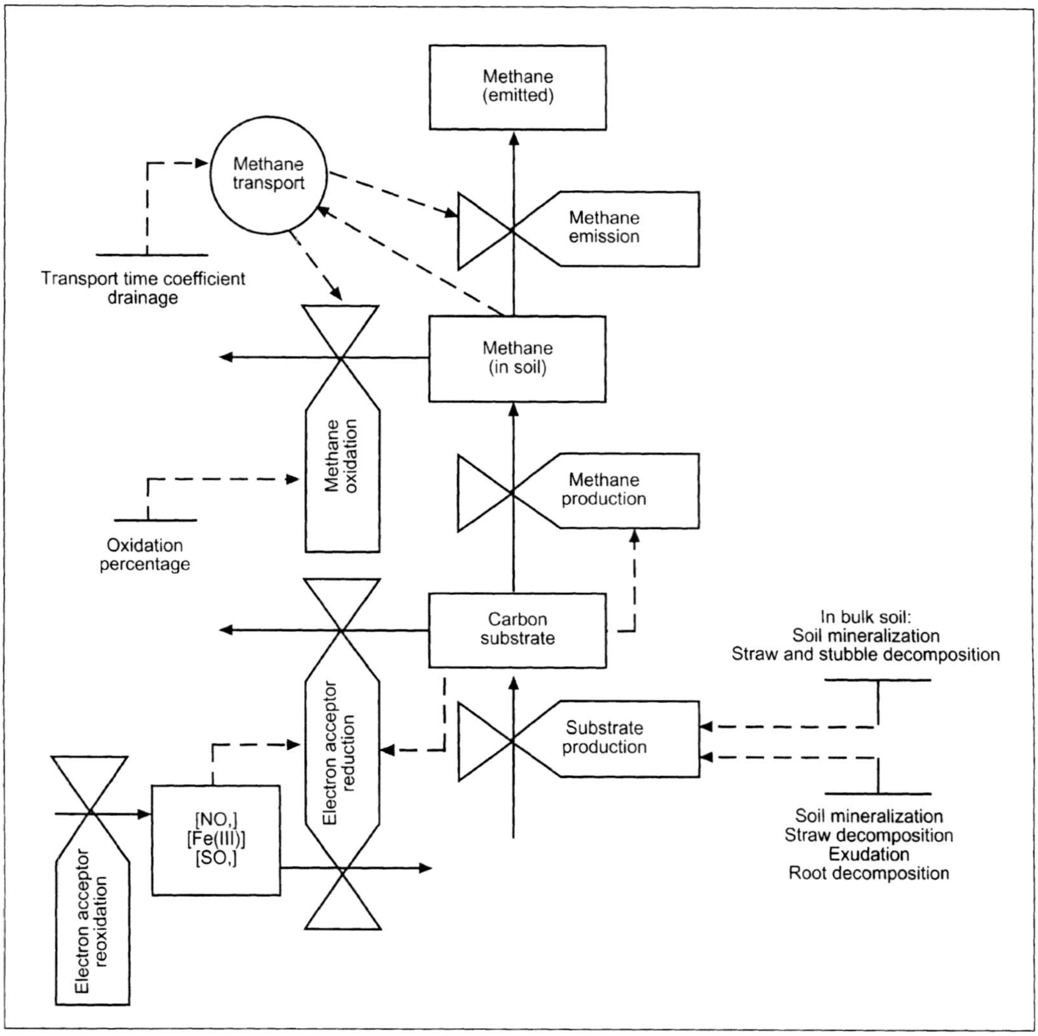

Figure 1. Relational diagram for carbon flow in each compartment of the model

biomass, based on data by Drenth et al. (1991), Tanaka et al. (1995), and Teo et al. (1995).

From the actual root length density, the distance between roots (*root_dist*) (in m) is calculated, assuming that all roots exchange gases and that roots are randomly distributed through the puddled soil by (based on Ogston, 1958):

$$root\_dist = \sqrt{\frac{\ln(2)}{\pi \cdot RLD}} \qquad (2)$$

The fraction of the soil dominated by the rhizosphere (*F_rhizosphere*) is equal to

$$F\_rhizosphere = \left(\frac{rhizo\_dist}{root\_dist}\right)^2 \qquad (3)$$

in which *rhizo_dist* is the estimated extent of the rhizosphere around a single root (Table 2). *F_rhizosphere* cannot become larger than one and is zero in the absence of plants. The fraction of the bulk soil is one minus *F_rhizosphere*. This description of

Table 2. Model parameter values (for an explanation on the kind of parameter, see main text)

Parameter	Value	Unit	Reference
K	85.5	-	*a*
rgr	13.3	-	*a*
k_{mor}	1.53	-	*a*
$time_{mor}$	0.6	-	*a*
rhizo_dist	2.10^{-3}	m	*b*
R_{min}	$1.25.10^{-4}$	$s^{-0.415}$	*a*
S_{min}	0.585	-	*a*
R_{fert}	$5.77.10^{-2}$	$s^{-0.623}$	*a*
S_{fert}	0.377	-	*a*
B_{exu}	$0.85.10^{-6}$	mol m^{-3} s^{-1}	*a*
A_{exu}	$4.41.10^{-6}$	mol m^{-3} s^{-1}	*a*
$time_{max}$	0.552	-	*a*
σ	0.14	-	*a*
$K_{d,root}$	$6.5.10^{-8}$	s^{-1}	*c*
$\tau_{rhizosphere}$	9.10^{3}	s	*a*
τ_{bulk}	$1.08.10^{6}$	s	*a*
B_{oxi}	0.10	-	*a*
A_{oxi}	0.63	-	*a*
k_{reox}(FeS)	$7.6.10^{-7}$	s^{-1}	*d*
k_{reox}(Fe(II))	$1.27.10^{-4}$	s^{-1}	*e*
k_{reox}(S^{2-})	$5.60.10^{-6}$	s^{-1}	*e*

[a]van Bodegom et al, 2000., [b]Kirk et al., 1993. [c]Saini, 1989. [d]Boudreau, 1996. [e]Ahmad & Nye, 1990; Cappellen & Wang, 1996; Murase & Kimura, 1997; Ratering & Conrad, 1998

the rhizosphere compartment is an extension of the model in van Bodegom et al. (2000) in which was assumed that optimal plant growth occurred, so that almost all CH_4 is emitted via the plant at RLD_{max}.

Process dynamics within the compartments

In both compartments, carbon substrates are produced by anaerobic mineralization, P_{min}, and fertilizer decomposition (from organic fertilizers and stubble incorporation), P_{fert}, both in mol C m^{-3} s^{-1}. The production rates are adapted from Yang (1996), assuming that substrates are consumed directly after release:

$$P_{min} = C_{min} \,.\, (1\text{-}S_{min}) \,.\, K_{dmin} \,.e^{-k_{dmin}.time} \text{ and } K_{dmin} = R_{min} \,.\, time^{-S_{min}} \tag{4}$$

$$P_{fert} = C_{fert} \,.\, (1\text{-}S_{fert}) \,.\, K_{dfert} \,.e^{-k_{dfert}.time} \text{ and } K_{dfert} = R_{fert} \,.\, time^{-S_{fert}} \tag{5}$$

in which C_{min} is the soil organic carbon pool and C_{fert} is the amount of organic fertilizer added or stubble incorporated (both in mol C m^{-3}). R_{min} (in s^{Smin-1}), R_{fert} (in $s^{Sfert-1}$), S_{fert} (-) and S_{min} (-) are empirical parameters (Table 2).

In the rhizosphere, additional substrates are provided by root exudation, P_{exu} (in mol C m^{-3} s^{-1}), described by a Gaussian curve:

$$P_{exu} = B_{exu} + A_{exu} \,.exp\,(-0.5\,.((time'-time'_{max})/\sigma)^2) \tag{6}$$

in which B_{exu} (baseline exudation), A_{exu} (maximum increase in exudation above the baseline), $time'_{max}$ (relative time of maximum exudation), and σ (spread of exudation in relative time) are empirical constants (Table 2). Root decomposition, P_{root}, described by a first-order decay rate (in mol m^{-3} s^{-1}) also only occurs in the rhizosphere:

$$P_{root} = K_{droot} \,.(pool\ of\ dead\ roots) \tag{7}$$

in which $K_{d,root}$ is the relative decomposition constant for roots (Table 2) and *pool of dead roots* (in mol m^{-3}) changes in time under influence of root mortality and root decomposition.

All available substrate is consumed directly either by methanogens or by other anaerobic bacteria using alternative electron acceptors. Oxygen concentrations in the rhizosphere are low (Frenzel et al., 1992) and it is assumed that these concentrations are too low to affect CH_4 production directly or to cause substantial electron acceptor reoxidation or aerobic respiration under flooded conditions. Therefore, NO_3^- is the first electron acceptor to be reduced:

$$\frac{d[NO_3^-]}{dtime} = -v_{NO3} \,.\Sigma P_x \tag{8a}$$

in which v_{NO3} is a stoichiometry factor for the carbon substrate needed to reduce NO_3^- and $\Sigma P_x=P_{min}+P_{fert}$ or $\Sigma P_x=P_{min}+P_{fert}+P_{exu}+P_{root}$ for the bulk soil and rhizosphere, respectively. After NO_3^-, Fe(III) is reduced:

$$\frac{d[Fe(III)]}{dtime} = -v_{Fe} \,.\Sigma P_x \tag{8b}$$

Methanogens and sulfate reducers are assumed to be completely outcompeted with respect to their carbon substrate by nitrate and iron-reducing bacteria, but they compete—after NO_3^- and Fe(III) disappearance—for available substrate. The competitive strength is proportional to $[SO_4^{2-}]$ and normalized for $[SO_4^{2-}]_{t=0}$:

$$\frac{d[SO_4^{2-}]}{dtime} = -v_{SO4} \,.\frac{[SO_4^{2-}]}{[SO_4^{2-}]_{t=0}} \,.\Sigma P_x \tag{8c}$$

$$\frac{d[CH_4]}{dtime} = v_{CH4} \,.\frac{[SO_4^{2-}]_{t=0}-[SO_4^{2-}]}{[SO_4^{2-}]_{t=0}} \,.\Sigma P_x - CH_4\_transport\_rate \tag{9a}$$

After all alternative electron acceptors have been reduced, all substrate is converted by methanogens:

$$\frac{d[CH_4]}{dtime} = v_{CH4} \cdot \Sigma P_x - CH_4\_transport\_rate \quad (9b)$$

Produced CH_4 is transported to an aerobic/anaerobic interface; root surface (rhizosphere compartment) or soil-water interface (bulk soil compartment):

$$CH_4\_transport\_rate = \frac{[CH_4]}{\tau} \quad (10)$$

The transport time coefficient, τ, the average time between production and the moment of reaching the interface, differs for the rhizosphere and the bulk soil (Table 2). If the field is dried (e.g., at the end of the season), gas transport via the soil matrix becomes increasingly faster, described by a decrease in the transport time coefficient (van Bodegom et al., 2000).

Part of the transported CH_4 is oxidized at the aerobic/anaerobic interface. The fraction that is oxidized at the soil-water interface is constant (37%, van Bodegom et al., 2000). In the rhizosphere, the oxidation fraction changes during the season as root activity—and thus root oxygen release—changes during the season. This change in activity is described by equation 6 with a different '*B*' and '*A*' value (Table 2). Non-oxidized CH_4 is released to the atmosphere.

For the purpose of the sensitivity analysis, a description for alternating aerobic/anaerobic periods was added to the model. The changes in transport of gases under influence of soil drying are already described above, but changes in CH_4 production and processes leading to CH_4 production occur as well if aeration changes. These changes were newly incorporated in the model. With drainage, reduced alternative electron acceptors (e^-acc_{red}) are reoxidized. Reoxidation rate (*reoxi_rate*) (in mol m^{-3} s^{-1}) is described by first-order kinetics:

$$reoxi\_rate = k_{reox} \cdot [e^-acc_{red}] \quad (11)$$

in which k_{reox} is the relative reoxidation constant (in s^{-1}). Equation 11 assumes that oxygen is not limiting the reoxidation rates at aeration. Not all e^-acc_{red} will be reoxidized. Nitrate is reduced to $NO/N_2O/N_2$ that is assumed to be emitted and thus not available for reoxidation. During the anaerobic phase, ammonia is formed by anaerobic mineralization. This ammonia and the nitrate formed by aerobic mineralization during drainage are, however, taken up by the plant and nitrogen concentrations are around zero during the rice cropping season (Witt et al., 1999). N reoxidation is thus neglected. Sulfate is reduced to sulfide that precipitates with ferrous iron. The precipitate first formed is amorphous FeS (Lord and Church, 1983; Rickard, 1975). In principle, this can react to pyrite, FeS_2, but the reaction rate frequently takes years at low reduced sulfur concentrations (Luther et al., 1982; Rickard, 1975) and this reaction was thus neglected. Reoxidation of FeS is described by first-order kinetics (Table 2). The amount of sulfide and ferrous iron that can be reoxidized is corrected for the precipitation of FeS, but emission of H_2S is neglected. An average k_{reox} for sulfide and ferrous iron was used in all simulations (Table 2).

Reoxidation is described independently of aerobic mineralization rates during drainage and CH_4 production stops at aerobic conditions. Aerobic mineralization rates are thus not important, in contrast to anaerobic mineralization rates upon reflooding. Anaerobic mineralization may be higher than before the aerobic conditions by an increased availability of organic substrates that are difficult to mineralize anaerobically (Cabrera, 1993; Inubushi & Wada, 1987), or lower by increased depletion of the organic matter pool during aerobic conditions. The sum of CO_2 and CH_4 release in a rice soil was hardly affected after reflooding (Ratering & Conrad, 1998). Therefore, no change in anaerobic mineralization rate was included.

With the onset of soil drying, not all soil is directly aerobic. The aerobic fraction of the soil (in which e^-acc_{red} can be reoxidized, while anaerobic processes continue in the anaerobic fraction) increases proportionally to the square root of time by evapotranspiration (Stroosnijder, 1982). It is estimated that the puddled layer is completely aerobic after 6 d (average from Kirchhof & So, 1996). This approach neglects variability between soils and heterogeneities due to soil structure and might overestimate drying and rewetting effects.

Sensitivity analysis of system and model

In this section, the model sensitivity is compared with that of the real system. In some cases, quantitative information on system sensitivity is known, so that a direct comparison can be made (on the influence of application of straw or sulfate fertilizer and of drainage). In other cases, only qualitative trends are known (on the change in contribution of different carbon substrates or on transport characteristics) and the quantitative model sensitivities are compared with these trends. Finally, there are cases that only model sensitivities can be calculated (on CH_4 oxidation, influence of yield and of reducible iron). If the model strongly reacts to these variables, experiments will be needed for verification. Confidence in the model is obtained if it reacts similarly as the real system in the first two cases. This helps to accept the nonverified model results from the third case. The relational diagram (Figure 1) forms the basis to organize this section.

Carbon substrate production

Methane can only be produced if carbon substrate is available. In rice soils, the most important carbon sources are soil organic matter mineralization, decomposition of organic fertilizers (like straw), stubble incorporation, root exudates, and root decay. The contribution of each of the sources changes during the season, but quantitative information on the different contributions is scarce. Figure 2 shows the modeled carbon production rates in case of a well-performing high-yielding variety (IR72) and a stubble incorporation of 15% of the aboveground plant biomass. In other field settings, the contributions may deviate from these findings as root development, and thus root exudation and root decay, depends on cultivar, nutrients, redox stress, and soil type. At the moment it is not possible to incorporate such changes more refinedly in a model, due to the lack of quantitative information. Organic fertilization (by rice straw) and soil organic matter mineralization contribute most to the available substrate pool, especially during the first half of the season (Figure 2). This was also found by Nugroho et al. (1997). We will thus focus on these two sources.

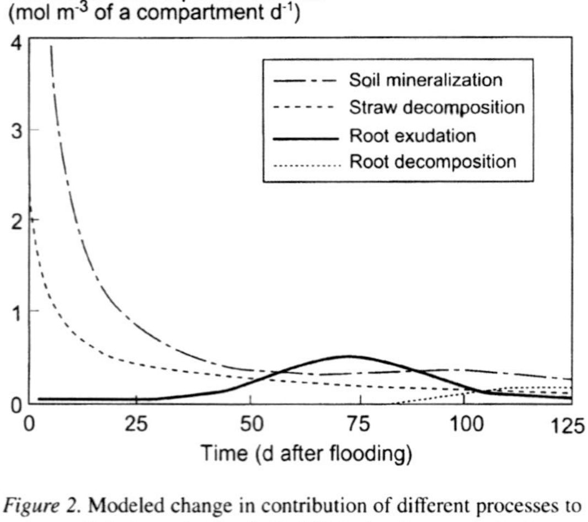

Figure 2. Modeled change in contribution of different processes to carbon substrate production in MA94 during the season. In the rhizosphere, all processes occur, while in the bulk soil compartment, only soil mineralization and straw decomposition occur

It is generally found that the application of rice straw leads to higher CH_4 emissions. The available data on the effects of rice straw addition are summarized in Figure 3a/b. Such data have been used to derive a logistic curve for the relative increase in CH_4 emissions vs straw application (Denier van der Gon & Neue, 1995a; Watanabe et al., 1995a), but a mechanistic explanation for such a curve was not given. The model (Figure 3c, default) produces a roughly linear increase from 0 to 10 t of rice straw, which means that a faster exhaustion of alternative electron acceptors, causing the site differences, only has a minor influence. Straw application will also affect other processes than organic matter supply. These other effects contributed considerably to the overall effects of straw application (Watanabe et al., 1998) and include the influence of straw on rice crop performance. Nugroho et al. (1994; 1996; 1997) found positive biomass effects at low straw additions of 5 t ha^{-1}, while Sass and Fisher (1995) and Kludze and Delaune (1995a) reported rice biomass decreases at straw additions of 11-22 t ha^{-1}. The negative effects might be explained by an inhibition of crop growth due to the accumulation of fermentative products (Bedford & Bouldin, 1994; Drenth et al., 1991) and N immobilization. If we include effects on rice biomass changes in the sensitivity analysis (Figure 3c, indirect) — simplified to a parabolic curve with a maximum at 5 t ha^{-1} and no change at 10 t ha^{-1}—then modeled data are still in the upper range of the experimental data (Figure 3b). This means that there are clearly more adverse interactions between rice plant and straw than were accounted for, especially if more than 10 t straw ha^{-1} is applied. Possible other interactions are changes in root oxygen release, plant carbon supply, or root morphology.

The characteristics of the soil itself also influence the amount of CH_4 emission. The important factors are 1) the amount of alternative electron acceptors, 2) the rate of transport within the soil, and 3) the amount of available substrate (soil organic C content, Table 1; C_{min} in equation 4). The (hypothetical) influence of total soil organic matter contents is presented in Figure 4a by imposing different levels of this parameter. If all other parameters remain constant, the influence of this parameter on the model outcome is very large and depends on the amount of alternative electron acceptors present. The finding, however, may be directly related to the anaerobic mineralization model itself, although

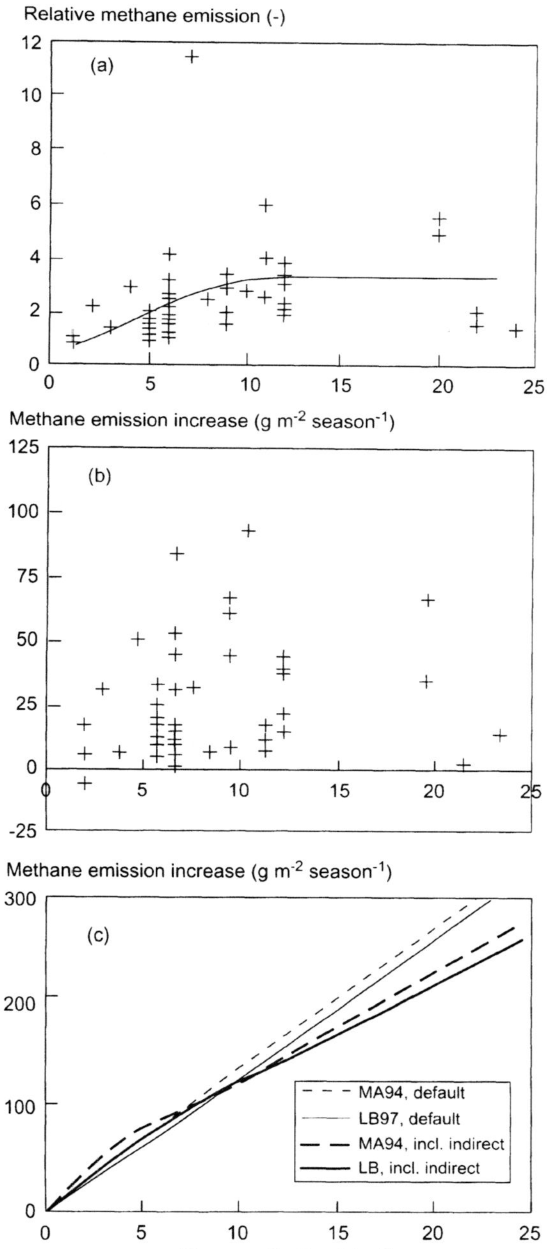

Figure 3. Measured increase in seasonal CH_4 emissions relative to controls without straw addition (a) and measured absolute increase in CH_4 emissions (b). Data are from Denier van der Gon & Neue (1995a), Kimura et al. (1991,1993), Kludze & Delaune (1995a), Lindau & Bollich (1993), Minoda & Kimura (1994), Nouchi et al. (1994), Nugroho et al. (1994,1996,1997), Sass et al. (1991a), Sass & Fisher (1995), Schütz et al. (1989a), Watanabe et al. (1993,1994,1998) and Yagi & Minami (1990). Modeled effects of rice straw application on CH_4 emissions (c) were calculated with and without ('default') the incorporation of an indirect effect of straw on rice yields

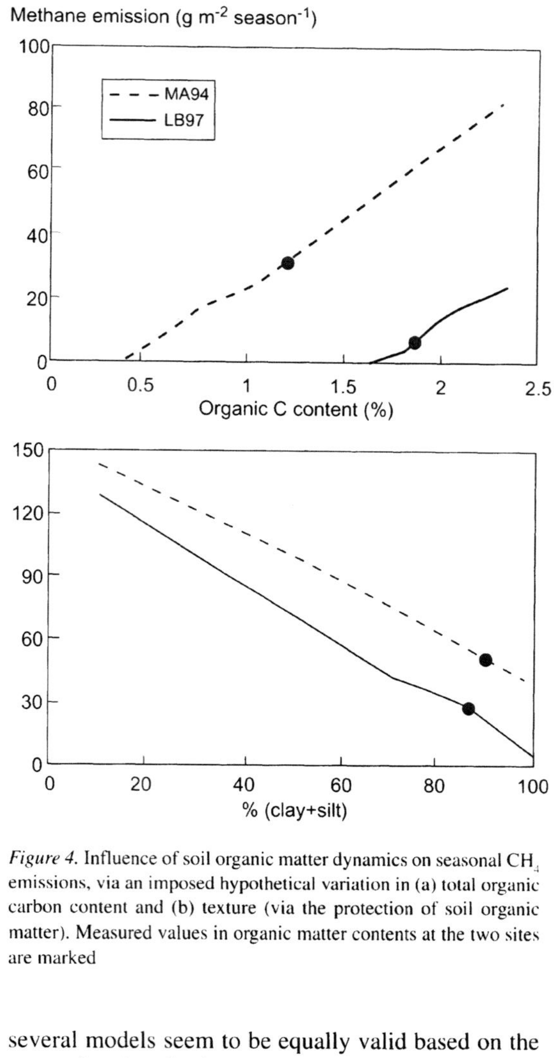

Figure 4. Influence of soil organic matter dynamics on seasonal CH_4 emissions, via an imposed hypothetical variation in (a) total organic carbon content and (b) texture (via the protection of soil organic matter). Measured values in organic matter contents at the two sites are marked

several models seem to be equally valid based on the scarce data (van Bodegom et al., 2000). Anaerobic mineralization processes are thus a very important uncertainty for predictive CH_4 emission models.

In some models (Huang et al., 1997; 1998), soil texture is also taken into account. Texture may affect diffusion of CH_4 (which will be addressed below) or of carbon substrates. Diffusion limitations would ultimately lead to substrate accumulation, which has never been found in field studies. Soil texture, in particular clay particles, may also protect soil organic matter against breakdown (e.g., Hassink & Whitmore, 1997). Quantitative descriptions on the influence of increased protection on R_{min} and S_{min} are unknown. Therefore, this texture influence on CH_4 emissions was tested with a different mineralization model, which leads to a new equation 4 for soil organic matter mineralization:

$$P_{min} = C_{min} \cdot (F_{fast} \cdot K_{fast} \cdot e^{-K_{fast} \cdot time} + (1-F_{fast}) \cdot K_{slow} \cdot e^{-K_{slow} \cdot time}) \quad (4')$$

in which F_{fast} is the fraction of the organic matter pool that is assigned to the fast pool (-) and K_{fast} and K_{slow} are the decomposition constants (s^{-1}) of the fast pool and the slow pool, respectively.

An increased protection will lead to a decrease in mineralization rates for the slow pool. This effect of texture was estimated from Parton et al. (1987), assuming that the slow pool in the two-compartment model is the same as the recalcitrant and lignin material pool in Parton et al. (1987):

$$K_{slow} = K_{default} \cdot (1-0.75^* \, (fraction_{clay+silt})) \quad (12)$$

K_{fast} and $K_{default}$ were calibrated using texture and soil mineralization data from anaerobic soil incubations (van Bodegom et al., 2000). K_{slow} for other textures can thus be calculated using equation 12 and means that the higher clay+silt content, the smaller becomes K_{slow}. This influence of texture via soil organic matter dynamics on CH_4 emission estimates is very large, as presented in Figure 4b by imposing different percentage of (clay+silt). The trends in Figure 4a,b are similar to the ones found in a correlative study (Huang et al., 1997).

Methane production

Methanogens, the bacteria producing CH_4, mainly use acetate as a carbon substrate, but other substrates like H_2/CO_2 and formate contribute 10-30% to CH_4 production (Achtnich et al., 1995a; Chin & Conrad, 1995; Rothfuss & Conrad, 1993). This contribution is less than the theoretical 33% valid for methanogenic systems (Gujer & Zehnder, 1983). Homoacetogens, converting H_2/CO_2 to acetate, might thus play a role in modifying the carbon flow. Besides the carbon substrate production, other conditions have to be fulfilled to produce CH_4.

The methanogens have to compete for the available substrates with other anaerobic bacteria, namely nitrate, manganese, ferric iron, and sulfate reducers. Bacteria using organic electron acceptors (Lovley et al., 1996) do not seem important in mineral rice soils (van Bodegom & Stams, 1999). The competition for carbon substrates in general follows thermodynamic rules: nitrate reducers outcompete the other anaerobic bacteria for the substrates. In practice, nitrate reducers are of minor importance, however, because nitrate concentrations are low in rice soils. All nitrate is thus reduced within a few hours (van Bodegom & Stams, 1999; Achtnich et al., 1995b; Westermann & Ahring, 1987). Ferric iron reducers are also able to maintain acetate and H_2 concentrations below concentrations that can be metabolized by sulfate reducers or methanogens (Lovley & Phillips, 1987). These bacteria suppress sulfate reduction (Jakobsen et al., 1981) unless the amount of carbon substrate is not limiting (Lovley & Phillips, 1986; 1987). The thermodynamic characteristics of sulfate reduction are not very different from CH_4 production. The affinity of sulfate reducers for H_2 is higher than the affinity of methanogens (Kristjansson et al., 1982) suppressing methanogens (Achtnich et al., 1995b). The differences in affinity for acetate are much smaller (Oude Elferink et al., 1994) and CH_4 production and sulfate reduction can occur simultaneously (Achtnich et al., 1995a). Other anaerobic bacteria can influence CH_4 production also through specific inhibitors such as NO, N_2O, or H_2S. The inhibition by NO and N_2O occurs already at low concentrations (Balderston & Payne, 1976; Klüber & Conrad, 1998), while the effects of inhibition by sulfide are small (Kristjansson et al., 1982; Winfrey & Zeikus, 1977).

All these interactions were expressed in the model by an outcompetition of methanogens by nitrate and ferric iron reducers and a competition with sulfate reducers (eq. 8,9), which is a close approximation for the competition for acetate. The influences of initial ferric iron (determined by dithionite extractable iron, Table 1) and sulfate concentrations (mainly determined by fertilization, i.e. ammonium sulfate) on CH_4 emissions are presented in Figure 5. As in Figure 4, we impose a fictive variation of one soil parameter, for conditions of both field experiments. Iron reduction, the dominating reduction process in soils (Inubushi et al., 1984), inhibits CH_4 production severely. At a given iron content, CH_4 emissions are higher for the soil with the higher soil mineralization (MA) (Figure 5a). Decreasing the anaerobic phase in rice soils, e.g. through dry seeding, decreases the period over which CH_4 can be produced, while increasing the relative importance of iron reduction. With a large effect of iron on CH_4 emission, one can also explain some very high Q_{10} values found for CH_4 production (Segers, 1998). With an increase in temperature, soil mineralization and thus CH_4 production are stimulated not only directly, but alternative electron acceptors are depleted faster as well. If one corrects for this indirect effect, the temperature effects on CH_4 production come in a normal range for

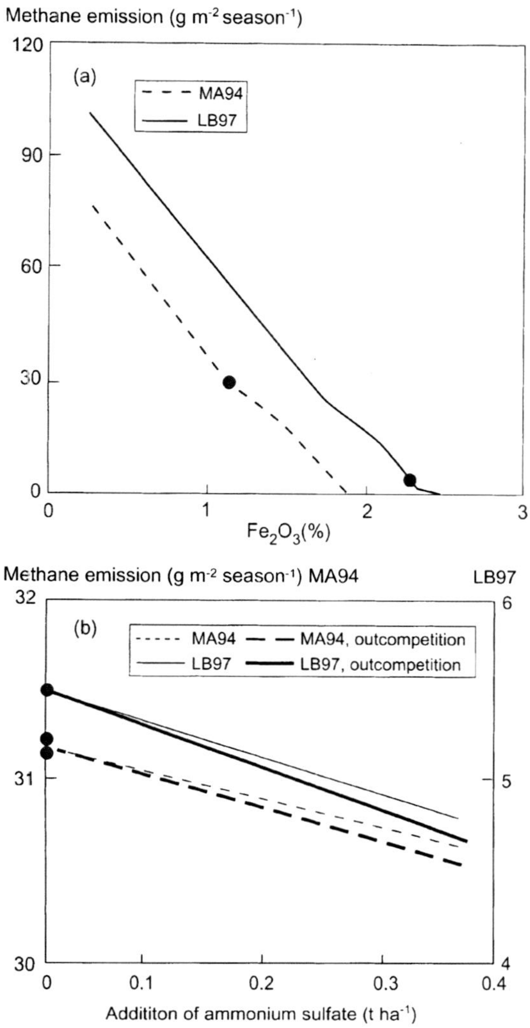

Figure 5. Influence of initial ferric iron (a) and sulphate concentrations (b) on the modeled CH_4 emissions. The influence of sulfate (b) was modeled with different model assumptions on the competition for carbon substrates. Measured values in sulfate contents at the two sites are marked. Note the different y-axes

biological processes. The effects of sulfate additions are much smaller. This is consistent with field data that show no CH_4 emission reduction (Wassmann et al., 1993) or a reduction of 20-30% (Schütz et al., 1989a). If it is assumed that H_2 is the dominating substrate for methanogens, then sulfate reducers will also outcompete the methanogens. This alternative assumption in the model hardly changes the outcome (Figure 5b).

Some models (Cao et al., 1995) relate CH_4 production to redox potential (Eh) and pH, which are in reality highly correlated (Tsutsuki and Ponnamperuma, 1987). Eh and pH do not seem good parameters for process-based models, as discussed elsewhere (van Bodegom et al., 2000). Only if pH<6.0, pH effects may occur. This might explain why urea application normally has no effect on CH_4 emissions (Nugroho et al., 1994; Wassmann et al., 1993), while urea application decreased CH_4 emissions in incubation experiments at application rates higher than 500 mg N kg^{-1} soil (Yang & Chang, 1998). Extreme salinity may also lead to a decreased CH_4 production (Denier van der Gon & Neue, 1995b), but this is not accounted for in any model.

Methane transport

Produced CH_4 is transported via aerobic interfaces, where CH_4 oxidation takes place, to the atmosphere. There are four ways to transport CH_4: leaching, diffusion through the soil, transport via the plant and ebullition. High percolation rates reduce CH_4, emissions significantly (Yagi et al., 1998) and will have to be considered in future models. Methane diffusion through the soil is a very slow process and hardly contributes to CH_4 emissions (Rothfuss & Conrad, 1993; Schütz et al., 1989b). The diffusion of CH_4 via the plant (in the rhizosphere compartment), which depends on root density, is the most important transport pathway to the atmosphere. On average, ebullition (in the bulk soil compartment) only contributes 10-20% to the seasonal CH_4 emission (Byrnes et al., 1995; Nouchi et al., 1994; Schütz et al., 1989b). In case CH_4 production is high at the start of the season (e.g. due to organic fertilization), the seasonal contribution of ebullition can be up to 60% (Denier van der Gon & Neue, 1995a; Wassmann et al., 1996). This difference can be understood from the conceptual ideas presented in the model.

Gas transport through rice plants is, contrary to other wetland plants, by diffusion and not by convection. In turn, CH_4 production does not show a short-term influence of photosynthetic activity (Denier van de Gon & Neue, 1995a; Wassmann et al., 1994), wind speed, humidity, light (Frenzel et al., 1992), transpiration (Byrnes et al., 1995), or radiation (Lee et al., 1981). Gases (both CH_4 and oxygen) exchange with the soil at the tips of roots (Flessa & Fischer, 1992; Kumazawa, 1984), but quantitatively little is known about the fraction of the root surface that is active in gas exchange. The gases are transported via the aerenchyma of root and shoot (affected by the porosity) and exchange with the atmosphere through special micropores in the shoot (Nouchi & Mariko, 1993). For the quantitative understanding of the flow, it is more important to know the largest resistance, which is probably at the root-shoot transition (Butterbach-Bahl et al., 1997). Quantitative data on this resistance are scarce, but probably this resistance will change during the season as root oxygen release and root morphology change during the season. The mechanism of transport through this transition is not known nor it is known if there is a transport interaction between tillers of one rice plant. The effects of those uncertainties might be small as the model showed hardly any influence of the transport time coefficient in the rhizosphere on seasonal CH_4 emissions (Figure 6a). In this simulation, it was assumed, however, that CH_4 oxidation in the rhizosphere was independent of transport. In reality this may not be the case as both processes are diffusion-related. The influence of transport rates in the rhizosphere on CH_4 oxidation is much larger in models that link these processes (e.g., Arah & Stephen, 1998).

The mechanisms of ebullition, gas transport via gas bubbles, are even less understood. Qualitatively, one might think of a mechanism in which there is always an equilibrium between the concentration in the soil solution and partial pressure of the gas in a bubble (Watanabe & Kimura, 1995). If the concentration in the soil increases, gas will be captured in bubbles as the concentration in the soil solution is limited (depending on temperature). If the pressure of the bubbles is larger than the combined pressure of overlying soil structure, root network and atmosphere, then bubble release will be triggered. From this conceptual idea, it can be understood why Mattson and Likens (1990) found influences of solar radiation, water temperature, air pressure, and local water table on ebullition and why ebullition was hardly found at cloudy or rainy days (Nouchi et al., 1994). Quantitative models on this process are not known. Again the effects of these uncertainties in the mechanism on the prediction of seasonal CH_4 emissions are small, as (hypothetical) changes in the transport time coefficient in the bulk soil (Figure 6b), e.g., caused by differences in soil texture or root density, hardly influence seasonal CH_4 emissions.

Transport time coefficients in bulk soil exceed those in the rhizosphere by several orders of magnitude (Table 2). From the combination of transport times per compartment and the seasonal changes in contribution of the compartments, the trends in conductance (Hosono & Nouchi, 1997) and in CH_4 residence times (Kimura & Minami, 1995; Watanabe & Kimura, 1995) during the season can be calculated and understood.

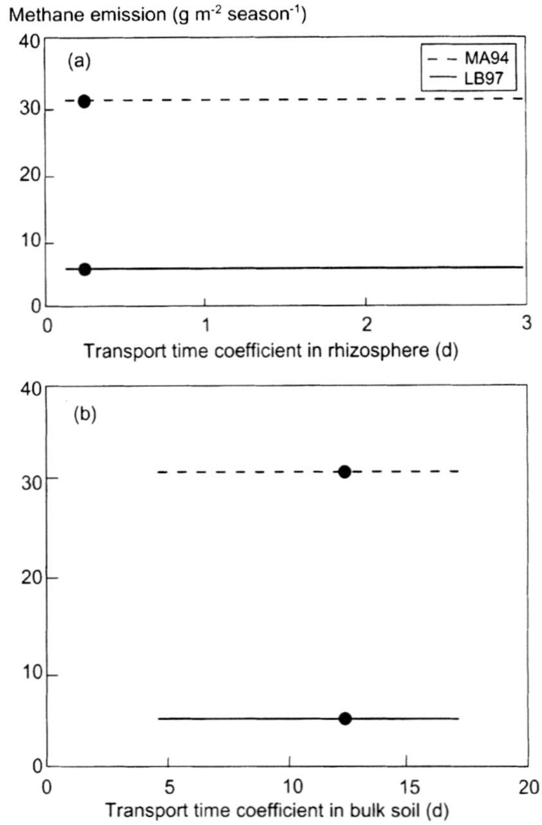

Figure 6. Influence of variation in modeled transport time coefficient of (a) the rhizosphere and (b) the bulk soil compartment on estimates of CH_4 emissions. Default parameter values are marked

The transport time coefficient hardly influences seasonal CH_4 emissions (Figure 6), but it changes the variation of emissions within the season (results not shown). Diurnal patterns may be related to fluctuations in ebullition and root oxygen release(results not shown). Ebullition may be the main factor, because 1) the magnitude of the diurnal fluctuations is highest at the start of the season, when rice plants are small (Denier van der Gon & Neue, 1995a; Husin et al., 1995); 2) diurnal amplitudes are much higher in unvegetated plots than in vegetated plots (Nouchi et al., 1994); and 3) diurnal patterns of CH_4 emissions are correlated with temperature (Sass et al., 1991b) and ebullition is triggered by temperature changes, while plant-mediated transport is hardly influenced by temperature. The influence of temperature might also be indirect: at a higher temperature, CH_4 production is stimulated, leading to an increase in CH_4 concentration in the soil. This increased concentration might again trigger ebullition. Neue et al. (1997) identified CH_4 concentration as a controlling factor for the diurnal patterns. If this indirect mechanism is indeed important, then some additional influence due to the plant might be expected, e.g. by diurnal changes in root exudation and root oxygen release. Unfortunately, there are no data available on these effects, but they may explain the differences in diurnal pattern between rice varieties found by Husin et al. (1995). As ebullition is not modeled mechanistically in CH_4 emission models, quantification of the factors determining diurnal patterns remains difficult.

When floodwater recedes and the soil falls dry, all CH_4 captured in the soil is released via the air-filled pores that are formed in the drying process (Denier van der Gon et al., 1996; Wassmann et al., 1994). The flush of methane is larger after a longer period of CH_4 production, as more CH_4 has been stored (Watanabe & Kimura, 1995). Similar effects occur by physical disturbances like cultural practices (Neue et al., 1997). Due to aerobic conditions developed by the disappearance of floodwater, the soil (and its electron acceptors) reoxidizes as well, resulting in suppressed CH_4 production after reflooding the soil. These negative effects on CH_4 emissions are larger than the flushing effects if the period of drainage is long enough. Midseason drainage has therefore become an effective mitigation option to decrease CH_4 emissions. A good timing of (hypothetical) drainage is important to obtain an optimal result (Figure 7), whereas the number of dry periods appears to be less important (results not shown). The modeled effects of drainage (Figure 7) are similar to what has been encountered experimentally (Sass et al., 1992; Nugroho et al., 1994; Yagi et al., 1996). The simplifications made in the model to describe reoxidation processes hardly influenced CH_4 emission estimates, as can be seen from the small effects of neglecting FeS formation and oxidation (Figure 7).

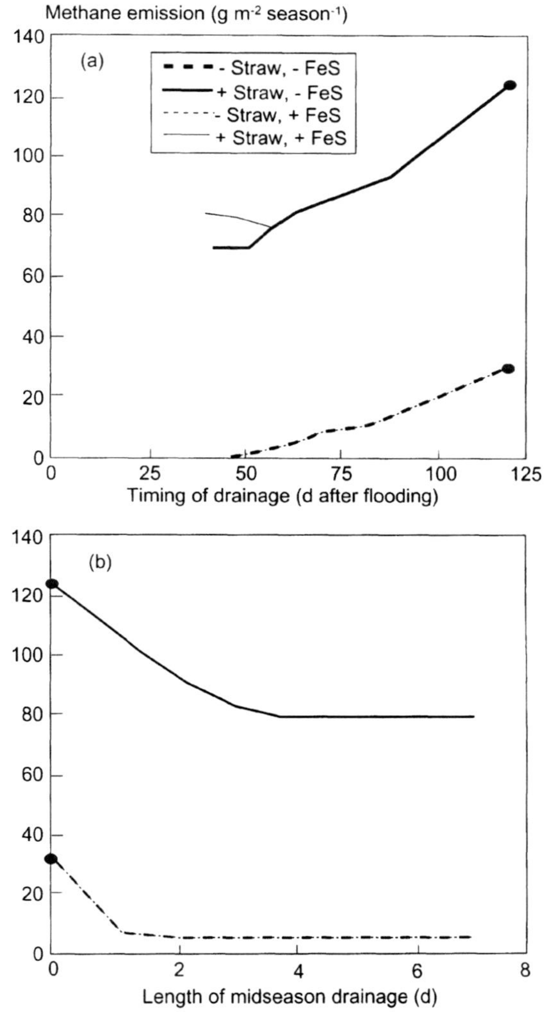

Figure 7. Effects of (a) timing of midseason drainage and (b) length of midseason drainage at 64 DAT, on top of a final drainage on modeled CH_4 emissions, using the data set of MA94 only. Calculations were carried out with and without straw application and with and without a correction for the formation and oxidation of FeS. The situation with no intermediate drainage is marked

Methane oxidation

At the aerobic interfaces, CH_4 can be oxidized in the soil by CH_4-oxidizing bacteria, methanotrophs. There are two types of CH_4 oxidizing activity: high affinity (at low CH_4 concentrations) and low affinity (at high CH_4 concentrations) (Bender & Conrad, 1992). For the study of CH_4 oxidation in wetlands, high affinity CH_4 oxidation does not need to be considered (Segers, 1998).

Low affinity CH_4 oxidation may in principle occur anaerobically and aerobically. In the first case, CH_4 oxidation may be coupled to nitrate, ferric iron or sulfate reduction. However, there is no evidence available that CH_4 oxidation coupled to nitrate reduction occurs in wetlands. Nedwell and Watson (1995) could not show sulfate reduction to be coupled to CH_4 oxidation in wetlands. Murase and Kimura (1994) and Miura et al. (1992) found a concurrence of a depletion of CH_4 and an accumulation in ferrous iron in rice subsoil and interpreted this as a coupled ferric iron reduction/CH_4 oxidation. Other interpretations are however also possible. No enrichments or kinetics of anaerobic methane oxidizers in rice fields are known.

If we restrict our considerations to aerobic CH_4 oxidation, then two sites for oxidation can be distinguished: the rhizosphere and the soil-water interface. At the soil-water interface, CH_4 oxidation is confined to 70-95% of the produced CH_4 (e.g., Banker et al., 1995; Schütz et al., 1989b). This small range indicates that the presence or absence of the oxygen produced by algae does not have a large influence on CH_4 oxidation. Oxidation at the soil-water interface is bypassed by ebullition.

The rhizosphere represents a far more dynamic system with many more uncertainties. Oxygen is released into the rhizosphere by root oxygen release (ROL), which is again influenced by root respiration and root transport resistances. The ROL changes diurnally (Satpathy et al., 1997), during the season (Satpathy et al., 1997), with cultivar (Wang et al., 1997; Kludze et al., 1994), with nutrient conditions (Kludze & Delaune, 1995a, b) and with Eh (Kludze et al., 1993). Moreover, the estimate for ROL highly depends on the used methodology (Sorrell & Armstrong, 1994). The oxygen input is thus very variable and uncertain.

The released oxygen is not only used by the methanotrophs. Part of the oxygen is used for the chemical and bacterial reoxidation of reduced compounds and for heterotrophic respiration of low-molecular organic compounds (Ponnamperuma, 1972; Watson et al., 1997). The contribution (and its dynamics) of the different processes to oxygen consumption is not known, but it is known that methanotrophic activity is affected by salinity (Denier van der Gon & Neue, 1995b), NH_4^+ (Conrad & Rothfuss, 1991), and elevated pH and $CaCO_3$ (King et al., 1990). Another complicating factor is that the aerobic zone moves through the soil (due to the combination of root growth and oxygen consumption). Bacterial activity will have to cope with this dynamics. This may result in growth of methanotrophs as the number of methanotrophs is higher in the rhizosphere than in the bulk soil (Gilbert & Frenzel, 1995; Kumaraswany et al., 1997) and increases during the growing season (Gilbert & Frenzel, 1995; Watanabe et al., 1997). It also may result in a limited mortality as mortality rates of methanotrophs are low at small oxygen and CH_4 availability (Roslev & King, 1994; Le Mer et al., 1996). Quantitative data on such adaptations are scarce. Finally, it is not certain whether CH_4 oxidation takes place in the rhizosphere or in the roots of the rice plant (as methanotrophs were found inside the roots (Gilbert et al., 1998)). Apart from mechanistic uncertainties, there are several uncertainties in the measurement of CH_4 oxidation, as discussed by Frenzel and Bosse (1996) and King (1996). All these uncertainties make the prediction of CH_4 oxidation rates extremely difficult.

The effects of all these uncertainties on the estimation of CH_4 emissions can be quite considerable as is shown by the model sensitivity of CH_4 emissions to hypothetical variation of this estimate (Figure 8). This clearly needs further attention. The sensitivity on CH_4 oxidation moreover depends on the time in the season that most CH_4 release occurs.

Rice plant influence on the processes

The above analysis shows the major importance of rice plants for CH_4 emissions, via its root system, exudation, oxygen release, and root-shoot resistance. These effects have been integrated in correlative models between CH_4 emissions and plant parameters, namely yield, total rice biomass, root density, plant height and shoot length. The results are however ambiguous. Watanabe et al. (1994) found a correlation between emission and shoot length, while Lindau et al. (1995) found no correlation between plant height and CH_4 emission. Sass et al. (1990) correlated CH_4 emission and aboveground biomass, while such a correlation was absent in the study of Watanabe et al. (1995b). Nouchi (1990) found a correlation between the number of tillers and methane emissions, while Denier van der Gon and Neue (1996) did not find such a correlation. The reason for these different results is that there are different influences of the plant on CH_4 release. Those influences will lead to nonlinear results and will moreover change during the season and with different conditions. This model can investigate some of those influences. Other interactions can better be explained by a fully mechanistic approach, like the one presented by Arah and Stephen (1998). An example of an interaction that changes with the conditions is given in Figure 9. In the first scenario, it is assumed that a constant aboveground biomass fraction equivalent to 30% of the yield (Table 1) is incorporated into the soil, which is a common, but unrealistic, assumption in global CH_4 emission esti-

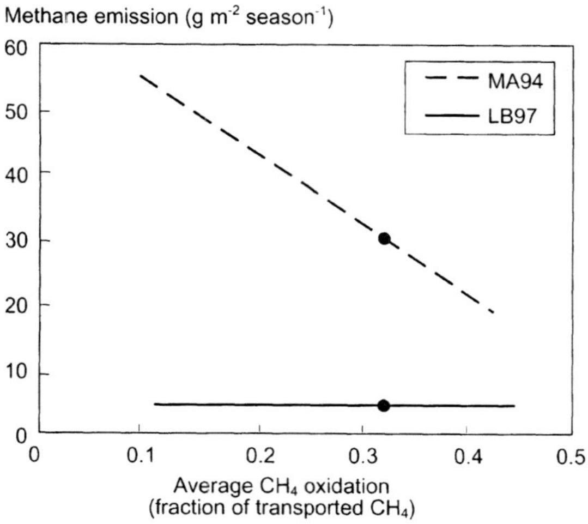

Figure 8. Effects of a hypothetical change in average seasonal methane oxidation in the rhizosphere on methane emissions. Default parameter values are marked.

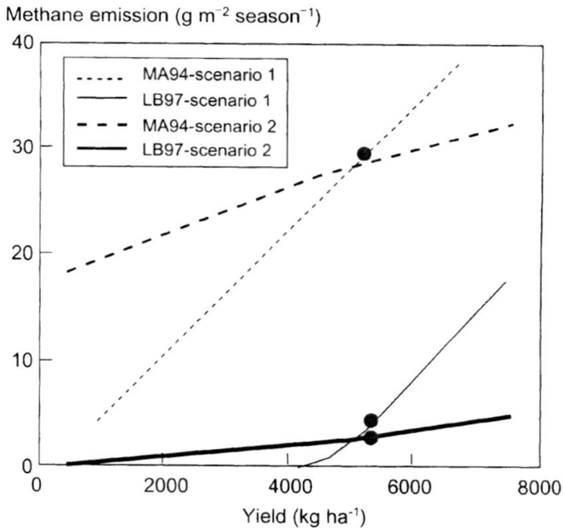

Figure 9. Influence of yield differences on methane emission for two scenarios. The first scenario assumes that the amount of stubble that is incorporated in the soil is equal to 30% of the yield. The second scenario assumes that the amount of stubble that is incorporated into the soil is 1.5 t ha^{-1}, independent of yield. Measured yields at the two sites are marked.

mates. In such a scenario, the organic matter supply dominates CH_4 emission changes, leading to an almost linear response with yield. In the second scenario, it is assumed that the farmer incorporates the same amount of stubble (e.g., by cutting the rice at a certain constant height) independent of the yield obtained. In both scenarios, the presence of rice plants stimulates CH_4 emissions (by providing a substrate for methanogens), but the response is quite different for the two scenarios.

Plant variables do not only change during the season and with conditions but also vary between varieties. Differences have been found in the root oxygen release (Kludze et al., 1994; Kludze & Delaune, 1995a; Wang et al., 1997), in gas permeability (Butterbach-Bahl et al., 1997), and root exudation (Kludze et al., 1999). This leads to large effects of rice varieties on CH_4 emission (Husin et al., 1995; Lindau et al., 1995; Nugroho et al., 1997; Sass & Fisher, 1995; Watanabe et al., 1995b). The combined effects have been incorporated in the model of Huang et al. (1998) by an empirical variable, the variety index. Lumping the various effects in a single variable leads to a loss of a mechanistic basis and hence to a reduction of extrapolation beyond the range of calibration. The plant physiological differences in gas permeability (influencing both root oxygen release and CH_4 transport) and root exudation (important in soils with a low carbon content) open possibilities for directed variety screening.

Concluding remarks

Uncertainties and variability in the knowledge on underlying processes leading to CH_4 emissions from rice cropping systems were reviewed. Sensitivity of these uncertainties and variabilities in processes on CH_4 emissions were investigated with a process-based model. Model sensitivities were compared with system sensitivities, as far as these were known, i.e. for the effects of organic matter supply, drainage, and sulfate additions. In those cases, the model behaved similarly and with a similar sensitivity as the real systems. For situations for which only trends are known (for the transport characteristics and the contribution of different carbon substrates), the model also behaved similarly. The model thus fairly reproduces the real variability in CH_4 emissions caused by the variability in underlying parameter values.

By plotting the relative change in CH_4 emission vs the relative change of a variable within its plausible range, all model sensitivities can be compared (Figure 10). The figure shows great differences in sensitivities and a large variety of linear and nonlinear responses. The responses were different for the two soils. Due to the high amount of reducible ferric iron in the Los Baños soil, this soil is more sensitive to variables influencing CH_4 production (Figure 10b). In the Maligaya soil, variables influencing carbon substrate production and CH_4 production are the most important variables as well, but CH_4 oxidation is also a sensitive variable (Figure 10a). Other well-known uncertainties, like mechanisms of CH_4 production or the pathway of gas transport through the plant do not seem to be important for the estimation of CH_4 emissions.

This analysis has two main implications. 1) The influence of straw application on soil-plant responses and the mechanisms of anaerobic soil organic matter mineralization belong to the main uncertainties, and also strongly influence CH_4 emissions (as is indicated by the influence of texture, organic C soil and straw application). As long as these processes are not well understood, the predictability and extrapolation of modeled CH_4 emissions will be limited at a field scale level and thus at a global scale level. The uncertainty in the range of CH_4 emissions will thus remain large. 2) CH_4 emissions react nonlinearly to variables describing the underlying processes, especially if interactions between underlying variables are taken into account (as can be seen from the different responses of the two soils). This means that global emission estimates based on average parameter values over large regions may deviate considerably from the real CH_4 emission. For a better global prediction of CH_4 emission, methodologies that account for spatial variability in sensitive parameters (like management and organic matter supply) will have to be developed.

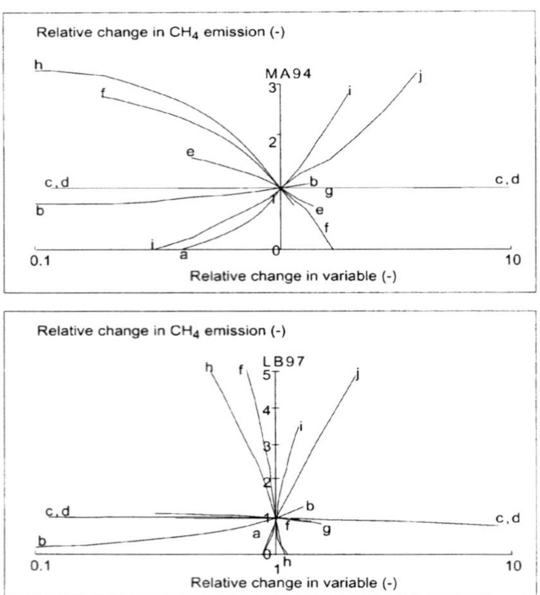

Figure 10. Sensitivity of modeled CH_4 emissions to changes in the underlying parameters (both relative to site input parameter values) for (a) Maligaya and (b) Los Baños, Philippines. The tested parameters are: 'a' moment of drainage (relative to drainage at the end of the season), 'b' yield, 'c' and 'd' transport time coefficient in the rhizosphere and bulk soil compartment, respectively, 'e' percentage oxidation, 'f' iron content of the soil, 'g' application of $(NH_4)_2SO_4$ (relative to initial soil sulfate concentration), 'h' texture (influencing soil mineralization), 'i' soil organic matter content and 'j' straw application. Note that the y-axis is linear and the x-axis is logarithmic

Acknowledgements

The authors thank Reinoud Segers, Jan Goudriaan, Hugo Denier van der Gon, Pieter Raats, Cor Langeveld, and Laura van Schöll for useful comments. This research was supported financially by the Dutch National Research Program on Global Air Pollution and Climate Change.

References

Achtnich C, Schuhmann A, Wind T & Conrad R (1995a) Role of interspecies H_2 transfer to sulfate and ferric iron-reducing bacteria in anoxic paddy soil. FEMS Microbiol Ecol 16: 61-70

Achtnich C, Bak F & Conrad R (1995b) Competition for electron donors among nitrate reducers, ferric iron reducers, sulfate reducers and methanogens in anoxic paddy soil. Biol Fertil Soils 19: 65-72

Ahmad AR & Nye PH (1990) Coupled diffusion and oxidation of ferrous iron in soils. I. Kinetics of oxygenation of ferrous iron in soil suspension. J. Soil Sci 41: 395-409

Arah JRM & Stephen KD (1998) A model of the processes leading to CH_4 emission from peatland. Atmos Environ 32: 3257-3264

Balderston WL & Payne WJ (1976) Inhibition of methanogenesis in salt marsh sediments and whole-cell suspensions of methanogenic bacteria by nitrogen oxides. Appl Environ Microbiol 32: 264-269

Banker BC, Kludze HK, Alford DP, Delaune RD & Lindau CW (1995) Methane sources and sinks in paddy rice soils: relationship to emissions. Agric Ecosyst Environ 53: 247-251

Bedford BL & Bouldin DR (1994) Response to the paper 'on the difficulties of measuring oxygen release by root systems of wetland plants. J Ecol 82: 185-186

Bender M & Conrad R (1992) Kinetics of CH_4 oxidation in oxic soils exposed to ambient air and or high CH_4 mixing ratios. FEMS Microbiol Ecol 101: 261-270

Beyrouty CA, Wells BR, Norman RJ, Marvel JN & Pillow JA (1990) Root growth dynamics of a rice cultivar grown at two locations. Agron J 80: 1001-1004

Boudreau BP (1996) A method-of-lines code for carbon and nutrient diagenesis in aquatic sediments. Computers Geosci 22: 479-496

Butterbach-Bahl K, Papen H & Rennenberg H (1997) Impact of gas transport through rice cultivars on CH_4 emission from rice paddy fields. Plant Cell Environ 20: 1175-1183

Byrnes BH, Austin ER & Tays BK (1995) CH_4 emissions from flooded rice soils and plants under controlled conditions. Soil Biol Biochem 27: 331-339

Cabrera ML (1993) Modeling the flush of nitrogen mineralization caused by drying and rewetting soils. Soil Sci Soc Am J 57: 63-66

Cao M, Dent JB & Heal OW (1995) Modeling CH_4 emissions from rice paddies. Global Biogeochem Cycles 9: 183-195

van Cappellen P & Wang Y (1996) Cycling of iron and manganese in surface sediments: a general theory for the coupled transport and reaction of carbon, oxygen, nitrogen, sulfur, iron and manganese. Am J Sci 296: 197-213

Chin K-J & Conrad R (1995) Intermediary metabolism in methanogenic paddy soil and the influence of temperature. FEMS Microbiol Ecol 18: 85-102

Conrad R & Rothfuss F (1991) Methane oxidation in the soil surface layer of a flooded rice field and the effect of ammonium. Biol Fertil Soils 12: 28-32

Corton TM, Bajita JB, Asis Jr CA & Pamplona RR (1995) CH_4 emission from an irrigated Philippine paddy field subjected to several fertilizer treatments. Phillip J Crop Sci 20: 39-55

Denier van der Gon HAC & Neue H-U (1995a) Influence of organic matter incorporation on the CH_4 emission from a wetland rice field. Global Biogeochem. Cycles 9: 11-22

Denier van der Gon HAC & Neue H-U (1995b) Methane emission from a wetland rice field as affected by salinity. Plant Soil 170: 307-313

Denier van der Gon HAC & Neue H-U (1996) Oxidation of CH_4 in the rhizosphere of rice plants. Biol Fertil Soils 22: 359-366

Denier van der Gon HAC, van Breemen N, Neue H-U, Lantin RS, Aduna JB, Alberto MCR & Wassmann R (1996) Release of entrapped CH4 from wetland rice fields upon soil drying. Global Biogeochem Cycles 10: 1-7

Drenth H, ten Berge HFM & Meijboom FW (1991) Effects of growth medium on porosity and branching of rice roots (*Oryza sativa* L.). In: Penning de Vries FWT, van Laar HH & Kropff MJ (eds) Simulation and Systems Analysis for Rice Production (SARP). Wageningen(NL): PUDOC. p 162-175

Flessa H & Fisher WR (1992) Plant-induced changes in the redox potentials of rice rhizospheres. Plant Soil 143: 55-60

Frenzel P & Bosse U (1996) Methyl fluoride, an inhibitor of CH_4 oxidation and CH_4 production. FEMS Microbiol Ecol 21: 25-36

Frenzel P, Rothfuss F & Conrad R (1992) Oxygen profiles and CH4 turnover in a flooded rice microcosm. Biol Fertil Soils 14: 84-89

Gilbert B & Frenzel P (1995) Methanotrophic bacteria in the rhizosphere of rice microcosms and their effect on porewater CH_4 concentration and CH_4 emissions. Biol Fertil Soils 20: 93-100

Gilbert B, Assmus B, Hartmann A & Frenzel P (1998) In situ localization of two methanotrophic strains in the rhizosphere of rice plants. FEMS Microbiol Ecol 25: 117-128

Gujer W & Zehnder AJB (1983) Conversion processes in anaerobic digestion. Water Sci Technol 15: 127-167

Hassink J & Whitmore AP (1997) A model of the physical protection of organic matter in soils. Soil Sci Soc Am J 61: 131-139

Hosono T & Nouchi I (1997) The dependence of CH_4 transport in rice plants on the root zone temperature. Plant Soil 191: 233-240

Houghton JT, Meira Filho LG, Callander BA, Harris N, Kattenberg A & Marskell K (1996) Climate change 1995. The science of climate change, 1st ed. Cambridge: University Press

Huang Y, Sass RL & Fisher Jr FM (1997) Methane emission from Texas rice paddy soils. 1. Quantitative multi-year dependence of CH_4 emission on soil, cultivar and grain yield. Global Change Biol 3: 479-489

Huang Y, Sass RL & Fisher Jr FM (1998) A semi-empirical model of CH_4 emission from flooded rice paddy soils. Global Change Biol 4: 247-268

Husin YA, Murdiyarso D, Khalil MAK, Rasmussen RA, Shearer MJ, Sabiham S, Sunar A & Adijuwana H (1995) CH_4 flux from indonesian wetland rice: the effects of water management and rice variety. Chemosphere 31: 3153-3180

Inubushi K & Wada H (1987) Easily decomposable organic matter in paddy soils: VII. Effect of various pretreatments on N mineralization in submerged soils. Soil Sci Plant Nutr 33: 567-576

Inubushi K, Wada H & Takai Y (1984) Easily decomposable organic matter in paddy soil. IV. Relationship between reduction process and organic matter decomposition. Soil Sci Plant Nutr 30: 189-198

Jakobsen P, Patrick Jr WH & Williams BG (1981) Sulfide and CH_4 formation in soils and sediments. Soil Sci 132: 279-287

Kang S-Y, Morita S & Yamazaki K (1994) Root growth and distribution in some Japonica-Indica hybrid and Japonica type rice cultivars under field conditions. Jpn J Crop Sci 63: 118-124

Kimura M & Minami K (1995) Dynamics of CH_4 in rice fields. In: Peng S et al. (eds) Climate change and rice. Berlin: Springer Verlag, p 33-45

Kimura M, Miura Y, Watanabe A, Katoh K & Haraguchi H (1991) Methane emission from paddy field, part I. Effect of fertilization, growth stage and midsummer drainage: Pot experiment. Environ Sci 4: 265-271

Kimura M, Minoda T & Murase J (1993) Water-soluble organic materials in paddy soil ecosystem, part 2. Effects of temperature on contents of total organic materials, organic acids, and CH_4 in leachate from submerged paddy soils amended with rice straw. Soil Sci Plant Nutr 39: 713-724

King GM (1996) In situ analyses of CH4 oxidation associated with the roots and rhizomes of a bur reed, Sparganium eurycarpum, in a Maine wetland. Appl Environ Microbiol 62: 4548-4555

King GM, Roslev P & Skovgaard H (1990) Distribution and rate of CH_4 oxidation in sediments of the Florida Everglades. Appl Environ Microbiol 56: 2902-2911

Kirchhof G & So HB (1996) The effect of puddling intensity and compaction on soil properties, rice and mungbean growth: a mini-ricebed study. In: Kirchhof G & So HB (eds) Management of clay soils for rainfed lowland rice-based cropping systems. Canberra: ACIAR, p 51-70

Kirk GJD, Begg CBM & Solivas JL (1993) The chemistry of the lowland rice rhizosphere. Plant Soil 155/156: 83-86

Klüber HD & Conrad R (1998) Effects of nitrate, nitrite, NO and N_2O on methanogenesis and other redox processes in anoxic rice field soil. FEMS Microbiol Ecol 25: 301-318

Kludze HK, Delaune RD & Patrick Jr WH (1993) Aerenchyma formation and CH_4 and oxygen exchange in rice. Soil Sci Soc Am J 57: 386-391

Kludze HK, Delaune RD & Patrick Jr WH (1994) A colorimetric method for assaying dissolved oxygen loss from container-grown rice roots. Agron J 86: 483-487

Kludze HK, Neue H-U, Llenaresas D & Lantin RS (1999) Rice root exudation and its impact on CH_4 production. Soil Sci Soc Am J (in press)

Kludze HK & Delaune RD (1995a) Straw application effects on CH_4 and oxygen exchange and growth in rice. Soil Sci Soc Am J 59: 824-830

Kludze HK & Delaune RD (1995b) Gaseous exchange and wetland plant response to soil redox intensity and capacity. Soil Sci Soc Am J 59: 939-945

Kristjansson JK, Schönheit P & Thauer RK (1982) Different Ks values for hydrogen of methanogenic bacteria and sulfate reducing bacteria: An explanation for the apparent inhibition of methanogenesis by sulfate. Arch Microbiol 131: 278-282

Kumaraswany S, Ramakrishan B, Satpathy SN, Rath AK, Misra S, Rao VR & Sethunathan N (1997) Spatial distribution of CH_4-oxidizing activity in a flooded rice soil. Plant Soil 191: 241-248

Kumazawa K (1984) Physiological specificity of rice root in relation to oxidizing power and nutrient uptake. In: Tsunoda S and Takahashi N (eds) Biology of rice. Tokyo/Amsterdam: Japan Sci Soc Press/ Elsevier. p 117-131

Lee KK, Holst RW, Watanabe I & App A (1981) Gas transport through rice. Soil Sci Plant Nutr 27: 151-158

Lelieveld J, Crutzen PJ & Dentener FJ (1998) Changing concentration, lifetime and climate forcing of atmospheric CH_4. Tellus 50B: 128-150

Lindau CW & Bollich PK (1993) Methane emissions from Lousiana 1st and ratoon crop rice. Soil Sci 156: 42-48

Lindau CW, Bollich PK & Delaune RD (1995) Effect of rice variety on CH_4 emission from Louisiana rice. Agric Ecosyst Environ 54: 109-114

Lord CJ III & Church TM (1983) The geochemistry of salt marshes: sedimentary ion diffusion, sulfate reduction and pyritization. Geochim Cosmochim Acta 47: 1381-1391

Lovley DR, Coates JD, Blunt-Harriss EL, Phillips EJP & Woodward JC (1996) Humic substances as electron acceptors for microbial respiration. Nature 382: 445-448

Lovley DR & Phillips EJP (1986) Organic matter mineralization with reduction of ferric iron in anaerobic sediments. Appl Environ Microbiol 51: 683-689

Lovley DR & Phillips EJP (1987) Competitive mechanisms for inhibition of sulfate reduction and CH_4 production in the zone of ferric iron reduction in sediments. Appl Environ Microbiol 53: 2636-2641

Luther GW III, Giblin A, Howarth RW & Ryans RA (1982) Pyrite and oxidized iron mineral phases formed from pyrite oxidation in salt marshes and estuarine sediments. Geochim Cosmochim Acta 46: 2665-2669

Mattson MD & Likens GE (1990) Air pressure and CH_4 fluxes. Nature 347: 718-719

Matthews E, Fung I & Learner G (1991) CH_4 emission from rice cultivation: geographic and seasonal distribution of cultivated areas and emissions. Global Biogeochem Cycles 5: 3-24

le Mer J, Escoffer S, Chessel C & Roger PA (1996) Microbiological aspects of CH_4 emission in a ricefield soil from the Camargue (France): 2. Methanotrophy and related microfauna. Eur J Soil Biol 32: 71-80

Minoda T & Kimura M (1994) Contribution of photosynthesized carbon to the CH_4 emitted from paddy fields. Geophys Res Lett 21: 2007-2010

Miura Y, Watanabe A, Murase J & Kimura M (1992) Methane production and its fate in paddy fields. II. Oxidation of CH_4 and its coupled ferric oxide reduction in subsoil. Soil Sci Plant Nutr 38: 673-679.

Murase J & Kimura M (1994) Methane production and its fate in paddy fields VI. Anaerobic oxidation of CH_4 in plow layer soil. Soil Sci Plant Nutr 40: 404-514.

Murase J & Kimura M (1997) Anaerobic reoxidation of Mn^{2+}, Fe^{2+}, S^0 and S^{2-} in submerged paddy soils. Biol Fertil Soils 25: 302-306

Nedwell DB & Watson A (1995) Methane production, oxidation and emission in a U.K. ombrotrophic peat bog: influence of SO_4 from acid rain. Soil Biol Biochem 27: 893-903

Neue H-U (1997) Fluxes of CH_4 from rice fields and potential for mitigation. Soil Use Manage 13: 258-267

Neue H-U, Wassmann R, Kludze HK, Wang B & Lantin RS (1997) Factors and processes controlling CH_4 emissions from rice fields. Nutr Cycling Agroecosyst 49: 111-117

Nouchi I (1990) Mechanisms of CH_4 transport through rice plants. In: CH_4 and N_2O: Global Emissions and Controls from Rice Fields and other Agricultural and Industrial Sources. NIAES. p 87-104

Nouchi I, Hosono T, Aoki K & Minami K (1994) Seasonal variation in CH_4 flux from rice paddies associated with CH_4 concentration in soil water, rice biomass and temperature and its modelling. Plant Soil 161: 195-208

Nouchi I & Mariko S (1993) Mechanisms of CH_4 transport by rice plants. In: Oremland RS (ed) Biogeochemistry of global change. Radiatively active trace gases. Chapman & Hall, p 336-352

Nugroho SG, Lumbanraja J, Suprapto H, Sunyoto, Ardjasa WS, Haraguchi H & Kimura M (1994) Effect of intermittent irrigation on CH_4 emission from an Indonesian paddy field. Soil Sci Plant Nutr 40: 609-615

Nugroho SG, Lumbaranaja J, Suprapto H, Sunyoto, Ardjasa WS, Haraguchi H & Kimura M (1996) Three-year measurement of CH_4 emission from an Indonesian paddy field. Plant Soil 181: 287-293

Nugroho SG, Sunyoto, Lumbanraja J, Suprapto H, Ardjasa WS & Kimura M (1997) Effect of rice variety on CH4 emission from an Indonesian paddy field. Soil Sci Plant Nutr 43: 799-809.

Ogston AG (1958) The spaces in a uniform random suspension of fibres. Trans Faraday Soc 54: 1754-1757

Oude-Elferink SJWH, Visser A, Hulshoff Pol LW & Stams AJM (1994) Sulfate reduction in methanogenic bioreactors. FEMS Microbiol Rev 15: 119-136

Parton WJ, Schimel DS, Cole CV & Ojima DS (1987) Analysis of factors controlling soil organic matter levels in Great Plains grasslands. Soil Sci Soc Am J 51: 1173-1179

Ponnamperuma FN (1972) The chemistry of submerged soils. Adv Agron 24: 29-96

Ratering S & Conrad R (1998) Effects of short-term drainage and aeration on the production of CH_4 in submerged rice soil. Global Change Bio 4: 397-407

Rickard DT (1975) Kinetics and mechanisms of pyrite formation at low temperatures. Am J Sci 275: 636-652

Roslev P & King GM (1994) Survival and recovery of methanotrophic bacteria starved under oxic and anoxic conditions. Appl Environ Microbiol 60: 2602-2608

Rothfuss F & Conrad R (1993) Vertical profiles of CH_4 concentrations, dissolved substrates and processes involved in CH_4 production in a flooded Italian rice field. Biogeochemistry 18: 137-152

Saini RC (1989) Mass loss and nitrogen concentration changes during the decomposition of rice residues under field conditions. Pedobiologica 33: 229-235

Sass RL, Fisher FM, Harcombe PA & Turner FT (1990) CH_4 production and emission in a Texas rice field. Global Biogeochem Cycles 4: 47-68

Sass RL, Fisher FM, Harcombe PA & Turner FT (1991a) Mitigation of CH_4 emissions from rice fields: possible adverse effects of incorporated rice straw. Global Biogeochem Cycles 5: 275-287

Sass RL, Fisher FM, Turner FT & Jund MF (1991b) Methane emission from rice fields as influenced by solar radiation, temperature and straw application. Global Biogeochem Cycles 5: 335-350

Sass RL, Fisher FM, Wang YB, Turner FT & Jund MF (1992) Methane emission from rice fields: the effect of floodwater management. Global Biogeochem Cycles 6: 249-262

Sass RL & Fisher FM (1995) Methane emissions from Texas rice fields: a five-year study. In: Peng S et al., (eds) Climate change and Rice. Berlin: Springer-Verlag. p 46-59

Satpathy SN, Rath AK, Ramakrishan B, Rao VR, Adhya TK & Sethunathan N (1997) Diurnal variation in CH_4 efflux at different growth stages of tropical rice. Plant Soil 195: 267-271

Schütz H, Holzapfel-Pschorn A, Conrad R, Rennenberg H & Seiler W (1989a) A 3-year continuous record on the influence of daytime, season, and fertilizer treatment on CH_4 emission rates from an Italian rice paddy. J Geophys Res 94: 16405-16416

Schütz H, Seiler W & Conrad R (1989b) Processes involved in formation and emission of CH_4 in rice paddies. Biogeochemistry 7: 33-53

Segers R (1998) Methane production and CH_4 consumption: a review of processes underlying wetland CH_4 fluxes. Biogeochemistry 41: 23-51

Slaton NA, Beyrouty CA, Wells BR, Norman RJ & Gbur EE (1990) Root growth and distribution of two short-season rice genotypes. Plant Soil 121: 269-278

Sorrell BK & Armstrong W (1994) On the difficulties of measuring oxygen release by root systems of wetland plants. J Ecol 82: 177-183.

Stroosnijder L (1982) Simulation of the soil water balance. In: Penning de Vries FWT and van Laar HH (eds) Simulation of plant growth and crop production. Wageningen(NL): PUDOC simulation monographs. p 175-193

Tanaka S, Yamuchi A & Kono Y (1995) Root system morphology of four rice cultivars: response of different component roots to nitrogen. Jpn J Crop Sci 64: 148-155

Teo YH, Beyrouty CA, Norman RJ & Gbur EE (1995) Nutrient uptake relationship to root characteristics of rice. Plant Soil 171: 297-302

Tsutsuki K & Ponnamperuma FN (1987) Behavior of anaerobic decomposition products in submerged soils. Effects of organic material amendment, soil properties, and temperature. Soil Sci Plant Nutr 33: 13-33

van Bodegom PM, Wassmann R & Corton TM (2000) A process-based model for CH_4 emission predictions from flooded rice paddies. Global Biogeochem Cycles (in press)

van Bodegom PM & Stams AJM (1999) Effects of alternative electron acceptors and temperature on methanogenesis in rice paddy soils. Chemosphere 39: 167-182

Walter BP, Heimann M, Shannon RD & White JR (1996) A process-based model to derive CH_4 emissions from natural wetlands. Geophys Res Lett 23: 3731-3734

Wang B, Neue H-U & Samonte HP (1997) Effect of cultivar difference ('IR72', 'IR65598' and 'Dular') on CH_4 emission. Agric Ecosyst Environ 62: 31-40

Wassmann R, Schütz H, Papen H, Rennenberg H, Seiler W, Aiguo D, Shen RX, Wang YS, Shangguan XJ & Wang MX (1993) Quantification of CH_4 emissions from Chinese rice fields Zhejiang province as influenced by fertilizer treatment. Biogeochemistry 20: 83-101

Wassmann R, Neue H-U, Lantin RS, Aduna JB, Alberto MCR, Andales MJ, Tan MJ, Denier van der Gon HAC, Hoffmann H, Papen H, Rennenberg H & Seiler W (1994) Temporal patterns of CH_4 emissions from wetland rice fields treated by different modes of N application. J Geophys Res 99: 16457-16462

Wassmann R, Neue H-U, Alberto MCR, Lantin RS, Bueno C, Llenaresas D, Arah JRM, Papen H, Seiler W & Rennenberg H (1996) Fluxes and pools of CH_4 in wetlands rice soils with varying organic inputs. Environ Monit Assess 42: 163-173.

Watanabe A & Kimura M (1995) Methane production and its fate in paddy fields. VIII. Seasonal variations in the amount of CH_4 retained in soil. Soil Sci Plant Nutr 41: 225-233.

Watanabe A, Katoh K & Kimura M (1993) Effect of rice straw application on CH_4 emission from paddy fields. I. Effect of weathering of rice straw in the field during off-crop season. Soil Sci Plant Nutr 39: 707-712

Watanabe A, Katoh K & Kimura M (1994) Effects of rice straw application on CH_4 emission from paddy fields. III. Effect of incorporation site of rice straw on CH_4 emission rates and their variation among shoots of a rice plant. Soil Sci Plant Nutr 40: 497-504

Watanabe A, Satoh Y & Kimura M (1995a) Estimation of the increase in CH_4 emission from paddy soils by rice straw application. Plant Soil 173: 225-231

Watanabe A, Kajiwara M, Tashiro T & Kimura M (1995b) Influence of rice cultivar on CH_4 emission from paddy fields. Plant Soil 176: 51-56

Watanabe I, Hashimoto T & Shimoyama A (1997) Methane-oxidizing activities and methanotrophic populations associated with wetland rice plants. Biol Fertil Soils 24: 261-265

Watanabe A, Yoshida M & Kimura M (1998) Contribution of rice straw carbon to CH_4 emission from rice paddies using ^{13}C-enriched rice straw. J Geophys Res D 103: 8237-8242

Watson A, Stephen KD, Nedwell DB & Arah JRM. 1997. Oxidation of CH_4 in peat: Kinetics of CH_4 and O_2 removal, and the role of plant roots. Soil Biol Biochem 29: 1257-1267

Westermann P & Ahring BK (1987) Dynamics of CH4 production, sulfate reduction and denitrification in a permanently waterlogged alder swamp. Appl Environ Microbiol 53: 2554-2559

Winfrey MR & Zeikus JG (1977) Effect of sulfate on carbon and electron flow during microbial methanogenesis in freshwater sediments. Appl Environ Microbiol 33: 275-281

Witt C, Cassmann KG, Ottow JCG & Biker U (1999) Soil microbial biomass and nitrogen supply in an irrigated lowland rice soil as affected by crop rotation and residue management. Biol Fertil Soils (in press)

Yagi K & Minami K (1990) Effect of organic matter application on CH4 emission from some Japanese paddy fields. Soil Sci Plant Nutr 36: 599-610

Yagi K, Minami K & Ogawa Y (1998) Effects of water percolation on CH_4 emission from rice paddies: a lysimeter experiment. Plant Soil 198: 193-200

Yagi K, Tsuruta H, Kanda K & Minami K (1996) Effect of water management on CH_4 emission from a Japanese rice paddy field: automated CH_4 monitoring. Global Biogeochem Cycles 10: 255-267

Yang HS (1996) Modelling organic matter mineralization and exploring options for organic matter management in arable farming in Northern China. PhD thesis, Wageningen (NL): Wageningen Agricultural University

Yang S-S & Chang H-L (1998) Effect of environmental conditions on CH_4 production and emission from paddy soil. Agric Ecosyst Environ 69: 69-80

Nutrient Cycling in Agroecosystems **58:** 249–258, 2000.

A process-based model for methane emissions from irrigated rice fields: experimental basis and assumptions

R.L. Sass[1], F.M. Fisher, Jr.[1] & Y. Huang[2]

[1]*Department of Ecology and Evolutionary Biology, Rice University, 6100 Main, Houston, Texas 77005, USA;* [2]*College of Resources and Environmental Sciences, Nanjing Agricultural University, 6 Tongwei Road, Wei Gang, Nanjing, Jiangsu 210095, People's Republic of China*

Key words: methane emission model, irrigated rice, atmospheric trace gases

Abstract

In this paper, we review the process-level studies that the authors have performed in rice fields of Texas since 1989 and the development of a semi-empirical model based on these studies. In this model, it is hypothesized that methanogenic substrates are primarily derived from rice plants and added organic matter. Rates of methane (CH_4) production in flooded rice soils are determined by the availability of methanogenic substrates and the influence of climate, soil, and agronomic factors. Rice plant growth and added carbon control the fraction of CH_4 emitted. The amount of CH_4 transported from the soil to the atmosphere is determined by the rates of production and the emitted fraction. Model calibration against observations from a single rice-growing season in Texas, USA, without organic amendments and with continuous irrigation demonstrated that the seasonal variation of CH_4 emission is regulated by rice biomass and cultivar type. A further validation of the model against measurements from irrigated rice paddy soils in various regions of the world, including Italy, China, Indonesia, Philippines, and the United States, suggests that CH_4 emission can be predicted from rice net productivity, cultivar character, soil texture, temperature, and organic matter amendments.

Introduction

Atmospheric methane (CH_4) is recognized as one of the most important greenhouse gases. Methane, with some 15-30 times greater infrared-absorbing capability than CO_2 on a mass basis, may account for 20% of anticipated global warming (Rodhe, 1990). The concentration of atmospheric CH_4, currently at 1.73 ppm, has been increasing at a rate of about 1% yr^{-1} but recently has slowed to approximately 0.5% yr^{-1} (Steele et al., 1992) and may be approaching a near steady state (Dlugokencky et al., 1998). The current burden of CH_4 in the atmosphere is approximately 4,700 teragrams (1 $Tg=10^{12}$ g). Recent estimates suggest an annual global CH_4 emission of approximately 550 Tg with 375 Tg from anthropogenic sources. The contribution from rice agriculture is estimated to range from 20 to 100 Tg with an average of 60 Tg (Denier van der Gon, 1996).

Many reports over the past decade have given the magnitudes of the sinks and sources for CH_4. Natural and agricultural wetlands have received particular attention because of their importance in global balances, inverse modeling, and tracer studies. Studies of the last several years have provided a wealth of information on the in situ processes and environmental controls of trace gas production and exchange, but they have done little to reduce the uncertainty in regional and country estimates of the exchange. Advances are needed in how to meaningfully scale measurements from point sources to a regional or larger scale. A first step in scaling field measurements to a regional or global scale is the development of predictive models based on process and environmental factors. In this paper, we review the process-level studies that the authors have performed in rice fields of Texas since 1989 (Sass & Fisher, 1997) and the development of a semi-empirical model based on these studies (Huang et al., 1998a). Rice fields, rather than natural wetlands, were studied because they provide an appropriate system to begin to address these ends. They are primarily composed of a single plant variety; can be tightly managed with respect to key variables such as planting times, flooding, and fertilization, and further, rice agroecosystems are widely distributed throughout many of the world's climate zones.

Model rationale and hypotheses

The processes involved in CH_4 emission from flooded rice fields to the atmosphere include CH_4 production in the soil by methanogens, CH_4 oxidation within oxic zones of the soil and floodwater by methanotrophs, and vertical transport of the gas from the soil to the atmosphere.

Methane is produced in the terminal step of several anaerobic microbial degradation chains. The metabolic pathways leading to CH_4 production include fermentation of methylated compounds and CO_2 reduction with molecular hydrogen (Takai, 1970; Conrad, 1989; Ferry, 1993). Acetate fermentation has been estimated to account for 50-90% of the CH_4 produced in rice fields (Burke & Sackett, 1986; Schütz et al., 1989a; Thebrath et al., 1992; Rothfuss & Conrad, 1993). The amount of CH_4 produced in flooded rice soils is primarily determined by the availability of methanogenic substrates and the influence of environmental factors. The sources of organic carbon for methanogenic substrates are primarily rice plants via root exudation, root senescence, and plant litter (Holzapfel-Pschorn et al., 1986; Schütz et al., 1991; Kludze et al., 1996) or added organic matter for fertilization (Schütz et al., 1989a; Yagi & Minami, 1990; Sass et al., 1991; Cicerone et al., 1992; Denier van der Gon & Neue, 1995). Emissions from soil organic carbon mineralization have been reported from other studies (Holzapfel-Pshorn, et al., 1986) but were essentially unobserved in Texas studies (Sass et al., 1990; Tyler et al., 1997). In these studies, control plots in unplanted fields generally showed little or no CH_4 emissions until short-term bursts of CH_4 were observed late in the season. These emissions were attributed to carbon sources from weeds and/or algal blooms in the floodwater developing at that time. The total seasonal CH_4 emissions from unplanted plots averaged less than 4% of emissions from plots planted to rice. The lack of emissions from soil organic carbon may be due to the management of the Texas fields. In general, they were fallow the season before experimental use and were kept fairly aerated during that time by plowing and disking to reduce weed crop formation. Also, these experimental soils are low in organic carbon (approximately 1.5%). In applying the model to emissions from China (Huang et al., 1998b), the model was modified in terms of emission calculations from the late crop of double cropping situations. This modification was done to take into account residual soil organic carbon remaining from the first crop. The analysis is essentially the same as that which would be required in the general case of soil organic carbon from areas of high soil organic carbon.

The environmental factors affecting CH_4 production include soil texture (Neue et al., 1994; Sass et al., 1994), climate (Schütz et al., 1990; Sass et al., 1991), and agricultural practices, such as water regime and management (Inubushi et al., 1990a,b; Sass et al., 1992; Lewis, 1996; Yagi et al., 1996).

Plant-mediated transport is the primary mechanism for the emission of CH_4 from rice fields, with approximately 90% of CH_4 transported to the atmosphere through the aerenchymal system of the rice plants (Cicerone & Shetter, 1981; Holzapfel-Pschorn et al., 1986; Schütz et al., 1989b). Under high organic fertilization, ebullition can play a significant role in CH_4 transport. Although ebullition does not appear to be significant in Texas soils, the model is not dependent on the specific mode of CH_4 transport.

The rice aerenchymal system not only transports CH_4 from the flooded rice to the atmosphere but also promotes the movement of atmospheric oxygen into the rhizosphere supporting root respiration and CH_4 oxidation (De Bont et al., 1978; Conrad & Rothfuss, 1991; Gerard & Chanton, 1993).

Experimental basis for the model

Simulation model equations

With an understanding of the processes of CH_4 production, oxidation, and emission, it is hypothesized that the rate-determining step in the process is that of CH_4 production with a time lag between production and emission of less than 3 h (Sass et al., 1991). Daily rates of CH_4 production in flooded rice soils are primarily dependent upon the availability of carbon substrates from rice plants and added organic amendments and influenced by the temperature, texture, and redox state of the soil. The emitted fraction of CH_4 is then determined by the extent of bacterial oxidation of the produced CH_4 (Huang et al., 1998a).

In the absence of other organic inputs, the daily amount of carbohydrate derived from rice plants, C_R ($g\ m^{-2}\ d^{-1}$), is postulated to be dependent on the rice cultivar and biomass represented by the allometric function:

$$C_R = \alpha \times VI \times SI \times TI \times W^{1.25} \quad (1)$$

where α ($g^{-.25}$ d^{-1}) is an empirical constant, VI (dimensionless) identifies the dependence on rice variety, SI (dimensionless) characterizes the effect of soil texture, TI (dimensionless) is a soil temperature index, and W (g m^{-2}) is the rice aboveground biomass on a given day (Huang et al., 1998a). The factors SI and TI are explained below. The exponential factor of 1.25 relating carbon substrate to biomass was obtained from a best-fit analysis as an empirical parameter (Huang et al., 1998a).

When organic inputs are present, the additional daily amount of carbohydrates is represented by

$$C_{OM} = SI \times TI \times (k_1 \times OM_N + k_2 \times OM_S) \quad (2)$$

where C_{OM} (g m^{-2} d^{-1}) is the daily amount of carbohydrate degraded from organic amendments, OM_N and OM_S (g m^{-2}) represent the amount of nonstructural and structural components, respectively, and k_1 and k_2 (d^{-1}) represent the first-order decay rates of the two components (Huang et al., 1998a). If the model is applied to situations where an appreciable amount of soil organic carbon is present and is mineralized during the season, this source could possibly be handled by this same treatment since organic amendments are the ultimate source of this carbon. Different values of k_1 and k_2 may need to be applied in these cases.

The daily production rate of CH_4 by methanogenic bacteria, P (g m^{-2} d^{-1}) is then represented by

$$P = 0.27 \times F_{Eh} \times (C_R + C_{OM}) \quad (3)$$

where F_{Eh} (dimensionless) describes the time development of the soil redox potential and 0.27 assumes that three moles of CH_4 is derived from one carbohydrate unit and is the ratio of their molecular weights (0.27 = 3 $CH_4/C_6H_{12}O_6$) (Huang et al., 1998a).

Having determined the daily CH_4 production rate, the emission rate, E (g m^{-2} d^{-1}), is given by

$$E = P \times E_f \quad (4)$$

where E_f is the emitted fraction of CH_4 determined by the rate of CH_4 oxidation and is simulated by

$$E_f = 0.55 \times (1 - W/W_{max})^{0.25} \quad (5)$$

where W_{max} (g m^{-2}) is the seasonal maximum aboveground biomass. The constant 0.55 represents the initial fraction of produced CH_4 which is emitted (Huang et al., 1998a).

Data needed to use the model

Emission values are calculated on a daily basis and summed over the season to give a seasonal estimate of CH_4 emission. To evaluate the model, one needs daily estimates of rice crop aboveground biomass and soil temperature; the relative emission potential of the rice cultivar used; the percent sand in the field soil; and the amount, timing, and composition of the organic amendments. Huang et al. (1998a) suggest that daily biomass, W, can be approximated by using the logistic growth equation:

$$dW/dt = r \times W \times (W_{max} - W)/ W_{max} \quad (6)$$

where r is the intrinsic growth rate for aboveground biomass and W_{max} can be approximated from the grain yield, GY, by the equation (Huang et al., 1997b):

$$W_{max} = 9.46 \times GY^{0.76} \quad (7)$$

The intrinsic growth rate, r, was experimentally determined to be 0.08 ± 0.02 d^{-1} based on 17 cases from four different cultivars and with 10–13 biomass measurements in each case (Huang et al., 1998a).

A simplified version of the model is also presented (Huang et al., 1998a) in which seasonal emission values can be estimated using integrated or average values of the time-dependent parameters.

Explicit and implicit assumptions in the model

Several assumptions have been incorporated into this model, both explicit and implicit. The explicit assumptions are easily recognized in that they appear as factors in the above equations. The implicit assumptions are less easily recognized but nevertheless are quite important in understanding how the model can be constructed based on experimental evidence.

Methane is produced by bacterial activity in a highly reduced soil environment. The primary driving force assumed in the model for the production of CH_4 is the availability and quantity of organic substrate supplied by the rice plant and other organic additions. A part of the produced CH_4 is reoxidized in oxidizing zones of the soil while the rest is transported to the atmosphere, mainly via the rice plants (Nouchi et al., 1990) with a lesser amount emitted by diffusion and ebullition through the soil-water system except in systems with very high or very decomposable organic amendments.

Equations 1, 2, and 3 assert that under conditions of constant soil temperature, soil composition, and soil redox potential, daily CH_4 production is proportional to the daily carbon substrate production derived from two sources: rice plants and added organic amendments. Implicit in this statement is the assumption that the conversion time from substrate formation to CH_4 production and emission is less than 1 d. In our studies, we have measured soil acetate turnover times ranging up to 7–10 h during the first 5 wk of the season, dropping to less than 1 h during the later half of the growing season (Sigren et al., 1997a). These values are less than the 10-16 h estimated by Schütz et al. (1989a) and 16 h estimated by Krumböck & Conrad (1991). However, all three estimates suggest that soil substrate pools in rice fields are turned over in less than 1 d. Temperature studies of CH_4 production and emission indicate that CH_4 production is the rate-determining step and that emission through the rice plant occurs effectively instantaneously (Sass et al., 1991).

The model also assumes that acetate is the major precursor of CH_4 in rice fields. Stable isotope measurements suggest that in our fields the percentage of CH_4 produced from acetate fermentation ranges from 57 to 80% (Tyler et al., 1997). Schütz et al. (1989a) estimated that acetate accounted for 50–70% of CH_4 production, whereas Thebrath et al. (1992) said it accounted for 80–90%. Regardless of the magnitude of this fraction, the model results will be valid if the ratio of CH_4 production from acetate to that from carbon dioxide reduction remains constant in all rice fields and during the whole season. The similarity of these three findings from different areas of the world suggests that this may be a reasonable assumption.

In Equation 1, the daily amount of carbon substrate and hence the daily amount of carbon substrate derived from rice plants of a particular variety is indicated to be directly related to the current aboveground biomass. This assumption has been evaluated and validated from several studies (Huang et al., 1997b).

In Equation 4, daily CH_4 emission is related to CH_4 production by multiplying by a time-dependent factor defining the fraction of CH_4 not oxidized. This assumption is discussed at length in Huang et al. (1998a). In the model, oxidation is assumed to range from 55% early in the season to approximately 80% during the late season. Some research suggests that more than 50% of the generated CH_4 is oxidized during the early phase of the vegetation period, whereas up to 90% may be consumed during the late season of rice maturation (Schütz et al., 1989b; Sass et al., 1992; Sigren et al., 1997a). Other studies suggest a lower amount of oxidation. Epp and Chanton (1993) reported that CH_4 oxidation in the rhizosphere of 3-mo-old rice plants ranged from 14 to 52%. A good review of the difficulties inherent in measuring the extent of methanotrophy in rice ecosystems is presented by Denier van der Gon (1996).

Correlations between CH_4 emission and aboveground biomass have been reported in subtropical sawgrass system (Whiting et al., 1991) and across a variety of agricultural and subarctic natural wetland ecosystems (Whiting & Chanton, 1993). Seasonal CH_4 emissions over a 5-yr period have been quantitatively described over a wide range of conditions (Huang et al., 1997a,b). In experiments carried out in Texas in 1994 and 1995, Huang et al. (1997b) showed that, over a 10-wk period after permanent flooding, total seasonal CH_4 emission was positively correlated with rice aboveground biomass ($r^2 = 0.845$, $n = 11$). A very strong dependence of daily CH_4 emission on aboveground vegetative biomass ($r^2 = 0.887$, $n = 93$) and on root biomass ($r^2 = 0.816$, $n = 33$) was also observed. Calculation from three developmental periods (vegetative, reproductive, and ripening) of rice plants indicated that more than 75% of total seasonal CH_4 was emitted during the last 5-wk period in concert with reproductive and ripening stages, while rice biomass production during the same period amounted to approximately 50% of the seasonal total. Carbon released as CH_4 was found to be approximately equivalent to 3% and 4.5% of photosynthetically fixed carbon in the biomass for low- and high-emitting cultivars.

Little attention has been paid to the relationship between CH_4 production and aboveground biomass. Sass et al. (1990) reported that daily CH_4 emissions from a flooded rice soil is highly correlated with rice aboveground biomass ($r^2 = 0.92$) and that CH_4 production is correlated with root biomass ($r^2 = 0.56$). A reanalysis of the data from the 1990 study shows a correlation between CH_4 emission and aboveground biomass with $r^2 = 0.79$. During an extended study of the effects of soil redox potential on CH_4 production and emission (Lewis, 1996), extensive data were collected in 1994 on CH_4 production levels as a function of soil depth. These data have been examined against aboveground biomass data collected concurrently from

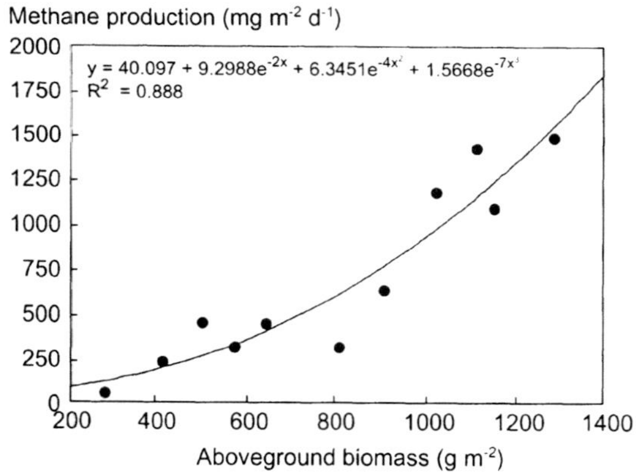

Figure 1. Correlation between CH_4 production and aboveground biomass data collected in 1994. Solid circles represent experimental measurements. The curve shown is a best-fit third-order polynomial of these data with accompanying equation

the same field plots (Huang et al., 1997b). The results, presented in Figure 1, indicate a good correlation between CH_4 production and aboveground biomass. A linear best-fit correlation results in an r^2 of 0.86. The curve shown is a best-fit third-order polynomial (r^2= 0.89). The model postulates a relationship between daily CH_4 production and aboveground biomass raised to the 1.25 power, which closely resembles the shape of the polynomial shown in Figure 1 and results in an r^2 value of 0.89.

Although a strong correlation can be shown to exist between CH_4 production and biomass for a single cultivar, the absolute relationship varies from cultivar to cultivar. That is, some cultivars appear to allocate more of the products of photosynthesis to root exudation than others do. In 1993, CH_4 emissions from 10 cultivars commonly used in Texas were investigated (Sass & Fisher, 1997). The period of maturation ranged from 114 d (Labelle) to 140 d (Jasmine). Semidwarf and conventional cultivars are represented with plant heights ranging from 90 cm (Lemont) to 140 cm (Dawn). Cultivars with yield potentials from medium to high as well as medium and long grain length are represented. Seasonal CH_4 emissions were found to vary from 17.95 to 41.05 g m^{-2}. A nonparametric test of medians was performed on the seasonal emissions of the 10 cultivars. The cultivars were sorted into three groups with the low emission group (Labelle and IR36) significantly different from the high emission group (Mars and Della), but not from the intermediate emission group (Lemont, Lebonnet, Dawn, Katy, Brazos, and Jasmine).

In 1994, the CH_4 emission from three of these cultivars were again measured (Sass & Fisher, 1997), one from each group: Mars, Labelle, and Lemont. The emission data were very similar to the 1993 study. The integrated seasonal emissions in 1994 vs 1993, respectively, were 34.26 g vs 34.06 g for Mars; 15.95 g vs 17.95 g for Labelle; and 17.97g vs 24.52 g for Lemont.

Other studies of CH_4 emissions from different cultivars have been reported. Methane emissions from eight different cultivars grown under similar conditions near New Delhi, India, differed by as much as an order of magnitude (Parashar et al., 1991). A study of five rice cultivars in irrigated fields near Beijing, China, indicated that CH_4 emission during the tillering-flowering stage varied by a factor of two (Lin, 1993).

Organic amendments such as rice straw or green manure increase CH_4 production and emission (Neue & Sass, 1994) by enhancing the reduction of soils and providing additional carbon sources. Different organic amendments vary considerably in their effectiveness in the production of CH_4 (Cicerone et al., 1992; Watanabe et al., 1993). Yagi & Minami (1990) show that the effectiveness of various organic amendments in producing CH_4 depends on the percentage of readily mineralized carbon (RMC). As shown in Equation 2, the model accounts for differences among various added amendments by dividing the available carbon substrate into two components in a first-order decay: a faster decomposing (k_1 = 0.027 d^{-1}) portion of "nonstructural" or RMC and a slower decomposing (k_2 = 0.002 d^{-1}) portion of "structural" carbon (see Murayama, 1984). In field studies (Sass, unpubl.), we have investigated the decomposition of rice straw during an entire flooded rice-growing season. Decomposition was measured by weighing soil-submerged nylon mesh bags of rice straw at various intervals during the season. Comparison of decomposition rates measured in this study with the rates given in Equation 2 results in a strong correlation (r^2 = 0.96) by assuming a rapidly decomposing straw fraction of 16%.

The bacterial processes involved in the processes leading to CH_4 emission should be temperature- and soil structure-dependent. These dependencies are represented in the model (Equations 1 and 2) by a temperature index, TI, and a soil index, SI.

The model accounts for soil temperature through TI, defined by the Arrhenius relationship:

$$TI = Q_{10}^{(T_{soil}-30/10)} \text{ with } T_{soil} = 30 \quad \text{for } 30 \leq T_{soil} \leq 40\ °C \qquad (8)$$

Values of Q_{10} for methanogenesis range widely in various wetland ecosystems (Segers, 1998). Field measurements in irrigated rice systems suggest a Q_{10} range from 2 (Khalil et al., 1991) to 4 (Schütz et al., 1989a). A model value of 3 was assigned to Q_{10} (Huang et al., 1998a) based on field and incubation measurements (Sass et al., 1991). In this study, it was shown that both CH_4 production (anaerobic laboratory incubations) and CH_4 emission (diel field experiments) followed the same temperature relationships with good agreement with the Arrhenius relationship. In the same study, diel soil temperatures varied by as much as 4 °C before canopy closure and by 3 °C after canopy closure later in the season. There was no observable time shift between trends in the measured soil temperature and CH_4 emission, indicating a rapid CH_4 production and emission response to temperature. Daily mean soil temperatures ranged by approximately the same amount throughout the season, but daily CH_4 emission values did not directly correlate with daily mean soil temperature, possibly due to the influence of other overriding factors such as plant growth and development.

Soil bacterial activity and hence CH_4 production, oxidation, and emission are found to be influenced by soil substrate conditions, mainly texture. Sass et al. (1994) compared a variety of CH_4 emission data sets obtained over a 4-yr period from three adjacent different soil types at the Texas Agricultural Experiment Station near Beaumont, Texas. A variety of physical and chemical properties of the soils were compared with

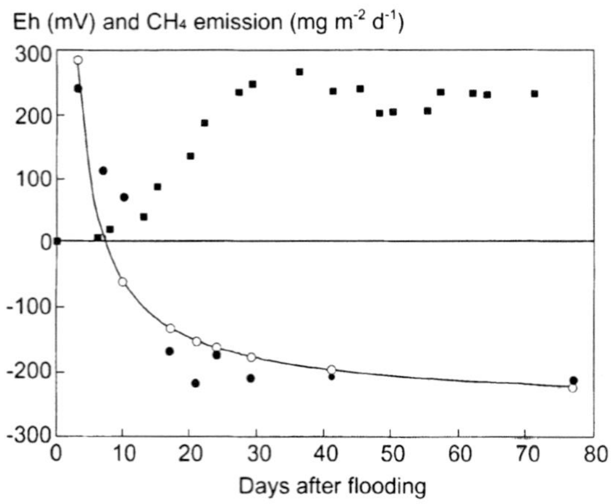

Figure 2. Methane emission in mg m^{-2} d^{-1} (solid squares) and soil Eh in mV (solid circles) measured in a Texas rice field in 1994. The Eh values are compared with the analytical expression Eh = 1390 $t^{-0.87}$-250 (see text) represented by the open circles and corresponding solid line

CH_4 emissions from fields planted with a single rice cultivar. It was observed that seasonal CH_4 emissions directly correlated with the percent sand in the soils. Soil percent sand ranged from 4.3 to 32.5%, while seasonal CH_4 emission values ranged from 13.6 to 36.3 g m^{-2}. The results of this study were directly incorporated into the model (Huang et al., 1998a) through the soil index, SI, as

$$SI = 0.3225 + 0.0225*\text{sand }\% \qquad (9)$$

This relationship has been modified in the model to scale the effect of soil texture to be unity when the soil sand percentage is 30%. Although the experimental evidence for this effect was based on CH_4 emission studies (Sass et al., 1994; Huang et al., 1997a), it is applied in the model in calculating CH_4 production. This application is justified by the observation that production and emission are very tightly coupled, with production being the rate-determining step in the process (Sass et al., 1991).

The temporal development of CH_4 production and emission is dependent on the reducing condition of the bacterial soil environment. The flooding of rice fields begins a series of events that lead to reduced soil conditions in which methanogenic activity can occur, beginning with the consumption of molecular oxygen by aerobic soil bacteria (Bohn et al., 1985). After oxygen depletion, a series of other terminal electron acceptors (NO_3^-, Mn^{+4}, Fe^{+3}, and SO_4^{-2}) are bacterially reduced, lowering the soil Eh from +250 to −100 mV. The critical soil Eh for the initiation of CH_4 production in laboratory incubations has been reported to be between −150 and −160 mV (Wang et al., 1993). Field soils are more heterogeneous than slurries due to the presence of microsites and soil aggregate structures; therefore in situ critical Eh values may be higher and CH_4 emissions may be observed even though the measured soil Eh has not reached a critical value. At any rate, as seen in Figure 2, initial CH_4 emission and critical soil Eh both develop over approximately the same time interval; approximately 2–3 wk after permanent flooding (Sigren et al., 1997b). The observed Eh is represented analytically by the best-fit equation

$$Eh = 1390\, t^{-0.87} - 250 \qquad (10)$$

where t is the time in days after flooding and the constant 250 represents the normal Eh in mV at the time of flooding (Huang et al., 1998a). This function is compared with experimental values in Figure 2. The development of redox conditions appropriate for methanogenesis depends on the amounts of other terminal electron acceptors in the soils such as iron and manganese. Equation 10 was able to describe the Eh development in soils which contained between 6,570 and 11,348 μg g^{-1} dw soil of total iron and between 905 and 1697 μg g^{-1} dw soil of manganese. During the rice-growing season, the concentration of ferrous iron in these submerged soil increased to steady-state values ranging from 500 to 3,000 μg g^{-1} dw soil (Lewis, 1996). These values compare with studies by Ponnamperuma (1981) in which ferrous concentrations increased to values as high as 600 μg g^{-1} within 1–3 wk of flooding and by Patrick (1981) in which ferrous ion concentrations increased to values greater than 2,000 μg g^{-1}.

The critical effect of the soil redox condition on CH_4 production and emission is thus during the early season. Once the critical value is reached, CH_4 production is dependent on other factors. This effect is treated in the model by a factor F_{Eh} where

$$F_{Eh} = \exp[-1.7\,(150 - Eh)/Eh] \quad \text{with } Eh = -150 \text{ for } Eh < -150 \qquad (11)$$

which ranges from 0 to 1 in the early season and equals 1 after a critical value of −150 has been reached or exceeded (Huang et al., 1998a).

In the model, daily CH_4 emission rates are calculated by multiplying production rates by E_f, the emitted fraction of produced CH_4 (Equation 4). If one knows

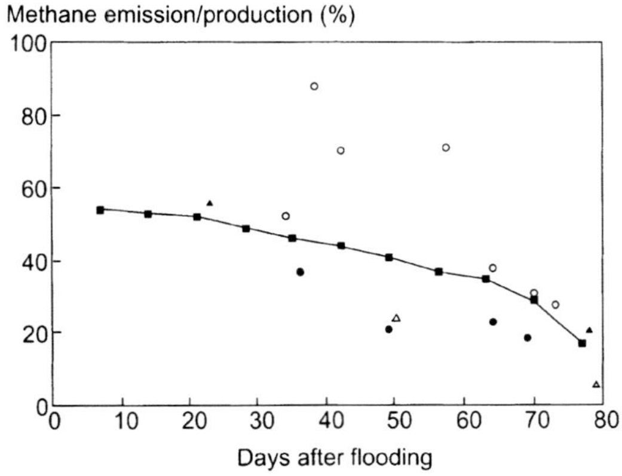

Figure 3. Experimentally determined ratios of CH_4 emission/production (%) determined at various times during the growing season in Vercelli, Italy, 1985 (closed triangles) and 1986 (open triangles) and Beaumont, Texas, 1991 (closed circles) and 1994 (open circles). The same ratio ($E/P = E_f$), calculated by the model equation $E_f = 0.55 \times (1-W/W_{max})^{0.25}$ using biomass data collected in Texas in 1994 is depicted by the line (closed squares)

the daily fraction of the produced CH_4 which is oxidized, then E_f would simply be equal to [1 − (fraction oxidized)]. In the model, E_f is approximated by a function of the daily and maximum aboveground biomass (Equation 5). The rationale behind this hypothesis lies in the assumption that soil bacterial activity, including both CH_4 production and oxidation, are coupled to rice plant development. Evidence of the validity of this assumption is presented in Figure 3. Experimentally determined ratios of CH_4 production (laboratory incubations) and emission (in situ field measurements) determined at various times during the growing season are presented from two locations and during four seasons: Vercelli, Italy, 1985 and 1986 (Schütz et al., 1989a) and Beaumont, Texas, 1991 and 1994 (Sass et al., 1992; Lewis, 1996). The same ratio ($E/P = E_f$) was calculated by Equation 5 using biomass data for the Beaumont, Texas 1994 field. Although there is considerable spread in the experimental ratios, there is generally good agreement between them and with calculated values. A gradual decrease with time is noted in the ratio, indicating that the fraction of CH_4 that is oxidized increases during the season. Since the model-calculated E_f for 1994 is in reasonable agreement with all four data sets, it may be reasonable to assume that, in the absence of reliable biomass data, general E_f values may be used in calculating CH_4 emission values. Conversely, if one knows the grain yield, one can calculate the biomass using Equation 7 to obtain the maximum biomass and then Equation 6. The validity of these relationships has been documented by Huang et al. (1997b).

Model usage

The model was tested by comparing calculated and reported observed values of seasonal CH_4 emissions from 20 studies in Texas and Louisiana, USA; Vercelli, Italy; Nanjing, Beijing, Sichuan, and Hangzhou, China; Taman Bogo, Indonesia; and IRRI, Philippines (Huang et al., 1998a) with considerable success. These studies were used because literature reports were available which contained the necessary model parameters of soil percent sand, average temperature, and grain yield. The variety used was generally not characterized, so the variety index was set to 1. The average calculated CH_4 emission value was 312 ± 138 mg m^{-2} d^{-1} while the average observed value was 322 ± 144. In a subsequent paper (Huang et al., 1998b), the model was used to calculate CH_4 emission values from China on a provincial scale. The resulting total calculated country emission value was reported to be 9.66 Tg with a range from

7.19 to 13.62 Tg, based on estimates of uncertainties in available data on soils, temperature, grain yields, and rice cultivars. To test the model using readily available data, we have calculated daily and seasonal emissions from a field in Texas and compared the results with data collected in 1994 (Sigren, 1996). The only parameters used were average soil temperature (25.1 °C), variety index (1.0), soil sand content (27.9%), and grain yield (570 g m^{-2}). Calculated and observed daily CH_4 emission values are shown in Figure 4. The calculated seasonal CH_4 emission was 17.50 g m^{-2} and the observed seasonal CH_4 emission was 17.97 g m^{-2} (Sigren, 1996). Before the model can be used with confidence in other regions of the world, it will be necessary to compare daily as well as seasonal calculated and observed emission values. This should be done as more complete information becomes available in the literature or as other scientists attempt to apply this and other models to their data.

Future extensions of the model

The current state of the model makes it particularly applicable to the simulation of CH_4 emissions from irrigated rice fields with a minimal amount of available data on climate, soil texture, rice cultivar, and grain yields. Modifications will be required to account for the effects of field drainage, a normal management practice used by farmers in many parts of the world and a potential strategy for the mitigation of CH_4 emissions (Sass et al., 1992). Also, systems of variable floodwater application such as in rainfed rice agriculture will need to be more carefully characterized before modeling of the process can be accomplished. The model dependence of CH_4 production and emission on rice cultivar as well as biomass is problematic in applying it on a large scale. Recent work in our laboratory indicates that plant height or certain aspects of the rice canopy geometry may be an indicator of the variety index, which would allow the model to be more easily applied in cases where varietal data are lacking. In cases where organic amendments have been applied or where indigenous soil organic carbon is an important source of carbon, CH_4 emissions are very dependent on specific composition and decomposition properties as well as on field management. More work is necessary to be able to simulate CH_4 emissions from such fields, particularly with respect to the pre-treatment (such as composting) the timing of such application (early or late treatment leading to possible partial aerobic decomposition), and the use of animal wastes (which have a much different rate of decomposition than plant matter). The ultimate goal of this type of model is to be able to accurately calculate CH_4 emissions on a regional or larger scale based on available geographic information system data sets and remotely sensed data. This model offers a solid beginning to this goal and a base for future development.

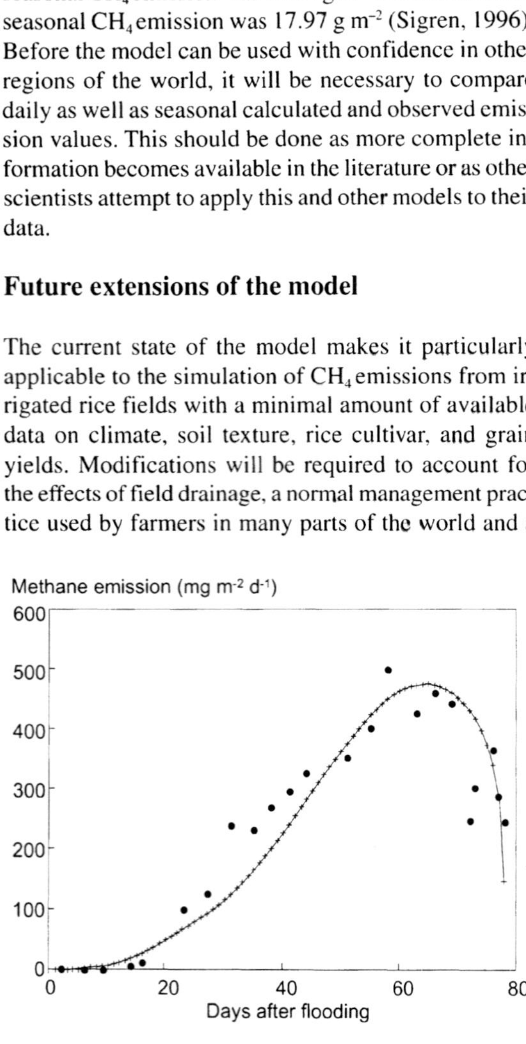

Figure 4. Comparison between observed (closed circles) and calculated (cross and corresponding line) CH_4 emission values from data collected in Texas in 1994

Acknowledgements

This work was supported by grants from the National Aeronautics and Space Agency, the US Department of Agriculture, the U.S. Department of Energy National Institute for Global Environmental Change, and by funding from The Houston Lighting and Power Company (Reliant Energy Inc.). We thank the many Rice University undergraduate and graduate students who have aided in the collection of data and the International Rice Research Institute and the United Nations Development Programme Global Environment Facility for the opportunity to present this study at the "Workshop of the Interregional Research Program on Methane Emission from Rice Fields" in Beijing, China, 10–15 Aug 1998.

References

Bohn HL, McNeal BL & O'Conner GA (1985) Soil Chemistry, 2nd edition. John Wiley & Sons, New York

Burke RA & Sackett WM (1986) Stable hydrogen & carbon isotopic compositions of biogenic CH_4 from several shallow aquatic environments. In: ML Sohn (ed) Organic Marine Chemistry, pp 297–313, American Chemical Society, Washington, DC

Cicerone RJ & Shetter JD (1981) Sources of atmospheric methane: measurements in rice paddies & a discussion. J Geophys Res 86:7203–7209

Cicerone RJ Delwiche CC, Tyler SC & Zimmermann PR (1992) Methane emissions from California rice paddies with varied treatments. Global Biogeochem Cycles 6:233–248

Conrad R (1989) Control of methane production in terrestrial ecosystems. In: Andreae MO & Schimel DS (eds) Exchanges of Trace Gases between Terrestrial Ecosystems & the Atmosphere, pp 39–58, John Wiley & Sons, Chichester

Conrad R & Rothfuss (1991) Methane oxidation in the soil surface layer of a flooded rice field and the effect of ammonium. Biol Fertil Soils 12:28–32

De Bont JAM, Lee K, Bouldin DF (1978) Bacterial oxidation of methane in a rice paddy. Ecol Bull 26:91–96

Denier van der Gon HAC (1996) Methane emission from wetland rice fields, Doctoral thesis, Agricultural University Wageningen, The Netherlands

Denier van der Gon HAC & Neue HU (1995) Influence of organic matter incorporation on the methane emission from a wetland rice field. Global Biogeochem Cycles 9:11–22

Dlugokencky EJ, Masarie KA, Lang PM & Tans PP (1998) Continuing decline in the growth rate of the atmospheric methane burden. Nature 393:447–450

Epp MA & Chanton JP (1993) Rhizospheric methane oxidation determined via the methyl fluoride inhibition technique. J Geophys Res 98:18413–18422

Ferry JG (1993) Fermentation of acetate. In: Ferry JG(ed) Methanogenesis, pp 305–324, Chapman & Hall, New York

Gerard G & Chanton J (1993) Quantification of methane oxidation in the rhizosphere of emergent aquatic macrophytes: defining upper limits. Biogeochemistry 23:79–97

Holzapfel-Pschorn A, Conrad R & Seiler W (1986) Effects of vegetation on the emission of methane by submerged paddy soil. Plant Soil 92:223–233

Huang Y, Sass RL & Fisher FM (1997a) Methane emission from Texas rice paddy soils. 1. Quantitative multi-year dependence of CH_4 emission on soil, cultivar and grain yield. Global Change Biol 3:479–489

Huang Y, Sass RL & Fisher FM (1997b) Methane emission from Texas rice paddy soils. 2. Seasonal contribution of rice biomass production to CH4 emission. Global Change Biol 3:491–500

Huang Y, Sass RL & Fisher FM (1998a) A semi-empirical model of methane emission from flooded rice paddy soils. Global Change Biol 4:247–268

Huang Y, Sass RL & Fisher FM (1998b) Model estimates of methane emission from irrigated rice cultivation of China. Global Change Biol 4:809–822

Inubushi KM, Umebayashi & Wada H (1990a) Control of methane emission from paddy soil. Paper presented at the 14th International Congress of Soil Science, Kyoto, August 1990

Inubushi KM, Umebayashi & Wada H (1990b) Methane emission from paddy fields. Paper presented at the 14th International Congress of Soil Science, Kyoto, August 1990

Khalil MAK, Rasmussen RA, Wang MX & Ren LX (1991) Methane emission from rice fields in China. Environ Sci Technol 25:979–981

Kludze HK, Neue HU & Llenaresas DL (1996) Rice root exudation and its impact on methane production. Paper presented at the Fourth Symposium on Biogeochemistry of Wetlands, 4-6 Mar 1996, New Orleans, Louisiana

Krumböck M & Conrad R (1991) Metabolism of position-labeled glucose in anoxic methanogenic paddy soil and lake sediment. FEMS Microbiol Ecol 85:247–256

Lewis ST (1996) The use of redox measurements to study methane mitigation options in Texas rice paddies. PhD thesis, Rice University, Houston, Texas, USA

Lin E (1993) Agricultural techniques: factors controlling methane emission. In: Gao L, Wu L, Zheng D & Han X (eds) Proceedings of the International Symposium on Climate Change, Natural Disasters and Agricultural Strategies, 26-29 May 1993, Beijing, China, pp 120-126, China Meteorological Press, Beijing

Murayama S (1984) Decomposition kinetics of straw saccharides and synthesis of microbial saccharides under field conditions. J Soil Sci 35:231–242

Neue HU & Sass RL (1994) Trace Gas Emissions from Rice Fields. In: Prinn RG (ed) Global Atmospheric-Biospheric Chemistry, pp 119-147. Plenum Press, New York

Neue HU, Lantin RS, Wassmann R, Aduna JB, Alberto CR, & Andales MJF (1994) Methane emission from rice soils of the Philippines. In: Minami K, Mosier A, Sass R (eds) CH_4 & N_2O: Global Emissions and Controls from Rice Fields and Other Agricultural and Industrial Sources, pp 55-63, Yokendo Publishers, Tokyo

Nouchi I, Mariko S & Aoki K (1990) Mechanism of methane transport from the rhizosphere to the atmosphere through rice plants. Plant Physiol 94:59–66

Parashar DC, Rai J, Gupta PK & Singh N (1991) Parameters affecting methane emission from paddy fields. Indian J Radio Space Physics 20:12–17

Patrick WH (1981) The role of inorganic redox systems in controlling reduction in paddy soils. In: Proceedings of Symposium on Paddy Soil, pp 107–138, Institute of Soil Science, Academia Sinica

Ponnamperuma FN (1981) some aspects of the physical chemistry of paddy soils. In: Proceedings of Symposium on Paddy Soil, pp 59-94, Institute of Soil Science, Academia Sinica

Rodhe H (1990) A comparison of the contribution of various gases to the greenhouse effect. Science 248: 1217–1219

Rothfuss F & Conrad R (1993) Vertical profiles of CH_4 concentrations, dissolved substrates, and processes involved in CH_4 production in a flooded Italian rice field. Biogeochemistry 18(3):137–152

Sass RL, Fisher FM & Harcombe PA (1990) Methane production and emission in a Texas rice field. Global Biogeochem Cycles 4:47–68

Sass RL, Fisher FM, Harcombe PA & Turner FT (1991) Mitigation of methane emissions from rice fields: possible adverse effects of incorporated rice straw. Global Biogeochem Cycles 5:275–287

Sass RL, Fisher FM, Wang YB, Turner FT & Jund MF (1992) Methane emission from rice fields: the effect of floodwater management. Global Biogeochem Cycles 6:249–262

Sass RL, Fisher FM, Lewis ST, Jund MF & Turner FT (1994) Methane emission from rice fields: effect of soil properties. Global Biogeochem Cycles 8:135–140

Sass RL & FM Fisher (1997) Methane emissions from rice paddies: a process study. Nutr Cycling Agroecosyst 49: 119–127

Schütz H, Seiler W & Conrad R (1989a) Processes involved in formation and emission of methane in rice paddies. Biogeochemistry 7:33–53

Schütz H, Holzapfel-Pschorn A, Conrad R, Rennenberg H & Seiler W (1989b) A 3-year continuous record on the influence of daytime, season, and fertilizer treatment on methane emission rates from an Italian rice paddy. J Geophys Res 94:16405–16416

Schütz H, Seiler W & Conrad R (1990) Influence of soil temperature on methane emission from rice paddy fields. Biogeochemistry 11:77–95

Schütz H, Schröder P & Rennenberg R (1991) Role of plants in regulating the methane flux to the atmosphere. In: Sharkey TD, Holl EA, Mooney HA (eds) Trace Gas Emissions by Plants, pp 29–63, Academic Press, Inc., New York

Segers R (1998) Methane production and methane consumption: a review of processes underlying wetland methane fluxes. Biogeochemistry 41:23

Sigren LK (1996) Soil acetate and methane emissions from irrigated rice: the effects of field drainage and cultivar choice. PhD thesis, Rice University, Houston, Texas, USA

Sigren LK, Byrd GT, Fisher FM & Sass RL (1997a) Comparison of acetate concentrations and methane production, transport, and emission in two rice cultivars. Global Biogeochem Cycles 11:1–14

Sigren LK, Lewis ST, Fisher FM & Sass RL (1997b) Effects of field drainage on soil parameters related to methane production and emission from rice paddies. Global Biogeochem Cycles 11:151–162

Steele P, Dlugokencky EJ, Lang PM, Tans PP, Martin RC & Masarie KA (1992) Slowing down of the global accumulation of atmospheric methane during the 1980's. Nature 358:313–316

Takai Y (1970) The mechanism of methane fermentation in flooded paddy soil. Soil Sci Plant Nutr 16:238–244

Thebrath B, Mayer HP & Conrad R (1992) Bicarbonate-dependent production and methanogenic consumption of acetate in an oxic paddy soil. FEMS Microbiol-Ecol 86: 295–302

Tyler SC, Bilek RS, Sass RL & Fisher FM (1997) Methane oxidation and pathways of production in a Texas paddy field deduced from measurements of flux, ^{13}C, and ?D of CH_4. Global Biogeochem Cycles 11:323–348

Wang ZP, Delaune RD, Masschelyn PH & Patrick WH (1993) Soil redox and pH effects on methane production in a flooded rice soil. Soil Sci Soc Am J 57:382–385

Watanabe A, Katoh Y & Kimura M (1993) Effect of rice straw application of CH_4 emission from paddy fields. 2. Contribution of organic constituents in rice straw. Soil Sci Plant Nutr 39:708–712

Whiting GJ, Chanton JP, Bartlett D & Happell (1991) Relationship between CH_4 emission, biomass, and CO_2 exchange in a subtropical grassland. J Geophys Res 96: 13067–13071

Whiting GK & Chanton JP (1993) Plant-dependent CH_4 emission in a subarctic Canadian fen. Global Biogeochem Cycles 8:325–331

Yagi K & Minami K (1990) Effect of organic matter application on methane emission from some Japanese paddy fields. Soil Sci Plant Nutr 36:599–619

Yagi K, Tsuruta H, Kanda K & Minami K (1996) Effect of water management on methane emission from a Japanese rice paddy field: automated methane monitoring. Global Biogeochem Cycles 10:255–267

Nutrient Cycling in Agroecosystems 58: 259–276, 2000.

Modeling trace gas emissions from agricultural ecosystems

C.S. Li

Institute for the Study of Earth, Oceans, and Space, University of New Hampshire, Durham, NH 03824, USA

Key words: agroecosystem, trace gas, modeling

Abstract

A computer simulation model was developed for predicting trace gas emissions from agricultural ecosystems. The denitrification-decomposition (DNDC) model consists of two components. The first component, consisting of the soil climate, crop growth, and decomposition submodels, predicts soil temperature, moisture, pH, Eh, and substrate concentration profiles based on ecological drivers (e.g., climate, soil, vegetation, and anthropogenic activity). The second component, consisting of the nitrification, denitrification, and fermentation submodels, predicts NH_3, NO, N_2O, and CH_4 fluxes based on the soil environmental variables. Classical laws of physics, chemistry, or biology or empirical equations generated from laboratory observations were used in the model to parameterize each specific reaction. The entire model links trace gas emissions to basic ecological drivers. Through validation against data sets of NO, N_2O, CH_4, and NH_3 emissions measured at four agricultural sites, the model showed its ability to capture patterns and magnitudes of trace gas emissions.

Introduction

In the context of global climate change, several trace gases, such as methane (CH_4), nitrous oxide (N_2O), nitric oxide (NO), and ammonia (NH_3), are drawing attention because of their radiative or chemical effects in the atmosphere. Field measurement campaigns were launched for quantifying gas fluxes at site scale. Meanwhile, models were developed to extrapolate results from the site scale to the regional or global scale. Soil is one of the major sources of the four trace gases. Under cultivated conditions, agricultural soils are subject to a great deal of anthropogenic disturbance including tillage, fertilization, irrigation, manure amendment, weeding, and liming. Anthropogenic activities elevate soil trace gas emissions and, hence, play an important role in the atmospheric balance of the trace gases. Various models, such as CASA (Potter et al., 1993), CENTURY (Parton et al., 1996), ExpertN (Baldioli et al., 1994), Hole-in-the-Pipe (Firestone and Davidson, 1989), NLOOS (Riley & Matson 1989), and others were developed for scaling up gas emission estimates. Each of the models has its own strategy or philosophy. Some models tried to use the least number of input parameters and more empirical equations to capture basic patterns of gas fluxes so that these models could be easily used at the regional or global scale. Some models tried to include more mechanisms to better track processes affecting gas production/consumption. To join the modeling efforts, a University of New Hampshire-based biogeochemical research group developed a process-oriented model to predict NO, N_2O, CH_4, and NH_3 emissions from agricultural ecosystems. Several papers have reported on the early development of the model, focusing only on N_2O and CO_2 (Li et al., 1992a; 1994). This paper discusses the latest research progress including simulations of NO, CH_4, and NH_3.

Model framework

Emissions of NO, N_2O, CH_4 and NH_3 are highly variable in space and time. The challenges of modeling the trace gas emissions come from three aspects: (1) some of the gases (e.g., NO and N_2O) have multiple sources (e.g., nitrification, denitrification, and chemo-denitrification); (2) all the gases are produced and consumed simultaneously in the soils, controlled by the kinetics of a series of geochemical or biochemical reactions; and (3) there are a large number of environmental variables driving the biogeochemical reactions.

To construct a process model of soil trace gases, all the factors including ecological drivers, soil environmental variables, and biogeochemical reactions should be integrated into one framework. To handle such a complex system, we adopted the concept of a biogeochemical field for our modeling practice. Paralleling the concept of biogeochemical cycle which describes the transport and transformation of the chemical elements, biogeochemical field answers what controls the elements' behavior. A biogeochemical field is an assembly of the spatially and temporally differentiated environmental forces that drive biogeochemical reactions in an ecosystem. For example, the biogeochemical field driving NO, N_2O, CH_4 and NH_3-relevant reactions consists of the environmental forces deriving from soil temperature, moisture, pH, Eh, substrate concentration, and other soil environmental factors. All the soil environmental factors are further controlled by several ecological drivers including climate, soil physical properties, vegetation, and anthropogenic activity. All the impacts in the system can be categorized into two groups. The first group includes the impacts of ecological drivers on soil environmental variables; the second includes the impacts of the soil environmental variables on trace gas-related geochemical or biochemical reactions (Figure 1). The goal of our modeling efforts was to build the two groups of impacts in a model framework.

The denitrification-decomposition (DNDC) model was constructed with two components. The first component, consisting of the soil climate, crop growth and decomposition submodels, predicts soil temperature, moisture, pH, redox potential (Eh), and substrate concentration profiles based on ecological drivers (e.g., climate, soil, vegetation, and anthropogenic activity). The second component, consisting of the nitrification, denitrification and fermentation submodels, predicts NO, N_2O, CH_4, and NH_3 fluxes based on the soil environmental variables. Classical laws of physics, chemistry, or biology or empirical equations generated from laboratory observations were used in the model to parameterize each specific reaction. The entire model forms a bridge between trace gas emissions and basic ecological drivers (Figure 2).

Linking ecological drivers to soil environmental variables

The first task in model development was to set links between ecological drivers and soil environmental variables. Usually, ecological drivers collectively affect soil environmental variables. Since the combination of ecological drivers in each ecosystem is unique, DNDC needs site-specific input data of climate, soil, vegetation, and farming practices for the simulated agricultural land. DNDC integrates the ecological drivers in the three submodels to generate their collective effects on soil temperature, moisture, pH, Eh, and substrate concentrations. The soil climate submodel calculates soil temperature, moisture, and Eh profiles by integrating air temperature, precipitation, soil thermal and hydraulic properties, and oxygen status. By integrating crop characters, climate, soil properties, and farming practices, the plant growth submodel simulates plant

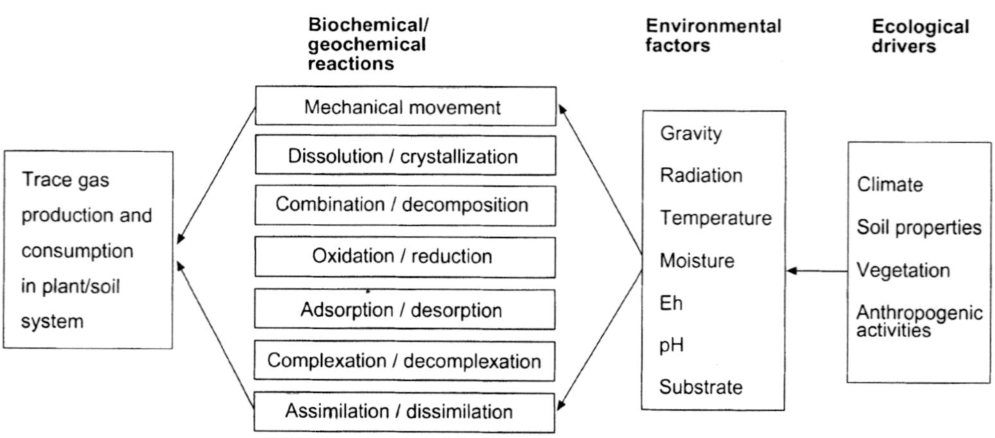

Figure 1. A biogeochemical model is a mathematical expression of biogeochemical field which consists of spatially and temporally differentiated environmental forces driving a series of biogeochemical reactions in ecosystems. Fluxes of NO, N_2O, CH_4, and NH_3 are regulated by directions and rates of the relevant biogeochemical reactions

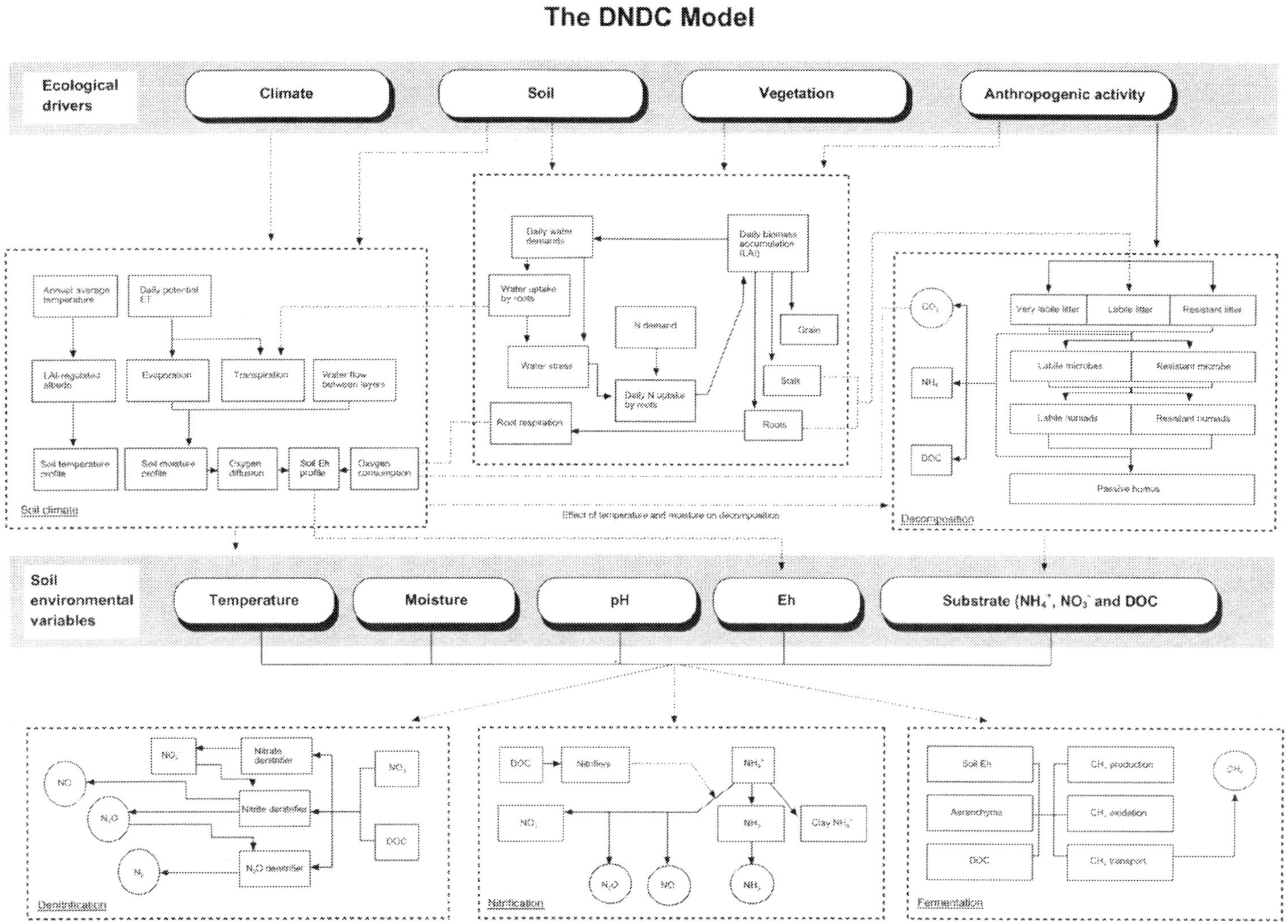

Figure 2. The DNDC model includes two components. The first component, consisting of the soil climate, plant growth, and decomposition submodels, predicts effects of climate, soil physical properties, vegetation, and anthropogenic activity on soil temperature, moisture, pH, Eh, and substrate concentration profiles. The second component, consisting of the nitrification, denitrification, and fermentation submodels, predicts NO, N_2O, CH_4 and NH_3 fluxes through simulating impacts of soil environmental conditions on the relevant geochemical and biochemical reactions

growth and its effects on soil temperature, moisture, pH, Eh, dissolved organic carbon (DOC), and available N concentrations. The decomposition submodel simulates concentrations of substrates (e.g., DOC, NH_4^+, and NO_3^-) by integrating climate, soil properties, plant effect, and farming practices. The three submodels interact with each other to finally determine soil temperature, moisture, pH, Eh, and substrate concentrations in the soil profile at a daily time step. Most of the equations used in this component have been reported in previous papers (see details in Li et al., 1992a; 1994; 1999).

Linking soil environmental factors to trace gases

As the second step for developing the DNDC model, we linked soil environmental variables to production and consumption rates of trace gases. The links were set up based on either the basic physical, chemical, or biological laws, or equations obtained from the experiments under controlled conditions so that the effect of each soil variable could be distinguished.

NO and N_2O

Biological oxidation/reduction dominates NO and N_2O evolution in soils. Nitrification (i.e., microbial oxidation of ammonium) has been observed to be the main source of NO and N_2O under aerobic conditions (Equation 1). Based on the observations reported by Hooper & Terry (1979), Bremner et al. (1980), Chalk & Smith (1983), Tiedje (1988), Sexstone et al. (1985), Anderson & Levine (1986), Papen et al. (1982), Davidson (1992), Hutchinson & Davidson (1993), and Bollmann & Conrad (1998), N_2O or NO production is proportional to nitrification rates, although the pathways remain unknown. The factors controlling nitrification have been determined to be soil temperature, moisture, pH, and NH_4^+ concentration (Johansson & Granat, 1984; Johansson, 1984; Slemr & Seiler, 1984; Williams et al., 1987; Anderson & Levine, 1987; Anderson & Poth, 1989; Valente & Thornton, 1993; Martin et al., 1998; Alexander, 1977; Saad & Conrad, 1993; Ingwerson et al., 1998; Davidson, 1992a, Bock et al., 1986; Ward, 1987). Relationships between environmental factors and nitrification rates were generalized from the observations and employed in the DNDC model. The model predicts nitrification rate by tracking nitrifier activity and NH_4^+ concentration (see equations 1.1-1.6 in the Appendix). Following Blagodatsky & Richter (1998) and Blagodatsky et al. (1998), growth and death rates of NH_4^+ oxidizers are calculated based on DOC concentration, temperature, and moisture. Many observations indicated that nitrification-induced NO or N_2O was a fraction of nitrification rate (Van Niel, 1991; Baumgartner & Conrad, 1992), and the fraction was related to temperature (Johansson & Granat, 1984; Johansson, 1984; Slemr & Seiler, 1984; Williams et al., 1987; Anderson & Levine, 1987; Anderson and Poth, 1989; Slemr & Seiler, 1991; Valente & Thornton, 1993; Martin et al., 1998). DNDC calculates nitrification-induced NO or N_2O production as a function of the predicted nitrification rate and temperature (equations 1.7 and 1.8 in the Appendix).

$$\text{Nitrification: } NH_4^+ \rightarrow \underset{\underset{NO}{\downarrow}}{H_2NOH} \rightarrow \underset{\underset{N_2O}{\downarrow}}{NOH} \rightarrow NO_2^- \rightarrow NO_3^- \quad (1)$$

Denitrification is another main source of N_2O and NO from soils. Denitrification includes a sequential reduction of nitrate to dinitrogen (N_2) driven by denitrifying bacteria under anaerobic conditions (equation 2) (Firestone et al., 1980; Payne, 1981; Anderson & Levine, 1986; Poth & Focht, 1985; SSSA, 1987). Based on field and laboratory observations, denitrification rates are controlled by soil moisture and Eh (Matsubara, 1971; Payne, 1973; Payne et al., 1971; Goreau et al., 1980; Knowles, 1982; Smith, 1980, 1990; Davidson & Schimel, 1995; Stevens et al., 1998), temperature (Nömmik,1956; Stanford, 1975; Bailey & Beauchamp, 1973; Dawson & Murphy, 1972), pH (Wijler & Delwiche, 1954; Khan & Moore, 1968; Focht, 1974; Klemedtsoon et al., 1988; Blackmer & Bremner, 1978; Firestone et al., 1980; Leffelaar & Wessel, 1988; Ashby et al., 1998), and substrate (e.g., DOC, NO_3^-, NO_2^-, NO, and N_2O) concentrations.

$$\text{Denitrification: } NO_3^- \rightarrow NO_2^- \rightarrow NO \rightarrow N_2O \rightarrow N_2 \quad (2)$$

The DNDC model simulates relative growth rates of nitrate, nitrite, NO, and N_2O denitrifiers based on soil Eh, concentrations of DOC, and nitrogen oxides. A simple scheme of "anaerobic balloon" was developed in the model to divide the soil matrix into aerobic and anaerobic parts. Tracking oxygen diffusion and consumption in the soil profile, DNDC simulates swelling and shrinking of the "anaerobic balloon." Only the substrates allocated in the anaerobic part are involved in denitrification (see details in Li et al., 1999). Following Bader (1978), a simple function describing multinutrient-dependent growth has been set in the model to calculate relative growth rates of the denitrifiers (equations 2.1, 2.2, and 2.3 in the Appendix). Death rate of denitrifiers is simply a constant fraction of the total denitrifier biomass (equation 2.3 in the Appendix). Following Leffelaar and Wessel (1988), we assume that the relative growth rates for denitrifiers with different substrates are independent, and competition among the bacteria takes place via the common DOC substrate. The Pirt equation is used to calculate consumption rates of the substrates (equations 2.4 and 2.5 in the Appendix). Since denitrification is a typical sequential reaction, we followed the basic laws of sequential chemical kinetic reactions to calculate NO, N_2O, and N_2 fluxes. As an intermediate of the reactions, NO or N_2O flux is determined by the rates of its production, consumption, and escape from the reacting system. A simplified equation was set in DNDC to calculate diffusion rates of NO and N_2O in the soil matrix. The predicted diffusion rate is a function of soil porosity, moisture, temperature, and clay content (equation 2.6 in the Appendix).

CH_4

Methane is an end product of the biological reduction of carbon dioxide (CO_2) or organic carbon under anaerobic conditions (equation 3) (Wassmann et al., 1993; Cleemput & El-Seboay, 1985; Zeikus, 1977; Yagi & Minami, 1990; Watanabe et al., 1993; Holland & Schimel, 1994; Zhou et al., 1994; Nouchi et al., 1994; Takai, 1970; Kimura et al., 1992; Kludze & Delaune, 1995; Li et al., 1993). According to the observations obtained from field or laboratory studies, CH_4 fluxes were strongly controlled by soil available carbon (i.e., DOC) content (Tao et al., 1994; Shangguan, 1994; Chen et al., 1992; Cicerone et al., 1992; Cai et al., 1995; Schütz et al., 1989; Wassmann et al., 1993; De Groot & Vermoessen, 1991; Inubushi et al., 1984; Sass et al., 1991; Van Vee & Paul, 1981), soil Eh (Takai, 1956; Oremland, 1988; Schipper & Reddy, 1996; Kludze & DeLaune, 1995; Masscheleyn et al., 1993), and soil temperature (Conrad et al., 1987; Vogels et al., 1988; Conrad, 1989; Yagi et al., 1990; Parashar et al., 1993; Wang et al., 1993). The reduction of available carbon to CH_4 is mediated by anaerobic microbes (e.g., methanogens) that are only active when the soil redox potential is low enough (Wassmann et al., 1993, Sass et al., 1991). According to field observations by Kludze & DeLaune (1994), Wang et al. (1993), and Masscheleyn et al. (1993), CH_4 production increased exponentially with decreasing Eh with a threshold range of –150 to –200 mV. Methane production increased with increasing temperature, with an optimum range of 30–40 °C. Based on the observations, DNDC calculates CH_4 production rate as a function of DOC content and temperature as soon as the predicted soil Eh reaches –150 mV or lower (equation 3.1 in Appendix).

$$\text{Methane production: } CO_2 + 8\,H^+ \rightarrow CH_4 + 2\,H_2O \quad \text{or} \quad \text{Organic C} + 4\,H^+ \rightarrow CH_4 \qquad (3)$$

Methane is oxidized by aerobic methanotrophs in the soil. Several researchers reported that 50–80% of CH_4 produced was oxidized in the same soil (Schütz et al., 1989; Holzapfel-Pschorn et al., 1985; Sass et al., 1991; Shangguan et al., 1993; Schipper & Reddy, 1996). Researchers assumed that CH_4 produced at low Eh soil microsites could diffuse into high Eh microsites (e.g., the topsoil or the soil around roots), and hence be oxidized rapidly under higher redox conditions (DeBont et al., 1978; Holzapfel-Pschorn et al., 1985; Schütz et al., 1989; Schipper & Reddy, 1994, 1996). DNDC calculates CH_4 oxidation rate as a function of soil CH_4 concentration and Eh (equation 3.2 in the Appendix). A highly simplified scheme was employed in DNDC to model CH_4 diffusion between soil layers based on CH_4 concentration gradients, temperature, and porosity in the soil (equation 3.5 in the Appendix).

Many researchers reported that plant-mediated transport dominated CH_4 emissions from the soil into the atmosphere (Kludze & DeLaune, 1995; Schütz et al., 1989; Nouchi et al., 1994; Cicerone & Shetter, 1981). Linear relationships between CH_4 emissions and crop aboveground biomass during the growing season have been observed by Sass et al. (1990) and Whiting et al. (1991). DNDC predicts plant-transported CH_4 flux as a function of CH_4 concentration and plant aerenchyma (equation 3.3 in the Appendix). If the soil is unvegetated or the plant aerechyma is not well developed yet, ebullition plays a major role in CH_4 emissions (Nouchi, 1994; Schütz et al., 1989; Chanton et al., 1989; Kelley et al., 1990; Byrnes et al., 1995). In DNDC, we assume that ebullition only occurs at the surface layer, and ebullition rate is regulated by soil CH_4 concentration, temperature, porosity, and plant aerenchyma (equation 3.4 in the Appendix).

NH_3

Soil NH_3 concentration is directly regulated by a chemical reaction occurring in the soil liquid phase:

$$[NH_4^+] + [OH^-] = [NH_{3\,(liquid)}] + H_2O \quad (4)$$

where $[NH_4^+]$ is ammonium concentration, $[OH^-]$ is hydroxide ion concentration, and $[NH_{3\,(liquid)}]$ is ammonia concentration in soil water.

DNDC calculates $NH_{3\,(liquid)}$ concentration based on NH_4^+ and OH^- concentrations (equation 4.1 in the Appendix). NH_4^+ concentration in the soil profile is calculated by the decomposition submodel. The submodel calculates turnover rates of soil organic matter at a daily time step (Li et al., 1992a). OH^- concentration is determined by soil pH and temperature based on Stumm and Morgan (1981). The concentration of NH_3 in the soil gas phase is proportional to the NH_3 concentration in the liquid phase as well as soil temperature (Glasstone, 1946; Sutton et al., 1993). We assume that daily emitted fraction of the gas phase NH_3 is related to the soil air-filled porosity and clay content due to their effects on NH_3 gas diffusion (equation 4.2 in the Appendix).

Based on field observations by Hooker et al. (1980) and Parton et al. (1988), ambient NH_3 can be absorbed and metabolized by the plants. Plant absorption rates of NH_3 have been observed to be related to NH_3 concentration in the air around the leaves (Hutchinson, 1972; Hutchinson et al., 1972; Meyer, 1973, Farquhar et al., 1979, 1980; Lockyer & Whitehead, 1986), N shortage in the crops (Harper et al., 1987), leaf surface moisture (Dabney & Bouldin, 1985; Harper et al., 1987; Sutton et al., 1993), and plant-growing stage (Farquhar et al., 1979; Hooker et al., 1980; Schjorring, 1991). A linear relationship between dry NH_3 deposition rates and air NH_3 concentrations was observed by Hutchinson (1972), Meyer (1973), Cowling & Lockyer (1981), Aneja et al. (1986), and Sommer & Jenson (1991). Based on their observations, the concept of N deposition velocity can be represented by the ratio of NH_3 absorption rate ($\mu g\ m^{-2}\ s^{-1}$) to air NH_3 concentration ($\mu g\ m^{-3}$). Reported velocity values range from 0.003 to 0.034 $m\ s^{-1}$ (Cowling & Lockyer, 1981; Aneja et al., 1986; Sommer & Jenson, 1991) for different crops such as grass, maize, snap bean, soybean, oats, and fescue. The maximum value of the range (i.e., 0.034 $m\ s^{-1}$) was adopted in DNDC for calculating NH_3 absorption rate by crops. In addition, factors such as plant N status and leaf surface moisture were also included in the calculation (equation 4.3 in the Appendix). A highly simplified scheme was included in DNDC to calculate NH_3 concentrations in the air between the ground and the top of the canopy, based on the predicted soil NH_3 flux, atmospheric background NH_3 concentration (0.06 ppm, based on Ayers & Gras [1980] and Tsunogai & Ikeuchi [1986]), and degree of closure of the canopy. Farquhar et al. (1979) and Harper et al. (1987) observed NH_3 release from the leaves during the late stages of crop growth. DNDC tracks total N content in the crops during the whole growing season (Li et al., 1994). When the model detects a decrease in the total plant N content, the reduced part will be regarded as the NH_3 flux released from the plants.

The equations describing the effects of soil environmental factors on NO, N_2O, CH_4, and NH_3 were organized into three submodels. The fermentation submodel contains all the CH_4-related equations. This submodel calculates production, oxidation, and transport of CH_4 under submerged conditions. The denitrification submodel contains all the denitrification equations. This submodel calculates production, consumption, and diffusion of N_2O and NO during rainfall, irrigation, or flooding events. Nitrification-related equations are included in the nitrification submodel. As a logical extension of the NH_4^+/ $NH_{3\,(liquid)}$/ $NH_{3\,(gas)}$ equilibrium, functions for NH_3 production and volatilization are also included in the nitrification submodel. The three submodels compose the second component of the DNDC model.

Input and output

Input parameters required by DNDC include daily temperature and precipitation, soil bulk density, texture, organic carbon content, pH, and farming practices (e.g., crop type and rotation, tillage, fertilization, manure amendment, irrigation, flooding, grazing, and weeding). Profiles of soil environmental variables as well as trace gas fluxes are calculated based on the input data. When DNDC is used for regional estimates of trace gas emissions, the model needs the spatially and temporally differentiated input data stored in geographic information system (GIS)-type databases in advance (Li et al., 1996). Based on the input parameters of the ecological drivers, DNDC first predicts daily soil temperature, moisture, Eh, pH, and substrate concentration, and then uses the environmental parameters to drive nitrification, denitrification, CH_4 production/oxidation, and other relevant geochemical or biochemical reactions. Daily emissions of NO, N_2O, CH_4, and NH_3 are finally calculated as their daily net fluxes. Most parts of the model run at a daily time step except the soil climate and denitrification submodels which run at an hourly time step. Output parameters from the model runs are daily soil profiles of temperature, moisture, Eh, pH, and concentrations of total soil organic carbon, nitrate, nitrite, ammonium, urea, ammonia, as well as daily fluxes of CO_2, NO, N_2O, CH_4 and NH_3. All the daily and annual output data are recorded for future use. For the regional version of DNDC, the simulated results are recorded as geographically explicit data in a GIS database.

Model tests

The DNDC model has been tested against several field studies. The old results related to N_2O and soil organic carbon have been published (e.g., Li et al., 1992b; Li et al., 1994; Li, 1997; Frolking, 1998). Here are reported four new cases that were examined recently for NO, N_2O, CH_4 and NH_3, respectively. The characteristics of the four agricultural sites are listed in Table 1.

NO

Fluxes of NO were measured at a winter wheat field in Wu County, Jiangsu Province, China, from 1 Nov 1996 to 9 Feb 1997 by Xunhua Zheng and her colleagues (1998). Urea and farmyard manure (equivalent to 114 kg N ha^{-1}) were applied on 1 Nov 1996. During the first 40 d following the application, high NO fluxes were observed in the fertilized plot but not in the control plot (Figure 3). Predicted results agreed with observed data and indicated that high fluxes were mainly caused by elevated nitrification rates following fertilizer application. In addition, the relatively high temperature in the early days of the experimental period also enhanced urea hydrolysis and nitrifier activity. The temperature effect can also be seen in the control plot.

N_2O

Nitrous oxide was measured by Crill et al. (1998) for two plots, fertilized and unfertilized, in a maize field at La Selva Biological Station in Costa Rica from November 1994 to March 1995. Ammonium nitrate and urea (30-90 kg N ha^{-1}) were applied on 25 Nov and 6 Dec 1994 on the fertilized plot. During the two maize-growing seasons, N_2O fluxes were consistently low in the unfertilized plot. In contrast, in the fertilized plot, high peaks were observed immediately after fertilizer application. Simulation results agreed with observed results showing the same two high peaks as observed in the field (Figure 4), indicating that the surges of N_2O

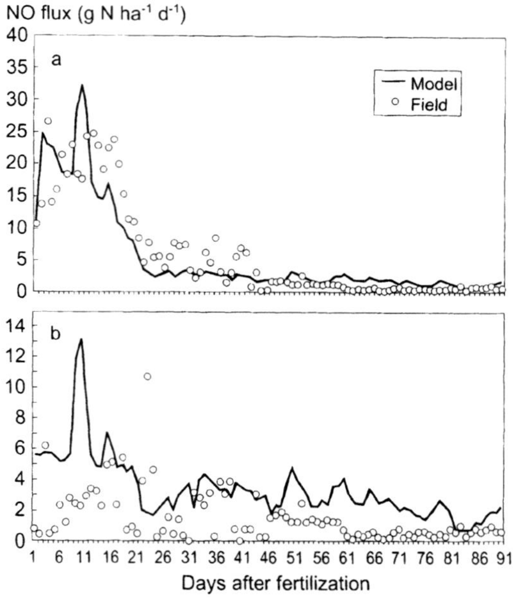

Figure 3. Measured and predicted NO fluxes from fertilized (a) and control (b) plots in a winter wheat field at Wu County, Jiangsu, China, 1 Nov 1996-9 Feb 1997

Table 1. Characteristics of four field sites for model validation tests.

Site	Crop type	Annual average temperature (°C)	Annual precipitation (cm)	Gas measured	Soil texture	Soil organic C	Soil pH
Wu, Jiangsu, China	Winter wheat	17.0	115.6	NO	Clay loam	0.01	7.0
La Selva, Costa Rica	Maize	24.7	438.2	N_2O	Clay	0.028	6.8
Texas, USA	Rice	18.7	99.7	CH_4	Loam	0.02	6.5
Fengqiu, Henan, China	Rice	14.6	64.2	NH_3	Sandy loam	0.0035	8.8

emissions were mainly caused by denitrification. Since soil temperature, moisture, and DOC did not limit denitrifier activity in the plot, additions of inorganic N immediately stimulated denitrification and N_2O emissions. Simulated results indicated that N was a limiting factor in the soil, although nitrification rates were high due to the rapid turnover of soil organic matter. The high demand of plants and soil microbes for N, as well as the strong leaching effect, did not allow NO_3^- or NH_4^+ to accumulate in the topsoil.

CH_4

Ron Sass and his colleagues (1991) measured CH_4 fluxes from two rice field plots, with and without straw amended, at Beaumont, Texas. The measured CH_4 fluxes from the straw-amended plot were almost twice higher than that from the control plot. Field observations indicated that the higher CH_4 production in the amended plot was mainly due to additional available C produced from straw decomposition. Model simulations

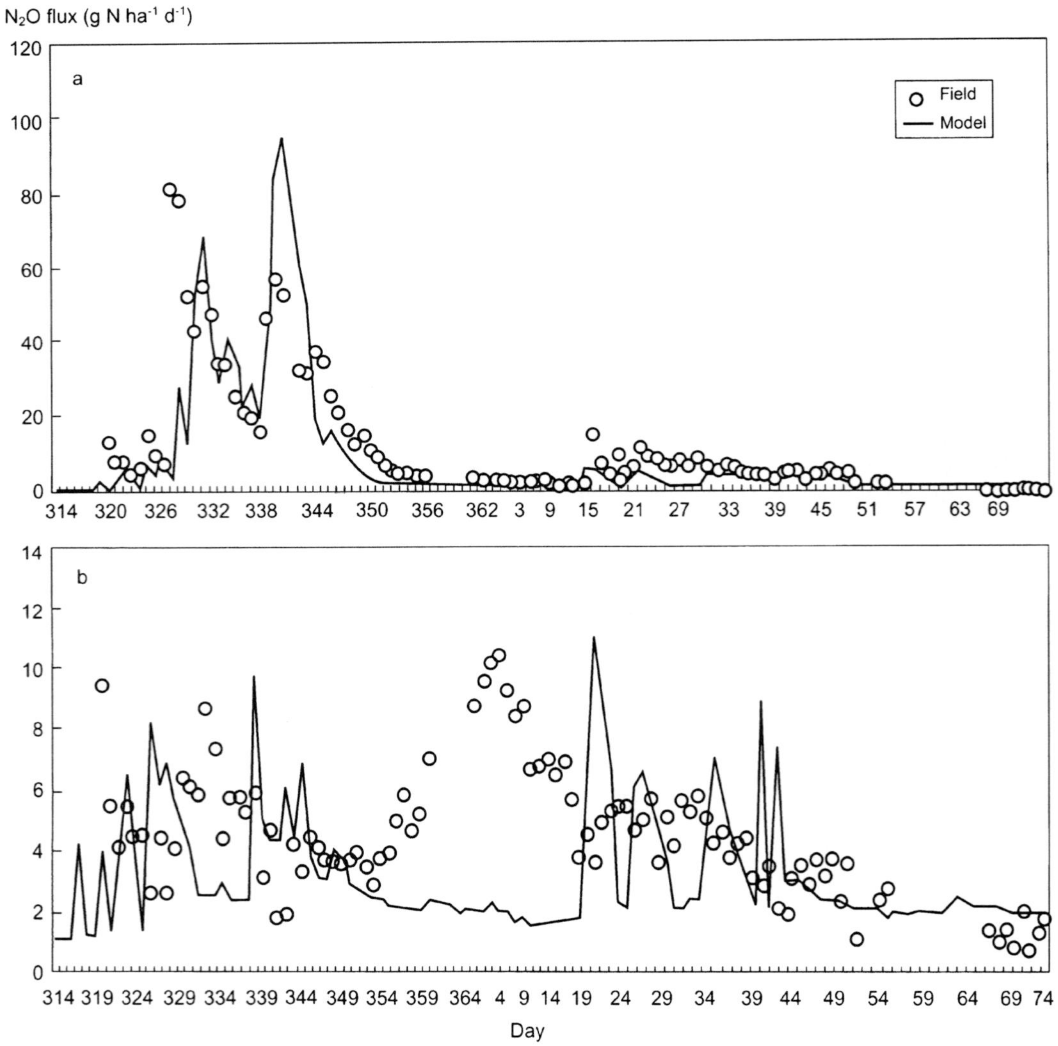

Figure 4. Measured and predicted N_2O fluxes from fertilized (a) and control (b) plots in a maize field at La Selva Biological Station in Costa Rica, November 1994-March 1995

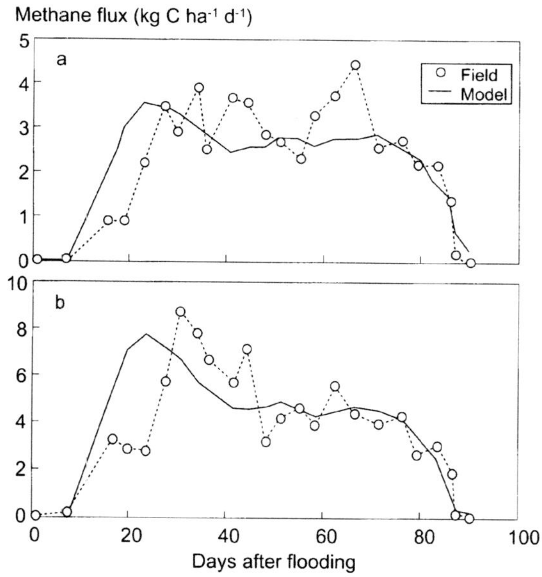

Figure 5. Measured and predicted CH_4 fluxes from control (a) and straw-amended (b) plots in a rice field at Texas A&M University Agricultural Center near Beaumont in Texas, USA, 1989-90

showed similar results (Figure 5). Both measured and model data showed a slight depression of CH_4 emissions in the middle of the growing season. Predicted data showed that the depression was caused by depletion of the labile straw and the undeveloped rice aerenchyma at that time.

NH_3

At a rice field in Fengqiu County, Henan Provice, China, Cai and Zhou (1995) measured NH_3 fluxes from the rice soils. Ammonium bicarbonate and urea were applied at the same rate (90 kg N ha^{-1}) to two plots to test the effect of different fertilizer types on NH_3 emissions. Field measurements were conducted at 4-h intervals for 9 d following fertilizer applications. NH_3 fluxes measured at the ammonium bicarbonate-applied plot were initially very high, and then rapidly decreased to almost zero in the 4 d after fertilizer application. In contrast, at the urea-fertilized plot, NH_3 fluxes were initially low, and gradually increased to a maximum value on the fifth day, and then decreased to a low level 8 d after application. Patterns of NH_3 fluxes observed in the field were simulated by the model (Figure 6). Simulation results showed that the applied ammonium bicarbonate immediately increased NH_3 concentration in the rice field water due to the equilibrium between NH_4^+ and NH_3 in the soil liquid phase. High soil pH (8.8) enhanced NH_3 volatilization from the rice soil. In contrast, it took 4 d for the applied urea to be gradually hydrolyzed. The hydrolysis slowed down NH_3 volatilization in the urea plot.

Simulated results from the four data sets showed that (1) DNDC was able to simulate the basic patterns of NO, N_2O, CH_4, and NH_3 fluxes under various farming conditions; (2) predicted total emissions during the experimental span agreed with the measurements (Table 2); and (3) measured temporal variations in gas

Table 2. Comparison between measured and predicted trace gas emissions

Site	Gas tested	Treatment	Experimental days (no.)	Total flux during experimental span: Measured	Predicted	Unit
Wheat field at Wu County, Jiangsu, China	NO	Fertilized	95	0.53	0.51	kg N ha^{-1}
		Control	95	0.14	0.31	kg N ha^{-1}
Maize field at La selva, Costa Rica	N_2O	Fertilized	125	1.25-1.40	1.17	kg N ha^{-1}
		Control	125	0.29-0.46	0.39	kg N ha^{-1}
Rice field at Texas, USA	CH_4	Amended with straw	90	98.9	93.8	kg C ha^{-1}
Control			90	54.7	53.9	kg C ha^{-1}
Rice field at Fengqiu County, Henan, China	NH_3	Fertilized with ammonium bicarbonate	9	48.3	55.2	kg N ha^{-1}
		Fertilized with urea	9	31.0	31.9	kg N ha^{-1}

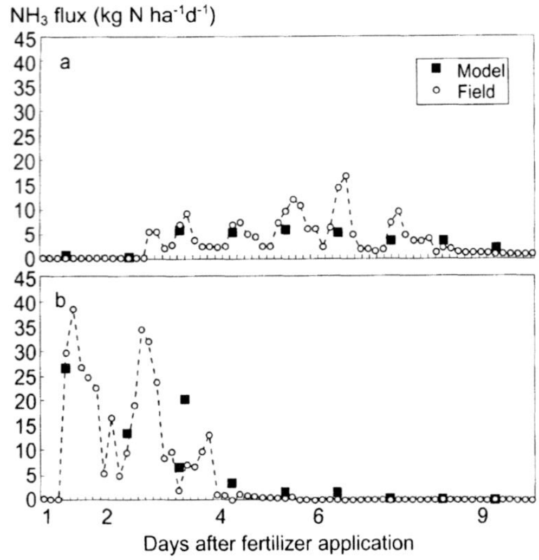

Figure 6. Measured and predicted NH_3 fluxes from urea-fertilized (a) and ammonium bicarbonate-fertilized (b) plots in a rice field at Fengqiu County, Henan, China. The NH_3 fluxes were measured in the field with 4-h intervals although DNDC only predicts daily NH_3 emissions

fluxes can be explained with the equations built in the model.

Discussion

The DNDC model reported in this paper is the result of a 10-yr effort to predict trace gas emissions from agricultural ecosystems. By linking ecological drivers to soil environmental variables, and further, to trace gas-related biogeochemical reactions, DNDC acts as a bridge between ecological drivers and the chemical elements' behavior. During development of the model, we made every effort to incorporate the basic mechanisms or processes into the model, although gaps still exist in almost every component of the model. For example, the highly simplified diffusion equations could have brought large uncertainties to the simulated results. Nevertheless, we hope that we have established a useful tool that can be used not only for synthesizing existing observations obtained by hundreds of researchers during the last several decades but also for testing new hypotheses for future studies. In comparison with other models focusing on a couple of trace gases, DNDC has the advantage of predicting CO_2, NO, N_2O, CH_4 and NH_3 simultaneously. This feature could be valuable in assessing the net effect of the changing climate or alternative agricultural management on either the atmosphere or agriculture. Linked to GIS databases of climate, soil, vegetation, and farming practices, DNDC is ready for regional estimation of trace gas emissions.

Methodology development is also one of the motivations for this modeling effort. Since V.I. Vernatski initiated the concept of biogeochemistry in his famous book *La Geochimie* in 1924, 75 yr have passed. During the first 50 yr of this time period, biogeochemistry, as a scientific discipline, did not develop very fast due to the lack of social demands. Only during the last two decades, when global climate change provided new challenges to the scientific community, did people rediscover the potential of biogeochemistry in integrating the macro processes occurring at the ecosystem level with the micro processes at the molecular or atomic scale. To meet the new demand, we need to develop new methodologies based on biogeochemical concepts or principles. The modeling effort reported in this paper is a continuation of our long-term biogeochemical studies. The strategy and methodologies used in this modeling study have been successfully used in several ecological studies including human health (Li & Yu, 1973) and environmental pollution (BEARG, 1997). The author hopes this paper will fuel more interest in the methodology studies in this interdisciplinary realm.

Acknowledgements

The author thanks Dr. R. Wassmann and Ms. R. S. Lantin for inviting this presentation in the UNDP/GEF Workshop of the Interregional Research Programme on Methane Emission from Rice Fields, 10-15 Aug1998, in Beijing. Thanks are also given to Drs. Zhaohua Dai, Xunhua Zheng, Klaus Butterbach-Bahl, Hans Papen, and Mr. F. Stange for their collaborations and contributions in developing the methane- and nitrification-related equations described in the paper. The author appreciates the kindness of Mr. Andrew Munton and Dr. Arvin Mosier for proofreading the manuscript. The development of the DNDC model was funded by the U.S. Environmental Protection Agency, the National Aeronautics and Space Agency, and the National Science Foundation through the Global Climate Change and the Terrestrial Ecosystem projects from 1989 to 1999.

References

Alexander M (1977) Introduction to Soil Microbiology. Wiley, New York

Anderson IC & Levine JS (1986) Relative rates of nitric oxide and nitrous oxide production by nitrifiers, denitrifiers, and nitrate respirers. Appl Environ Microbiol 51:938–945

Anderson IC & Levine JS (1987) Simultaneous field measurements of biogenic emissions of nitric oxide and nitrous oxide. J Geophys Res 92:965–976

Anderson IC & Poth MA (1989) Semiannual losses of nitrogen as NO and N_2O fron unburned and burned chaparral. Global Biogeochem Cycle 3:121–135

Aneja VP, Rogers HH & Stahel EP (1986) Dry deposition of ammonia at environmental concentrations on selected plant species. J. Air Pollut Contr Assoc 36:1338–1341

Ashby JA, Bowden WB & Murdoch PS (1998) Controls on denitrification in riparian soils in headwater catchments of a hardwood forest in the Catskill Mountains, USA. Soil Biol Biochem 30:853–864

Ayers GP & Gras JL (1980) Ammonia gas concentrations over the Southern Ocean. Nature 284:539–540

Bader FG (1978) Analysis of double-substrate limited growth. Biotechnol Bioeng 20:183–202

Bailey LD & Beauchamp EG (1973) Effects of temperature on NO_3^- and NO_2^- reduction, nitrogeneous gas production, and redox potential in a saturated soil. Can J Soil Sci 53:213–218

Baldioli M, Engel T, Klocking B, Priesack E, Schaaf T, Sperr C & Wang E (1994) xpert-N, ein Baukasten zür Simulation der Stickstoffdynamik in Boden und Pflanze. Prototyp, Benutzerhandbuch, Lehreinheit für Ackerbau und Informatik im Pflanzenbau, TU Munchen, Freising, 1–106

Baumgartner D & Conrad R (1992) Effects of soil variables and season on the production and consumption of nitric oxide in oxic soils. In: Fertility of Soils, pp 166–174, Springer-Verlag

BEARG (Beijing Environmental Assessment Research Group) (1977) Assessment of Environmental Quality in Beijing Western Suburbs. 465 p

Blackmer AM & Bremner JM (1978) Inhibitory effect on nitrate on reduction of N_2O to N_2 by soil microorganisms. Soil Biol Biochem 10:187–191

Blagodatsky SA & Richter O (1998) Microbial growth in soil and nitrogen turnover: a theoretical model considering the activity state of microorganisms. Soil Biol Biochem 30(13):1743–1755

Blagodatsky SA, Vevdokimov IV, Larionova AA & Richter J (1998) Microbial growth in soil and nitrogen turnover: model calibration with laboratory data. Soil Biol Biochem 30(13):1757–1764

Bock E, Koops HP & Harms H (1986) Cell biology of nitrifying bacteria. In: Nitrification: Special Publications of the General Microbiology. Vol. 20. Prosser JI (ed) pp 17–38, IRL Press, Oxford

Bollmann A & Conrad R (1998) Influence of O_2 availability on NO and N_2O release by nitrification and denitrification in soils. Global Change Biol 4:387–396

Bremner JM, Blackmer AM & Waring SA (1980) Formation of nitrous oxide and dinitrogen by chemical decomposition of hydroxylamine in soils. Soil Biol Biochem 12: 263–269

Byrnes BH, Austin ER & Tays BK (1995) Methane emission from flooded rice soils and plants under controlled conditions. Soil Biol Biochem 27:332–339

Cai GX & Zhou ZL (1995) Gaseous loss of fertilizer nitrogen from rice fields. Acta Pedol Sin 32:128–135

Cai ZC, Yan XY, Xu H, Tsuruta H, Yagi K & Minami K (1995) Effect of nitrogen form on CH_4 emission from rice paddy field. Acta Pedol Sin 32:136–143

Chalk PM & Smith CJ (1983) Chemodinitrification. In: Freney JF & Simpson JR (eds) Gaseous Loss of Nitrogen from Plant-Soil Systems, pp 65-89, Martinus Nijhoff Publishers, The Hague

Chanton JP, Martens CS & Kelley CA (1989) Gas transport from methane-saturated tidal freshwater and wetland sediments. Limnol Oceanogr 34:807–819

Chen ZL, Gao JH, Yuan Y, Shao KS, Zhang JB, Yu ZJ & Wang BJ (1992) Study on the effects of different cultivation methods on methane emission from rice fields in Beijing. Res Environ Sci 5:1–7

Cicerone RJ, Delwiche CC, Tyler SC & Zimmerman PR (1992) Methane emission from California rice paddies with varied treatments. Global Biogeochem Cycles 6: 233–248

Crill P, Keller M & Weizt A (1998) Measuring N_2O fluxes from tropical agricultural soils. Global Biogeochem Cycles (in press)

Conrad R (1989) Control methane production in terrestrial ecosystems. Exchange of trace gases between terrestrial ecosystems and atmosphere. Andereae MO & Schimel DS (eds) pp 39-58, J. Wiley & Sons, Chichester

Conrad R, Schütz H & Babbel M (1987) Temperature limitation of hydrogen turnover and methanogenesis in anoxic paddy soil. FEMS Microbiol Ecol 45:281–289

Cowling DW & Lockyer DR (1981) Increased growth of ryegrass exposed to ammonia. Nature 292:337–338

Dabney SM & Bouldin DR (1985) Fluxes of ammonia over an alfalfa field. Agron J 77:572–578

Davidson EA & Schimel JP (1995) Microbial processes of production and consumption of nitric oxide, nitrous oxide, and methane. In: Matson PA & Harriss RC (eds) Biogenic Trace Gases: Measuring Emissions from Soil and Water, pp 327–357, Cambridge University Press, Cambridge

Davidson EA (1992a) Sources of nitric oxide and nitrous oxide following wetting of dry soil. Soil Sci Soc Am J 56:95–102

Davidson EA (1992b) Pulses of nitric oxide and nitrous oxide flux following wetting of dry soil: an assessment of probable sources and importance relative to annual fluxes. Ecol Bull 42:1–7

Dawson RN & Murphy KL (1972) The temperature dependence of biological denitrification. Water Res 6:71–83

De Groot CJ, Vermoesen A & van Cleemput O (1994) Laboratory study of the emission of N_2O and CH_4 from a calcareous soil. Soil Sci 158:355–364

DeBont JAM, Lee KK & Bouldin DF (1978) Bacterial oxidation of methane in a rice paddy. Ecol Bull 26:91–96

Farquhar GD, Firth PM, Wetselaar R & Wier B (1980) On the gaseous exchange of ammonia between leaves and the environment: determination of the ammonia compensation point. Plant Physiol 66:710–714

Farquhar GD, Wetselaar R & Firth PM (1979) Ammonia volatilization from senescing leaves of maize. Science 203:1275–1258

Firestone MK, Firestone RB & Tiedje JM (1980) Nitrous oxide from soil denitrification: factors controlling its biological production. Science 208:749–751

Firestone MK & Davidson EA (1989) Microbiological basis of NO and N_2O production and consumption in soil. In: Andreae MO & Schimel DS (eds) Exchange of Trace Gases between Terrestrial Ecosystems and the Atmosphere, pp 7-21, John Wiley & Sons Ltd., Chichester, UK

Focht DD (1974) The effect of temperature, pH and aeration on the production of nitrous oxide and gaseous nitrogen – a zero-order kinetic model. Soil Sci 118:173–179

Frolking SE, Mosier AR, Ojima DS, Li C, Parton WJ, Potter CS, Priesack E, Stenger R, Haberbosch C, Dorsch P, Flessa H & Smith KA (1998) Comparison of N_2O emissions from soils at three temperate agricultural sites: simulations of year-round measurements by four models. Nutr Cycling Agroecosyst (in press)

Glasstone S (1946) Textbook of Physical Chemistry. 2d ed. Van Nostrand Reinhold, New York, 1320 p

Goreau TJ, Kaplan WA, Wofsy SC, McElroy MB, Valios FW & Watson SW (1980) Production of NO_2^- and N_2O by nitrifying bacteria at reduced concentrations of oxygen. Appl Environ Microbiol 40:526–532

Harper LA, Sharpe RR, Langdale GW & Giddens JE (1987) Nitrogen cycling in a wheat crop: soil. plant and aerial nitrogen transport. Agron J 79:965–973

Hartel PG & Alexander M (1987) Effect of growth of bacteria in freshly moistened soil. Soil Sci Am J 51:93-96

Holland EA & Schimel DS (1994) Ecosystem and physiological controls over methane production in northern wetlands. J Geophys Res 99:1563–1571

Holzapfel-Pschorn A, Conrad R & Seiler W (1985) Production, oxidation, and emission of methane in rice paddies. FEMS Microbiol Ecol 31:343–351

Hooker ML, Sanders DH, Peterson GA & Daigger LA (1980) Gaseous N losses from winter wheat. Agron J 72:789–792

Hooper AB & Terry KR (1979) Hydroxylamine oxidoreductase of *Nitrosomonas*: production of nitric oxide from hydroxylamine. Biochim Biophys Acta 571:12–22

Hutchinson GL & Davidson EA (1993) Processes for production and consumption of gaseous nitrogen oxides in soil. In: Harper LA, Mosier AR & Duxbury JM (eds) Agricultural Ecosystems on Trace Gases and Global Climate Change, pp 79–93

Hutchinson GL (1972) Air containing nitrogen-15 ammonia: foliar absorption by corn seedings. Science 175:759–761

Hutchinson GL, Millington RJ & Peters DB (1972) Atmospheric ammonia: absorption by plant leaves. Science 175: 771–772

Ingwerson J, Butterbach-Bahl K, Gasche R, Richter O & Papen H (1998) Barometric process separation (BaPS): new method for quantifying nitrification, denitrification, and N_2O sources in soils (submitted)

Inubushi K, Wada H & Takai Y (1984) Easily decomposable organic matter in paddy soil. IV. Relationship between reduction and organic matter decomposition. Soil Sci Plant Nutr 30:189–198

Johansson C (1984) Field measurements of emission of nitric oxide from fertilized and unfertilized forest soils in Sweden. J Atmos Chem 1:429–442

Johansson C & Granat L (1984) Emission of nitric oxide from arable land. Tellus 36B:26–37

Kelley CA, Martens CS, Chanton JP (1990) Variations in sedimentary carbon remineralization rates in the White Oak River estuary. NC Limnol Oceanogr 35:372–383

Khan MFA & Moore AW (1968) Denitrification capacity of some Alberta soils. Can J Soil Sci 48:89–91

Kimura M, Minoda T & Murase J (1993) Water-soluble organic materials in paddy soil ecosystem. II. Effects of temperature on contents of total organic materials, organic acids, and methane in leachate from submerged paddy soils amended with rice straw. Soil Sci Plant Nutr 39:713–724

Klemedtsson L, Svensson B & Rosswal T (1988) A method of selective inhibition to distinguish between nitrification and denitrification as a source of nitrous oxide in soil. Biol Fertil Soils 6:112–119

Kludze HK & DeLaune RD (1995) Gaseous exchange and wetland plant response to soil redox intensity and capacity. Soil Sci Soc Am J 59:939–945

Knowles R (1982) Denirification. Microbiol Rev 46:43–70

Leffelaar PA & Wessel WW (1988) Denitrification in a homogeneous closed system: experiment and simulation. Soil Sci 146:335–349

Li C & Yu Z (1973) An environmental model and its application for an endemic human heart failure Keshan Disease in Heilongjiang Province, China. Environ Geol Health 1:9–26

Li C, Frolking S & Frolking TA (1992a) A model of nitrous oxide evolution from soil driven by rainfall events: 1. Model structure and sensitivity. J Geophys Res 97:9759–9776

Li C, Frolking S & Frolking TA (1992b) A model of nitrous oxide evolution from soil driven by rainfall events. 2. Model applications. J Geophys Res 97:9777–9783

Li C, Frolking S & Harriss R (1994) Modeling carbon biogeochemistry in agricultural soils. Global Biogeochem Cycles 8:237–254

Li. C, Narayanan V & Harriss R (1996) Model estimates of nitrous oxide emissions from agricultural lands in the United States. Global Biogeochem Cycles 10:297–306

Li C, Frolking S, Crocker GJ, Grace RR, Klir J, Korchens M & Poulton PR (1997) Simulating trends in soil organic carbon in long-term experiments using the DNDC model. Geoderma 81:45–60

Li C, Aber J, Stange F, Butterbach-Bahl K & Papen H (1999) A process oriented model of N_2O and NO emissions from forest soils. 1. Model development. J Geophys Res (submitted)

Li DB, Zhang JW, Li WX, Yue ZT, Yu F & Gu JN (1993) Effect of agricultural methods on methane emission from rice paddies. Rural Eco-Environ 9:13–18

Lockyer DR & Whitehead DC (1986) The uptake of gaseous ammonia by the leaves of Italian Ryegrass. J Exp Bot 37:919–927

Martin RE, Scholes MC, Mosier AR, Pjima DS, Holland EA & Parton WJ (1998) Controls on annual emissions of nitric oxide from soils of the Colorado shortgrass steppe. Global Biogeochem Cycles 12:81–91

Masscheleyn PH, DeLaune RD & Patrick Jr WH (1993) Methane and nitrous oxide emission from laboratory measurements of rice soil suspension: effect of soil oxidation-reduction status. Chemosphere 26:251–260

Matsubara T (1971) Studies on denitrification. XIII. Some properties of the N_2O -anaerobically grown cell. J Biochem (Tokyo) 69:991–1001

Meyer MW (1973) Absorption and release of ammonia from and to the atmosphere by plants. Dissertation directed by J.H. Axley, University of Maryland, College Park

Nömmik H (1956) Investigations on denitrification in soil. Acta Agric Scand 6:195–228

Nouchi I, Hosono T, Aoki K & Minami K (1994) Seasonal variation in methane flux from rice paddies associated with methane concentration in soil water, rice biomass and temperature, and its modelling. Plant Soil 161:195–208

Oremland RS (1988) Biogeochemistry of methanogenic bacteria. In: Zehnder AJB (ed) Biology of Anaerobic Microorganisms. John Wiley & Sons, pp 641–705

Papen H, von Berg R, Hinkel I, Thoene B & Rennenberg H (1982) Heterotrophic nitrification by *Alcaligenes faecalis*: NO_2^-/ NO_3^-. and NO production in exponentially growing cultures. Appl Environ Microbiol 55:2068–2072

Parashar DC, Gupta PK, Rai J, Sharma RC & Singh N (1993) Effect of soil temperature on methane emission from paddy fields. Chemosphere 26:247–250

Parton WJ, Morgan JA, Altenhofen JM & Harper LA (1988) Ammonia volatilization from spring wheat plants. Agron J 80:419–425

Parton WJ, Mosier AR, Ojima DS, Valentine DW, Schimel DS, Weier K & Kulmala AE (1996) Generalized model for N_2 and N_2O production from nitrification and denitrification. Global Biogeochem Cycles 10:401–412

Payne WJ (1973) Reduction of nitrogenous oxides by microorganisms. Bacteriol Rev 37:409–452

Payne WJ (1981) The status of nitric oxide and nitrous oxide as intermediates in denitrification. In: Delwiche CC (ed) Denitrification, Nitrification and Atmospheric Nitrous Oxide, pp 85–103, Wiley, New York

Payne WJ, Riley PS & Cox CD (1971) Separate nitrite, nitric oxide, and nitrous oxide-reducing fractions from *Pseudomonas perfectomarinus*. J Bacteriol 106:356–361

Poth M & Focht DD (1985) 15-N kinetic analysis of N_2O production by *Nitrosomonas europaea*: an examination of nitrifier denitrification. Appl Environ Microbiol 49: 1134–1141

Potter CS, Randerson JT, Field CB, Matson PA, Vitousek PM, Mooney HA & Klooster SA (1993) Terrestrial ecosystem production: a process model based on global satellite and surface data. Global Biogeochem Cycles 7:811–841

Riley WJ & Matson PA (1989) The NLOSS model. 1998 Fall AGU meeting. Poster A41B-25. AGU volume 79, Number 45, 10 Nov 1998

Saad OALO & Conrad R (1993) Temperature dependence of nitrification. denitrification. and turnover of nitric oxide in different soils. Biol Fertil Soils 15:21–27

Sass RL, Fisher FM, Harcombe PA & Turner FT (1990) Methane production and emission from a Texas rice field. Global Biogeochem Cycles 4:47–68

Sass RL, Fisher FM, Harcombe PA & Turner FT (1991) Mitigation of methane emissions from rice fields: possible adverse effects of incorporated rice straw. Global Biogeochem Cycles 5: 275–287

Schütz H, Seiler W & Conrad R (1989) Processes involved in formation and emission of methane in rice paddies. Biogeochemistry 7:33–53

Schipper LA & Reddy KR (1994) Methane production and emission from four reclaimed and pristine wetlands of southeastern United States. Soil Sci Soc Am J 58:1270–1275

Schipper LA & Reddy KR (1996) Determination of methane oxidation in the rhizosphere of *Sagittaria lancifolia* using methyl fluoride. Soil Sci Soc Am J 60: 611–616

Schjorring JK (1991) Ammonia emission from the foliage of growing plants. In: Sharkey TD, Holland EA & Mooney HA (eds) Trace Gas Emissions by Plants, pp 267–292, Academic Press, San Diego, CA

Sexstone AJ, Revsbech NP, Parking TB & Tiedie JM (1985) Direct measurement of oxygen profiles and denitrification rates in soil aggregates. Soil Sci Soc Am J 49: 645–651

Shah DB & Coulman GA (1978) Kinetics of nitrification and denitrification reactions. Biotechnol Bioeng 20:43-72

Shangguan XJ, Wang MX, Shen RX, Wang YS, Wassmann R, Rennenberg H, Seiler, W, Xie XL, Wang WD & Xie KH (1994) The feature of methane emission from a paddy field in the central China region. Sci Atmos Sin 18:358-365

Shangguan XJ, Wang MX, Shen RX, Wang YS, Wassmann R, Rennenberg H, Seiler, W, Xie XL, Wang WD & Xie KH (1994) The feature of methane emission from a paddy field in the central China region. Sci Atmos Sin 18:358–365

Slemr F & Seiler W (1984) Field measurements of NO and NO_2 emissions from fertilized and unfertilized soils. J Atmos Chem 2:1–24

Slemr F & Seiler W (1991) Field study of environmental variables controlling the NO and N_2O emissions from soil and of the NO and N_2O compensation points. J Geophys Res 96:13017–13031

Smith KA (1980) A model of the extent of anaerobic zones in aggregated soils and its potential application to estimates of denitrification. J Soil Sci 31:263–277

Smith KA (1990) Anaerobic zones and denitrification in soil: modelling and measurement. In: Revsbech NP & Sorensen J (eds) Denitrification in Soil and Sediment, pp 229-243, Plenum Press, New York

SSSA (Soil Science Society of America) (1987) Glossary of soil science terms. SSSA, Madison, WI, USA, 11 p

Sommer SG & Jeneson ES (1991) Foliar absorption of atmospheric ammonia by ryegrass in the field. J Environ Qual 20:153–156

Stanford G, Dzienia S & Vander Pol RA (1975) Effect of temperature on denitrification rate in soils. Soil Sci Soc Am Proc 39:867–870

Stevens RJ, Laughlin RJ & Malone JP (1998) Soil pH affects reducing nitrate to nitrous oxide and di-nitrogen. Soil Biol Biochem 30:1119–1126

Stumm W & Morgan JJ (1981) Aquatic Chemistry: An Introduction Emphasizing Chemical Equilibia in Natural Waters. 2d ed, pp 418-503, John Wiley & Sons, New York

Sutton MA (1990) The surface/atmosphere exchange of ammonia. PhD thesis, Institute of Ecology and Resource Management, University of Edinburgh, Edinburgh

Sutton MA, Pitcairn CER & Fowler D (1993) The exchange of ammonia between the atmosphere and plant communities. In: Advances in Ecological Research. Vol. 24, pp 301-393, Academic Press Limited.

Takai Y (1970) The mechanism of methane formation in flooded paddy soil. Soil Sci Plant Nutr 16:238–244

Tao Z, Du DD, Zhou Y, Mai GX & Liu XW (1995) Effect of different agricultural methods on methane emission from rice paddy. Agro-Environ Prot 14:101–104

Tiedje JM (1988) Ecology of denitrification and dissmilatory nitrate reduction to ammonium. In: Zehnder AJB (ed) Biology of Anaerobic Microorganisms. pp 179-244, Wiley, New York

Tsunogai S & Ikeuchi K (1986) Ammonia in the atmosphere. Geochemistry J 2:157–166

Valente RJ & Thornton FC (1993) Emissions of NO from soil at a rural site in Central Tennessee. J Geophys Res 98:16745–16753

Van Cleemput O & El-Sebaag AS (1985) Gaseous hydrocarbons in soil. Adv Agron 38:181–195

Van Niel EWJ (1991) Nitrification by heterotrophic denitrifiers and its relationship to autotrophic nitrification. PhD thesis, Delft University of Technology, Delft

Van Veen JA & Paul EA (1981) Organic carbon dynamics in grassland soils. 1. Background information and computer simulation. Can J Soil Sci 61:185–201

Van Verseveld HW, Meijer EM & Stouthamer AH (1977) Energy conservation during nitrate respiration in *Paracoccus denitrificans*. Arch Microbiol 112:17-23

Van Verseveld HW & Stouthamer AH (1978) Growth yields and the efficiency of oxidative phosphorylation during autotrophic growth of *Paracoccus denitrificans* on methanol and formate. Arch Microbiol 118:21-26

Vernatski VI (1924) La Geochimie, P Revue scient

Vogels GD, Keltjens JT & van der Drift C (1988) Biochemistry of methane production. In: Zehnder AJB (ed) Biology of Anaerobic Microorganisms, pp 707-770. John Wiley & Sons, New York

Wang ZP, DeLaune RD, Masscheleyn PH & Patrick Jr WH (1993) Soil redox and pH effects on methane production in a flooded rice soil. Soil Sci Soc Am J 51:382–385

Ward BB (1987) Kinetic studies on ammonia and methane oxidation by *Nitrosococcus oceanus*. Arch Microbiol 147:126–133

Wassmann R, Papen H & Rennenberg H (1993) Methane emission from rice paddies and possible mitigation strategies. Chemosphere 26:201–217

Watanabe A, Katoh K & Kimura M (1993) Effect of rice straw application on CH_4 emission from paddy fields. I. Effect of weathering of rice straw in the field during off-crop season. Soil Sci Plant Nutr 39:701–706

Whiting GJ, Chanton JP, Barlett DS & Happell JD (1991) Relationships between CH_4 emission. biomass and CO_2 exchange in a subtropical grassland. J Geophys Res 96: 13067–13071

Wijler J & Delwiche CC (1954) Investigations on the denitrifying process in soil. Plant Soil 5:155–169

Williams EJ, Parrish DD & Fehsenfeld FC (1987) Determination of nitrogen oxide emissions from soils: results from a grassland site in Colorado, United States. J Geophys Res 92:2173–2179

Yagi K, Minami K & Breitenbeck GA (1990) Emission and production of methane from paddy fields. Transactions of the 14th International Congress of Soil Science, pp 238–243, International Society of Soil Science, Kyoto

Zeikus JG (1977) The biology of methanogenic bacteria. Bacteriol Rev 41:514–541

Zheng XH, Wang YS, Wang MX, Shen RX, Li LT & Jin JS (1998) Automated measurements of NO emissions from cropland. Environ Sci 19(2) (in press)

Zhou Y, Tao Z & Du DD (1994) Agricultural measure options for reducing methane emission from rice field. Rural Eco-Environ 10:6–8

Appendix: Equations and parameters

1. Nitrification

Equation 1.1. Relative growth rate of nitrifiers

$dG/dt = 0.0166 * (DOC / (1.0 + DOC) + Fm / (1.0 + Fm));$

Equation 1.2. Relative death rate of nitrifiers

$dD/dt = 0.008 * BIO * 1.0 / (1.0 + DOC) / (1.0 + Fm);$

Equation 1.3. Net increase in nitrifier biomass

$dBIO/dt = (dG/dt - dD/dt) * BIO * Ft * Fm;$

$BIO = \int^{24} dBIO/dt;$

Equation 1.4. Nitrification rate

$Rn = [NH_4+] * (0.005 * BIO) * pH,$ (kg N/ha/day);

Equation 1.5. Temperature factor

$Ft = 3.503^{(60.0-T/(60.0-34.22)} * e^{3.503*(T-34.22)/(60.0-34.22)};$

Equation 1.6. Moisture factor

$Fm = 0.8 + 0.21 * (1.0 - wfps)$, if $wfps > 0.05$;

$Fm = 0$, if $wfps <= 0.05$;

Equation 1.7. Nitrification-induced NO (kg N/ha/d)

$NO = 0.0025 * Rn * Ft;$

Equation 1.8. Nitrification-induced N_2O (kg N/ha/d)

$N_2O = 0.0024 * Rn;$

DOC – Concentration of dissolved organic C, kg C/ha;

BIO – Nitrifier biomass, kg C/ha;

$[NH_4^+]$ – Concentration of ammonium, kg N/ha;

pH – Soil pH.

2. Denitrification

Equation 2.1. Relative growth rate of NOx denitrifiers (1/h)

$GR_{NOx} = GR_{NO_x(max)} * [C / (Kc + C)] * [NOx / (Kn + NOx)];$

Equation 2.2. Relative growth rate of total denitrifiers (1/h)

$GR = Ft * (GR_{NO_3} * PH1 + GR_{NO_2} * PH2 + GR_{NO} * PH3 + GR_{N_2O} * PH4);$

$Ft = a*2^{(T - 22.5) / 10.0};$

Equation 2.3. Denitrifier growth/death and consumption of soluble carbon (kg C/m^3/h)

Growth rate : $(dBIO/dt)_g = GR * BIO(t);$

Death rate: $(dBIO/dt)_d = Mc * Yc * BIO(t);$

Carbon consumption rate: $dC/dt = (GR / Yc + Mc) * BIO(t);$

Equation 2.4. Consumption rates of N oxides (kg N/m^3/h)

$d(Nox)/dt = (GR_{NOx} / Y_{NOx} + M_{NOx} * NOx / N) * BIO(t);$

Equation 2.5. Nitrogen assimilation rate (kg N/m^3/h)

$(dN/dt)_a = (dBIO/dt)_g / CN;$

Equation 2.6. NO, N_2O and N_2 diffusion rates (%)

NO and N_2O: diffuse = (0.0006+0.0013*AD)+(0.013-0.005*AD)*PA*(1-anvf);

N_2: diffuse 0 0.017+((0.025-0.0013*AD)*PA*(1-anvf);

$GR_{NO_3(max)}$ – Maximum growth rate of NO_3^- denitrifiers, 0.67 1/h (Hartel & Alexander, 1987);

$GR_{NO_2(max)}$ – Maximum growth rate of NO_2^- denitrifiers, 0.67 1/h (Hartel & Alexander, 1987);

$GR_{NO(max)}$ – Maximum growth rate of NO denitrifiers, 0.34 1/h (Hartel & Alexander, 1987);

$GR_{N_2O(max)}$ – Maximum growth rate of N_2O denitrifiers, 0.34 1/h (Hartel & Alexander, 1987);

Kc – Half-saturation value of soluble carbon, 0.017 kg C/m^3 (Shah & Coulman, 1978);

Kn – Half-saturation value of N oxides, 0.083 kg N/m^3 (Shah & Coulman, 1978);

C - Soluble C concentration, kg C/m^3 (calculated by DNDC);

NO_x - Concentration of NO_3^-, NO_2^-, NO or N_2O, kg N/m^3 (calculated by DNDC);
PH1, PH2, PH3 = a * (soil pH – b), a=0.4, and b=2.5, 3.0, or 3.5, respectively; factors of impact of pH on NO_3^-, NO_2^-, and NO and N_2O denitrifiers (Focht,1974);
BIO(t)– Denitrifier biomass at time t, kg C /m^3 (calculated by DNDC);
Mc – Maintainance coefficient on carbon, 0.0076 kg C /kg/h (Van Verseveld et al., 1977);
Yc – Maximum growth rate on soluble carbon, 0.503 kg C /kg C (Van Verseveld et al., 1977);
Y_{NO_3} Y_{NO_2} Y_{NO} Y_{N_2O} – Maximum growth rate on NO_3^-, NO_2^-, NO and N_2O, respectively, 0.401, 0.428, 0.151, 0.151 kg C /kg N (Van Verseveld et al., 1977);
$M_{(NO_3)}$ $M_{(NO_2)}$ $M_{(N_2O)}$ $M_{(NO)}$– Maintainance coefficient on NO_3^-, NO_2^-, N_2O and NO, respectively, 0.09, 0.035, 0.079, 0.079 kg N /kg/h (Van Verseveld et al., 1977);
CN – C/N in denitrifiers, 3.45 (Van Verseveld & Stouthamer, 1978).

3. CH_4 Production and Oxidation

Equation 3.1. CH_4 production rate (kg C/ha/d)
$CH_4p = a * AC * Ft$;
$Ft = b * e^{(0.2424 * T)}$; (factor of temperature)

Equation 3.2. CH_4 oxidation rate (kg C/ha/d)
$CH_4o = CH_4[l] * e^{(8.6711 * Eh[l] / 1000)}$;

Equation 3.3. CH_4 flux through plant aerenchyma (kg C/ha/d)
$CH_{4(aere)} = 0.5 * CH_4[l] * AERE$;
$AERE = -0.0009*PGI^5+0.0047*PGI^4 -0.883*PGI^3+1.9863*PGI^2-0.3795*PGI+0.0251$;
PGI = (days since planting) / (season days); (plant growth index)
Function 3.4. CH_4 flux through ebullition (kg C/ha/d)
$CH_{4(ebullition)} = 0.025 * CH_4[l] * PORO * Ft * (1 - AERE)$;
$Ft = -0.1687*(0.1*T[l])^3 +1.167*(0.1*T[l])^2 -2.0303*(0.1*T[l])+1.042$;
Function 3.5. CH_4 diffusion rate (kg C/ha/d)
$Rd = 0.01 * (CH_4[l] - CH_4[l+1]) * T[l] * PORO$;
AC – Available C concentration, kg C/ha;
T – soil temperature, °C;
l – soil layer number;
AERE – plant aerenchyma;
FloodDay – flooding days;
PORO – soil porosity;
$CH_4[l]$ – CH_4 concentration at layer l, kg C/ha.

4. NH_3 Volatilization

Equation 4.1. NH_3 concentration in liquid phase (mol/l)
$[NH3(l)] = [NH_4^+][OH-] / Ka$;
NH_4^+/NH3 equilibrium constatnt: $Ka = (1.416 + 0.01357 * T) * 10^{-5}$;
$[OH-] = Kw / [H+]$, mol/l;
$[H+] = 10^{-pH}$, mol/l;
$Kw = 10\wedge(0.08946 + 0.03605 * T) * 10^{-15}$; (water dissociation constant)

Equation 4.2. NH_3 concentration in gas phase and flux (kg N/ha)
$NH_3(g) = [NH_3(l)] * (T/T_0)^2$;
$Flux(NH_3) = NH_3(g) * AFPS * (1\text{-}Clay)$, kg N/ha/d;

Equation 4.3. NH_3 deposit (kg N/ha/d)
Vg = MaxVg * F(plant-N) * F(lsm);
F(plant-N) = Plant-N(act) / Plant-N(opt);
F(lsm) = LSM(act) / LSM(max);
$PlantUp(NH_3) = Vg * Air(NH_3) * LAI * 0.864$;
$Air(NH_3) = Base(NH_3) + Flux(NH_3) * 10\wedge9 / V(canopy) * LAI / (LAI + k2) * k3$;
V(canopy) = Height * 10000;
T_0 – reference temperature, 45°C;
T – soil temperature, °C;
PH – soil pH;
AFPS – soil air-filled porosity;
Clay – soil clay content;

MaxVg – maximum NH_3 deposit velocity , 0.05 m/s;
Plant-N(act) – crop N content, kg N/ha;
Plant-N(opt) – crop optimum N content, kg N/ha;
LSM(act) – water content on leaf surface, cm;
LSM(max) – maximum water content on leaf surface, cm;
Base(NH_3) – background NH_3 concentration, 0.06 ug/m^3;
V(canopy) – volume of the room from ground to the top of canopy, m^3/ha;
Height – maximum height of plant, m;
LAI – crop leaf area index;
MaxLAI – maximum crop leaf area index;
K2, k3 – constant coefficients;

Nutrient Cycling in Agroecosystems **58:** 277–284, 2000.

Simulation of methane production in anaerobic rice soils by a simple two-pool model

Y. Lu[1,2], J.R.M. Arah[1,3], R. Wassmann[1,4] & H.U. Neue[1,5]

[1]*International Rice Research Institute, MCPO 3127, 1271 Makati City, Philippines;* [2]*China National Rice Research Institute, Hangzhou, Zhejiang 310006, People's Republic of China;* [3]*AAT Consultants, 15 Clerk Street, Edinburgh EH8 9JH, United Kingdom;* [4]*Fraunhofer Institute for Atmospheric Environmental Research, Garmisch-Partenkirchen, Germany;* [5]*UFZ-Center for Environmental Research, Department of Soil Sciences, Theodor-Lieser Strasse 4, D-06120 Halle, Germany*

Key words: methane production, exogenous substrates, simulation, two-pool model, rice soils

Abstract

Methane (CH_4) is produced in flooded rice fields by anaerobic decomposition of applied organic residues, root-derived materials and native soil organic matter (SOM). Since CH_4 is an important greenhouse gas it is important to understand, and to be able to model, the processes which produce it. Anoxic incubation of soils employed in the cultivation of irrigated rice, with and without the addition of various potentially-available organic substrates, provides information on potential CH_4 emissions which can be incorporated into process-based models. In this study, a simple two-pool model is employed to simulate the CH_4 production of a number of anaerobically-incubated rice soils, and their responses to amendment with a variety of organic substrates. The model differs from most accounts of SOM transformation in that kinetics are microbially-mediated rather than first-order. Simulation yields a reproduction of the general trends of CH_4 production in response to amendments of acetate, glucose and rice straw.

Introduction

Increasing atmospheric concentrations of methane (CH_4) contribute to global warming and affect the photochemistry of the atmosphere (Cicerone & Oremland, 1988). Wetland rice soils have been shown to be an important CH_4 source at the global scale (Bartlett & Harriss, 1993; IPCC, 1995). Estimates of the strength of this source, however, are uncertain, varying from 20 to 100 Tg yr^{-1} (Sass & Fisher, 1997; Neue et al., 1997). Much of the uncertainty is due to the large spatial and temporal variability of the factors and processes involved in CH_4 emission. Modeling the underlying processes is necessary in order to predict CH_4 productions and emissions from flooded rice fields.

Methane emission from rice fields is the result of production and oxidation in the soil and transport from soil to the atmosphere. The field-scale modeling of CH_4 emission requires calculations of these basic processes as precise as possible. In recent years, a number of models on CH_4 production and emissions from rice paddies have been developed. In these models, the rate of CH_4 production was generally described as function of the availability of organic substrates and preferential oxidants (Segers & Kengen 1998; van Bodegom et al., 2000; Matthews et al., 2000), and modifiers such as soil Eh and pH, soil texture and temperature, crop growth and variety (Cao et al., 1995; Huang et al., 1998). Methane oxidation occurs in the rhizosphere and soil surface layer where O_2 is available. The rate of CH_4 oxidation was empirically described as a function of the aboveground biomass of the crop (Cao et al., 1995; Huang et al., 1998), and more mechanistically as a function of the depth distribution of root transmissivity (Arah & Kirk, 2000). Inevitably, however, the predictive power and the extrapolation reliability of the models depend on the mechanistic understanding and submodeling on the individual processes and driving

Table 1. Characteristics of the soils used in the experiments

Soil characteristic	Maahas	Pangil	Luisiana	Pila
pH (1:1 water)	6.40	4.40	4.50	7.4
OC (%)	1.57	3.96	1.84	2.08
Active Mn (%)	0.119	0.0340	0.109	0.058
Active Fe (%)	2.27	5.91	4.63	0.800
Inherent capacity (μmol g^{-1} soil)[a]	0.171	16.0	3.44	30.5
Texture	Clay	Clay	Clay	Silt
Soil order	Mollisol	Inceptisol	Entisol	Alfisol

[a]Inherent capacity is the cumulative CH_4 production within 28 d incubation without organic amendment

variables involved. This paper specifically focuses on simulation of the CH_4-production response of rice soils to additions of various exogenous substrates under anaerobic conditions.

Materials and methods

Methane production in anaerobic incubation

Much of the data employed in the development of the model described here is derived from a previously reported incubation experiment (Lu et al., 2000). A full description of the experimental methodology is given in that paper, but a brief summary is included here for convenience. The four soils—Maahas, Pangil, Luisiana and Pila—were collected from rice fields in the Philippines (Table 1). These soils represent a broad range in terms of soil pH (4.5 to 7.8), organic matter content (1.57 to 3.76%), and inherent CH_4 production potentials (0.171 to 30.5 μmol g^{-1} soil). Two experiments were conducted with these soils.

In experiment I, Maahas and Pangil soils were amended with acetate (100 μg C g^{-1} soil), glucose (100 μg C g^{-1} soil) and three types of root exudates (6.7, 9.6, and 16.0 100 μg C g^{-1} soil for exudate A, B, and C, respectively) [exudates were collected in a separate experiment in which IR72 rice was grown under different levels of P supply (Lu et al., 1999)]. The incubation procedure consisted of the following steps: 10 g of air-dried soils (Maahas and Pangil) were mixed with 16 ml deionized water in incubation vessels; vessels were sealed with rubber stopper and flushed with N_2; soil suspensions were then preincubated at 30 °C for 15 d to ensure the development of anaerobic condition (Wassmann et al., 1998); 4 ml of stock substrate solution was spiked into each vessel; the incubation was continued for another 14 d while CH_4 production rates were determined at 1, 2, 4, 6, 9, 11, 13, and 15 d after spiking. At the sampling date, soil suspensions were flushed with N_2 for 3 min and incubated for exactly 12 h. One ml of gas was then taken by syringe from the headspace and analyzed for CH_4 concentration.

In experiment II, Luisiana, Pila and Maahas soils were amended with rice straw (1% by weight). Straw was incorporated into the soil at the start of the incubation without preincubation. Soils were flushed with N_2 and incubated at temperatures of 25 °C, 30 °C, and 35 °C. Methane was measured at 7, 14, 21, 28, 35, 42, 49, and 56 d after incubation with an identical protocol as described above.

Model description

In the simulation of soil carbon dynamics (Molina et al., 1983; Parton et al., 1987; Paustian et al., 1992), SOM is generally partitioned into several components with each fraction having a defined turnover rate reflecting its resistance to mineralization. It has been suggested that CH_4 production in rice soils is mainly related to decomposition of the labile portions of SOM (Gaunt et al., 1997). We assume that only the active fractions of SOM are responsible for CH_4 production in the anaerobic incubation, and we further divide them into two functional pools: pool *F* in which decomposition is rapid, and pool *S* with a slower rate of decomposition. The decomposition of both pools is mediated by microorganisms. The model structure is shown in Figure 1.

At constant temperature, the decomposition rates of the two pools are

$$C_F = k_F F \mu \tag{1}$$
$$C_S = k_S S \mu \tag{2}$$

where C_F and C_S are the decomposition rates of pool *F* and pool *S* (μg C d^{-1} g^{-1} soil); *F* and *S* are the concentrations of pool *F* and pool *S* (μg C g^{-1} soil); k_F and k_S are reaction rate constants of pool *F* and *S* (μg^{-1} biomass C d^{-1}); and μ is the microbial biomass (μg C g^{-1} soil).

It is assumed that the production of methanogenic substrate is directly coupled to the anaerobic organic-matter decomposition. Intermediate fermentation reactions and hydrolysis are not explicitly taken into account (Segers & Kengen, 1998). The methanogenic substrate is then converted into CH_4 and CO_2, with a factor of 0.5 to produce 0.5 CO_2 and 0.5 CH_4 for each

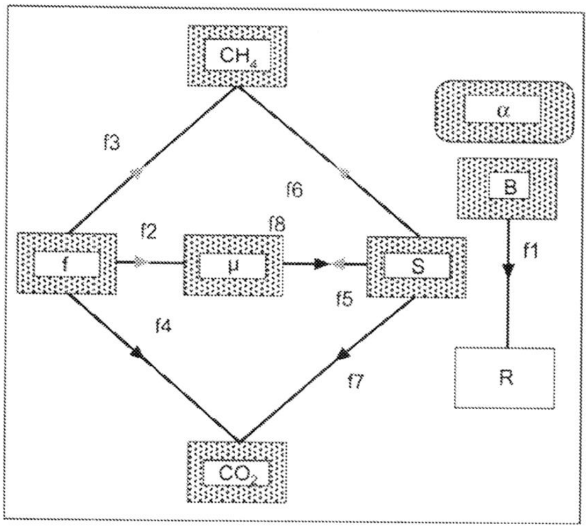

Figure 1. Material flow diagram for the two-pool model

carbon. Under anaerobic conditions, the decomposition of both pools produces microbial biomass (μ), CO_2 and CH_4. The initiation of CH_4 production, however, is hampered by a preferentially reduced buffer (*B*), which represents a pool of alternative electron acceptors in the soil (e.g., NO_3^-, Mn^{4+}, Fe^{3+}, and SO_4^{2-}). For simplicity, we specified the quantity of *B* (μg C eq g^{-1} soil) for all species. Methane production will not occur until most of this pool has been reduced and become reduced form (*R*).

Methane production was calculated by

$$f_3 = 0.5\alpha(1 - \eta_F)C_F \quad (3)$$
$$f_6 = 0.5\alpha(1 - \eta_S)C_S \quad (4)$$
$$\alpha = 1/(1 + \sigma B) \quad (5)$$

where f_3 is the CH_4 production rate derived from pool *F* (μg d^{-1} g^{-1}); f_6 is the CH_4 production rate from pool *S* (μg d^{-1} g^{-1}); and η_F and η_S are the growth constants of microbial biomass from pool *F* and *S*, respectively (g biomass g^{-1} C); α and σ are the inhibition factors of preferentially-reduced buffer (*B*). When *B* is zero, α is equal to 1.

The other units in the model are defined by equations 6 to 11:

$$f_1 = (C_F + C_S)B \quad (6)$$
$$f_2 = \eta_F C_F \quad (7)$$
$$f_4 = C_F B + \alpha C_F - F_2 - F_3 \quad (8)$$
$$f_5 = \eta_S C_S \quad (9)$$
$$f_7 = C_S B + \alpha C_S - F_6 - F_5 \quad (10)$$
$$f_8 = k_\mu \mu \quad (11)$$

where f_1 is the rate of reaction between substrate C and buffer *B*; f_2 and f_5 are growth rates of microbial biomass derived from pool *F* and pool *S*, respectively (μg d^{-1} g^{-1}); f_4 and f_7 are the CO_2 production rate from pool *F* and pool *S* (μg d^{-1} g^{-1}); f_8 is the death rate of microbial population; and k_μ is the biomass mortality constant.

Model parameters were summarized in Table 2. Statistical analysis and optimization of parameters were accomplished using ModelMaker program (Version 3.0, Cherwell Scientific Publishing Ltd, 1997).

Results

Experiment I: addition of acetate, glucose, and exudates

In experiment I, soils were preincubated under N_2 for 14 d before substrate spiking. Soil Eh was below −150 mV, according to previous observations (Wassmann et al., 1998). It is therefore assumed that the soils were

Table 2. Model parameters

Symbol	Meaning	Unit
B_0	Initial redox buffer concentration	μg C eq g^{-1}
F_0	Initial fast pool concentration	μg C g^{-1}
S_0	Initial slow pool concentration	μg C g^{-1}
μ_0	Initial microbial biomass concentration	μg C g^{-1}
η_F	Fast pool biomass production efficiency	g biomass g^{-1} C
η_S	Slow pool biomass production efficiency	g biomass g^{-1} C
k_F	Fast pool reaction constant	$μg^{-1}$ biomass C d^{-1}
k_S	Slow pool reaction constant	$μg^{-1}$ biomass C d^{-1}
k_μ	Biomass mortality constant	d^{-1}
σ	Methanogenesis sensitivity	g μg C eq^{-1}

fully reduced at the time of substrate spiking—i.e., the initial concentration of preferentially reduced buffer (B_0) is set to zero at the start of substrate spiking.

Acetate and glucose were added to Maahas and Pangil soils at the rate of 100 µg C g^{-1} soil. It is assumed that the initial fast pool F is zero and becomes equal to 100 µg C g^{-1} soil upon addition of substrate.

The other parameters were optimized with iterative numerical methods to obtain minimized values of the weighted sum of squares (expressed as χ^2). Initially, optimization was performed with data from the acetate treatment. Optimized parameters include initial concentrations of pool S and microbial biomass, biomass mortality constant, biomass growth constants, and reaction rate constants of pools F and S. Subsequently for the glucose treatment, the previously optimized initial microbial biomass, initial pool S concentration, biomass mortality constant, biomass growth constants, and reaction rate constant of pool S were introduced as known parameters. The reaction rate constant and microbial growth constant of pool F became the only parameters optimized. For the control soil, the same approach was applied while pool F was set to zero.

The results of the optimization are presented in Table 3. The initial pool S concentration is 3000 µg C g^{-1} soil for Maahas and 1,400 µg C g^{-1} soil for Pangil. Reaction rate constant of pool S was optimized at 0.001 g^{-1} biomass d^{-1} and 0.003 g^{-1} biomass d^{-1} for Maahas and Pangil, respectively. Initial biomass was 5 µg C g^{-1} soil for both soils.

Table 3. Optimized values of parameters for treatments of acetate and glucose

Parameter	Unit	Pangil	Maahas
B_0	µg C eq g^{-1}	0	0
μ_0	µg C g^{-1}	5	5
σ	g µg C eq^{-1}	100	100
S_0	µg C g^{-1}	1400	3000
k_S	g^{-1} biomass C d^{-1}	0.003	0.001
η_S	g biomass g^{-1} C	0.055	0.070
k_μ	d^{-1}	0.185	0.615
Acetate			
k_F	g^{-1} biomass C d^{-1}	0.680	0.022
η_F	g biomass g^{-1} C	0.359	0.426
Glucose			
k_F	g^{-1} biomass C d^{-1}	0.006	0.001
η_F	g biomass g^{-1} C	1	1

A comparison between the experimental and simulated kinetics is shown in Figure 2. Although discrepancies between simulated and mean values of experimental data are evident, the trend of simulated response to substrate amendments agrees well with that of the measured data ($r^2 = 0.84$ for Pangil and 0.78 for Maahas, both significant at $P < 0.01$).

For the treatment of root exudates, a similar simulation approach as for glucose treatment was applied. Pool F was set to zero before substrate addition and became equal to the amounts of added exudate upon the point of addition. Simulations were performed with the reaction rate constant and microbial growth constant of pool F being the only parameters to be optimized. The simulation, however, did not result in a good fit. Simulation was then performed with all the parameters for pools S and F, and microbial biomass set as unknown to permit the optimization program to search for low χ^2 values. However, the model again showed poor performance ($r^2 = 0.3$ and 0.1 respectively,

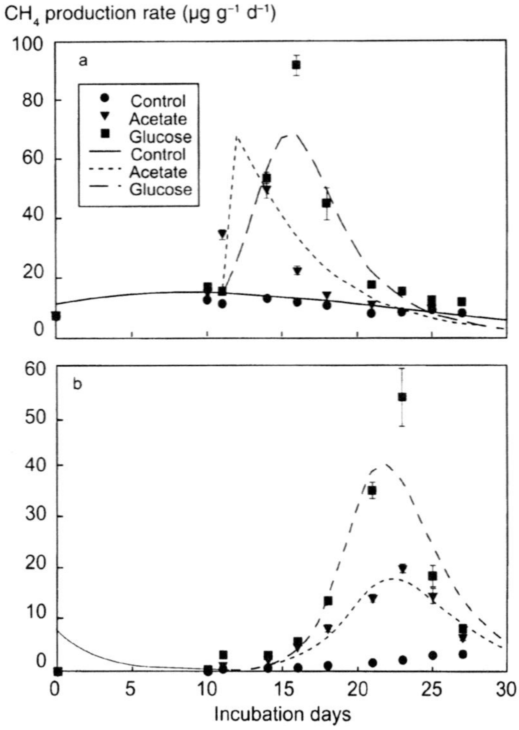

Figure 2. Measured and simulated response of CH_4 production to addition of acetate and glucose: (a) Pangil and (b) Maahas. Measured data = symbols; model output = lines; bars = standard errors of measured data

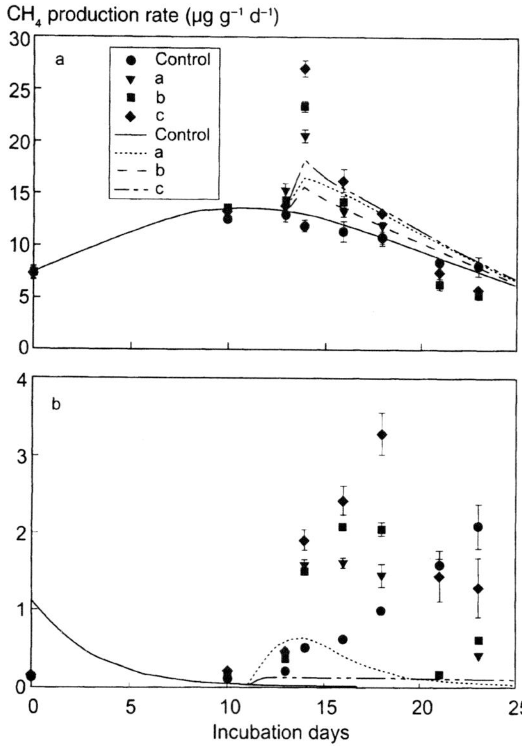

Figure 3. Measured and simulated response of CH_4 production to addition of root exudates a, b, and c: (A) Pangil and (B) Maahas. Measured data = symbols; model output = lines; bars = standard errors of measured data

for Pangil and Maahas) (Figure 3). Possible reasons for this are discussed later.

Experiment II: amendments of straw and temperature responses

In these experiments, incubation conditions differed from those of experiment I in that (a) rice straw was incorporated at the start of the experiment without pre-incubation; (b) batches of anaerobic incubations were conducted under 25, 30, and 35 °C.

The simple two pool model was slightly modified: (i) reaction rate constants k_S and k_F at 30 and 35°C were taken to be proportional to those at 25°C, with proportionality constants Q_5 and Q_{10}; and (ii) straw addition at time t_0 added C to both pool F (taken to be zero in the absence of straw addition) and pool S. The values of buffer (B_0) and temperature constants (Q_5 and Q_{10}) were optimized together with initial concentrations of pool S and pool F, reaction rate constants, microbial

Table 4. Optimized values of parameters for treatments of straw addition with temperature effect

Parameter	Unit	Luisiana	Pila	Maahas
B_0	μg C eq g^{-1}	3000	0	0
μ_0	μg C g^{-1}	1	1	1
σ	g μg C eq^{-1}	10	10	10
k_μ	d^{-1}	0.074	0.11	0.063
Q_5		1.26	1.35	1.06
Q_{10}		1.84	1.87	1.75
Control				
S_0	μg C g^{-1}	560	17400	0
k_S	g^{-1} biomass C d^{-1}	0.0015	0.004	0.0002
η_S	g biomass g^{-1} C	0.16	0.02	0.07
Straw				
ΔS_0	μg C g^{-1}	540	1360	3200
F_0	μg C g^{-1}	2400	61	200
k_F	g^{-1} biomass C d^{-1}	0.0005	0.012	0.0008
η_F	g biomass g^{-1} C	0	0	2.1

biomass, biomass growth constants, and mortality constant. The results of the optimization are presented in Table 4. The initial pool F concentrations derived from straw amendments were 200, 2400, and 61 μg C g^{-1} soil for Maahas, Luisiana, and Pila, respectively. The corresponding values for pool S were 3200, 1360, and 540 μg C g^{-1}. Reaction rate constants of pool F were optimized at 0.0008, 0.0005, and 0.012 g^{-1} biomass C d^{-1} for Maahas, Luisiana, and Pila, respectively, and at 0.0002, 0.00015, and 0.004 g^{-1} biomass C d^{-1} for pool S, respectively.

Figure 4 shows a comparison between experimental and simulated kinetics. The trend of simulated response to straw amendments at three temperatures agreed well with that of the measured data (r^2 = 0.88, 0.61, and 0.91 respectively for Luisiana, Pila, and Maahas, all significant at $P < 0.01$).

Discussion

Allowing for its simplicity and the number of arbitrary assumptions it involves, the simple two-pool model simulates CH_4 production in anaerobic soils quite well. Simulation yields a reproduction of the general trends of CH_4 production in response to amendments of acetate, glucose, and rice straw.

In this model, the production of substrate for methanogenesis is directly coupled to anaerobic carbon decomposition. The rate of carbon decomposition depends on reaction rate constant, substrate pool concentration, and microbial biomass. These kinetics differ from the first-order form found in most multiple-

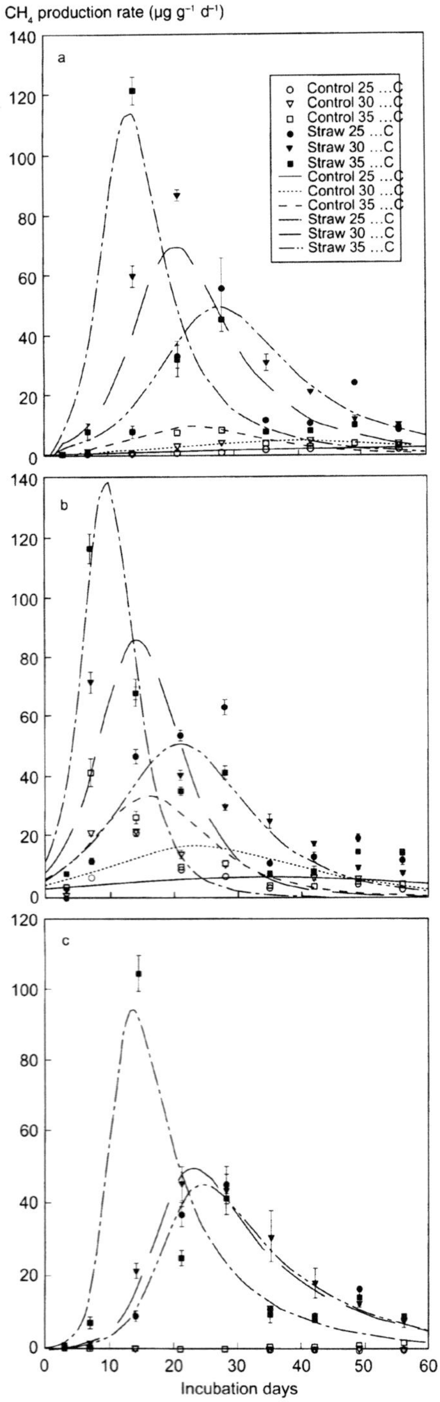

Figure 4. Measured and simulated response of CH_4 production to temperature and rice straw addition: (a) Luisiana, (b) Pila, and (c) Maahas. Measured data = symbols; model output = lines; bars = standard errors of measured data.

pool models of SOM decomposition, in which the decomposition rate of SOM from each pool is usually a function of substrate quality (lignin content and C-N ratio) and external factors such as temperature and moisture (Paustian et al., 1992), but not of microbial biomass. The microbial biomass is usually taken to be a most active pool which participates in the carbon cycling (Parton et al., 1987; Paustian et al., 1992; Nicolardot et al., 1994), but which does not in itself influence the decomposition rates of other pools. This implies that the microbial population is always ready to consume readily metabolized substrates and that dead biomass is rapidly decomposed. This might not be true if soils are amended with large amounts of readily decomposable organic materials, where the development of microbial population may lag behind the supply of substrate. Segers and Kengen (1998) indicated that in the initial phase, the rate of CH_4 production was limited by methanogenic biomass. Under the conditions of substrate-enriched incubation as in this study, it appears necessary to include microbial biomass in the model and to adopt microbially mediated kinetics.

The active SOM pool is partitioned into two pools in the model. Acetate and glucose, which are immediately converted to CH_4 under anaerobic conditions, belong to fast pool F, while cellulose and the like correspond to slow pool S. The decomposition of pool F accounts for the initial phase and the peak of CH_4 production, while the decomposition of pool S contributes most to the later phase of CH_4 production.

Model parameters F_0, S_0, and B_0 are the initial concentrations of the active organic carbon and buffer pools; k_F and k_S are reaction rate constants. It appears that reaction rate constant is more important than total pool concentration, as would be expected over the short term. For example, in experiment I, although Pangil showed lower S_0, the higher reaction rate constant led to a higher CH_4 production rate than in Maahas (Figure 3). Similarly, in experiment II, although the fast pool concentration of Pila was lower than Maahas and Luisiana, the reaction rate constant was much higher and CH_4 production was faster in Pila (Table 4, Figure 4). It should be stressed that F, S, and B are functional pools, not measurable fractions. They are defined exclusively by their role in the model, and no extraction procedure can be expected to measure them. There may be loose correlations between F and dissolved C, and between B and "active Fe", for example, but it would be misleading to expect (or assume) equivalence.

The performance of the model is poor ($r^2 = 0.3$ and 0.1) in simulating the responses to additions of exudate (Figure 3). Two reasons may be advanced: the first is the relative size of the amendments and their concomitant effects: between 6 and 15 times more C was added in the acetate and glucose experiments than in the exudate experiment; trends apparent in Figure 3 are little more than noise in Figure 2. The model may simply be too insensitive to reproduce the subtle effects apparent in Figure 3. Another possible explanation concerns the C/N. Acetate and glucose contain no N. Their addition cannot enhance the efficiency (η_F and η_S) of microbial biomass production under N-limited conditions. However, root exudate has a low C/N (lower than that of rice straw), and may thus promote biomass growth as well as SOM mineralization even under N-limited circumstances. Introducing such considerations into the simple model may improve the model performance for CH_4 production with various organic inputs.

In conclusion, a simple two-pool model for the prediction of CH_4 production under anaerobic incubation was developed. The model incorporated the effect of microbial biomass, which we thought necessary under conditions of high organic input in the paddy soils. In the model, the active soil organic phase was divided into a fast pool and a slow pool and methanogenic substrate and CH_4 production were directly coupled to the decomposition of these pools. Methane production was delayed in the presence of preferentially reduced oxidants. With exceptions for the treatments of root exudates, the simple model simulated trends of CH_4 production in response to organic amendments ($r^2 = 0.61$ to 0.91) well. However, it should be indicated that the model represents only a routine of CH_4 production under controlled substrate supply and anaerobic condition. To be suitable to simulate field-scale CH_4 production, it should be integrated with subroutines on substrate production, soil aeration, and electron-acceptor reoxidation and incorporated with modifiers such as soil texture and temperature and others.

References

Arah JRM & Kirk GJD (2000) Modelling rice-plant-mediated methane emission. Nutr Cycling Agroecosyst (this issue)

Bartlett KB & Harriss RC (1993) Review and assessment of methane emissions from wetlands. Chemosphere 26(1-4): 261-320

Cao M, Dent JB & Heal OW (1995) Modelling methane emissions from rice paddies. Global Biogeochem Cycles 9:183-195

Cicerone RJ & Oremland RS (1988) Biogeochemical aspects of atmospheric methane. Global Biogeochem Cycles 2: 299-327

Gaunt JL, Neue HU, Bragais J, Grant IF & Giller KE (1997) Soil characteristics that regulate soil reduction and methane production in wetland rice soils. Soil Sci Soc Am J 61:1526-1531

Huang Y, Sass RL & Fisher FM Jr (1998) A semi-empirical model of methane emission from flooded rice paddy soils. Global Change Biol 4(3):247-268

IPCC – Intergovernmental Panel on Climate Change (1995) Climate Change: The Supplementary Report on the IPCC Scientific Assessment. Cambridge University Press, New York.

Lu Y, Wassmann R, Neue HU & Huang C (1999) Impact of phosphorus supply on root exudation, aerenchyma formation and methane emission of rice plants. Biogeochemistry 47:203-218

Lu Y, Wassmann R, Neue HU, Huang C & Bueno CS (2000) Response of methanogenesis in anaerobic rice soils to exogenous substrates. Soil Biol Biochem 32:1683-1690

Matthews R, Wassmann R & Arah J (2000) Using a crop/soil simulation model and GIS techniques to assess methane emissions from rice fields in Asia. I. Model development. Nutr Cycling Agroecosyst (this issue)

Molina JAE, Clapp CE, Shaffer MJ, Chichester FW & Larson WE (1983) NCSOIL, a model of nitrogen and carbon transformations in soil: description, calibration, and behavior. Soil Sci Soc Am J 47:85-91

Neue HU, Wassmann R, Kludze HK, Bujun W & Lantin RS (1997) Factors and processes controlling methane emissions from rice fields. Nutr Cycling Agroecosyst 49: 153-161

Nicolardot B, Molina JAE & Allard MR (1994) C and N fluxes between pools of soil organic matter: model calibration with long-term incubation data. Soil Biol Biochem 26:235-243

Parton WJ, Schimel DS, Cole CV & Ojima DS (1987) Analysis of factors controlling soil organic matter levels in great plains grasslands. Soil Sci Soc Am J 51:1173-1179

Paustian K, Parton WJ & Persson J (1992) Modeling soil organic matter in organic-amended and nitrogen-fertilized long-term plots. Soil Sci Soc Am J 56:476-488

Sass RL &Fisher FM (1997) Methane emissions from rice paddies: a process study summary. Nutr Cycling Agroecosyst 49:119-127

Segers R & Kengen SWM (1998) Methane production as a function of anaerobic carbon mineralization: a process model. Soil Biol Biochem 30:1107-1117 van Bodegom PM, Wassmann R & Corton TM (2000) A process-based model for methane emission predictions from flooded rice paddies. Nutr Cycling Agroecosyst (this issue)

Wassmann R, Neue HU, Bueno C, Lantin RS, Alberto MCR, Buendia LV, Bronson K, Papen H & Rennenberg H (1998) Methane production capacities of different rice soils derived from inherent and exogenous substrates. Plant Soil 203:227-237

Nutrient Cycling in Agroecosystems **58:** 285–301, 2000.

Combining upscaling and downscaling of methane emissions from rice fields: methodologies and preliminary results

H.A.C. Denier van der Gon[1], P. M. van Bodegom[2], S. Houweling[3], P. H.Verburg[1] & N. van Breemen[1]

[1]*Laboratory of Soil Science and Geology, Wageningen University, P.O. Box 37, 6700 AA, Wageningen, The Netherlands;* [2]*Free University Amsterdam, Department of Systems Ecology, de Boelelaan 1087, The Netherlands;* [3]*Institute for Marine and Atmospheric Research, Utrecht University, The Netherlands*

Key words: methane emission, rice fields, temporal scaling, spatial scaling, regional estimates, inverse modeling

Abstract

The uncertainty in the methane (CH_4) source strength of rice fields is among the highest of all sources in the global CH_4 budget. Methods to estimate the source strength of rice fields can be divided into two scaling categories: bottom-up (upscaling) and top-down (downscaling). A brief review of upscaling and downscaling methodologies is presented. The combination of upscaling and downscaling methodologies is proposed as a potential method to reduce the uncertainty in the regional CH_4 source strength of rice fields. Some preliminary results based on upscaling and downscaling are presented and the limitations of the approaches are discussed. The first case study focuses on upscaling by using a field-scale model in combination with spatial databases to calculate CH_4 emissions for the island of Java. The reliability of upscaling results is limited by the uncertainty in model input parameters such as soil properties and organic carbon management. Because controlling variables such as harvested rice area may change on relatively short time scales, a land use change model (CLUE) was used to quantify the potential land use changes on Java in the period 1994-2010. The predicted changes were evaluated using the CH_4 emission model. Temporal scaling by coupling land use change models and emission models is necessary to answer policy-related questions on future greenhouse gas emissions. In a downscaling case study, we investigate if inverse modeling can constrain the emissions from rice fields by testing a standard CH_4 from rice scenario and a low CH_4 from rice scenario (80 and 30 Tg CH_4 yr^{-1}, respectively). The results of this study are not yet conclusive; to obtain fine-resolution CH_4 emission estimates over the Southeast Asian continent, the monitoring network atmospheric mixing ratios need to be extended and located closer to the continental sources.

Introduction

Wetland rice fields are an important source of methane (CH_4), a potent greenhouse gas (Wang, 1976; IPCC, 1994). The first field measurements were done in California (Cicerone & Shetter, 1981; Cicerone et al., 1983), followed by extensive studies in Spain (Seiler et al., 1984) and Italy (Holzapfel-Pschorn & Seiler, 1986; Schütz et al., 1989). From the 1980s to the 1990s, CH_4 emissions from rice fields were measured at numerous locations. For an overview by country, we refer to Minami et al. (1994). Since the late 1980s, the database of flux measurements from rice fields has expanded and the combined field and laboratory studies have greatly increased our understanding of the processes controlling CH_4 emission from rice fields. However, the newly available field results revealed a huge variation of flux rates and methods to select which flux rates are "representative" of the world's rice fields are lacking. Hence, the uncertainty in the global CH_4 source strength of rice fields of about ~65% is among the highest of all CH_4 sources, for example 60 ± 40 Tg yr^{-1} (IPCC, 1994) or 80 ± 50 Tg yr^{-1} (Lelieveld et al., 1998).

Techniques used for extrapolating measurements and constraining results between different spatial and temporal scales are generally referred to as "scaling." Two approaches to scaling of the CH_4 source strength of rice fields can be distinguished: (1) bottom-up scaling methodologies and (2) top-down scaling methodologies, often referred to as "upscaling" and "downscaling," respectively. Upscaling typically uses small scale (~ 1 m^2) flux measurements that are extrapolated to the regional or global scale. Downscaling typically uses atmospheric transport and chemistry to deduce information on CH_4 sources and sinks from the temporal and spatial variation of atmospheric CH_4 mixing ratios as measured by global air sampling networks. In this paper, we give a condensed chronological review of upscaling and downscaling methodologies used to estimate the CH_4 source strength of rice fields. The advantage of combining upscaling and downscaling to reduce the uncertainty in the CH_4 source strength of rice fields is discussed and some preliminary results based on upscaling and downscaling are presented.

Some variables controlling CH_4 emissions from rice fields are quite stable over time—e.g., soil type and climate. By contrast, other controlling variables may change drastically on time scales >~5 yr, e.g., harvested rice area, cropping index, fertilizer use, rice varieties, and water management. For these variables, the rate of change and its impacts on emissions have to be quantified. This asks for temporal scaling, estimating past or future emissions based on current emissions. Temporal scaling is necessary to answer policy-related questions on future greenhouse gas emissions but also to avoid comparing incompatible results from upscaling and downscaling methodologies—e.g., if data from the 1980s are combined with those from the 1990s, the potential impact on the calculations should be carefully considered.

Upscaling of CH_4 emissions from rice fields

Estimates of the global CH_4 source strength of rice fields have been made using various bottom-up scaling approaches, further referred to as upscaling methodologies. Table 1 describes, in chronological order, the major categories of upscaling methods, without aiming for completeness. The first attempt to scale up was published by Koyama (1963). Koyama measured CH_4 production of nine Japanese rice field soils upon anaerobic incubation as a function of temperature in the range of 5–40 °C. By assuming that all rice soils are similar in nature to Japanese rice soils and deriving rice field areas and average soil temperature from statistics, Koyama (1963) estimated the global CH_4 source strength of rice paddies as 190 Tg yr^{-1}.

In the 1980s, the first measurements of CH_4 emission from rice fields were published. To estimate the global CH_4 source strength from these measurements, the harvested area of rice is multiplied by the average CH_4 emission per day times the length of the growing season (method 2, Table 1). This method is characterized by the use of a uniform emission factor. The geographical location and local management practices are not taken into account. Various amendments on this method have been made. For example, upland rice, which contributes about 12% of the world harvested rice area and is characterized by no flooding for any significant amount of time, was not excluded in early calculations (e.g., Holzapfel-Pschorn & Seiler, 1986; Schütz et al., 1989) and IPCC (1995) proposed a correction for growing-season average temperature. Nevertheless, in essence, a uniform emission factor is used. To deal with the huge variation in measured emissions, measurements are averaged to yield the uniform emission factor and the standard deviation is used to calculate the range in the emission estimate. This frequently used methodology is the basis of most rice field source strengths in global CH_4 budgets. For example, the EDGAR database (Olivier et al., 1996) assumes a uniform emission of 350 mg CH_4 m^{-2} d^{-1} and a fixed number of 130 flooding days for each rice crop, except upland rice which has no flooding for any significant amount of time. Multiplication of the uniform emission factor with the number of flooding days and harvested area of non-upland rice results in a global source strength of ~60 Tg yr^{-1} (Kreileman & Bouwman, 1994; Olivier et al., 1996).

The third methodology used for estimating the global CH_4 source strength of rice fields acknowledges that differences in rice-growing environments will result in different levels of emission. A certain ecosystem process or function, which is better known than CH_4 emission from the specific rice environments, is assumed to be proportional to CH_4 emission. This process is then used as a so-called proxy for CH_4 emission from rice fields. Examples of processes that can be used as a proxy are net primary production (NPP, method 3a, Table 1) or the amount of carbon returned to the rice soil during a full rice crop cycle (method 3b, Table 1). The fundamental difference with method 2 (Table 1) is the absence of a uniform emission factor. If the proxy value, e.g. NPP, varies in an ecosystem, CH_4 emission will vary proportionally. The proxy method is sensitive to the transfer function between the proxy and CH_4 emission. This transfer function is often uncertain or variable. For example, the proposed fraction of rice NPP emitted as CH_4 ranges from 1.5 to 7% (Aselman & Crutzen, 1989; Taylor, 1991; Huang et al., 1997). So, if it is assumed that 6% instead of 3% of the net primary production is emitted as CH_4, the estimated

Table 1. Bottom-up scaling (upscaling) methodologies that have been used for estimating the CH_4 source strength of rice agriculture in chronological order. Numbers indicate distinctly different methodologies, bullets indicate a modification on a methodology

	Upscaling methodologies for CH_4 emission from rice fields	Source strength (Tg yr^{-1})
1	CH_4 production in incubated rice soil samples multiplied by the estimated amount of rice soil (Koyama, 1963)	190
2	Uniform emission factor based on flux measurements multiplied by harvested area of rice (Cicerone & Shetter, 1981; Holzapfel-Pschorn & Seiler, 1986; Schütz et al., 1989, respectively)	59, 70-170, 50-150
	• Excluding upland rice area because no potential for CH_4 emission	-12%
	• Growing season average temperature correction (IPCC, 1995)	60-105[a]
3a	CH_4 emission proportional to net primary production (NPP), e.g., 3-7% (Aselman & Crutzen, 1989), 5% (Taylor, 1991) of NPP emitted as CH_4	60-140
	• Including soil CH_4 emission potential (Bachelet & Neue, 1993)	47[b]
3b	CH_4 emission proportional to carbon returned to the soil: 30% of the soil returned carbon emitted as CH_4 (Neue et al., 1990)	63
	• Including soil CH_4 emission potential (Bachelet & Neue, 1993)	52[b]
4	Specific emission factors for specific ecosystems, regions and/or management (IPCC, 1997)	
	• Rice ecosystem-specific emission factors (Neue & Sass, 1998)	30-50
	• Country-specific emission factors (Neue & Sass, 1998)	32[c]
	• National rice regionalization (Yao et al., 1996)	15 (China only)
5	Empirical (regression) models using input from national statistics and / or geographical information system (GIS)	
	Kern et al. (1997)	10 ± 3 (China only)
6	Simulation models for CH_4 emission from rice fields linked to a geographical information system	
	Cao et al. (1996)	53
	Huang et al. (1998b)	7.2 - 13.6 (China only)

[a]Proposed for national communications by IPCC (1995), not applied on a global scale. Estimated source strength here is based on growing season average temperature between 25 and 32 °C. [b]Original calculation by Bachelet & Neue (1993) was for Asian rice fields only (~90% of world harvested area). For comparison, the source strength is increased proportionally to cover the world rice area. [c]32 Tg yr^{-1} is presented as median, 104 Tg yr^{-1} as maximum.

global source strength using this methodology obviously doubles.

A fourth method was introduced to make better use of newly reported CH_4 emission data from rice fields and account for the observed emission differences from different rice cropping systems (method 4, Table 1). Based on reported observed CH_4 emissions, proposed OECD/IPCC default guidelines discriminate rice fields and respective CH_4 emissions according to rice ecology and introduce factors for organic amendments and water regimes (IPCC, 1997). A default seasonally integrated CH_4 emission of 20 g m^{-2} is recommended for continuously irrigated and continuously flooded lowland rice ecosystems without organic amendments with proportionately lower values for other rice ecosystems and a multiplier factor of 2 (range 2-5) for emissions for the corresponding rice ecosystems with organic amendments. For an extensive discussion of this method, we refer to Sass (1999). Method 4 can be further expanded by replacing the default seasonally integrated CH_4 emission factor with national or regional emission factors where available (IPCC, 1997; Neue & Sass, 1998; Sass, 1999).

Calculations using emission factors introduce unquantifiable measures of uncertainty, mainly because of two reasons. First, the highly dynamic and nonlinear interactions between processes underlying CH_4 emissions make it difficult to relate CH_4 emissions to single environmental variables. Second, the local variations in biotic and abiotic parameters controlling CH_4 emission ask for a spatial explicit approach. In recent years, various geo-referenced databases and digital maps relevant to CH_4 emissions from rice fields have been published. For example, rice by type of culture (Huke and Huke, 1997) and the digital FAO soil map of the world (FAO, 1995). A geographic information system (GIS) can be used to overlay, integrate, and analyze the relevant data sets to derive a new, spatial explicit database with controlling variables of CH_4 emission. The newly derived database contains controlling variables of CH_4 emission of a rice field that is assumed to be representative of a particular spatial resolution and is used as model input for empirical, often regression-based, models (method 5, Table 1) or process-based models (method 6, Table 1). In theory, the lowest resolution database or map in the GIS determines the spatial resolution of the derived database. However, the classification scheme and the model input parameters can be optimized with sensitivity analysis of the model used. Therefore, in practice, the resolution of the derived database is determined by the lowest resolution database of a critical parameter. The emissions from the 'representative' rice field of the smallest spatial unit multiplied by their hectarage can be aggregated at the regional level using the GIS. The major advantage of spatial explicit upscaling methodologies is the ability to build up regional profiles of CH_4 emissions from detailed (process) studies. Intricate feedback mechanisms, adaptation strategies, mitigation strategies, and predicted changes in controlling variables can be tested. An empirical regression-based model combined with a GIS was used by Kern et al. (1997) to make a spatial analysis of CH_4 emission from Chinese rice fields and to evaluate potential mitigation strategies (method 5, Table 1). Unfortunately, regression-based models are only valid within their domain. However, the regression is usually based on a few observations or sites and extrapolated to numerous locations with combinations of controlling variables not covered by the observations used to build the regression model.

To obtain regional CH_4 emission estimates with minimized uncertainty, the use of process-based models to simulate CH_4 emissions using GIS-derived model input is preferable (method 6, Table 1). Recently, several models were developed to predict field-scale CH_4 emissions under varying conditions (Cao et al., 1995; Hosono & Nouchi, 1997; Huang et al., 1998a; van Bodegom et al., 2000). Field-scale models designed for larger scale emission estimation should, in anticipation of difficulties with obtaining input parameters, minimize their demand of input parameters. An example of how such a field-scale model can be used in combination with a GIS is presented further in this paper.

Downscaling of CH_4 emissions with inverse modeling

As sources and sinks of trace gases are also reflected in the spatial distribution and temporal variation of their atmospheric mixing ratio, an alternative approach consists of inverting observed atmospheric mixing ratios into a spatial and temporal resolution of the trace gas sources (Heimann & Kaminski, 1999). To do this, the atmospheric transport from the source regions to the observation sites has to be described using simulation models of atmospheric transport and, depending on the trace gas studied, atmospheric chemistry because the atmospheric mixing ratio may change during the atmospheric transport from the source region to the observation site. Atmospheric trace gases for which global- or regional-scale sources and sinks have been estimated from observational data using inverse approaches are CO_2, CH_4, N_2O, halocarbons, and CO. The reasons for restriction to these trace gases are (1) their life time is longer than ~1 mo, (2) chemical transformations are either absent or relatively well understood, and (3) the mathematical inverse problem of these gases is linear or may be assumed to be linear in the relevant concentration range (Heimann & Kaminski, 1999). For a comprehensive review of global approaches to infer surface trace gas fluxes from observed atmospheric mixing ratios using inverse modeling, we refer to Heimann & Kaminski (1999).

The first applications of inverse modeling techniques to atmospheric problems appeared during the 1980s, for example, investigations of CO_2 sources and sinks (Enting, 1985; Enting & Mansbridge, 1989). Brown (1993) applied similar mathematical techniques to study sources of CFC-11 ($CFCl_3$), methyl chloroform (CH_3CCl_3), and CH_4. Later, this CH_4 inversion was extended with measurements of isotopic ratios (Brown, 1995). Initially, global-scale, zonally averaged, two-dimensional (2D) models were used. The two dimensions in atmospheric transport models are height and latitude. Therefore, the number of distinguishable unknown sources and sinks is limited in the 2-D model studies. A source is not defined as an activity that causes trace gas emission (e.g., rice agriculture, animal husbandry) but as the integrated emission over, for example, a latitudinal band (Brown, 1993). Three dimensional (3-D) atmospheric transport models, where also longitude is included, allow a much better geographical definition of source location and a relatively large number of sources and sinks can be distinguished. Hartley & Prinn (1993) were the first to publish a global 3-D inverse modeling study dealing with sources of $CFCl_3$. To retain uniqueness, Hartley and Prinn (1993) aggregated their sources to a few geographical units, such as countries and continents. Hein et al. (1997) applied a 3-D model to sources and sinks of CH_4 and included measurements of CH_4 isotopes. By introducing a priori information on temporal and spatial distribution of sources, Hein et al. (1997) could distinguish a large number of individual CH_4-emitting activities, such as rice agriculture or biomass burning. So, in principle, the global atmospheric CH_4 concentration distribution can be used to constrain the regional emission. This is referred to as downscaling from the global to the regional scale. The inverse modeling method optimizes the agreement between model-calculated and observed CH_4 mixing ratios by adjusting the magnitudes of the various CH_4 sources and sinks. Often, the adjustment is constrained by specified a priori information on source distributions, seasonal variations, and associated uncertainty ranges.

Combining upscaling and downscaling: a novel approach to reduce uncertainties in the CH_4 source strength of rice fields

When scaling CH_4 emissions from one scale to another using models, it is crucial to validate the model with an observational data set at the target scale to limit uncertainty and obtain high reliability. For example, if flux measurements at a scale of 1 m^{-2} are extrapolated to the local (100-1,000 ha) or regional (≥10^4 ha) scale, flux measurements at the local or regional scale are required. Measurements of CH_4 fluxes at the local scale may be achieved using micrometeorological methods (Fowler et al., 1995; Liu et al., 2000). However, both the increased technical complexity compared with a closed chamber method and the prerequisite of a rather homogeneous landscape cause limited availability and applicability of such local-scale flux measurements. Measurement of CH_4 fluxes at the regional scale is feasible using an aircraft (Choularton et al., 1995; Thohjima et al., 1997), but this is very expensive. Moreover, this is not expected to be done in the tropics, where most of the rice emissions occur, and not likely to cover a semi-continuous period such as a full rice cropping cycle. Thus there is a need for alternative approaches that can be used to obtain a quantitative evaluation of the precision and reliability of the calculated source strength estimates at the target (regional) scale.

The major limitation in narrowing the uncertainty of regional and global CH_4 source strength estimates from rice agriculture is the lack of a constraint on the regional source strength. In theory, a major improvement could result from constraining the regional source strength obtained by upscaling with the regional source strength derived from an independent downscaling approach and vice versa. For example, upscaling from the soil-rice ecosystem perspective by using monitored CH_4 fluxes and a field-scale model in combination with spatial databases to calculate CH_4 emissions and downscaling from the global atmosphere perspective to a smaller scale by inverse modeling of sources and sinks of atmospheric CH_4. Comparison between CH_4 budgets for the same geographical region, derived from independent upscaling and downscaling approaches, could result in a reduction of the uncertainty in the magnitude of individual CH_4 sources. Moreover, regions where the discrepancy between the source strength estimates based on upscaling and downscaling is found to be large could be given higher priority in future research. This would help attain a cost-effective reduction of uncertainties in greenhouse gas budgets at the national, continental, and global scales. The feasibility of combining upscaling and downscaling approaches to reduce the uncertainty in regional source strength estimates is assessed through case studies for the island of Java (Indonesia) and China. We present preliminary results for upscaling CH_4 emissions from rice fields on Java and, in a second case study, investigate, using downscaling from the global to the regional scale, to what extent CH_4 sources of Southeast Asia are constrained by available measurements over this region.

A case study on upscaling: CH_4 emissions from Java

The upscaling approach followed consists of using a process-based, field-scale model to simulate CH_4 emissions and using GIS-derived model input (method 6, Table 1).

Methane emission model description and model input

The CH_4 emission model (MEM) used in this case study is described by van Bodegom et al. (1999). Two compartments, the rhizosphere and the bulk soil, are distinguished in this model. To simulate CH_4 emissions, the MEM contains simplified process-based descriptions of CH_4 production, transport, and oxidation for each compartment. The model was validated with experimental data collected in the Philippines (van Bodegom et al., 2000). Simulated seasonal CH_4 emissions did not significantly differ from measured seasonal emissions (with a coefficient of variation of 7%). To calculate CH_4 emissions, the MEM requires data on reducible soil iron content (Fe) and soil organic carbon content (OC), rice variety and rice yield, inorganic and organic fertilizer input, length of growing season and temperature. District-level data on harvested rice area (irrigated or rainfed), rice yield, inorganic fertilizer, and temperature were based on an integrated set collected from land use maps and agricultural statistics by Verburg et al. (1999b) and converted to a gridded map of Java consisting of 329 grids of 20 × 20 km. The rice variety grown was assumed to be a high-yielding IR variety. Rice fields of Java are largely planted to one variety, IR64, which covers 50–70%, depending on the province (BPS, 1996b). The length of the growing season was 119 d (BPS, 1996b). To be able to calculate emissions from rainfed areas, we assumed rainfed rice to be grown only in the wet season and the flooding pattern of rainfed rice on Java was based on expert judgement (Setyanto, pers. commun. 1998). We assumed that part of the straw (equivalent to 30% of the yield [Neue et al., 1990]) was returned to the soil and that no other organic amendments are applied. The soil property data were derived by overlaying the 20- × 20-km grid map of Java on to the digital soil map of the world (FAO, 1995) and extracting the fraction of each soil association within the 20- × 20-km grids. Next, the FAO soil associations were broken down into the FAO_74 soil units (FAO, 1988). The soil units that are presumably used for growing rice on Java based on Soepraptohardjo & Suhardjo (1978) and Batjes (pers. commun., 1998) were selected. The WISE database (Batjes, 1995) was checked for presence of one or more Asian soil profiles describing the selected soil units; if no Asian profile was found, the full database was used. Because the WISE soil profile database (Batjes, 1995) contains no data on soil Fe, a subset of the ISIS database (Van de Ven & Tempel, 1994) was added. A weighted average of the OC and reducible Fe for the top 20 cm of the profiles for each soil unit was calculated. Next, an average soil OC and Fe was calculated by the proportion of each soil unit suitable for rice growing in the 20- × 2-km grid cell (block). A similar approach is described in more detail by Knox et al. (2000).

Upscaling of CH_4 emissions from rice for the Island of Java

The MEM was used to calculate the CH_4 emission in g m^{-2} in each of the 20- × 20-km blocks for irrigated rice and rainfed rice. The CH_4 emission per 20- × 20-km block was calculated using the GIS (Figure 1). Aggregated to the Island of Java, a total CH_4 emission from rice agriculture of 0.91 Tg yr^{-1} is calculated. The results presented here are preliminary and need to be carefully evaluated. However, Figure 1 illustrates the potential of this methodology. Clear spatial patterns can be recognized and these can largely be explained by the variation in rice harvested area and soil properties.

A case study on downscaling: CH_4 emissions from rice fields

By applying the inverse model, Hein et al. (1997) computed average magnitude of rice field emissions close to the a priori value (~70 Tg yr^{-1}) but reduced the uncertainty in this estimate significantly (from –50 to ±20 Tg CH_4 yr^{-1}). However, no observations in or close to the rice-growing regions of Asia were used and it is questionable how well the rice source strength can be

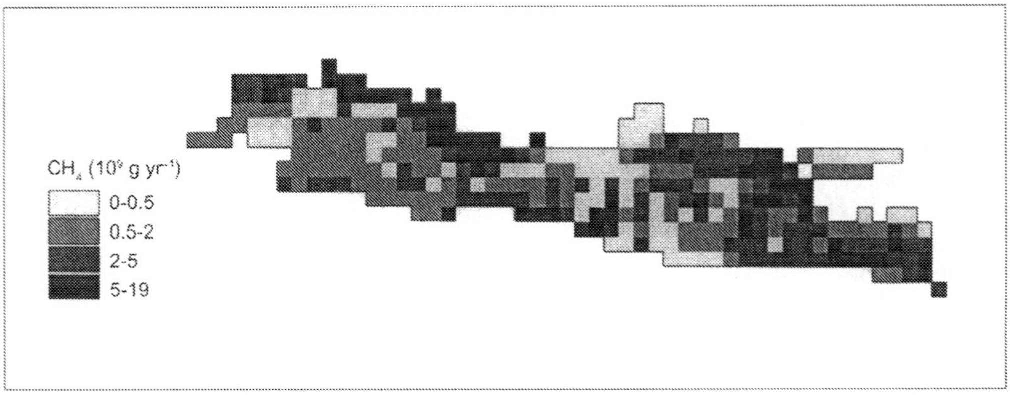

Figure 1. Annual CH_4 emission from rice fields for Java in 1994 using van Bodegom et al. (1999) with gridded 20- × 20-km land use data from Verburg et al. (1999b) and soil data derived from FAO (1995) and Batjes (1995) (see text for detailed description of model input data)

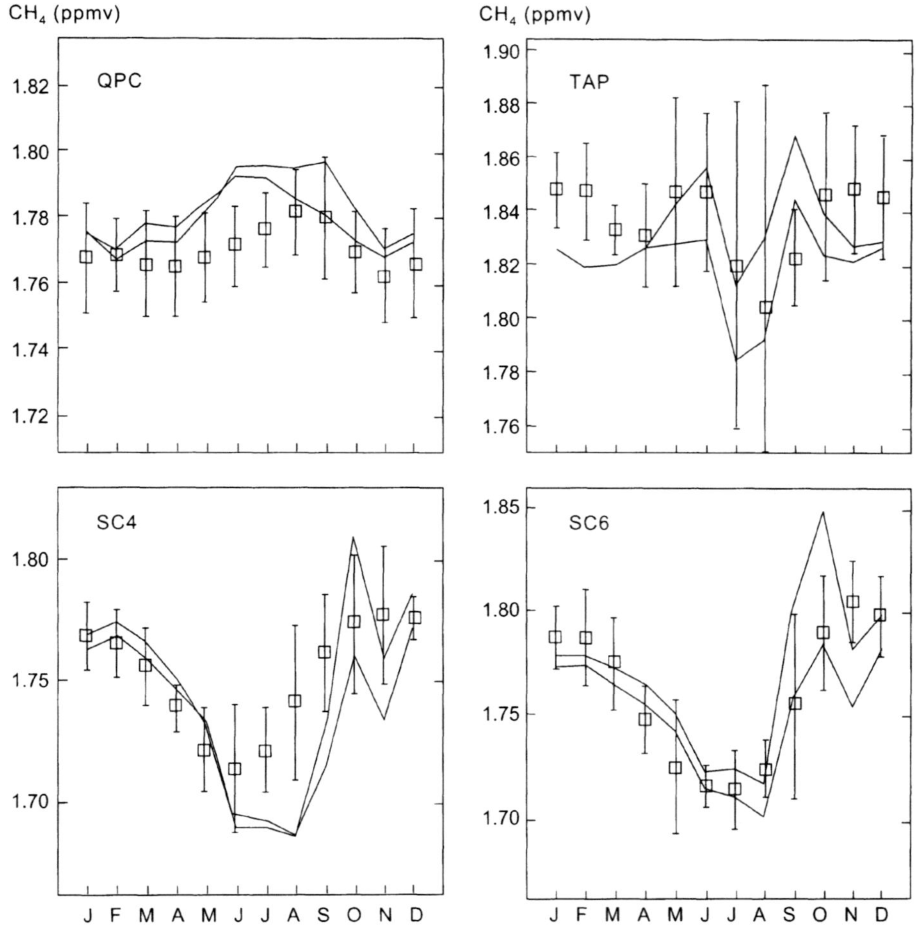

Figure 2. Comparison of measurements and model-derived seasonal cycles in atmospheric CH_4 mixing ratio at stations Quinghai Province (QPC), Tae-ahn Peninsula (TAP), and the South China Sea, 12 °N (SC4) and 18 °N (SC6). All concentrations are representative of the year 1994. Model calculations are based on 'standard' (solid line) and 'low-rice'(dashed line) emission scenarios

constrained without such observations. In recent years, the NOAA/CMDL global network of trace gas-monitoring stations (Dlugokencky, 1994) expanded and we were able to add 12 stations, three of them located in Asia, to the 22 used by Hein et al. (1997). However, the resulting number of stations, where samples are taken in weekly intervals only, proved still insufficient to resolve the complex concentration distribution over the Southeast Asian continent. Further improvement was achieved by incorporating the air sampling cruises on the Pacific and South China Sea (Lang, 1992). This added another 11 observational points, bringing the number of locations with atmospheric CH_4 mixing ratios suitable for use in our inverse model calculations to 45. The seasonal variation at four of the most relevant stations for the rice-growing regions of China is shown in Figure 2. The Tae-ahn Peninsula, Korea (TAP) station is influenced by many different sources, which is reflected in the large standard deviation. The Quinghai Province station (QPC), located on a high mountain in China (3810 m), mainly samples free tropospheric air and therefore shows little seasonal variation with small standard deviation. The other two locations are derived from the South China Sea cruises, SC4 (12NB) and SC6 (18NB), receiving air masses from the Chinese continent from August/September through February/March. Low-resolution inversions, in combination with sparse nonhomogeneous measurement networks, yield biased a posteriori estimates (Trampert & Snieder, 1996). This bias can be reduced by increasing the resolution or by defining regions such that the emission distributions over these regions are well known. Unfortunately, in case of CH_4, this prerequisite is not satisfied for most parts of the globe. In our inversion, we define the sources at the scale of the model grid (8° × 10°), in contrast to Hein et al (1997) who applied global scale source distributions. As a consequence, the number of unknowns (the sources to be estimated) is much larger than the number of measurements. To obtain a unique solution for the resulting underdetermined inverse problem, a priori information is introduced (for detailed description, we refer to Houweling et al. [2000]).

A priori assumptions

Initially, a priori emission distributions were taken from Hein et al. (1997). These distributions have been verified and updated. In addition, minor CH_4 sources have been accounted for—e.g., termites, oceans, continental shelves, permafrosts, volcanoes, and wild animals. Anthropogenic CH_4 emissions were derived from Olivier et al. (1996), except for rice agriculture which have been derived from distribution estimates by Matthews et al. (1991). Annual totals were adjusted in agreement with Lelieveld et al. (1998). Local uncertainties were derived from global-scale uncertainty estimates, under the assumption that all fluxes are uncorrelated.

Constraining the magnitude of rice field emissions

To investigate how well inverse modeling can constrain the emissions from rice fields, we tested two distinctly different a priori estimates of rice field emissions: 80–50 Tg CH_4 yr^{-1} (Lelieveld et al., 1998) and our own "best guess" estimate of 30–15 Tg CH_4 yr^{-1}, hereafter referred to as "standard" and "low-rice" scenario, respectively. The low-rice scenario is backed by recent emission estimates for Chinese rice fields ranging from 9 to 16 Tg yr^{-1} (e.g., Dong et al. [2000], Yao et al. [1996]), reassessment of previously published empirical methods (Denier van der Gon, 2000a) and assessments based on the revised OECD/IPCC guidelines (IPCC, 1997) combined with recently reported emission measurements (Neue & Sass, 1998). The aim of this exercise is to determine whether atmospheric CH_4 measurements favor one or both emission scenarios. We assumed the same globally integrated CH_4 budget for the two scenarios, which is crucial because otherwise differences between the standard and low-rice scenario are mainly explained by the difference in the a priori assumed global budgets. A constant global budget was achieved by keeping the sum of rice field and tropical wetland emissions constant, meaning that the low-rice scenario differs from the standard scenario in that rice field emissions were substituted for natural wetland emissions. To analyze the difference between both scenarios, we first look at the global integrated emissions and gradually zoom in to a region where rice is relatively important (10°N, 75°W to 40°N -135°W square). The a posteriori integrated emissions appear to be quite insensitive to the applied a priori scenario (Table 2). Globally, both scenarios show a small decrease of a posteriori totals, compared with the first guesses. In the standard scenario, the decrease over the region of intensive rice cultivation is large (–31%) compared with the global emission change (16%), which can be interpreted as a regional decrease superimposed on a global scale change. To compare the two a priori scenarios in more detail, we can look at individual measurement

Table 2. A priori assumed and a posteriori model-calculated emissions for the 'standard' and 'low-rice' scenario, integrated over the globe, northern hemisphere (NH), southern hemisphere (SH), the 10° N–40° N latitudinal band (zone) and the 75° W - 135° W part of the zone (region). Numbers in Tg CH_4 yr^{-1}

Scenario	Globe	NH	SH	Zone	Region
			A priori		
Standard	528 (±90)[a]	405 (±81)	123 (±40)	212 (±66)	111 (±56)
Low rice	528 (±77)	384 (±66)	143 (±38)	185 (±47)	74 (±31)
			A posteriori		
Standard	505 (±24)	340 (±19)	165 (±18)	169 (±25)	77 (±23)
Low rice	508 (±24)	342 (±18)	166 (±17)	164 (±23)	68 (±18)

[a]95% confidence interval (±2 sigma).

stations close to a large rice-growing region (Figure 2). Comparison of the model-calculated mixing ratios at SC4 and SC6 shows that a peak around September-October is associated with rice emissions from the Chinese continent; this peak is much higher in the standard scenario than in the low-rice scenario. Surprisingly, the observations at South China Sea do not show this peak, although the concentrations do increase after July/August due to a change in wind direction bringing air masses from the continent toward the stations. It is not yet possible to exclude either the standard or the low-rice scenario but, in line with Table 2, the results for the individual stations show that the low rice scenario is certainly not less realistic.

Temporal scaling of CH_4 emissions from rice fields

Various controlling variables of CH_4 emission from rice may change drastically on time scales >~5 yr—e.g., harvested rice area, cropping index, fertilizer use, rice varieties, and water management. The resulting temporal variations in CH_4 emissions due to land use change or new rice technology are expected to be considerable (Denier van der Gon, 1999, 2000). However, temporal scaling of CH_4 emissions from rice has not received much attention yet. The changes are driven by socio-economic developments and technological advances, and also depend on biophysical conditions, all of these are not uniform across Asia. Therefore, global or continental generalizations are not adequate to capture the temporal trends in CH_4 emissions from rice fields or its controlling variables. For example, CH_4 emissions are strongly enhanced by organic amendments (Denier van der Gon & Neue, 1995, Wassmann et al., 1996). The amount of organic manure applied in Chinese rice agriculture had doubled between 1952 and the early 1980s (Wen, 1984), whereas in Japan the use of organic manure declined sharply over the same period (Kanazawa, 1984). Such national trends in rice agricultural management significantly influence the national emission from rice fields, in the order of 10–40% (Denier van der Gon, 1999, 2000). Land use change, a process with a clear temporal dimension, can also significantly change the magnitude of CH_4 emissions from rice agriculture. For example, a future change to a rice–wheat rotation instead of double rice cropping would significantly reduce the harvested rice area and therewith, CH_4 emissions from rice agriculture.

Land use change for the case of Java 1994–2010

The impact of land use change on CH_4 emission is studied in a case study for Java with the CLUE modeling framework. The CLUE modeling framework is a dynamic spatial simulation methodology that uses actual and historical land use patterns in relation to biophysical and socioeconomic determining factors for the exploration of realistic land use changes in the near future (Veldkamp & Fresco, 1996; Verburg et al., 1999a). The CLUE methodology uses a multiscale approach to determine the competitive power of the different land use types at a certain location. The model can calculate the changes in land use pattern given a scenario of land use change at the national level (Verburg et al., 1999b). Such scenarios can be based on expected changes in consumption patterns, urbanization, and others. As an example, a scenario, based on a study by the World Bank (1992), is evaluated. The major land use change represented in this scenario, which is assumed to be representative of realistic future land use changes in Java, is caused by an increasing demand for nonagricultural land (e.g., land for urban and manufacturing development). Based on demand-supply studies, it is expected that within agriculture, there will be shifts away from rice toward horticultural crops and other cash crops. Model predictions for land use changes for the period 1994–2010 indicate 'hot-spots' of land use change (Figure 3). Land use dynamics in the uplands are generally low. Along the northern coast of Java, large decreases in rice area are expected. The model is spatially explicit and it can be seen that the decrease in rice area of Java takes place in the most productive and not in marginal (rainfed) ones. This is important for rice production predictions but also for CH_4 emissions from rice on Java because emissions depend on soil and management factors. For accurate temporal scaling of CH_4 emissions from rice fields, both land cover and land management change have to be included, but incorporation of management aspects in the model is not yet accomplished.

Land use change and CH_4 emission

The output of the land use change model for a chosen scenario can be used as model input for the CH_4 emission model described earlier. This is feasible because the areas of change are known and biophysical input parameters can be derived from the spatial databases or maps. Here we used the results presented in Figure 3

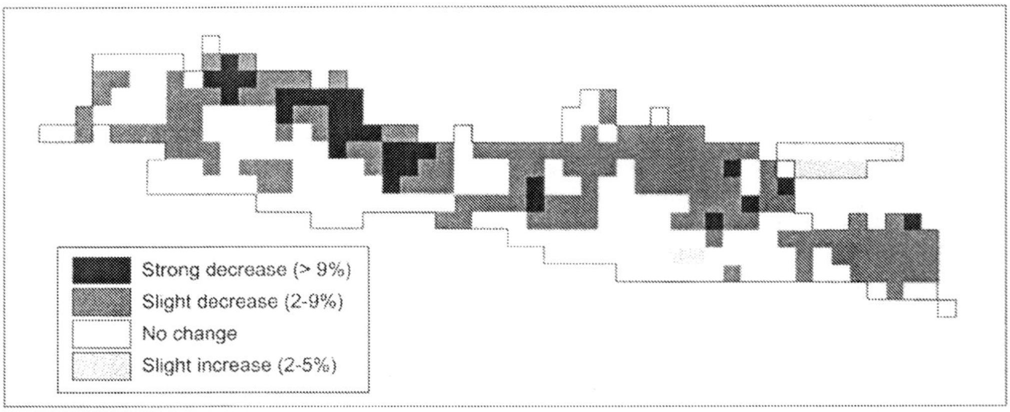

Figure 3. Predicted changes in rice area for Java from 1994 to 2010 with the CLUE modeling framework (see Verburg et al. [1999b] for a detailed description of the selected scenario)

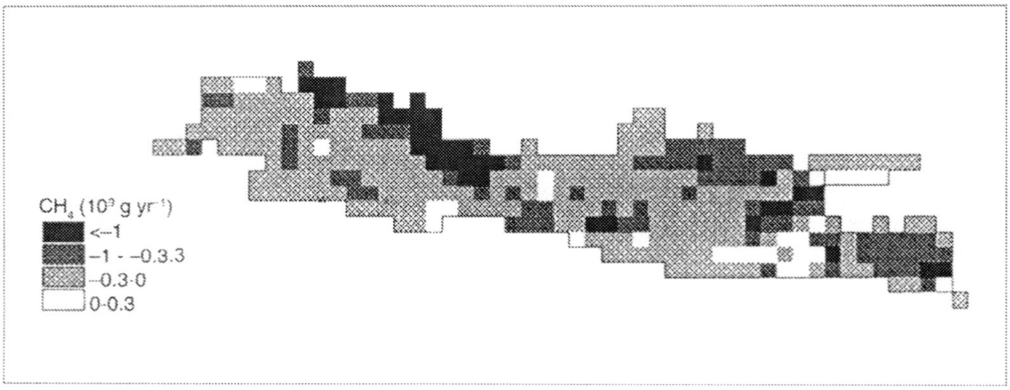

Figure 4. Predicted change in annual CH_4 emission from rice fields for Java by comparing predicted 2010 emissions with 1994 emissions using van Bodegom et al. (1999) with gridded 20- × 20-km land use data from Verburg et al. (1999b), soil data derived from FAO (1995) and Batjes (1995), land use change as predicted in Figure 2 and assuming no change in cropping index and ratio of irrigated rice to rainfed rice in 2010 as compared with 1994

as model input, resulting in a CH_4 emission map for 2010 comparable with Figure 1 (results not shown). The change in annual CH_4 emission can be calculated by subtracting the 1994 level emission per 20- × 20-km block from the calculated 2010 emissions. The spatial explicit visualization of where changes in CH_4 emission are expected to occur, given the scenario studied, may help in understanding the overall calculated change (Figure 4). In this particular case, an emission of 0.80 Tg yr^{-1} for rice agriculture on Java was calculated, an overall decrease of 0.12 Tg yr^{-1} when compared with 1994. In the past, a change in CH_4 emission may have occurred due to abandonment of marginal lands and making more intensive use of the fertile and easily accessible lowlands. For the 2010 scenario, the change in CH_4 emission from rice agriculture is mainly caused by competition between agriculture and housing or infrastructure, resulting in a loss of fertile, intensively managed rice soils.

Discussion

Limitations of the application of CH_4 emission models in spatial upscaling

A major problem in upscaling methodologies as used here for Java is that essential data on spatial distribution of one or more crucial variables may be lacking.

Since the model cannot be applied without such data, the missing data will be replaced by 'best guesses,' expert judgement, derived data, or assumed to be homogeneous for the study area. This causes unquantifiable variability within the 'representative' rice field, leading to an also unquantifiable uncertainty in the final emission estimates. A related complication is that field-scale models are validated with data from field studies where the model-input parameters are accurately measured or estimation of these parameters is relatively easy and accurate. When applying the model to larger areas, the input parameters are not measured but derived from other sources such as local statistics, maps, etc. The parameter estimation to be used as model input is a critical process, which greatly affects the reliability of model-calculated emissions as is illustrated with two examples.

Carbon availability for CH_4 production

The amount of carbon available for microbial decomposition is a key factor in process-based models predicting CH_4 emission (Cao et al., 1995; Huang et al., 1998a; van Bodegom et al., 2000). Sources of decomposable carbon are soil organic matter (SOM), organic amendments, root exudates, turnover of roots, incorporated weeds, and remains of previous crops. Although SOM may be derived from soil maps or surveys, spatial explicit data on all the other sources of decomposable carbon are scarce. When data on carbon sources are missing, assumptions have to be made explicitly or implicitly. The effect of these assumptions on calculated emissions is usually not evaluated because quantification is very difficult.

For example, even if no organic amendments are used in a particular region, local farmer's management may significantly affect the amount of carbon returned to the soil. Incorporation of residues of the previous crop, further referred to as stubble, before the new crop is planted is a common farmer's practice and not considered a special treatment. Stubble in our definition is aboveground biomass left in the field after the straw has been cut off plus the underground roots. The straw is the part of the plant that is cut with the panicle and it generally starts from 30 to 40 cm above the soil in a country such as the Philippines (R.S. Lantin, pers. commun., 1998). The height of cutting will change if there is local use for straw such as fuel or animal fodder and whether harvest is done mechanically or manually. But fields with large stubble may be burned to ease plowing and puddling. In field experiments used for model validation, straw is usually cut close to the soil because researchers want to know how much straw was produced.

The consequences of assumptions on stubble management for the model of van Bodegom et al. (2000) were explored. The first scenario assumes that stubble is 15% of the aboveground biomass of the previous crop (rel. stubble, Figure 5). Because the aboveground biomass is estimated from yield data, this scenario results in an almost linear response of CH_4 emission with yield. The second scenario assumes a fixed stubble incorporation of 1.5 $t.ha^{-1}$ independent of yield obtained (fixed stubble, Figure 5). The third, rather extreme, scenario assumes all stubble (so including the belowground remains of the previous crop), removed from the field (no stubble, Figure 5). This results in strongly reduced or negligible CH_4 emissions. According to the model, without organic amendments and no stubble incorporation, the CH_4 production in some soils may be very low due to limited substrate supply combined with high contents of alternative electron acceptors, such as reducible iron in the case of the Maahas soil of Los Banos. In such cases, minimal CH_4 emissions are predicted. Calculated CH_4 emissions differed considerably, depending on stubble management. This illustrates the importance of crop residue management for CH_4 emission in the following growing season. In all scenarios, the presence of rice plants stimulates CH_4 emissions because of substrate supply caused by other plant parameters such as root exudates and root turnover.

Soil parameter estimation for CH_4 emission modeling

Various soil properties such as SOM, texture, or reducible Fe are important input parameters for field-scale CH_4 emission models (Huang et al., 1998a; van Bodegom et al., 2000). For regional studies, these properties are not measured but mostly derived from soil maps. This introduces several complications that are generally ignored because good alternatives are lacking. First, classification of soil maps in rice-growing regions is often heavily based on (geo)morphological criteria determined in the field rather than laboratory analysis, and soil classification is usually not based on topsoil properties since these are considered to be too variable. Second, when a soil is used for long-term rice cultivation, some important topsoil parameters may change due to physical processes—e.g., terrace building, puddling and plowing, alternate reduction-oxidation cycles, or cultivation practices (IRRI, 1978; Suzuki et al., 1990). However, rice soils are generally not a

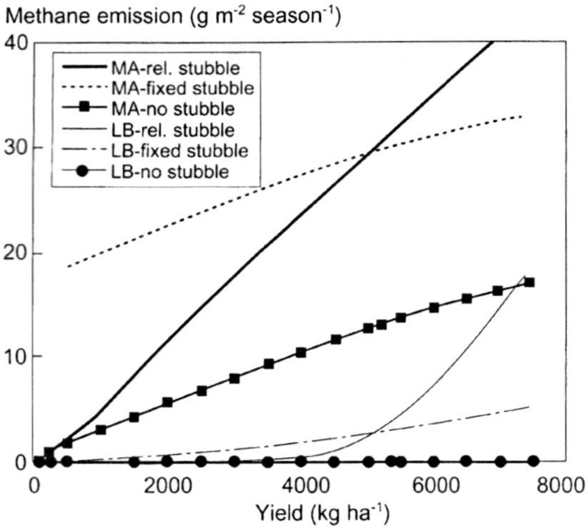

Figure 5. Methane emission calculated using van Bodegom et al. (1999) for two sites in the Philippines, Maligaya (MA) and Los Baños (LB) for three different stubble management scenarios without organic amendments. Stubble is defined here as the belowground biomass of the previous crop plus the aboveground biomass of the previous crop left in the field after the straw has been cut

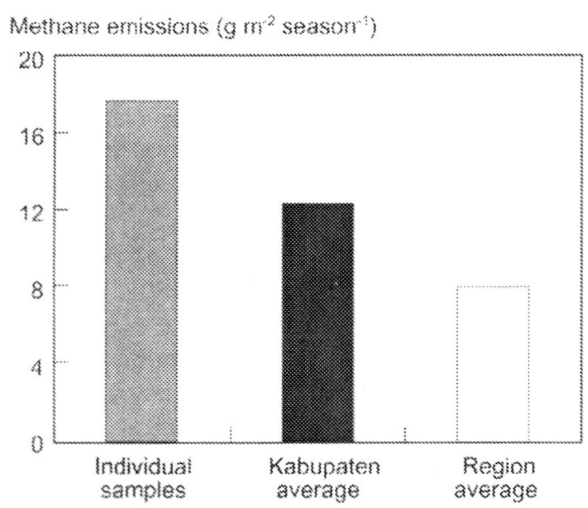

Figure 6. An example of the effecs of averaging soil data on he model-calculated average regional CH_4 emission for a rainfed rice region in Central Java using van Bodegom et al. (1999)

soil unit. As a result, topsoil properties of the fraction of a soil unit used for rice cultivation will differ significantly from the average topsoil properties of that soil unit. Analysis of reconnaissance soil maps in the Philippines indicated that the value of these maps, when used for quantitative spatial modeling, is questionable. The existing soil maps could only explain 0–40% of the variance for 14 agronomically important soil properties and large within–map unit variabilities were found (Oberthuer et al., 1996). It should be realized that soil maps were produced to bring a certain systematic order in soil formation, not to delineate mapping units for spatial CH_4 emission simulations.

Aggregation error

Independent of the accuracy in parameter estimation, a spatial explicit upscaling approach using field-scale models is hindered by scale discontinuities. Related phenomena at different spatial scales respond to completely different sets of causal factors (Clarke, 1985). It is therefore questionable whether CH_4 emission on larger scales is still controlled by variations at the field scale. Furthermore, the nonlinear relationship between controlling variables and CH_4 emission may cause an aggregation error. For a detailed description of aggregation errors made when modeling large-scale attributes of ecosystems, we refer to Rastetter et al. (1992). To illustrate the aggregation error, CH_4 emissions were calculated using the model of van Bodegom et al. (2000) with SOM and reducible Fe as model input from individual soil samples, average values for an administrative unit (*kabupaten*) and average values for a region (encompassing three kabupatens) in Central Java (Figure 6). In this particular case, a factor of 2 in final estimated emission was found, depending on whether a fine-scale resolution or average values were used because CH_4 emissions react nonlinearly to parameter changes.

Limitations of inverse modeling approaches to infer regional CH_4 source strength of rice fields

In top-down studies of the global CH_4 budget, the magnitude of rice agriculture as a CH_4 source is estimated at 70-100 Tg yr^{-1} (e.g., Fung et al., 1991; Hein et al., 1997). A regional top-down study using atmospheric CH_4 measurements in Korea confirmed the estimate of rice agriculture as a global CH_4 source of ~100 Tg yr^{-1} (Dlugokencky et al., 1993). However, the inverse modeling method as used by Hein et al. (1997) was not designed to study regional-scale sources. Global-scale constraints may well be insufficient to study the complex heterogeneous source signature at smaller scales. Moreover, the large variation observed in monitoring CH_4 from rice fields and the known dependence of emissions on say, management, irrigation, and soil type, indicate that large regional differences in source strengths of rice fields per unit area are to be expected.

This is not accounted for in current top-down approaches. Therefore, at present, results of inverse modeling at the global scale should not be interpreted as "proof" that emissions from rice fields have to be in the range of 70–100 Tg yr^{-1}. An alternative low rice scenario, with rice emissions at 30 Tg yr^{-1}, while keeping the sum of rice and tropical wetland emissions fixed, explained the variation in atmospheric CH_4 equally or, depending on the station, slightly better than a standard scenario. Because in the low-rice scenario the rice plus natural wetland emissions were kept constant, it was also an "enhanced tropical wetlands" scenario. Indeed, higher source strength for tropical wetlands may be realistic. In the standard scenario, the natural wetland emissions were estimated at 145 Tg yr^{-1} (Lelieveld et al., 1998) but Hein et al. (1997), using the inverse modeling method, estimated natural wetland emissions as 232 Tg yr^{-1}. Walter (1998), using a process-based model to derive CH_4 emissions, estimated the source strength of natural wetlands at 263 Tg yr^{-1} and suggested that especially the source strength of tropical wetlands was much higher than previously reported. The inverse model method needs further improvements to more precisely answer questions concerning the regional CH_4 budget of Southeast Asia. Possible improvements to be made are improving the a priori source distribution and fine-tuning of the interhemispheric exchange time using tracers with well-defined budgets such as F-11 and SF_6.

There are other top-down approaches aiming at quantification of regional CH_4 budgets than the methodology followed in the case study presented in this paper. Recently, emissions for the European continent have been estimated by Vermeulen et al. (1999) by means of a trajectory model and measurements at a relatively high sampling frequency (~200 samples d^{-1}) taken at Cabauw, The Netherlands. Results of this study show reasonable agreement with emission inventories such as EDGAR (Olivier, 1996), indicating that inverse modeling of regional-scale sources is indeed feasible. The methodology of Vermeulen et al. (1999) could be applied to other target regions. However, the number of available measurement sites and the sampling frequency are critical and at present too low to apply this technique to Southeast Asia.

Temporal scaling of CH_4 emissions from rice

The case study on land use change in Java indicated that the rice production capacity of the fertile lowlands may be reduced in the near future (Figure 3). As a result of these land use changes, our preliminary calculations indicate a small decrease in the magnitude of CH_4 emissions from Java. It is of interest to speculate how detailed information about one region such as Java may help to understand developments in other related regions. Despite a decline in rice area on Java, food demands of a growing population will have to be fulfilled. The average rice yield on Java, 5.2 t ha^{-1}, is ~40% higher than the average yield of the other Indonesian islands of 3.7 t ha^{-1} (BPS, 1996a). So, merely substituting rice produced on Java with rice produced on the outer islands asks for a considerably larger harvested area than is lost on Java. Moreover, to realize a growth in Indonesian rice production, while harvested area on Java is stable or declining, even larger areas on the outer islands have to be converted to rice fields. So, CH_4 emissions on Java are expected to decline. But considering the whole of Indonesia, an increase may be expected due to more than proportional rice area increases outside of Java to compensate for area losses on Java to cover future rice demand. This type of information may be highly valuable for predicting future emissions and designing efficient greenhouse gas mitigation policies.

Conclusions

Independent of the scaling methodology used, validation of regional CH_4 source strength estimates derived from scaling are severely hampered by the lack of independent regional-scale emission measurements that could constrain or be used to validate the scaling results. For example, as in the case of Java, a monthly measurement of the CH_4 emission of the whole Java would be extremely useful to validate the output predicted by the process-based emission model coupled to the GIS (Figure 1). The comparison of CH_4 budgets based on independent upscaling and downscaling methods may be a feasible methodology to reduce the uncertainty in the magnitude of regional CH_4 sources if the selected region can be chosen in such a way that the number of sources contributing to the regional CH_4 budget is small. This is because the downscaling approach is not source-specific, although, based on isotopic composition, a distinction between biogenic and fossil fuel-related sources can be made. However, another constraint of the downscaling approach is observational data of atmospheric CH_4 mixing ratios. Java is an example of a place where such observations are not present. To include such observational data, the size of the geographic region has to increase, which, in turn, has consequences for upscaling. China may be an example of a region where surface fluxes from rice agriculture can be inferred from inverse modeling of atmospheric mixing ratios, but the results are not yet conclusive. In general, quantification of regional flux estimates using inversion techniques for verification of upscaling estimates, national greenhouse gas budgets, or reduction targets in the Asian region calls for a considerable extension of the monitoring networks. Moreover, the current observational networks are heavily biased toward oceanic areas. A better and more detailed regional determination of continental sources requires observations closer to these sources (Heimann & Kaminski, 1999).

The reliability of upscaling results using spatially distributed data and a CH_4 emission model are limited by the uncertainty surrounding the model input parameters. Soil property estimation and local organic carbon management significantly influence the calculated emissions. Increasing the input and accessibility of local information and expertise may be an important improvement. Moreover, the upscaling methodology used here for Java allows, in principle, the use of different models as well as the use of different data input sources. This is highly recommended to reduce and better understand the uncertainty of the calculated regional emission estimate.

Given the limitations, developing other independent approaches to verify or constrain the regional source strength estimates should be encouraged. Apparent feasible alternative options include additional experiments and/or literature reviews to improve the proxy methods (method 3, Table 1). For example, Huang et al. (1997) measured the fraction of NPP emitted as CH_4 in Texas rice fields and found a range of 1.2–5.4% of NPP emitted as CH_4. Similar measurements could be done in other rice-growing regions.

Methane emission models as well as land use change studies have to be developed in such a way that they can be linked to each other and an integrated assessment of the effects of land use change can be made. The preliminary results presented here show that such a coupling is feasible. This may be essential if we aim at accurately predicting future CH_4 emissions from rice. Easterling (1997) convincingly argued that regional studies are essential in support of integrated assessment modeling of global change processes. To a large extent, Easterling's (1997) arguments also apply to the assessment of the global CH_4 source strength of rice fields. National or subnational policymakers will need regional studies for mitigation strategies and global change policy in general because global emission factors are not reliable at the (sub)national scale. The composition of regional greenhouse gas budgets, a multiple source approach, may further reduce uncertainties in estimates of individual sources. The combination of upscaling and downscaling approaches may be a future tool to reduce uncertainties in greenhouse gas budgets but at present the problems to successfully apply upscaling and downscaling approaches, at a resolution where these approaches match, present a major scientific challenge.

Acknowledgements

We thank the United Nations Development Programme Global Environment Facility, Interregional Research Programme on Methane Emissions from Rice Fields (GLO/91/G31) and especially R. Wassmann, R.S. Lantin, A.K. Makarim, T. Corton, L. Buendia, Lu Wanfang, J. Bajita, & P. Setyanto for their cooperation, discussion, and sharing of unpublished results. We also thank two referees for their constructive comments. We are grateful to N.H. Batjes (ISRIC) for making the WISE database available and supplying a subset of the ISIS database containing soil Fe data. The research was funded by the Dutch National Research Programme on Climate Change and Global Air Pollution, Project 951202.

References

Aselman I & Crutzen PJ (1989) Global distribution of natural freshwater wetlands and rice paddies: their net primary productivity, seasonality, and possible methane emissions. J Atmos Chem 8:307–358

Bachelet D & Neue HU (1993) Methane emission from wetland rice areas of Asia. Chemosphere 26:219–237

Batjes NH (ed) (1995) A homogenized soil data file for global environmental research: a subset of FAO, ISRIC and NRCS profiles (Version 1.0). Working Paper 95/10, International Soil Reference and Information Centre (ISRIC), Wageningen

BPS – Biro Pusat Statistik (1996a) Statistical yearbooks Biro Pusat Statistik Indonesia. In: Arif M, Madiadipura H & Zulfikar H (eds), CGPRT crops in Indonesia: a statistical profile. Working Paper 18. The CGPRT Centre, Bogor, Indonesia

BPS – Biro Pusat Statistik (1996b) Survei luas tanaman padi dengan pendekatan rumah tangga di Jawa. Jakarta

Brown M (1993) Deduction of emissions of source gases using an objective inversion algorithm and a chemical transport model. J Geophys Res 98:12,639–12,660

Brown M (1995) The singular value decomposition method applied to the deduction of the emissions and the isotopic composition of atmospheric methane. J Geophys Res 100: 11,425–11,446

Cao MK, Dent JB & Heal OW (1995) Modeling methane emissions from rice paddies. Global Biogeochem Cycles 9:183–195

Cao MK, Gregson K, Marshall S, Dent JB, & Heal OW (1996) Global methane emissions from rice paddies. Chemosphere 33:879–897

Cicerone RJ & Shetter JD (1981) Sources of atmospheric methane: measurements in rice paddies and a discussion. J Geophys Res 86:7203–7209

Cicerone, RJ, Shetter JD & Delwiche CC (1983) Seasonal variation of methane flux from a California rice paddy. J Geophys Res 88:11022–11024

Choularton TW, Gallagher MW, Bower KN, Fowler D, Zahniser M & Kaye A (1995) Trace gas flux measurements at the landscape scale using boundary-layer budgets. Phil Trans Royal Soc London 351:357–368

Clark WC (1985) Scales of climate impacts. Clim Change 7: 5–27

Denier van der Gon HAC & Neue HU (1995) Influence of organic matter incorporation on the methane emission from a wetland rice field. Global Biogeochem Cycles 9: 11–22

Denier van der Gon HAC (2000) Changes in CH_4 emissions from rice fields from 1960 to the 1990s. I. Impacts of modern rice technology. Global Biogeochem Cycles 14:61-72

Denier van der Gon HAC (1999) Changes in CH_4 emissions from rice fields from 1960 to the 1990s. II. The declining use of organic inputs in rice farming. Global Biogeochem Cycles 13:1053-1062

Dlugokencky EJ, Harris JM, Chung YS, Tans PP & Fung I (1993) The relationship between the methane seasonal cycle and regional sources and sinks at Tae-Ahn Peninsula, Korea. Atmos Environ 27:2115–2120

Dlugokencky EJ, Steele LP, Lang PM & Masarie KA (1994) The growth rate and distribution of atmospheric methane. J Geophys Res 99:17021–17043

Dong HM, Erda L, Li Y, Rao MJ & Yang QC (1996) An estimation of methane emissions from agricultural activities in China. Ambio 25:292–296

Easterling WE (1997) Why regional studies are needed in the development of full-scale integrated assessment modelling of global change processes. Global Environ Change 7: 337–356

Enting IG (1985) A classification of some inverse problems in geochemical modelling. Tellus 37B: 216–229

Enting IG & Mansbridge JV (1989) Seasonal sources and sinks of atmospheric CO_2, direct inversion of filtered data. Tellus 41B:111–126

FAO – Food and Agriculture Organization (1988) Soil map of the world, revised legend. FAO/UNESCO World Soil Resources Report 60. Rome, Italy

FAO – Food and Agriculture Organization (1995) The digital soil map of the world. Notes, version 3.5. Rome, Italy

Fowler D, Hargreaves KJ, Skiba U, Milne R, Zahniser MS, Moncrieff JB, Beverland IJ & Gallagher MW (1995) Measurements of CH_4 and N_2O fluxes at the landscape scale using micrometeorological methods. Phil Trans Royal Soc London A - Physical Sciences and Engineering 351:339–355

Fung I, John J, Lerner J, Matthews E, Prather M, Steele LP & Fraser PJ (1991) Three-dimensional model synthesis of the global methane cycle. J Geophys Res 96(D): 13033–13065

Hartley DE & Prinn R (1993) Feasability of determining surface emissions of trace gases using an inverse method in a three-dimensional chemical transport model. J Geophys Res 98:5183–5197

Heimann M & Kaminski T (1999) Inverse modelling approaches to infer surface trace gas fluxes from observed atmospheric mixing ratios. In: Bouwman AF (ed) Approaches to Scaling of Trace Gas Fluxes in Ecosystems. Developments in Atmospheric Science 24, Elsevier, Amsterdam

Hein R, Crutzen PJ & Heimann M (1997) An inverse modeling approach to investigate the global atmospheric methane cycle. Global Biogeochem Cycles 11:43–76

Holzapfel-Pschorn A & Seiler W (1986) Methane emission during a cultivation period from an Italian rice paddy. J Geophys Res 91:11803–11814

Hosono T & Nouchi I (1997) The dependence of methane transport in rice plants on the root zone temperature. Plant Soil 191: 233–240

Houweling S, Dentener FJ, Lelieveld J, Kaminski T & Heimann M (2000) Inverse modelling of the sources and sinks of methane using the adjoint of a global chemical transport model. J Geophys Res (in press)

Huang Y, Sass RL & Fisher FM (1997) Methane emission from Texas rice paddy soils. 2 Seasonal contribution of rice biomass production to CH_4 emission. Global Change Biol 3:491–500

Huang Y, Sass RL & Fisher FM (1998a) A semi-empirical model of methane emission from flooded rice paddy soils. Global Change Biol 4:247–268

Huang Y, Sass RL & Fisher Jr FM (1998b) Model estimates of methane emission from irrigated rice cultivation of China. Global Change Biol 4:809–822

Huke RE & Huke EH (1997) Rice area by type of culture: South, Southeast, and East Asia. A revised and updated data base. International Rice Research Institute, Los Banos, Philippines

IPCC – Intergovernmental Panel on Climate Change (1994) Climate Change. Radiative forcing of climate change and an evaluation of the IPCC IS92 emission scenarios. Cambridge University Press, New York, USA

IPCC – Intergovernmental Panel on Climate Change (1995) IPCC Guidelines for National Inventories of Sources and Sinks of Greenhouse Gases. Workbook. OECD, Paris

IPCC – Intergovernmental Panel on Climate Change (1997) Revised IPCC Guidelines for National Greenhouse Gas Inventories. OECD, Paris

IRRI – International Rice Research Institute (1978) Soils & Rice. Manila, Philippines

Kanazawa N (1984) Trends and economic factors affecting organic manures in Japan. In: Organic Matter and Rice, pp 557–569, International Rice Research Institute, Manila, Philippines

Kern JS, Gong ZT, Zhang GL, Zhuo HZ & Luo GB (1997) Spatial analysis of methane emissions from paddy soils in China and the potential for emission reduction. Nutr Cycling Agroecosyst 49:181–195

Knox J, Matthews RB & Wassmann R (2000) Using a crop/soil simulation model and GIS techniques to assess methane emissions from rice fields in Asia. II. Databases. Nutr Cycling Agroecosyst (this issue)

Koyama T (1963) Gaseous metabolism in lake sediments and paddy soils and the production of atmospheric methane and hydrogen. J Geophys Res 68:3971–3973

Kreileman GJJ & Bouwman AF (1994 Computing land use emissions of greenhouse gases. Water Air Soil Pollut 76: 231–258

Lang PM, Steele LP, Waterman LS, Martin RC, Masarie KA & Dlugokencky EJ (1992) CMDL atmospheric methane data for the period 1983-1990 from shipboard flask sampling. NOAA Tech Mem ERL CMDL-4 NOAA Environ Res Lab, Boulder, Colorado

Lelieveld J, Crutzen PJ & Dentener FJ (1998) Changing concentration lifetime and climate forcing of atmospheric methane. Tellus 50B:128–150

Liu J, Zhang Y, Shao K, Shao M, Zeng L, Lu S, Slanina J & Denier van der Gon HAC (2000) Regional methane emission from rice fields by atmospheric diffusion model. Nutr Cycling Agroecosyst (this issue)

Matthews E, Fung I & Lerner J (1991) Methane emission from rice cultivation: geographic and seasonal distribution of cultivated areas and emissions. Global Biogeochem Cycles 5:3–24

Minami K, Mosier A & Sass R (eds) (1994) CH_4 and N_2O: global emissions and controls from rice fields and other agricultural and Industrial sources. National Institute of Agro-Environmental Science, Tsukuba, Japan

Neue HU, Becker-Heidmann P & Scharpenseel HW (1990) Organic matter dynamics, soil properties, and cultural practices in rice lands and their relationship to methane production. In: Bouwman AF (ed) Soils and the Greenhouse Effect, pp 457–466, Wiley and Sons, New York

Neue HU & Sass RL (1998) The budget of methane from rice fields. IGACtivities Newsl 12:3–11

Oberthür T, Dobermann A & Neue HU (1996) How good is a soil map for agronomic purposes? Soil Use Manage 12: 33–43

Olivier JGJ et al (1996) Description of EDGAR Version 2.0. A set of global emission inventories of greenhouse gases and ozone-depleting substances for all anthropogenic and most natural sources on a per country basis and on 1° ∞ 1° grid. RIVM report 771060 002/TNO-MEP report R96, RIVM Bilthoven, The Netherlands

Rastetter EB, King AW, Cosby BJ, Hornberger GM, O'neil RV & Hobbie JE (1992) Aggregating fine-scale ecological knowledge to model coarser-scale attributes of ecosystems. Ecol Appl 2:55–70

Sass RL (1999) Methane from rice agriculture. Paper presented at the IPCC/OECD/IEA Programme on National Greenhouse Gas Inventories, Expert Group Meeting on Good Practice in Inventory Preparation. Agricultural Sources of Methane and Nitrous Oxide, 24-26 February 1999, Wageningen, The Netherlands

Schütz H, Holzapfel-Pschorn A, Conrad R, Rennenberg H & Seiler W (1989) A 3-year continuous record on the influence of daytime, season and fertilizer treatment on methane emission rates from an Italian rice paddy. J Geophys Res 94:16405–16416

Seiler W, Holzapfel-Pschorn A, Conrad R & Scharffe D (1984) Methane emission from rice paddies. J Atmos Chem 1: 241–268

Soepraptohardjo M & Suhardjo H (1978) Rice soils of Indonesia. In: Soils and Rice, pp 99–115, International Rice Research Institute, Manila, Philippines

Suzuki M, Kamekawa K, Sekiya S & Shiga H (1990) Effect of continuous application of organic or inorganic fertilizer for sixty years on soil fertility and rice yield in paddy field. In: Proceedings of the 14th International Congress of Soil Science, IV. pp 14–19, Kyoto, Japan

Taylor JA, Brasseur GP, Zimmerman PR & Cicerone RJ (1991) A study of the source and sinks of methane and methyl chloroform using a global three-dimensional lagrangian tropospheric tracer transport model. J Geophys Res 96 (D):3013–3044

Thohjima Y, Wakita H, Maksuytov S, Machida T, Inoue G, Vinnichenko N & Khattatov V (1997) Distribution of tropospheric methane over Siberia in July 1993. J Geophys Res 102:25,371–25,382

Trampert J & Snieder R (1996) Model estimations biased by truncated expansions: possible artifacts in seismic tomography. Science 271:1257–1260

Van Bodegom PM, Wassmann R & Corton TM (2000) A process-based model for methane emission predictions from flooded rice paddies. Global Biogeochem Cycles (in press)

Van de Ven T & Tempel P (1994) ISIS 4.0, ISRIC Soil Information System: User Manual (rev ed). Technical Paper 14. International Soil Reference and Information Centre (ISRIC), Wageningen

Veldkamp A & Fresco LO (1996) CLUE: a conceptual model to study the conversion of land use and its effects. Ecol Model 85:253–270

Verburg PH, de Koning GHJ, Kok K, Veldkamp A & Bouma J (1999a) A spatial explicit allocation procedure for modelling the pattern of land use change based upon actual land use. Ecol Model 116:45–61

Verburg PH, Veldkamp A & Bouma J (1999b) Land use change under conditions of high population pressure: the case of Java. Global Environ Change 9:47–56

Vermeulen AT, Eisma R, Hensen A & Slanina J (1999) Transport model calculations of NW-Europe methane emissions. Environ Sci Policy 2:315–324

Walter B (1998) Development of a process-based model to derive methane emissions from natural wetlands for climate studies. PhD thesis, Max Planck Institut fur Meteorologie, Hamburg

Wassmann R, Neue HU, Alberto MCR, Lantin RS, Bueno C, Llenaresas D, Arah JRM, Papen H, Seiler W & Rennenberg H (1996) Fluxes and pools of methane in wetland rice soils with varying organic inputs. Environ Monitor Assess 42:163–173

Wen QX (1984) Utilization of organic materials in rice production in China. In: Organic Matter and Rice. pp 45–56, International Rice Research Institute, Manila, Philippines

World Bank (1992) Indonesia: agricultural transformation challenges and opportunities. Report no. 1054-IND. Washington, USA

Yao H, Zhuang YB & Chen ZL (1996) Estimation of methane emission from rice paddies in mainland China. Global Biogeochem Cycles 10:641–649

Nutrient Cycling in Agroecosystems **58:** 303–310, 2000.

Estimation of regional methane emission from rice fields using simple atmospheric diffusion models

J.G. Liu[1], Y.H. Zhang[1], K.S. Shao[1], M. Shao[1], L.M. Zeng[1], S.H. Lu[1], S. Slanina[2] & H.A.C. Denier van der Gon[3]

[1]*Center of Environmental Sciences, Beijing University, Beijing 100871, China;* [2]*Netherlands Energy Research Foundation ECN, Westerduinweg 3, P.O.Box 1, 1755 ZG Petten (N.H.), The Netherlands;* [3]*Laboratory of Soil Science and Geology, Wageningen Agricultural University, P.O.Box 37, 6700 AA, Wageningen, The Netherlands*

Key words: closed chamber method, box model, ATDL model, CH_4, China

Abstract

Two atmospheric diffusion models, the box model and the ATDL (Atmospheric Turbulent and Diffusion Laboratory) model, were used to calculate regional methane (CH_4) emissions of rice fields in the Beijing area. Compared with conventional closed chamber measurements, the box model overestimated CH_4 emission because of meteorological conditions—the ground inverse layer was not favorable for the application of the model during the rice-growing season. The ATDL model, on the other hand, handled this unfavorable meteorological condition and gave reasonable CH_4 emission estimates (about 6.1–8.5 mg m^{-2} h^{-1}) close to conventional measurements (about 0.3–14.3 mg m^{-2} h^{-1}) in June, a period generally characterized by significant CH_4 emission from rice fields. In September, CH_4 emission as measured with closed chambers was negligible (about 0–0.3 mg $m^{-2}h^{-1}$), but the ATDL model still calculated it to be about 2.8–5.3 mg m^{-2} h^{-1}, albeit at a low level and considerably below the June emission level. This discrepancy cannot be explained at present and needs further study. Most likely causes are measurement artifacts and/or the presence of minor local CH_4 sources (ditches, field depressions) in the study area. The application of atmospheric diffusion models for regional CH_4 emission estimation depends greatly on meteorological conditions. Moreover, the models tend to give much more reliable results during periods of rather high CH_4 emission. This coincides with the time that such regional CH_4 emission estimates are most valuable. The atmospheric diffusion models complement the closed chamber method by providing integrated CH_4 emission estimates from 1–100-km^2 rice areas. Detailed information about agricultural management of rice fields and other potential CH_4 sources within the study region are necessary to better understand the integrated regional emission estimates.

Introduction

As a radiatively active trace gas and an important reactant in the atmospheric chemical system, the increase in atmospheric methane (CH_4) can significantly influence global climate and atmospheric chemistry (Cicerone & Oremland, 1988). However, still large uncertainties in the global CH_4 budget exist, especially for some major CH_4 sources such as rice fields. The closed chamber method, generally used to measure CH_4 fluxes from rice fields, introduces a large uncertainty when used to calculate fluxes from large areas because the chamber covers only about 1 m^2, despite the large spatial variations in CH_4 emission. To reduce the uncertainties extrapolated by conventional methods, a comparison can be made with estimates obtained using other measurement methods, such as micro-meteorological techniques or models. Micrometeorological techniques (e.g., gradient or eddy-correlation method) need highly sensitive and high-frequency CH_4 concentration responder (Shurpali et al., 1993; Simpon et al., 1995), while models can give the estimation by relatively ordinary CH_4 concentration measurement. The models used for this purpose include the atmospheric tracer model (Czpiel et al., 1996; Shorter et al., 1996), regression models (Aselmann & Crutzen, 1990; Bachelet & Neue, 1993), process-based models (Cao et al., 1995; Walter et al., 1996), and trajectory models (Veltkamp et al., 1995).

China is one of the most important rice-producing countries in the world. Methane emission from rice fields has always been estimated in China by using the conventional static chamber method (Shao et al., 1993; Wang et al., 1993; Khalil & Rasmussen, 1993). This pilot study aims to develop a methodology for regional validation of CH_4 emission by using atmospheric diffusion models to calculate spatial average CH_4 emission from rice fields over a large area.

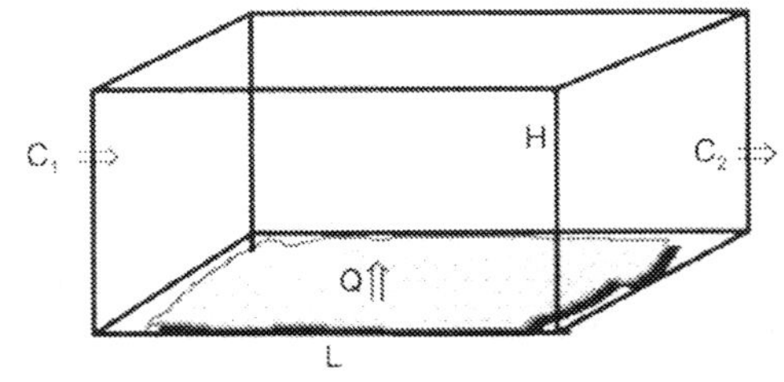

Figure 1. Methodology of the box model for estimating CH_4 emission from rice fields

Methodology

Rice fields can be regarded as an area source of CH_4. Under specific meteorological conditions—i.e., mixing layer formed completely and advection wind as dominant transport process for the CH_4 emitted—the CH_4 emission rate from rice fields can be calculated through a box model using this formula (Hanna et al., 1982, Figure 1):

$$Q = \frac{(C_2 - C_1)\, UH}{L} \tag{1}$$

where Q is emission rate (mg m^{-2} s^{-1}); C_1 and C_2 are concentrations (mg m^{-3}) upwind and downwind of the source area, respectively; U is windspeed (m s^{-1}), H is the height of mixing layer (m), and L is distance (m) in the wind direction of the virtual box.

The Atmospheric Turbulent and Diffusion Laboratory (ATDL) model is essentially an improved box model (Hanna et al., 1982). The model divides the whole area source into a number of small units and substitutes H (the height of the mixing layer) for the vertical dispersion parameter σ_z used in the Gaussian diffusion models. Consequently, the ATDL model can handle the unfavorable meteorological condition—e.g., the mixing layer was not formed or is too high. Methane is thus not well mixed vertically as assumed by the box model. The emission can be calculated by this ATDL model formula:

$$Q = \frac{(C_2 - C_1)\, Ua\,(1 - b)}{\sqrt{\frac{\pi}{2}}\,(\frac{2}{3}L)^{1-b}} \tag{2}$$

where Q, C_1, C_2, U, and L represent the same variables in Equation 1; a and b are parameters relating to the formula $\sigma_z = ax^b$, which is in common use in the Gaussian dispersion model.

The models depend on the measurement of concentration difference. Therefore, a numerical simulation was made prior to the experiment and thus indicated (data not shown) to identify unfavorable conditions (e.g., emission rate too small, height of mixing layer too high, or windspeed too high) that prohibit obtaining an effective concentration difference. Thus, the meteorological conditions set for the application of the box model are strict to some degree, though the model itself is simple. Moreover, the analytical precision of CH_4 concentration measurement in our study is 50 ppb (GC-FID). As a result, measured concentration differences less than 50 ppb are not suitable for model calculation.

The experiment

Experimental site

The rice area selected for the experiment was a 50-ha area, located about 60 km northwest of Beijing. It is a rather large and homogeneous area, rectangular in shape, with a length of 2.9 km in the north-south direction. A small village and a brook can be found in this region; no other significant CH_4 sources are nearby. In summer and autumn, the dominant wind in the Beijing area follows the north-south direction. The prevailing wind direction in the daytime is southern and windspeed is often high (about 2.0–4.5 m s^{-1}). At night, the prevailing wind direction is toward the north and windspeed is usually low (about 1.0–2.0 m s^{-1}). Based on numerical simulation, 5:00–9:00 in the morning of June-September was selected as the best time to carry out the experiment since windspeed and mixing layer height were low during this time.

Monitoring items and instruments

Methane concentrations were measured at the upwind and downwind sampling sites simultaneously in 15-min intervals. The air, at a height of 2 m above ground, was sampled with a 30-ml syringe. Air samples were carried back to the laboratory and CH_4 concentration was measured by gas chromatography (GC-FID) as soon as possible. Random flux measurements were made in the rice field by using the closed chamber method (Shao, 1993) to get an approximate estimate and comparison with our model results. Five to six plots at different locations within the experimental rice area were measured during the experiment with three replicates each. The Beijing Meteorological Science Research Institute made meteorological observations. Height of mixing layer, wind fields and their vertical profile, atmospheric stability, air temperature and humidity were measured by sounding radar, theodolite, sounding balloons, teleanemometer, and other instruments.

Results

Meteorological conditions and CH_4 concentration differences

The experiment was carried out on 27–29 Jun and 12–13 Sep. Figures 2 and 3 show typical results of the experiment in June and September. Ground inversion layers were generally observed during early morning in both months. The inverse layers at the lower altitude disappeared very rapidly at about 6:00~7:00 am in June. Then the mixing layer formed and became very high in a short time. The ground inverse layer was maintained longer in September, not disappearing completely until 8:00-9:00 am. The ground temperature increases were generally faster in June than in September.

Wind direction above 100 m was always different in June but it became homogeneous in September. Ground windspeed was generally higher in September (about 2.5–8.0 m s^{-1} under 500 m) than in June (about 0.5-3.0 m s^{-1} under 500 m). The dominant ground wind direction was northern both in June and September; Windspeed was always very low in June (about 0.5–1.6 m s^{-1}), becoming relatively higher in September (about 1.5–3.0 m s^{-1}).

Generally, the concentration difference measured between the upwind and downwind sites of the rice fields was small at early morning. It increased gradually and reached a maximum at around 8:00 am, and then decreased gradually to values smaller than 50 ppb. Semicontinuous closed-chamber measurements have shown that CH_4 emission from rice fields is lowest around 5:00 am, gradually increasing to a peak at noon or early afternoon (Denier van der Gon & Neue, 1995). The height of the inverse layer and windspeed were low in the morning, making the concentration difference between upwind and downwind sample location relatively high. When the mixing layer is formed and as it rises gradually, CH_4 concentration will be lower due to dilution and so will the CH_4 concentration difference.

Based on measurements for many days, it was clear that CH_4 concentrations at the measurement sites were always lower in September than in June and CH_4 emissions became much smaller in autumn. As a result, meteorological conditions and other local CH_4 sources have a more adverse influence on the concentration difference during autumn than during summer. Therefore, small or even negative concentration differences were often observed in September.

Model results

The emission rates calculated by the box and ATDL models are shown in 'Qbm' and 'Qam' column, respectively (Table 1). The CH_4 emission rate from the rice fields calculated with the box model was within the range of 13.2–30.4 mg m^{-2} h^{-1} in June and 10.4–20.0 mg m^{-2} h^{-1} in September. The corresponding values calculated with the ATDL model were 6.1–8.5 mg m^{-2} h^{-1} in June and 2.8–5.3 mg m^{-2} h^{-1} in September.

Flux measurements with the closed chamber method

The random measured CH_4 flux was in the range of 0.4–4.8 mg m^{-2} h^{-1} in June (Table 2). In September, the closed chamber measurements revealed that CH_4 emissions were close to zero from different plots. After the first 3 d, flux measurements were stopped since the fields were dry and CH_4 emission was expected to remain low. Our experimental area was divided into about 40 portions belonging to some 100 farmer families and quite different agricultural practices were carried out in the selected area. As a result, large spatial variations in emission rate were found in measurements from different rice plots within the experimental rice field in June, which shows the problem in using the closed chamber method reported earlier (Folorunso & Rolston, 1984). However, a full-scale flux measurement by box chambers in a 50-ha rice area was unpractical. The limited random flux measurements could only give us an approximate emission range for the selected experimental rice area. To give a more reliable comparison with model estimates, mapping of agricultural practices and field practices in the selected study area is needed and, if variability is high, more flux measurements are needed.

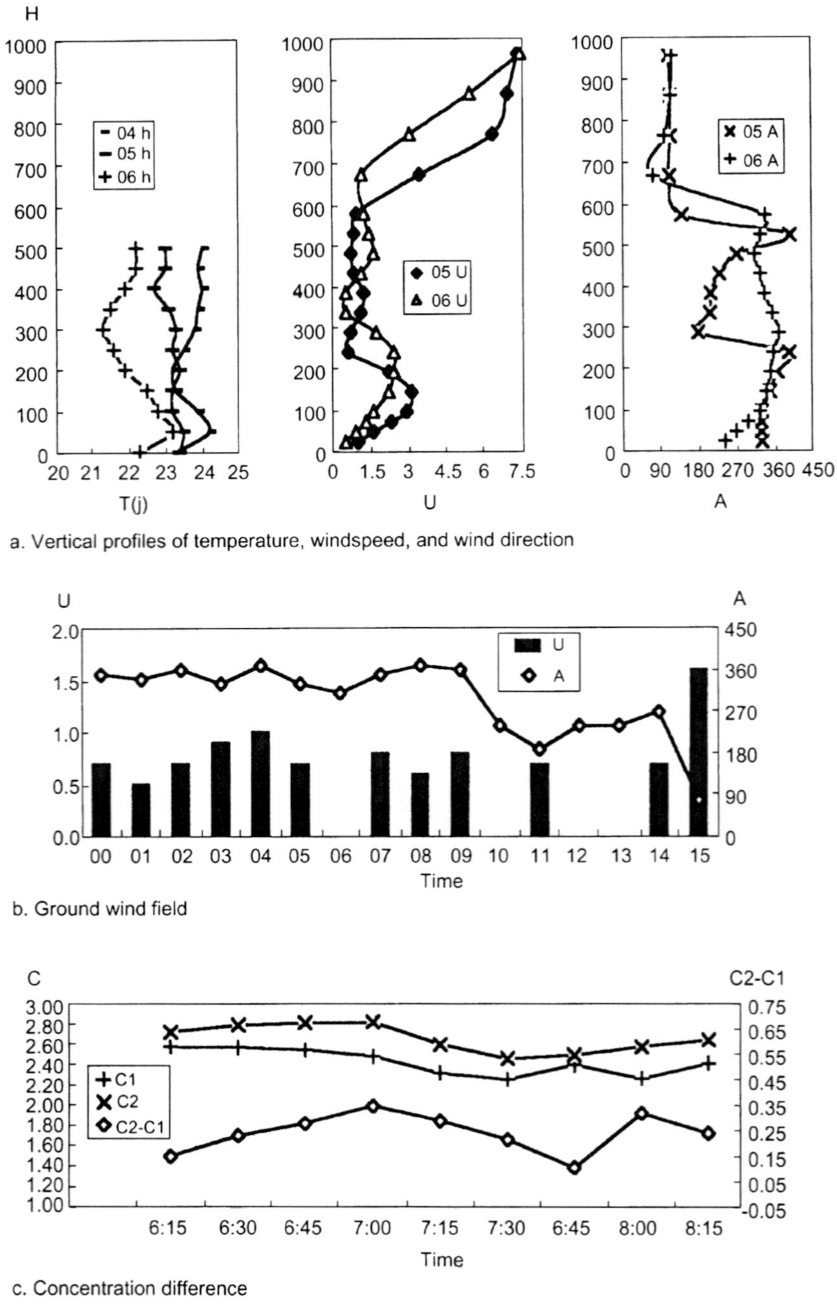

Figure 2. Typical results of the experiment in June. T = air temperature (°C), U = windspeed (m s^{-1}), A = wind direction, H = altitude (m), and C = CH_4 concentration (mg m^{-3}). "05" represents experimental time during 5:00-6:00 am, and so on.

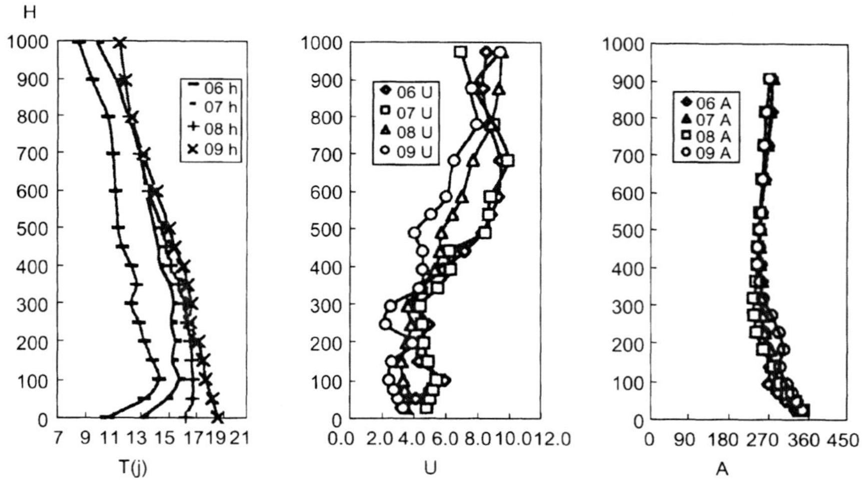

a. Vertical profiles of temperature, windspeed, and wind direction

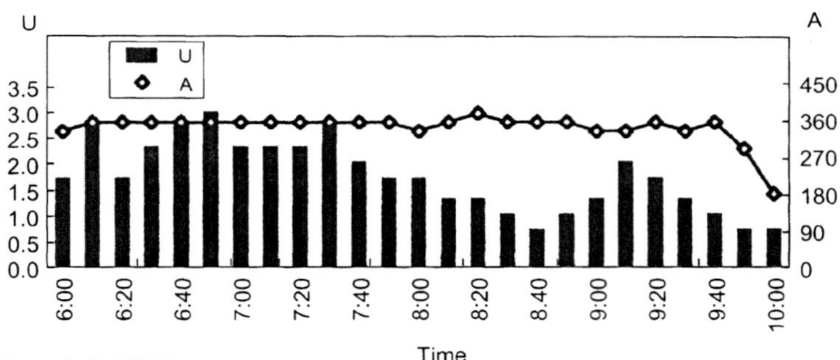

b. Ground wind field

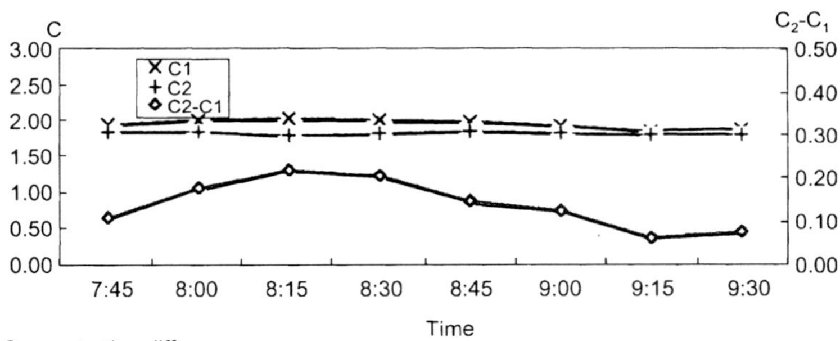

c. Concentration difference

Figure 3. Typical results of the experiment in September. (See Figure 2 for explanation of abbreviations.)

Table 1. Methane emissions of experimental rice fields calculated by models

Time	Wind direction A^a degree	Wind speed U^a (ms^{-1})	Inverse layer height H^b(m)	Stability[c]	Distance L^d (m)	Concentration difference $(C_2\text{-}C_1)^e$ (mg m^{-3})	Calculated emission rate (mg $m^{-2}h^{-1}$)	
							Qbm	Qam
6/27/5:00-6:00	250	0.90	200	E	4101	0.19	30.4	7.8
6/28/6:00-7:00	310	1	150	D	4101	0.13	17.6	6.1
6/29/6:00-7:00	330	1	75	D	3349	0.16	13.2	7.8
6/29/7:15-8:15	350	0.8	100	C	2945	0.15	14.6	8.5
Av							19.0	7.6
9/13/7:30-8:45	190	0.80	75	E	2945	0.14	10.4	2.8
9/18/7:45-8:30	360	1.40	50	E	2900	0.14	11.8	4.7
9/18/8:30-9:30	350	1.30	150	D	2945	0.08	20.0	5.3
Av							14.1	4.3

[a]Av of measured values at the time of calculation. [b]Detected by sounding balloon at the time of calculation. [c]Pasquill category : A = extremely unstable, B = moderately unstable, C = slightly unstable, D = neutral, E = slightly stable, F = moderately stable (Pasquill, 1961). [d]Calculated by size of experimental area, location of sample sites, and wind direction. [e]Av of effective concentration difference (i.e., value larger than 50 ppb) measured at the time of calculation

Table 2. Methane emissions (mg $m^{-2}h^{-1}$) by random flux measurements using closed chambers in the experimental rice fields

	Jun 26-29	Sep 12-18
Flux	0.4-8.5	0
Av	4.7	0

Table 3. Methane emission rates (mg $m^{-2}h^{-1}$) from rice fields under two kinds of fertilization conditions, CAAS

Fertilizer treatment	Jun 26-29	Av	Sep 12-18	Av
Rice straw input	12.6-16.3	14.3	0-0.5	0.3
Chemical fertilizer only	0.2-0.6	0.3	0-0.1	0

The field investigation shows that water regime and rice cultivar were similar throughout the experimental region on the whole. However, fertilization practices, such as use of organic amendment, are different because different farmers have access to different facilities and fertilizer sources. At least two distinctly different fertilizer practices were identified in our study area: mineral fertilizer plus rice straw or mineral fertilizer only. At the same time and close to our experimental rice area, a project conducted by the China Academy of Agricultural Sciences (CAAS) in cooperation with UNDP was proceeding. Methane emissions from rice fields with different fertilization practices were measured using their auto-monitoring box chamber system (Wang et al., 1997). Table 3 shows that the CH_4 emission measured from experimental rice fields with fertilization schemes similar to those in our study (data from CAAS-UNDP project, Wang et al., this issue), were in the range found by our random flux measurements (see Tables 2 and 3). In June, the emission from plots with straw amendment was much higher than from plots receiving mineral fertilizer only (14.3 and 0.3 mg $m^{-2}h^{-1}$, respectively). To give an exact estimate, the CH_4 emission of our 50-ha rice area using the closed chamber results, an exact mapping of plots with and without rice straw amendments is necessary. This information is not available and we can only conclude that the average emission on Jun 26–29 is between 0.3 and 14.3 mg $m^{-2}h^{-1}$. The September period is different because most rice fields in the Beijing area are drained by that time, and emissions measured with closed chambers are almost zero.

There are typically two CH_4 emission peaks during the rice-growing season in the Beijing area (Shao, 1993). The first peak coincides with the tillering stage and is generally much higher than the second one, which is observed at the heading stage. In the selected rice fields, the experimental period, Jun 25-29, coincided with late tillering, toward the end of the first CH_4 emission peak. The Sep 12-18 period corresponds to the milk grain or early ripening stage, usually characterized by baseline CH_4 emissions. Emission rate during the experiment was expected to be relatively low, especially in September.

Discussion

In comparison with the closed chamber method, the box model calculations resulted in much higher emission rates both for June and September (Table 4). However,

Table 4. Comparison between emission rates (mg $m^{-2}h^{-1}$) measured by the closed chamber method and the atmospheric diffusion models

Month	Closed chamber	Box model	ATDL model
June	0.3-14.3	13.2-30.4	6.1-8.5
September	0-0.3	10.4-20.0	2.8-5.3

CH_4 emission rates for June calculated by the ATDL model were in agreement with the closed chamber data. Emission rates for September calculated by the ATDL model were higher than those measured by the closed chamber techniques.

The box model assumes that the emitted CH_4 has been mixed well under the inverse layer. However, the ground inverse layer existing during the experiment was unfavorable for vertical mixing of the emitted CH_4. Therefore, the emitted CH_4 accumulated near the ground. Another factor not favorable for mixing was the low windspeed. Therefore, the measured concentration difference near the ground would be higher than that assumed in the box model, resulting in high emission rates calculated by the box model in both months. The ATDL model could handle this "not-so-ideal" condition because it incorporated the dispersion theory of the Gaussian diffusion models (instead of the complete mix theory used in the box model). Reliable results were thus obtained in June. The discrepancy between closed chamber results and ATDL model in September can be explained by 1) the ATDL model overestimating the CH_4 emissions at very low emission levels such as in September and/or 2) there were still local emissions in the area — e.g., from other minor sources like ditches, reservoirs, depressions in the fields, or slow release of soil-entrapped CH_4. In future experiments, a detailed survey of the study area during periods of low emissions is necessary to explain this discrepancy. In addition, analytical precision has to be improved in order to detect concentration differences smaller than 50 ppb and to identify emissions close to zero with the ATDL model.

Conclusion

The meteorological conditions in the Beijing area were unfavorable for CH_4 emission estimation from rice fields with the box model, but the ATDL model gave results in agreement with those from the conventional closed chamber method. However, use of the ATDL model requires high analytical precision and favorable meteorological conditions. It can therefore be used only during a certain time of day and not on a day-to-day basis. The methodology is therefore not suited for identifying diel emission patterns and only crude estimation of total seasonal emission is possible due to incomplete season coverage. However, the ATDL model can measure integrated regional CH_4 emissions from rice fields. The method is therefore complementary to the closed chamber method and may be a simple way of answering the pressing question of larger scale emission estimates (Khalil et al., 1998). Information on soil and agricultural practices employed in rice fields at the regional scale are essential in understanding these integrated emission estimates.

Acknowledgement

We especially thank Dr. Wang Zhengyuan of the China Academy of Agricultural Sciences who generously provided valuable data.

References

Aselmann I & Crutzen PJ (1990) A global inventory of wetland distribution and seasonality, net productivity, and estimations. In: Bouwman AF (ed) Soils and the Greenhouse Effect, pp 441-449, John Wiley, New York

Bachelet D & Neue HU (1993) Methane emission from wetland rice areas of Asia. Chemosphere 26:219-237

Cao MK & Dent JB (1995) Modeling CH_4 emissions from rice paddies. Global Biogeochem Cycles 9 (2):183-195

Cicerone RJ & Oremland RS (1988) Biogeochemical aspects of atmospheric methane. Global Biochem Cycles 2(4): 299-327,

Czpiel PM et al (1996) Landfill methane emissions measured by enclosure and atmospheric tracer methods. J Geophys Res 101:D11, 16711-16719

Denier van der Gon HAC & Neue HU (1995) Influence of organic matter incorporation on the methane emission from a wetland rice field. Global Biogeochem Cycles 9: 11-22

Folorunso OA & Rolston DE (1984) Spatial variablity of field measured denitrification gas fluxes. Soil Sci Soc Am J 43:89-95

Hanna SR & Steven R (1982) Handbook on atmospheric diffusion. Technical Information Center, U.S. Department of Energy

Khalil MAK & Rasmussen RA (1993) The methane sources in China, historical and current emissions. Chemosphere 26(1-4):127-142

Khalil MAK, Rasmussen RA, Shearer MJ, Dalluge R W, Ren Lixin & Chang-lin Duan (1998) Factors affecting methane emissions from rice fields. J Geophys Res 103:D19, 25219-25231

Pasquill F (1961) The estimation of the dispersion of windborn material. Meteorol Mag 90:33-49

Shao KS (1993) Preliminary study on the relationship of agricultural practices on mehane emission flux from rice paddy near Beijing area. Rural Eco-Environ 35 (in Chinese)

Shorter JH et al. (1996) Methane emission measurements in urban areas in eastern Germany. J Atmos Chem 24:121-140

Shurpali NJ, Verma SB, Clement RJ & Billesbach DP (1993) Seasonal distribution of methane flux in a Minnesota peatland measured by eddy correlation. J Geophys Res 98:D11, 20649-20655

Simpson IJ, Thurtell GW, Kidd G E, Lin M, Demetriades-Shah TH, Flitcroft ID, Kanemasu ET, Nie D, Bronson KF & Neue HU (1995) Tunable diode laser measurements of CH_4 fluxes from an irrigated rice paddy field in the Philippines. J Geophys Res 100:D4, 7283-7290

Veltkamp AC, Vermeulen AT, Kieskamp WM & Van Den Bulk WCM (1995) Source apportionment studies of atmospheric methane at two monitoring stations in the Netherlands (Progress Report 1990-1994), ECN-C—95-034

Walter BP, Heimann M, Shannon RD & White JR (1996) A process-based model to derive methane emissions from natural wetlands. Geophys Res Lett 23 (25):3731-3734

Wang ZY et al. (1997) Integrated report of IRRI-UNDP methane projects in Beijing. Paper presented at the In-house Review and Evaluation of the Methane Project in China, 08-10 Dec 1997

Wang MX, Shangguan WQ & Shen RX (1993) Sources of methane in China: rice fields, agricultural waste treatment, cattle, coal mines, and other minor sources. Atmos Sci 17:5 (in Chinese)

Nutrient Cycling in Agroecosystems 58: 311–314, 2000.

The effects of cultural practices on methane emission from rice fields

J.-Y. Ko & H.-W. Kang

National Yeongnam Agricultural Experiment Station, Rural Development Administration, Milyang, Kyweongnam, 627-130, Korea

Key words: direct seeding, transplanted seedling, plowing time

Abstract

A field experiment was conducted in a clayey soil to determine the effects of cultural practices on methane (CH_4) emissions from rice fields. The factors evaluated were a) direct seeding on dry vs wet soil, b) age of transplanted seedlings (8 d old and 30 d old), and c) fall vs spring plowing. Methane emissions were measured weekly throughout the rice-growing season using a closed static chamber technique. Transplanted 8-d-old seedlings showed the highest emission of 42.4 g CH_4 m^{-2} $season^{-1}$, followed by transplanted 30-d-seedlings (40.3 g CH_4 m^{-2} $season^{-1}$), and direct seeding on wet soil (37.1 g CH_4 m^{-2} $season^{-1}$). Direct seeding on dry soil registered the least emission of 26.9 g CH_4 m^{-2} $season^{-1}$. Thus transplanting 30-d-old seedlings, direct seeding on wet soil, and direct seeding on dry soil reduced CH_4 emission by 5%, 13%, and 37%, respectively, when compared with transplanting 8-d-old seedlings. Methane emission under spring plowing was 42.0 g CH_4 m^{-2} $season^{-1}$ and that under fall plowing was 31.3 g CH_4 m^{-2} $season^{-1}$. The 26% lower emission in the field plowed in spring was caused by degradation of organic matter over the winter.

Introduction

Irrigated rice fields are known as an important source of methane (CH_4), one of the greenhouse gases. They are estimated to contribute between 25.4 and 54 t yr^{-1} (Cole, 1996) of the total 410 to 660 million t yr^{-1} emitted globally (Houghton et al., 1996).

Methane is the decomposed product of organic matter under highly anaerobic condition and its production is, therefore, closely related to the soil redox potential. Takai et al. (1956) demonstrated that the redox potential of soils must be below –200 mv to produce CH_4. Wang et al. (1993) also reported that the critical initial Eh of methanogenesis was –150 to –160mv. Thus, the effect of cultural practices on CH_4 emission should be studied inasmuch as these practices differ according to duration of soil submergence during the cropping season. In addition, root growth and activity, which may affect CH_4 emission because the rice plant is an important transport medium of CH_4 from the rice fields to the atmosphere, would be diverse under different cultural practices. Plowing time is another cultural method that may influence CH_4 emission because it changes the chemical and physical properties of the soil. It may eventually affect the decomposition of organic matter in the soil.

Rice is the major staple food of Korea, and rice cultivation is necessary for food security. Therefore, it is important to decrease CH_4 emissions from rice fields without reducing the cultivated area. In this study, we investigated the effects on CH_4 emission of age of transplanted seedlings and time of plowing.

Materials and methods

The experimental site was in southeastern Korea. Mean temperature during cropping season (June to September) is 22.8 °C and precipitation during the period is 800 mm, which is two-thirds of total annual precipitation. The soil at this site is silty clay loam, which has a good water-holding capacity. Selected soil properties are presented in Table 1.

Treatment and field management

The experimental design was a randomized complete block, strip-plot experiment with three replicates. The main plot treatments were plowing times (two levels)

Table 1. Chemical and physical properties of soil used

Soil depth (cm)	pH (1:5)	OM (g kg^{-1})	P (mg kg^{-1})	Exch. cations (cmol^{+} kg^{-1})			Soil texture
				Ca	Mg	K	
0-15	5.7	29.4	40.6	3.5	0.8	0.3	SiCl
15-30	6.0	28.1	38.8	3.6	0.7	0.5	SiCl

and subplot treatments were cultural practices (four levels). The area of each plot was 670 m^2 and it had two chambers in it.

To measure the CH_4 fluxes as affected by cultural practices (particularly direct seeding and age of transplanted seedlings), this experiment used the following treatments: a) direct seeding on dry soil, b) direct seeding on wet soil, c) 8-d-old seedlings transplanted, and d) 30-d-old seedlings. (The most common cultural practice in South Korea is transplanting 30-d-old seedlings [67%], followed by transplanting 8-d-old seedlings [21%], and direct seeding [11%]). Direct seeding on dry and wet soil was done on 6 May 1996. Eight-day-old and 30-d-old seedlings were transplanted on 25 May and 6 Jun 1996, respectively. Fertilization rates were 150 kg N ha^{-1}, 30 kg P ha^{-1}, and 80 kg K ha^{-1} in direct-seeded plots. In transplanted plots, fertilizers were 110 kg N ha^{-1}, 30.5 kg P ha^{-1}, and 80 kg K ha^{-1}.

To measure the effects of plowing time on CH_4 fluxes, plowing was done in the fall (23 Nov 1995). Spring plowing to a depth of 15 cm was performed on 22 Apr 1996. Rice straw was applied at a rate of 5 t ha^{-1} just before plowing.

The rice cultivar used in all experimental treatments was Hwanambye, which is a japonica-type rice requiring about 120 d of growing period.

Fields were continuously flooded until harvest in all treatments.

Sampling and analysis

Methane emissions were measured with a closed static chamber of polyacrylic plastic (60 × 60 × 100 cm)with a fan (DC 12volt, 220 mA) to mix the air (Shin, 1996). The chamber has a top that opens and closes: the top remained opened during the cropping season except when air samples were collected. Wooden footbridges were installed beside the chamber to prevent CH_4 emission due to soil disturbance during the process of chamber installation and gas sampling.

The samples were collected once a week at 10 am with a stopcock-fitted PP syringe from 15 May to 5 Oct. The collected air samples were analyzed for CH_4 concentration by a gas chromatograph (HP 5890 Series) equipped with a flame ionization detector, using a Porapak N stainless steel column (80/100 mesh, 0.3 cm × 2 m) at 40 °C. Nitrogen was used as the carrier gas at a flow rate of 30 mL min^{-1} and a CH_4 standard of 0.93 mL L^{-1} was employed.

Results and discussion

Effects of cultural practices on CH_4 flux from rice fields

The CH_4 fluxes with different seedling age and seeding on wet and dry soils are shown in Figure 1.

Methane emissions increased until 22 Aug and decreased thereafter in all treatments. During tillering (12 Jun-11 Jul), CH_4 flux from direct seeding on wet soil was highest, followed by 8-d-old and 30-d-old transplanted seedlings. Direct seeding on dry soil emitted the least. The higher CH_4 emission from direct seeding on wet soil as compared with the other treatments may be attributed to soil redox potential being reduced sufficiently to an anaerobic condition suitable for CH_4 production.

From panicle formation to heading (16 Jul to 22 Aug), there was a large difference in CH_4 emissions between plots, and the highest CH_4 flux, 42 mg m^{-2} h^{-1}, was observed in the transplanted 30-d-old seedlings. This high flux value was due to a coincidence in the rise of temperature of both air and floodwater. When soil is submerged, as temperature increases, rice straw decomposes rapidly to produce CH_4 under anaerobic condition. The CH_4 fluxes obtained from direct seeding on dry soil were least among treatments. The differences among cultural practices were negligible at harvesting stage.

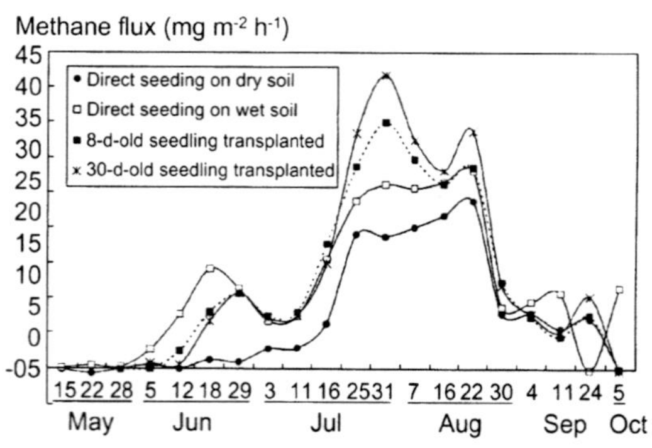

Figure 1. Variations in CH_4 emission as affected by different cultural practices

Table 2. Methane emissions as affected by cultural practices

Cultural practice	Methane emission ($g\ m^{-2} d^{-1}$)			Total emission ($g\ m^{-2} season^{-1}$)	Grain yield ($t\ ha^{-1}$)
	Min	Max	Av		
Direct seeding					
On dry soil	- 0.031	0.59	0.17	26.9 a[a]	5.28
On wet soil	0.003	0.66	0.24	37.1 b	5.38
Transplanting					
8-d-old seedling	0.001	0.70	0.31	42.4 b	5.39
30-d-old seedling	0.011	0.76	0.31	40.3 b	5.32

[a]Means within a column followed by the same letter are not significantly different at P= .05 LSD level. LSD(5%) = 5.25

Methane emissions due to cultural practices throughout the cultivation period are presented in Table 2. Average daily CH_4 emissions from direct seeding on dry soil was least (0.17 g CH_4 m^{-2}) among the treatments, followed by direct seeding on wet soil (0.24 g CH_4 m^{-2}). The transplanted treatments had the highest emission (0.31g CH_4 m^{-2}). However, based on total quantity of CH_4 emissions over the season, the 8-d-old seedling treatment gave values a little higher (42.4g m^{-2}) than the 30-d-old seedling treatment (40.3 g m^{-2}) because the cultivation period of the former was longer than that of the latter. But statistically, these were not different. So CH_4 emissions from 30-d-old transplanted seedlings, direct seeding on wet soil, and direct seeding on dry soil were reduced by 5%, 13%, and 37%, respectively, with respect to that from 8-d-old transplanted seedlings. Grain yield trend was similar to that of CH_4 emission. But there were no statistical differences among treatments.

The reason for the low CH_4 emission from direct seeding on dry soil plot was the aerobic condition during the early growth stages, resulting in small CH_4 production from applied organic matter such as rice straw. The other treatments, on the other hand, were flooded. In addition, the plant root system, which may affect the oxidation of soil-entrapped CH_4, was better developed here than in any other treatments because the soil was not submerged in the early growth stage. In a study of CH_4 emission from direct seeding on dry soil in China, the practice reduced CH_4 emission by 59-74% compared with the use of 30-d-old seedlings and application of pig manure (Liang, 1995).

The negative emission observed in direct seeding on dry soil may be brought about by the activity of methanotrophic bacteria, which oxidized the CH_4 under aerobic condition. This result indicates that the soil, which is not flooded, can act as a CH_4 sink. Similar CH_4 uptake patterns were seen in unflooded rice soils (Thurlow et al., 1995) and in Indian rice fields (Parashar et al., 1994).

Effects of plowing time on CH_4 flux from rice fields

Seasonal changes in CH_4 fluxes due to plowing times are shown in Figure 2. Three CH_4 peaks were observed during cultivation, regardless of plowing time. With both spring and fall plowing treatments, the first peak occurred at 4 wk after transplanting (WAT). The second peak occurred at 8 WAT and the last at 12 WAT. After the third peak, CH_4 fluxes were reduced rapidly. This type of seasonal change in CH_4 emission was typical of flooded rice fields were rice straw was applied. This result confirms the findings of Minami (1993) and Neue and Sass (1994). We observed another large flux at 16 WAT when the floodwater receded. The flux at

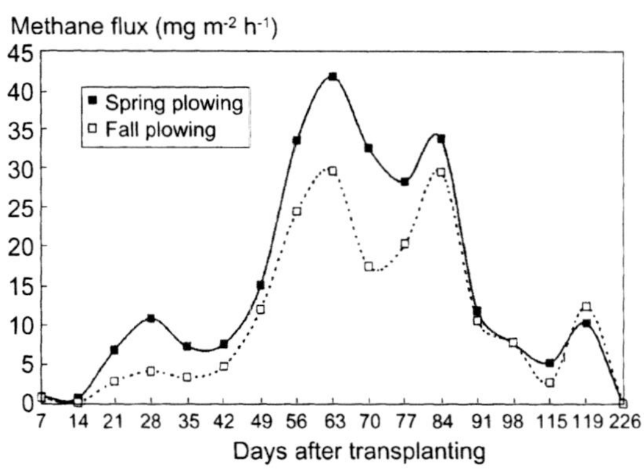

Figure 2. Seasonal changes in methane emissions as affected by plowing time

Table 3. Methane emissions as affected by plowing time

Plowing time	Methane emission($g\ m^{-2}\ d^{-1}$)			Total emission ($g\ m^{-2}\ season^{-1}$)	Grain yield ($t\ ha^{-1}$)
	Min	Max	Av		
Spring	0.002	0.69	0.29	42.0 a[a]	5.30
Fall	0.001	0.66	0.22	31.3 b	5.36

[a]Means within a column followed by the same letter are not significantly different at P= .05 LSD level. LSD(5%) = 6.19

that time could probably be direct soil emission of entrapped CH_4 after the water receded from the macropores. Neue et al. (1994) reported a similar observation of high emission of 90 mg $m^{-2}\ h^{-1}$ at 6 - 8 d after the floodwater receded.

Methane emissions following spring plowing were much greater than those following fall plowing. The fall treatments emitted 3-12 mg $m^{-2}\ h^{-1}$ less than the spring treatment during ripening. However, after ripening, the difference between treatments became small. These results indicate that the effect of plowing time on CH_4 emission was related to the amount of decomposition of the applied straw, which is a carbon source for methanogenic bacteria and causes the redox conditions to become more anaerobic as organic matter is consumed. Inubushi et al.(1992) reported that rice straw application at 1 and 2 mo before transplanting, compared with application just before transplanting, reduced CH_4 emissions by 50% and 63%, respectively.

Average daily CH_4 emissions following spring plowing was 0.29 g CH_4 m^{-2} and that following fall plowing was 0.22 g CH_4 m^{-2} (Table 3). In terms of total quantity of CH_4 emission during cultivation, the spring plot had 42.0 g CH_4 m^{-2} and the fall plot had 31.3 g CH_4 m^{-2}. The grain yield of fall plowing plot was a little higher even though there was no statistical difference.

Conclusion

Among the cultural practices tested, direct seeding on dry soil was the most effective in reducing CH_4 emission. Moreover, this method also decreased labor for transplanting. This cultural method is recommended in situations where the weed control problem could be resolved.

As to plowing time, CH_4 emissions following fall plowing were 26% less than those following spring plowing. In addition, fall plowing promoted early crop growth because the readily mineralizable nutrients in the soil increased as the organic matter decomposed during winter. Therefore, fall plowing is a more effective way of mitigating CH_4 emission from rice fields when organic amendment is required.

References

Cole V et al. (1996) Agricultural options for mitigation of greenhouse gas emissions. In: Watson RT, Zinyowera MC, Moss RH (eds) Climate Change 1995 Impacts. p 745-771. Adaptations and Mitigation of Climate Change: Scientific Technical Analyses. Cambridge University Press, New York

Houghton JT, Meria Filho LG, Callander BA, Harris N, Kattenberg A & Maskell K (eds) (1996) Climate Change 1995. The Science of Climate Change. Cambridge University Press, Cambridge, UK

Inubushi K, Muramatsu Y & Umebayashi M (1992) Influence of percolation on methane emission from flooded paddy soil. Jpn J Soil Sci Plant Nutr 63:184-189

Liang Sicui & Gen Yang (1994) Possibility for reducing methane emission from rice fields in China. Int Rice Res Notes 19(3):39-40

Minami K (1993) Methane from rice production. Res Rep Div Environ Planning 9: 243-258

Neue HU & Sass R (1994) Trace gas emissions from rice fields. In: Prinn R (ed) Global Atmospheric-Biospheric Chemistry. Plenum Press, New York, USA, pp 119-148

Neue HU, Wassmann R, Lantin RS, Alberto MC & Aduna JB (1994) Effect of fertilization on methane emission. Int Rice Res Notes 19(3):33-34

Parashar DC, Mitra AP, Shinha SK, Gupta PK, Sharma RC, Singh N, Kaul S, Lal G, Chaudhay A, Ray HS, Das SN, Parida KM, Rao SB, Kanung SP, Ramasami T, Nair BU, Swamy M, Gupta SK, Singh AR, Saikia BK, Barua AKS, Pathak MG, Iyer CPS, Gopalakrishnan M, Sane PV, Singh SN, Banerjee R, Sethunathan N, Adhya TK, Rao VR, Palit P, Saha AK, Purkait NN, Chaturvedi GS, Sen SP, Sen M, Sarkar B, Banik A, Subbaraya BH, Lal S & Venkatramani (1994) Methane budget from Indian paddy fields. In: CH_4 and N_2O- Global Emissions and Controls from Rice Fields and Other Agricultural and Industrial Sources. NIAES Series 2: p 27-39

Shin YK (1996) Mitigation options for methane emission from rice fields in Korea. Ambio 25(4):289-291

Takai Y, Koyama T & Kamura T(1956) Microbial metabolism in reduction process of paddy soils (Part 1). Soil Plant Food 2:63-66

Thurlow M, Kanda K, Tsuruta H & Minami K (1995) Methane uptake by unflooded paddy soils. Soil Sci Plant Nutr 41:371-375

Wang ZP, DeLaune RD, Masscheleyn PH & Patrick Jr WH (1993) Soil redox and pH effects on methane production in a flooded rice soil. Soil Sci Soc Am J 57:382 - 385

Nutrient Cycling in Agroecosystems **58:** 315–319, 2000.

Varietal differences in methane emission from Korean rice cultivars

Y.-K. Shin & S.-H. Yun

National Institute of Agricultural Science and Technology, Rural Development Administration, Suwon 441-701, Korea

Abstract

Methane (CH_4) emission from eight cultivars planted under uniform field conditions was measured by the closed static chamber method. Mean daily CH_4 emission and seasonally integrated CH_4 flux followed similar trends among the different varieties, irrespective of growth duration. The CH_4 flux (g CH_4 m^{-2}) among the varieties was in the order of Dasanbyeo (36.9) < Ilpumbyeo (42.9) < Gyehwabyeo (47.8) < Daeanbyeo (50.9) < Dongjinbyeo (58.8) < Hwaseongbyeo (59.7) < Odaebyeo (62.9) < Mangeumbyeo (76.0). No significant correlation was observed between CH_4 emission factor and root distribution in the 0-5 cm soil profile and dry matter weight in the canopy at heading stage.

Key words: methane, rice cultivars, root distribution, dry matter weight

Introduction

Methane (CH_4) is one of the important greenhouse gases emitted from both biological and industrial processes (Houghton et al., 1996). The atmospheric concentration of CH_4 has increased approximately to 246% of preindustrial concentration (Houghton et al., 1996).

Wetland rice cultivation is a major anthropogenic source of CH_4, contributing 15-35% of total CH_4 released. World rice requirements are predicted to increase at the compounded rate of 1.75% yr^{-1} between 1990 and 2025 (IRRI, 1997). Owing to the conversion of upland soils for wetland rice cultivation, global CH_4 emissions to the atmosphere may increase by 20% over the next decade (US-EPA, 1991).

Methane has a relatively short atmospheric lifetime (10 yr) compared with CO_2, N_2O, and chlorofluorocarbons (CFCs) (50–200 yr) and reduction from wetland rice in the future could help stabilize or reduce the global warming potential (Batjes & Bridges, 1992). It is estimated that a 10% reduction in anthropogenic emission would stabilize CH_4 at current concentrations, whereas CO_2 (60%), N_2O (70–80%), and CFCs (70–85% reduction) would require much higher levels (Lelieveld et al., 1993).

To maintain or increase rice yield and to reduce CH_4 emission, new management practices must be developed for wetland rice agriculture. Suggested mitigation options include cultivar selection and breeding, altered water management, and addition of chemicals or soil amendments (Neue, 1993; Wassmann et al., 1993). Early-maturing varieties, intermediate maturing varieties and late-maturing varieties occupy 10, 30, and 60% of rice areas respectively, in Korea (RDA, 1998).

Some data on varietal effects are currently available, but information on Korean rice varieties are not relatively known. The objective of this field study was to assess the CH_4 emission potential of eight rice cultivars over the cropping seasons.

Materials and methods

Cultivation of rice

A field experiment was carried out at the rice farm of the National Institute of Agricultural Science and Technology (Suwon, Korea) in 1997. The soil belongs to the Hwadong series of fine clayey, mixed, mesic, Aquic Hapludalfs. Soil pH was 5.9 (1:5 soil/water), soil organic matter content was 10 g kg^{-1}, content of available phosphorus was 20 mg kg^{-1}, and exchangeable potassium was 0.26 cmol kg^{-1}. All the plots were given 110-30.6–66.4 kg NPK ha^{-1} and rice straw (5 t ha^{-1}). Basal dressing was done just before transplanting and topdressing was applied at 14 and 48 d after transplanting (DAT). Rice straw was applied on the surface in the fall of 1996 and incorporated in the spring of 1997.

Seven rice japonica-type cultivars—Odaebyeo, Hwaseongbyeo, Ilpumbyeo, Daeanbyeo, Gyehwabyeo, Dongjinbyeo, and Mangeumbyeo—and one Tongil-type cultivar, Dasanbyeo, were cultivated in a 7.5- × 7-m field. Odaebyeo belongs to the early-maturing group. Dasanbyeo and Hwaseongbyeo are intermediate-maturing rice varieties. Ilpumbyeo, Daeanbyeo, Gyehwabyeo, Dongjinbyeo, and Mangeumbyeo are late-maturing rice varieties. Four seedlings of each cultivar were transplanted at 15- × 30-cm plant spacing on 28 May. Plots of 3.1- × 7-m were prepared for each cultivar in triplicates.

Water in the rice plots was supplied by intermittent irrigation. All plots were flooded until 30 DAT. Thereafter, the field was intermittently flooded until 2 wk before harvest; intermittently irrigated plots remained without any irrigation until small cracks were noticed on the soil surface.

Collection and analyses of gas samples

Gas samples were collected using the closed static chamber method (Shin et al., 1995, 1996), in which eight rice plants were enclosed in a transparent polyacrylic plastic chamber with internal dimensions of 60 × 60 × 110 cm. One chamber was installed in each experimental plot. Gas samples were collected between 9 am and noon at 7-d intervals from the day of transplanting until maturity. Gas samples were taken using a 60-mL polypropylene syringe fitted with a Mininert valve. A Varian Star 3400 gas chromatograph fitted with a flame ionization detector and in-board data handling was used to determine CH_4 concentration. Gas samples (2 mL were injected into a stainless steel column (3 mm outside diameter × 2 m) packed with Porapak N (80/100 mesh). The temperatures of the column, injector, and detector were 45, 80, and 200 °C, respectively. Gas samples were injected using an air-actuated six-port valve (Valco valves, Houston, TX, USA) with the aid of a mass flow controller unit (Tylan Inc., CA, USA). Calibration gas (15.1 ppmv CH_4) was purchased from MG Industries (Malvern, PA, USA). All collected gas samples were analyzed within 3 h of field collection. Flux data were subjected to analysis of variance and Duncan's multiple range test (P = 0.05) using the statistical analysis system (SAS, 1988).

Distribution profile of rice roots

Rice roots were collected with PVC samplers (inside diameter of 19.5 cm) in a series of four at heading stage. Soils inside the samplers were subdivided into 5-cm sections and sieved (2 mm) under running water. Roots left on the sieve were taken and dried at 70 °C in a drying oven until a constant weight was obtained.

Dry matter weight of canopy

The canopy of rice was collected at heading. It was dried at 70 °C in a drying oven to constant weight.

Results and discussion

Change in CH_4 emission among Korean rice cultivars

The seasonal change in CH_4 emission rates is shown in Figures 1 and 2. The CH_4 emission rates increased from 3 wk after transplanting and showed a maximum value at the end of July, or 62 DAT, which corresponded to the end of the vegetative stage. Methane emission rates decreased twice on 8 Jul and 22 Jul, which was due to midsummer drainage. Small peaks were observed on 18 Jun, 2 Jul, 15 Jul, and 19 Aug. The peak observed on 19 Aug corresponded to the tillering stage. After 25 Aug, CH_4 flux declined in all varieties.

Rice cultivar did not influence the pattern of seasonal variation in CH_4 emission rates. The same pattern mentioned above was observed in 24 plots of eight varieties. However, the amount of CH_4 emitted differed among the cultivars. The largest CH_4 emission was recorded in plots planted to Mangeumbyeo (japonica), while the smallest was in plots planted to Dasanbyeo (japonica). Methane emission rates from Mangeumbyeo and Dasanbyeo plots differed significantly ($p<0.05$). As shown in Table 1, different values of CH_4 emission factor (g CH_4 m^{-2} d^{-1}) were observed among the eight different varieties: Dasanbyeo (0.298), Ilpumbyeo (0.33), Gyehwabyeo (0.379) < Daeanbyeo (0.391) < Dongjinbyeo (0.452), Hwaseongbyeo (0.482) < Odaebyeo (0.566), Mangeumbyeo (0.603). On the basis of CH_4 emission, these eight rice varieties can be grouped as follows: low CH_4 emission (Dasanbyeo, Ilpumbyeo, Gyehwabyeo, and Daeanbyeo), intermediate CH_4 emission (Hwaseongbyeo and Dongjinbyeo), and high CH_4 emission (Odaebyeo and Mangeumbyeo).

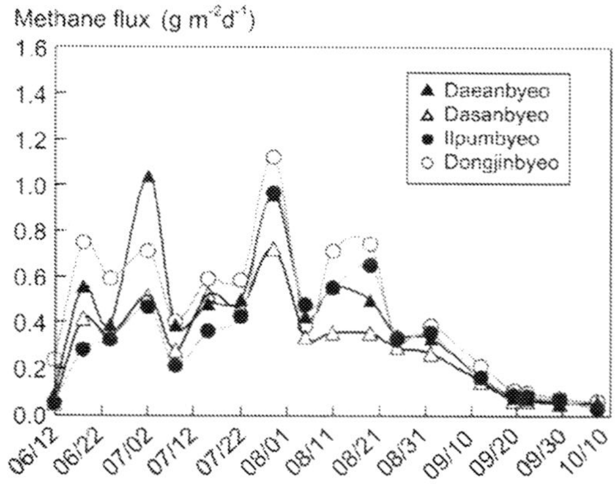

Figure 1. Seasonal changes in CH_4 emission of different rice cultivars

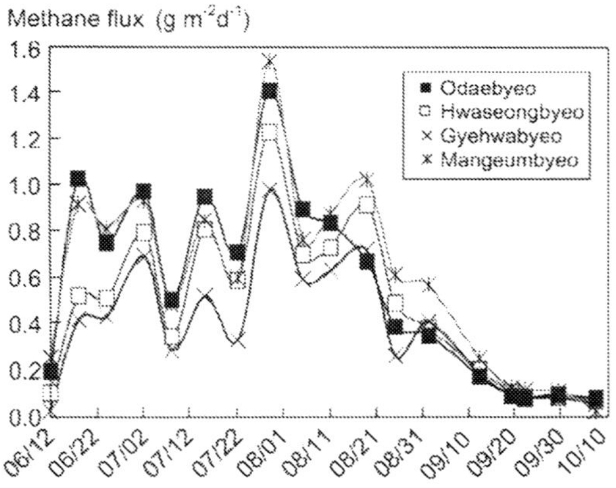

Figure 2. Seasonal changes in CH_4 emission of different rice cultivars

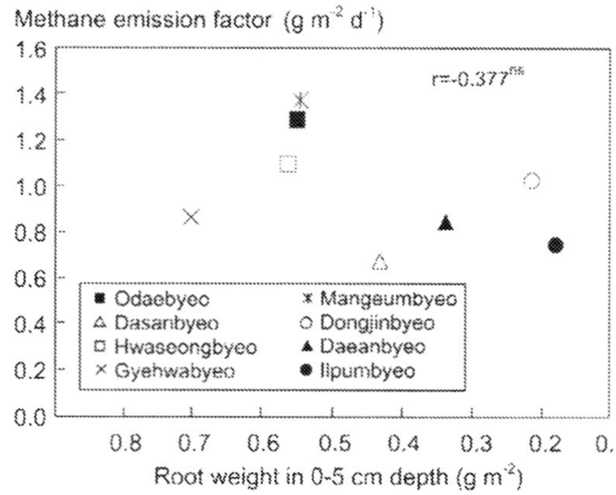

Figure 3. Relationship between root weight at 0-5 cm soil depth at heading and CH_4 emission factor

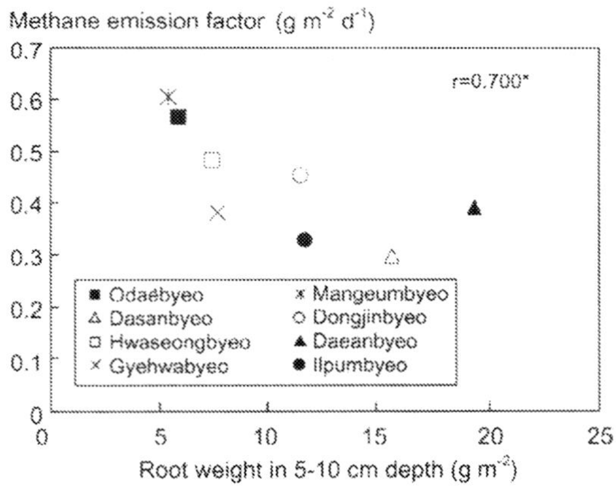

Figure 4. Relationship between root weight at 5–10 cm soil depth at heading and CH_4 emission factor

Table 1. Methane emission factor and integrated emission factor

Cultivar	Growth duration[a] (d)	CH4 emission (EF)[b]* ($g\ CH_4\ m^{-2}\ d^{-1}$)	Integrated emission factor= EF × growth duration ($g\ CH_4\ m^{-2}$)
Ilpumbyeo	130	0.330 d	42.898
Dasanbyeo	124	0.298 d	36.969
gyehwabyeo	126	0.379 d	47.751
Daeanbyeo	130	0.391 cd	50.877
Hwaseongbyeo	124	0.482 bc	59.725
Donhjinbyeo	130	0.452 bc	58.822
Odaebyeo	111	0.566 ab	62.867
Mangeumbyeo	126	0.603 a	76.022

[a]Values not followed by the same letter differ significantly at $p<0.05$ (Duncan's multiple range test). [b]Transplanting –45 d after heading.

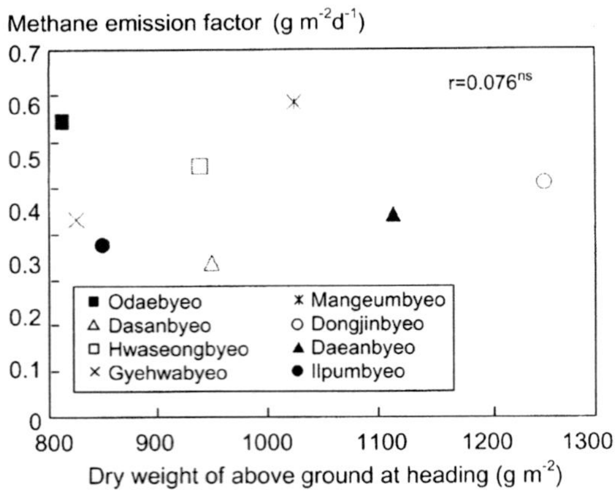

Figure 5. Relationship between dry weight of above ground at heading and CH_4 emission factor

The integrated CH_4 emission flux (g CH_4 m^{-2} season^{-1}) showed a similar trend with CH_4 emission factor despite differences in growth period—Dasanbyeo (36.9) < Ilpumbyeo (42.9) < Gyehwabyeo (47.8) < Daeanbyeo (50.9) < Dongjinbyeo (58.8) < Hwaseongbyeo (59.7)< Odaebyeo (62.9) < Mangeumbyeo (76.0).

Relation between root distribution and CH_4 flux

The relationship between dry weight of roots at the 0–5 and 5–10 cm depths and CH_4 emission factors were determined (Figure 3 and 4). Rice roots are assumed to be associated with the collection, production, and oxidation of CH_4. However, the CH_4 emission rates were not correlated with root weight at 0–5 cm depth (Figure 3). Similar results were reported by Watanabe et al. (1995). Methane emission rates were negatively related with root weight at 5–10 cm depth (Figure 4).

Armstrong (1969) and Kludze et al. (1994) reported some differences in cultivar rhizosphere oxygenation (per unit area of root and per plant). The differences in the amount of oxygen or exudates released per unit weight of root among cultivars may have more influence than the total weight of roots.

Relationship between dry weight of canopy and CH_4 flux

Dry weight of canopy at heading stage was not correlated with CH_4 emission factors (Figure 5). Sass et al. (1990) found a positive correlation between aboveground biomass and CH_4 emission rates in two fields using one cultivar. On the contrary, aboveground biomass in the present study was not correlated with CH_4 emission rates among plots with different cultivars. These results are similar to those reported by Watanabe et al. (1995). This indicates that cultivars with large biomass are not necessarily related to higher CH_4 emission.

Our results indicate that the big difference between CH_4 emission factor and the integrated emission factor among rice cultivars tested shows a potential mitigation option. Rice cultivars low in CH_4 emission may be selected. It was shown in this study that there is no significant correlation between CH_4 emission factor and dry matter weight of canopy and root at heading stage. Thus, further studies must be conducted to determine other factors that could affect varietal differences in CH_4 emission.

References

Armstrong W (1969) Rhizosphere oxidation in rice: an analysis of intervarietal difference in oxygen flux from the roots. Physiol Plant 22:296-303

Batjes NH & Bridges EM (1992) Introduction. In: Batjes NH and Bridges EM (eds) A Review of Soil Fcators and Processes that Control Fluxes of Heat, Moisture and Greenhouse Gases, Technical Paper 23, pp 1-7, International Soil Reference and Information Center, Wageningen

Houghton JT, Meira Filho LG, Callander BA, Harris N, Kattenberg A, Maskell K (eds) (1996) Climate Change 1995. The Science of Climate Change. Casmbridge University Press, Cambridge, UK.

IRRI – International Rice Research Institute (1997) Rice almanac, 2nd ed. 181 p, IRRI, Manila, Philippines.

Kludze HK, DeLaune RD & Patrick Jr WH (1994) A colorimetric method for assaying dissolved oxygen loss from container-grown rice roots. Agron J 86: 483-487

Lelieveld J, Crutzen PJ & Bruehl C (1993) Climate effects of atmospheric methane. Chemosphere 26:739-768

Neue HU (1993) Methane emission from rice fields. Bioscience 43:466-474

RDA – Rural Development Administration (1998) Reference data for rice cultivation in Korea. 200 p, Korea

SAS Institute (1988) SAS/STAT users' guide. Release 6.03 SAS Inst, Cary, NC

Sass RL, Fisher FM, Harcombe PA & Turner FT (1990) Methane production and emission in a Texas rice field. Global Biogeochem Cycles 4:47-68

Shin YK, Lee YS, Yun SH & Park ME (1995) A simplified closed static chamber method for measuring methane flux in paddy soils. J Korean Soc Soil Sci 28:183-190

Shin YK, Yun SH, Park ME & Lee BY (1996) Mitigation options for methane emission from rice fields in Korea. Ambio 25:289-291

US-EPA – Environmental Protection Agency (1991) Introduction. In: Braatz BV & Hogan KB (eds) Sustainable Rice Productivity and CH_4 Reduction Research Plan. pp 12-18, U.S. EPA/Office of Air and Radiation, Washington, DC

Wassmann R, Papen H & Rennenberg H (1993) Methane emission from rice paddies and possible mitigation strategies. Chemosphere 26:201-217

Watanabe A, Kajiwara M, Tashiro T & Kimura M (1995) Influence of rice cultivar on CH_4 emission from rice fields. Plant Soil 176:51-56

Nutrient Cycling in Agroecosystems **58:** 321–326, 2000.

Influence of *Azolla* on CH_4 emission from rice fields

Z. Ying[1], P. Boeckx[2], G.X. Chen[1] & O. Van Cleemput[2]

[1]*Institute of Applied Ecology, Academia Sinica, Shenyang, People's Republic of China;* [2]*Ghent University, Coupure 653, B-9000 Gent, Belgium*

Keywords: Azolla, rice field, China, methane emission, methane production, methane oxidation

Abstract

Azolla is an aquatic fern that has been used successfully as a dual crop with wetland rice. Rice fields are a major source of atmospheric CH_4, which is an important greenhouse gas. In this study, field and laboratory experiments showed that growing *Azolla* as a dual crop could enhance CH_4 emission from rice fields. In pot experiments, indications showed that *Azolla* could mediate CH_4 transport from the floodwater of a rice soil into the atmosphere. It was also found that due to the presence of *Azolla*, chemical soil properties could be developed, stimulating CH_4 production and decreasing in situ CH_4 removal.

Introduction

Azolla is a genus of aquatic ferns found floating in swamps, ditches, lakes, and rivers. Because of its aquatic nature, rapid growth, ability to fix N_2 (due to symbiosis with *Anabaena*, a blue-green algae), and high N content, *Azolla* has been used as a green manure or a dual crop in rice cultivation for many years (Wagner, 1997). It has been shown that incorporation of *Azolla* as green manure is beneficial for rice production both in terms of rice yield and N uptake. Incorporation of *Azolla* appears to be equivalent to using urea as a source of N (Galal, 1997). *Azolla* also increases N recovery by the soil and therefore improves soil fertility in the long term (Kumarasinghe & Eskew, 1993). The use of *Azolla* as a floating cover in rice fields is also effective in reducing NH_3 volatilization from applied urea (Vlek et al., 1995). As a result, *Azolla* is most beneficial as a sustainable natural source of N. In addition, it is also useful in reducing weed growth and improving the soil structure and water economy. The major fundamental constraints are limitations of water supply and phosphorus and its susceptibility to temperature changes, pests, and pathogens (Kulasooriya, 1991).

Rice fields have to be considered as a significant source of greenhouse gases (CH_4 and N_2O) (Bronson et al., 1997a,b). They account for about 60 Tg CH_4 per year, or about 12% of the global annual CH_4 emission (IPCC, 1996). Much attention has been paid to the influence of fertilization, organic matter amendment, water management, and rice varieties on CH_4 and N_2O emissions from rice fields. Recently, it has been reported that growing *Azolla* in rice fields could enhance CH_4 and N_2O emissions (Chen et al., 1997). Therefore, it is important to collect more information on the role of *Azolla* on greenhouse gas emissions from rice soils. This paper gives evidence suggesting that *Azolla* could enhance CH_4 emission from flooded rice soils at the level of CH_4 transport, production, and oxidation.

Materials and methods

Pot experiments

Pot experiments were conducted in a greenhouse at the Ghent University (Belgium). The objectives were to evaluate the influence of *Azolla* on soil properties, rice growth, and CH_4 emission from a flooded rice soil. The soil used originated from a maize field and had the following physicochemical characteristics: pH 7.3, total C 1.4%, total N 0.14%, 54% sand, 31% silt, and 15% clay. Plastic pots (20 cm in height and 16 cm in diameter) were filled with 3.3 kg of soil amended with 0.5% wheat straw. The soil was mixed with de-ionized water until it was flooded to 2 cm depth. The soil was preincubated at constant water level for 20 d. During the entire experiment, the water content was kept constant and average temperature in the greenhouse was 25 °C.

The four treatments were the flooded soil without rice or *Azolla* (P1), without rice but with *Azolla* (P2), with rice and without *Azolla* (P3), and with both rice and *Azolla* (P4). Each treatment was replicated five times. Four 18 d-old rice seedlings (variety: Liao Kai 79) were planted in the center of each pot of treatments P3 and P4. Seven days after planting, urea was added to the overlying water of all treatments at a rate of 100 mg urea N kg^{-1} soil. *Azolla fuliculoides* was cultivated (Watanabe et al., 1977), and 14 d after planting of the rice seedlings, 2 g fresh *Azolla* was spread on top of the water layer of treatments P2 and P4.

The CH_4 emission from the rice soil microcosms was determined using the static chamber technique (IAEA, 1992). A chamber (50 cm height and 15 cm diameter) equipped with a septum to sample the gas phase was put over the plastic pots. Methane emissions were collected weekly between day 21 and day 92 after planting the rice seedlings. At day 50, the O_2 concentration in the water layer and the NH_4^+ content of the soil (0–20 cm) were determined. The O_2 concentration was measured with an oxygen electrode (Oxi320/CellOx325, WTW, Weilheim, Germany). Ammonium N was determined acidimetrically after distillation of an extract (shaking time: 60 min) of the soil with 1 N KCl (soil/KCl = 1/2) (Keeney & Nelson, 1982).

In a second pot experiment, the transport capacity of *Azolla* for CH_4 was investigated. Pots with rice were prepared as described for treatment P3. In this experiment, *Azolla* was put as a complete cover onto the water layer 1 d before CH_4 measurements were conducted. When the CH_4 measurements were finished, *Azolla* was removed from the pot. As such, two treatments were handled: one with rice but without *Azolla* (P3) and one with rice and temporally *Azolla* (P4'). Five replicates were used. A split chamber (Figure 1) was used to distinguish between CH_4 emitted from the overlying water and CH_4 escaping via the rice plants. The rice plants were separated from the soil chamber via an air-tight plastic tube and modeling clay (Figure 1). As such, it was possible to determine separately the amount of CH_4 emitted via diffusion through the water layer, eventually mediated by *Azolla* (soil chamber), and the amount of CH_4 emitted via the rice plants (rice chamber). The CH_4 emission was determined on day 53 and day 78 after transplanting the rice. As such, CH_4 emission was measured during a period of high flux and during a period of low flux.

The concentrations of CH_4 in rice and soil chambers were measured with a Chrompack CP 9000 gas chromatograph (GC) (Chrompack, Delft, The Netherlands). After injection, part of the gas was directed through a 1.8 m × 3 mm activated aluminum column (100-120 mesh). Methane was measured using a flame ionization detector (FID). Helium was used as carrier gas (46 mL min^{-1}). The analyses were carried out under the following conditions: injector temperature 65 °C, oven temperature 55 °C, and detector temperature 200 °C. The CH_4 concentrations were calculated from the peak area. As standard gas, 50.3 ± 1.5 ppmv CH_4 in argon was used (L' air Liquide, Belgium). The chromatograms were registered and analyzed using "Winner on Windows" (Thermo Separation System, Fremont, California).

Field experiments

A field experiment was carried out at the experimental station of the Institute of Applied Ecology (Shenyang, China). The objective was to find out whether there was a difference between CH_4 emissions from a rice field without *Azolla* (F1), a rice field where *Azolla* had been grown for 1 yr (F2), and a rice field where *Azolla* had been grown for 5 consecutive years (F3). The experimental site is characterized in Table 1. The rice and *Azolla* species were similar to those used in the pot experiments.

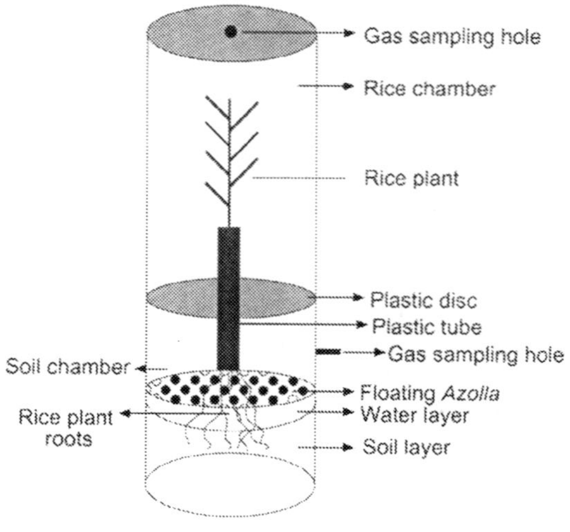

Figure 1. Split chamber for measuring CH_4 emission from the overlying water (soil chamber) and CH_4 escaping through the rice plants (rice chamber)

Table 1. Characterization of the experimental site in Shenyang, China

Latitude and longitude	10° 32' N, 123° 23' E
Soil temperature during the growing season	19.5°C (mean) 9 - 24°C (range)
Annual precipitation	570 - 680 mm
Cropping system	Wetland rice, single harvest per year
Rice variety	Liao Kai 79
Soil type	Meadow brown soils sand 54%, clay 22%, silt 24%
pH (H_2O)	6.5
Organic matter (g kg^{-1})	16.2
Total N (g kg^{-1})	0.8
CEC (cmol kg^{-1})	18

The soil was flooded on 18 May 1997 and rice seedlings were planted on 21 May and harvested on 16 Oct. *Azolla* inoculum (156 g m^{-2}) was spread on fields F2 and F3 on 11 Jun. Pig manure was applied as a basal dressing at a rate of 37.5 t ha^{-1} (± 112 kg N ha^{-1}) on 20 May. The rice fields were also fertilized with 170 kg urea-N ha^{-1} (60 kg N ha^{-1} on 28 May and 27 Jul, and 50 kg N ha^{-1} on 25 Aug). There was no difference in soil temperature between the different fields. The average soil temperature during the growing season was 19.5 °C.

The CH_4 emission was determined weekly between 28 May and 15 Oct, using the static chamber technique (0.8 × 0.8 × 1.0 m^3) (IAEA ,1992). Two chambers were used per field. The chambers were closed during 40 min and gas samples were collected at 0 and 40 min. The CH_4 concentration in the headspace of the chambers was determined using a Shimadzu GC-14B GC (Shimadzu, Tokyo, Japan). After injection, the gas was directed through a packed column (molecular sieve 5 Å). Methane was measured using a FID. Helium was used as carrier gas. The analyses were carried out under the following conditions: injector temperature 100 °C, oven temperature 100 °C and detector temperature 200 °C. The redox potential (Eh), pH, water soluble organic carbon (WSOC) (McCardell & Fuhrmann, 1992) and the NH_4^+-N content (Keeney & Nelson, 1982) of the soil (0-20 cm) were also monitored during the rice-growing season.

Results and discussion

Pot experiments

For an entire period of 70 d, the presence of *Azolla* enhanced total CH_4 emission from a flooded soil without rice by 75% (P1 and P2, Table 2a). The increase in CH_4 emission could be explained by a significant decrease in dissolved O_2 in the overlying water and an increase in the NH_4^+-N content of the soil in the presence of *Azolla* (Table 2a). A decrease in O_2 in the overlying water could result in more reduced soil conditions (not measured), leading to an enhanced CH_4 production (Patrick & DeLaune, 1977). The effect of *Azolla* on the redox potential (Eh) of the soil was clearly shown in the field experiments (Table 3a). In situ oxidation of indigenously produced CH_4 mitigates CH_4 emission from wetland soils (Boeckx & Van Cleemput, 1996; van der Gon & Neue, 1996). However, NH_4^+ can inhibit the biological oxidation of CH_4 (King & Schnell, 1994). As a result, an increase in the NH_4^+-N content of the rice soil could decrease its CH_4-oxidizing capacity. Thus, based on the O_2 and NH_4^+ -N data in Table 2a, CH_4 production may be higher and in situ CH_4 oxidation may be lower in treatment P2 than in treatment P1. This results in an enhanced amount of CH_4 available for transport to the atmosphere in treatment P2 com-

Table 2a. Total CH_4 emission during 70 d; O_2 concentration in the floodwater, NH_4^+ -N content of the soil (all measured on day 50) and dry weight (dw) of rice shoots and roots; values between parentheses are standard errors

Treatment[a]	CH_4 flux (g CH_4 m^{-2})	Dissolved O_2 (Mg L^{-1})	NH_4^+ -N content (mg N kg^{-1} dw)	Rice shoots (g dw)	Rice roots (g dw)
P1	123a (10)	14.0a (1.0)	25.9a (1.1)		
P2	211b (35)	4.1b (0.2)	33.7b (0.6)		
P3	144a (19)	11.5a (0.6)	2.4c (0.4)	14.0a (1.3)	5.5a (0.5)
P4	138a (7)	5.8b (0.4)	0.8c (0.1)	23.8b (1.6)	11.5b (1.5)

[a]See text for treatment description. Treatments followed by the same letter in each column are not significantly different (P <0.05) - one way Anova test with Student – Newman – Keuls comparison of means

Table 2b. Effect of *Azolla* on CH_4 emission via the overlying water (water) and CH_4 emission via the rice plants (rice); for treatments, see text P4' *Azolla* covered the water only 1 d prior to CH_4 measurements; values between parentheses are standard errors

Treatment	High CH_4 flux (day 53)[a] ($mg\ CH_4\ m^{-2}\ h^{-1}$)			Low CH_4 flux (day 78) ($mg\ CH_4\ m^{-2}\ h^{-1}$)		
	Water	Rice	Total	Water	Rice	Total
P3	4.1a (0.2)	16.3a (0.1)	20.4	2.9a (0.1)	3.1a (0.1)	6.0
P4'	10.4b (0.2)	18.6a (0.3)	29.0	2.9a (< 0.0)	2.2a (0.1)	5.1

[a]Treatments followed by the same letter in each column are not significantly different ($P < 0.05$) - one way Anova test with Student – Newman – Keuls comparison of means

Table 3. Total CH_4 emission (during 147 d); water soluble organic carbon (WSOC) and NH_4^+-N content (data shown here are integrated values (during 147 d) of the WSOC and NH_4^+-N contents that were determined each time CH_4 emissions were measured, see Fig. 2); av pH, av redox potential (Eh) and porosity of the rice soil of the field experiment; F1 = rice field without *Azolla* (control). F2 = rice field with first year *Azolla*, F3 = rice field with fifth year *Azolla*

Treatment	CH_4 flux[a] ($g\ CH_4\ m^{-2}$)	WSOC ($Mg\ C\ g^{-1}$ dw)	NH_4^+-N content ($mg\ N\ kg^{-1}$ dw)	pH (H_2O)	Eh (mV)	Porosity (%)
F1	12.7 – 15.2	16.8	408	6.8	–50	52
F2	22.0 – 23.9	11.9	650	7.0	–80	54
F3	22.2 – 25.2	9.4	880	7.0	–100	55

[a]Flux range measured in both static chambers

pared with P1. This explains the elevated CH_4 emission from treatment P2.

However, *Azolla* did not increase the total CH_4 emission from the soils grown with rice (P3 and P4, Table 2a), although the O_2 concentration of the water was also markedly lower in the presence of *Azolla* (P4). The NH_4^+ levels of treatment P3 and P4 were low, probably because N has been taken up by the rice plants. Therefore, inhibition of CH_4 oxidation will be of minor importance in these treatments. Thus, based on the observations of the O_2 concentration of the overlying water, one could also expect a higher CH_4 emission from treatment P4 compared with P3. This was not the case because of some influence of *Azolla* on the development of the rice plants. The formation of NH_4^+ through N_2 fixation by the *Azolla-Anabaena* association (supplementary to NH_4^+ produced via urea hydrolysis) increased the dry weight of the rice roots and shoots (P3 and P4, Table 2a). This rhizosphere was mainly found in the subsurface layer of the soil. In general, it has been observed that O_2 transport through the rice plants results in in situ CH_4 oxidation in the rhizosphere (Gilbert et al., 1998; van der Gon & Neue, 1996). Methane oxidation may be higher in treatment P4 than in treatment P3 because the volume of the rhizosphere of treatment P4 was twofold that of treatment P3. Thus, due to the presence of *Azolla,* chemical soil properties could be developed, stimulating CH_4 production in the deeper soil layer and at the same time in situ CH_4 removal in the rhizosphere. Methane produced in the deeper soil layers is oxidized while diffusing through the rhizosphere. As a result, an increased oxidizing capacity is offsetting the enhanced CH_4 production in treatment P4, resulting in less CH_4 available to be transported to the atmosphere.

From a second pot experiment (Table 2b), it could be deduced that *Azolla* can mediate transport of CH_4 from the overlying water into the atmosphere. In this pot experiment, *Azolla* was put onto the water layer 1 d before CH_4 emissions were determined. Thereafter it was removed again. Thus, the presence of Azolla could not have affected the rice plant roots or the soil properties. *Azolla* served here only as a possible, additional pathway for CH_4 to escape from the soil-water interface. When the total (= water + rice) CH_4 emission from the soil-rice microcosm was relatively high, *Azolla* significantly affected "total" (and "water") CH_4 emission (day 53, Table 2b). Apparently *Azolla* could transport CH_4, which was released from the soil and dissolved in the water layer, into the atmosphere. When the total CH_4 emission was relatively low, "total" and "water" CH_4 emission was not affected by *Azolla* (day

78, Table 2b). Thus, the transport ability of *Azolla* probably also depended on the concentration of dissolved CH_4. However, further experimental evidence is required to confirm this observation.

Field experiments

It has been shown that the use of *Azolla* for dual cropping with rice can improve N fertilizer efficiency and rice yield and can reduce NH_3 volatilization from rice fields (Kumarasinghe & Eskew, 1993). However, recently, it was shown that an *Azolla* cover increased CH_4 and N_2O emissions from rice fields (Chen et al., 1997). From the above pot experiments, it was clear that *Azolla* could mediate CH_4 transport. However, due to the development of a larger subsurface CH_4-oxidizing rhizosphere, the effect of *Azolla* on CH_4 emission could not be shown in microcosms, wherein *Azolla* and rice plants were grown as dual crops (P4). Therefore, the effect of A*zolla* on CH_4 emission was also investigated in the field.

Methane emission was measured from rice fields with and without an *Azolla* cover. The presence of *Azolla* appears to increase CH_4 emission (Figure 2). The total CH_4 emission from a rice field grown with *Azolla* for 1 yr (F2, Table 3) was 65% higher than emission from a rice field without *Azolla* (F1, Table 3). When the floodwater had been inoculated with *Azolla* for 5 consecutive years (F3, Table 3), CH_4 emission was 70% higher than the control (F1). This finding suggests an effect of *Azolla* on CH_4 emission. The effect of a successive growth of *Azolla* (F3) on CH_4 emission seems to be minimal (F3). However, lack of repetitive measurements does not allow proving the latter statistically.

Here, *Azolla* also showed some important effects on chemical soil properties, which could affect CH_4 emission. In general, emission of CH_4 from rice soils is controlled by the balance of three processes: CH_4 production, oxidation, and transport (both from the soil into the water layer and from the water layer into the atmosphere). The presence of *Azolla* (F2 and F3, see Table 3) appeared to depress WSOC and Eh and to increase NH_4^+-N content and porosity of the rice soil (Table 3). Insufficient treatment replication, however, did not allow, comparing F1, F2, and F3 statistically. Nevertheless, both F2 and F3 showed a possibly higher CH_4 emission. The WSOC tended to be lower in treatments F2 and F3 than in the control treatment (F1). This result suggests that the C substrate was not the limiting factor for CH_4 emission. The presence of *Azolla*

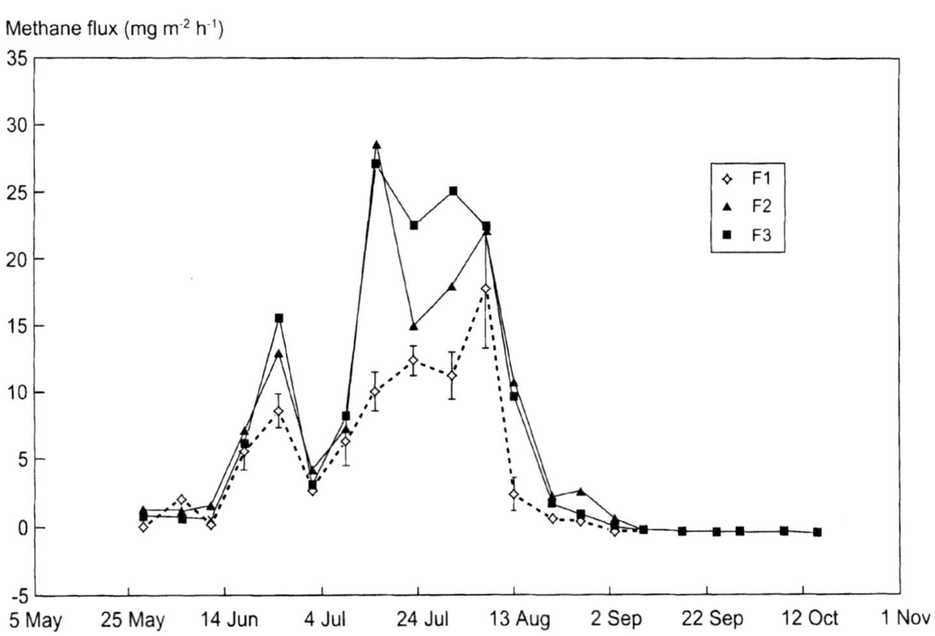

Figure 2. Av CH_4 emission from field experiments in Shenyang during an entire growing season in 1997; F1-rice field without *Azolla* (control), F2-rice with first year *Azolla*, F3-rice field with fifth year Azolla

appeared to decrease Eh and increase overall NH_4^+-N content. The Eh decrease could result in an increased CH_4 production (Patrick & DeLaune, 1977) and the NH_4^+ increase could result in a reduced biological CH_4 oxidation (King & Schnell, 1994). These two processes may result in an enhanced net amount of CH_4 available for transport (diffusion) into the overlying water. In addition, the presence of *Azolla* seemed to slightly increase soil porosity, thereby improving diffusion of CH_4 from the soil into the overlying water. Thus, the presence of *Azolla* may enhance all three processes controlling CH_4 emission from the rice soil into the overlying water. The second pot experiment indicated that *Azolla* could also mediate CH_4 transport from the overlying water into the atmosphere.

Thus, the observed increase in CH_4 emission in the presence of *Azolla* could be explained by soil conditions promoting CH_4 availability and CH_4 diffusion into the overlying water and by the fact that *Azolla* served as an additional pathway for CH_4 transport into the atmosphere. However, more such field experiments are needed to provide statistical evidence of these results. Finally, it is also worth mentioning that in the field experiments, root formation was not concentrated in the subsurface layer (as observed in the pot experiment). Therefore, CH_4 oxidation in the rhizosphere was probably less intense than that in the pot experiment.

Acknowledgement

The financial support by ABOS (Ministry of Foreign Affairs, Belgium) and NSFC (National Natural Science Foundation of China) is highly appreciated.

References

Boeckx P & Van Cleemput O (1997) Methane emission from a freshwater wetland in Belgium. Soil Sci Soc Am J 61: 1250-1256

Bronson KF, Neue HU, Singh U & Abao Jr EB (1997a) Automated chamber measurements of methane and nitrous oxide flux in a flooded rice soil. I. Residue nitrogen and water management. Soil Sci Soc Am J 61:981-987

Bronson KF, Singh U, Neue HU & Abao Jr EB (1997b) Automated chamber measurements of methane and nitrous oxide flux in a flooded rice soil. II. Fallow periods. Soil Sci Soc Am J 61:988-993

Chen GX, Huang GH, Huang B, Yu KW & Xu H (1997) Nitrous oxide emissions from soil-plant systems. Nutr Cycling Agroecosyst 49:41-45

Denier van der Gon HAC & Neue HU (1996) Oxidation of methane in the rhizosphere of rice plants. Biol Fertil Soils 22:359-366

Galal YGM (1997) Estimation of nitrogen fixation in an *Azolla* - rice association using the nitrogen-15 isotope dilution technique. Biol Fertil Soils 24:76-80

Gilbert B, Assmus B, Hartmann A & Frenzel P (1998) In situ localization of two methanotrophic strains in the rhizosphere of rice plants. FEMS Microbial Ecol 25:117-128

IAEA – International Atomic Energy Agency (1992) Manual on measurement of methane and nitrous oxide emission from agriculture. IAEA - TECDOC – 674 Vienna: IAEA. 91 p

IPCC – Intergovernmental Panel on Climate Change (1996) Climate Change 1995. The Science of Climate Change. Cambridge: Cambridge University Press, 572 p

Keeney DR & Nelson DW (1982) Nitrogen - Inorganic forms. In: Page et al. (eds) Methods of Soil Analysis. Madison: American Society of Agronomy; Soil Science Society of America Agronomy No. 9, Part 2, 2nd edition, p 643-698

King GM & Schnell S (1994) Ammonium and nitrite inhibition of methane oxidation by *Methylobacter albus* BG8 and *Methylosinus trichosporum* OB3b at low methane concentrations. Appl Environ Microbiol 60:3508-2513

Kulasooriya SA (1991) Constraints on the widespread use of Azolla in rice production. In: Polsinelli M, Materassi R & Vincenzini M (eds) Nitrogen fixation. Proceedings of the fifth International Symposium on Nitrogen Fixation with Non-legumes, 10-14 Sep 1990, Florence, Italy, 145 p

Kumarasinghe KS & Eskew DL (1993) Isotopic studies of Azolla and nitrogen fertilization of rice. Kluwer, Dordrecht. 145 p

McCardell A &Fuhrmann JJ (1992) Determination of persulfate oxidizable carbon by gas chromatography. Soil Biol Biochem 24:615-616

Patrick WH Jr & DeLaune RD (1977) Chemical and biological redox systems affecting nutrient availability in coastal wetland. Geosci Manage 18:131-137

Vlek PLG, Diakite MY, Mueller H (1995) The role of Azolla in curbing volatilization from flooded rice systems. Fert Res 42:165-174

Wagner GM (1997) Azolla: a review of its biology and utilization. Bot Rev 63:1-26

Watanabe I, Espinas CR, Berta NS & Alimango BV. (1977). Utilization of *Azolla-Anabaena* complex as a nitrogen fertilizer for lowland rice. Int Rice Res Pap Ser II

Nutrient Cycling in Agroecosystems **58:** 327–332, 2000.

Effect of land management in winter crop season on CH_4 emission during the following flooded and rice-growing period

H. Xu[1], Z.C. Cai[1], Z.J. Jia[1] & H. Tsuruta[2]

[1]*Institute of Soil Science, Chinese Academy of Sciences, Nanjing 210008, China;* [2]*National Institute of Agro-Environmental Sciences, Kannondai, 3-1-1, Tsukuba 301, Japan*

Key words: methane emission, rice soil, land management, winter crop season

Abstract

A greenhouse pot experiment was carried out to study the effect of land management during the winter crop season on methane (CH_4) emissions during the following flooded and rice-growing period. Three land management patterns, including water management, cropping system, and rice straw application time were evaluated. Land management in the winter crop season significantly influenced CH_4 fluxes during the following flooded and rice-growing period. Methane flux from plots planted to alfalfa (ALE) in the winter crop season was significantly higher than those obtained with treatments involving winter wheat (WWE) or dry fallow (DFE). Mean CH_4 fluxes of treatments ALE, WWE, and DFE were 28.6, 4.7, and 4.1 mg CH_4 m^{-2} h^{-1} in 1996 and 38.2, 5.6, and 3.2 mg CH_4 m^{-2} h^{-1} in 1997, respectively. The corresponding values noted with continuously flooded fallow (FFE) treatment were 6.1 and 5.2 times higher than that of the dry fallow treatment in 1996 and 1997, respectively. Applying rice straw just before flooding the soil (DFL) significantly enhanced CH_4 flux by 386% in 1996 and by 1,017% in 1997 compared with rice straw application before alfalfa seed sowing (DFE). Land management in the winter crop season also affected temporal variation patterns of CH_4 fluxes and soil Eh after flooding. A great deal of CH_4 was emitted to the atmosphere during the period from flooding to the early stage of the rice-growing season; and CH_4 fluxes were still relatively high in the middle and late stages of the rice-growing period for treatments ALE, DFL, and FFE. However, for treatments DFE and WWE, almost no CH_4 emission was observed until the middle stage, and CH_4 fluxes in the middle and late stages of the rice-growing period were also very small. Soil Eh of treatments ALE and DFL decreased quickly to a low value suitable for CH_4 production. Once Eh below −150 mV was established, the small changes in Eh did not correlate to changes in CH_4 emissions. The soil Eh of treatments DFE and WWE did not decrease to a negative value until the middle stage of the rice-growing period, and it correlated significantly with the simultaneously measured CH_4 fluxes during the flooded and rice-growing period.

Introduction

Methane (CH_4) is an important greenhouse gas and a key factor in tropospheric and stratospheric chemistry (IPCC, 1992). Rice fields are one of the major contributors to the increasing atmospheric CH_4 concentration (Schutz et al., 1989). Since the first field study of CH_4 emission from a rice field was made in California (Cicerone & Shetter, 1981), research has focused on studying CH_4 emissions from rice fields and exploring strategies for mitigating them (Schutz et al., 1989; Sass et al., 1991; Yagi et al., 1994; Wassmann et al., 1993). Unfortunately, almost all experimental treatments in previous studies concentrated on the rice-growing period. Many questions concerning variables within annual field management have not been answered. For example, does land management during a nonrice growth period affect CH_4 emission in the following rice-growing period? If this is the case, CH_4 emissions from rice fields may be reduced with appropriate land management techniques in the winter crop season.

China is one of the major rice-producing countries in the world, occupying 22.6% of the total rice-growing area and contributing 36.3% of the total rice grain production (IRRI, 1991). Rice soils are diversely managed in the winter crop season in China: they can be left fallow or cultivated with different kinds of crop, given a wet or dry water regime, or amended with rice straw at different times. The more popular cropping systems involve fallow and cultivation of green manure and winter wheat. There are 2.7-4.0 × 10^6 ha of rice fields in China which are continuously flooded in the winter crop season (Cai, 1995). Rice fields, which maintain a water layer during this season, seem to emit particularly large amounts of CH_4 (Cai, 1997). Chinese farmers now use more rice straw as organic manure rather than as daily fuel, the consequence of standards of living and decreasing cost of coal gas. Rice straw was applied to fields at the beginning of the winter crop season and before rice transplanting. Different water management schemes, cropping systems, and rice straw application times may result in different soil methanogenic populations and activities that influence CH_4 emission during the following flooded rice-growing period. Due to their diversity and feasibility, land management options in the winter crop season may provide more effective mitigation strategies than those recommended by studies which focused on the rice-growing period. In China, early rice field measurements revealed very high CH_4 emission, the highest recorded throughout the rice-growing period in the world (Khalil et al., 1991). Thereafter, the majority of mean CH_4 fluxes measured from Chinese rice fields were much lower (Cai, 1997). Until today, research that aims to explain this unusually high CH_4 emission in the early years in China is scanty. It is worthwhile to look into the underlying mechanisms that resulted in the unusual high CH_4 emission in order to map out strategies that will reduce CH_4 emission from rice fields. In China, rice-growing soils are exposed to three kinds of water management: intermittent irrigation, continuous flooding during rice-growing period but dry in the winter crop season, and flooding all year-round (Cai, 1997). The highest mean CH_4 flux was recorded in a rice field flooded year-round (Khalil et al., 1991). This suggests that water management in the winter crop season may be a very important factor influencing CH_4 emission during the following rice-growing period. To evaluate the effect of land management, especially water management in the winter crop season, a greenhouse pot experiment was conducted from October 1995 to October 1997. This paper presents the results of the 2-yr study.

Materials and methods

Soil and experimental design

The experiment was conducted in a greenhouse. The soil was collected from the experimental farm of Jurong Agricultural College, Jiangsu Province, immediately after rice harvest in 1995. It was derived from Xiashu loess and classified into Typic Haplaquepts (USDA, 1975). Before treatment, the soil was air-dried and passed through a 5-mm sieve. The soil has 9.87 g organic carbon kg^{-1}; 1.18 g total N kg^{-1}, and a pH of 6.3. Experimental pots, 20 cm inner diameter and 30 cm height, were filled with 6 kg of soil. At the beginning of the winter crop season, the prepared soils were treated as follows: dry fallow (DFE and DFL), flooded fallow (FFE) with more than 2 cm floodwater layer, soil planted to alfalfa (ALE), and soil planted to winter wheat (WWE). All treatments had three replications. Thirty grams of rice straw containing organic carbon (413 g kg^{-1}, 1995; 451 g kg^{-1}, 1996) was incorporated into the surface soil in all treatments (except for DFL) before the alfalfa seed was sowed. Soils in all 15 pots were flooded on 1 Jun in 1996 and 1997. Twenty-two grams of air-dried alfalfa containing organic carbon (378 g kg^{-1}, 1996; 401 g kg^{-1}, 1997) and the same amount of rice straw were incorporated into the surface soil in treatments ALE and DFL just before flooding. Rice was transplanted on 14 Jun and 17 Jun and harvested on 12 Oct and 7 Oct in 1996 and 1997, respectively. The rice stubble remained in the pot after the 1996 rice harvest.

Water management of rice pots

A water layer of more than 2 cm was maintained during the rice-growing period for all treatments in 1996 and 1997.

Gas sampling and CH_4 measurement

Gas samples were collected with plexiglass chambers (51 × 51 ×100(h) cm) at 3-7 d intervals after the rice pots were placed on specially designed wood tables. Methane concentration in the gas samples was determined with a gas chromatograph (Shimadzu GC-12A) equipped with a flame ionization detector.

Soil Eh measurement

When the CH_4 flux was measured, soil Eh was also simultaneously determined by using Pt-tipped electrodes (Hirose Rika Co., Ltd.) and an ORP meter (Toa RM-1K). The electrodes were inserted into the soil at a depth of 10 cm and kept in place throughout the rice-growing period. All soil Eh measurements were made in triplicate.

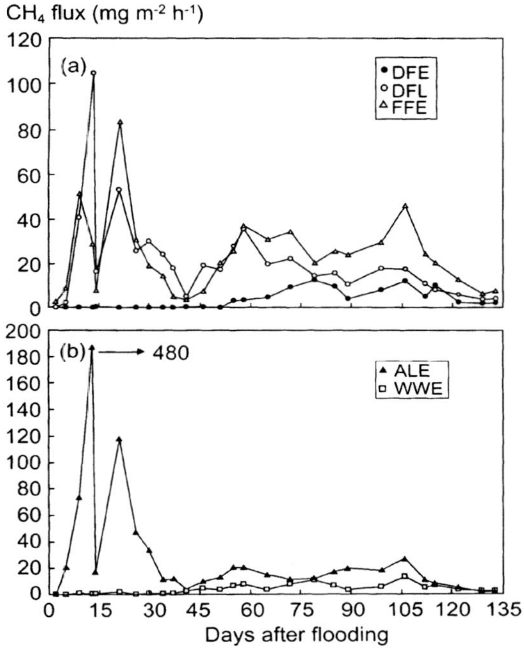

Figure 1. Temporal variations in CH_4 fluxes during the period from flooding to rice harvest for treatments with different water management levels, cropping systems, and rice straw application times in the 1996 winter crop season. (a) Treatments DFE, DFL and FFE; (b) Treatments ALE and WWE. DFE, dry fallow to which rice straw was applied just before the winter crop season; DFL, dry fallow to which rice straw was applied just before rice transplanting; FFE, flooded fallow to which rice straw was applied just before the winter crop season; ALE, alfalfa to which rice straw was applied just before the winter crop season; WWE, wheat to which rice straw was applied just before the winter crop season

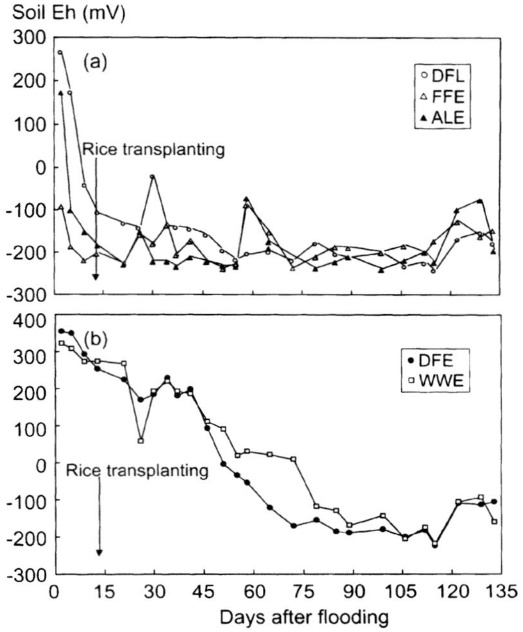

Figure 2. Temporal variations in soil Eh during the period from flooding to rice harvest for treatments with different water management levels, cropping systems, and rice straw application times in the 1996 winter crop season. (a) Treatments FFE, DFL, and ALE; (b) Treatments DFE and WWE. FFE, flooded fallow to which rice straw was applied just before the winter crop season; DFL, dry fallow to which rice straw was applied just before rice transplanting; ALE, alfalfa to which rice straw was applied just before the winter crop season; DFE, dry fallow to which rice straw was applied just before the winter crop season; WWE, wheat to which rice straw was applied just before the winter crop season.

Results and discussion

Figure 1a,b illustrates temporal variations in CH_4 fluxes among treatments with different water management schemes, cropping systems, and rice straw application times in the 1996 winter crop season. The patterns of temporal variations from flooding to rice harvest could be clearly divided into two sections. For treatments DFL, ALE, and FFE, CH_4 fluxes were substantial during the first 21 d after flooding, and CH_4 emissions measured thereafter were still relatively high. On the other hand, for treatments WWE and DFE, almost no CH_4 emission were observed up to 55 d after flooding; and CH_4 fluxes during the following period were also very small.

Land management in the winter crop season affected not only the temporal variation pattern of CH_4 flux but also the pattern of soil Eh change after flooding (Figure 2a,b). Soil Eh of treatment FFE was very low and within the active range of methanogenic bacteria all the time after flooding. Soil Eh values of the other four treatments were very high just after flooding, but the patterns of soil Eh change after flooding differed among treatments ALE and DFL and treatments WWE and DFE. In ALE and DFL, soil Eh decreased quickly after flooding, approximating that of FFE 5 d and 13 d after flooding, respectively. Meanwhile, it took 65 d (DFE) and 79 d (WWE) after flooding for soil Eh of treatments WWE and DFE to drop to within the active range of methanogenic bacteria. The results indicated that the maintenance of soil Eh at a high level for more than 2 mo after flooding was the main reason for the almost negligible CH_4 fluxes in treatments WWE and DFE. Some of the results support Trolldenier's find-

Table 1. Mean CH_4 fluxes[a] ($mg\ m^{-2}\ h^{-1}$) of different treatments during the period from flooding to rice harvest in 1996 and 1997

Treatment[b]	CH_4 fluxes in 1996	CH_4 fluxes in 1997
ALE	28.60 ± 5.60a	38.17 ± 14.39a
FFE	24.59 ± 2.96a	16.21 ± 1.05b
DFL	19.73 ± 0.83a	35.20 ± 12.18ac
WWE	4.73 ± 1.37b	5.62 ± 1.88bd
DFE	4.06 ± 0.62b	3.15 ± 0.74d

[a]CH_4 fluxes followed by the same letter are not significantly different at P = 0.05. [b]DFE, dry fallow to which rice straw was applied just before the winter crop season; FFE, flooded fallow to which rice straw was applied just before the winter crop season; DFL, dry fallow to which rice straw was applied just before rice transplanting; ALE, alfalfa to which rice straw was applied just before the winter crop season; WWE, wheat to which rice straw was applied just before the winter crop season

ings (1995) that the soil Eh of a rice pot with dry fallow in the previous crop season decreased very slowly, and that CH_4 flux was very low until soil Eh dropped to within the active range of methanogenic bacteria about 90 d after flooding.

Land management in winter also significantly influenced CH_4 fluxes during the period from flooding to rice harvest. The mean CH_4 fluxes of different treatments during the period from flooding to rice harvest in 1996 and 1997 were shown in Table 1. As affected by cropping system, the CH_4 flux of the treatment ALE was significantly higher than those of treatments WWE and DFE. The mean CH_4 fluxes of treatments ALE, WWE, and DFE were 28.6, 4.7, and 4.1 mg CH_4 m^{-2} h^{-1} in 1996 and 38.2, 5.6, and 3.2 mg CH_4 m^{-2} h^{-1} in 1997, respectively. As affected by water management, mean CH_4 flux of treatment FFE was 6.1 and 5.2 times higher than that of treatment DFE in 1996 and 1997, respectively (Table 1). Applying rice straw just before flooding of the soil for rice growth (DFL) significantly enhanced CH_4 flux by 386% in 1996 and by 1,017% in 1997 compared with rice straw application before alfalfa seed sowing (DFE) (Table 1).

Although water and crop residue management during the winter cropping period can directly influence soil Eh during the following rice-growing season, soil Eh is not always a good predictor of CH_4 emissions. Methane production occurs when soil suspension Eh is below −150 mV (Masscheleyn et al., 1993; Wang et al., 1993). Wang et al (1993) found that CH_4 production increased exponentially with decrease in soil Eh from −150 to −250 mV. As noted above, for prerice planting treatments ALE, FFE, and DFL, soil Eh rapidly decreased into the CH_4 production zone and CH_4 emissions increased (Figure 2a, Figure 1a,b). Once Eh below −150 mV was established in the puddled soil, the small changes in Eh observed up to rice harvest did not correlate to changes in CH_4 emissions (r = 0.11, 0.32, and 0.36 for ALE, DFL, and FFE treatments, respectively, using the equation $y = ax^2 + bx + c$). In contrast, simultaneously measured CH_4 flux and soil Eh for treatments WWE and EFE were significantly correlated (r = 0.69; $y = -1E\text{-}05x^2 - 0.0126x + 4.96$ for WWE; r = 0.87; $y = 6E\text{-}05x^2 - 0.023x + 1.56$ for DFE). In these prerice crop treatments, reducing conditions appropriate for CH_4 production were not reached until >60 d after flooding (Figure 2b). As a result, soil Eh and CH_4 emissions were highly correlated throughout the rice-growing period.

Soil Eh status and its change in direction depend on the relative concentration of electron donors and receptors in the soil. Easily decomposable organic carbon is a main electron donor; and NO_3^-, reducible Mn^{4+} and Fe^{3+}, and SO_4^{2-} are electron receptors (Yagi et al., 1994). Laboratory anaerobic incubation demonstrated a good relationship between CH_4 production and soil organic carbon content (Crozier et al., 1995). Green manure or rice straw which was applied into soil just before the rice-growing season played a role as soil organic matter in affecting soil Eh direction. Although the soils were not flooded under treatments ALE and DFL in the winter crop season, addition of green manure or rice straw before flooding provided the soils with extra electron donors, energy and carbon sources; therefore soil Eh of treatments ALE and DFL decreased faster after flooding and their temporal variations were close to that of treatment FFE (Figure 2a). On the other hand, applying rice straw before the winter crop season allowed the rice straw to decompose during the whole winter season under aerobic condition when the soil was planted to winter wheat (WWE) or dry fallow (DFE) (Figure 2b). This explains the greater mean CH_4 fluxes of treatments ALE and DFL than those of treatments WWE and DFE (Table 1).

Water management during the rice-growing season had a strong influence on CH_4 emissions from rice fields (Sass et al., 1992). Our results showed that water management in the nonrice growing period also played an important role. Mean CH_4 flux of treatment with continuously flooded fallow (FFE) in the winter crop season was 6.1 and 5.2 (1996) and 5.2 and 2.9 (1997) times higher than those of treatments with dry fallow and winter wheat (Table 1). Cai et al. (1998) also found that CH_4 flux from rice fields continuously flooded in the previous crop season was 2.8 times higher than fields previously planted to winter wheat. In China, rice fields flooded year-round accounted for 8-12% of total rice-cultivating area. Mainly distributed in southwest China, they were the dominant contributor to total CH_4 emissions from Chinese rice fields (Cai, 1997). If irrigation and drainage facilities for the year-round flooded rice fields could be improved substantially and if floodwater could be drained completely during winter, total CH_4 emissions would be significantly reduced.

Methane fluxes from rice fields were strongly enhanced by incorporation of green manure or rice straw (Denier van der Gon & Neue, 1995; Yagi & Minami, 1990). The unusually high mean CH_4 fluxes of treatments ALE and DFL mainly resulted from the addition of green manure or rice straw before flooding (Table 1). To mitigate CH_4 emission from rice fields, organic amendments should be minimized. However, this may conflict with soil fertility aspects, as well as local availability of fertilizers. In this experiment, rice straw incorporation with the surface soil before the winter crop season, whether soil was fallow or planted to winter wheat, resulted in very low CH_4 fluxes during the following flooded and rice-growing period (Table 1). This suggests that application time is an important factor that should be taken into account in evaluating the effect of rice straw application on CH_4 emissions from rice fields.

Conclusions

Land management in the winter crop season significantly affected CH_4 emission and soil Eh during the following flooded and rice growth period. The difference in soil Eh and temporal variation patterns as a result of land management in the previous crop season explains why CH_4 fluxes and the temporal variation patterns under different treatments were not alike.

Water management in the preceding crop season was a very important factor that influenced CH_4 emissions from rice fields. Compared with the management of flooded fallow in the winter crop season, a practice mainly adopted in southwest China, planting winter wheat or dry fallow, which is rather popular in rice-growing areas in China, could result in significantly reduced CH_4 emissions during the following flooded and rice-growing period.

Rice straw, which undergoes aerobic decomposition in the winter crop season after being incorporated into the soil, had a greatly decreased effect on CH_4 emission during the following flooded and rice-growing period. Rice straw and possibly green manure application at a suitable application time not only could sustain soil fertility and meet the needs of sustainable agriculture but also could prevent large amounts of CH_4 being emitted to the atmosphere.

Acknowledgements

The authors wish to thank Dr A.R. Mosier and the other anonymous editors and reviewers for corrections to and suggestions for the manuscript. Financial support was provided by the National Natural Science Foundation of China (49371039), National Key Basic Research Development Foundation (G1999011805) and the Red Soil Ecological Experimental Station, Chinese Academy of Sciences.

References

Cai ZC, Xu Hua, Lu WS & Wei CF (1998) Influence of water management in winter crop season on CH_4 emission during rice-growing season. Chinese J Appl Ecol 9(2):171-175

Cai ZC (1995) Estimation of methane emission from Chinese rice fields. In: Chen Yiyu (ed) Study of Wetlands in China, Jilin Science Technology press. p 79-85

Cai ZC (1997) A category for estimate of CH_4 emission from rice paddy fields in China. Nutr Cycling Agroecosyst 49:171-179

Cicerone RJ & Shetter JD (1981) Sources of atmospheric methane: measurements in rice paddies and a discussion. J Geophys Res 86:7203-7209

Crozier CR, Devai & Delaune RD (1995) Methane and reduced sulfur gas production by fresh and dried wetland soils. Soil Sci Soc Am J 59:277-284

Denier van der Gon HAC & Neue HU (1995) Influence of organic matter incorporation on the methane emission from a wetland rice field. Global Biogeochem Cycles 9:11-22

IPCC – Intergovernmental Panel on Climate Change (1992) Climate Change, the supplementary report to the IPCC scientific assessment. Cambridge (UK): Cambridge University Press. XII+200 p

IRRI – International Rice Research Institute (1991) World Rice Statistics 1990. Manila (Philippines)

Khalil MAK, Rasmussen RA, Wang MX & Ren L (1991) Methane emissions from rice fields in China. Environ Sci Technol 25:979-981

Masscheleyn PH, Delaune RD & Patrick Jr WH (1993) Methane and nitrous oxide emission from laboratory measurements of rice soil suspension: effect of soil oxidation-reduction status. Chemosphere 26:251-260

Sass RL, Fisher FM, Harcombe PA & Turner FT (1991) Methane production and emission in a Texas rice field. Global Bigeochem Cycles 4:47-68

Sass RL, Fisher FM, Wang YB, Turner FT & Jund MF (1992) Methane emission from rice fields: the effect of flood water management. Global Biogeochem Cycles 6:249-262

Schutz H, Seiler W & Conrad R (1989) Processes involved in formation and emission of methane of rice paddies. Biogeochemistry 7:33-53

Soil survey staff (1975) Soil Taxonomy. U. S. Dep. Agricultural Handbook. No. 436. Washington, D.C.

Trolldenier G (1995) Methanogenesis during rice growth as related to the water regime between crop seasons. Biol Fertil Soils 19:84-86

Wang ZP, Delaune RD, Masschelegn PH & Patrick Jr WH (1993) Soil redox and pH effects on methane production in flooded rice soils. Soil Sci Soc Am J 57:382-385

Wassmann R, Papen H &Rennenberg H (1993) Methane emission from rice paddies and possible mitigation strategies. Chemosphere 26:201-217

Yagi K & Minami K (1990) Effect of organic matter application on methane emission from some Japanese paddy fields. Soil Sci Plant Nutr 36:599-610

Yagi K, Tsuruta H, Minami K, Chairoj P & Cholitkul W (1994) Methane emission from Japanese and Thai paddy fields. In: Minami K, Mosier A & Sass R (eds) CH_4 and N_2O. Proceedings of an International Workshop; 25-26 March 1992; Tsukuba, Japan. p 41-53

Nutrient Cycling in Agroecosystems **58:** 333–338, 2000.

Effects of organic and N fertilizers on methane production potential in a Chinese rice soil and its microbiological aspect

A.X. Hou[1,2], Z.P. Wang[2], G.X. Chen[1] & W.H. Patrick, Jr.[2]

[1]*Institute of Applied Ecology, Chinese Academy of Sciences, Shenyang 110015, China;* [2]*Wetland Biogeochemistry Institute, Louisiana State University, Baton Rouge, LA 70803, USA*

Key words: rice straw, organic manure, related microbial groups, zymogenic bacteria

Abstract

An incubation experiment to determine the effects of organic and chemical N fertilizers on methane (CH_4) production potential in a Chinese flooded rice soil was conducted. Organic matter, added as rice straw and organic manure, increased CH_4 production rate significantly. Chemical N fertilizers such as ammonium bicarbonate (AB), modified ammonium bicarbonate (MAB), and urea (U) did not show a clear effect when they were applied with rice straw. Field results may be very different because of the involvement of rice plants. Organic manure showed different promoting effects on CH_4 production rate. Pig manure stimulated the production rate most, followed by chicken and cattle manure. This difference in organic manure was not related to either total C added to the system or to C/N. The study on bacteria groups related to CH_4 production indicated that the different effects of organic matter may be closely related to content of easily decomposable organic matter. A significant linear relationship between CH_4 production and the logarithm of the number of zymogenic bacteria was found with an *r* value of 0.96. This finding suggests that the number of zymogenic bacteria may be used as an index to predict CH_4 production potential in flooded rice fields and other wetlands.

Introduction

Methane (CH_4) production in flooded rice soils is a microbiological process affected by many biochemical and physical factors in the environment. Organic carbon (C) source supply to the soil, whether it comes from the disposal of crop residues or as organic fertilizer, appears to be the most important factor in controlling the CH_4 production rate. Nitrogen (N) fertilization, which is essential for high rice yield, is also likely to influence CH_4 production by changing the properties of the soil and the litter and root exudates from the rice plants. Therefore, studying the effects of organic and N fertilizers, which are widely applied in Asian rice cropping systems, on CH_4 production potential would be helpful in estimating global CH_4 production and explaining the complexity of field results. Changes in CH_4 emission with rice straw application have been observed in both laboratory and field experiments (Yagi & Minami, 1990; Wang et al., 1992). Much attention has also been paid to the influence of urea and ammonium sulfate application on CH_4 emission rate, but results obtained by different researchers were contradictory (Yagi & Minami, 1991; Schütz et al., 1989a).

Most studies on the microbiological aspect of CH_4 production in flooded rice soil have focused on methanogens (Asakawa & Hayano, 1995; Asakawa et al., 1996, 1998). In addition to methanogens, the degradation of organic matter to its most reduced status (i.e., CH_4), however, involves at least two other kinds of nonmethanogens, the zymogenic bacteria and the acetic acid- and hydrogen-producing bacteria. Methanogenic bacteria are strictly anaerobic autotrophs that catalyze the terminal step in the anaerobic decomposition of organic matter. Only a limited number of small molecules can be used as their substrates, but these small molecules are mainly supplied by the metabolic activities of zymogenic bacteria and acetic acid- and hydrogen-producing bacteria in most environments such as flooded rice soil and natural wetland soil. Thus, zymogenic bacteria and acetic acid- and hydrogen-producing bacteria should have some relationships with CH_4 production.

In this study, we compared the effects of chemical N fertilizers (ammonium bicarbonate [AB], modified ammonium bicarbonate [MAB], and urea [U]) and organic fertilizers (including rice straw, pig, chicken, and cattle manure) on CH_4 production potential in Beijing rice soil incubated in the laboratory. The number of zymogenic bacteria, acetic acid- and hydrogen-producing bacteria and methanogens were estimated by the most probable number (MPN) method.

Materials and methods

Soil sample

The soil used in the study was a sandy soil obtained from a rice field located in Yongfeng County, Haidian District, Beijing, China. It contained 15.9 g organic matter kg^{-1} soil and 0.78 g total N kg^{-1} soil and had a pH of 8.1 (1:1, soil/water). Bulk samples from the surface 15 cm of the soil were passed through a 2-mm sieve, air-dried, and stored at room temperature.

Fertilizers

The organic fertilizers tested in this experiment included rice straw and pig, chicken, and cattle manure. These are widely applied to rice in Asia. Their main characteristics are shown in Table 1.

Table 1. Main chemical characteristics of organic manure and rice straw

Type	Total C (%)	Total N (%)	C/N
Pig manure	46.62	3.56	13.10
Chicken manure	34.75	3.30	10.53
Cattle manure	48.18	3.25	14.82
Rice straw	67.14	1.13	59.61

Three chemical N fertilizers, which are extensively used in Chinese rice fields, were tested. The traditional AB and U are commercially available. The MAB, AB co-crystallized with dicyandiamide during production, was obtained from the Fengcheng Fertilizer Manufacturing Co. It has an N content of 17% and a water content of 3.5%.

Effect of fertilization on CH_4 production potential incubation experiment

In this experiment, 11 treatments were designed as follows: (1) control, (2) rice straw, (3) AB with rice straw, (4) AB without rice straw, (5) MAB with rice straw, (6) MAB without rice straw, (7) U with rice straw, (8) U without rice straw, (9) pig manure, (10) chicken manure, and (11) cattle manure. Each treatment was replicated three times.

Twenty g of air-dried soil were placed into 100-ml incubation bottles. The bottles were sealed with rubber stopper with an inlet and outlet for gas. Forty ml of distilled water or chemical N fertilizer solution was added to bring the ratio of water to soil to 2:1. Four mg of fertilizer N was applied. Organic manure was added to attain the same N concentration as the chemical N fertilizer treatment. The amount of rice straw applied was 0.1% (dry soil basis).

The soil suspensions were incubated at 30 °C after they were homogenized. Methane production rate was measured at 0, 3, 6, 10, 15, 20, 25, 30, 35, and 40 d after incubation. Meanwhile, soil pH and redox potential values were also determined. During measurement of CH_4 production rate, the soil suspension was stirred by a magnetic stirrer and purged with O_2-free N_2 (250 ml min^{-1}) for 3 min to clear the originally existing gases away 24 h before the gas sampling time. Just before gas sampling, the soil suspension was stirred again by a magnetic stirrer for 3 min to release the CH_4 trapped into the soil suspension. A gas sample was then taken from the headspace of the bottle by using a 1-ml plastic syringe. After gas sampling, the incubation bottle was again purged with O_2-free N_2 for 3 min, resealed, and set aside until the next measurement.

Gas analysis

Methane concentration was analyzed using a Shimadzu gas chromatograph (GC) fitted with a flame ionization detector. Standard CH_4 gas was provided by the National Institute of Standard Material, China. The rate of CH_4 production was expressed as μg CH_4 g^{-1} soil d^{-1}.

Separation and measurement of microorganisms

Five ml of soil suspension was collected by a 5-ml sterilized plastic syringe after 15 d of incubation. Samples were inoculated immediately to determine the number of zymogenic bacteria, acetic acid- and hydrogen-producing bacteria, and methanogens. The medium preparations and procedures for separating and cultivating these three related bacteria groups have been described by Hou et al. (1997). The items tested were optical density (OD) for zymogenic bacteria by colorimetric analysis, H_2 for acetic acid- and hydrogen-producing bacteria by GC, and CH_4 for methanogens by GC, respectively. The MPN was used as the enumeration method of bacteria.

Statistical analysis

The SPSS 6.0 software package from SPSS Inc. (17 Jun 1993) was used to calculate correlation coefficients between different variations and to compare differences in total amounts of CH_4 production among treatments by analysis of variance at the 0.05 probability level.

Results and discussion

Effect of rice straw application on CH_4 production potential

The effect of rice straw application on CH_4 production potential is shown in Figure 1. Methane production in the treatment without rice straw supplement was at a much lower rate during the whole period of incubation, in which the highest production rate was less than 40 μg CH_4 kg^{-1} soil d^{-1}. In situ results might differ because of the involvement of rice plants. It has been proved that leaf litter and root exudates from growing rice plants could enhance CH_4 emission by providing substrates for methanogenesis (Raimbault et al., 1977; Kludze et al., 1993; Holzapfel-Pschorn & Seiler, 1986; Schütz et al., 1991). After the application of rice straw, CH_4 production rate increased substantially. Methane production rate quickly reached a maximum value at 6 d after the start of the incubation. This peak lasted for around 2 wk before it started to decrease; it was almost not detectable after 40 d of incubation. This confirms that the exogenous supply of organic C is an important contributor to CH_4 production. Wang et al. (1992) reported a linear relationship between CH_4 production rate and rice straw addition rate in Crowley rice soil. A field study (Yagi & Minami, 1990) also showed that rice straw applied at rates of 6-9 t ha^{-1} enhanced CH_4 emission rates by 1.8-3.5 times.

Effect of chemical N fertilizers on CH_4 production potential

Chemical N fertilizers—AB, MAB, and U—had a slight inhibiting effect on CH_4 production when they were applied without the supplement of organic matter (Figure 2). The CH_4 production rates were much lower compared with treatments with rice straw (Figure 3); and no significant differences existed among the AB, MAB,

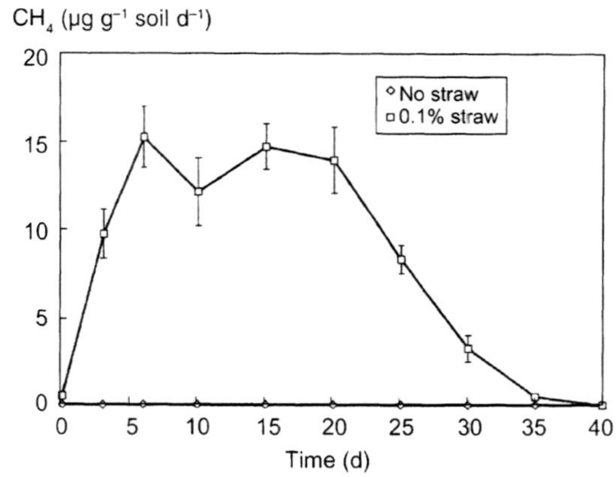

Figure 1. Effect of rice straw application on CH_4 production potential (n=3, mean ± SE)

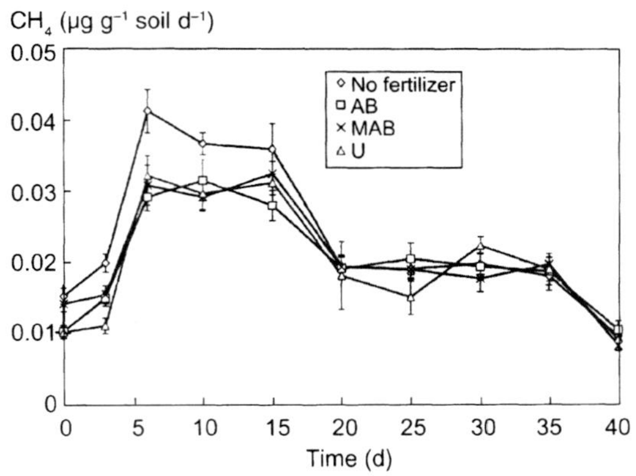

Figure 2. Effect of chemical fertilizer application on CH_4 production potential (without rice straw; n=3, mean ± SE)

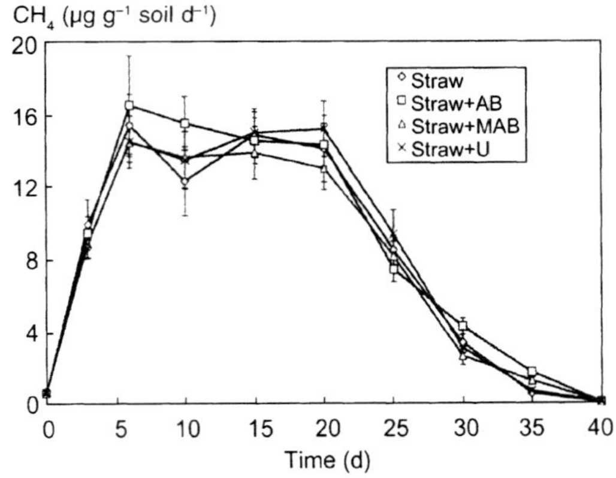

Figure 3. Effect of chemical fertilizer application on CH_4 production potential (with rice straw; n=3, mean ± SE)

and U treatments. This suppression might be mainly attributed to the shift (around 0.5 to 1 unit increase) in pH value of the tested soil sample away from the range for CH_4 production after application of AB, MAB, and U. Some studies showed that most methanogens preferred to grow at the relatively narrow pH range of 6-8 and optimal pH was around 7 (Alexander, 1977; Oremland, 1988), although there also exist a few acidophilic and alkaliphilic methanogen strains (Crawford, 1984; Oremland et al., 1982). Studies by Wang et al. (1995) suggested that addition of U in most acidic soils enhanced CH_4 production, but in all nonacidic and alkaline soils, CH_4 production was inhibited probably because of an increase in soil pH by U. AB, MAB, and U did not show a clear effect on CH_4 production when they were applied with straw. The total amounts of CH_4 production in straw, AB with straw, MAB with straw, and U with straw were 344, 365, 331, and 352 μg CH_4 g^{-1} soil, respectively. There were no significant differences among these treatments. The results from both treatments with and without rice straw showed that chemical N fertilizers had only slight effect on CH_4 production rate in nonacidic rice soil, which could be masked by the importance of C source to CH_4 production. This further suggests that the exogenous organic carbon is the key factor to control CH_4 production rate in rice soils. The contradictory field results obtained by different researchers (Yagi & Minami, 1991; Schütz et al., 1991a) might be due to changes in soil chemical characteristics (such as pH) and plant litter and root exudates following chemical fertilizer application.

Effect of organic manure application on CH_4 production potential

As shown in Figure 4, CH_4 production rates quickly increased following the application of pig, chicken, and cattle manure. The activities of methanogens and related bacteria groups existing in air-dried soils were rapidly restored shortly after incubation. The results also showed that these three kinds of organic manure had different promoting effects, with pig manure increasing the CH_4 production rate most, followed by chicken and cattle manure.

The results from correlation analyses (Table 2) indicate that the difference in CH_4 production potential caused by organic manure seemed neither closely related to total C added to the system nor to the C/N of the materials.

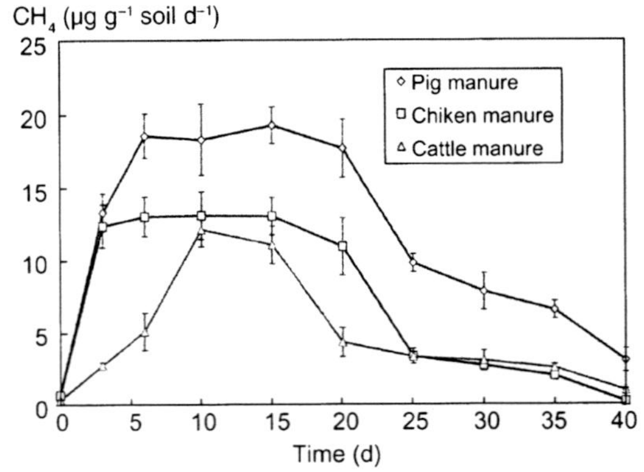

Figure 4. Effect of organic manure application on CH_4 production potential (n=3, mean ± SE)

Table 2. Relationship between CH_4 production and amount and C/N of organic matter added to the system

Item	Total CH_4 (μg g^{-1})	Total C added (g kg^{-1})	C/N
Pig manure	499	2.60	13.10
Chicken manure	302	2.10	10.53
Cattle manure	204	2.95	14.82
Rice straw	344	1.00	59.61
r[a]		-0.13	0.019
P[b]		0.87	0.98

[a] *r* stands for correlation coefficient between CH_4 production and amount and C/N of organic matter added to the system. [b]P stands for significance of correlation between CH_4 production and amount and C/N of organic matter added to the system; P>0.05 means no significant correlation

Microbiological aspects of production potential following application of various organic fertilizers

To understand the microbiological mechanism behind the influence of organic matter on CH_4 production, three related microbial groups involved in degrading organic matter under strictly anaerobic conditions—zymogenic bacteria, acetic acid- and hydrogen-producing bacteria, and methanogenic bacteria—were evaluated. The cell numbers in treatments with organic fertilizers after 15 d of incubation are shown in Table 3. The maximum cell number of these three bacteria was observed in the pig manure treatment, whereas the minimum was found in the cattle manure treatment. These results were consistent with the amount of CH_4 produced in the treatments. The correlation between CH_4 production potential and number of related microbial groups showed a significant relationship between CH_4 production and the logarithm of the number of zymogenic bacteria (Table 3). The correlation between CH_4 production and acetic acid- and hydrogen-producing bacteria and methanogenic bacteria was lower. These results suggest that the cell number of zymogenic bacteria was most sensitive to changes in environmental conditions. We also found this linear relationship between CH_4 emission and zymogenic bacteria in a field experiment (Hou et al., 2000). This means that the number of zymogenic bacteria may be used as an index to predict CH_4 production potential in flooded rice fields (and wetland). Further studies should be done to test its practicability.

The dependence of methanogens on their microbial partners is due to the fact that nonmethanogens release fermentation products, which are the catabolic substrates for methanogens. The nonmethanogenic bacteria can hydrolyze and ferment a wide range of complex organic molecules into small molecular weight substrates for methanogenic bacteria. The zymogenic bacteria function at the first step in the anaerobic food chain. In this laboratory study, all incubation conditions, except organic matter, were the same. Our results showed that pig, chicken, and cattle manure had essentially the same effect on soil pH and Eh. The pH of all treatments was approximately 7 and the Eh value decreased to approximately –250 mV. Thus, one reasonable explanation for the significant correlation between CH_4 production and the logarithm of the number of zymogenic bacteria should be the larger amount of organic matter easily decomposed by zymogenic bacteria in the pig treatment compared with those in the chicken and cattle manure treatments. Accordingly, more precursors were supplied for methanogenesis. This suggests, from the point of view of microbiological ecology, that the different effects of various organic fertilizers on CH_4 production potential might be closely related to amount of easily decomposable organic matter.

Table 3. Number of related microbial groups (no. g^{-1} dry soil) in treatments with organic fertilizer and correlation analysis

Item	Zymogenic bacteria	Acetic acid- and hydrogen-producing bacteria	Methanogens
Pig manure	6.0×10^8	7.5×10^4	2.3×10^5
Chicken manure	6.0×10^7	2.9×10^4	6.0×10^4
Cattle manure	2.3×10^7	2.9×10^3	4.5×10^4
Rice manure	4.5×10^7	2.9×10^3	1.4×10^5
r[a]	0.96	0.73	0.94
P[b]	0.045	0.27	0.058

[a]*r* stands for correlation efficient between logarithm number of related microbial groups and CH_4 production potential in treatments with organic fertilizer. [b]P stands for significance of correlation between number of related microbial groups and CH_4 production potential in treatments with organic fertilizer; $P<0.05$ means significant correlation

Conclusions

Organic fertilizers including rice straw and organic manure substantially increased CH_4 production potential in flooded rice soil showing that organic matter applied in rice cropping systems makes a big contribution to CH_4 emission from rice fields. Chemical N fertilizers had no significant effect on CH_4 production potential in nonacidic flooded rice soil.

Various organic fertilizers had different promoting effects on CH_4 production, and these differences appeared to be closely related to composition of organic matter instead of total C or C/N of the materials. When studying the contribution of rice cropping systems to CH_4 production, the easily decomposable C content of the organic material added to the soil must be considered.

The linear correlation between CH_4 production and logarithm of the number of zymogenic bacteria suggests that the number of zymogenic bacteria may predict CH_4 production potential in rice fields and possibly other wetland ecosystems.

References

Alexander M (1977) Introduction to soil microbiology, 2nd ed, John Wiley & Sons, Inc., New York

Asakawa S, Akagawa-Matsushita M, Koga Y & Hayano K (1998) Communities of methanogenic bacteria in paddy field soils with long-term application of organic matter. Soil Biol Biochem 30:299-303

Asakawa S & Hayano K (1995) Populations of methanogenic bacteria in paddy field soil under double cropping conditions (rice-wheat). Biol Fertil Soils 20:113-117

Asakawa S, Koga Y & Hayano K (1996) Enumeration of methanogenic bacteria in paddy field soil by the most probable number (MPN) method. Soil Microorg 47:31-36

Crawford W (1984) Methane production in Minnesota peat lands. Appl Environ Microbiol 47:1266-1271

Holzapfel-Pschorn A & Seiler W (1986) Methane emission during a cultivation period from an Italian rice paddy. J Geophy Res 91:11803-11814

Hou AX, Chen GX, Wu J & Cleemput O (1997) Relationship between CH_4 and N_2O emissions from rice field and its microbiological mechanism and impacting factors. Chin J Appl Ecol 8: 270-274

Hou AX, Chen GX, Wang ZP, Yu K, Cleemput O & Patrick WH Jr (2000) CH_4 and N_2O emissions from a rice field in relation to soil redox and microbiological processes. Soil Sci Soc Am J 64: (in press)

Kludze HK, Delaune RD & Patrick WH Jr (1993) Aerenchyma formation and methane and oxygen exchange in rice. Soil Sci Soc Am J 57:386-391

Oremland RS, Marsh LM & Polcin S (1982) Methane production and simultaneous sulfate reduction in anoxic, salt marsh sediments. Nature 296:143-145

Oremland RS (1988) Biogeochemistry of methanogenic bacteria. In: Zehnder AJB (ed) Biology of anaerobic microorganisms, p 641-706, John Wiley and Sons, Inc., New York

Raimbault M, Rinaudo G, Garcia JL & Boureau M (1977) A device to study metabolic gases in the rice rhizosphere. Soil Biol Biochem 2:193-196

Schütz H, Holzapfel-Pschorn A, Conrad R, Rennenberg H & Seiler W (1989) A 3-year continuous record on the influence of daytime, season, and fertilizer treatment on methane emission rates from an Italian rice paddy. J Geophys Res 94:16405-16416

Schütz H, Schröder P & Rennenberg H (1991) Role of plants in regulating the methane flux to the atmosphere. In: Trace Gas Emissions by Plants, pp 29-63, Academic Press, New York

Wang Z, Delaune RD, Lindau CW & Patrick WH Jr (1992) Methane production from anaerobic soil amended with rice straw and nitrogen fertilizers. Fert Res 33:115-121

Wang Z, Kludze HK, Crozier CR & Patrick Jr WH (1995) Soil characteristics affecting methane production and emission in flooded rice. In: Pen S et al (eds) Climate Change and Rice, p 81-90, Springer-Verlag, Berlin

Yagi K & Minami K (1990) Effect of organic matter application on methane emission from some Japanese paddy fields. Soil Sci Plant Nutr 36:599-610

Yagi K & Minami K (1991) Emission and production of methane in the paddy fields of Japan. JARQ 25:165-171

Nutrient Cycling in Agroecosystems **58**: 339–347, 2000.

Effects of elevated CO_2 and temperature on methane production and emission from submerged soil microcosm

W. Cheng[1], K. Chander[2] & K. Inubushi[1]

[1]*Department of Bioproduction Sciences, Faculty of Horticulture, Chiba University, 648 Matsudo, Chiba 271-8510, Japan;* [2]*Guru Jambheshwar University, Hisar 125001, India*

Key words: elevated CO_2 and temperature, methane emission, methane production, rice soil

Abstract

Incubation experiments were conducted under controlled laboratory conditions to study the interactive effects of elevated carbon dioxide (CO_2) and temperature on the production and emission of methane (CH_4) from a submerged rice soil microcosm. Soil samples (unamended soil; soil + straw; soil + straw + N fertilizer) were placed in four growth chambers specifically designed for a combination of two levels of temperature (25 °C or 35 °C) and two levels of CO_2 concentration (400 or 800 µmol mol^{-1}) with light intensity of about 3000 Lx for 16 h d^{-1}. At 7, 15, 30, and 45 d after incubation, CH_4 flux, CH_4 dissolved in floodwater, subsurface soil-entrapped CH_4, and CH_4 production potential of the subsurface soil were determined. The results are summarized as follows: 1) The amendment with rice straw led to a severalfold increase in CH_4 emission rates, especially at 35 °C. However, the CH_4 flux tended to decrease considerably after 15 d of incubation under elevated CO_2. 2) The amount of entrapped CH_4 in subsurface soil and the CH_4 production potential of the subsurface soil were appreciably larger in the soil samples incubated under elevated CO_2 and temperature during the early incubation period. However, after 15 d, they were similar in the soil samples incubated under elevated or ambient CO_2 levels. These results clearly indicated that elevated CO_2 and temperature accelerated CH_4 formation by the addition of rice straw, while elevated CO_2 reduced CH_4 emission at both temperatures.

Introduction

Recent anthropogenic emissions of key atmospheric trace gases (e.g., CO_2 and CH_4) which absorb infrared radiation may lead to an increase in mean surface temperatures and potential changes in climate. Continuous increases in greenhouse gases in the atmosphere have been attributed to population growth, global reliance on burning fossil fuel for energy, and changes in land use practices. The concentration of CO_2 in the atmosphere may double during the next century (Bolin, 1986). The concentration of atmospheric methane (CH_4) has been increasing at about 1% yr^{-1}. Most of the atmospheric CH_4 is produced by the bacterial activities in extremely anaerobic ecosystems such as natural and cultivated wetlands, sediments, sewage, landfills, and the rumen of herbivorous animals (IPCC, 1995). Rice fields are considered as one of the most important sources of CH_4, taking into account the recent increase in harvested rice area in the world. Methane emission from flooded rice soils occurs through plant-mediated transport, ebullition, and diffusion (Kimura et al., 1996). Hitherto, research has been focused on plant-mediated CH_4 emission from rice fields (Inubushi et al., 1989, 1994; Chidthaisong et al., 1996; Kimura et al., 1996, Huang et al., 1998). Some researchers (Ziska et al., 1998) have monitored CH_4 emission at elevated CO_2 concentration and temperature conditions in tropical rice field using open-top chambers. However, there is a lack of information on the release pattern of CH_4 by ebullition and diffusion, albeit their possible dominance particularly during the early stage of flooding, and when rice plants are small (Crill et al., 1988; Takai & Wada, 1990). There is a need to accurately predict and elucidate fully the impact of changing climatic factors on CH_4 production and emission from flooded rice soils without the rice plant. In this context, one may hypothesize that increasing atmospheric CO_2 and/or temperature may stimulate growth and photosynthetic activity of algae which is commonly present on submerged soil surface and floodwater (Wang et al., 1994). Algae may cause increased inhibition of CH_4 flux by providing a physical barrier to ebullition or by releasing O_2 and thereby stimulating the methanotrophs. Therefore, in this work, an incubation experiment was conducted under controlled conditions with the main objective to study the interactive effects of elevated CO_2 and temperature on CH_4 production and emission through ebullition and diffusion (in the absence of plant) from a flooded soil microcosm.

Materials and methods

Soil and experimental treatments

The soil used for this study was a sandy soil collected from the plow layer (0-10 cm) of a nonexperimental farmer's rice field at Kuju-Kuri, Chiba Prefecture, Japan. It contained 7.4 g organic C kg^{-1}, 0.8 g total N kg^{-1} and had a pH of 6.4. The soil was air-dried, then sieved (<2 mm) before use. Portions (0.35 kg) of air-dried soil were weighed into plastic pots (10 cm diam, 10 cm height) covered with aluminum foil, except at the mouth. Each pot was then fertilized with 269 mg of P-K fertilizer (equivalent to each of 100 kg P_2O_5 and K_2O ha^{-1}). The experiment consisted of three treatments: 1) unamended soil; 2) soil amended with rice straw at 1% w/w (equivalent to 8 t ha^{-1}); rice straw dried at 35 °C for 2 d and pulverized with small electric mill before use; contained 400 g organic C kg^{-1} and 8 g total N kg^{-1}; and 3) soil amended with rice straw at 1% w/w and supplemented with N fertilizer as $(NH_4)_2SO_4$ at the rate of 100 mg N kg^{-1}(equivalent to 100 kg N ha^{-1}).

Design of experimental equipment

Four growth chambers were used in this study. Of two growth chambers receiving a continuous flow of elevated CO_2 (equivalent to 800 µmol mol^{-1}), one was set at 25 °C and the other at 35 °C. The two remaining growth chambers were maintained at near ambient CO_2 level (400 µmol mol^{-1}) with one set at 25 °C and the other at 35 °C. The growth chambers were provided with light at an intensity of about 3000 Lx for 16 h d^{-1}. Soil samples in 32 pots for each treatment were transferred to the respective growth chambers. Each pot was flooded with 250 ml of distilled water. The pots were weighed every day and weight loss was compensated for by adding water to maintain a constant water level throughout the incubation period.

Analyses

Duplicate pots of each treatment were withdrawn from each of the incubator after 7, 15, 30, and 45 d of incubation and the following measurements were made as per procedures explained below.

Measurement of CH_4 emission. Methane flux from pots was estimated using the method described by Inubushi et al. (1989). At every sampling, each pot was transferred into a closed chamber, and after an initial settling period (< 2 min), the amount of CH_4 emitted from the pots during the next 30 min were measured by taking 1 ml of the gas in the closed chamber and injecting it into a gas chromatograph (GC) (Shimadzu GC-7A) with a flame ionization detector.

Measurement of dissolved CH_4 in floodwater. From each pot, 5 ml of the aliquot supernatant water was directly transferred gently into a 30-ml erlenmeyer flask using an autopipette. The flask was then sealed immediately with a butyl stopper and shaken vigorously for 2 min to mobilize the dissolved gas to the headspace. Methane was measured as per procedure described above.

Measurement of entrapped CH_4 in subsurface soil. Immediately after siphoning off the floodwater, surface (0-1 cm) soil sample from each pot was removed gently by a spatula. A truncated syringe (6 ml, 12 mm inner diam) was inserted into the undisturbed subsurface soil (below 1 cm) with fixing head of plunger at 1 cm depth. The contents of the soil were transferred gently into a 30-ml erlenmeyer flask which was then sealed immediately with butyl stopper. The flasks were then shaken vigorously for 2 min to mobilize the trapped gas to the headspace. Methane collected in the headspace was measured in a GC as described above.

Measurement of CH_4 production potential and soluble C. From each pot, 20-g portions of wet subsurface soil was put into a 100-ml glass bottle, to which 20 ml of oxygen-free water was added. The headspace gas in the bottle was then replaced with nitrogen gas before sealing tightly with a butyl stopper. The bottles were then incubated either at 25 °C or 35 °C (i.e., at temperatures similar to those in their respective pots) under dark condition. Methane concentration in the headspace over 7 d of incubation was determined in a GC as described above. The amount of soluble carbon in soil samples was determined by extracting with 0.5 M K_2SO_4 and dichromate digestion (Inubushi et al., 1991).

Results and discussion

Methane flux

The CH_4 emission rates obtained from various treatments at different intervals during incubation were calculated on a per pot basis and the results are presented in Figure 1. The amount of CH_4 emitted from the unamended (control) soil samples was low, ranging between 0 and 4 µg C pot^{-1} h^{-1} and thus all the values were comparable, irrespective of incubation condition. The addition of rice straw led to a severalfold increase in CH_4 emission rates over that of the respective unamended soil samples under all incubation conditions. However, the rate of increase varied greatly among the different incubation conditions. The maximum emission rates were observed in the soil samples incubated under elevated CO_2 conditions at 35 °C at day 7; under ambient CO_2 at 35 °C and under elevated CO_2 conditions at 25 °C at day 15; in soil samples incubated under 25 °C ambient CO_2 condition at day 30. At 45 d, the emission rates decreased in all the amended soils, irrespective of incubation condition. These results indicate that higher incubation temperature had caused faster decomposition of organic matter, leading to its increased conversion to CH_4.

Averaged over both temperatures, the CH_4 emission rates were about 20-50% less in soil samples incubated under elevated CO_2 than in those incubated under ambient CO_2. The emission rates in soil samples amended with straw + N were generally less pronounced than in those amended with straw alone at an incubation temperature of 35 °C. Increasing atmospheric CO_2

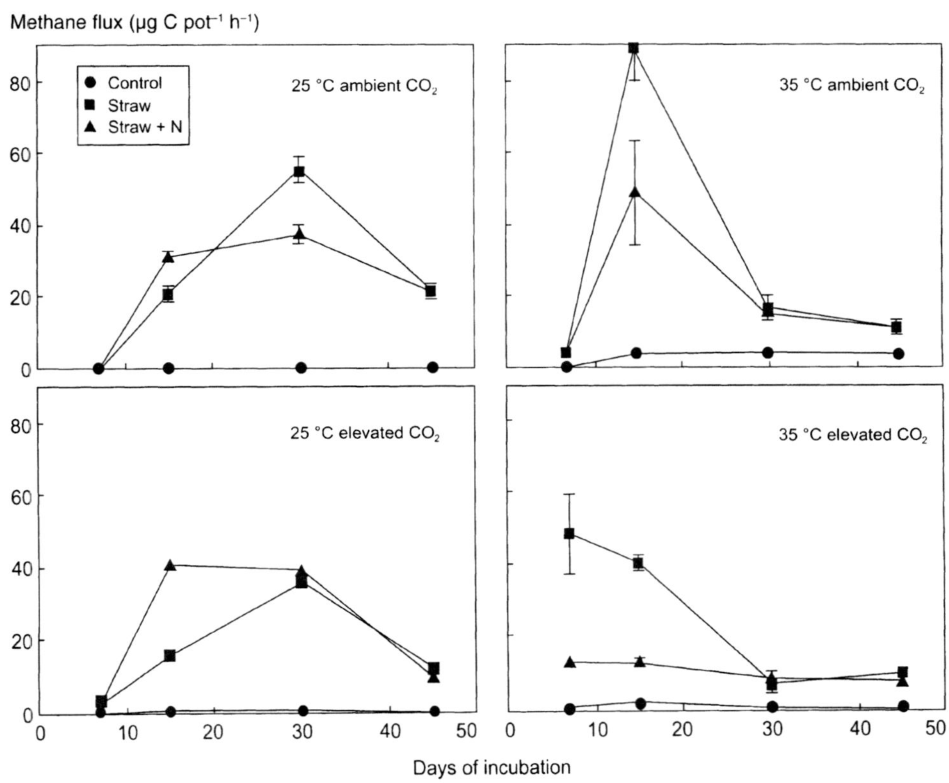

Figure 1. Methane emission from three treatments under four different conditions

concentration tended to decrease CH_4 emission from soil samples amended with straw after 15-d incubation at 35 °C, indicating faster conversion of straw carbon to CH_4 under elevated CO_2. However, this effect was not clear at 25 °C.

Amounts of CH_4 dissolved in floodwater during incubation

It is known that CH_4 emission in nonplanted submerged soils occur primarily by ebullition and/or diffusion. But the diffusion of gases in water is about 10,000 times slower than in air; therefore the diffusive exchange of gases drastically slows down when the soils are waterlogged. Thus, the depth of water layer over the soil may control CH_4 fluxes. Sebacher et al. (1986) reported that CH_4 emission rates were linearly related to water depth up to about 10 cm; depths greater than this did not promote CH_4 emission. Therefore, in this study, similar water depth (3 cm) was maintained throughout the incubation.

The data in Figure 2 revealed no specific pattern of the effects of soil amendments or incubation conditions on the amount of water-dissolved CH_4. For example, the amount of CH_4 dissolved in floodwater (MDFW) of all the control pots generally was very low and similar throughout the incubation, irrespective of incubation conditions (i.e., temperature or CO_2 levels). However, in the amended pots, the amount of MDFW varied considerably with respect to both soil treatments and incubation conditions. In all the amended soils and at all incubation conditions, the amount of MDFW correlated with CH_4 emission. At 15 d , the amount of MDFW in the straw-amended soil incubated at 25 °C was about threefold higher than in pots incubated at 35 °C under both levels of CO_2. However, the amount of MDFW in soil samples incubated under elevated CO_2 was about twice as large as that in soil samples under ambient CO_2, irrespective of temperature. On the other hand, soil samples amended with both straw and N showed no specific trend in the amount of MDFW with respect to incubation period. At 15 d, in the case of

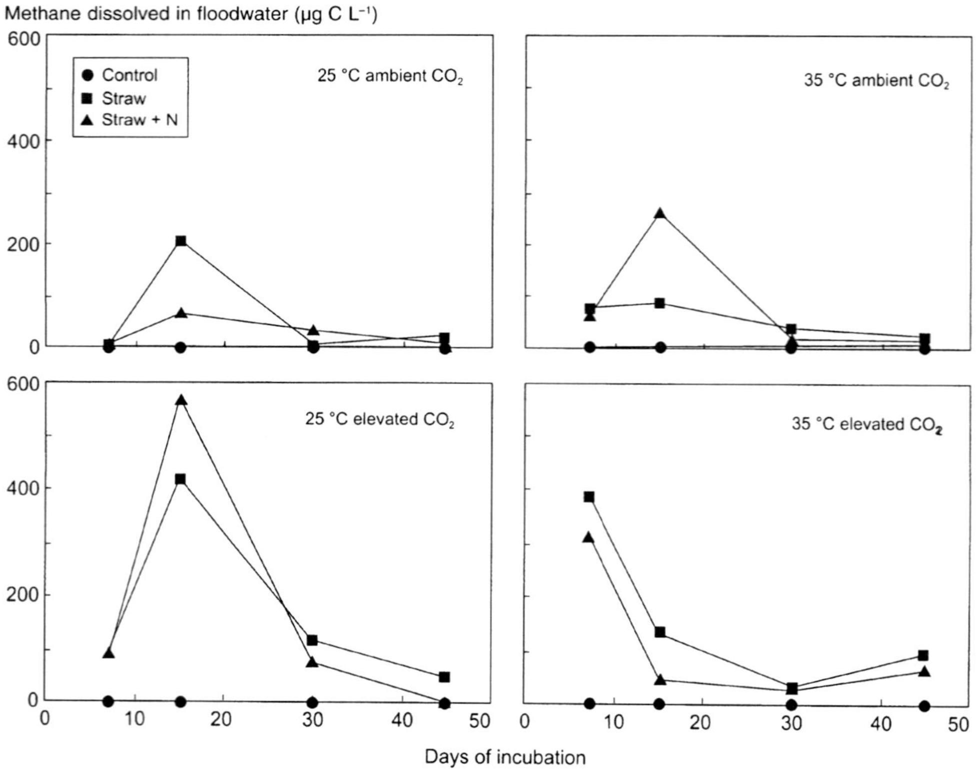

Figure 2. Amount of CH_4 dissolved in floodwater

ambient CO_2, the amount of MDFW was larger in soils incubated at 35 °C than at 25 °C. However, in the case of elevated CO_2, the reverse was true. After 15 d, MDFW decreased in all treatments at all conditions.

Averaged over the whole incubation period, the amount of MDFW in the amended (rice straw or rice straw + N) soil samples incubated at 25 °C under elevated CO_2 was about twice as high as that in soil samples similarly incubated but under ambient CO_2. However, no such effects of elevated CO_2 were observed at 35 °C, although the amount of MDFW was generally larger in soil samples incubated at 25 °C than that at 35 °C. These results suggest that the elevated level of CO_2 increased the concentration of MDFW, whereas the increase in temperature from 25 to 35 °C decreased the MDFW pool. This implies that at high temperature, relatively more CH_4 would either have escaped to the atmosphere from the floodwater or have oxidized.

Amount of entrapped CH_4 in the subsoil

The amount of entrapped CH_4 in subsoils (MES) varied considerably among different amendments and incubation conditions (Figure 3). In most cases, the amount of MES in the soil samples amended with straw with and without added N was remarkably similar. Therefore, the results have been discussed with only one treatment from the amended soils.

The amount of MES in the control soils incubated at 25 °C under ambient CO_2 was very small (approximately 1 μg g^{-1} soil). However, the amount of MES in the straw-amended soil samples incubated similarly as above (25 °C, ambient CO_2) increased steadily with incubation. On an average, the amount of MES was about 15 times greater in the amended soil samples than in the control soil. The pattern of change in the amount of MES in soil samples incubated under elevated CO_2 but at 25 °C was almost similar to that in soil samples incubated under ambient CO_2.

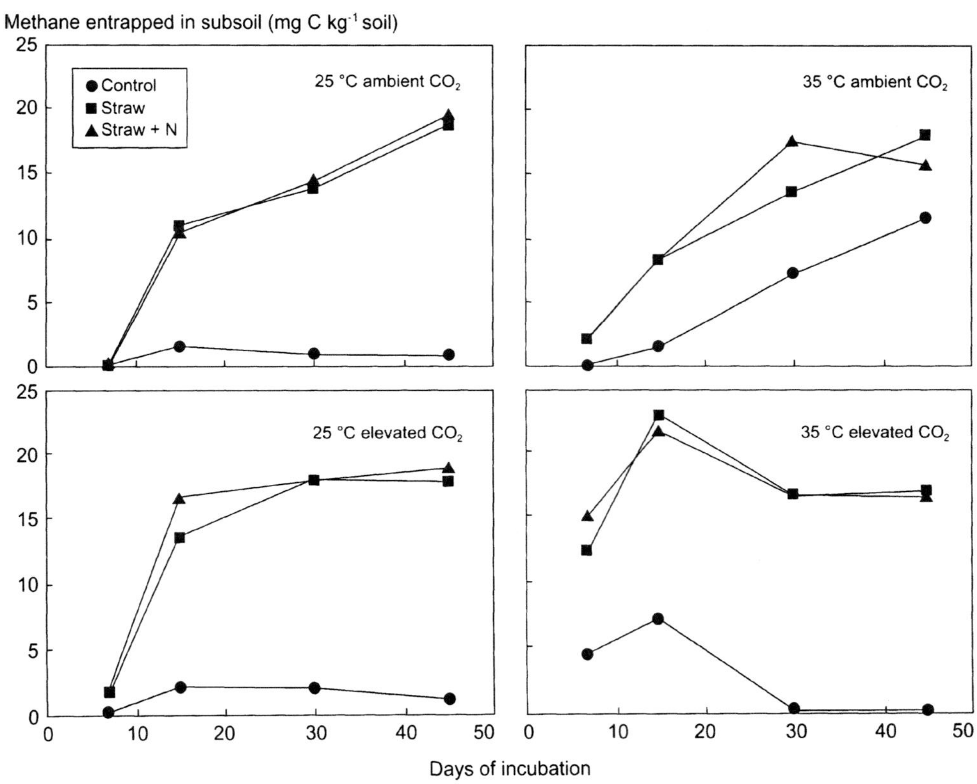

Figure 3. Amount of CH_4 entrapped in subsurface soil

However, a different pattern was observed in soil samples incubated under elevated CO_2 and at high temperature (35 °C). The amount of MES both in the control and amended soil samples incubated under elevated CO_2 at 35 °C was highest at 15 d, and it then decreased. On the other hand, the amount of MES in both control and amended soil samples incubated under ambient CO_2 at 35 °C continuously increased until the end of the incubation. These results indicate that the amount of MES was greatly affected by incubation conditions. However, it is interesting to note that at the end of the incubation (45 d), the amounts of MES were similar in all the amended soil samples, irrespective of incubation temperature or CO_2 level. These results indicate that a significant amount of the CH_4 produced in the soils during anaerobic decomposition of native or added organic material was held by the soil itself and that the different incubation conditions tested in this study seemed to have different effects on the amount of MES.

Methane production potential

Methane generation is considered to be the terminating step during anaerobic microbial decomposition of organic matter and any parameter affecting the biological, chemical, or physical characteristics of the flooded soil environment will influence CH_4 production and eventual emission (Bouwman, 1990; Inubushi et al., 1994; Wang et al., 1996). There is circumstantial evidence that methanogens, which can metabolize only a limited number of substrates, are dependent upon associated microorganisms for supply of substrates. Therefore, CH_4 production is a function of the collective activities of a broad group of obligate and facultative anaerobes, which are sensitive to changes in the soil environment (Wang et al., 1996).

Our results also support the above observations. In this study, CH_4 production of control soil samples incubated at 25 °C was very low and did not differ

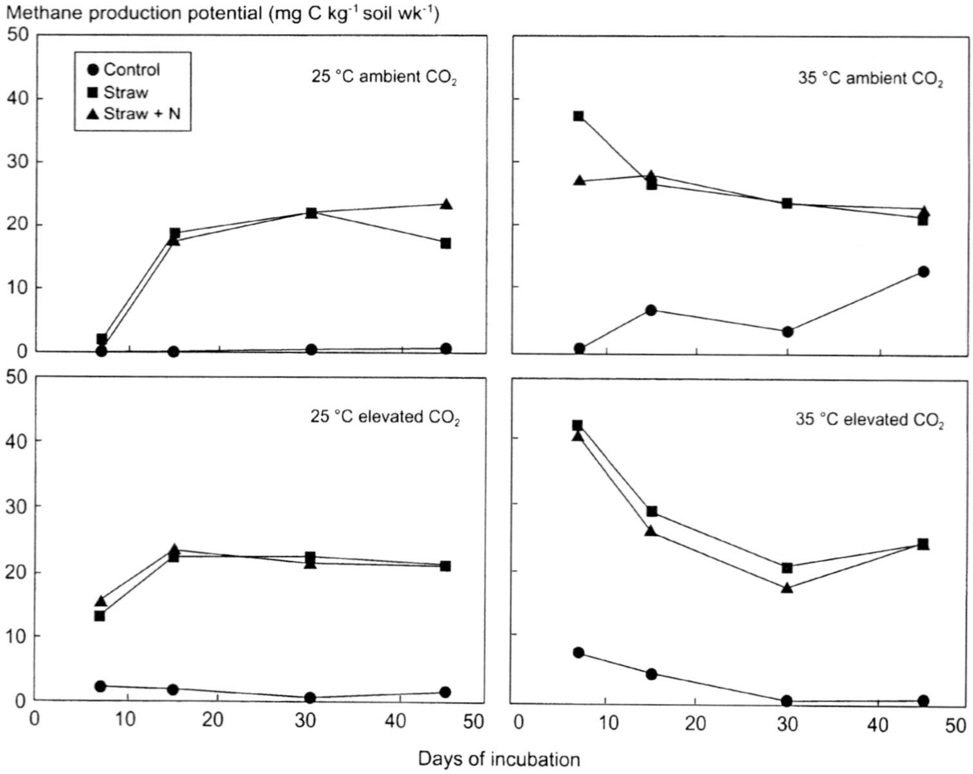

Figure 4. Methane production potential of subsurface soil under different conditions

greatly between soil samples incubated under elevated or ambient CO_2 levels throughout the incubation period (Figure 4). However, the control soil samples incubated at 35 °C produced significantly more CH_4 under ambient CO_2 than under elevated CO_2. It is interesting to note that the CH_4 production potential (MPP) increased by about twofold at 45 d as compared with that at 15 d of incubation in control soil incubated at 35 °C under ambient CO_2. In most cases, the amounts of MPP in soil samples amended with straw with and without added N were similar. The results indicate that CH_4 entrapped in subsoil was related to MPP.

Incorporation of straw stimulated MPP in all amended soil samples, irrespective of temperature and CO_2 level. On an average, the MPP of the straw-amended soil samples was 15-20 times more than that of unamended soils. However, the extent of stimulation varied considerably with temperature. At 25 °C, the MPP increased gradually up to 15 d and then stabilized until the end of the incubation, irrespective of CO_2 level. In contrast, at 35 °C, the MPP sharply increased and reached the maximum at 7 d and then declined gradually. This flush of CH_4 production during early incubation at higher temperature indicates that the conversion efficiency from added straw carbon to CH_4 by the methanogenic and associated bacteria was more pronounced at 35 °C than at 25 °C. Averaged over the whole incubation period and at both levels of CO_2, the MPP of the straw-amended soil samples was about 25% greater at 35 °C than at 25 °C. On the other hand, the MPP of the soil samples incubated under ambient and elevated CO_2 levels was almost identical. There was a highly significant correlation (r= 0.957**) between soluble organic C and MPP of the soil samples from elevated CO_2 condition at 35 °C after 1 wk of incubation (Figure 5). These results clearly show that the MPP of the flooded soil depended primarily on the availability of labile organic carbon and that incubation temperature only affected the pattern of CH_4 production. This finding may imply that methanogens and other associated anaerobes could adapt better at higher temperature (37 °C), but their activities largely remain un-

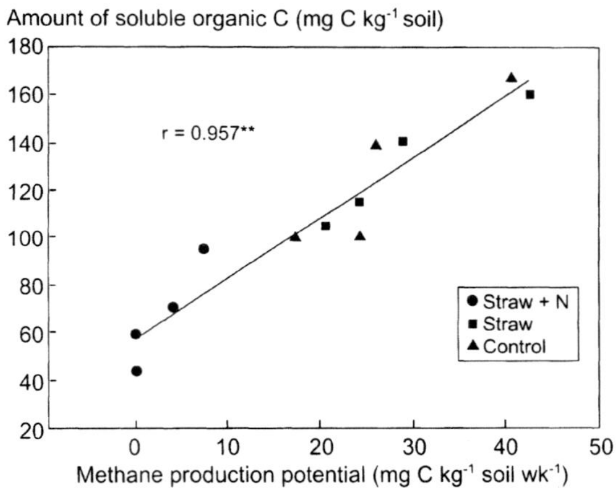

Figure 5. The relationship between amount of soluble C and CH_4 production potential of soil samples under elevated CO_2 condition at 35 °C after 1 wk of incubation

Table 1. Estimated CH_4 budget (mg C pot^{-1}) in pots of two treatments under four conditions

Condition	Treatment	MP	MES	MDFW*	MF	MO	MO/MP
Ambient CO_2 25 °C	Control	0.92	0.219	0.0	0.22	0.48	0.52
	Straw	40.36	5.616	2.0	31.20	3.55	0.09
Ambient CO_2 35 °C	Control	15.60	3.462	0.2	3.23	8.91	0.57
	Straw	49.01	5.382	1.7	39.95	3.68	0.08
Elevated CO_2 25 °C	Control	2.73	0.366	0.1	0.49	1.87	0.68
	Straw	45.86	5.367	5.2	20.57	19.91	0.43
Elevated CO_2 35 °C	Control	3.14	0.069	0.0	0.86	2.21	0.71
	Straw	51.75	5.133	9.4	18.61	28.00	0.54

MP: amount of CH_4 produced during 45 d; MES: amount of CH_4 entrapped in subsoil at 45 d; MDFW: amount of CH_4 dissolved in floodwater at 45 d; MF: amount of CH_4 emitted during 45 d; MO: amount of CH_4 oxidized during 45 d, MO=MP-MES-MDFW-MF. Straw C: 1.40 g pot^{-1}; soil C: 2.59 g pot^{-1}. *Unit: μg pot^{-1}.

affected with increase in atmospheric CO_2 level. Several investigators have reported the stimulatory effects of added organic matter on CH_4 production in flooded rice soils (Inubushi et al., 1989, 1994; Wang et al., 1992, 1996; Chidthaisong et al., 1996; Huang et al., 1998; Singh et al., 1998). Similarly, there are numerous reports that CH_4 emission in submerged rice soils increased when soil temperature increased up to 37 °C; however, optimum temperature for both production and consumption was 25 °C (Holzapfel- Pschorn et al., 1986; Bouwman, 1990; Chapman et al., 1996; Huang et al., 1998). Combining these results, one might hypothesize that the microbes involved with production of CH_4 preferred the easily decomposable organic materials than the more resistant native soil organic matter.

Estimated CH_4 budget and CH_4 oxidation

To summarize various forms of CH_4 in the pots, we estimated CH_4 oxidation (MO) calculated from produced CH_4 (MP) and CH_4 emitted (MF) both during 45 d of incubation and MES and MDFW both at 45 d as

$$MO = MP - MES - MDFW - MF$$

in control and straw-amended soil samples. MO was about 50-70% of MP in control soil samples and increased by elevated temperature and straw amendment. In straw-amended soil samples, MO was severalfold higher in CO_2 elevated than in ambient soil samples, indicating again that MO was enhanced by elevated CO_2 in this experiment. Overall, about 40% of MP was oxidized in straw-amended soil samples by rising CO_2 concentration. However, this estimation, especially for MP, needs further investigation.

In conclusion, our results clearly demonstrate that elevated CO_2 and temperature accelerated CH_4 production in rice straw-amended soil samples, while elevated CO_2 reduced CH_4 emission at both temperatures.

Acknowledgements

Part of this experiment was financially supported by the Research Institute of Innovative Technology for the Earth through a fellowship for K. Chander.

References

Bolin B (1986) How much CO_2 will remain in the atmosphere? In: Bolin B, Doos BR, Jagar J& Warrick RA (eds) The Greenhouse Effect, Climatic Change, and Ecosystems, pp 93–155, Wiley, New York

Bouwman AF (1990) Soils and the Greenhouse Effect. John Wiley and Sons, New York

Chapman SJ, Kanda K, Tsuruta H & Minami K (1996) Influence of temperature and oxygen availability on the flux of CH_4 and carbon dioxide from wetlands: a comparison of peat and rice soils. Soil Sci Plant Nutr 42: 269–277

Chidthaisong A, Inubushi K, Muramatsu Y & Watanabe I (1996) Production potential and emission of CH_4 in flooded rice soil microcosms after continuous application of straws. Microb Environ 11: 73–86

Crill PM, Barlett KB, Harris RC, Gorham E, Very ES, Sebacher DL, Madsar L & Sanner W (1988) Methane flux from Minnesota peat-lands. Global Biogeochem Cycl 2: 371–384

Gaunt JL, Neue HU, Bragais J, Grant IF & Giller KE (1997) Soil characteristics that regulate soil reduction and methane production in wetland rice soils. Soil Sci Soc Am J 61: 1526–1531

Holzapfel-Pschorn A, Conrad R & Seiler W (1986) Effects of vegetation on the emission of methane from submerged rice soil. Plant Soil 92: 223–233

Huang Y, Sass RL & Fisher FM (1998) A semi-empirical model of methane emission from flooded rice paddy soils. Global Change Biol 4: 247–268

Inubushi K, Hori K, Matsumoto S, Umebayashi M & Wada H (1989) Methane emission from the flooded rice soil to the atmosphere through rice plant. Jpn J Soil Sci Plant Nutr 60: 318–324

Inubushi K, Brookes PC & Jenkinson DS (1991) Soil microbial biomass C, N and ninhydrin-N in aerobic and anaerobic soils measured by the fumigation-extraction method. Soil Biol Biochem 23: 737–741

Inubushi K, Muramatsu Y & Umebayashi M (1994) Effect of incorporation-timing of rice straw on methane emissions from rice soil. Jpn J Soil Sci Plant Nutr 65: 22–26

IPCC (Intergovernmental Panel on Climate Change) (1995) Climate Change 1994, pp 7–34, Cambridge University Press, Cambridge, UK

Kimura M, Miura A, Watanabe A, Katoh T & Haraguchi H (1996) Methane emission from paddy field (part 1). Effect of fertilization, growth stage and midsummer drainage: pot experiment. Environ Sci 4: 265–271

Sebacher DJ, Harriss KB, Bartlett SM & Grice SS (1986) Atmospheric methane sources: Alaskan tundra bogs, an alpine fen, and a subarctic boreal march. Tellus 38B: 1–10

Singh JS, Raghubanshi AS, Reddy VS, Singh S & Kashyap AK (1998) Methane flux from irrigated rice and dryland rice fields, and from seasonally dry tropical forest and savanna soils of India. Soil Biol Biochem 30: 135–139

Takai Y & Wada E (1990) Methane formation in waterlogged paddy soils and its controlling factor. In: Soils in a Warmer Earth, Scharpenseel WH, Schomaker M & Ayoub A (eds), pp 101–107. Elsevier, Amsterdam

Wang ZP, DeLaune R, Lindau CW & Patrick WH Jr (1992) Methane production from anaerobic soil amended with rice straw and nitrogen fertilizers. Fert Res 33: 115–121.

Wang ZP, DeLaune RD, Masscheleyn P & Patrick WH Jr (1993) Soil redox and pH effects on methane production in flooded rice soil. Soil Sci Soc Am J 57: 382–385

Wang M, Shangguan X & Ding A (1996) Methane in the rice agriculture. In: From Atmospheric Circulation to Global Change. The Institute of Atmospheric Physics, Chinese Academic of Sciences, pp 647–659, China Meteorological Press, Beijing

Ziska LH, Moya TB, Wassmann R, Namuco OS, Lantin RS, Aduna JB, Abao E Jr, Bronson KF, Neue HU & Olszyk D (1998) Long-term growth at elevated carbon dioxide stimulates methane emission in tropical paddy rice. Global Change Biol 4: 657–665

Nutrient Cycling in Agroecosystems **58:** 349–356, 2000.

Differences among rice cultivars in root exudation, methane oxidation, and populations of methanogenic and methanotrophic bacteria in relation to methane emission

B. Wang[1,2] & K. Adachi[2]

[1]*Institute of Crop Breeding and Cultivation, Beijing 100081, China;* [2]*Japan International Research Center for Agricultural Sciences, Okinawa Subtropical Station, Ishigaki, Okinawa 907-0002, Japan*

Key words: control measure, methane emission, methane oxidation, rice cultivar, root exudation

Abstract

Greenhouse experiments were conducted under subtropical conditions to understand the mechanism of rice cultivar differences in methane (CH_4) emission. Three rice cultivars were studied. Differences in CH_4 emission rates among the three rice cultivars became evident in the middle and late growth stages. Rice root exudates per plant measured as total released C were significantly different among rice cultivars. The effect of root exudates on CH_4 production in soil slurry differed accordingly. The amount of root exudates was not significantly different among rice cultivars when computed on a dry matter basis, indicating that it is positively correlated to root dry matter production. The root CH_4-oxidizing activity differed among rice cultivars. IR65598 had a higher oxidative activity than IR72 and Chiyonishiki. Root air space was not significantly different among rice cultivars at the late growth stage, indicating that it is probably not a factor contributing to cultivar differences in CH_4 emission. The population level of methanogenic bacteria differed significantly in soil grown to different rice cultivars, but not in roots, at booting stage and ripening stage. Methanotrophic bacteria population differed significantly in roots among rice cultivars at ripening. Rice cultivars with few unproductive tillers, small root system, high root oxidative activity, and high harvest index are ideal for mitigating CH_4 emission in rice fields.

Introduction

Rice fields are one of the most important methane (CH_4) sources. Estimated annual CH_4 emission from rice fields range from 57 to 82 Tg yr^{-1} (Bachelet & Neue, 1993) and may contribute 10-15% to global CH_4 emissions (Neue, 1993). Research on measures to control CH_4 emission from rice fields to the atmosphere has focused on rice cultivar, irrigation water management, organic matter management, and fertilization (Neue et al., 1995; Neue, 1997; Yagi et al., 1997; Wang et al., 1999). Rice cultivars received high research priority because high-yielding rice cultivars with low CH_4 emission rates can be easily extended to farmers' fields without any additional input and management. Rice cultivars showed significantly different effects on CH_4 emissions from rice fields (Neue et al., 1994; Wang et al., 1997a; Wang et al., 1999).

Methanogens utilize H_2 and CO_2, methanol, acetate, and methylamides as C sources to produce CH_4 (Takai 1970; Vogels et al., 1988) and methanotrophs use CH_4 as substrate in flooded soils. The activities of both methanogens and methanotrophs result in accumulation of CH_4 that escapes to the atmosphere mainly by diffusion through rice plant aerenchyma (Schütz et al., 1989; Nouchi et a.l, 1990; Byrnes et al., 1995; Wang et al., 1997b). Compared with studies concerning the process and controlling factors of CH_4 emission from rice fields, studies dealing with mechanisms of rice cultivar differences on CH_4 emission rate and the ideal rice plant types for mitigating CH_4 emissions from rice fields are scanty. Yet, knowledge on microbial activities and distributions of methanogens and methanotrophs as affected by rice cultivars is still limited (Watanabe et al., 1997).

This study was conducted under greenhouse condition at the Okinawa Subtropical Station, Japan International Research Center for Agricultural Sciences. The aim of this study was to understand cultivar differences in CH_4 emission rate and to describe the ideal rice plant type for mitigating CH_4 emissions.

Materials and methods

Soil and rice cultivars used in the experiment

The alluvial soil used in the experiments was collected from a rice field in Sandabaru, Ishigaki Island, Japan (latitude 24° 23' N, longitude 124° 12' E) during the fallow season. The soil was air-dried, ground, and passed through a 2.0-mm mesh sieve before it was used. It has a light clay texture (coarse sand 16%, fine sand 24%, silt 20%, clay 27%, and $CaCO_3$ 5%) with 3.8% total organic C, 6.6% total organic matter, and pH 7.3 (soil:water, 1:1). Three rice cultivars—IR72 (indica), IR65598-112-2 (IR65598, tropical japonica), and Chiyonishiki (japonica)—were planted.

Growth of rice plants

Rice seeds were germinated and grown on a nylon screen framed with styrofoam floating on culture solution for 2 wk. A 2-wk-old rice seedling was transplanted to a 4-L pot that contained 2.5 kg of soil. The soil in pots was submerged 1 wk before transplanting and was kept flooded throughout the rice-growing season. Combined fertilizer (NPK, 14-6-12) was applied at a rate of 2 g pot^{-1} (equivalent to 1,000 kg ha^{-1}) before transplanting as basal which was mixed well with soil. An additional 2 g of the combined fertilizer was topdressed in two splits: half at tillering stage and half before flowering.

Methane sampling and analysis

Methane emission rates from pots planted to rice were measured in a greenhouse at tillering (4 wk after transplanting, [WAT]), booting (8 WAT), flowering (11 WAT), and ripening stages (14 WAT). Methane samples were taken with closed chambers (30.5 cm in diameter and 129 cm in height). A small electric fan was fixed inside each chamber to homogenize the air before sampling. The experimental pots were put in big containers that were filled with water to seal the bottom of the chambers during the CH_4 flux measurements. Two to three minutes after placing the pots inside the chambers, air samples were taken four times with syringes at 10-30-min intervals. Methane concentrations of the air samples were determined with a Hewlett-Packard 5890 A/II gas chromatograph equipped with a Porapak R polymer (50-80 mesh) column and a flame ionization detector. The temperature settings were 60, 100, and 150 °C for column, injector, and detector, respectively. Methane fluxes from pots were determined by measuring the temporal increase of CH_4 concentration of air within the chambers. The CH_4 emission rates were determined from the increasing rate of CH_4 concentration in the chambers by using the following equation:

$$F = 60 \times 10^{-6} \times (AH) \times (PM/RT) \times (dc/dt)$$

where F is CH_4 emission rate, in mg pot^{-1} h^{-1}; A is bottom area of the chamber, in cm^2; H is effective height of the chamber, in cm; P is pressure, in atm; M is CH_4 molecular weight in g; R is gas constant; T is absolute temperature, in K; dc/dt is increasing rate of CH_4 in the chamber, in μL^{-1} min^{-1}.

Collection and analysis of root exudation

A two-week-old rice seedling was planted in pots with 4-L culture solution. The pots were covered with styrofoam to support the plant and to prevent algal growth. Roots were inserted through small openings into the culture solution. The culture solution was prepared according to Yoshida et al. (1976). The solution was changed weekly. The culture solution was collected at the fourth and eighth week after the rice plant had grown in it for 1 wk. About 50 mL of solution collected from each pot was filtered through filter paper #1 and 0.45-μm membrane filter to remove root detritus and microbial cells. The filtrates were kept in a refrigerator (0 °C) until analysis.

The amount of exudates in each cultivar was determined by the anthrone colorimetric method (Brink et al., 1960) as the total amount of water-soluble C released by the rice roots. Two g of anthrone was dissolved in 1 L of 95% H_2SO_4 to form anthrone reagent. Five mL of exudate sample was mixed thoroughly with 10 mL of the anthrone reagent and the absorbance of the mixture was measured at 625 nm on a Hitachi U-

2000 spectrophotometer. A standard curve was obtained from the absorbance of glucose standards in which the C contents were 0.113, 0.592, 1.184, 1.776, and 2.367 mg C mL^{-1}. Exudation rates were calculated as mg C $plant^{-1}$ d^{-1} and mg C g^{-1} root d^{-1}.

Effect of root exudation on CH_4 production of soil slurry

Five mL of soil slurry (water content, 177.5%) incubated in a greenhouse for 3 mo was placed in 31.5-mL test tubes, and the test tubes were closed with W-shape butyl stoppers. Five mL of root exudates collected from three rice cultivars at tillering and heading stages was injected into the tubes. Headspace of the tubes was flushed with N_2 for about 10 s, then the tubes were shaken for 1 min on a vortex shaker and incubated at 30 °C in the dark. Methane produced in the tubes was sampled for analysis after 3, 7, 14, 21, 28, 35, 42, and 49 d of incubation. The tubes were shaken on a vortex shaker for 1 min before gas samples were obtained and 1 mL N_2 was injected into each of the tubes after each sampling to maintain normal pressure inside the tubes. Methane produced was computed using the following equation:

$$CH_4\ (\mu g\ g^{-1}\ soil) = 3.272 \times 10^{-5} \times Hs \times Mc \times Ds^{-1}$$

where Hs is the volume of headspace, in cm^3; Mc is the CH_4 concentration of air sample, in μL L^{-1}; Ds is the dry weight of the soil slurry, in g.

Oxidation of dissolved CH_4 by rice roots

About 2 g of fresh rice roots was sampled from three rice cultivars at three growth stages and placed in 31.5 -mL test tubes. Five mL of previously prepared CH_4 solution was transferred with a 5-mL Gilson pipette to each of the tubes to immerse the rice roots. The test tubes were immediately closed with W-shape butyl stoppers and were placed under laboratory condition for 1 d. A reference solution was prepared by the same procedure without placing fresh rice roots. Tubes were shaken for 2 min on a vortex shaker to release CH_4 from the solution to the headspace after 1 d of incubation. The CH_4 concentrations of the air in the headspace were analyzed with a Hewlett-Packard 5890 A/II gas chromatograph. The decrease of CH_4 in the CH_4 solution during incubation was calculated as amount of CH_4 oxidized by the fresh roots in a given time. Methane solution was prepared as follows: degas tap water in a flask under vacuum for 30 min, inject 50 mL pure CH_4 gas into the evacuated flask, stir the water inside the flask gently, and leave the flask overnight.

Root air space

Rice roots from three cultivars at four growth stages (4, 8, 11, and 14 WAT) were sampled for measurement of root air space. Root porosity was measured by the pycnometer method (Jensen et al., 1969). Roots were rinsed with tap water. A 50-mL pycnometer was filled with water and weighed. About 1-2 g fresh roots were sampled and gently blotted dry on tissue paper. The roots were then introduced into the water-filled pycnometer and reweighed. The roots were later retrieved, ground into a paste with mortar and pestle, and returned quantitatively to the pycnometer for reweighing. The porosity of the roots was determined, using the formula

$$RAS = ((p\&gr) - (p\&r))/((r + p) - (p\&r)) \times 100$$

where RAS is root air-space, in %; r is weight of root, in g; p is weight of water-filled pycnometer, in g; p&r is weight of pycnometer with roots and water, in g; and p&gr is weight of pycnometer with ground roots and water, in g.

Enumeration of methanogenic and methanotrophic in flooded soil and roots

Flooded soil in pots was transferred to a container with about 2,000 mL water. The soil was stirred vigorously to form a soil suspension. Ten mL of the soil suspension was sampled and placed in an oven at 105 °C for 1 d for measuring water content of the soil suspension. For enumeration of methanogenic bacteria (MGB), 1 mL of the soil suspension was diluted to 10^{-2} to 10^{-7} levels anaerobically in tubes under N_2 gas, using anaerobic dilution fluid (0.5 g of cysteine-hydrochloride, 0.5 g of $Na_2S{\cdot}9H_2O$, and 1 mL of 1 g L^{-1} resazurin solution in 1 L distilled water; pH=7.0). One mL of the suspension at 10^{-3} to 10^{-7} dilution was inoculated to the tubes containing 5 ml of MGB medium with sterile syringes. Top gas phase in the tubes was replaced with H_2-CO_2 (4:1) after inoculation. There were five replications for each dilution. The tubes were incubated at 30 °C in the dark for 60 d. The gas phase in the tubes was assayed with a gas chromatograph for CH_4 production. The proportion of the positive and negative tubes in CH_4 production indicates the most probable number. For enumeration of methanotrophic bacteria (i.e., methane-oxidizing bacteria [MOB]), one mL of the soil suspension was diluted to 10^{-2} to 10^{-7} levels in tubes under ambient air condition, using distilled water. One mL of the suspension at dilution of 10^{-3} to 10^{-7} was inoculated to the tubes containing 5 mL of MOB medium with sterile syringes. Six mL of CH_4 passed through 0.2 μm filter was injected into each tube, leading to about 18% CH_4 in the headspace of each tube after inoculation. Control tubes were prepared without inoculation. There were five replications for each dilution. The tubes were incubated at 30 °C in the dark for 60 d. The gas phase in the tubes was assayed with a gas chromatograph for CH_4 consumption. The proportion of positive and negative tubes in CH_4 consumption indicates the most probable number. The medium for MGB was prepared according to Adachi et al. (1996) and the medium for MOB was prepared according to Graham et al. (1992).

About 2 g of fresh roots was sampled and ground in 20-mL water into paste with mortar and pestle. The paste was diluted and inoculated in the same way as the flooded soil for enumeration of MGB and MOB. The paste was oven dried at 80 °C for 1 d and its dry weight measured.

Results and discussion

Methane emission rates of three rice cultivars

Methane emission rates were low and not significantly different among the three rice cultivars at tillering stage (Table 1). This was probably related to high soil redox potential and small plant size. Methane emission rates per pot increased at the late growth stages and differed significantly among the three rice cultivars. Methane emission in IR72 was significantly higher than in IR65598, but not Chiyonishiki, at the booting stage. IR72 and Chiyonishiki had significantly higher emission than IR65598 at flowering and ripening. Methane emission rates per g of plant dry matter among 3 rice cultivars differed only at ripening stage. Chiyonishiki had significantly higher emission rate per gram of plant dry matter than IR65598, but not IR72, at ripening stage. Pot-based CH_4 emission rates differed among 3 rice cultivars at late growth stages. This finding is consistent with reports (Wang et al., 1997a) indicating significant differences in pot-based CH_4 emission rate among different rice cultivars and growth stages. In this study dry matter-based CH_4 emission rates were not significantly different among rice cultivars before ripening, confirming earlier observations that CH_4 emission rates were closely related to dry matter production without discrimination of rice cultivars and growth stages (Wang et al., 1997a). However, in contrast to Wang's findings (Wang et al., 1997a) the dry matter-based CH_4 emission rate of IR65598 was significantly lower than that of Chiyonishiki at ripening stage, although dry matter production levels are the same.

Root exudations of three cultivars and their effect on CH_4 production of soil slurry

IR72 had significantly higher plant weight and root weight than IR65598 and Chiyonishiki (Table 2). IR72 released more C per plant than did IR65598 and

Table 1. Plant dry weight and CH_4 emission rate of three rice cultivars at four growth stages. Values are means of three replicates ± SD[a]

Growth stage	Cultivar	Plant dry wt (g plant^{-1})	CH_4 emission rate (mg pot^{-1} h^{-1})	CH_4 emission rate (mg g^{-1} plant h^{-1})
Tillering	IR72	9.33±1.16a	0.380±0.107a	0.042±0.017a
	IR65598	7.60±1.08ab	0.304±0.157a	0.040±0.022a
	Chiyonishiki	6.59±0.50b	0.239±0.015a	0.036±0.001a
Booting	IR72	13.32±0.39a	1.268±0.402a	0.095±0.031a
	IR65598	11.60±1.31a	0.707±0.113b	0.061±0.005a
	Chiyonishiki	11.91±1.28a	1.161±0.208ab	0.097±0.007a
Flowering	IR72	20.82±2.54a	1.648±0.186a	0.080±0.014a
	IR65598	15.65±2.28a	0.979±0.279b	0.065±0.029a
	Chiyonishiki	17.24±2.54a	1.826±0.209a	0.108±0.019a
Ripening	IR72	29.45±2.72a	2.252±0.461a	0.077±0.014ab
	IR65598	20.50±4.42b	0.664±0.252b	0.032±0.010b
	Chiyonishiki	15.82±3.20b	1.775±0.517a	0.119±0.057a

[a]Data in a column at a growth stage followed by a common letter are not significantly different.

Table 2. Plant dry weight, root dry weight, and root exudation of three rice cultivars at tillering and heading stages in hydroponic rice cultivation. Values are means of 3 replicates ± SD[a]

Growth stage	Cultivar	Plant dry wt ($g\ plant^{-1}$)	Root dry wt ($g\ plant^{-1}$)	C released ($mg\ plant^{-1}\ d^{-1}$)	C released ($mg\ g^{-1}$ root d^{-1})
Tillering	IR72	9.42±1.68a	1.19±0.31a	6.62±1.45a	5.67±1.19a
	IR65598	4.47±0.98b	0.40±0.07b	2.09±0.40b	5.45±1.56a
	Chiyonishiki	4.42±0.26b	0.49±0.06b	3.07±0.48b	6.36±1.33a
Heading	IR72	37.05±3.33a	4.61±0.26a	23.15±4.00a	5.01±0.76a
	IR65598	15.42±3.81b	1.29±0.31b	8.62±2.43b	6.67±0.45a
	Chiyonishiki	17.22±2.54b	1.79±0.24b	10.66±1.99b	6.05±1.44a

[a]Data in a column followed by a common letter are not significantly different.

Chiyonishiki at both tillering and heading stages. However, there was no difference in released C among the three rice cultivars when the released C was computed on dry root basis. This finding is consistent with the observation that the amount of C released from root is closely related to root dry weight (Wang et al., 1997a). No difference in root exudation on dry matter basis indicates that root weight discriminates rice cultivars in root exudation that provides C source for methane formation. Cultivars with small roots are ideal for mitigating methane emissions.

At the heading stage, addition of IR72 root exudates to soil slurry gave highest CH_4 production, followed by addition of IR65598 root exudates. Addition of Chiyonishiki root exudates to the soil slurry gave the lowest CH_4 production (Figure 1). The effect of addition of IR65598 root exudates was more pronounced than that of adding Chiyonishiki root oxidates although both varieties were not significantly different in terms of root exudation. This suggests that the root exudates of IR65598 were probably preferred by methanogens. At tillering, the effect of addition of IR72 root exudates on CH_4 production in the soil slurry was more pronounced than that of IR65598 and Chiyonishiki, the latter 2 cultivars showed similar effect on CH_4 production of the soil slurry. In general, the effect of adding root exudates on CH_4 production of the soil slurry was in accordance with the amount of C released from the roots.

Oxidation of dissolved CH_4 by rice roots

Root CH_4-oxidizing activity of IR65598 was significantly higher than that of Chiyonishiki and slightly higher than that of IR72 at tillering stage; it was slightly higher than that of IR72 and Chiyonishiki at flowering stage (Table 3). The root CH_4-oxidizing activity of IR65598 was slightly higher than that of Chiyonishiki and that of both IR65598 and Chiyonishiki were significantly higher than those of IR72 at ripening stage. IR65598 showed the highest root CH_4 oxidative activity among the three rice cultivars in all growth stages. This may partly explain its low CH_4 emission rate. IR72 was higher than Chiyonishiki in root CH_4-oxidizing capacity at tillering and flowering stages, but lower than Chiyonishiki at ripening stage, indicating that root senescence of IR72 may have started early. Root oxidation power as measured by oxidation of α-naphthylamine decreased when the roots grew older

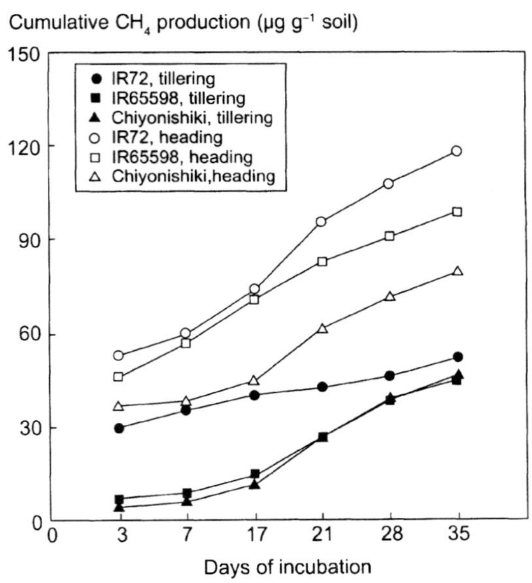

Figure 1. Effect of root exudates collected in hydroponic rice cultivation on CH_4 production in soil slurry. Data are means of six replicates

Table 3. Root dry weight and oxidative activity of three rice cultivars at three growth stages. Values are means of 6-10 replicates ± SD[a]

Growth stage	Cultivar	Root dry wt (g $plant^{-1}$)	CH_4 oxidized (μg g^{-1} root d^{-1})
Tillering	IR72	2.70±0.54a	18.78±6.42a
	IR65598	2.32±0.40ab	24.85±6.30a
	Chiyonishiki	1.54±0.07b	9.55±8.60b
Flowering	IR72	4.15±0.96a	15.59±4.33a
	IR65598	2.83±0.54b	16.51±6.77a
	Chiyonishiki	1.79±0.24b	13.38±3.78a
Ripening	IR72	3.63±0.54a	11.11±3.39b
	IR65598	3.06±0.84ab	15.99±3.65a
	Chiyonishiki	2.23±0.43b	13.85±2.52a

[a]Data in a column followed by a common letter are not significantly different. The reference CH_4 concentration was 10.85±0.35 μg mL^{-1} H_2O at tillering stage, 8.60±0.28 μg mL^{-1} H_2O at flowering stage, and 8.79±0.36 μg mL^{-1} H_2O at ripening stage.

Table 4. Root air space of three rice cultivars at four growth stages. Values are means of three replicates ± SD[a]

Growth stage	Cultivar	Root air space	
		(%)	(cm^3 g^{-1} dry root)
Tillering	IR72	27.76±4.83a	4.63±0.85a
	IR65598	22.26±5.06a	3.16±0.79b
	Chiyonishiki	24.72±2.59a	3.42±0.55b
Booting	IR72	23.69±2.30a	3.47±0.19a
	IR65598	13.16±4.83b	1.70±0.66b
	Chiyonishiki	18.40±5.78ab	2.61±0.88ab
Flowering	IR72	21.73±5.10a	2.90±0.78a
	IR65598	26.66±4.76a	3.27±0.60a
	Chiyonishiki	20.33±2.48a	2.93±0.41a
Ripening	IR72	16.91±2.82a	2.55±0.57a
	IR65598	14.97±3.32a	2.31±0.54a
	Chiyonishiki	12.01±3.89a	1.96±0.51a

[a]Data in a column followed by a common letter are not significantly different.

Table 5. Population level of MGB in flooded soil and rice roots as influenced by rice cultivars[a]

Growth stage	Cultivar	MGB in soil (no. g^{-1} dry soil)	MGB in roots (no. g^{-1} dry roots)
Tillering	IR72	7.0×10^4	4.9×10^3
	IR65598	7.0×10^4	3.4×10^3
	Chiyonishiki	1.2×10^5	2.9×10^3
Booting	IR72	7.1×10^4	-[b]
	IR65598	1.2×10^5	-
	Chiyonishiki	8.4×10^5	-
Flowering	IR72	2.1×10^5	-
	IR65598	1.4×10^5	-
	Chiyonishiki	1.1×10^6	-
Ripening	IR72	6.2×10^5	1.9×10^6
	IR65598	5.2×10^4	8.9×10^5
	Chiyonishiki	8.6×10^5	6.5×10^6

[a]Data with 10.9-fold difference are significant at 5% level (Alexander, 1982).
[b]Not enumerated.

(Wang et al., 1997c). Root CH_4-oxidizing capacity differs greatly among rice cultivars. If rice cultivars have similar root weights, those with high oxidative capacity are ideal for mitigating CH_4 emission. The significant difference in root CH_4-oxidizing capacity opens a chance for screening and breeding cultivars with low CH_4 emission rates.

Root air space of three rice cultivars

IR72 was significantly higher than IR65598 and Chiyonishiki in terms of root air space on dry root basis, but not on percentage, at tillering stage (Table 4). IR72 was significantly higher than IR65598 in root air space both on dry root basis and on percentage at booting stage. Root air space both on dry root basis and percentage was not significantly different at flowering and ripening among the three rice cultivars. Root air space facilitates CH_4 emission from the soil to the atmosphere. It also facilitates transport of oxygen from the air to the rhizosphere. It is not understood that root air space should be large or small for mitigating CH_4 emission in rice fields. Root air space is probably not an important factor contributing to cultivar differences in CH_4 emission rate since it was not significantly different among the three rice cultivars at late growth stages.

Population levels of MGB and MOB in flooded soil and in rice roots

The population level of MGB depends on the availability of C sources, assuming other conditions remain the same. The population level of MGB in flooded soil planted to different cultivars was significantly different at booting and ripening stages (Table 5), indicating that cultivars supplied different amounts of C sources for methanogens or that redox of the flooded soil was altered. No significant difference in the population level of MGB in roots at tillering and ripening stages was observed probably due to the oxidative condition in the roots. Oxygen diffuses from aboveground shoots via roots to the rhizosphere. The high concentration of oxygen in the roots depresses MGB growth and may narrow the differences among rice cultivars. The higher population level of MGB in the flooded soil planted to Chiyonishiki (compared with that in flooded soil planted to IR72) at booting stage may be attributed to

Chiyonishiki's lower oxidative capacity. Chiyonishiki had smaller root system than IR72 and IR65598 (Table 3). Oxygen release from Chiyonishiki roots might be less than that from IR72 roots. More MGB in the flooded soil planted to IR72 at ripening stage may be attributed to IR72's higher root exudation. The significantly higher population level of MGB in roots at ripening stage than at tillering stage may be raised by the availability of C sources from root exudation and root senescence.

The population level of MOB among rice cultivars differed only in roots at ripening stage (Table 6). MOB in the roots of IR65598 were significantly more than those in the roots of IR72 and Chiyonishiki. MOB in flooded soil at booting and flowering were more than those at tillering and ripening. MOB in roots of IR65598 at ripening were more than those at tillering stage. The rice plant at booting and flowering stages grew fast and its size was large at these rice growth stages. More oxygen may be transported from the air to the rhizosphere in these growth stages compared with other growth stages. This may explain why more MOB were observed at the booting and flowering stages. More MOB in the roots of IR65598 at ripening stage indicate that these roots had greater oxidizing capacity than the roots of IR72 and Chiyonishiki.

In summary, the three rice cultivars studied had significantly different CH_4 emission rates on a single-plant basis. The differences became evident at late growth stages. IR65598 gave the lowest CH_4 emission rate. Dry matter-based CH_4 emission rates among rice cultivars differed only at ripening stage, indicating that root weight is closely related to the amount of root exudates. Rice cultivars with small roots are ideal for mitigating CH_4 emissions. The small root weight results in few ineffective tillers and high harvest index, since rice root weight is closely related to rice dry matter production (Wang et al., 1997a). The large difference in root CH_4- oxidizing capacity indicates that if rice cultivars produce the same root weight, then those with higher CH_4-oxidizing capacity will have lower CH_4 emission rates. The population level of MGB increased in flooded soil planted to rice cultivars giving high root exudation, while the population level of MOB increased in the roots of rice cultivars giving higher root CH_4-oxidizing capacity. Rice cultivars with few unproductive (ineffective) tillers, higher harvest index, smaller root system, and higher oxidative capacity are ideal for mitigating CH_4 emissions in rice fields.

Table 6. Population level of MOB in flooded soil and rice roots as influenced by rice cultivars[a]

Growth stage	Cultivar	MOB in soil (no. g^{-1} dry soil)	MOB in roots (no. g^{-1} dry roots)
Tillering	IR72	3.1×10^5	5.8×10^5
	IR65598	2.0×10^5	3.4×10^5
	Chiyonishiki	3.2×10^5	4.5×10^5
Booting	IR72	$>10^7$	-[b]
	IR65598	$>10^7$	-
	Chiyonishiki	$>10^7$	-
Flowering	IR72	4.6×10^6	-
	IR65598	1.3×10^6	-
	Chiyonishiki	3.5×10^6	-
Ripening	IR72	2.8×10^5	4.5×10^6
	IR65598	3.6×10^4	6.5×10^7
	Chiyonishiki	3.7×10^4	4.2×10^6

[a]Data with 10.9-fold difference are significant at 5% level (Alexander, 1982).
[b]Not enumerated.

Acknowledgements

We thank Dr. S. Hirabayashi for providing us seeds of IR65598-112-2. This research was funded by and conducted at the Japan International Research Center for Agricultural Sciences, Okinawa Subtropical Station, Ishigaki, Okinawa 907-0002, Japan.

References

Adachi K, Oyediran G & Senboku T (1996) Effect of application of rice straw and cellulose on methane emission and biological nitrogen fixation in a subtropical paddy field. II Enumeration of populations of methanogenic bacteria by most probable number method and roll tube method. Soil Sci Plant Nutr 42:713-723

Alexander M (1982) Most probable number method for microbial populations. In: Page AL, Miller RH, Keeney DR (eds) Method of Soil Analysis, Part 2, Chemical and Microbiological Properties, pp 815-820, ASA-SSSA, Madison, Wisconsin

Bachelet D & Neue HU (1993) Methane emissions from wetland rice areas of Asia. Chemosphere 26:219-237

Brink RH Jr, Dubach P & Lynch DL (1960) Measurement of carbohydrates in soil hydrolyzates with anthrone. Soil Sci 89:157-166

Byrnes BH, Austin ER & Tays BK (1995) Methane emissions from flooded soils and plants under controlled conditions. Soil Biol Biochem 27:331-339

Graham DW, Korich DG, LeBlanc RP, Sinclair NA & Arnold RG (1992) Application of a colorimetric plate assay for soluble methane monooxygenase activity. Appl Environ Microbiol 58:2231-2236

Jensen CR, Luxmoor RG, van Gundy SD & Stolzy LH (1969) Root airspace measurements by a pycnometer method. Agron J 61:474-475

Neue HU (1993) Methane emission from rice fields. Bioscience 43:466-475

Neue HU (1997) Fluxes of methane from rice fields and potential for mitigation. Soil Use Manage 13:258-267

Neue HU, Wassmann R & Lantin RS (1995) Mitigation options for methane emissions from rice fields. In: Peng S, Ingram KT, Neue HU, Ziska LH (eds) Climate Change and Rice, pp 136-144, Springer-Verlag, Berlin

Neue HU, Wassmann R, Lantin RS, Alberto MCR & Aduna JB (1994) Effect of rice cultivars on methane emission. Int. Rice Res News 19:32

Nouchi I, Mariko S & Aoki K (1990) Mechanism of methane transport from the rhizosphere to the atmosphere through rice plants. Plant Physiol 94:59-66

Schütz H, Seiler W & Conrad R (1989) Processes involved in formation and emission of methane in rice paddies. Biogeochemistry 7:33-35

Takai Y (1970) The mechanism of methane fermentation in flooded paddy soil. Soil Sci Plant Nutr 16:238-244

Vogels GD, Keltjens JT & Van der Drift C (1988) Biochemistry of methane production. In: Zehuder AJB (ed) Biology of Anaerobic Microorganisms, pp 707-770, John Wiley & Sons, New York

Wang B, Neue HU & Samonte HP (1997a) Effect of cultivar difference ('IR72', 'IR65598' and 'Dular') on methane emission. Agric Ecosyst Environ 62:31-40

Wang B, Neue HU & Samonte HP (1997b) Role of rice in mediating methane emission. Plant Soil 189: 107-115

Wang B, Neue HU, & Samonte HP (1997c) Effect of rice plant on seasonal methane emission pattern. Acta Agron Sin 23:170-179

Wang B, Xu Y, Wang Z, Li Z, Guo Y, Shao K & Chen Z (1999) Methane emissions from ricefields as affected by organic amendment, water regime, crop establishment, and rice cultivar. Environ Monitor Assess 57:213-228

Watanabe I, Hashimoto T & Shimoyama A (1997) Methane-oxidizing activities and methanotrophic populations associated with wetland rice plants. Biol Fertil Soils 24: 261-265

Yagi K, Tsuruta H & Minami K (1997) Possible options for mitigating methane emission from rice cultivation. Nutr Cycling Agroecosyst 49:213-220

Yoshida S, Forno DA, Cock JH & Gomez KA (1976) Laboratory manual for physiological studies of rice, (3rd ed), International Rice Research Institute, Manila, Philippines

Nutrient Cycling in Agroecosystems 58: 357–366, 2000.

Methane transport capacity of rice plants. I. Influence of methane concentration and growth stage analyzed with an automated measuring system

M.S. Aulakh[1,2], J. Bodenbender[3], R. Wassmann[1,3] & H. Rennenberg[3]

[1]*International Rice Research Institute, MCPO Box 3127, Makati City 1271, Philippines;* [2]*Present address: Department of Soils, Punjab Agricultural University, Ludhiana 141004, Punjab, India;* [3]*Fraunhofer Institute for Atmospheric Environmental Research, Kreuzeckbahnstrasse 19 D-82467 Garmisch-Partenkirchen, Germany*

Key words: plant-mediated gas transfer, methane emissions, rice cultivars, rhizosphere, automated methane measurements, plant growth stages, global warming, greenhouse effects

Abstract

A major portion (60–90 %) of the methane (CH_4) emitted from rice fields to the atmosphere is transported through the aerenchyma of the rice plants. However, a rapid and accurate method to study the CH_4 transport capacity (MTC) of rice plants is not available. We developed a gas sampling and analytical device based on a closed two-compartment chamber technique and analyzed the enrichment of the CH_4 mixing ratio inside the shoot compartment of cylindrical cuvettes enclosing individual rice plants under ambient conditions. The computer-controlled analytical system consists of a gas chromatograph (GC) and a pressure-controlled autosampler for eight cuvettes (seven for plants and one for CH_4-calibration gas). The system automates closure and opening of plant cuvettes using pneumatic pressure, air sample collection and injection into the GC, and CH_4 analysis. It minimizes sources of error during air sampling by continuously mixing headspace air of each cuvette, maintaining pressure and composition of the headspace inside the cuvettes, purging the dead volumes between the sampler induction tube and GC, and running a reference CH_4-calibration gas sample in each cycle. Tests showed that the automated system is a useful tool for accurate sampling of headspace air of cylindrical cuvettes enclosing individual rice plants and enables rapid and accurate fully automated analysis of CH_4 in the headspace air samples. A linear relationship was obtained between CH_4 transported by rice plants of two cultivars (IR72, a high-yielding dwarf, and Dular, a traditional tall cultivar) and concentration of CH_4 up to 7,500 ppm used for purging the nutrient culture solution surrounding the roots in the root compartment of the chamber. Further increase in CH_4 emission by shoots was not observed at 10,000 ppm CH_4 concentration in the root compartment of the chamber. The MTC of IR72 was measured at six development stages; it was lowest at seedling stage, increasing gradually until panicle initiation. There was no further change at flowering, but a marked decrease at maturity was noted. These results suggest that the plants have 45–246% greater potential to transport CH_4 than the highest CH_4 emission rates reported under field conditions, and plants would not emit CH_4 at early growth and at a reduced rate close to ripening.

Introduction

Tropospheric methane (CH_4) concentrations have more than doubled over the past 300 yr (Blake & Rowland, 1988; Etheridge et al., 1992), with more recent data indicating a rate of increase of 0.50-75% a year over the last decade (Dlugokencky et al., 1994). On the basis of ice core data, the rapid increase of CH_4 concentrations is unprecedented during the industrial period (Raynaud et al., 1993) and is presumably linked to human activity. As CH_4 is one of the main greenhouse gases accounting for about 19% of the direct radiative forces of climate (IPCC, 1995), there is a growing interest in the sources and sinks of CH_4 and in the influence of anthropogenic activities on its exchange rates.

Rice cultivation is one of the most important sources of atmospheric CH_4, with a global emission ranging from 20 to 150 Tg CH_4 yr^{-1} according to IPCC (1992) and from 30 to 50 Tg CH_4 yr^{-1} according to recent estimates (Neue & Sass, 1998). The development of methods and strategies to reduce the emission of CH_4 from rice fields is a central issue of ongoing efforts to protect the earth's atmosphere and to avert possible climatic changes. Rice plants act in three key functions regulating the CH_4 budget: (i) as a source of methanogenic substrate, (ii) as a conduit for CH_4 through a well-developed system of intercellular air spaces (aerenchyma), and (iii) as an active CH_4-oxidizing site in the rice rhizosphere by supporting O_2 counter transport through the aerenchyma system. Several studies showed that the CH_4 emitted from rice fields to the atmosphere is transported mostly (60–90%) through the aerenchyma of the rice plants rather than by molecular diffusion across water-air interfaces or release of gas bubbles (Cicerone & Shetter, 1981; Holzapfel-Pschorn & Seiler, 1986; Holzapfel-Pschorn et al., 1986; Wassmann et al., 1996; Wassmann & Aulakh, 2000). Thus, one promising strategy to reduce CH_4 emissions from rice fields is to select and cultivate high-yielding rice cultivars with a reduced CH_4 transport capacity (MTC) (Butterbach-Bahl et al., 1997).

Nouchi et al. (1990) measured CH_4 transport through rice plants using a manual sampling system where a plant was enclosed in an acrylic cylindrical chamber with its roots sealed in a glass vessel that contained nutrient culture solution saturated with a high CH_4 concentration. Then inlet and outlet air samples were collected from the chamber in Tedlar bags and were analyzed for CH_4 by gas chromatography with manual injection. The studies by Nouchi et al. (1990) and Nouchi and Mariko (1993) indicated that the plant-mediated transport of CH_4 is influenced by the concentration of CH_4 in the soil solution around plant roots and the size of the plant. However, keeping in view the high variability in CH_4 emission rates both in time and space, there is a need for a rapid and accurate method that can minimize sources of error during air sampling and CH_4 analysis and can handle a large number of plant-mediated CH_4 flux measurements for screening rice cultivars for reduced MTC. The present study was undertaken (a) to develop, test, and optimize a reliable automated sampling and analysis system for the determination of MTC of different rice cultivars; (b) to analyze the influence of different concentrations of CH_4 in rhizosphere solution on CH_4 emission by rice plants; and (c) to determine MTC of IR72, a high-yielding dwarf cultivar, at six developmental stages.

Materials and methods

Design of the automated measuring system

The design of the fully automated system for measuring the gas transport capacity of rice plants was based on the closed chamber technique for single plants. The main components of the automated system are (a) two-compartment cuvettes, (b) an autosampler connected to valve-control and data-logging system, and (c) a gas chromatograph (GC) connected to a GC-control and data-logging system (Figure 1). The measuring system comprises a total of 8 two-compartment cuvettes, seven for enclosing individual plants and one for calibration. For the sake of simplicity, only one cuvette is shown in Figure 1. All the cuvettes were placed in a cage in the ambient environment in the vicinity of the greenhouse. The cage had a wooden roof for protecting the electrical connections from rainwater. All other components of the automated system were installed inside the greenhouse laboratory.

Two-compartment cuvettes. Each cuvette can hold a single plant with its roots in the lower and the shoots in the upper compartment (Figure 2). Both compartments are made of plexiglass tubes to accommodate plants of different heights and tillers, shoot compartments of three sizes (5 cm id × 60 cm long; 9 cm id × 77 cm long; 9 cm id × 117 cm long) were fabricated. A root chamber of only one size (9 cm id × 18 cm long) was found to be appropriate for enclosing roots of rice plants of all sizes. A connector made of plexiglass was used for fastening the plant and for separating the two chambers. The rice plant was held in place by sealing the base of culm with modeling clay (Plastic-fermit,

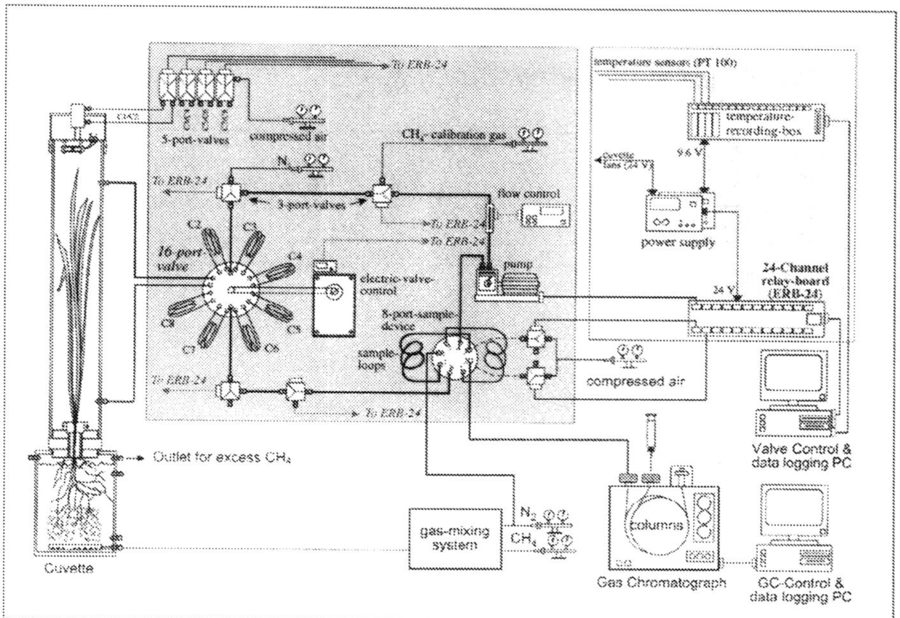

Figure 1. Schematic drawing of the automated measuring system for determining CH_4 transport capacity of rice plants. Only one of eight cuvettes is shown here. Arrows labeled as 'To ERB-24' indicate the connection with the 24-channel relay board

Nissen and Volk, Hamburg, Germany) and a rubber stopper divided into two pieces. The two compartments were sealed from each other by filling agar-agar jelly (2% agar-agar in water) into the case of the connector. To ensure complete sealing, a 3-cm water seal was provided over the agar-agar layer in the shoot chamber. The root compartment was filled with nutrient culture solution saturated with a desirable concentration of CH_4 by purging CH_4 through a gas sieve (air stone) placed at the bottom. After passing through the culture solution, excessive CH_4 gas could escape to the atmosphere through two outlets located at the upper edge of the root chamber (Figure 2). This design ensured the maintenance of ambient pressure inside the root chamber. The CH_4 gas injected and dissolved in the culture solution in the root compartment can escape to the shoot compartment only via the rice plant. Thus plant-mediated CH_4 transport can be measured from the increase in CH_4 concentration inside the closed shoot compartment. The cover of the shoot compartment was opened and closed automatically by a pneumatic pressure device. A fan was mounted on the inner side of the shoot compartment near its upper end to ensure (i) rapid replacement of the air inside the shoot compartment by ambient air when the cuvette is open, and (ii) thorough mixing of the headspace air of the shoot compartment to avoid vertical CH_4 gradients within the shoot compartment when the cuvette is closed.

To monitor temperature inside and outside the cuvettes, one temperature sensor (PT-100) is installed inside the shoot chamber of one of the cuvettes and another inside the cage. The sensors were connected to a personal computer (PC-1) equipped with DAS-1600/1400/1200 Series Board software and hardware package (Keithley Instruments, Taunton, MA, USA) for recording temperature data continuously during the operation of the automated system.

Autosampler, valve control, and data-logging system. The autosampler comprises an automatic, valve-controlled CH_4 sampling and calibration complex (Figure 1). The automated, valve-controlled CH_4 sampler mediates air sample transfer from the shoot compartment of each cuvette to the sample loops and a direct injection of each air sample onto the GC column. A membrane pump provides a circular airflow from one cuvette to the sample loop and back to the same cuvette equilibrating the headspace air of the shoot compartment with that of the sample loop. An electrically driven 16-port valve is connected to the shoot compartment of the eight cuvettes of the measuring system. The valve sequentially opens one connection to one cuvette and switches to the next cuvette in a fixed-

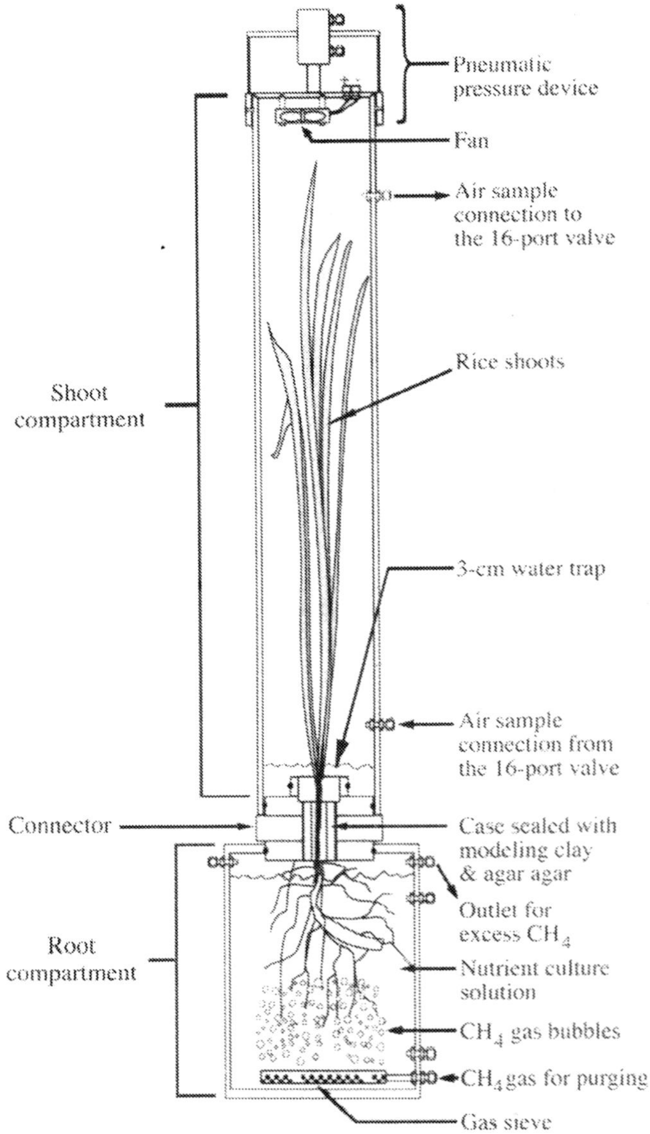

Figure 2. Components of a two-compartment cuvette used for enclosing individual rice plants

time pattern. By switching the eight-port sample device, the air samples are transferred by the carrier N_2 gas stream to the GC column. The tubes connecting the cuvettes with the valves, pump, sampling loops, and GC are made of stainless steel. In between the measurements of each cuvette, the tubes are flushed with N_2 gas.

The magnetic valves controlling the gas fluxes in the system and the 24-channel relay box (ERB-24, Keithley Instruments, Taunton, MA, USA) were operated by the PC-1 equipped with DAS-1600/1400/1200 Series Board software and hardware, which also recorded temperature as mentioned above.

Gas chromatograph, GC control and data-logging system. The gas chromatograph (GC-14B, Shimadzu Corporation, Kyoto, Japan) attached to the autosampler was equipped with a flame ionization detector (FID) and porapak N column. The column oven and injection port temperatures were maintained at 80 and 140 °C, respectively. The operating temperature for the FID was 140 °C. Hydrogen as fuel gas and synthetic air as supporting gas were used with flow rates of 30 and 50 mL min^{-1}, respectively. Pure N_2 was used as a carrier gas with a flow rate of 25 mL min^{-1}.

Another computer, PC-2 equipped with Shimazu Class-VP Chromatography Data system (Shimazu Scientific Instruments, Columbia, MD, USA) software, controlled the operation of the GC. This software also analyzed and stored the data acquired from the GC.

Both valve- and GC control data-logging systems were designed to handle up to 45 consecutive runs each of 120 min (90 h).

Pattern of the measurement cycle

After several preliminary tests using different durations for the measuring cycle, the final pattern of the measurement cycle used during the investigations was of 24-min duration. It started with a 3-min calibration of the GC with CH_4 calibration gas (100 ppm CH_4) followed by a 21-min period for sampling the headspace air in the shoot compartments of the seven cuvettes. During sampling of each cuvette, the system was switched consecutively for 3 min in the sample gas stream by the 16-port valve. An eight-port sampling device was switched on every 1.5-min for transferring CH_4 calibration gas and air samples to the GC column with carrier N_2 gas stream. To avoid pressure-induced errors in the sampling volume, the air in the sampling loop was recompressed to atmospheric pressure by switching on the pump shortly before and again after the operation of sampling device. Between the measurements of each cuvette, flushing the valves with N_2 cleaned the tubes to avoid contamination of the subsequent sample with the residual sample. After the first sampling cycle, the cuvettes were closed by pneumatic pressure cylinders and remained closed for another four cycles (96 min). After a complete run of 120 min (24 + 96 min), the cuvettes were opened again to reset and equilibrate their headspace CH_4 concentration with ambient air.

Rice cultivation and CH_4 transport measurements

Wooden frames (25 cm × 30 cm) covered with nylon mesh were prepared and about 100 healthy seeds of IR72 (a high-yielding dwarf rice cultivar) and Dular (a traditional tall cultivar) were uniformly distributed on the mesh frames. Each mesh frame was floated on a nutrient culture solution (3 cm deep) in a plastic tray. As the seedlings grew, the roots passed through the nylon mesh and were submerged in nutrient culture solution, whereas the base part and shoots remained outside the solution. This procedure facilitated uniform germination of the seeds and growth of seedlings. The nutrient solution contained 40 mg N L^{-1} (as NH_4NO_3), 10 mg P L^{-1} (as $NaH_2PO_4{\cdot}H_2O$), 40 mg K L^{-1} (as KCl), 40 mg Ca L^{-1} (as $CaCl_2$), 40 mg Mg L^{-1} (as $MgCl_2$), 0.5 mg Mn L^{-1} (as $MnCl_2{\cdot}4\ H_2O$), 0.05 mg Mo L^{-1} (as $(NH_4)_6MoO_{24}{\cdot}4\ H_2O$), 0.2 mg B L^{-1} (as H_3BO_4), 0.01 mg Zn L^{-1} (as $ZnSO_4{\cdot}7\ H_2O$), 0.01 mg Cu L^{-1} (as $CuSO_4{\cdot}5\ H_2O$), and 0.01 mg Fe L^{-1} (as $FeSO_4{\cdot}7\ H_2O$ + EDTA in 1 N KOH) and was adjusted at pH 5.5. Potassium was supplied as KCl instead of K_2SO_4 and Mg as $MgCl_2$ instead of $MgSO_4$ in order to avoid inhibitory effects of sulfate on CH_4 production (Westermann & Ahring, 1987; Achtnich et al., 1995).

Two weeks after seeding, the plants were transplanted individually into plastic pots (16 cm id × 15 cm length) each lined with a polyethylene bag and filled with crushed 2.3 kg Maahas clay soil collected from a rice field of IRRI. The air-dried Maahas clay had pH 6.4, CEC 34.1 cmole kg^{-1}, 19 mg available P kg^{-1}, 15.7 g organic C kg^{-1} and 1.9 g total N kg^{-1}. Each pot was irrigated with water maintaining a 5-cm water layer overlying the soil surface. Fertilizer N (150 kg ha^{-1}) as urea in three splits was applied as basal, at maximum tillering, and at flowering.

For MTC measurements, rice plants were taken out of the pots with the intact soil and plastic bag. The plastic bag was cut open and the soil around the roots was washed off with a gentle water spray. This procedure facilitated soil removal without injuring the roots. An individual plant was placed in the root compartment of each cuvette containing nutrient culture solution. The shoot compartments were connected to the root compartments and sealed. The root compartments were purged with CH_4 and the measuring system was operated for at least three consecutive runs of 120 min each.

Calculation of CH_4 transport rates through rice plants

During each measuring run of 120 min, a linear increase in the CH_4 mixing ratio was usually observed inside the closed shoot compartment of each cuvette. Nonlinear results originated from an experimental setup that was not gas-tight and therefore were discarded. The CH_4 transport rate through the rice plant was calculated from the slope of the linear increase of the CH_4 concentration [ppmv min^{-1}] in the headspace of the closed shoot compartments of each cuvette by equations (1) and (2):

$$CH_4 \text{ transport rate (mole } CH_4 \text{ plant}^{-1} \text{ min}^{-1}) = CH_4 \text{ [ppmv min}^{-1}]^* \, 10^{-6} \text{ [1 ppmv}^{-1}] * V * 1/(R * T) \quad (1)$$

$$CH_4 \text{ transport rate (mg } CH_4 \text{ plant}^{-1} \text{ d}^{-1}) = (\text{mole } CH_4 \text{ plant}^{-1} \text{ min}^{-1}) * 16000 \text{ [mg } CH_4 \text{ mole}^{-1}] * 1440 \text{ [min d}^{-1}] \quad (2)$$

where V = total volume of shoot compartment (L), R = the universal gas constant equal to 0.08205 liter-atm $°K^{-1}$ $mole^{-1}$, and T = measured temperature in Kelvin scale (°K).

Experiments

Three experiments were conducted (a) to determine the accuracy of the measuring system for sampling and analysis of CH_4, (b) to test the influence of CH_4 concentration in rhizosphere solution on CH_4 transport by rice plants, and (c) to measure MTC of the rice plants of cultivar IR72 at different stages of development.

Accuracy of measuring system for sampling and analysis of CH_4. The precision of the automated system in sampling the headspace of shoot chamber and analysis of CH_4 was tested in two ways. The cuvettes were installed without enclosing a rice plant and nutrient culture solution. Treatments, in triplicate, included (A) ambient air, (B) 10 ppm CH_4, (C) 580 ppm CH_4, and (D) continuous flushing with 580 ppm CH_4 for 360 min. In treatments B and C, the air space in the cuvettes was flushed for 5 min with 10 and 580 ppm CH_4, respectively, and then the cuvettes were closed. In treatment D, cuvettes were flushed with 580 ppm CH_4 continuously during the period of testing. The measuring system was then operated for three conjunctive runs (360 min) and CH_4 concentration was measured. Simultaneously, air samples were drawn manually from the shoot chamber of cuvettes with airtight syringes and analyzed on the GC for CH_4 concentration.

Influence of CH_4 concentration on CH_4 transport by the rice plants. To study the relationship between CH_4 emission rate of rice plants and CH_4 concentration in soil water surrounding the plant roots, nutrient culture solution was purged with ambient air containing (A) 1.8 ppm CH_4, (B) 2,500 ppm Ch_4, (C) 5,000 ppm CH_4, (D) 7,500 ppm CH_4, and (E) 10,000 ppm CH_4. Treatment F with 10,000 ppm CH_4 without plant was included to check the scaling between the root compartment and the shoot compartment. Plants of IR72 and Dular cultivars at panicle initiation stage were chosen for this experiment because a preliminary experiment indicated maximum MTC of plants at this stage. In treatments B to F, nutrient culture solution in the root chamber was saturated with CH_4 by bubbling CH_4 of different concentrations as per treatment at flow rate of 2 L min^{-1} continuously during the course of the experiment. Nouchi and Mariko (1993) found that the rate of CH_4 emission by plants began to increase within 10 min and reached maximum values within 25-40 min after the start of bubbling CH_4 through the culture solution. Our preliminary experiments with bubbling CH_4 of a particular concentration at 2 L min^{-1} flow rate showed that a period of 40-50 min was sufficient to create an equilibrium between CH_4 in culture solution and CH_4 emitted by plants. Therefore, before initiating actual measurements, the culture solution was purged with CH_4 of a particular concentration for 1 h. The measuring system was then operated for three conjunctive runs (3 × 120 min) and CH_4 concentrations in the shoot-compartments were measured. All treatments were performed in three replicates with three different plants.

Methane transport capacity of rice at different stages of plant development. Methane transport capacity of rice plants of IR72 was studied at six stages of development—i.e., seedling, early tillering, maximum tillering, panicle initiation, flowering, and maturity. At each growth stage, three plants were enclosed in cuvettes with nutrient culture solution. Culture solution was purged with 10,000 ppm CH_4 gas 1 h before and continuously during the actual MTC measurements. The measuring system was operated for three conjunctive runs (360 min) and CH_4 concentration was measured.

Statistical analysis

The data presented are means ± standard deviation of three different plants with each plant analyzed in triplicate. Statistical analysis of experimental data was accomplished by standard analysis of variance in completely randomized design (Cochran & Cox, 1950) using IRRISTAT statistical software (Bartolome et al., 1999). Mean separation for different treatments in each experiment was performed using the least significant difference (LSD) test for significance at the 0.05 level of probability.

Results and discussion

Accuracy of the measuring system for sampling and analysis of CH_4

The precision of the automated sampling and analysis system was tested by filling the cuvettes with ambient air, 10 or 580 ppm CH_4, or by continuously flushing the cuvettes with 580 ppm CH_4 for 360 min. Fluctuations in CH_4 sampling and analysis by the automated system ranged from negligible to 11% over a period of 360 min (Figure 3). Relatively higher deviations were associated with the analysis of low concentration of 10 ppm CH_4, presumably due to the very wide range of CH_4 concentrations tested. The CH_4 concentrations measured from one-time addition of 580 ppm CH_4 and continuous flushing with 580 ppm CH_4 for 360 min were comparable with deviations within $\pm$ 5%. The CH_4 concentrations measured by the automated system differed from samples collected manually from the same treatments with airtight syringes by less than 5% (data not shown). These results confirm that the system was reliable in sampling the headspace air of the upper cuvettes and analysis of CH_4 in these samples.

Influence of CH_4 concentration on CH_4 transport by rice plants

Methane concentration in the shoot compartment of the cuvettes closed without a plant did not increase over a period of 360 min despite continuous purging of culture solution in the root compartment with 10,000 ppm CH_4 (Figure 4). On the other hand, a linear increase in the CH_4 concentration of the shoot compartment was observed when a rice plant was included in the cuvette. Nouchi (1994) observed unavoidable leakage of CH_4 through gaps between the rice plant and the modeling clay. In the present study, after sealing the two compartments from each other with modeling clay, the case of the connector between the compartments was filled with agar-agar jelly and, in addition, a 3-cm water trap was provided. This method ensured perfect sealing.

Figure 5 shows the CH_4 transported by IR72 and Dular cultivars purged with four concentrations of CH_4 in the root compartment. There was a linear relation-

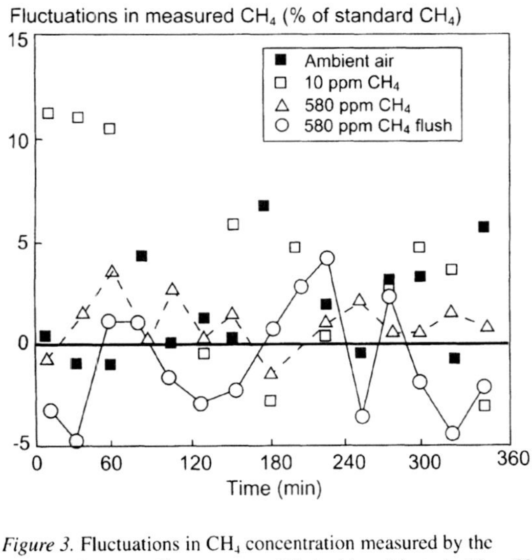

Figure 3. Fluctuations in CH_4 concentration measured by the automated system. Cuvettes were filled with ambient air, 10 or 580 ppm CH_4 at zero time, or were continuously flushed with 580 ppm CH_4 for 360 min. The deviation in % of the respective CH_4 standard added into the cuvette is shown

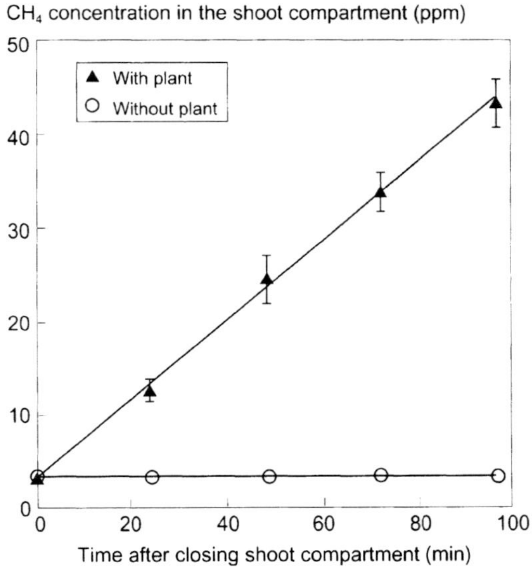

Figure 4. Methane concentration in the shoot compartment of the cuvette with or without rice plant of cultivar IR72 at panicle initiation. The nutrient culture solution in the root compartment was continuously purged with 10,000 ppm CH_4. Vertical bars indicate standard deviations with three replicate plants each measured in triplicate

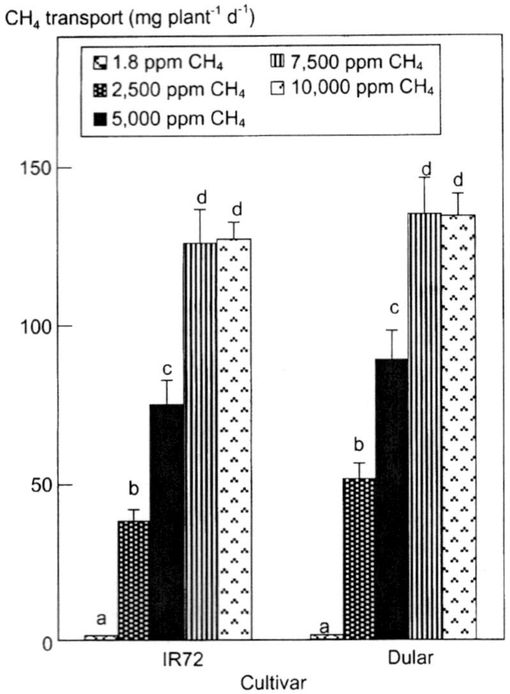

Figure 5. Influence of the CH_4 concentration used for purging the nutrient culture solution surrounding roots on CH_4 transport by rice plants of cultivars IR72 and Dular at panicle initiation. Data shown are means ± SD of three replicate plants each measured in triplicate. Different letters indicate significant differences between CH_4 concentrations ($p \geq 0.05$).

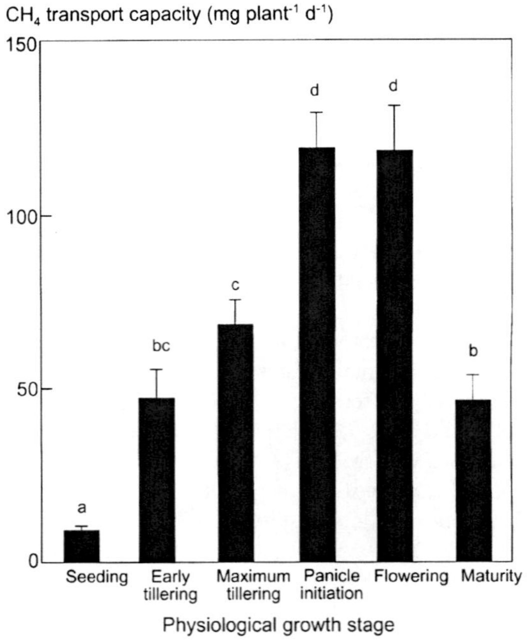

Figure 6. Methane transport capacity of rice plants of cultivar IR72 at seedling, early tillering, maximum tillering, panicle initiation, flowering and maturity. Data shown are means ± SD of 3 replicate plants each measured in triplicate. Different letters indicate significant differences ($p \geq 0.05$).

ship between the amount of CH_4 transported by the plants and the concentration of purging CH_4 up to 7,500 ppm. Further increase in CH_4 concentration to 10,000 ppm did not affect CH_4 transport through the rice plants, suggesting that maximum transport of CH_4 was reached at 7,500 ppm. This finding illustrates that a CH_4 concentration of 10,000 ppm in the purging gas is adequate to determine the maximum CH_4 transport through the plants, irrespective of rice cultivar.

Methane transport capacity of rice plants at different physiological growth stages

The MTC of IR72 plants was determined at six growth stages using 10,000 ppm CH_4 for purging the rhizosphere solution. At the seedling stage (plant age 25 d), MTC was lowest with mean values of 8 ± 1 mg CH_4 plant^{-1} d^{-1}; it increased by a factor of about 6 and 8 at the early tillering stage (35 d old) and maximum tillering (50 d old), respectively (Figure 6). Plants at panicle initiation (60 d old) showed maximum MTC (120 mg CH_4 plant^{-1} d^{-1}), and further growth to the flowering stage (80 d old) did not change the MTC. However, there was a significant decrease in MTC at maturity. In an earlier study, using manual gas collection, Butterbach-Bahl et al. (1997) also observed a substantial increase in MTC of rice plants of two Italian varieties (Lido and Roma) from young seedlings with an age of 22 d to an age between 35 and 40 d and no further increase during subsequent growth until 60 d of plant age. The authors did not report a marked decrease in MTC at maturity as observed in the present study with plants of IR72.

The MTC of 120 mg CH_4 plant^{-1} d^{-1} observed at panicle initiation in our study corresponds to 4,500 mg CH_4 m^{-2} d^{-1}: 1.5 (plants hill^{-1}) × 25 (hills m^{-2}) × 120 (mg CH_4 plant^{-1} d^{-1}). The density factor of 1.5 plants hill^{-1} was calculated from the difference between shoot biomass of 2 plants hill^{-1} of IR72 grown in the field during the same season on the same soil and a single plant grown in an individual pot in the greenhouse (based on 10 random replicated measurements). The density of 25 hills m^{-2} is based on a commonly followed 20- × 20-cm plant spacing under field conditions. The highest CH_4 emission rates reported from field studies conducted with organic inputs ranged from about 1,300 mg CH_4 m^{-2} d^{-1} (Buendia et al., 1997), 2,000 mg CH_4 m^{-2} d^{-1} (Denier van der Gon & Neue, 1995) and 3,100 mg CH_4 m^{-2} d^{-1} (Wassmann et al., 2000). The results of the present study suggest that the rice plants have 45–246% greater potential to transport CH_4 than the highest emission rates observed under field conditions. Furthermore, plants would not emit CH_4 at early growth and, to a reduced extent, close to ripening. If CH_4 is produced at a high rate during early growth of the rice crop, as often found in fields treated with crop residues or green manure, CH_4 will have to move to the atmosphere through other pathways, such as molecular diffusion across the water-air interfaces or release of gas bubbles. This observation is consistent with the findings of earlier field studies (Schütz et al., 1989; Wassmann et al., 1996) showing CH_4 emission mainly by bubbling during the first few weeks after transplanting. However, keeping in view the enormous genotypic and phenotypic variations among different rice cultivars, more detailed investigations are needed to assess the role of plant-mediated transport of CH_4 in CH_4 emissions from rice agriculture, e.g., by analysis of commonly used and new high-yielding cultivars. The automated measuring system developed and used for analyzing MTC of cultivar IR72 in this study was proven to be a useful tool for such an approach.

Acknowledgements

We thank Sonny Pantoja for his help with the greenhouse experiments. This research was funded by the German BMZ/GTZ for the project "Reduction of methane emission from rice fields by screening for low methane transport capacity (No. 95.7860.0-001.05)."

References

Achtnich C, Bak F & Conrad R (1995) Competition for electron donors among nitrate reducers, ferric iron reducers, sulfate producers and methanogens in anoxic paddy soil. Biol Fertil Soils 19:65-72

Bartolome VI, Casumpang RM, Ynalvez MAH, Olea AB & McLaren CG (1999) IRRISTAT for Windows - statistical software for agricultural research. Biometrics, International Rice Research Institute, Makati City, Philippines

Blake DR & Rowland FS (1988) Continuing worldwide increase in tropospheric methane, 1978 to 1987. Science 239:1129-1131

Buendia LV, Neue HU, Wassmann R, Lantin RS, Javellana AM, Yuchang X, Makarim AK, Corton TM & Charoensilp N (1997) Understanding the nature of methane emission from rice ecosystems as basis of mitigation strategies. Appl Energy 56:433-444

Butterbach-Bahl K, Papen H & Rennenberg H (1997) Impact of gas transport through rice cultivars on methane emission from paddy fields. Plant Cell Environ 20: 1175-1183

Cicerone RJ & Shetter JD (1981) Sources of atmospheric methane: measurements in rice paddies and a discussion. J Geophys Res 86:7203-7209

Cochran WG & Cox GM (1950) Experimental Designs. John Wiley and Sons, Inc, New York

Denier van der Gon HAC & Neue HU (1995) Influence of organic matter incorporation on the methane emission from a wetland rice field. Global Biogeochem Cycles 9: 11-22

Dlugokencky E, Steele P, Lang PM, Tans P & Masaire K (1994) The growth rate and distribution of atmospheric methane. J Geophys Res 99:17021-17043

Etheridge DM, Pearman GI & Fraser PJ (1992) Changes in tropospheric methane between 1841 and 1978 from a high accumulation rate Antarctic ice core. Tellus 44:282-294

Holzapfel-Pschorn A, Conrad R & Seiler W (1986) Effects of vegetation on the emission of methane from submerged paddy soil. Plant Soil 92:223-391

Holzapfel-Pschorn A & Seiler W (1986) Methane emission during a cultivation period from an Italian rice paddy. J Geophy Res 91:11803-11814

IPCC – Intergovernmental Panel on Climate Change (1992) Climate Change: The Supplementary Report to the IPCC Scientific Assessment, Houghton JT, Callander BA & Varney SK (eds) IPCC, Cambridge University Press, Cambridge

IPCC – Intergovernmental Panel on Climate Change (1995) Climate Change: The Science of Climate Change, Houghton JT, Meira Filho LG, Callander BA, Harris N, Kattenberg A, Maskell K (eds) IPCC, Cambridge University Press, Cambridge

Neue HU & Sass R (1998) The budget of methane from rice fields. IGAC tivities 17:3-11

Nouchi I, Mariko S & Aoki K (1990) Mechanism of methane transport from the rhizosphere to the atmosphere through rice plants. Plant Physiol 94:59-66

Nouchi I & Mariko S (1993) Mechanism of methane transport by rice plants. In: Oremland RS (ed). Biogeochemistry of Global Change, Radiatively Active Trace Gases, pp 336-352, Chapman Hall, New York

Nouchi I (1994) Mechanisms of methane transport through rice plants. In: Minami K, Mosier A, Sass R (eds) CH_4 and N_2O: Global Emissions and Controls from Rice Fields and Other Agricultural and Industrial Sources, pp 87-104, National Institute of Agro-Environmental Sciences, Tsukuba, Japan..

Raynaud D, Jouzel J, Barnola JM, Chappellaz J, Delmas RJ & Lorius C (1993) The ice record of greenhouse gases. Science 259:926-933

Schütz H, Seiler W & Conrad R (1989) Processes involved in formation and emission of methane in rice paddies. Biogeochem 7:33-53

Wassmann R & Aulakh MS (2000) The role of rice plants in regulating mechanisms of methane emissions: A review. Biol Fertil Soils 31:20-29

Wassmann R, Neue HU, Alberto MCR, Lantin RS, Bueno C, Llenaresas D, Arah JRM, Papen H, Seiler W & Rennenberg H (1996) Fluxes and pools of methane in wetland rice soils with varying organic inputs. Environ Monitor Assess 42:163-173

Wassmann R, Buendia LV, Lantin RS, Bueno CS, Lubigan LA, Umali A, Nocon NN, Javellana AM & Neue HU (2000) Mechanisms of crop management impact on methane emissions from rice fields in Los Baños, Philippines. Nutr Cycling Agroecosyst, this issue

Westermann P & Ahring BK (1987) Dynamics of methane production, sulfate reduction and denitrification in a permanently waterlogged alder swamp. Appl Environ Microbiol 53:2554-2559

Nutrient Cycling in Agroecosystems **58:** 367–375, 2000.

Methane transport capacity of rice plants. II. Variations among different rice cultivars and relationship with morphological characteristics

M.S. Aulakh[1,2], J. Bodenbender[3], R. Wassmann[1,3], & H. Rennenberg[3]

[1]*International Rice Research Institute, MCPO Box 3127, Makati City 1271, Philippines;* [2]*Present address: Department of Soils, Punjab Agricultural University, Ludhiana 141004, Punjab, India;* [3]*Fraunhofer Institute for Atmospheric Environmental Research, Kreuzeckbahnstrasse 19 D-82467 Garmisch-Partenkirchen, Germany*

Key words: plant-mediated gas transfer, methane emissions, rice cultivars, rhizosphere, automated methane measurements, plant growth stages, global warming, greenhouse effects, plant biomass, plant tillers

Abstract

Of the total methane (CH4) emitted from a rice field during the growing season, 60-90% is emitted through the rice plants. We determined the methane transport capacity (MTC) of rice plants at different physiological growth stages using an automatic measuring system under greenhouse conditions. A total of 12 cultivars (10 inbred varieties and 2 hybrids) were studied in sets of two experiments and was distinguished into three groups according to the patterns of MTC development. MTC is generally increasing from seedling stage to panicle initiation (PI), but differs in the development from PI to maturity. While the hybrid showed a gradual increase in MTC, the inbred cultivars showed either minor changes in MTC or a drastic decrease from flowering to maturity. Among tall cultivars, Dular showed the highest MTC, followed by B40; the lowest MTC was found in Intan. High-yielding dwarf cultivars showed MTC in the descending order of IR72 > IR52 > IR64 > PSBRc 20. New plant type cultivars showed very low MTC with IR65600 exhibiting the smallest MTC at PI, flowering, and maturity. Hybrids (Magat and APHR 2) showed the largest MTC that continued to increase with plant growth. The MTC patterns were attributed to growth parameters and the development of morphological characteristics of the aerenchyma. These results suggest that in tall, dwarf, and NPT cultivars, increase in root or aboveground biomass during initial growth determines a corresponding increase in MTC. Once aerenchyma has fully developed, further increase in plant biomass would not influence MTC. However, in the case of hybrids, a positive relationship of MTC with root + shoot biomass ($r = 0.672$, $p \geq 0.05$) and a total plant biomass including grain ($r = 0.849$, $p \geq 0.01$) indicate continuous development of aerenchyma with plant growth, resulting in enhanced MTC. In all cultivars, tiller number, but not height, was linearly related to MTC, indicating that the number of outlets/channels rather than plant size/biomass determines the transport of CH_4. These results clearly demonstrate that rice cultivars differ significantly in MTC. Therefore, the use of high-yielding cultivars with low MTC (for example, PSBRc 20, IR65598, and IR65600) could be an economically feasible, environmentally sound, and promising approach to mitigate CH_4 emissions from rice fields.

Introduction

Methane (CH_4) is one of the main greenhouse gases, accounting for about 19% of the direct radiative forces of climate, with atmospheric concentrations increasing at a current rate of about 8 ppbv yr^{-1} (IPCC, 1995). Since this increase is expected to alter the earth's climate, there is a growing interest in the sources and sinks of CH_4 and in the development of mitigation options. Rice cultivation is one of the most important sources of atmospheric CH_4 with a global emission ranging from 20 to 150 Tg CH_4 yr^{-1} according to IPCC (1992) and 30 to 50 Tg CH_4 yr^{-1} according to recent estimates (Neue & Sass, 1998). Increased land use for rice cultivation and multiple cropping have increased the strength of this source of atmospheric CH_4 during the last century. In view of the future rice demand for feeding the increasing world population, the traits of high-yielding rice cultivars will further affect the CH_4 source strength of rice cultivation. Therefore, the high contribution of rice cultivation to the global CH_4 budget demands strategies to reduce CH_4 emissions from rice fields.

Of the total CH_4 emitted from a rice field during the growing season, 60–90% is emitted through the rice plants (Cicerone & Shetter, 1981; Holzapfel-Pschorn & Seiler, 1986; Holzapfel-Pschorn et al., 1986; Wassmann et al., 1996; Wassmann & Aulakh, 2000). Field studies from China (Lin, 1993; Kesheng & Zhen, 1997), India (Adhya et al., 1994; Shalini et al., 1997; Mitra et al., 1999), Italy (Butterbach-Bahl et al., 1997), Japan (Watanabe et al., 1995), and Texas, USA (Sigren et al., 1997) have indicated substantial differences in the rate of CH_4 emission between different rice cultivars. These differences in CH_4 flux rates could be attributed to differences in CH_4 production, oxidation, and gas transport capacities of different cultivars. Comparative studies on different rice cultivars are therefore crucial for the development of mitigation options. One promising strategy to reduce CH_4 emissions from rice fields, for example, could be to select and grow high-yielding rice cultivars with a reduced CH_4 transport capacity (MTC). Considering the enormous genotypic and phenotypic variations in the genus *Oryza* (Leon & Carpena, 1995) that comprises approximately 80,000 known cultivars, a thorough understanding of the mechanisms involved in CH_4 production, oxidation, and gas transport capacities would help in the selection and breeding for the traits of high yield and low CH_4 emission potential.

We developed an automated system for accurate sampling of headspace air of cylindrical cuvettes each enclosing individual rice plants that enables the rapid and fully automated analysis of CH_4 in the air samples (Aulakh et al., 1999). The present study was undertaken (a) to determine the MTC of 12 cultivars (10 inbred varieties and 2 hybrids) at different growth stages using this automatic system, (b) to identify cultivars with low MTC, and (c) to investigate the relationships between MTC of the rice plants and growth parameters or morphological characteristics.

Materials and methods

A detailed description of the fully automated measuring system that was used for determining the MTC of rice plants and the procedure for MTC measurements are reported in an accompanying paper (Aulakh et al., 1999).

Methane transport measurements of rice plants during vegetative growth

A first set of experiments was conducted in the greenhouse of the Fraunhofer Institute for Atmospheric Environmental Research (IFU) at Garmisch-Partenkirchen, Germany, with seven rice cultivars selected from four categories: (a) traditional tall cultivars (Dular, B40, and Intan), (b) high-yielding dwarf cultivars developed by the International Rice Research Institute (IRRI), Los Banos, Philippines (IR72 and IR64), (c) a new plant type (NPT) from IRRI (IR65597), and (d) a hybrid from IRRI (Magat). Seeds were sown on a framed nylon mesh that was floated on a nutrient culture solution (3 cm deep) in a plastic tray. The nutrient solution contained 40 mg N L^{-1} (as NH_4NO_3), 10 mg P L^{-1} (as $NaH_2PO_4{\cdot}H_2O$), 40 mg K L^{-1} (as KCl), 40 mg Ca L^{-1} (as $CaCl_2$), 40 mg Mg L^{-1} (as $MgCl_2$), 0.5 mg Mn L^{-1} (as $MnCl_2{\cdot}4H_2O$), 0.05 mg Mo L^{-1} [(as $(NH_4)_6MoO_{24}{\cdot}4H_2O$)], 0.2 mg B L^{-1} (as H_3BO_4), 0.01 mg Zn L^{-1} (as $ZnSO_4{\cdot}7H_2O$), 0.01 mg Cu L^{-1} (as $CuSO_4{\cdot}5H_2O$), and 0.01 mg Fe L^{-1} (as $FeSO_4$ $7H_2O$ + EDTA in 1 N KOH) and was adjusted at pH 5.5. Two weeks after seeding, the plants were transplanted individually into plastic pots (16 cm id × 15 cm length) each lined with a polyethylene bag and filled with crushed 2.3 kg soil collected from rice fields of the Italian Rice Research Institute in Vercelli, Italy. The soil was a sandy loam and had a pH 6.0, 13.5 g organic C kg^{-1}, and 1.5 g total N kg^{-1}. Each pot was irrigated with water maintaining a 5-cm water layer overlying the soil surface. Fertilizer N (150 kg N ha^{-1}) as urea in three splits was applied as basal, at maximum tillering, and flowering. Four physiological growth stages—seedling, early tillering, maximum tillering, and panicle initiation (PI)—were selected for measuring MTC of the rice plants by a procedure reported by Aulakh et al. (2000). After enclosing individual plants in cylindrical cuvettes, the nutrient culture solution in the root compartment was purged with 10,000 ppm CH_4 1 h prior to and continuously during the MTC measurements. The automated system was operated for three conjunctive runs (360 min) for MTC measurements. At the end of each MTC measurement, the plants were divided into roots and shoots (aboveground portions) and dried in an oven at 110 °C for biomass determination.

Methane transport measurements of rice plants during reproduction and maturity

In a second set of experiments in the IRRI greenhouse at Los Banos, Philippines, 10 cultivars were selected from four categories: (a) traditional tall cultivars (Dular, B40, and Intan), (b) high-yielding dwarf cultivars from the Philippines (PSBRc 20) and IRRI (IR72, IR52, and IR64), (c) NPT cultivars (IR65598 and IR65600), and (d) a hybrid from India (APHR 2). Maahas clay soil collected from a rice field of IRRI was used for the experiment. The air-dried Maahas clay had a pH 6.4, 15.7 g organic C kg^{-1}, and 1.9 g total N kg^{-1}. Following the procedure explained above, MTC was studied at three reproductive growth stages starting with PI through flowering and maturity. At the end of each MTC measurement, developed and underdeveloped tillers were counted and tiller height was measured using methods by Gomez (1972). Developed tillers are defined as the tallest tillers that were productive at the end. Underdeveloped tillers comprise remaining tillers having variable height and were often unproductive. Tiller height is the distance from ground level to the tip of the tallest leaf (at vegetative growth stages) or of the highest panicle (at maturity). Mean height of developed and underdeveloped tillers in each plant was then calculated. Biomass of roots, shoots, and grains (at maturity only) was determined after drying the samples in an oven at 110 °C for 3 d.

To test further the relationship between tiller number and MTC, plants from B40 and IR72 cultivars with 4, 6, and 8 developed tillers were selected at PI for MTC measurement.

Statistical analysis

The data presented are means $\pm$ standard deviation (SD) of three different plants with each plant analyzed in triplicate. Statistical analysis of experimental data was accomplished by standard analysis of variance in completely randomized design (Cochran & Cox, 1950) using IRRISTAT statistical software (Bartolome et al., 1999). Mean separation for different treatments in each experiment was performed using the least significant difference (LSD) test for significance at the 0.05 level of probability. Correlation coefficients (r values) between MTC and different growth parameters (root, shoot, grain biomass) and morphological characteristics (tiller number and height) were calculated.

Results and discussion

Effect of physiological growth stage on CH_4 transport capacity

The MTC of seven rice cultivars during vegetative growth (seedling to PI) are summarized in Table 1. In general, MTC was lowest at the seedling stage (plant age 25-27 d), with mean values ranging between 2 mg CH_4 plant^{-1} d^{-1} (Dular) and 15 mg CH_4 plant^{-1} d^{-1} (IR65597). At early tillering stage (2 wk later), MTC increased by a factor of about 5 in B40 (tall), 16 in IR64 (high-yielding dwarf), and 14 in Magat (hybrid). Measurements at maximum tillering and PI indicated insignificant differences in MTC of B40 but a continuous increase in IR64 and Magat. In an earlier study using a manual gas collection method, Butterbach-Bahl et al. (1997) observed a substantial increase in MTC of rice plants of two Italian varieties (Lido and Roma) from young seedlings (from 22 d of age to between 35 and 40 d) and no further increase during subsequent growth until 60 d of plant age. Aulakh et al. (2000), using an automated system, reported the lowest MTC in plants of cultivar IR72 at the seedling stage and a predominant increase in MTC with increasing plant growth until PI, no further change at flowering, but a marked decrease at maturity. The results of this experiment revealed that MTC is generally increasing from seedling stage to PI.

In a second set of experiments, the MTC of 10 rice cultivars was determined during reproductive stage and maturity (ripening stage). The results for tall, dwarf, NPT and hybrid plants are presented in Figure 1. Among tall cultivars, Dular showed the highest MTC, followed by B40; the lowest MTC was found in Intan. High-yielding cultivars showed MTC in the descending order of IR72 > IR52 > IR64 > PSBRc 20. NPT cultivars showed very low MTC, with IR65600 having the smallest value at all three stages (PI, flowering, and maturity) studied. Interestingly, hybrid APHR 2 showed very high MTC as did hybrid Magat in the first set of experiments. The low MTC of NPT cultivars explains the lower amounts of CH_4 emitted by IR65598 as compared with those of Dular and IR72 throughout the growing season in an earlier study of Wang et al. (1997).

The variability in MTC between the different replicates of the same variety at the same growth stage ranged from 6% observed in Intan up to more than 100% in Dular at the seedling stage (Table 1) and from 3% (IR65598) to 35% (Dular) at maturity stage (Figure 1). In general, the variability among replicates of each cultivar was high at the early growth stage when MTC values were minimum, but it decreased with plant growth. Part of the variability in MTC between the different replicates could be due to the sensitivity of the automated measuring system that showed ± 5 % deviation in analysis (Aulakh et al., 2000). Despite plant-specific variations, the present results show significant differences in MTC between individual cultivars at a given growth stage. However, no uniform developmental pattern of differences in MTC could be observed.

Table 1. Methane transport capacity (mg CH_4 plant^{-1} d^{-1}) of rice plants of traditional tall cultivars (Dular, B40,, and Intan), high-yielding dwarf cultivars (IR72 and IR64), new plant type cultivar (IR65597), and a hybrid (Magat) at seedling to panicle initiation growth stages[a]

Cultivar	Growth stage			
	Seedling	Early tillering	Maximum tillering	Panicle initiation
Dular	2 ± 2[b]	[c]	9 ± 2	66 ± 2
Intan	7 ± 0	[c]	20 ± 6	30 ± 8
B40	14 ± 10	65 ± 10	80 ± 10	88 ± 6
IR64	4 ± 2	69 ± 3	68 ± 3	101 ± 6
IR72	7 ± 5	[c]	54 ± 10	102 ± 4
IR65597	15 ± 4	[c]	33 ± 8	70 ± 7
Magat	3 ± 1	42 ± 8	137 ± 10	252 ± 9
Overall LSD (0.05)	22			

[a]MTC of plants grown under greenhouse conditions were studied with a fully automated measuring system applying the closed chamber technique. The chamber was divided into a shoot and a root compartment. The root compartment was purged with CH_4 and accumulation of CH_4 in the shoot compartment was determined. [b]Standard deviation. [c]Measurements were not made

For example, IR65597, which showed the largest MTC at the seedling stage, was only at position 3 at maximum tillering and at PI out of the seven cultivars tested. On the other hand, Magat, which had a very small MTC at the seedling stage, showed the largest MTC at PI, 3–8 times higher than those of the six other cultivars.

Based on the results of both sets of experiments, the patterns of MTC development with plant growth could be divided into three types. All cultivars showed an increase in MTC from seedling to PI but differed in the succeeding growth stages until maturity. Only the hybrid (APHR2) showed a gradual increase in MTC. The inbred varieties showed either minor changes in MTC (Intan, PSBRc20, IR65600) or a drastic decrease (2-4 times) from flowering to maturity (Dular, B40, IR72, IR52, IR64, IR65598). The decrease in MTC at maturity in eight out of nine inbred cultivars may be due to dying of root cells, root-stem intersection (base) and tillers that caused the aerenchyma to collapse, consequently the aerenchyma channels were blocked. On the other hand, a continuous increase in MTC with plant growth in the hybrids studied may be due to the simultaneous development of the aerenchyma. These assumptions, however, need to be tested by microscopic analysis of aerenchyma sections and measurement of the aerenchyma areas.

Effect of growth parameters on CH_4 transport capacity

The growth parameters of the 10 cultivars tested (Table 2) revealed large differences among cultivars. During vegetative growth (seedling to PI), a strong correlation was found between MTC and plant root or total plant biomass in all the cultivars studied, but the nature of this relationship varied among cultivars (Figure 2). For instance, Magat showed a linear relationship between MTC and plant biomass, whereas B40 and IR64 exhibited a logarithmic relationship. On the other hand, increase in root and shoot biomass during the reproductive period and maturity (Table 2) did not affect MTC as evident from the very poor correlations in all cultivars except hybrid APHR 2 (Table 3). Combining the biomass of root, shoot, and grain resulted in a negative relationship that was significant in five out of nine cultivars. These results suggest that in tall, dwarf, and NPT cultivars, the increase in root or aboveground biomass during initial growth would determine the corresponding increase in MTC. Once aerenchyma has fully developed, further increase in plant biomass does not affect MTC. The positive correlation between

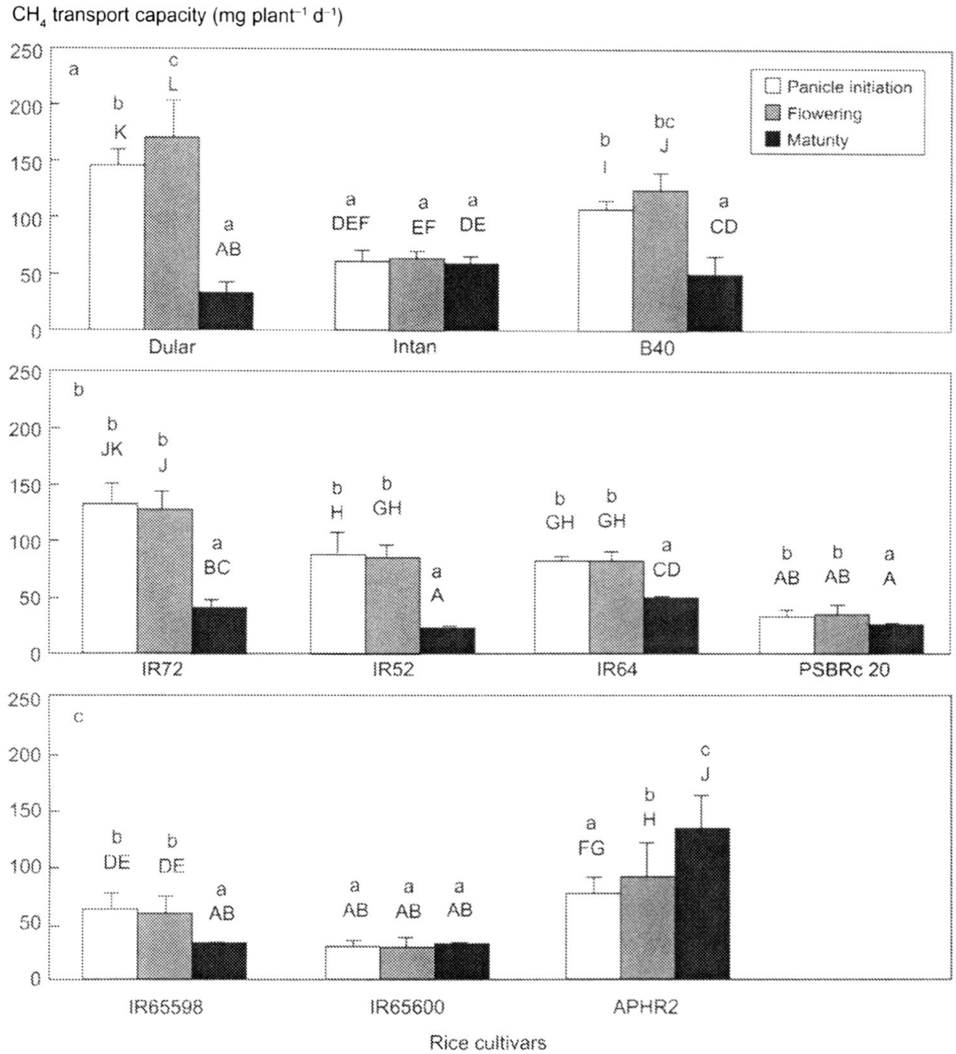

Figure 1. Methane transport capacity (MTC) of (a) traditional tall cultivars (Dular, B40, and Intan), (b) high-yielding dwarf cultivars (IR72, IR52, IR64, and PSBRc 20), (c) new plant type cultivars (IR65598 and IR65600) and a hybrid (APHR 2) at three growth stages during reproductive period. Data shown are means ± SD of three replicate plants each measured in triplicate. Different small letters indicate significant differences between growth stages ($p \geq 0.05$). Different capital letters indicate significant differences between cultivars with a particular growth stage ($p \geq 0.05$). MTC determined as outlined in Table 1

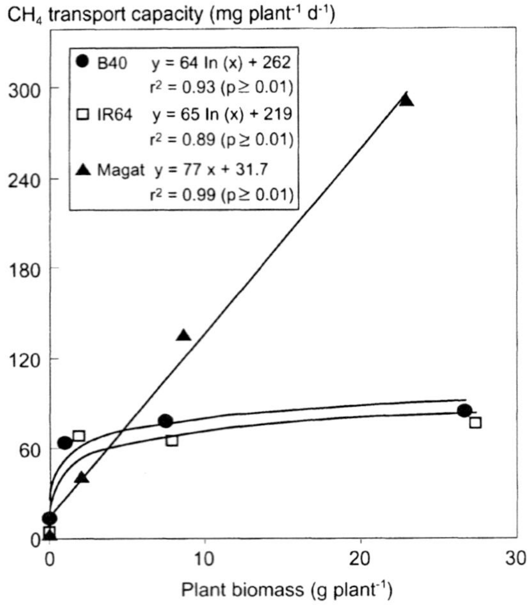

Figure 2. Relationship between plant biomass and CH_4 transport capacity (MTC) of B40 (tall), IR64 (high-yielding dwarf), and Magat (hybrid) rice cultivars (composite data from four growth stages). MTC determined as outlined in Table 1

aboveground biomass and CH_4 emission by rice plants observed under field conditions (Sass et al., 1991; Shalini et al., 1997) may be due to the increase in MTC of plants during the early vegetative growth and possibly determined by plant-derived CH_4 production during later growth (Minoda & Kimura, 1994; Wang et al., 1997).

In the case of the hybrids APHR 2 and Magat, a strong positive relationship between MTC and root + shoot biomass (r =0.672, $p \geq 0.05$) and total plant biomass including grain (r =0.849, $p \geq 0.01$) was observed. It may therefore be concluded that radial development of aerenchyma continues with plant growth in these hybrids, resulting in enhanced MTC.

Effect of morphological characteristics on CH_4 transport capacity

The number of fully developed tillers was largest in APHR 2, medium in traditional tall and dwarf cultivars, and lowest in NPT cultivars (Table 2). Traditional cultivars were tallest (124-136 cm), whereas the tiller height of all other cultivars ranged from 77 to 95 cm at maturity. The number of developed tillers at three growth stages (PI, flowering, and maturity) showed a significantly positive relationship with MTC ($p \geq 0.05$) in all 10 cultivars with r values ranging from 0.690 to 0.763 (Table 3). Underdeveloped tillers, which keep on emerging at different times during plant growth and often end up unproductive, failed to show any relationship with MTC in any of the cultivars including APHR 2. Consequently, the combined number of developed and underdeveloped tillers did not exhibit a significant relationship with MTC in all cultivars. When data from different cultivars were pooled in terms of tall, dwarf, or NPT plants, the relationship between tiller number and MTC was insignificant, indicating that each cultivar has its own typical size and pattern of aerenchyma formation that develops as plant growth proceeds.

MTC measurements of plants from B40 and IR72 cultivars with four, six, and eight tillers at PI (Figure 3) further showed that the influence of tiller density on MTC is not only due to radial growth. It is presumably due to the proportional enhancement in channels/out-

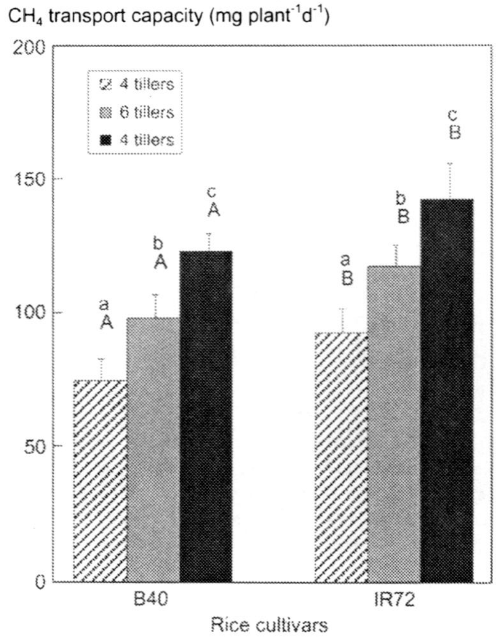

Figure 3. Methane transport capacity (MTC) of plants of B40 (tall) and IR72 (high-yielding dwarf) rice cultivars with different tiller number at panicle initiation. Data shown are means + SD of three replicate plants each measured in triplicate. Different small letters indicate significant differences between tillers ($p > 0.05$). Different capital letters indicate significant differences between cultivars with a particular tiller number ($p > 0.05$). MTC was determined as outlined in Table 1

Table 2. Growth parameters and morphological characteristics of traditional tall cultivars (Dular, B40, and Intan), high-yielding dwarf cultivars (IR72, IR52, IR64, and PSBRc 20), new plant type cultivars (IR65598 and IR65600), and a hybrid (APHR2) at panicle initiation (PI), flowering (Fl) and maturity (Mt)

Cultivar	Growth stage	Root biomass (g plant^{-1})	Shoot biomass (g plant^{-1})	Grain yield (g plant^{-1})	Developed tillers (no. plant^{-1})	Under-developed tillers (no. plant^{-1})	Developed tiller height (cm)	Under-developed tiller height (cm)
Dular	PI	5.2 ± 0.9[a]	12.9 ± 1.2	-	5.3 ± 1.2	2.3 ± 0.6	115 ± 4	68 ± 25
	Fl	5.6 ± 1.4	18.0 ± 1.9	-	6.3 ± 0.6	1.3 ± 0.6	123 ± 0	71 ± 15
	Mt	6.1 ± 1.2	16.4 ± 2.3	13.8 ± 0.8	4.7 ± 0.6	2.0 ± 0.0	125 ± 7	86 ± 5
Intan	PI	13.5 ± 2.6	12.6 ± 0.3	-	9.3 ± 1.5	2.3 ± 1.5	93 ± 7	53 ± 2
	Fl	15.2 ± 3.7	26.4 ± 1.1	-	7.0 ± 2.6	2.0 ± 0.0	124 ± 1	78 ± 0
	Mt	15.9 ± 1.0	34.4 ± 1.3	18.3 ± 0.8	6.7 ± 1.5	0.0 ± 0.0	136 ± 9	111 ± 6
B40	PI	3.2 ± 1.1	12.6 ± 0.7	-	6.0 ± 1.0	2.3 ± 0.6	101 ± 9	53 ± 8
	Fl	4.6 ± 0.6	18.9 ± 0.9	-	7.0 ± 1.0	1.3 ± 0.6	121 ± 2	59 ± 9
	Mt	4.8 ± 0.7	17.5 ± 1.3	12.1 ± 1.3	5.3 ± 1.2	1.7 ± 0.6	124 ± 2	68 ± 7
IR72	PI	5.4 ± 2.0	10.5 ± 1.2	-	7.0 ± 1.0	2.3 ± 0.6	74 ± 4	48 ± 1
	Fl	7.1 ± 1.9	24.7 ± 2.3	-	6.3 ± 1.5	2.3 ± 1.5	91 ± 4	63 ± 7
	Mt	7.0 ± 1.0	16.7 ± 2.2	16.9 ± 1.0	5.0 ± 1.0	2.7 ± 0.6	87 ± 1	65 ± 3
IR52	PI	6.7 ± 1.4	10.9 ± 2.1	-	7.0 ± 1.0	3.7 ± 0.6	74 ± 2	50 ± 2
	Fl	6.3 ± 0.5	21.5 ± 3.6	-	6.7 ± 1.2	1.7 ± 0.6	94 ± 3	52 ± 2
	Mt	6.1 ± 1.5	14.9 ±0.6	13.0 ± 2.2	6.0 ± 0.0	2.3 ± 0.6	91 ± 4	55 ± 8
IR64	PI	5.9 ± 0.5	10.3 ± 0.6	-	8.7 ± 1.2	2.0 ± 1.0	69 ± 1	42 ± 7
	Fl	7.5 ± 2.8	22.2 ± 0.9	-	8.0 ± 1.7	2.0 ± 1.7	87 ± 1	52 ± 7
	Mt	7.6 ± 1.3	16.3 ± 1.7	14.6 ± 1.8	6.3 ± 1.5	2.7 ± 1.2	82 ± 4	59 ± 9
PSBRc20	PI	3.2 ± 1.5	7.2 ± 0.8	-	7.0 ± 1.0	2.7 ± 0.6	75 ± 7	44 ± 5
	Fl	5.6 ± 0.4	19.9 ± 1.1	-	5.7 ± 1.5	1.7 ± 0.6	87 ± 2	62 ± 10
	Mt	4.8 ± 0.2	11.9 ± 0.8	13.9 ± 1.7	5.7 ± 0.6	1.0 ± 1.0	85 ± 5	56 ± 7
IR65598	PI	4.7 ± 0.5	8.2 ± 1.8	-	3.5 ± 2.1	1.5 ± 0.7	74 ± 0	53 ± 8
	Fl	6.1 ± 0.8	15.3 ± 2.8	-	3.7 ± 1.5	0.0 ± 0.0	89 ± 1	0 ± 0
	Mt	5.4 ± 0.4	13.5 ± 2.8	13.9 ± 1.7	3.3 ± 0.6	0.3 ± 0.6	93 ± 3	60 ± 0
IR65600	PI	3.2 ± 0.5	7.7 ± 0.5	-	4.7 ± 1.5	3.3 ± 0.6	80 ± 3	45 ± 4
	Fl	7.0 ± 0.4	22.5 ± 3.6	-	4.7 ± 1.2	0.7 ± 0.6	94 ± 2	60 ± 0
	Mt	9.5 ± 3.0	21.0 ± 2.9	10.0 ± 1.1	4.0 ± 0.0	0.3 ± 0.6	95 ± 2	78 ± 0
APHR2	PI	5.1 ± 0.8	12.3 ± 2.6	-	8.3 ± 2.1	6.7 ± 1.2	62 ± 2	51 ± 1
	Fl	6.4 ± 1.6	14.7 ± 6.4	-	11.7 ± 1.5	4.0 ± 3.5	72 ± 1	55 ± 0
	Mt	6.3 ± 0.5	16.3 ± 0.9	18.3 ± 2.3	12.3 ± 5.9	4.7 ± 1.5	77 ± 1	55 ± 5
LSD (0.05)		1.4	2.1	2.3	1.3	1.0	4	18

[a]Standard deviation.

Table 3. Correlation coefficients of CH_4 transport capacity of rice plants of traditional tall cultivars (Dular, B40, and Intan), high-yielding dwarf cultivars (IR72, IR52, IR64, and PSBRc 20), new plant type cultivars (IR65598 and IR65600), and a hybrid (APHR2) with growth parameters and morphological characteristics

Cultivar	Root biomass	Shoot biomass	Root + shoot biomass	Root + shoot + grain biomass	Developed tiller number	Under-developed tiller number	Total tiller number	Developed tiller height	Under-developed tiller height
Dular[a]	-0.125	-0.122	-0.172	-0.857[b]	0.749[c]	-0.127	0.611	-0.359	-0.614
Intan	0.348	-0.083	0.008	-0.110	0.718[c]	-0.068	0.519	-0.001	-0.355
B40	-0.128	-0.037	-0.064	-0.715[c]	0.763[c]	-0.073	0.798[c]	-0.399	-0.364
Pooled[d]	-0.397[c]	-0.287	-0.368[a]	-0.549[b]	0.092	0.046	0.106	-0.108	-0.144
IR72	-0.007	0.039	0.032	-0.700[c]	0.747[c]	-0.028	0.549	-0.325	-0.452
IR52	0.250	0.089	0.140	-0.691[c]	0.713[c]	0.156	0.542	-0.333	-0.359
IR64	-0.083	-0.050	-0.067	-0.745[c]	0.690[c]	-0.097	0.466	-0.243	-0.506
PSBRc20	0.238	0.147	0.170	-0.304	0.751[c]	0.108	0.580	0.103	0.282
Pooled	0.312	0.212	0.260	-0.255	0.439[b]	0.165	0.416[c]	-0.167	0.062
IR65598	0.240	-0.189	-0.129	-0.653	0.716[c]	0.215	0.68[c]	-0.584	-0.085
IR65600	0.167	-0.101	-0.022	0.075	0.744[c]	-0.200	0.230	0.092	-0.570
Pooled	-0.086	-0.337	-0.282	-0.391	0.215	-0.234	-0.026	-0.479	-0.336
APHR 2	0.322	0.652	0.672[c]	0.849[b]	0.738[c]	-0.037	0.701[c]	0.723[c]	0.160

[a] In each cultivar, data for panicle initiation, flowering, and maturity were used (n = 9 plants). [b] Significant at 0.01 probability level. [c] Significant at 0.05 probability level. [d] Pooled data from each category of cultivars (n = 36, 36, and 18 plants for tall, high-yielding dwarf, and new plant type cultivars, respectively)

lets of aerenchyma for the upstream transport of CH_4 from the base to the sites of release to the atmosphere. In an earlier field study, Watanabe et al. (1995) could not find any relationship between CH_4 emission rates and tiller number. However, under greenhouse conditions with constant supply of CH_4 to plant roots, Wang et al. (1997) reported that tiller number is positively related to CH_4 emission rates. The present results show that tiller number can become a major controlling factor of plant-mediated CH_4 transport in widely different cultivars. Therefore, plants with less number of tillers would minimize CH_4 emission from the soil to the atmosphere. For example, NPT cultivars that had the minimum number of tillers, a high proportion of productive tillers, large panicles on each tiller, and strong stems exhibited low MTC.

Tiller height showed neither relationship nor negative nonsignificant relationship with MTC in all cultivars except APHR 2 (Table 3). Underdeveloped tiller height also did not show any significant relationship with MTC in all cultivars. These observations are consistent with earlier findings of no correlation between CH_4 emission rates and shoot length (Watanabe et al., 1995). The significantly positive relationship between developed tiller height and MTC in APHR 2 may suggest a proportionally enhanced continuity of aerenchyma channels with increasing plant height in this cultivar. At maturity, hybrid plants were very sturdy and upright, whereas plants of other cultivars exhibited weak and bent shapes. Further investigations, such as a study of anatomical appearance of pattern and distribution of aerenchyma in different parts of rice plants and microscopic analysis including measurement of aerenchyma areas are needed to enhance our understanding of the differences in aerenchyma development in different cultivars.

In summary, there are large differences in MTC of rice plants during different growth stages and among cultivars. Root and aboveground biomass determines MTC during initial vegetative growth in all cultivars, except in hybrids where it is directly related to growth during the entire plant development. Tiller number is a major controlling factor of plant-mediated CH_4 transport rates in widely different cultivars. Therefore, plants with less biomass and fewer tillers could minimize CH_4 emission. Identification of these plant traits could help efforts in breeding for high-yielding rice plants with low CH_4 emission potential. For example, cultivation of NPT cultivars that have the minimum tiller number, higher proportion of productive tillers and larger panicles (more grains) on each tiller, and that can transport less amounts of CH_4 seems to be an an economically feasible, environmentally sound, and promising approach to mitigate CH_4 emissions from rice fields.

Acknowledgements

We thank Sonny Pantoja for his help with the greenhouse experiments. This research was funded by the German BMZ/GTZ under the project "Reduction of methane emission from rice fields by screening for low methane transport capacity (No. 95.7860.0-001.05)."

References

Adhya TK, Rath AK, Gupta PK, Rao VR, Das SN, Parida KM, Parashar DC & Sethunathan N (1994) Methane emission from flooded rice fields under irrigated conditions. Biol Fertil Soils 18: 245-248.

Aulakh MS, Bodenbender J, Wassmann R & Rennenberg H (2000) Methane transport capacity of rice plants. I. Influence of CH_4 concentration and growth stage analyzed with an automated measuring system. Nutr Cycling Agroecosyst (this issue)

Bartolome VI, Casumpang RM, Ynalvez MAH, Olea AB & McLaren CG (1999) IRRISTAT for Windows - statistical software for agricultural research. Biometrics, International Rice Research Institute, Makati City, Philippines.

Butterbach-Bahl K, Papen H & Rennenberg H (1997) Impact of gas transport through rice cultivars on methane emission from paddy fields. Plant Cell Environ 20: 1175-1183

Cicerone RJ, Shetter JD (1981) Sources of atmospheric methane: measurements in rice paddies and a discussion. J Geophys Res 86: 7203-7209

Cochran WG & Cox GM (1950). Experimental Designs. John Wiley and Sons, Inc, New York

Gomez KA (1972) Techniques for field experiments with rice. International Rice Research Institute, Makati City, Philippines

Holzapfel-Pschorn A, Conrad R & Seiler W (1986) Effects of vegetation on the emission of methane from submerged paddy soil. Plant Soil 92: 223-391

Holzapfel-Pschorn A, Seiler W (1986) Methane emission during a cultivation period from an Italian rice paddy. J Geophy Res 91: 11803-11814

IPCC (Intergovernmental Panel on Climate Change) (1992) Climate Change: The Supplementary Report to the IPCC Scientific Assessment Houghton JT, Callander BA, Varney SK (eds) IPCC, Cambridge University Press, Cambridge

IPCC (Intergovernmental Panel on Climate Change) (1995) Climate Change: The Science of Climate Change Houghton JT, Meira Filho LG, Callander BA, Harris N, Kattenberg A & Maskell K (eds) IPCC, Cambridge University Press, Cambridge

Kesheng S & Zhen L (1997) Effect of rice cultivars and fertilizer management on methane emission in a rice paddy in Beijing. Nutr Cycling Agroecosyst 49: 139-146

Leon JC & Carpena AL (1995) Morphology-based diversity analysis of improved irrigated lowland rice (*Oryza sativa* L.) varieties in the Philippines. Philipp J Crop Sci 20(2): 113-121

Lin E (1993) Agricultural techniques: factors controlling methane emissions. In: Gao L, Wu L, Zheng D & Ham X (eds) Proceedings of the International Symposium on Climate Change, Natural Disasters, and Agricultural Strategies, p120-126. China Meteorological Press, Beijing

Minoda T & Kimura M (1994) Contribution of photosynthesized carbon to the methane emitted from paddy fields. Geophys Res Lett 21: 2007-2010.

Mitra S, Jain MC, Kumar S, Bandyopadhya SK & Kalra N (1999) Effect of rice cultivars on methane emission. Agric Ecosyst Environ 73: 177-183

Neue HU & Sass R (1998) The budget of methane from rice fields. IGACtivities 17: 3-11

Sass RL, Fisher FM, Harcombe PA & Turner FT (1991) Mitigation of methane emission from rice fields: effect of incorporated rice straw. Global Biogeochem Cycles 5: 275-288

Shalini S, Kumar S & Jain MC (1997) Methane emission from two Indian soils planted with different rice cultivars. Biol Fertil Soils 25: 285-289

Sigren LK, Byrd GT, Fisher FM & Sass RL (1997) Comparison of soil acetate concentrations and methane production, transport, and emission in two rice cultivars. Global Biogeochem Cycles 11: 1-14

Wang B, Neue HU & Samonte HP (1997) Effect of cultivar difference ('IR72', IR65598' and 'Dular') on methane emission. Agric Ecosyst Environ 62: 31-40

Wassmann R & Aulakh MS (2000) The role of rice plants in regulating mechanisms of methane emissions. Biol Fertil Soils 31:20-29

Wassmann R, Neue HU, Alberto MCR, Lantin RS, Bueno C, Llenaresas D, Arah JRM, Papen H, Seiler W & Rennenberg H (1996) Fluxes and pools of methane in wetland rice soils with varying organic inputs. Environ Monitor Assess 42: 163-173

Watanabe A, Kajiwara M, Tashiro T & Kimura M (1995) Influence of rice cultivar on methane emission from paddy fields. Plant Soil 17: 51-56

Nutrient Cycling in Agroecosystems 58: 377–388, 2000.

Methane production, oxidation, and emission from Indian rice soils

N. Sethunathan[1*], S. Kumaraswamy[1], A. K. Rath[1], B. Ramakrishnan[2], S. N. Satpathy[3], T. K. Adhya[3] & V. R. Rao[3]

[1]*Division of Microbiology, Indian Agricultural Research Institute, New Delhi 110012, India;* [2]*Max-Planck-Institut für Terrestrische Mikrobiologie, Karl-von-Frisch-Str., D-35043 Marburg, Germany;* [3]*Central Rice Research Institute, Cuttack 753006, India (*Present address: CSIRO Land and Water, PMB2, Glen Osmond, SA 5064, Australia)*

Abstract

Experiments were conducted to investigate methane (CH_4) production, oxidation, and emission from flooded rice soils. Incorporation of green manure (*Sesbania rostrata*) into rice fields led to a several-fold increase in CH_4 emission. A stimulatory effect of organic sources on CH_4 production in soil samples was noticed even under nonflooded conditions. Addition of rice straw at 1% (w/w) to nonflooded soil samples held at -1.5 MPa effected a 230-fold increase in CH_4 production over that in corresponding unamended soil samples at 35 d, as compared with a threefold increase in rice straw-amended soil over that in unamended soil under flooded conditions. In a study involving two experimental field sites differing in water regimes but planted to the same rice cultivar (cv Gayatri) and fertilized with prilled urea at 60 kg N ha^{-1}, the field plots with deep submergence of around 30 cm (site I) emitted distinctly more CH_4 than did the plots with continuous water depth of 3-6 cm (site II). Likewise, in another incubation study, CH_4 production in flooded soil samples increased with a progressive increase in standing water column from 5 mm to 20 mm. Application of carbamate insecticide, carbofuran, at 2 kg ai ha^{-1} to rice fields retarded CH_4 emission through enhanced CH_4 oxidation. Hexachlorocyclohexane was found to inhibit CH_4 emission. The results suggest the need for extensive research efforts to develop technologies with dual objectives of environmental protection and crop productivity.

Introduction

The increasing concentrations of trace gases such as carbon dioxide, methane (CH_4), and nitrous oxide in the earth's atmosphere are of global concern because of their potential influence on atmospheric chemistry and climate pattern (Houghton et al., 1996). Studies conducted between 1978 and 1988 indicate that atmospheric CH_4 concentration is increasing at about 1.0% yr^{-1} (Crutzen, 1991). However, the recent trend in atmospheric CH_4 concentration shows that the rate of increase has slowed down, with an annual increase of 0.7% (Steele et al., 1992; Khalil & Rasmussen, 1993). The anthropogenic sources of CH_4 include rice fields, domestic ruminants, biomass burning, landfills, coal mining, oil and natural gas flaring, animal wastes and domestic sewage (Crutzen, 1991; Khalil & Rasmussen, 1991). Rice fields alone may account for about 15–20% of global atmospheric CH_4 budget (US-EPA 1990; Minami & Neue, 1994; Neue et al., 1995). Currently, global research is intended at estimating the source strength of rice fields and identifying as well as developing mitigation technologies for CH_4 emission from flooded rice fields.

Extensive field measurements in Spain (Seiler et al., 1984), Italy (Holzapfel-Pschorn & Seiler, 1986; Schütz et al., 1989), Japan (Yagi & Minami, 1990; Yagi et al., 1994), the United States of America (Sass et al., 1984), India (Mitra, 1992; Adhya et al., 1994; Parashar et al., 1996; Sethunathan et al., 1998), China (Chen et al., 1993; Wassmann et al., 1993a; Shao & Li, 1997), the Philippines (Neue et al., 1994), and Thailand (Minami, 1994; Yagi et al., 1994) indicate that there are large temporal variations in CH_4 fluxes and that the flux is critically dependent upon several factors including climate, characteristics of soils, and agricultural practices. The estimates of global rice field CH_4 emission remain rather uncertain. The Intergovernmental Panel on Climate Change estimated the global emission rate from rice fields at 60 ± 40 Tg CH_4 yr^{-1} (Houghton et al., 1996). Lelieveld et al. (1998) estimated a total CH_4 emission of agricultural origin to be

260 ± 115 Tg yr^{-1} and rice fields of 80 ± 50 Tg yr^{-1}. The world's annual rough rice production must increase to meet the demand of the human population, from the present 520 million t to at least 880 million t by 2025 as rice is a staple food for more than half of the world's population (Lampe, 1995). Such intensification of rice cultivation by adopting new cultivation practices may increase CH_4 emission (Anastasi et al., 1992; Neue et al., 1995). Although a flooded soil planted to rice is predominantly anaerobic, surface soil, rhizosphere and standing water are in oxidized state (Ponnamperuma, 1972). Not all CH_4 produced in the anaerobic zones is emitted to the atmosphere. In the oxic zones of flooded soils, as much as 80% of CH_4 produced in the anaerobic soil layers is oxidized (Conrad & Rothfuss, 1991). Methane-oxidizing activity decreases with increasing depth in flooded soil (Kumaraswamy et al., 1997a). The production and oxidation of CH_4 in flooded rice soils are controlled by many soil, plant, and microbial factors. Once the soil is disturbed, CH_4 oxidation is inhibited for months or years (Mosier et al., 1991). Undisturbed soils have higher CH_4 uptake capacity than agricultural soils (Boeckx et al., 1998). There are, however, less information on the influence of commonly used cultural practices on CH_4 production, oxidation, and emission from tropical Indian rice fields. Studies were initiated in our Institute to identify the cultural practices that show mitigation potential in flooded rice fields. Some of the results are summarized in this report and more details of this work together with the experimental procedures and the analytical methods used are provided elsewhere (Sethunathan, 1997; Rao, 1998; Kumaraswamy, 1998; Rath, 1998).

Organic matter application

It has been reported that addition of fresh organic sources to the rice soil increases the availability of methanogenic substrates and thereby enhances CH_4 production and emission (Neue, 1993). Application of organic sources such as rice straw, *Azolla*, blue-green algae, green manure (leguminous and nonleguminous), animal and human excreta to rice soils is one of the common cultural practices. In a laboratory incubation experiment, the effect of various organic sources on CH_4 production in soil samples was determined under flooded conditions (Satpathy, 1997). The organic sources used were rice straw, cellulose, *Azolla* (a water fern harboring a nitrogen-fixing blue-green alga, *Anabaena*) compost, blue-green algae (BGA) compost, farmyard manure (FYM) and green manure (GM). The rice straw (aboveground parts only), *Azolla* compost (*Azolla pinnata* and *Azolla mexicana* were composted in a pit for 7 d after harvest), BGA compost (composted in a pit for 7 d from BGA mats with dominant species of *Nostoc, Anabaena, Aulosira* and *Tolypothrix*), FYM (compost of cattle wastes), and GM (leaves and tender twigs of leguminous plant *Sesbania rostrata*) were air-dried in shade, crushed, and sieved to pass through a 2-mm mesh and stored in polyethylene bags at room temperature. At regular intervals, net production in organic-amended and unamended samples during 40-d incubation was monitored by gas chromatography as described earlier by Ramakrishnan et al. (1995). Organic amendments effected a several-fold increase in CH_4 production in alluvial soil under flooded conditions over that of corresponding unamended controls during 40-d incubation, irrespective of organic sources used (Table 1). In general, the stimulatory effect of organic sources on CH_4 production was evident even at 10–15 d of submergence. The stimulation of CH_4 production by organic sources under flooded conditions generally followed the order: GM> cellulose > rice straw > BGA compost > *Azolla* compost > FYM> unamended control. It is interesting to note that composted organic sources (*Azolla* compost, BGA compost and FYM) were less effective than the fresh organic sources (rice straw, cellulose, and GM) in stimulating the production of CH_4 in alluvial soil samples. Composted organic sources are known to support low production of CH_4 when compared with the fresh organic sources (Debnath et al., 1996).

In a followup field study, the effect of certain organic manures (GM, FYM, or poultry manure) applied in combination with inorganic N fertilizer (urea) on CH_4 emission from flooded rice field plots was determined. Nitrogen fertilizers were applied at 60 kg N ha^{-1}, 50% as basal and 50% at the tillering stage of the crop. Basal (50%) application of fertilizers included GM, FYM, poultry manure, and prilled urea. Green manure (*S. rostrata*), with organic carbon, 42.22%; total N, 4.284%; grown in neighboring plots, were harvested, chopped, and incorporated into the field soil. The organic carbon and total N in FYM were 14.6% and 1.7%, respectively. Poultry manure collected from a poultry farm contained organic carbon of 26.1% and total N of 1.7%. Organic manures were incorporated into the field soil 1 d before transplanting and prilled urea was broadcast onto the standing water of the flooded rice fields. The remaining 50% N was applied as prilled urea by surface-

Table 1. Effect of added organic sources on CH_4 production (mg CH_4 g^{-1} soil) in flooded alluvial soil samples under laboratory incubation (Satpathy, 1997)[a]

Source	Days after flooding						
	10	15	20	25	30	35	40
Control	-	0.3c	2d	21e	31f	32f	20f
Rice straw	6a	84b	124b	257a	269a	182c	169c
Cellulose	7a	18c	125b	212b	229b	280b	213b
Azolla compost	5a	23c	82c	85d	117d	103d	109d
BGA compost	7a	69b	120b	162c	177c	176c	147c
FYM	1a	10c	22d	24e	72e	77e	82e
GM	1a	181a	216a	243a	282a	338e	303e

[a]Organic sources were added to the soil at 1% (w/w) level just before flooding. Mean of five replicates. BGA = blue green algae; FYM = farmyard manure; GM = green manure (*Sesbania rostrata*). In a column, means followed by a common letter are not significantly different at the 5% level by Duncan's multiple range test (DMRT)

Table 2. Effect of organic manure and urea application on CH_4 emission (mg m^{-2} h^{-1})[a] from flooded field plots planted to rice (cv CR 1002) (Satpathy, 1997)

Treatment	Days after transplanting					
	30	40	60	80	95	115
Control	4.1b	7.8d	20.9a	8.7c	5.8c	42.4b
Urea (30+30 kg N ha^{-1})	12.0a	9.1d	24.6a	4.1d	6.9c	82.5a
GM (30 kg N ha^{-1}) + urea (30 kg N ha^{-1})	7.4a	32.5a	26.7a	61.7a	29.1a	109.7a
FYM (30 kg N ha^{-1}) + urea (30 kg N ha^{-1})	10.4a	17.8b	18.4a	29.6b	18.3b	7.3d
PM (30 kg N ha^{-1}) + urea (30 kg N ha^{-1})	7.7a	13.9bc	20.2a	6.3c	6.9c	15.5c

[a]Mean of four replicate values. In combined application of organic manure and urea, organic manures were applied to the field as basal and urea was applied at tillering stage (45 d after transplanting) of the crop. In treatment with urea alone, urea (60 kg N ha^{-1}) was applied in two equal splits, 50% as basal and 50% at tillering stage (45 d after transplanting) of the crop. GM = green manure (*Sesbania rostrata*); FYM = farmyard manure; PM = poultry manure. Growth stages: 60, 80, 95 and 115 d of sampling correspond to maximum tillering, panicle initiation, flowering and maturity stages of the crop, respectively. In a column, means followed by a common letter are not significantly different at the 5% level by DMRT

broadcast to all the plots at the tillering stage of the rice crop (45 d after transplanting). Methane emission from the flooded rice fields was monitored in the morning for 30 min and in the afternoon for 30 min, at different growth stages of the rice plant, employing the manual closed chamber method as described earlier (Adhya et al.,1994). Results indicated that integrated use of organic (30 kg N ha^{-1}) and chemical (30 kg N ha^{-1}) fertilizers led to a distinct increase in CH_4 emission from rice fields over that of control (Table 2). Urea, applied alone at 60 kg N ha^{-1}, also effected a significant increase in CH_4 emission. Among the organic sources used in combination with urea, GM was the most stimulatory to CH_4 emission almost throughout the cropping season. The stimulatory effect of GM was noticed even at 115 d after transplanting. Denier van der Gon and Neue (1995) also reported higher CH_4 efflux from GM-amended field plots than in field plots treated with urea. Farmyard manure effected only a marginal increase in CH_4 emission over that of control. Interestingly, despite higher organic carbon content (26.2%), poultry manure inhibited CH_4 emission. Methane emission reached the peak during the maturity stage of the crop in almost all treatments. It may be mentioned that also in laboratory incubation studies, GM effected a more pronounced production of CH_4 than did other organic sources (Table 1). The low emission of CH_4 with poultry manure might be due to its high sulfur content (1.3% total sulfur) when compared with the relatively low sulfur content of GM (0.5% total sulfur), FYM (0.6% total sulfur), and urea (0.001% sulfate). There is evidence that methanogenesis is inhibited in sulfate-rich anaerobic environments because of the competition of sulfate-reducers with methanogens for common

Table 3. Methane production (mg g^{-1} soil) in alluvial soil samples with different water potentials under laboratory incubation conditions (Rath et al., 1999a)[a]

Water potential (MPa)	Days of incubation							
	5	10	15	20	25	30	35	40
-1.5	3b	3d	9d	11d	14d	23d	49d	205d
-0.01	3b	16c	20c	27[c]	32c	48c	352c	2537c
0 (saturated)	3b	28b	44b	610b	1838b	6311b	7603b	8764b
Flooded	12a	43a	294a	1187a	8055a	32852a	37400a	43413a

[a]Mean of three replicate observations. In a column means followed by a common letter are not significantly different at the 5% level by DMRT

Table 4. Methane production (mg g^{-1} soil) in alluvial soil samples under different water potential, amended with rice straw (1% w/w) and incubated under laboratory conditions (Rath et al., 1999a)[a]

Water potential (MPa)	Days of incubation							
	5	10	15	20	25	30	35	40
-1.5	0.1c	1.3c	25.4c	83.5b	94.4c	52.5c	50.0b	46.5[b]
-0.01	0.3b	4.9b	30.8bc	94.1b	125.4bc	57.0c	55.7b	48.9b
0	0.4b	5.8b	38.4ab	76.5c	175.0ab	116.1b	70.1b	61.4b
Flooded	1.0a	9.9a	47.7a	175.7a	21.8a	172.2a	150.4a	128.3a

[a]Mean of three replicate observations. Rice straw added at 1% (w/w). In a column, means followed by a common letter are not significantly different at the 5% level by DMRT.

substrates such as H_2 and acetate (Lovley & Klug, 1983; Achtnich et al., 1995) and the toxicity of sulfide formed during anaerobiosis to methanogenesis (Winfrey & Zeikus, 1979). There is also report of low production of CH_4 in laboratory-incubated soil amended with chicken manure (Wang & Patrick, 1995).

Influence of floodwater regimes

Methane production in an alluvial soil, unamended or amended with rice straw (1% w/w), was examined under nonflooded and flooded conditions during a 40-d incubation in closed vacutainer tubes (Rath et al., 1999a). Methane production in the alluvial soil samples, not amended with rice straw, was negligible at –1.5 MPa during the 40-d incubation period but increased progressively with an increase in water potential to –0.01 MPa, 0 MPa (saturated), and flooded (1:1.25 soil-water ratio) conditions (Table 3). Methane production in unamended soil increased, relative to that at –1.5 MPa, 12-fold at -0.01 MPa, 40-fold at 0 Mpa, and 200-fold under flooded conditions. According to Ramakrishnan et al. (1995), CH_4 production in alluvial soil (same as that used in this study) held under nonflooded conditions (60% water-holding capacity) was less than that in flooded soil. In general, CH_4 production is low in nonflooded soils as the redox status of nonflooded soils is not favorable for methanogenic activities (van Cleemput et al., 1983). Addition of rice straw (1% w/w) to soil samples effected a several-fold increase in CH_4 production at all water potential levels relative to that of the respective unamended soil (Table 4). Net CH_4 production in rice straw-amended soil, as in unamended soil, distinctly increased with an increase in the soil moisture level. What is particularly interesting is the substantial increase in the CH_4 production in nonflooded rice straw-amended soil, even when held at –1.5 MPa. Thus, for rice straw-amended soil held at –1.5 MPa, a 230-fold increase in CH_4 production relative to that of the corresponding unamended soil at 25 d was recorded, as compared with a threefold increase in the rice straw-amended soil relative to that of unamended soil under flooded conditions during the corresponding period. However, CH_4 production in rice straw-amended soil was three times higher under flooded conditions than at -1.5 MPa.

Table 5. Methane emission (mg pot^{-1} d^{-1})from nonflooded and flooded soil (in pots) planted to rice (cv IR72), unamended or amended with rice straw (Rath, 1998)[a]

Treatment	Days after transplanting (DAT)							
	20	25	40	50	65	75	85	95
Nonflooded	0.2c	0.3c	0.3b	0.4b	0.9b	0.5d	0.2d	0.5c
Nonflooded + rice straw (1% w/w)	3.4b	1.1b	0.6b	1.0b	1.8a	2.5b	1.6c	3.2b
Flooded	0.4c	0.7b	0.4b	0.9b	0.9b	1.5c	7.2b	3.1b
Flooded + rice straw (1% w/w)	22.2a	27.9a	16.1a	4.3a	2.6a	22.9a	26.8a	9.5a

[a]Mean of four replicate observations. In a column, means followed by a common letter are not significantly different at the 5% level by DMRT

In a followup greenhouse experiment, CH_4 efflux from nonflooded and flooded alluvial soil samples in pots, with or without rice straw addition, planted to rice was examined (Rath, 1998). Addition of rice straw to potted soil planted to rice enhanced the emission of CH_4 under both nonflooded and flooded conditions by 7-8 fold over that of the respective unamended control (Table 5). Flooded soils, rice straw-amended or unamended, emitted distinctly more CH_4 than corresponding nonflooded pots. Cumulative CH_4 emission followed the order: flooded rice straw-amended (1,040 mg CH_4 pot^{-1}) > nonflooded rice straw-amended (112 mg CH_4 pot^{-1}) ≥ flooded unamended (104 mg CH_4 pot^{-1}) > nonflooded unamended (24 mg CH_4 pot^{-1}). Interestingly, cumulative CH_4 emission from nonflooded soil amended with rice straw was almost on a par with that of flooded soil not amended with rice straw. There was a distinct increase in CH_4 flux from rice straw-amended and flooded soils as compared with that of other treatments. Decomposition of rice straw in predominantly anaerobic flooded soil can lead to the accumulation of acetate, a major, but a transitory intermediate (Rao & Mikkelsen, 1977). Acetate is the important substrate for methanogens in the flooded soils (Takai, 1970). About 80% of CH_4 is formed from acetic acid in rice soils (Achtnich et al., 1995). This would explain the substantial accumulation of CH_4 in rice straw-amended pots. What is particularly interesting is the fact that addition of rice straw distinctly enhanced CH_4 efflux even under greenhouse conditions not only from flooded soils but also from nonflooded soils. These results from pot culture experiments support the data generated under laboratory incubation (Rath et al., 1999a). Substantial production of CH_4 in nonflooded soils amended with rice straw probably occurs at anaerobic microsites that can be abundant in nonflooded soils (Sextone et al., 1985) and more so in nonflooded soil amended with organic sources. Thus, application of organic amendments, a conventional practice in rice culture, will have a significant influence on CH_4 emission from both flooded and nonflooded soils. Continuous flooding was found to emit more CH_4 than alternate flooding and drying in a greenhouse experiment, and single or multiple drainage retarded CH_4 emission from pots planted to rice (Mishra et al., 1997). A single midseason drainage reduces seasonal CH_4 emission rates by about 50% (Kimura, 1992; Sass et al., 1992). Thus, floodwater management is one of the important mitigation strategies. Intermittent irrigation and mid-season drainage retard CH_4 emission from rice fields but increase the emission of nitrous oxide, another important greenhouse gas (Neue, 1993; Wassmann et al., 1993b). Moreover, in rainfed lowland rice as in eastern and northeastern India, drainage of water from rice fields is virtually impossible due to high water table. Hence, there is a need to evaluate these mitigation strategies before adopting them as technologies.

Effect of chemical fertilizers and floodwater depth

The effects of fertilizer management and water regime on CH_4 emission were studied in two sets of field plot experiments (Rath et al., 1999b). Table 6 presents the physicochemical properties of the soil at both sites. The experimental plots of rice field with 30-cm water depth (site I) were treated with prilled urea (60 kg N ha^{-1}), prilled urea (60 kg N ha^{-1}) coated with Nimin (a nitrification inhibitor; neem triterpenes, Godrej Agrovet Limited, Bombay), and urea supergranules (60 kg N ha^{-1}).

Table 6. Physicochemical characteristics of soil samples from rice field plots (sites I & II) of the Institute's experimental farm (Rath et al., 1999b)

Soil characteristic	Site I	Site II
pH (1:2 soil : water ratio)	6.40	7.63
Electrical conductivity (dS m^{-1}, 1:2 soil : water ratio)	0.78	2.03
Water-holding capacity (%)	50.0	47.1
Organic carbon (%)	0.57	0.51
Total nitrogen (%)	0.089	0.082
Cation exchange capacity (cmol (+) kg^{-1})	14.0	11.4
Bulk density (g cm^{-3})	1.2	1.3
Particle size distribution		
Clay (%)	22	9
Silt (%)	12	10
Sand (%)	66	81

Table 7. Effect of fertilizer management practices on CH_4 efflux (mg m^{-2} h^{-1}) from rice field plots with 30-cm water depth (site I), planted to cv Gayatri (Rath et al., 1999b)[a]

Treatment	Days after transplanting								
	30	50	70	85	100	110	125	130	140
Control	8.3a	21.0a	40.0a	90.7a	62.8a	75.1a	102.9a	58.7a	8.2a
Prilled urea	5.7a	13.1a	26.8a	67.2ab	71.6a	85.2a	94.3a	28.5b	7.8a
Prilled urea + Nimin	5.2a	17.7a	27.1a	48.0b	51.0a	64.8a	77.2a	21.7b	7.4a
Urea supergranule	6.1a	13.2a	30.7a	58.4b	57.6a	74.3a	90.2a	56.2a	12.3a

[a]Mean of four replicate observations. In a column, means followed by a common letter are not significantly different at the 5% level by DMRT

The rice field plots (site II) were treated with prilled urea (60 kg N ha^{-1}), GM (60 kg N ha^{-1}), and prilled urea (30 kg N ha^{-1}) combined with GM (30 kg N ha^{-1}). In the first experiment in site I, CH_4 emission peaked 100-125 d after transplanting followed by a decline in all plots (Table 7). Methane emission from rice fields in site I was little affected by broadcast application of prilled urea. Subsurface application of urea supergranules was marginally effective in reducing the CH_4 flux over that of control. Evidently, the mode of application of the fertilizer compounds might have direct effects on CH_4 emission in a rice field with 30 cm water depth. Methane emission was less pronounced in plots treated with the mixture of urea and Nimin than in plots with no fertilizer control and prilled urea alone. Nimin is known to inhibit autotrophic oxidation of NH_4^+ to NO_2^- (Sahrawat & Parmar, 1975).

In another field plot experiment, the application of prilled urea and GM (*S. rostrata*) to plots with water depth of 4-6 cm (site II) significantly enhanced CH_4 emission over that of control (Table 8). In general, prilled urea or GM at 60 kg N ha^{-1} effected a 1.5- to 2-fold increase in net CH_4 emission over that in control. The application of prilled urea and GM stimulated CH_4 emission at the early stage of the crop. Application of GM in combination with prilled urea further enhanced CH_4 emission significantly over that in treatments with prilled urea and GM alone. The cumulative CH_4 emission was 1.8-, 1.9-, and 3-fold with prilled urea, GM, and prilled urea combined with GM, respectively, over that of control. Both the experimental plots (sites I and II with water depth of 30 cm and 4-6 cm, respectively) were planted to the same cultivar, cv Gayatri. Among the physicochemical properties of soil samples from the two sites, appreciable differences were detected only in pH and electrical conductivity, and clay, sand, and silt contents. The levels of total carbon and nitrogen were, however, similar at both sites. Interestingly, in control and prilled urea-treated plots, CH_4 emission from rice plots with water depth of 30 cm was 4-10 times higher than that of rice plots with water depth of 4-6 cm. Increased CH_4 emission from rice fields at site

Table 8. Effect of fertilizer management practices on CH_4 efflux (mg m^{-2} h^{-1}) from rice field plots with 4-6 cm water depth (site II), planted to cv Gayatri (Rath et al., 1999b)[a]

Treatment	Days after transplanting						
	25	40	60	75	90	105	120
Control	2.9c	9.5c	9.8d	8.9b	6.7c	0.9b	0.4b
Prilled urea	16.7a	16.5b	18.7b	14.7a	9.7b	3.2	0.4b
Green manure	19.8a	9.6c	13.7c	14.5a	11.7ab	2.2ab	2.2a
Prilled urea + green manure	7.6b	36.8a	32.7a	15.3a	14.1a	2.3ab	3.3a

[a]Mean of four replicate observations. In a column, means followed by a common letter are not significantly different at the 5% level by DMRT.

I with deeper water depth could be attributed to continuous deep submergence for prolonged period (about 30-cm water depth for 70 d, Figure 1) as compared with 4-6-cm water depth in rice fields (site II) (Figure 2). Data presented in Figures 1 and 2 also show a more rapid and sharper drop in redox potential in plots with 30-cm water depth than in rice field plots with water depth of 4-6 cm. Moreover, field plots at site I produced more plant biomass than did those at site II. Cumulative flux values showed that control plots with 30-cm water depth at site I emitted around ninefold more CH_4 than did rice field plots with water depth of 4-6 cm at site II. In plots at site I, maturity duration of the rice plants increased by about 20 d and the prolonged growth period due to high water table had also led to a significant increase in cumulative CH_4 emission.

In a followup laboratory incubation study, CH_4 production in flooded alluvial soil was monitored at different depths (5, 8, 10.5, 14, 17, and 20 mm) of standing water (Rath et al., 1999b). During the initial 20 d of incubation, there was no appreciable increase in the concentrations of CH_4 among treatments (Table 9). However, after 20 d, CH_4 production from soil samples was distinctly enhanced with progressive increase in water level. Following soil submergence, oxygen in the soil is rapidly consumed by the aerobic microorganisms and soil can soon be devoid of molecular oxygen. Moreover, oxygen is sparingly soluble (37.18 µg g^{-1}) in water and oxygen diffused to the soil can decrease with increase in standing water column. Therefore, oxygen-stress conditions in flooded soil may be more intense in situations with deeper water depth than with shallow water depth.

There are reports on the inhibitory effects of N fertilizers on methanotrophic microorganisms in soils. Application of ammonium sulfate and, to a lesser extent, urea to surface, rhizosphere, and subsurface soil samples from flooded field planted to rice inhibited CH_4 -oxidizing activity (Kumaraswamy et al., 1997a). This difference may be attributed to the competitive inhibition of CH_4 oxidation by the readily released ammonium from ammonium sulfate, while urea can be inhibitory only upon release of ammonium by hydrolysis. There are reports that ammonium sulfate decreases CH_4 from rice fields (Lindau et al., 1993). Urea addition enhances CH_4 production, probably due to the increase in soil pH following urea hydrolysis and the drop in redox potential which stimulates methanogenic activities (Wang et al., 1992). In spite of the significantly contrasting effects of these two compounds on CH_4 production/emission from rice fields, both had adverse effect on CH_4 oxidation. Yan-XiaoYuan et al. (1996) also showed that NH_4 and NO_3 inhibited CH_4 oxidation. Urea did not inhibit CH_4 oxidation initially, but strongly inhibited the process after a lag period of 2 d in a rice soil. In principle, three different causes have been suggested for the inhibitory effect of nitrogenous fertilizers, especially NH_4-N fertilizers on CH_4 oxidation : (i) an immediate inhibition of methanotrophic enzyme system (CH_4 monooxygenase - MMO) (Bedard & Knowles, 1989); (ii) secondary inhibition through NO_2^- production from methanotrophic ammonium oxidation (Megraw & Knowles, 1987); and (iii) dynamic alterations of microbial communities of soil (Adamsen & King, 1993).

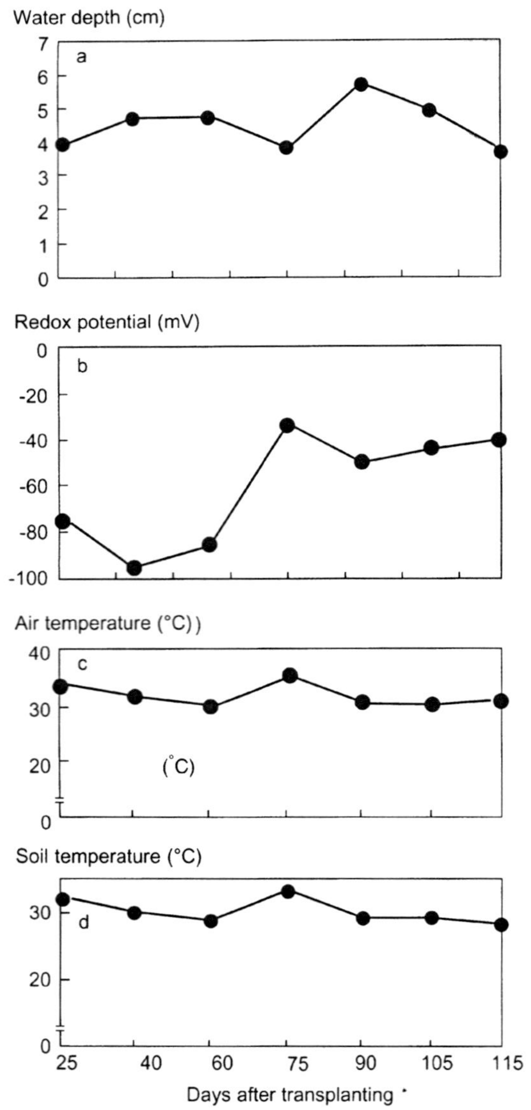

Figure 1. (a) Variations in water depth (cm), (b) redox potential (mV) of flooded soil, (c) ambient temperature, and (d) soil surface temperature of rice field plots of 30-cm water depth (site I)

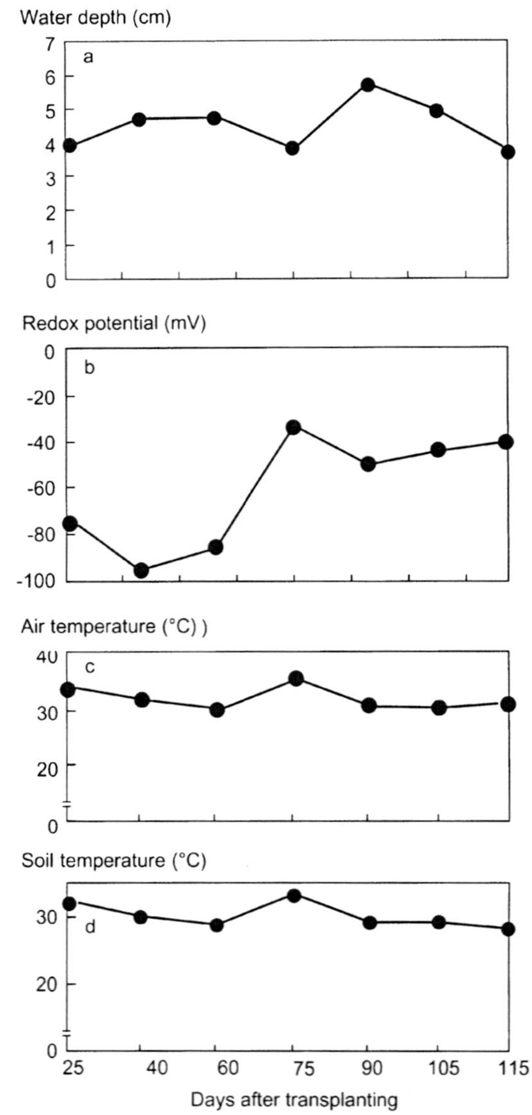

Figure 2. (a) Variations in water depth (cm), (b) redox potential (mV) of flooded soil, (c) ambient temperature, and (d) soil surface temperature of rice field plots of 4-6 cm water depth (site II)

Nitrification inhibitors are also known to inhibit CH_4 oxidation (Bronson & Mosier, 1994). Kumaraswamy et al. (1997a) also showed that nitrification inhibitors (thiourea, sodium thiosulfate, and dicyandiamide) inhibited CH_4-oxidizing activity of flooded rice field samples. These inhibitors had repressing effects on the population of CH_4 oxidizers with soluble CH_4 monooxygenase activity. A similar trend of decrease in population of ammonium oxidizers was also noticed. Nitrification inhibitors such as acetylene and nitrapyrin can inhibit the growth of nitrifiers, methanogens, and methanotrophs (Oremland & Capone, 1988; Bedard & Knowles, 1989). Bronson and Mosier (1991) reported significant reduction in CH_4 emission from rice fields following application of urea in combination with encapsulated calcium carbide.

Table 9. Methane production (mg g^{-1} soil) in soil samples, experimentally flooded to provide different water levels (Rath et al., 1999b)[a]

Water level (mm) above soil surface	Days of incubation 5	10	15	20	25	30	35	40
5.0	5a	32a	77a	235a	2381e	11077f	15034f	21345f
8.0	5a	41a	113a	365a	4111d	19809e	21973e	25180e
10.5	6a	33a	170a	496a	5849c	23979d	26465d	29488d
14.0	11a	24a	221a	547a	6545bc	26518c	29681c	36814c
17.0	4a	46a	284a	611a	7277ab	30733b	33709b	40242b
20.0	4a	44a	311a	1108a	8034a	32352a	37489a	43538a

[a]Mean of five replicate observations. In a column, means followed by a common letter are not significantly different at the 5% level by DMRT.

Table 10. Effect of carbofuran application on production and oxidation of CH_4 in soil samples, and CH_4 emission from flooded field plots planted with cv IR72 (Kumaraswamy et al., 1998)

Carbofuran	Methane production under flooded conditions [nmol of CH_4 g^{-1} soil d^{-1}]	Methane oxidation* [μmol of CH_4 oxidized g^{-1} soil d^{-1}] [*measured on 4th day of incubation]		Methane emission from field plots [μmol CH_4 efflux m^{-2} h^{-1}]
		Soil samples held at 60% water-holding capacity	Flooded soil samples	
0	20	239	249	-
5 μg g^{-1}	4	512	545	-
10 μg g^{-1}	3	550	526	-
50 μg g^{-1}	ND	301	292	-
100 μg g^{-1}	30	55	126	-
Treatment in field plots			-	
Control	-	-	-	945
+ 2 kg ai ha^{-1}	-	-	-	505
+ 12 kg ai ha^{-1}	-	-	-	445

ND = not determined

Lindau et al. (1993) found that CH_4 emission from rice fields decreased by 35% and 14% following application of encapsulated calcium carbide and dicyandiamide, respectively.

Effect of pesticide application

In modern rice culture, pesticides are increasingly used. There is little information available on the effects of pesticides on bacteria involved in the production or consumption of CH_4. Satpathy et al. (1997) found that application of a commercial formulation of a widely used organochlorine insecticide, hexachlorocyclohexane (HCH) to flooded rice fields or its technical grade isomers (α, β, and δ) to laboratory-incubated flooded soils retarded the production and emission of CH_4, even at the field application rate of 1-2 kg ai ha^{-1} to control insect pests. Hexachlorocyclohexane inhibited CH_4 oxidation, measured using the treated soil samples under laboratory incubation, significantly at 5 μg g^{-1} soil and almost completely at 10 μg g^{-1} soil (Kumaraswamy et al., 1997b). The commercial formulation of carbofuran, a carbamate insecticide, when applied at rates of 2 kg and 12 kg ai ha^{-1} to a flooded field planted to rice, resulted in significant inhibition of CH_4 emission (Kumaraswamy et al.,1998). On the 9th day after application of carbofuran (56 d after transplanting), CH_4 emission from untreated field plots was 1.60 mmol CH_4 m^{-2} h^{-1} as compared with 0.47 and 0.87 mmol CH_4 m^{-2} h^{-1} in plots treated with carbofuran at rates of 2 and 12 kg ai ha^{-1}, respectively. In the laboratory-incubation study on CH_4 production, the soil samples treated with carbofuran at rates of 5 and 10 μg g^{-1} soil accumulated substantially less CH_4 under flooded conditions than the control during a 30-d incubation period. In contrast, carbofuran at 100 μg g^{-1} soil effected a distinct stimulation of CH_4 production compared with that of control. Interestingly, CH_4 oxidation, measured using the soil samples incubated under laboratory conditions, proceeded more rapidly at low concentrations of carbofuran (5 μg g^{-1} soil) than in controls or soil samples amended with high concentrations of carbofuran (100 μg g^{-1} soil). Data presented in Table 10 show that when carbofuran was applied at a rate of 100 μg g^{-1} to the soil samples incubated under flooded conditions, the production of CH_4 was stimulated, but its oxidation was inhibited by this concentration of carbofuran. At low concentrations of carbofuran, CH_4 oxidation was stimulated, and this led to a decrease in net CH_4 production compared with that of control.

Conclusions

Methane emission from flooded rice fields differ markedly with climate, characteristics of soil and rice cultivar, application of organic matter and mineral fertilizer, and other agricultural practices. Composted organic sources (*Azolla* compost, BGA compost, and FYM) had less effect on the production of CH_4 than the fresh organic sources (rice straw, cellulose, and GM). Application of poultry manure, due to its high sulfur content, resulted in low emission of CH_4 from rice fields. Methane emission can be reduced significantly by adopting certain cultural practices which include floodwater management and choice of rice cultivars, fertilizers, and agrochemicals. However, the universal applicability of these mitigation options can depend on factors, such as soil characteristics, plant factors and the associated microbiological processes. Many of these mitigation options are location-specific, a major constraint to their universal adaptability in diverse rice ecologies. More research is needed to identify suitable and economically viable management practices for different rice-growing areas and socioeconomic situations.

References

Achtnich C, Bak F & Conrad R (1995) Competition for electron donors among nitrate reducers, ferric iron reducers, sulfate reducers and methanogens in anoxic paddy soil. Biol Fertil Soils 19:65-72

Adamsen APS & King GM (1993) Methane consumption in temperate and subarctic forest soils: rates, vertical zonation and responses to water and nitrogen. Appl Environ Microbiol 59:485-490

Adhya TK, Rath AK, Gupta PK, Rao VR, Das SN, Parida KM, Parashar DC & Sethunathan N (1994) Methane emission from flooded rice fields under irrigated conditions. Biol Fertil Soils 18:245-248

Anastasi C, Dowding M & Simpson VJ (1992) Future CH_4 emissions from rice production. J Geophys Res 97:7521-7525

Bedard C & Knowles R (1989) Physiology, biochemistry and specific inhibitors of CH_4, NH_4^+ and CO oxidation by methanotrophs and nitrifiers. Microbiol Rev 53:68-84

Boeckx P, van Cleemput O & Meyer T (1998) The influence of land use and pesticides on methane oxidation in some Belgian soils. Biol Fertil Soils 27:293-298

Bronson KF & Mosier AR (1991) Effect of encapsulated calcium carbide on dinitrogen, nitrous oxide, methane and carbon dioxide emissions from flooded rice. Biol Fertil Soils 11:116-120

Chen Z, Li D, Shao K & Wang B (1993) Features of CH_4 emission from rice paddy fields in Beijing and Nanjing. Chemosphere 26:239-245

van Cleemput O, El-Sebaay AS & Baert L (1983) Evolution of gaseous hydrocarbons from soil: effect of moisture content and nitrate level. Soil Biol Biochem 15:519-524

Conrad R & Rothfuss F (1991) Methane oxidation in the soil surface layer of a flooded rice field and the effect of ammonium. Biol Fertil Soils 12:28-32

Crutzen PJ (1991) Methane's sinks and sources. Nature 350:380-381

Crutzen PJ (1995) On the role of CH_4 in atmospheric chemistry: sources, sinks and possible reductions in anthropogenic sources. Ambio 24:52-55

Debnath G, Jain MC, Kumar S, Sarkar K & Sinha S (1996) Methane emissions from rice fields amended with biogas slurry and farmyard manure. Clim Change 33:97-109

Denier van der Gon HAC & Neue HU (1995) Influence of organic matter incorporation on the methane emission from a wetland rice field. Global Biogeochem Cycles 9:11-22

Holzapfel-Pschorn A & Seiler W (1986). Methane emission during a cultivation period from an Italian rice paddy. J Geophys Res 91:11803-11814

Houghton JT, Meria Filho LG, Callander BA, Harris N, Kattenberg A & Maskell K (1996) IPCC report on Climate Change 1995: The Science of Climate Change. WGI contribution to the IPCC second assessment report on methane emission from rice cultivation. London: Cambridge University Press

Khalil MAK & Rasmussen RA (1991) The global methane cycle. In: Abrol YP, Wattal PN, Gnanam A, Govindjee, Ort DR, Teramura (eds) Impact of Global Climate Changes on photosynthesis and Plant Productivity, pp 641-651, New Delhi: Oxford and IBH Publishing Co

Khalil MAK & Rasmussen RA (1993) Decreasing trend of methane: unpredictability of future concentrations. Chemosphere 26:803-814

Kimura M (1992) Methane emission from paddy soils in Japan and Thailand. In: Batjes NH & Bridges EM (eds). World Inventory of Soil Emission Potentials, WISE report 2. Wageningen: International Soil Reference and Information Centre, p 43-79

Kumaraswamy S (1998) Studies on methane production and its fate in rice paddy ecosystem. PhD thesis, Bhubaneswar : Utkal University

Kumaraswamy S, Ramakrishnan B, Satpathy SN, Rath AK, Mishra S, Rao VR & Sethunathan N (1997a) Spatial distribution of methane-oxidizing activity in a flooded rice soil. Plant Soil 191:241-248

Kumaraswamy S, Rath AK, Bharati K, Ramakrishnan B & Sethunathan N (1997b) Effect of pesticides on methane oxidation in a flooded tropical rice soil. Bull Environ Contam Toxicol 59:222-227

Kumaraswamy S, Rath AK, Satpathy SN, Ramakrishnan B, Adhya TK & Sethunathan N (1998) Influence of an insecticide carbofuran on the production and oxidation of methane in a flooded rice soil. Biol Fertil Soils 26: 362-366

Lampe K (1995) Rice research:food for 4 billion people. GeoJournal 35:253-259

Lelieveld J, Crutzen PJ & Dentener FJ (1998) Changing concentration, lifetime and climate forcing of atmospheric methane. Tellus 50B:128-150

Lindau CW, Bollich PK, DeLaune RD, Mosier AR & Bronson KF (1993) Methane mitigation in flooded Louisiana rice fields. Biol Fertil Soils 15:174-178

Lovely DR & Klug MJ (1983) Sulfate reducers can outcompete methanogens at freshwater sulfate concentrations. Appl Environ Microbiol 45:187-192

Megraw SR & Knowles R (1987) Methane consumption and production in a cultivated humisol. Biol Fertil Soils 5:56-60

Minami K (1994) Methane from rice production. Fert Res 37:169-179

Minami K & Neue HU (1994) Rice paddies as a methane source. Clim Change 27:13-26

Mishra S, Rath AK, Adhya TK, Rao VR & Sethunathan N (1997) Effect of continuous and alternate water regimes on methane efflux from rice under greenhouse conditions. Biol Fertil Soils 24:399-405

Mitra AP (1992) Greenhouse gas emission in India, 1991 Methane Campaign. Scientific Report No. 20. Council of Scientific and Industrial Research, and Ministry of Environment and Forest. New Delhi: National Physical Laboratory

Mosier A, Schimel D, Valentine D, Bronson K & Parton W (1991) Methane and nitrous oxide fluxes in native, fertilized and cultivated grasslands. Nature 350:330-332

Neue HU (1993) Methane emission from rice fields. BioScience 43:466-474

Neue HU, Lantin RS, Wassmann R, Aduna JB, Alberto MCR & Andales MJF (1994) Methane emission from rice soils of the Philippines. In: Minami K, Mosier AR & Sass RL (eds) CH_4 and N_2O : Global Emissions and Controls from Rice Fields and other Agricultural and Industrial Sources, pp 55-63, Tokyo, Japan : Yokendo Publishers

Neue HU, Ziska LH, Matthews RB & Dai Q (1995) Reducing global warming-the role of rice. GeoJournal 35:351-362

Oremland RS & Capone DG (1988) Use of specific inhibitors in biogeochemistry and microbial ecology. Adv Microbial Ecol 10:285-383

Parashar DC, Mitra AP, Gupta PK, Rai J, Sharma RC, Singh N, Koul S, Ray HS, Das SN, Parida KM, Rao SB, Kanungo SP, Ramasami T, Nair BU, Swamy M, Singh G, Gupta SK, Singh AR, Saikia BK, Barua AKS, Pathak MG, Iyer CSP, Gopalakrishnan M, Sane PV, Singh SN, Banerjee R, Sethunathan N, Adhya TK, Rao VR, Palit P, Saha AK, Purkait NN, Chaturvedi GS, Sen SP, Sen M, Sarkar B, Banik A, Subbaraya BH, Lal S, Venkatramani S, Lal G, Chaudhary A & Sinha SK (1996) Methane budget from paddy fields in India. Chemosphere 33:737-757

Ponnamperuma FN (1972) The chemistry of submerged soils. Adv Agron 24:29-96

Ramakrishnan B, Satpathy SN, Adhya TK, Rao VR & Sethunathan N (1995) Methane production in two Indian rice soils. Geomicrobiol J 13:193-199

Rao VR (1998) Methane emission from rice based cropping system. Final report on ICAR AP Cess Fund Project. Cuttack: Central Rice Research Institute

Rao DN & Mikkelsen DS (1977) Effect of rice straw additions on production of organic acids in a flooded soil. Plant Soil 47:303-311

Rath AK (1998) Studies on methane emission from tropical rice soils. PhD thesis, Bhubaneswar: Utkal University

Rath AK, Mohanty SR, Mishra S, Kumaraswamy S, Ramakrishnan B & Sethunathan N (1999a) Methane production in unamended and rice-straw amended soil at different moisture levels. Biol Fertil Soils 28:145-149

Rath AK, Swain B, Ramakrishnan B, Panda D, Adhya TK, Rao VR & Sethunathan N (1999b) Influence of fertilizer management and water regime on methane emission from rice fields. Agric Ecosyst Environ 76:99-107

Sahrawat KL & Parmar BS (1975) Alcohol extract of neem (*Azadirachta indica* L.) seed as nitrification inhibitor. J Indian Soc Soil Sci 23:131-134

Sass RL, Fisher FM, Harcombe PA &Turner FT (1990) Methane production and emission in a Texas rice field. Global Biogeochem Cycles 4:47-68

Sass RL, Fisher FM, Wang YB, Turner FT & Jund MF (1992) Methane emission from rice fields: the effect of floodwater management. Global Biogeochem Cycles 6:249-262

Satpathy SN (1997) Factors affecting methane emission in tropical rice soil. PhD thesis, Bhubaneswar: Utkal University

Satpathy SN, Rath AK, Mishra S, Kumaraswamy S, Ramakrishnan B, Adhya TK & Sethunathan N (1997) Effect of hexachlorocyclohexane on methane production and emission from flooded rice soils. Chemosphere 34:2663-2671

Schütz H, Holzapfel-Pschorn A, Conrad R, Rennenberg H, & Seiler W (1989) A 3-year continuous record on the influence of daytime, season and fertilizer treatment on methane emission rates from an Italian rice paddy. J Geophys Res 94 :16,405-16,416

Seiler W, Holzapfel-Pschorn A, Conrad R & Scharffe D (1984) Methane emission from rice paddies. J Atmos Chem 1:241-268

Sethunathan N (1997) Methane production from tropical Indian rice soils. Final report of the project. Department of Science & Technology, Ministry of Science and Technology. Cuttack : Central Rice Research Institute

Sethunathan N, Neue HU, Parashar DC, Wassmann R, Rao VR, Adhya TK & Ramakrishnan B (1998) Greenhouse effect and mitigation options. In: Mohanty SK et al. (eds) Rainfed Rice for Sustainable Food Security. Cuttack : Central Rice Research Institute and Association of Rice Research Workers. p 409-424

Sextone AJ, Revsbach NP, Parkin TB & Tiedje JM (1985) Direct measurement of oxygen profiles and denitrification rates in soil aggregates. Soil Sci Soc Am J 49:645-651

Shao KS & Li Z (1997) Effect of rice cultivars and fertilizer management on methane emission in a rice paddy in Beijing. Nutr Cycling Agroecosyst 49:139-146

Steele LP, Dlugokencky EJ, Lang PM, Tans PP, Martin RC & Masarie KA (1992) Slowing down of the global accumulation of atmospheric methane during the 1980's. Nature 358:313-316

Takai Y (1970) The mechanism of methane fermentation in flooded paddy soil. Soil Sci Plant Nutr 16:238-244

US-EPA –United States Environmental Protection Agency – (1990) Overview of methane's contribution to global warming. In: Methane Emissions and Opportunities for Control, EPA 1400/9-90/007. Washington : US-EPA/Air and Radiation. p 2-24

Wang ZP, DeLaune RD, Lindau CW & Patrick Jr WH (1992) Methane production from anaerobic soil amended with rice straw and nitrogen fertilizers. Fert Res 33:115-121

Wang ZP & Patrick Jr WH (1995) Fertilization effects on methane production potential in a Chinese flooded rice soil. Proceedings of the Fourth Annual IRRI-EPA-UNDP Planning Meeting of Methane Emission from Rice Fields, 19-25 Nov 1995, Chonburi, Thailand

Winfrey MR & Zeikus JG (1979) Microbial methanogenesis and acetate metabolism in a Meromictic lake. Appl Environ Microbiol 37:213-221

Wassmann R, Schütz H, Papen H, Rennenberg H, Seiler W, Dai A, Shen R, Shangguan X & Wang M (1993) Quantification of methane emission from Chinese rice fields (Zhejiang Province) as influenced by fertilizer treatment. Biogeochemistry 20:83-101

Wassmann R, Papen H & Rennenberg H (1993b) Methane emission from rice paddies and possible mitigation strategies. Chemosphere 26:201-217

Yagi K & Minami K (1990) Effect of organic matter applications on methane emissions from Japanese paddy fields. Soil Sci Plant Nutr 36:599-610

Yagi K, Tsuruta H, Minami K, Chairoj P & Cholitkul W (1994) Methane emission from Japanese and Thai paddy fields. In: Minami K, Mosier AR, Sass RL (eds) CH_4 and N_2O : Global Emissions and Controls from Rice Rields and other Agricultural and Industrial Sources, pp 41-53, Tokyo, Japan: Yokendo Publishers

Yan XY & Cai ZC (1996) Effects of nitrogen fertilizer, soil moisture and temperature on methane oxidation in paddy soil. Pedosphere 6: 175-181

Nutrient Cycling in Agroecosystems **58:** 389–394, 2000.

Influence of six nitrification inhibitors on methane production in a flooded alluvial soil

K. Bharati[1], S. R. Mohanty[1], P.V.L. Padmavathi[2], V. R. Rao[1] & T. K. Adhya[1]

[1]*Laboratory of Soil Microbiology, Division of Soil Science and Microbiology, Central Rice Research Institute, Cuttack 753006, Orissa, India;* [2]*Department of Botany and Microbiology, Nagarjuna University, Nagarjunanagar, Guntur 522510, Andhra Pradesh, India*

Keywords: incubation, dicyandiamide, methanogenic bacteria, redox potential, pH

Abstract

The influence of six nitrification inhibitors (NI) on CH_4 production in an alluvial soil under flooded condition was studied in a laboratory incubation experiment. The inhibition of CH_4 production followed the order of sodium azide > dicyandiamide (DCD) > pyridine > aminopurine > ammonium thiosulfate > thiourea. Inhibition of CH_4 production in DCD-amended soils was related to a high redox potential, low pH, low Fe^{2+} and lower readily mineralizable carbon content as well as lower population of methanogenic bacteria and their activity. In the presence of higher levels of urea N (40 μg), the inhibitory effect of DCD was only partially alleviated. Results indicate that several NIs can differentially regulate CH_4 production in a flooded alluvial soil.

Introduction

Methane (CH_4), one of the most abundant gaseous hydrocarbons in the environment, is an important greenhouse gas and a key factor in the tropospheric and stratospheric chemistry (Wang et al., 1976). Flooded rice soil, which contributes up to 20% or ~100 Tg CH_4 on an annual basis (Houghton et al., 1992), is one of the major anthropogenic sources of global CH_4. The projected increase in rice production during the coming decades (IRRI, 1999), is expected to result in further increase in CH_4 fluxes to the atmosphere if prevalent cultivation practices are continued (Anastasi et al., 1992).

Measurement of CH_4 fluxes from rice fields all over the world show large temporal variation. The flux differs markedly with climate, soil characteristics and application of organic materials and mineral fertilizers (Bouwman, 1990; Cicerone & Shetter, 1981; Lindau et al., 1990, 1991; Minami, 1995; Schutz et al., 1989; Yagi & Minami, 1990). Urea is the dominant form of N fertilizer applied to rice in Asia (Vlek & Byrnes, 1986), but it is subjected to various forms of loss including nitrification-denitrification (Prasad, 1998). Nitrification inhibitors (NI) are being increasingly recommended for rice agriculture to minimize fertilizer N losses (Prasad & Power, 1995) by limiting the formation of nitrate from ammonium. Many potential NIs like ammonium thiosulfate and thiourea (Bremner & Yeomans, 1986) are produced and used in Japan. Dicyandiamide (DCD), a commercially available NI, suitable for use with solid chemical fertilizer in rice cultivation (Gorelik et al., 1992), is produced and marketed in both Japan and Germany, while nitrapyrin is licensed for use in the USA (Hauck, 1984). In addition to their role in controlling various processes of N losses, NIs such as calcium carbide and nitrapyrin have been shown to inhibit CH_4 emission from flooded soil planted to rice (Bronson & Mosier, 1991; Keertisinghe et al., 1993). While NIs are well-recognized in inhibiting CH_4-oxidizing processes and CH_4-oxidizing microbial populations (Hanson & Hanson, 1996), their exact role in CH_4 production is not well investigated. In a laboratory incubation study, we studied the influence of six different NIs on CH_4 production in a tropical alluvial rice soil under flooded condition. In addition, effects of DCD alone or in combination with urea N on CH_4 production in an alluvial soil under flooded condition and the associated physical, chemical, and microbial changes were also investigated.

Materials and methods

Soil, treatments, and incubation setup

An alluvial soil (a typic Haplaquept), collected during fallow period from the experimental farm of the Central Rice Research Institute, Cuttack, India, was used in the study. Physicochemical parameters of the soil were determined according to the SSSA/ASA guidelines (Sparks, 1996) for soil analysis. The soil was sandy clay loam in texture (256 g clay kg^{-1}, 216 g silt kg^{-1}, 528 sand g kg^{-1}) with pH 6.2, cation exchange capacity 15 meq 100 g^{-1}, electrical conductivity 0.6 dS m^{-1}, organic C 9.3 g kg^{-1}, total N 1.1 g kg^{-1}, SO_4^{2-}-S 34.2 mg kg^{-1}, Olsen P 8 mg kg^{-1}. The soil, collected from the plow layer (0-15 cm), was air-dried, ground and sieved (< 2 mm) and stored at 4 °C until used in the study.

An incubation method, as described by Adhya et al. (1998), was used in studies on CH_4 production. In brief, individual 5-g portions of the air-dried soil samples were placed in B-D Vacutainer tubes (13 ml capacity) (Becton-Dickinson and Co., NJ, USA). Stock solutions (1,000 mg L^{-1}) of NIs (Table 1) were prepared in sterile distilled water immediately before use. The required amounts of the stock solutions were added separately to the soil to get a final concentration of 10 mg kg^{-1} soil. Similarly placed soils without any amendment served as control. All treatments were replicated five times. After amendments, the soils in tubes were flooded (1.5 cm standing water) with sterile distilled water, stoppered with a rubber septum, and incubated in a BOD incubator (30 ± 2°C) in the dark up to 40 d.

In a followup experiment on the effect of DCD on CH_4 production in the presence or absence of urea, the required quantity of DCD was added to similarly placed soil to provide a final concentration of 15 mg kg^{-1} soil. Urea was added at either 0, 20, or 40 mg N kg^{-1} soil as per treatment. Soils without any amendment served as control.

To estimate CH_4 production in the soil, tubes were shaken for 10 s on a vortex mixer to release soil-entrapped CH_4 (Wang et al., 1993), if any, and 5 mL of the headspace gas was collected for CH_4 analysis. On every sampling day, five soil tubes from each treatment were sacrificed for the estimation of CH_4.

Estimation of CH_4

Methane was estimated in a Shimadzu GC-8A gas chromatograph (GC) equipped with FID and a Porapak N column (Bharati et al. 1999). The column and detector were maintained at 70 and 110 °C, respectively. The gas samples were injected through a sample loop (3 mL) with the help of an on-column injector using a multiport valve (VICI AG, Schenkon, Switzerland). The GC was calibrated before and after each set of measurement using 5.38, 9.03 and 10.8 µL CH_4 mL^{-1} in N_2 (Scotty$^{(R)}$ II analyzed gases, M/s Altech Associates Inc., USA) as primary standard and 2.14 µL CH_4 mL^{-1} in air as secondary standard. Under these conditions, the retention time of CH_4 was 0.65 min and the minimum detectable limit was 500 µL L^{-1}.

Methane production (P) was calculated by

$$P = \frac{d_c}{d_t} \times \frac{V_H}{W_S} \times \frac{MW \cdot T_{st}}{MV \cdot (T_{st} + T)} \; (\mu g \; CH_4 \; kg_{\,(d.w.\,soil)}^{-1})$$

where d_c / d_t is the recorded change in the mixing ratio of CH_4 in the headspace over time (ppmv), V_H the volume of headspace, W_S the dry weight of soil, MW the molecular weight of CH_4, MV the molecular volume, T the temperature (K), and T_{st} the standard temperature.

Soil and microbiological analyses

Soil samples (40 g) placed in 100-mL beakers after amendment with either DCD or urea or both were flooded with sterile distilled water at 1:1.25 ratio. Soils without any amendment and flooded with only sterile distilled water served as control. Following flooding, the soil samples were incubated at room temperature (28 ± 2°C) in diffuse light. On 0, 5, 10, 20, 30, and 40 d of flooding, the redox potential of duplicate soil samples from each treatment was measured by inserting a combined platinum-calomel electrode (Barnant Co. IL, USA) into the reduced zone (about 1-2 cm below the oxidized zone) of the soil and measuring the potential difference in mV (Pal et al., 1979). All the values were corrected to that of a hydrogen electrode by adding +240 mV to the redox readings. Immediately after the measurement of the redox potential, the pH of the soil was measured with a portable pH meter (Philips model PW 9424, Philips Analytical, Cambridge, UK).

For measurement of extractable Fe^{2+}, another set of soil samples (10 g) was placed in sterile test tubes (150- × 20-mm) and after amendment with either urea or DCD or both, was flooded with sterile distilled water at 1:1.25 ratio. Soil samples, thus flooded, were incubated at room temperature (28 ± 2°C) and two replicates of each treatment were extracted with

Table 1. Names, chemical formulas, and sources of nitrification inhibitors used in the study

Nitrification inhibitor	Chemical formula	Pure/ commercial	Source
Aminopurine	$C_5H_5N_5$	Pure	Sigma, St. Louis
Ammonium thiosulfate	$(NH_4)_2SO_3$	Pure	E. Merck, Mumbai
Dicyandiamide	$NH_2C(:NH)NHCN$	Pure	Loba-chemie, Mumbai
Pyridine	C_5H_5N	Pure	E. Merck, Mumbai
Sodium azide	NaN_3	Pure	E. Merck, Mumbai
Thiourea	NH_2CSNH_2	Pure	E. Merck, Mumbai

Table 2. Effects of various nitrification inhibitors[a] on CH_4 production (μg kg^{-1}) in an alluvial soil under flooded condition[b]

Treatment	Days after flooding							
	5	10	15	20	25	30	40	Mean
Unamended control	47bc	126a	157a	168a	573a	2929a	4426c	1204
Sodium azide	40c	42c	48d	58d	135f	541g	1795g	380
Aminopurine	56bc	82b	133ab	140b	569a	2066b	3540c	941
Pyridine	52bc	87b	92c	130bc	346d	1558d	4094d	908
Dicyandiamide	39c	50c	55d	63d	276e	1112f	2844f	634
Thiourea	95a	103ab	117bc	110c	399c	1843c	4791b	1065
Ammonium thiosulfate	71b	90d	103c	125bc	491b	1443e	5098a	1060

[a]The nitrification inhibitors were added to the soil at 10 mg kg^{-1} soil. [b]Mean of five replicate observations. In a column, means followed by a common letter are not significantly different at the 5% level by DMRT.

NH_4OAC:HCl (pH 2.8) on 10, 20, 30 and 40 d of flooding. The soil extract was filtered and analyzed for Fe^{2+} by colorimetry after reacting with orthophenanthroline (Pal et al., 1979). Readily mineralizable carbon (RMC) was also measured by extracting another set of similarly incubated soil samples with 0.5 M K_2SO_4 and titrating the extract with ferrous ammonium sulfate after wet digestion with chromic acid (Vance et al., 1982).

Methanogenic bacterial population of flooded alluvial soil was enumerated using anaerobic culture tube technique by the most probable number (MPN) method (Kasper & Tiedje, 1982). Detection of CH_4 in the headspace of culture tubes was considered as evidence for the presence of methanogens and the population was counted (Alexander, 1982).

Statistical analyses

Data were statistically analyzed using statistical package (IRRISTAT, version 3.1: International Rice Research Institute, Philippines) and means of different treatments were seperated by Duncan's multiple range test (DMRT) at $P < 0.05$.

Results and discussion

Methane production in both untreated and treated flooded soils was low up to 20 d of incubation and increased enormously thereafter (Table 2). All the NIs used in the study significantly reduced CH_4 production over that of the unamended control following the order of sodium azide > DCD > pyridine > aminopurine > ammonium thiosulfate > thiourea. Sodium azide inhibited the mean CH_4 production by 75% over that of unamended control. Sodium azide, a potent NI, is also a respiratory inhibitor and a microbial inhibitor but is known to increase N_2O emission in soils (Aulakh & Rennie, 1985). Although ammonium thiosulfate and thiourea stimulated CH_4 production, especially at 40 d of incubation, the mean CH_4 production was lower than the unamended control. nitrogen-containing compounds are known to stimulate CH_4 production in flooded soils (Bollag & Czlonkowski, 1984). Possibly the N contained in the two inhibitors was released from the parent compound upon decomposition and stimulated CH_4 production. Methane production in DCD amended soil was significantly lower compared with unamended con-

Table 3. Effect of dicyandiamide (DCD) on CH_4 production (μg kg^{-1} soil) in an alluvial soil in the presence or absence of urea-N under flooded condition[a]

Treatment	Days after flooding						
	5	10	20	25	30	40	Mean
Unamended control	20a	116a	142a	576a	3024a	6549a	1738
Urea-N (20 mg kg^{-1})	21a	88a	194a	416a	2686a	6174a	1597
Urea-N (40 mg kg^{-1})	16a	74a	217a	285a	877b	1795c	544
DCD (15 mg kg^{-1})	21a	72a	114a	208a	654b	636d	2841
Urea-N (20 mg kg^{-1}) + DCD (15 mg kg^{-1})	21a	78a	93a	160a	389b	601d	224
Urea-N (40 mg kg^{-1}) + DCD (15 mg kg^{-1})	26a	80a	135a	476a	2682a	5534b	1489

[a]Mean of five replicate observations In a column, means followed by a common letter are not significantly different at the 5% level by DMRT.

Table 4. Changes in redox potential and pH of an alluvial soil under flooded condition amended with urea N and/or dicyandiamide (DCD)[a]

Treatment	Days after flooding											
	0		5		10		20		30		40	
	Eh	pH	Eh	pH	Eh	pH	Eh	pH	Eh	pH	Eh	pH
Unamended control	203a	6.46a	–180a	6.80c	–231b	7.01b	–300d	6.98c	–276bc	7.15a	–230b	6.96c
Urea-N (20 mg kg^{-1})	203a	6.46a	–182a	6.82bc	–234b	7.02b	–301d	7.03b	–287cd	7.16a	–258c	7.02b
Urea-N (40 mg kg^{-1})	203a	6.46a	–212b	7.15a	–259b	7.39a	–324e	7.30a	–304d	7.16a	–267c	7.11a
DCD (15 mg kg^{-1})	203a	6.46a	–177a	6.83bc	–216c	7.04b	–259b	6.98c	–261b	7.07b	–189a	6.83d
Urea-N (20 mg kg^{-1}) + DCD (15 mg kg^{-1})	203a	6.46a	–186a	6.82bc	–183a	6.96c	–236a	6.95d	–240a	6.98c	–178a	6.79e
Urea-N (40 mg kg^{-1}) + DCD (15 mg kg^{-1})	203a	6.46a	–195ab	6.85b	-235b	7.02b	–281c	6.96cd	–275bc	7.10b	–254c	6.98e

[a]Mean of duplicate observations. In a column, means followed by a common letter are not significantly different at the 5% level by DMRT.

trol and the inhibitory effect persisted throughout the incubation period of 40 d.

DCD, a commercially available NI suitable for flooded rice soil system, was used in followup experiments to understand the nature and extent of inhibition of CH_4 production in flooded soil. Application of urea at 20 mg N kg^{-1} soil led to a 26.4% increase in the mean CH_4 production compared with unamended soil (Table 3). In contrast, DCD applied alone at 15 mg kg^{-1} soil inhibited CH_4 production by about 46.03% over the control. Interestingly, the combined application of DCD (15 mg kg^{-1}) and urea (20 mg N kg^{-1}) resulted in the highest inhibition (54.9%) of CH_4 production. Although, higher levels of urea (40 mg N kg^{-1}), in combination with DCD, partially alleviated the inhibitory effect of DCD, CH_4 production did not exceed that of the unamended control. Increasing the level of DCD (30 mg kg^{-1}), however, did not further reduce CH_4 production (data not shown).

Methane production is linked to a decrease in redox potential (Eh) and an increase in pH of inundated soils. Redox potential was low in the unamended flooded soil (Table 4). Application of urea caused a further drop in redox potential of the soil and also resulted in higher CH_4 production. On the contrary, DCD-amended soil recorded a higher redox status and a corresponding lower CH_4 production. What is more interesting was that soils, supplemented with urea at 20 and 40 mg N kg^{-1} soil in combination with DCD, registered higher redox potential compared with that of urea N alone. Application of urea increased the pH of the soil, while DCD amendment with or without urea N registered a lower pH (Table 4). Fe^{2+} content was also high in urea-amended soil but decreased upon amendment

Table 5. Accumulation of Fe^{2+} (mg kg^{-1} soil) in alluvial soil treated with the nitrification inhibitor dicyandiamide (DCD) under flooded condition[a]

Treatment	Days after flooding			
	10	20	30	40
Unamended control	2035b	3425d	3785b	1905c
Urea-N (20 mg kg^{-1})	2069b	3795a	3885a	3400b
Urea-N (40 mg kg^{-1})	2047b	3596c	3703c	3532a
DCD 15 (mg kg^{-1})	2125a	3740b	3664d	1815e
Urea-N (20 mg kg^{-1}) + DCD (15 mg kg^{-1})	1870c	2805f	3171e	1555f
Urea-N (40 mg kg^{-1}) + DCD (15 mg kg^{-1})	2071b	3315e	3755b	2035c

[a]Mean of two replicates. In a column, means followed by a common letter are not significantly different at the 5% level by DMRT.

Table 6. Effect of dicyandiamide (DCD) on the concentration of K_2SO_4 extractable carbon (readily mineralizable carbon) content (g kg^{-1} soil) of flooded alluvial soil[a]

Treatment	Days after flooding			
	10	20	30	40
Unamended control	2035b	3425d	3785b	1905c
Unamended control	56ab	44b	204c	78d
Urea-N (20 mg kg^{-1})	55ab	88a	261a	143b
Urea-N (40 mg kg^{-1})	56ab	47b	237b	155a
DCD (15 mg kg^{-1})	63a	44b	152e	46e
Urea-N (20 mg kg^{-1}) + DCD (15 mg kg^{-1})	11c	22c	122f	34f
Urea-N (40 mg kg^{-1}) + DCD (15 mg kg^{-1})	53b	44b	182d	88c

[a]Mean of two replicate observations. In a column, means followed by a common letter are not significantly different at the 5% level by DMRT.

Table 7. Changes in methanogenic population (MPN × 10^3 g^{-1} soil) in a flooded alluvial soil treated with nitrification inhibitor dicyandiamide (DCD)

Treatment	Days after flooding	
	20	40
Unamended control	5.4c	7.5c
Urea-N (20 mg kg^{-1})	9.0b	9.2b
Urea-N (40 mg kg^{-1})	11.2a	10.9a
DCD (15 mg kg^{-1})	2.2d	2.5e
Urea-N (20 mg kg^{-1}) + DCD (15 mg kg^{-1})	2.8d	2.1e
Urea-N (40 mg kg^{-1}) + DCD (15 mg kg^{-1})	5.2c	5.3d

[a]MPN = most probable number. In a column, means followed by a common letter are not significantly different at the 5% level by DMRT.

with DCD (Table 5). Urea amendment, perhaps, caused a spurt in the heterotrophic activity resulting in higher Fe^{2+} content. Admittedly, higher Fe^{2+} content is indicative of low redox condition of a flooded soil, a status conducive to higher CH_4 production.

The RMC content of the soil, an indicator of substrates available to the methanogenic consortia (Mishra et al., 1997), reached a peak around 30 d after which it declined (Table 6) with a corresponding increase in CH_4 production. Amendment with urea alone at 20 and 40 mg N kg^{-1} soil resulted in a higher RMC content which was available in appreciable amounts even beyond 30 d of incubation. The RMC contents of the DCD-amended soil samples incubated with or without 20 or 40 mg urea N kg^{-1} soil were low. Thus, DCD amendment might have directly or indirectly influenced the RMC content, resulting in a low CH_4 production.

Strictly anaerobic methanogenic bacteria prevalent in the reduced flooded soil produce CH_4 (Conrad, 1996). The population of methanogenic bacteria was stimulated following application of urea N alone while it was inhibited in DCD-amended soil even in the presence of urea-N (Table 7). The inhibitory effect of DCD on MPN of methanogens was more pronounced in soils amended with low levels of urea N while higher levels of urea-N alleviated it to a certain extent.

Results of the present study reveal the role of several NIs in regulating CH_4 production in a flooded alluvial soil. The impact, however, varied among NIs. The inhibitory effect of DCD on CH_4 production in the alluvial soil studied, appears to be a combined result of higher redox status, lower pH, lower Fe^{2+}, and RMC contents that supported a lower population of methanogenic bacteria. Our study demonstrates that DCD, applied even in the presence of higher levels of urea N, exhibited substantial inhibitory effect on CH_4 production.

Acknowledgements

We thank Dr. K.C. Mathur, director, for permission to publish the results. This research was supported, in part, by the IRRI-UNDP Interregional Research Program on Methane Emission from Rice Soils (GLO/91/G31).

References

Adhya TK, Pattnaik P, Satpathy SN, Kumaraswamy S & Sethunathan N (1998) Influence of phosphorus application on methane emission and production in flooded paddy soils. Soil Biol Biochem 30:177-181

Alexander M (1982) Most probable number method for microbial population. In: Page AL, Miller RH & Keeney DR (eds) Methods of Soil Analysis, Part 2, Chapter 46, pp 815-1009, American Society of Agronomy, Madison, Wisconsin

Anastasi C, Dowding M & Simpson VJ (1992) Future CH_4 emissions from rice production. J Geophys Res 97:7521-7525

Aulakh MS & Rennie DA (1985) Azide effects upon N_2O emission and transformations of N in soils. Can J Soil Sci 65:205-212

Bollag JM & Czlonkowski ST (1973) Inhibition of methane formation in soil by various nitrogen-containing compounds. Soil Biol Biochem 5:673-678

Bouwman AF (1990) Exchange of greenhouse gases between terrestrial ecosystems and the atmosphere. In: Bouwman AF (ed) Soils and the Greenhouse Effect, pp 61-192, John Wiley & Sons, New York

Bharati K, Mohanty SR, Adhya TK, Banerjee A, Rao VR & Sethunathan N (1999) Influence of a commercial formulation of tridemorph on methane production and oxidation in a tropical rice soil. Chemosphere 39:933-943

Bremner JM & Yeomans JC (1986) Effects of nitrification inhibitors on denitrification of nitrate in soil. Biol Fertil Soils 2:173-179

Bronson KF & Mosier AR (1991) Effect of encapsulated calcium carbide on dinitrogen, nitrous oxide, methane and carbon dioxide emissions from flooded rice. Biol Fertil Soils 11:116-120

Cicerone RJ & Shetter JD (1981) Sources of atmospheric methane: measurements in rice paddies and a discussions. J Geophys Res 86:7203-7209

Conrad R (1996) Soil microorganisms as controllers of atmospheric trace gases (H_2, CO, CH_4, OCS, N_2O and NO). Microbiol Rev 60:28-32.

Gorelik LA, Yanishevskii FV, Podkolzina CV, Golov VG & Yashinevsky FV (1992) Efficiency of the nitrification inhibitor dicyandiamide in field experiments of the NIUIF geo-network. Agrokhimiya 9:14-18

Hanson RS & Hanson TE (1996) Methanotrophic bacteria. Microbiol Rev 60:439-471.

Hauck RD (1984) Technological approaches to improving the efficiency of nitrogen fertilizer use by crop plants. In : Hauck RD (ed) Nitrogen in Crop Production, pp 551-560, American Scociety of Agronomy, Madison, Wisconsin

Houghton JT, Callander BA & Varney SK (1992) Intergovernmental Panel on Climatic Change (IPCC) 1992 : Climate change. The supplementary report to the IPCC Scientific Assessment, pp 1-200, Cambridge University Press, New York

IRRI (International Rice Research Institute) (1999) Sustaining Food Security Beyond the Year 2000 - A Global Partnership for Rice Research: Medium-term Plan 2000-2002, p 130, International Rice Research Institute, Los Banos, Philippines

Kasper HF & Tiedje JM (1982) Anaerobic bacteria and processes. In: Page AL, Miller RH and Keeney DR (eds) Methods of Soil Analysis, Part 2, Chapter 46, pp 989-1009, American Scociety of Agronomy, Madison, Wisconsin

Keerthisinghe DG, Freney JR & Mosier AR (1993) Effect of wax-coated calcium carbide and nitrapyrin on nitrogen loss and methane emission from dry-seeded flooded rice. Biol Fertil Soils 16:71-75

Lindau CW, Bollich PK, DeLaune RD, Patrick WH, Jr. & Law VJ (1991) Effect of urea fertilizer and environmental factors on CH_4 emission from a Louisiana USA rice field. Plant Soil 136:195-203

Lindau CW, DeLaune RD, Patrick WH Jr. & Bollich PK (1990) Fertilizer effects on dinitrogen nitrous oxide and methane emission from lowland rice. Soil Sci Soc Am J 54:1789-1794

Minami K (1995) The effect of nitrogen fertilizer use and other practices on methane emission from flooded rice. Fert Res 40:71-84

Mishra S, Rath AK, Adhya TK, Rao VR & Sethunathan N (1997) Effect of continuous and alternate water regimes on methane efflux from rice under greenhouse conditions. Biol Fertil Soils 24:399-405

Pal SS, Barik S & Sethunathan N (1979) Effects of benomyl on iron and manganese reduction and redox potential in flooded soil. J Soil Sci 30:155-159

Prasad R (1998) Fertilizer urea, food security, health and the environment Curr Sci 75:677-683

Prasad R & Power JF (1995) Nitrification inhibitors for agriculture, health and environment. Adv Agron 54:233- 281

Schutz H, Seiler W & Conrad R (1989) Processes involved in the formation and emission of methane in rice paddies. Biogeochemistry 7:33-53

Sparks DL (1996) Methods of soil analysis. Part 3: Chemical Methods, SSSA Book Series No. 5, Soil Science Society of America / American Society of Agronomy, Madison, Wisconsin. 1390 p

Vance ED, Brookes PC & Jenkinson DS (1987) An extraction method for measuring soil microbial biomass. Soil Biol Biochem 19:703-707

Vlek PLG & Byrnes BH (1986) The efficacy and loss of fertilizer N in lowland rice. Fert Res 9: 131-147

Wang WC, Yung YL, Lacis AA, Mo T & Hansen JE (1976) Greenhouse effects due to man made perturbation of trace gases. Science 194:685-690

Wang ZP, Lindau CW, DeLaune RD & Patrick WH Jr (1993) Methane emission and entrapment in flooded rice soils as affected by soil properties. Biol Fertil Soils 16:163-168

Yagi K & Minami K (1990) Effect of organic matter application on methane emission from some Japanese paddy fields. Soil Sci Plant Nutr 36:599-610

Nutrient Cycling in Agroecosystems 58: 395, 2000.

Author index

Developments in Plant and Soil Sciences

68. O. van Cleemput, G. Hofman and A. Vermoesen (eds.): *Progress in Nitrogen Cycling.* 1996 ISBN 0-7923-3962-2
69. N. Ahmad (ed.): *Nitrogen Economy in Tropical Soils.* 1996 ISBN 0-7923-4094-9
70. M. Rahman, A. Kumar Podder, C. van Hove, Z.N. Tahmida Begum, T. Heulin and A. Hartmann (eds.): *Biological Nitrogen Fixation Associated with Rice Production.* 1996 ISBN 0-7923-4197-X
71. D. Rosen, E. Tel-Or, Y. Hadar and Y. Chen (eds.): *Modern Agriculture and the Environment.* 1997 ISBN 0-7923-4295-X
72. G.H. Elkan and R.G. Upchurch (eds.): *Current Issues in Symbiotic Nitrogen Fixation.* 1997 ISBN 0-7923-4367-0
73. H.M. Anderson, P.W. Barlow, D.T. Clarkson, M.B. Jackson and P.R. Shewry (eds.): *Plant Roots – From Cells to Systems.* 1997 ISBN 0-7923-4369-7
74. Zhu Zhao-liang, Wen Qi-xiao and J.R. Freney (eds.): *Nitrogen in Soils of China.* 1997 ISBN 0-7923-4372-7
75. J.K. Ladha, F.J. de Bruijn and K.A. Malik (eds.): *Opportunities for Biological Nitrogen Fixation in Rice and Other Non-Legumes* (2nd Working Group Meeting, Faisalabad, Pakistan, 1996). 1997 ISBN 0-7923-4514-2
76. R.W. Bell and B. Rerkasem (eds.): *Boron in Soils and Plants* (International Symposium, Chiang Mai, Thailand, 1997). 1997 ISBN 0-7923-4705-6
77. B. Dell, P.H. Brown and R. Bell (eds.): *Boron in Soils and Plants: Reviews.* 1997 ISBN 0-7923-4729-3
78. T. Ando, K. Fujita, T. Mae, H. Matsumoto, S. Mori and J. Sekiya (eds.): *Plant Nutrition for Sustainable Food Production and Environment.* 1997 ISBN 0-7923-4796-X
79. K.A. Malik, M.S. Mirza and J.K. Ladha (eds.): *Nitrogen Fixation with Non-Legumes.* 1998 ISBN 0-7923-4873-7
80. P.A. Finke, J. Bouma and M.R. Hoosbeek (eds.): *Soil and Water Quality at Different Scales.* 1998 ISBN 0-7923-4994-6
81. V. Balasubramanian, J.K. Ladha and G.L. Denning (eds.): *Resource Management in Rice Systems: Nutrients.* 1999 ISBN 0-7923-5199-1
82. J.E. Box, Jr. (ed.): *Root Demographics and Their Efficiencies in Sustainable Agriculture, Grasslands and Forest Ecosystems.* 1998 ISBN 0-7923-5230-0
83. G. Hardarson and W. Broughton (eds.): *Molecular Microbial Ecology of the Soil.* Results from an FAO/IAEA Co-ordinated Research Programme, 1992–1996. 1998 ISBN 0-7923-5252-1
84. C. Kirda, P. Moutonnet, C. Hera and D.R. Nielsen (eds.): *Crop Yield Response to Deficit Irrigation.* Report of an FAO/IAEA Co-ordinated Research Program by using Nuclear Techniques. 1999 ISBN 0-7923-5299-8
85. M.J. McLaughlin and B.R. Singh (eds.): *Cadmium in Soils and Plants.* 1999 ISBN 0-7923-5843-0
86. D. Anac and P. Martin-Prével (eds.): *Improved Crop Quality by Nutrient Management.* 2000 ISBN 0-7923-5850-3
87. A. Stokes (ed.): *The Supporting Roots of Trees and Woody Plants: Form, Function and Physiology.* 2000 ISBN 0-7923-5964-X
88. P.A. Finke, M.F.P. Bierkens and P. De Willigen: *Upscaling and Downscaling Methods for Environmental Research.* 2000 ISBN 0-7923-6339-6

Developments in Plant and Soil Sciences

89. C.J. Staiger, F. Baluska, D. Volkmann and P. Barlow (eds.): *Action: A Dynamic Framework for Multiple Plant Cell Functions*. 2000 ISBN 0-7923-6412-0
90. O. Gašparíková, M. Čiamporová, I. Mistrík and F. Baluška (eds.): *Recent Advances of Plant Root Structure and Function*. 2000 ISBN 0-7923-6658-1
91. R. Wassmann, R. Lantin and H.-U. Neue (eds.): *Methane Emissions from Major Rice Ecosystems in Asia*. 2000 ISBN Pb 0-7923-6765-0; Hb 0-7923-6759-6

Kluwer Academic Publishers – Dordrecht / Boston / London

Printed in the United States
120588LV00004B/79-80/A